THE LAW OF BIODIVERSITY AND ECOSYSTEM MANAGEMENT

by

JOHN COPELAND NAGLE
Professor of Law
University of Notre Dame

J.B. RUHL
Joseph Story Professor of Law
The Florida State University

NEW YORK, NEW YORK
FOUNDATION PRESS

2002

COPYRIGHT © 2002 By FOUNDATION PRESS

395 Hudson Street
New York, NY 10014
Phone Toll Free 1–877–888–1330
Fax (212) 367–6799
fdpress.com

All rights reserved
Printed in the United States of America

ISBN 1–58778–134–4

 TEXT IS PRINTED ON 10% POST CONSUMER RECYCLED PAPER

In memory of Patricia Campbell Nagle
J.C.N.

To Lisa, Grant, and Grayson
J.B.R.

*

PREFACE

It is unlikely that this book would have been published ten years ago. Biodiversity and ecosystem management were far from household words in the early 1990s, and had only begun to emerge in the sciences as relevant ways of thinking about the environment and human impacts on it. Law, always a straggler, was far behind in this case. A law school text on biodiversity and ecosystem management written in 1990, even by the most inventive of authors, would have been quite thin. Its market would have been even more slim, as few law schools had strayed in their curriculum beyond the traditional offerings of a survey course on environmental law and, maybe, another on natural resources law.

Much has changed since then. In 1999, the National Geographic Society declared biodiversity one of six subjects shaping our destiny and devoted an issue of *National Geographic* to the topic. Despite that sea change in the sciences, the law of biodiversity and ecosystem management has only begun to coalesce. After passing the 1990 Clean Air Act amendments, Congress has been unable to reach a consensus in support of any new major environmental law initiatives. That has left biodiversity far off Congress' radar screen. Nevertheless, as recent years have shown, Congress is not the only potent force of change in environmental law. Federal courts and agencies have interpreted federal laws—in many cases laws that have remained virtually unchanged for decades—in ways that have opened to door to the formation of policy on questions central to biodiversity conservation and ecosystem management. Many states have also lived up to the ideal of states as the laboratory of innovation when it comes to biodiversity and ecosystem management. Suffice it to say that, as much as has happened in the sciences, law has finally discovered biodiversity and ecosystem management as core subjects of environmental regulation and begun the process of catching the law up with the science.

We have attempted in this book to capture where the law at present stands in that regard. To do so, we begin in Part I with an overview of the topic that shows how law and science gradually are fusing. The materials introduce the student to the important threshold issues that face the development of a law of biodiversity, including defining what biodiversity is, choosing among different policy approaches for its protection, and finding the appropriate levels of administration—federal, tribal, state, and local—for implementing those policies.

Part II then covers the law that has emerged as the centerpiece of federal biodiversity policy—the Endangered Species Act. Famous for its lofty aspiration of conserving imperiled species and their ecosystems, and infamous to many for how it goes about doing so, perhaps no federal environmental law wields such power and produces as much controversy as does the Endangered Species Act. Our coverage in Part II walks the student through the mechanics and policy of the framework provisions of the law, covering the identification of threatened and endangered species and their critical habitat, the duties that apply to federal agencies once a species is "listed," and the strong prohibitions that apply to all persons and governments against harming protected species. Over time the latter of these themes–the so-called "take"prohibition that applies to private and public landowners alike—has moved the center of gravity for the Endangered Species Act from primarily a federal agency concern to having substantial impact on local land use decisions around the nation, which has in turn led the law to become engulfed in property rights issues. Part II thus also covers the emerging innovations in endangered species policy that may prove in the long run to promote both biodiversity conservation and flexibility for landowners.

Notwithstanding its broad scope and potent authorities, the Endangered Species Act falls far short of providing a comprehensive biodiversity conservation law. Yet, along with endangered species protection, a patchwork of existing state and federal laws can go a long way toward broad coverage of biodiversity conservation values. The difficulty is in how to describe that amalgam of laws. We have chosen the approach in Part III that seems most consistent with the way in which biodiversity is discussed at policy levels—as a core objective of ecosystem management. As then Secretary of the Interior Bruce Babbitt once quipped, in congressional testimony no less, an ecosystem can be anything from a raindrop to the planet. The point is that ecosystems, and their biological diversity, can be defined at very small, reductionist scales, or very large, holistic scales, or somewhere in between. For the convenience of presentation, we have chosen the middle ground, dividing coverage of ecosystem management according to different types of ecosystems we felt would be familiar to any student, but without violating sound scientific treatment of the topic.

Indeed, the science of ecology–the study of ecosystems–has forged bold new paths in recent years in tandem with the increased understanding of the importance of biological diversity. An important overarching theme of much of this new work in ecology is the dynamic, open nature of ecosystems and the processes which are in constant flux. We are increasingly appreciating that ecosystems involve complex interrelations, adaptive strategies for dealing with disturbance, and a sustainability that depends on change. Boundaries and parts have little meaning in such open systems. Yet humans have a difficult time managing chaos. Administrative convenience necessitates some division and compartmentalization of complex management problems, even if at the expense of a perfect description of the subject matter. Indeed, this, so far, has been the case for the web of laws oriented–more accurately, being reoriented—toward ecosystem management. So, while we recognize that forests and wetlands may overlap, it is useful, particularly given the structure of the law, to give them separate coverage in this text.

Consistent with that approach, the chapters in Part III are organized based on types of ecosystems. The first series of chapters covers conventional divisions of ecosystem types into forests, grasslands, freshwater, and coastal and marine. The bundles of laws associated with managing these ecosystem regimes are covered in each corresponding chapter. The theme in these chapters is the conservation of biodiversity values in ecosystems that have not been significantly degraded by human intrusion, or at least which can be recovered to that state. Some ecosystems, however, such as coral reefs, islands, and deserts, present dire circumstances as a result of their critical dynamics. There are also ecosystems no longer capable of being defined without reference to human presence. These human-dominated ecosystems include lands in agricultural, recreational, and urban uses. Far from unimportant in terms of biodiversity conservation, scientists increasingly are finding that these two extreme ecosystem types play an important role in overall biodiversity conservation goals. Accordingly, we cover them in the final chapter of Part III.

Parts I through III provide an ample framework for understanding the law of biodiversity conservation and ecosystem management in the United States. But, as we have observed, at a more holistic level the entire planet must be managed as an ecosystem. Biodiversity knows no political boundaries. Thus, in Part IV we turn to issue of global biodiversity. The law of global biodiversity is forming on two levels. First, recognizing that local events can have profound impacts on global biodiversity, many nations have started to shape their domestic laws to take transboundary effects into account. Simultaneously, the international community has begun to develop conventions for managing the "big picture" of biodiversity through international law in its customary, bilateral, and multilateral forms.

As unlikely as it is that this text would have been published ten years ago, it is equally likely that it will not resemble its present form in ten years. This field of law is changing rapidly. That we can even call it a field of law today suggests we have moved far and fast in the past decade. There is no reason to believe that this momentum will be lost, or that the field will be considered to have "matured," any time in the near future. But the materials covered in this book will always be known as the beginning of the law of biodiversity and ecosystem management, and are likely always to be important in that respect even if they fade from day-to-day prominence as the field evolves. Law students who use this first edition of the text thus can say they got into the topic of biodiversity and ecosystem management law at the ground floor. Where the law of biodiversity and ecosystem management goes from here is anyone's call, but wherever that may be, we plan to be there every step of the way and, through subsequent editions of this book, to record the journey.

Our road traveled thus far would not have been possible without the assistance of many. Numerous student research assistants contributed countless hours finding source materials: Ryan Carson, Vinette Godelia, Sean Moloney, Kyle Payne, Susan Lyndrup, Wes Wheeler, and Kristina Zurcher. Many colleagues at other institutions offered invaluable insight on drafts and on the topic in general. Chief among them was Rob Fischman, who meticulously reviewed many of the chapters and provided comments well beyond the call of duty. Also thanks in this department go to Jim Salzman, Oliver Houck, Marc Miller, Jim Salzman, and Buzz Thompson. Students in our respective biodiversity law seminars also endured early drafts and helped us immensely in refining the materials to provide what we hope other instructors and students will find a teachable, probing, and informative text. Dean Patricia O'Hara of the Notre Dame Law School and Dean Don Weidner of the Florida State University College of Law provided generous support of our research. We also thank our publisher, Steve Errick, for having the vision to chart new ground in the field of environmental law texts.

Of course, like all those before us who conceived law school texts with starry-eyed vision, jumped for joy upon receiving a publication agreement, and then found out just how much work it would be after all, we owe or deepest thanks to our families, who patiently supported us every page of the way, particularly our wives: Lisa Nagle and Lisa LeMaster.

We welcome comments on the text, on both its strengths and areas for improvement. They may be sent to John Nagle (nagle.8@nd.edu) or J. B. Ruhl (jruhl@law.fsu.edu).

JOHN COPELAND NAGLE
J.B. RUHL

ACKNOWLEDGMENTS

The authors gratefully acknowledge the permissions granted to reproduce the following materials.

Books and Articles

Adams, Jonathan S., Bruce A. Stein and Lynn S. Kutner, "Biodiversity: Our Precious Heritage," in "Precious Heritage: The Status of Biodiversity in the United States" (Bruce A. Stein, Lynn S. Kutner & Jonathan S. Adams eds., 2000). Reprinted with permission of Oxford University Press.

Barrera-Hernandez, Lila Katz de, and Alastair R. Lucas, "Environmental Law in Latin America and the Caribbean: Overview and Assessment," 12 Georgetown International Environmental Law Review 207 (1999). Reprinted with permission of the publisher, Georgetown International Environmental Law Journal © 1999, and the authors.

Babbitt, Bruce, "Between the Flood and the Rainbow: Our Covenant to Protect the Whole of Creation," 2 Animal Law 1 (1996). Reprinted with permission of Animal Law and the author.

Blair, John M., Scott L. Collins, and Alan K. Knapp, "Ecosystems As Functional Units In Nature," 14 Natural Resources & Env't 150 (2000). Reprinted by permission of the American Bar Association copyright © 2000.

Bowden, Matthew M., "An Overview of the National Estuary Program," 11 Natural Resources & Env't 35 (Fall 1996). Reprinted by permission of the American Bar Association copyright © 1996.

Dickson, Barnabas, "CITES in Harare: A Review of the Tenth Conference of the Parties," Colorado Journal of International Environmental Law, 1997 Year Book 55. Reprinted by permission of the author and the Colorado Journal of International Environmental Law.

Green, Martha Hodgkins, "How Green Is My Valley?," 47 Nature Conservancy 18 (Sept. / Oct. 1997). Reprinted by permission of the Nature Conservancy.

Grumbine, R. Edward, "What Is Ecosystem Management?," 8 Conservation Biology 27 (1994). Reprinted with permission by Blackwell Science, Inc.

Harrigan-Lum, June F. and Arnold L. Lum, "Hawaii's Total Maximum Daily Load Program: Legal Requirements versus Environmental Realities," 15 Natural Resources & Env't 12 (2000). Reprinted by permission of the American Bar Association copyright © 2000.

Hodas, David, NEPA, Ecosystem Management and Environmental Accounting, 14 Natural Resources & Env't 185 (2000). Reprinted by permission of the American Bar Association copyright © 2000.

Hsu, Shi-Ling and James E. Wilen, "Ecosystem Management and the 1996 Sustainable Fisheries Act," 24 Ecology L.Q. 799 (1997). Reprinted from Ecology Law Quarterly, vol. 24, no. 4, pp. 799-811, by permission of Ecology Law Quarterly and the authors. Copyright © 1997 by the Regents of the University of California.

Hughes, Harry R. and Thomas W. Burke, Jr., "The Cleanup of the Nation's Largest Estuary: A Test of Political Will," 11 Natural Resources & Env't 30 (Fall 1996). Reprinted by permission of the American Bar Association copyright © 1996.

"In re Fund for Animals Ltd.," Administrative Appeals Tribunal, 9 A.L.D. 622 (1986). Reprinted with permission of LexisNexis Butterworths.

Karkkainen, Bradley C., "Biodiversity and Land," 83 Cornell Law Review 1 (1997). Reprinted by permission of the author and the Cornell Law Review.

Kelly, Christopher R. and James A. Lodoen, "Federal Farm Program Conservation Initiatives: Past, Present, and Future," 9 Natural Resources & Env't 17 (Winter 1995). Reprinted by permission of the American Bar Association copyright © 1995.

Lorenzo, Mark, "Sizing Up Sprawl," 9 Wild Earth 72 (Fall 1999). Reprinted with permission of Wild Earth and the author.

Morell, Virginia, "New Mammals Discovered by Biology's New Explorers," 273 Science 1491 (1996). Reprinted with permission from Science, vol. 273, page 1491. Copyright © 1996 American Association for the Advancement of Science.

Postel, Sandra and Stephen Carpenter, "Freshwater Ecosystem Services," in "Nature's Services: Societal Dependence on Natural Ecosystems" (Gretchen Dailey ed. 1997). Reprinted with permission of Island Press copyright © 1997.

Ruhl, J.B. and R. Juge Gregg, "Integrating Ecosystem Services Into Environmental Law: A Case Study of Wetlands Mitigation Banking," 20 Stan. Envtl. L.J. 365 (2001). Reprinted by permission of the authors and the Stanford Environmental Law Journal.

Ruhl, J.B., "Ecosystem Management, The Endangered Species Act, and the Seven Degrees of Relevance," 14 Natural Resources & Env't 156 (2000). Reprinted by permission of the American Bar Association copyright ©.

Salzman, James, "Valuing Ecosystem Services," 24 Ecology L.Q. 887 (1997). Reprinted from Ecology Law Quarterly, vol. 4. no. 4, pp. 887-903, by permission of Ecology Law Quarterly and the author. Copyright © 1997 by Ecology Law Quarterly.

Taylor, William E. and Mark Gerath, "The Watershed Protection Approach: Is the Promise About to Be Realized?," 11 Natural Resources & Env't 16 (Fall 1996). Reprinted by permission of the American Bar Association copyright © 1996.

Thompson, Barton H., Jr., "Markets for Nature," 25 Wm. & Mary Envtl. Law & Pol'y Rev. 261 (2000). Reprinted with permission of the William and Mary Environmental Law and Policy Review.

Thompson, Rebecca, "'Ecosystem Management'– Great Idea, But What Is It, Will It Work, and Who Will Pay?," 9 Natural Resources & Env't 42 (Winter 1995). Reprinted by permission of the American Bar Association copyright © 1995.

United Nations Conference on Environment and Development: Convention on Biological Diversity, 31 International Legal Materials 818 (1992). Reprinted by permission of the Secretariat of the Convention on Biological Diversity.

United Nations Food and Agriculture Organization, Code of Conduct for Responsible Fisheries. Copyright (c) 2001 Food and Agriculture Organization of the United Nations. Reprinted with permission of the Food and Agriculture Organization of the United Nations.

Watermolen, D.J., C. Bleser, D. Zastrow, J. Christenson, D.R. Lentz and 20 co-authors. 2000. Executive Summary Pp. Iii-xviii in "Wisconsin Statewide Karner Blue Butterfly Habitat Conservation Plan and Environmental Impact Statement." 2 Vols. Madison: Wisconsin Department of Natural Resources. Reprinted by permission of the Wisconsin Department of Nature Resources.

World Trade Organization Appellate Body, "United States – Import Prohibition of Certain Shrimp and Shrimp Products," WT/DS58/AB/R; (98-3899); AB-1998-4, 1998 WTO DS LEXIS 13 (1998). Reprinted by permission of the World Trade Organization.

Illustrations

1990 Total Water Withdrawals illustration, United States Geological Survey.

Alabama red-bellied turtle, photograph by Robert H. Mount. Reproduced by permission of the Alabama Department of Archives and History.

Apalachicola-Chattahoochee-Flint River Basin map, United States Army Corps of Engineers.

Atlas of America's Polluted Waters, United States Environmental Protection Agency.

Blackburn's sphinx moth, photograph by W.P. Mull.

aimen yacare, photograph by David S. Kirshner.

CALFED Problem and Solution Areas map, CALFED.

California condor, photograph, collection of the Arizona Ecological Field Services Office of the United States Fish and Wildlife Service.

California gnatcatcher, photograph by Arnold Small, United States Fish and Wildlife Service.

Chequamegon-Nicolet National Forest, photograph by United States Forest Service.

Chesapeake Bay Watershed map, Chesapeake Bay Program.

Coastal Zone Management Program map, National Ocean Service.

Dead crab, photograph by United States Environmental Protection Agency.

Delhi Sands Flower-Loving Fly, photograph by Guy Bruyea.

Destin, Florida, photographs by J.B. Ruhl

Ecoregions map, United States Bureau of Land Management.

Ecosystem Units map, United States Fish and Wildlife Service.

Entry to Cimarron National Grassland, photograph by United States Forest Service.

Illinois cave amphipod, photograph by Steve Taylor, Center for Biodiversity, Illinois Natural History Survey.

Karner blue butterfly, photograph by Ann B. Swengel.

Key deer, photograph by Gerald Ludwig, United States Fish and Wildlife Service.

Loach minnow, photograph by Marty Jakle, United States Fish and Wildlife Service.

Lower Keys marsh rabbit, photograph by Beth Forys.

Marine Protected Areas map, National Ocean Service.

Mississippi River Basin with Gulf of Mexico Hypoxia map, United States Environmental Protection Agency.

National Forest Lands map, United States Forest Service.

New York City Water Supply map, New York City Department of Environmental Protection.

Overall National Coastal Condition chart, United States Environmental Protection Agency.

Percentage of wetland losses map, United States Geological Survey.

Phase I Inventory of Current EPA Efforts to Protect Ecosystems map, United States Environmental Protection Agency.

Piping plover, photograph by J.P. Mattson, United States Fish and Wildlife Service.

Principal Oxidized Nitrogen Airsheds map, United States Environmental Protection Agency and National Oceanic and Atmospheric Administration.

Proposed Darby Prairie National Wildlife Refuge map, United States Fish and Wildlife Service.

Public Lands Managed by the Bureau of Land Management map, Bureau of Land Management.

Shawnee National Forest map, United States Forest Service.

Shawnee National Forest, photograph by United States Forest Service.

Southeast Orlando land use map, Planning and Development Department, City of Orlando, Florida.

The White Marlin illustration, National Marine Fisheries Service.

Watershed Information Network Atlas map, United States Environmental Protection Agency.

Watersheds of National Estuaries map, United States Environmental Protection Agency.

White Marlin Open, photograph by Angel Bolinger, Maryland Department of Natural Resources.

*

SUMMARY OF CONTENTS

*

TABLE OF CONTENTS

TABLE OF CASES

Principal cases are in bold type. Non-principal cases are in roman type. References are to Pages.

*

THE LAW OF BIODIVERSITY AND ECOSYSTEM MANAGEMENT

*

PART ONE

BIODIVERSITY

CHAPTER 1

THE PROBLEM OF THE DELHI SANDS FLOWER–LOVING FLY

Long ago, the Santa Ana winds picked up sand from several creeks and deposited it over about 35,000 acres of land located about sixty miles east of the Pacific Ocean in southern California. Thus were born the Delhi sand dunes (also known as the Colton Dunes), the only inland sand dune system in the Los Angeles basin. At first glance, the dunes are inhabited only by blowing sand and scattered shrubs. But contrary to the popular image of deserts as barren of wildlife, desert ecosystems are in fact teeming with birds, insects, reptiles, mammals and plants. The Delhi Sands are a good example of what one finds in a coastal sage scrub ecosystem. Birds such as Western meadowlarks and burrowing owls frequent the area. The San Diego horned lizard and the legless lizard live in the dunes, as do insects such as the Delhi sands metalmark butterfly and the Delhi sands Jerusalem cricket. The onset of night entices the Los Angeles pocket mouse, the San Bernadino kangaroo rat and other small mammals to survey the land.

The Delhi sands

Primrose, goldfields and other wildflowers flourish after the winter rains, replaced later in the year by the wild buckwheat and the colorful butterflies that the plant attracts. The yellow flowers of telegraph weeds appear in the summer.

Increasingly, though, the Delhi Sands are home to many people, too. The area had long been inhabited by many different Native American peoples, including the Serrano, the Cahuilla, the Chemehuevi and the Mojave. Spanish and Mormon missionaries traveled across the land during the eighteenth and early nineteenth centuries, and the first European settlements in what is now western San Bernadino and Riverside Counties began after California became a state in 1850 and after the railroad reached the area in the early 1870's. The city of Colton, for example, was first settled in 1874 and named after a Civil War general who also served as the vice president of the Southern Pacific Railroad. The settlers immediately began planting citrus orchards despite concerns that the land was inadequate for farming. The citrus thrived in the warm climate once irrigated water was delivered from the nearby Santa Ana River, and much of the land was cultivated for grapes, oranges and other fruits by the late 1800's. Dairies, residential homes, and commercial and industrial development were the next to appear on the scene.

The results of the human settlements have not been especially attractive. The California Portland Cement Company mined Slover Mountain for over 100 years, leaving a pile of granite and no dunes in its wake. Similar enterprises have taken the sand for road fill and other purposes. Junk yards and petroleum tank farms abound. The Southern Pacific Railroad and Interstate 10 bisect the area. A landfill, a sewage treatment facility and many illegally dumped cars are also nearby. Off-road vehicle enthusiasts alter the terrain of the little bit of the dunes that remains.

Yet the land is in great demand. The population of Colton, in the heart of the dunes, is expected to grow from 45,000 in 1995 to 53,160 in 2000, and then to 69,710 in 2020. The entire western San Bernadino and western Riverside County region is expected to see even more explosive growth. The two counties are expected to add 128,000 people each year through 2010. The additional people bring additional demands for housing, shopping, offices, road, and other facilities to be built on the previously barren land. The attractiveness of the area has spiked land prices to as high as $160,000 per acre.

The human population of the dunes is as diverse as the wildlife population. Nearly sixty percent of the residents of Colton, for example, are of Hispanic origin. Another fifteen percent of Colton's residents are African–Americans, Asians, or Native Americans. The city's median family income is only slightly above $30,000, making it one of the poorest cities in California. The closure of many military bases and the loss of defense jobs in San Bernadino County caused the region to suffer a significant economic recession beginning in the 1980's. The economic plight of the area was illustrated by the creation of the Agua Mansa Enterprise Zone, which was established by San Bernadino and Riverside Counties and the cities of Colton, Rialto and Riverside in an effort to lure economic development to a

10,000 acre site in the region. The 1986 environmental study preceding the creation of the enterprise zone assured that there were no rare or endangered species living on the affected land.

The view of the dunes today, with I-10 in the foreground and the sand mining operation in the background.

The growth in the human population has produced a corresponding shrinkage of the original Delhi sands. Most of the original dunes were destroyed by the onset of agricultural uses at the end of the nineteenth century. Over the next one hundred years, commercial, industrial and residential development eliminated much of the remaining dunes. A shopping center replaced seventy acres of dunes in the early 1990s, and a county park split another segment of the dunes in 1998. Only about forty square miles of dunes—or about two percent of the original sands—exist in several patches stretching between the cities of Colton and Mira Loma.

As the Delhi sands have disappeared, so has the native wildlife. Pringle's monardella, a wildflower that once grew only in the Delhi sands, has already gone extinct. The number of meadowlarks and burrowing owls has diminished as their habitats have been converted into human uses, though both birds have displayed a surprising resiliency in the presence of bulldozers and landfills and the like. And the area is the still the only place on earth where the Delhi Sands Flower–Loving Fly clings to life.

The Fly—known to entomologists as *Rhaphiomidas terminatus abdominalis*—is colored orangish and brown, with dark brown oval spots on its abdomen and emerald green eyes. It is one inch long, much larger than a common house fly. The Fly undergoes a metamorphosis from egg to larva to pupa to adult fly over a three-year period. Once it emerges from the sand at the end of the three years, an individual Fly lives for about a week in

August and September. The Flies are active during the heat of the day, with females burrowing into the sand to deposit about forty eggs at one time. The Flies remain inactive during cloudy, rainy or windy conditions. There are twice as many males as females, suggesting that the males are more active as they investigate vegetation or sand where females can deposit their eggs. The Fly is only found in the Delhi sands.

The Delhi-Sands Flower Loving Fly

photo by Guy Bruyea

The Fly is one of nearly 85,000 species of flies that scientists have identified around the world so far. It is one of a dozen species that comprise the *Rhaphiomidas* genus. The Fly is distinguished from the other species within the *Rhaphiomidas* genus—all of which live in arid and semi-arid parts of California, Arizona, New Mexico and northern Mexico—by its bicolored abdomen and its widely separated eyes. The Fly is actually a subspecies of the *Rhaphiomidas terminatus* species. Its companion subspecies and closest relative—the El Segundo Flower–Loving Fly—lived in the sand dunes north of downtown Los Angeles until the dunes were replaced by runways upon the construction of Los Angeles International Airport in the 1960's.

As its name suggests, the Fly loves flowers. It hovers like a humming-bird as it removes nectar from the native buckwheat flowers with its long tubular proboscis. It flies low, preferring sparsely vegetated areas. The Fly is a pollinator. It is one of a small group of flies that are pollinators. Other flower-loving flies survive elsewhere. For example, the acton flower-loving fly pollinates an endangered Californian plant known as the Santa Ana wooly star.

Entomologists do not know a whole lot about the Fly. The first Fly was collected in 1888, but it was not until a century later that the Fly was identified as a separate subspecies. The Fly probably lived throughout the full historic extent of the Delhi Sands, though we will never be sure about that. Entomologists do not know what the larval flies-in-waiting eat during their three years underground. Nor is there any indication that the Fly provides any nutritional, medicinal or other tangible benefit to people.

The Fly survives in just five locations within an eight mile radius along the border of San Bernadino and Riverside Counties. No one knows for sure how many Flies are alive today, though estimates run from a couple hundred to less than a thousand. What everyone agrees upon, though, is that the number of Flies is shrinking and that the species may soon become extinct.

The Fly faces a variety of threats to its survival. Birds, reptiles, dragonflies, and the Argentine ant—an insect that is not native to the area—sometimes attack and kill a larval or adult Fly. Pesticides used for agricultural purposes eliminate the native vegetation upon which the Fly relies for its survival. Native plants have been smothered by local dairies that have dumped tons of cow manure on sections of the dunes—often without the landowner's permission—thus providing nutrient-rich soil for exotic plants. Mustard, cheeseweed, the Russian thistle, and other plants that are new to the area affect the soil in a way that is harmful to the Fly. The native vegetation is trampled by off-road vehicle riders and removed for fire control efforts. The Fly lives best in those few areas that have yet to be disturbed by human activities. Also, the fact that so few populations of the Fly still exist makes the entire species vulnerable to catastrophic events such as fires and droughts. The small, scattered populations reduce the genetic variability of the Fly—and thus, its ability to respond to environmental stresses—as well.

Mostly, though, the Fly is on the brink of extinction because the Delhi sands are disappearing in the wake of human development. An estimated 98% of the Fly's original habitat has been destroyed. By 1993, the Delhi sands that remained were threatened by a host of residential, commercial and industrial development projects. The most notable development to threaten the Fly was a hospital. Plans to replace San Bernadino's aging County Medical Center began in the late 1970s. County officials designed a large regional medical center that could resist earthquakes and satisfy the demands of federal health officials. The site of the hospital was a vacant piece of land just north of I–10 in Colton. By September 1993, the county was finally ready to break ground for its new Arrowhead Regional Medical Center.

Meanwhile, the Fly had attracted the attention of Greg Balmer, a graduate student in entomology at the University of California at Riverside. Balmer viewed the Fly as "spectacular," yet he quickly became concerned about its plight. The rapid residential, commercial and industrial development of the region posed a grave threat to the survival of the dunes, and thus to the survival of the Fly. So Balmer did what any smart entomologist would do: he filed a petition with the United States Fish & Wildlife Service (FWS) to list the Fly as "endangered" under the Endangered Species Act of 1973.

Congress had enacted the ESA in 1973 during the heyday of federal environmental legislation. The proponents of the law evoked images of bald eagles, grizzly bears, alligators and other national symbols that were on the brink of disappearing from this land. Almost immediately, though, the ESA was deployed to protect much less popular creatures. The listing of the snail darter as endangered just months after the ESA became law resulted in the Supreme Court's decision in Tennessee Valley Authority v. Hill, 437 U.S. 153 (1978), confirming that the multi-million dollar Tellico Dam project could not be completed because of the threat that the dam posed to the snail darter's survival. In more recent years, the law's application to the northern spotted owl became a focal point for broader debates between the timber industry and environmentalists in the Pacific Northwest.

Balmer had asked the FWS to list the Fly on an emergency basis because of the urgency of the development pressures on all of the Fly's remaining habitat. The agency did not act until September 1993, when it agreed to add the Fly to the permanent list of endangered species. That also happened to be the day before construction was to begin on San Bernadino County's new hospital project smack in the middle of some of the Fly's prime habitat. At the first meeting between local officials and the FWS, a FWS employee surveyed the scene and suggested that nearby I–10—the major east-west highway between the San Bernadino Valley and Los Angeles—would have to be closed two months each year when the Fly was above ground.

Local officials were stuck. They protested to Congress. And they tried to cut a deal with the FWS. At first, the parties agreed that the hospital could be built if it was moved three hundred feet to the north and if the county established a refuge for the Fly. The "refuge" was vacant land adjacent to the hospital that was bordered by orange plastic fencing. Happily, the Flies loved the fencing. Then the county realized that they would need to build a new electrical substation to power the hospital; that resulted in seven more acres for the refuge. But when the county sought permission to reconfigure the roads in the area surrounding the hospital, the FWS balked. The county sued, joined by local developers, claiming that the ESA could not be constitutionally applied to regulate construction projects involving a species like the Fly that lived in only one state and that was not involved in interstate commerce itself. The district court held that such an application of the ESA was constitutional, as did a divided D.C. Circuit, and any hope for a constitutional exit disappeared in June 1998 when the Supreme Court denied certiorari. *See* National Ass'n of Home Builders v. Babbit, 949 F.Supp. 1 (D.D.C.1996), *aff'd*, 130 F.3d 1041 (D.C.Cir.1997), *cert. denied*, 524 U.S. 937 (1998).

The Fly, thanks to the ESA, now occupied a position of great strength in future discussions about the development of the region. A host of developments were challenged because of their possible impact on the Fly and its remaining habitat. The projects included:

- A 2.8 million square foot WalMart distribution facility to be built in the dunes near Colton.

- A 27 hole golf course and accompanying 202 home development slated for Fontana which a city official defended because the sighting of a couple of Flies there over a two-year period is "just not enough science to put people's land at risk."

- A truck stop and industrial center to be built by Kaiser Ventures, which estimates that the project could create 5,300 jobs and $75 million per year to the San Bernadino County economy.

- A cement plant and a facility that produces sidewalk pavers in Rialto that was blocked by a federal court when the FWS claimed that the plant would wipe out a major portion of the Fly's habitat, but which the FWS approved in 1999 when the company agreed to set aside 30.5 acres of land for Fly habitat.

- A large project that would include new homes, theaters and restaurants in Fontana.

- The proposal of Viny Industries, a paper products company, to create 400 jobs by building on sixty acres of land in Colton.

- A $110 million plant to make fiberboard from recycled waste wood which opened in the Agua Mansa Industrial Center in May 1999 only after the industrial center contributed $450,000 to purchase other habitat for the Fly.

The hospital itself finally opened in March 1999 after the county set aside a total of twelve acres of land for a Fly refuge. The county estimated that moving the site of the hospital, establishing the Fly preserves, and otherwise accommodating the Fly cost the county nearly $3,000,000.

The Fly also interfered with environmental cleanup activities in the area. When petroleum leaking from a nearby tank farm contaminated the groundwater, the presence of the Fly underground so complicated any remediation work that the tank farm owner planned to wait for the plume of contamination to migrate past the Fly's habitat. Colton residents were prevented from cleaning up a vacant lot that was used as an illegal dump—thus causing some families to fear for their children—until the FWS approved any alterations to the property.

In early 1999, Fontana officials warned that the Fly could cause the city to default on $42 million in municipal bonds. The city had issued the bonds in 1991 to build streets, sewers, lighting, and other improvements on vacant land adjacent to a small shopping center. The possibility that the Fly lives on the land prevented the expected commercial development of the land, and when the landowner stopped paying taxes on the land, the city began to use its reserve funds to pay the bondholders. The Fly did facilitate one source of new employment: developers hired consultants to determine the extent of the presence of the Fly in the area. It was only when the landowner paid its taxes after one such survey failed to find any Flies on the property that Fontana barely avoided defaulting on the bonds in October 1999.

The Fly continued to block the proposed road construction projects that resulted in the commerce clause litigation. Colton officials and the FWS had not reached an agreement that would permit the realignment of roads near the new hospital despite meetings held throughout 1999. Similarly, when Riverside County asked the FWS for permission to build new ramps for I–15 in Mira Loma, the federal agency responded that the county would need to establish a 200 acre preserve for the Fly. The agency reasoned that although the ramps would only displace a little more than eight acres of Fly habitat, the effect on the Fly of the accompanying development and increased traffic justified a larger reserve. The purchase of that much land would cost the county as much as $32 million in an area where land sells for up to $160,000 an acre, which would make the Fly reserve more expensive than the highway ramps themselves. More generally, when officials representing Colton, Fontana, Rialto and other local cities met with the FWS in July 1999 to propose setting aside 850 acres of land for Fly habitat in exchange for permission to develop throughout the area, they were told that FWS biologists were seeking 2,100 acres that could cost $220 million to purchase. Much of that land—including a former dairy in Ontario—would have to be rehabilitated in order to serve as viable habitat for the Fly.

The Fly was vilified. Fontana Mayor David Eshleman complained that the Fly "is costing the Inland Empire thousands of jobs and millions of dollars. I think we should issue fly swatters to everyone." Colton's city manager estimated the stalled development, uncollected tax revenue, and lost jobs attributable to the Fly as totaling $661 million. Julie Biggs, the Colton city attorney, characterized the Fly's habitat as "a bunch of dirt and weeds." Jerry Eaves, the chairman of the San Bernadino County Board of Supervisors, stated that "the Endangered Species Act was intended to save eagles and bears. Personally, I don't think we should be spending this money to save cockroaches, snails and flies." Advocates for reform of the ESA seized on the controversy as an example of the kinds of problems that the law created, with the "people vs. flies" argument being voiced frequently.

The Fly has been featured on network television news shows, leading newspapers across the country, *National Geographic*, and other national media. CBS described it as "superfly, with the power to stop bulldozers." The *Los Angeles Times* reported that the Fly could become "the snail darter of the 1990s." Many portrayals of the Fly have been unsympathetic. The *Washington Post* described the Fly as "a creature that spends most of its life underground, living as a fat, clumsy, enigmatic maggot." The *Washington Times* editorialized that "one could build the flies their own mansion in Beverly Hills ... fill it up from top to bottom with leftover potato salad and other fly delicacies, and it would still be cheaper than the royal estate Fish & Wildlife has in mind for them."

By contrast, UCLA professor Rudy Mattoni described the Fly as "a national treasure in the middle of junkyards.... It's a fly you can love. It's beautiful." A FWS official told CNN that the Fly "isn't as charismatic as a panda bear or a sea otter, but that doesn't make it any less important." Another FWS official insisted that "the value of the fly to mankind is a very difficult thing to judge. It's much more of a moral issue. Do we have the right to destroy another creature when we, in our day-to-day activities, have the ability not to destroy a creature?" The statement of county supervisor Jerry Eaves that the ESA was not intended to save flies provoked a letter to the editor of the *Los Angeles Times* complaining that "developers and their minions in public office will go to any length to satisfy their corporate greed." Environmentalists also emphasized the importance of the dune ecosystem rather than the Fly. A FWS biologist reminded that "every ecosystem has its intrinsic value, and maybe we can't quite put a dollar value on it. But every time one disappears, it's an indication that something else is wrong." Dan Silver, the head of the Endangered Habitats League, asserted that the ESA "is saving Riverside County from itself, its own short-sightedness. It is forcing people to take a longer view."

Having lost in the courts, the local communities turned to Congress. They paid $48,000 for a Washington lobbyist to persuade Congress to remove the Fly from the list of protected species. Democratic state representative Joe Baca introduced a resolution in the California legislature calling for lifting of the ESA's protection of the Fly; the voters rewarded him by electing him to Congress in 1999. Republican Senate candidate Matt Fong was not so fortunate in 1998: he campaigned against the Fly's impact on development in the region, but he lost to incumbent Senator Barbara Boxer, a supporter of the ESA. The protected status of the Fly survived all of these efforts.

Another strategy involved the crafting of a habitat conservation plan (HCP) that would set aside some land for habitat for the Fly and other wildlife while allowing other land—including wildlife habitat—to be developed. Eleven local cities joined San Bernadino County in planning a HCP that would encompass over 300,000 acres of land comprised of eight different kinds of ecosystems and containing the Fly and other rare species such as the San Bernadino kangaroo rat and the Santa Ana River woolystar. Several years of negotiations failed to produce an agreeable plan, but those efforts have aided ongoing work by Riverside County to combine a multiple species HCP with comprehensive land use and transportation planning. The county hoped to complete its integrated plan by 2002, and environmentalists have testified before Congress that the work is "ambitious and visionary."

Meanwhile, both economic development and protection of the Fly proceeded on a piecemeal basis. The typical approach involved a landowner agreeing to set aside some of its property to serve as habitat for the Fly in exchange for FWS permission to build on another part of its property. In other instances a developer simply paid for the purchase of other land that could be used by the Fly. For example, in August 2000 the developer of a warehouse project agreed to pay $82,500 so that a community group could purchase habitat for the Fly. But neither side was really satisfied by such arrangements: environmentalists worried that the haphazard patches of protected land would not sustain a healthy population of the Fly, while developers watched as their proposed building sites remained vacant as the economic boom of the 1990's ended.

So the dawn of the twenty-first century finds government officials, developers, environmentalists, and other interested parties still debating the needs of the Fly, the dunes, and the people who live there. The FWS continues to meet with local county and city officials in an effort to resolve both specific proposed projects and the broader issues raised by the Fly. Scientists are trying to breed the Fly in captivity, though they have not succeeded yet. Private efforts to help the Fly have begun, such as the work of volunteers and local students to restore a four-acre right-of-way owned by Southern California Edison. Nonetheless, scientists and federal officials still fear that the Fly will go extinct early this century despite all of the efforts to save it.

Notes & Questions

1. Assume that you are a billionaire who wants to protect the Delhi Sands and the Fly at all costs. What would you do?

Assume that you are an attorney for an environmental group that wants to protect the Delhi Sands and the Fly. What would you advise?

Assume that you are the land use attorney for the City of Colton, and that you have been asked to report on how to protect the Delhi Sands and the Fly while still accounting for human needs in the economically depressed city. What would you recommend?

2. The saga of the dunes and the Fly raises nearly every imaginable question about the relationship between biodiversity and the law. How would you answer the following questions:

- Why should we preserve the dunes? Why should we preserve the Fly?
- What cost should we be willing to pay to save the dunes or the Fly?
- Should we be more concerned about saving ecosystems like the dunes, or individuals species like the Fly?
- Are some species more worth saving than others? Who should make that decision?
- What role should private efforts play in preserving the dunes and the Fly? What role should the law play?
- Are the dunes and the Fly best protected by international law, federal law, state law, or some combination of such laws?

• Who should pay the costs of protecting the Fly?

3. We do not know how the story of the Delhi Sands and the Fly will end. In that respect, the saga is typical of most modern controversies involving biodiversity, as will become obvious as we consider the stories of other ecosystems and species throughout this book. Meanwhile, the best sources of updates on the Fly and the related controversy are the reports of the *Press-Enterprise* (the Riverside newspaper available on LEXIS in the News library, PRSENT file). For additional material on the Fly's happenings to date, *see, e.g.,* Determination of Endangered Status for the Delhi Sands Flower-loving Fly, 58 Fed. Reg. 49881 (1993); U.S. FISH AND WILDLIFE SERVICE, PACIFIC REGION, FINAL RECOVERY PLAN FOR THE DELHI SANDS FLOWER-LOVING FLY (1997); National Ass'n of Home Builders v. Babbitt, 130 F.3d 1041 (D.C.Cir.1997), *cert. denied,* 524 U.S. 937 (1998); Kenneth J. Kingsley, *Behavior of the Delhi Sands Flower–Loving Fly (Diptera: Mydidae), A Little–Known Endangered Species,* 89 ANNALS OF THE ENTOMOLOGICAL SOC'Y OF AMERICA 883 (1996).

CHAPTER 2

AN OVERVIEW OF BIODIVERSITY

Chapter Outline:
A. Types of Biodiversity
B. Why We Care About Biodiversity
C. Threats to Biodiversity

People have long pondered the biological wealth of this world with a mixture of wonder, fear, awe, and concern. The term "biodiversity" itself was not coined until the 1980s, when the eminent Harvard biologist Edward O. Wilson popularized it. *See* EDWARD O. WILSON, THE DIVERSITY OF LIFE 15 (1992) (referring to "[b]iological diversity—'biodiversity' in the new parlance"). By 1992, the Convention on Biological Diversity defined biological diversity as "the variability among living organisms from all sources, including, *inter alia*, terrestrial, marine and other aquatic ecosystems and the ecological complexes of which they are a part; this includes diversity within species, between species and of ecosystems." That statement only begins to hint at the vast range of plant, animal, and other life encompassed by the idea of biodiversity. Biodiversity means many things to many people, though the core understanding of the breadth of life on this planet pervades all of the multiple definitions.

This chapter seeks to provide an elementary guide to the nature of biodiversity. It explains what biodiversity means, why it is important, and what threatens its existence today. Each of these topics has elicited far more thoughtful scientific study than can be captured in this overview. The best sources to consult for a sense of the vast scientific and popular literature that has examined the nature of biodiversity just in the past few years include PRECIOUS HERITAGE: THE STATUS OF BIODIVERSITY IN THE UNITED STATES 7–10 (Bruce A. Stein, Lynn S. Kutner & Jonathan S. Adams eds., 2000), a joint product of the Nature Conservancy and the Association for Biodiversity Information, NATIONAL RESEARCH COUNCIL, PERSPECTIVES ON BIODIVERSITY: VALUING ITS ROLE IN AN EVERCHANGING WORLD (1999); REED F. NOSS & ALLEN Y. COOPERRIDER, SAVING NATURE'S LEGACY: PROTECTING AND RESTORING BIODIVERSITY (1994); and WORLD CONSERVATION MONITORING CENTRE, GLOBAL BIODIVERSITY: STATUS OF THE EARTH'S LIVING RESOURCES (Brian Groombridge ed. 1992). And for an excellent interactive tour of biodiversity, conservation biology, and environmental science, try the CD–ROM *Conserving Earth's Biodiversity with E.O. Wilson*, prepared by Edward O. Wilson and Dan L. Perlman.

A.   TYPES OF BIODIVERSITY

Jonathan S. Adams, Bruce A. Stein & Lynn S. Kutner
Biodiversity: Our Precious Heritage

PRECIOUS HERITAGE: THE STATUS OF BIODIVERSITY IN THE UNITED STATES 7–10.
(Bruce A. Stein, Lynn S. Kutner & Jonathan S. Adams eds., 2000).

From the Devils Hole pupfish to the delicate spring ecosystems at Ash Meadows to the Mohave Desert, from genes to species to ecosystems to landscapes: Each is part of the fabric of life. Each is a component of biodiversity. But what is biodiversity? Although the term is now common, many people are bewildered by it. Still others use it in an all-encompassing way to refer to any and all nature.

Biodiversity is, in essence, the full array of life on Earth. The most tangible manifestations of this concept are the species of plants, animals, and microorganisms that surround us. Yet biodiversity is more than just the number and diversity of species, as immense as that might be. It also includes the genetic material that makes up those species. And at a higher level, it includes the natural communities, ecosystems, and landscapes of which species are a part. The concept of biodiversity includes both the variety of these things and the variability found within and among them. Biodiversity also encompasses the processes—both ecological and evolutionary—that allow life on Earth to continue adapting and evolving.

While the term *biodiversity* was coined and popularized only recently, the concept is as old as the human desire to know and name all the creatures of the earth. Nature's daunting complexity demands some method of differentiating among its various components. Four of the principal levels of biological organization are genes, species, ecosystems, and landscapes.

Genetic diversity refers to the unique combinations of genes found within and among organisms. Genes, composed of DNA sequences, are the fundamental building blocks of life. The complexes of genes found within individual organisms, and their frequencies of occurrence within a population, are the basic levels at which evolution occurs. Genetic variability is an important trait in assuring the long-term survival of most species, since it allows them to respond to unpredictable changes in their environment.

Species diversity encompasses the variety of living organisms inhabiting an area. This is most commonly gauged by the number of different of organisms—for instance, the number of different birds or plants in a state, country, or ecosystem. While species are the most widely understood aspect of biodiversity, it is actually the individual populations that together make up a species that are the focus of on-the-ground conservation efforts.

Ecological diversity refers to the higher-level organization of different species into natural communities, and the interplay between these communities and the physical environment that forms ecosystems. Interactions are key to ecological diversity. This includes interactions among different species—predators and prey, for instance, or pollinators and flowers—as well as interaction among these species and the physical processes, such as nutrient cycling or periodic fires, that are essential to maintain ecosystem functioning.

Landscape diversity refers to the geography of different ecosystems across a large area and the connection among them. Natural communities and ecosystems change across the landscape in response to environmental gradients such as climate, soils, or altitude and form characteristic mosaics. Understanding the patterns among these natural ecosystems and how they relate to other landscape features, such as farms, cities, and roads, is key to maintaining such regional diversity.

Conservation of biodiversity requires attention to each of these levels, because all contribute to the persistence of life on Earth. More than most people realize, humans rely on wild biological resources for food and shelter. Genes from wild plants, for instance, allow plant breeders to develop disease-resistant crops or increase crop yields, passing along the benefits of biodiversity to farmers and ultimately consumers. Similarly, medicines derived from plants, animals, and especially microbes are an established part of the Western pharmacy and include such widely used medications as aspirin, penicillin, and digitalis. The emerging biotechnology industry, perhaps more than any other, depends on such wild genetic resources. Indeed, a crucial piece of the technology that enables scientists and industry to easily multiply strands of DNA—and thereby create useful commercial products—derives from the bacterium *Thermus aquaticus*, first discovered in a hot spring in Yellowstone National Park.

The value of these biodiversity goods is enormous, but even so it is just a fraction of the value of the ecosystem services on which human life depends, such as waste assimilation, climate regulation, water supply and regulation, erosion control and sediment retention, soil formation, waste treatment, and pollination. Ecosystem services, however, are largely outside the financial markets and therefore are ignored or undervalued. By one rough estimate the value of ecosystem services for the entire biosphere is $33 trillion, nearly double the global gross national product.

When most people think about biodiversity, however, they think not about ecosystems and their services but rather about species. Yet scientists still don't know how many species share the planet with us. Estimates vary by an order of magnitude. A conservative guess is roughly 14 million species, only one-eighth supports far more species than previously believed, from tremendous numbers of beetles living in the canopy of tropical trees to bacteria inhabiting rocks more than a mile beneath the earth's surface.

Individual species, like the pupfishes of the desert Southwest, form threads in the lustrous ecological tapestry of the United States. Further examination reveals a dense weave of thousands of species, many found nowhere else. Together these threads spell out superlatives: tallest, largest, oldest. Topping out at more than 360 feet in height, northern California's redwoods (*Sequoia sempervivens*) are the tallest trees in the world. Their close relatives the giant sequoias (*Sequoiadendron giganteum*) rank among the most massive living things on Earth, and bristlecone pines (*Pinus longaeva*), overlooking the Owens Valley near the summit of eastern California's White Mountains, are the world's oldest living trees, some dating back nearly 5,000 years.

The difference between two species can be visually obvious, as with the Devils Hole and Owens pupfish, or so subtle that only sophisticated molecular techniques can reveal the distinctions. Nonetheless, scientists have documented more than 200,000 species from the United States, and the true number of species living here is probably at least double this figure. By any measure, the United States is home to an exceptionally diverse flora and fauna. On a global scale the nation is particularly noteworthy for certain groups of organisms, including salamanders, coniferous plants, and freshwater fishes, turtles, mussels, snails, and crayfishes. The United States harbors nearly 16,000 species of vascular plants, about 9% of the world's total mammal species, and about 10% of known freshwater fishes worldwide.

This wealth of life owes a great deal to the nation's size and location. While covering only about 6% of the earth's total land area, the United States spans nearly a third of the globe, extending more than 120 degrees of longitude from eastern Maine to the tip of the Aleutian chain, and 50 degrees in latitude from Point Barrow above the Arctic Circle to the southern tip of Hawaii below the tropic of Cancer. Together with this expanse of terrain comes a variety of topographic features and climates, from Death Valley to Mt. McKinley. This range of climates has given rise to a wide array of ecological types, from tundra and subarctic conifer forests called taiga, to deserts, prairie, boreal forest, deciduous forests, temperate rain forest, and even tropical rain forests.

While still far from complete, the process of documenting the nation's ecological diversity suggests that the United States is also extraordinary from an ecological perspective. For example, of the 14 biome types worldwide that represent major ecosystem groups, the United States contains 12, more than any other country. Three biomes—temperate broadleaf forests, temperate grasslands, and mixed mountain systems—are particularly well represented: At least 10% of their area occurs in the United States. Around the world and on a more detailed scale within the United States, ecologists have also identified relatively large areas, known as ecoregions, that in ecological terms function more or less as a unit. With 21 of 28 globally defined ecoregions, the United States is also the most diverse country in the world from an ecoregional perspective.

On a much finer ecological scale, natural heritage ecologists have identified more than 4,500 distinct vegetation communities in the United States. This figure is likely to grow as additional inventory and classification work proceeds, and we can project that on the order to 7,000 to 9,000 natural and seminatural vegetation associations ultimately will be documented from the United States.

Notes & Questions

1. What does the term biodiversity mean? Is the term biodiversity really just a synonym for "life on earth?" For a chart listing several other similar, though not identical, definitions, and an analysis of the conceptual problems presented by the idea of biological diversity, see Dan L. Perlman & Glenn Adelson, Biodiversity: Exploring Values and Priorities in Conservation (1997).

2. The different types of biodiversity can be viewed along a continuum from the characteristics of individual organisms to the features of vast tracts of land. The same term "biodiversity" is used in each instance because the extent of the biological resources is important at the micro level, the macro level, and everywhere in-between.

Genetic diversity tends to be treated as an issue that is distinct from the other kinds of biodiversity. The value of such diversity is its avoidance of the negative adaptive and reproductive consequences of limiting a species to a population of genetically similar individuals. Much variation in genes requires biological and chemical analysis to detect. Genetic diversity may also be seen in variation among individuals in anatomy and behavior. The genetic differences in Douglas fir trees, for example, are evident in their response to cold and to moisture. Most discussions of genetic diversity occur in either of two contexts: the preservation of seeds for agricultural purposes, and the prevention of inbreeding among endangered animals. Contemporary farming practices often rely upon genetically uniform crops, which present a threat of widespread crop failure if those crops cannot combat a certain pest or disease. Consequently, many environmentalists and agricultural reformers promote seed banks, variation in crops, and other means of achieving greater genetic diversity in agriculture. The importance of genetic diversity in preventing extinction is revealed in the extent of the southern California habitat of the arroyo toad that needs to be protected for the species to survive. "Populations on the periphery of the species range or in atypical ecological environments are important for maintaining the genetic diversity of the species which could be essential to evolutionary adaptation to changing climatic and environmental conditions." The protection of the arroyo toad thus required 22 discrete habitat units totaling over 182,000 acres. *See* Final Designation of Critical Habitat for the Arroyo Toad, 66 Fed. Reg. 9414, 9419–26 (2001). A similar challenge is presented by the genetic similarity of all cheetahs: "The fastest animal on land, an apparent model of evolutionary fitness, is also as inbred as the average lab mouse." Richard Conniff, *Cheetahs: Ghosts of the Grasslands*, Nat'l Geo., Dec. 1999, at 20.

Species diversity is the most familiar form of biodiversity. It is the only kind of biodiversity protected by its own federal law—the Endangered Species Act (ESA)—and by international treaties such as the Convention on the International Trade in Endangered Species (CITES). Yet it is surprisingly difficult to identify what constitutes a "species." Why are there eight distinct species of bears: Asiatic black bears, American black bears, brown bears, the giant panda, polar bears, sloth bears, spectacled bears, and sun bears? Why are the Siberian tiger, the South China tiger, the Indo–Chinese tiger, and the Bengal tiger viewed as members of the same species (albeit different subspecies)? Why was it not until 2001 that scientists decided that African elephants actually constitute two distinct species, African forest elephants and African savannah elephants? *See* Alfred L. Roca et al., *Genetic Evidence for Two Species of Elephant in Africa*, 293 Science 1473 (2001), Why not treat all tigers and leopards and jaguars as members of a single species of large, orange, carnivorous cats? Which characteristics should be relevant in determining the identity of a species? While there is no definitive answer to such questions, the most common understanding of what constitutes a species is attributed to Harvard evolutionary biologist Ernst Mayr: "A species is a reproductive community of populations, repro-ductively isolated from other populations, that occupies a specific niche in nature." Ernst Mayr, The Growth of Biological Thought 273 (1982). Mayr's explanation does not solve the definitional problem, though, as evidenced by the struggle to decide what constitutes a "species" for purposes of the ESA, recounted below at pages 122–29.

Ecosystem diversity has gained increasing attention in recent years. That attention is reflected in the countless scientific conferences examining many kinds of ecosystems, the popular concern about specific types of ecosystems such as wetlands and rainforests, and the attention that ecosys-tems have received pursuant to the collection of state and federal laws considered in Part III. There are still few laws that are specifically designed to protect a wide collection of different kinds of ecosystems, though many environmentalists favor the enactment of something like an "Endangered Ecosystem Act" modeled on the existing ESA. *See, e.g.*, Julie B. Bloch, *Preserving Biological Diversity in the United States: The Case for Moving to an Ecosystems Approach to Protect the Nation's Biological Wealth*, 10 Pace Envtl. L. Rev. 175, 217–22 (1992) (advocating a federal "Ecosystems Protection Act").

Yet the idea of an "ecosystem" is even more difficult to define than a "species." The general idea is of a community of organisms found in a particular geographical area. That community is dynamic and its borders cannot always be ascertained. Nonetheless, a variety of types of ecosystems have been recognized. For example, "[a]n ecosystem can be a vegetation type, a plant association, a natural community, or a habitat defined by floristics, structure, age, geography, condition, or other ecologically relevant factors. Thus, virgin and old-growth forests, pitcher plant (*Sarracenia* spp.) bogs, ungrazed sagebrush steppe, wetlands, (general or specific types), Midwestern oak (*Quercus* spp.) savanna, vernal pools, free-flowing rivers, and seagrass meadows are ecosystem types." Reed F. Noss, Edward T. LaRoe III & J. Michael Scott, *Endangered Ecosystems of the United States: A Preliminary Assessment of Loss and Degradation*, 28 Biological Rep. 1, 3 (1995), *available at* <http://biology.usgs.gov/pubs/ecosys.htm>.

Landscape diversity refers to the collection of different kinds of ecosys-tems. Many areas contain multiple ecosystems, such as grasslands, wet-lands and forests. The diversity of ecosystems, and the resulting diverse landscapes, becomes visible only from a regional perspective. The next category—ecoregions—describes "large landscapes that can be distinguished from other regions on the basis of climate, physiography, soils, species composition patterns (biogeography) and other variables." NOSS & COOPERRIDER, *supra*, at 11. Finally, a "biome" is a "major regional community of plants and animals with similar life forms and environmental conditions." M. Lynne Corn, *Ecosystems, Biomes, and Watersheds: Definitions and Use*, CRS Rep.No. 93–655 ENR, at 4 (1993). The fourteen biomes of the world are tropical humid forests, subtropical/temperate rain forests/woodlands, temperate needleleaf forests/woodlands, tropical dry forests/woodlands, temperate broadleaf forests, evergreen sclerophyllous forests, warm deserts/semideserts, tropical grasslands/savannas, temperate grasslands, mixed island systems, tundra communities, mixed mountain systems, cold-winter deserts, and lake systems. *See* Mark T. Bryer et al., *More Than the Sum of the Parts: Diversity and Status of Ecological Systems*, in PRECIOUS HERITAGE, *supra*, at 206 (providing a map of the world's biomes).

Virginia Morell, *New Mammals Discovered by Biology's New Explorers*

273 SCIENCE 1491 (1996).

Last July, evolutionary biologist James L. Patton took a brief reconnaissance trip to Colombia's central Andes, scouting a site for a student's doctoral thesis project. Just 2 weeks later, he rode out of the hills carrying six new species of mammals in his saddle bags: four mice, a shrew, and a marsupial. Six species in 2 weeks may be a record haul for mammalogists in the late 20th century. But although the biggest boom in biological exploration ended decades ago, there is now a renaissance in species discovery, not just of insects and microbes, but also of humans' closest relatives, mammals. "Because we're mammals, you'd think that we'd already know everything there is to know about other mammals." says Patton, a professor at the University of California, Berkeley. "But we don't."

In the last decade, partly because of a new round of exploratory field surveys, scientists have fumed up a surprising variety of new mammals, including a deer species, a wild ox, 10 new species of primates, several bats, a new genus of antelope, and several genera and species of rodents. When all of these new creatures are officially named and described, researchers estimate that the number of known mammals will jump by at least 15%. And the pace of discovery shows no signs of slowing. "Right now we're at a little more than 4600 mammalian species," says Lawrence R. Heaney, an evolutionary biogeographer at Chicago's Field Museum, who recently discovered 11 new mammals in the Philippine Islands. "But I think that number will ultimately go up to around 8000."

All that is more than just taxonomic scorekeeping. With each new mammal comes a set of associated organisms—parasites and pathogens— and new data for research in biogeography, evolution, and conservation. In particular, scientists say, the new species are giving them a far better understanding of mammalian diversity and distribution patterns. At the same time, each discovery is a victory in a race to get a relatively complete picture of mammalian diversity before it is swept away by habitat loss.

The new mammals have come to light in several ways. Some have emerged from detailed genetic analyses that split apart species once lumped together; others have been "found" in museum collections. One new species, the wild ox, was discovered in a Vietnamese marketplace—or at least its horns were; scientists have yet to see the animal in the wild. But many of the new mammals are "really new" species, as Patton calls them, which scientists have never seen before in any form. Most of these were found the old-fashioned way: on lengthy biological surveys, usually in the farther reaches of the globe. "The tropics are still so poorly known, even for mammals," says Heaney, "that just about anywhere you go, you'll find something new."

For example, the combination of a remote destination and a lengthy stay paid off handsomely for Patton in 1991, when he and a team of researchers from the Instituto Nacional de Pesquisas de Amazonia in Manaus, Brazil, spent an entire year surveying mammals along the Rio Jurua in the Amazon basin. Seven of the 52 species they collected were new. And one, a spiny mouse (recently named *Scolomys juaraense*), had no known relatives in this region: Mice of this genus previously had been known only from the Andean foothills in Ecuador, 1500 kilometers away. "It just shows how phenomenally little we know about mammalian geographical distributions," says Patton.

Similarly, Steven M. Goodman's survey of Madagascar has unveiled an unexpectedly diverse set of rodents. Goodman, a field biologist with the Field Museum, helped launch a Malagasy biodiversity survey with the World Wildlife Fund in 1991. Since then, he and his team have discovered several as-yet-undescribed species, including two new genera of rodents in the island's endemic subfamily Nesomyinae. In time, these finds may help "paint a whole new picture of how and when Madagascar was colonized" by rodents, he says, and give scientists a glimpse of the island's complicated history of biological invasions and radiations.

Even in the 20th century, uncovering new species, particularly tiny shrews and tenrecs (insectivores found only in Madagascar), is something of an art, researchers say. Goodman's technique is to bury 15–liter buckets in the ground, then wait for the littlest of the mammals to tumble in. "There's no other way to get these 3 to 4–gram tenrecs," each about the size of two grapes, he says. In South America, Patton and the Field Museum's curator emeritus, Philip Hershkovitz, place their traps in places such as rocky outcrops that they suspect may serve as microhabitats for more reclusive mammals. "I smell them out," says Hershkovitz, who has discovered two new genera and 16 new species of field mice in Brazil's Cerrado grasslands in the last 6 years.

Other new species come to light courtesy of local people; that was the case for several of the six new species of Brazilian primates (tamarins, marmosets, and a capuchin monkey) found in the last 6 years, says Russell A. Mittermeier, a primatologist and president of Conservation International in Washington, D.C. Mittermeier predicts that 10 more primate species will be found in the next decade.

Finding a new mammal has secondary benefits too, such as leading biologists to new parasites and diseases infecting the mammalian host. "For every new species of mammal we find, we also discover a whole community of other organisms associated with it," explains mammalian systematist Terry L. Yates of the University of New Mexico. And because mammals are close to human hearts, the unveiling of a new mammal can help rally efforts to preserve an entire area, says Mittermeier. For example, government officials and residents of Camiguin Island in the Philippines invited Heaney and his team to search for new mammals in 1995. They fumed up two new rodents, one a fluffy, long-haired moss mouse, the other resembling a deer mouse. Both are as yet undescribed, but already the government has recommended setting aside the island's remaining forests.

Such efforts aren't a moment too soon, says Heaney, for these mice, like many of the new mammals, are already on "the red list"—highly endangered, chiefly because of habitat loss. Even as scientists reach out to identify them, the world's mammals are vanishing.

Notes & Questions

1. We continue to discover new species at a rapid rate. Several new species of mammals have been found in the years since this report. Steven Goodman, the same Field Museum scientist who found the new rodents in Madagascar described above, has since helped discover three new species of mouse lemurs there. An expedition in China's Xinjiang province near the Tibetan mountains identified what may be a new species of camel that survives on salt water that bubbles up in a desert that had been used for Chinese nuclear weapons testing.

Vietnam has hosted some of the most remarkable finds in recent years, including the Vu Quang ox—also known as the soala—that was known only to villagers living near the mountainous rainforests in northern Vietnam until a British biologist made the ox the largest mammal discovered in nearly a century. The ox was not seen in the wild until after the *Science* article was written in 1996. It and other species in the area had remained hidden in part because war and trade embargoes prevented scientists from exploring the area. *See* Eugene Linden, *Ancient Creatures in a Lost World: In An Isolated Region that Divides Vietnam and Laos, Scientists Find a Trove of New Species*, Time, June 20, 1994, at 52.

Other species are discovered far more frequently than mammals. Entomologists are always finding new insects, such as the hundreds of beetles that Smithsonian Institution scientist Terry Erwin found in a single tree in the Panamanian rain forest. Nearly 1,000 new species of mollusks have been identified in the Florida Keys since 1995. In 1994, biologists found a tree that had been thought to have been extinct for 60 million years in a canyon in Wollemi National Park north of Sydney. Nor are such discoveries limited to remote, uninhabited parts of the world. Three new species of spiders were found by collectors in a forest preserve outside of Chicago in 2000. A new species of ant was discovered in a potted plant in a Washington, D.C. office. Besides these individual species, entirely new ecosystems continue to be discovered, especially in the virgin forests of the Amazon.

2.   The ongoing discovery of new species of wildlife and plants presents obvious challenges to any estimate of the total number of species in the world. Scientists believe that "[t]he majority of species on Earth have yet to be discovered," guessing that we have identified as few as ten percent of the species that live throughout the world. ANDREW BEATTIE & PAUL EHRILICH, WILD SOLUTIONS: HOW BIODIVERSITY IS MONEY IN THE BANK 9 (2001). Perhaps it should be surprising, then, that estimates regarding the amount of biodiversity vary greatly. Medieval scientists assumed that the total number of species in the world was constrained by the capacity of Noah's ark. Today's scientists guess that the total number of species ranges anywhere from one million to 100 million. The proper methodology for calculating the number of species is often contested as scientists disagree about how to extrapolate from the relatively small number of known species to the admittedly unknown total number of species. The methodology and contours of that debate are well summarized in Nigel E. Stork, *The Magnitude of Global Biodiversity and Its Decline*, in THE LIVING PLANET IN CRISIS 10–21 (1999); WORLD CONSERVATION MONITORING CENTRE, *supra*, at 17–38; NORMAN MYERS, THE SINKING ARK (1979). It is important to remember, though, that "biodiversity is not just a numbers game," and that "[w]hen we consider species richness at any scale smaller than the biosphere, quality is more important than quantity." NOSS & COOPERRIDER, *supra*, at 4.

3.   Whatever number is accurate, it is apparent that biodiversity is not distributed evenly throughout the world. On a local level, the number of species increases as the area of land increases. For example, if ten species live on one acre of land, then 20 species would be expected on 32 acres of land. But the amount of land is not the sole determiner of the extent of biodiversity. "Life does not lie evenly across the landscape. For a variety of reasons—evolution, geography, climate, or historical accident, to name a few—some places harbor more species than others." Stephen J. Chaplin, *The Geography of Imperilment: Targeting Conservation Toward Critical Biodiversity Areas*, in PRECIOUS HERITAGE, *supra*, at 187. Maps of the homes of endangered species in the United States display a concentration along the Pacific Coast, Florida, and the Appalachians, with relatively few rare species in the northeast or the central Great Plains. *See id.* at 162–83 (illustrating the distribution of biodiversity nationwide according to several different measures). The tendency of biodiversity to concentrate in certain areas has led conservationists to focus on so-called "hot spots" where the most biodiversity is found. One recent study identified six such hot spots in the United States:

- *Hawaii*, whose isolation has resulted in a remarkable collection of plant and animal life, including many striking forest birds;

- *The southern Appalachians*, where rivers teem with unique fish and mussels;

- *The San Francisco Bay area*, which features many endemic plants and vernal pool invertebrates;

- *Coastal and interior southern California*, where an array of birds, fish, plants, and insects—including the Delhi Sands Flower–Loving Fly—live in coastal sage scrub and grassland ecosystems;

- *The Florida panhandle*, whose woodlands, bays and rivers host an exceptional array of reptiles and amphibians; and

- *The Death Valley region*, where unusual fish, snails, and plants survive amidst the harsh desert environment.

Id. at 188–99. These hotspots only begin to show how much biodiversity exists close to home. As one recent study concluded, "By any measure, the United States is home to an exceptionally diverse flora and fauna. On a global scale the nation is particularly noteworthy for certain groups of organisms, including salamanders, coniferous plants, and freshwater fishes, turtles, mussels, snails, and crayfishes." Jonathan S. Adams, Bruce A. Stein & Lynn S. Kutner, *Biodiversity: Our Precious Heritage* 9, in PRECIOUS HERITAGE, *supra*, at 9; *see also* Bruce A. Stein et al., *A Remarkable Array: Species Diversity in the United States*, in PRECIOUS HERITAGE, *supra*, at 55–92 (detailing the number and kinds of species that live in the United States).

Of course, the United States is only one of a number of places blessed with abundant biodiversity. Russell Miettermeier, whose work for Conservation International is discussed in the *Science* article, has written a book with two of his colleagues documenting biological hotspots around the world. They describe 25 of those places, including the tropical Andes, the

Caribbean, Madagascar and Indian Ocean islands, the Caucasus, the mountains of south-central China, and southwest Australia. *See* RUSSELL A. MITTERMEIER, NORMAN MYERS & CRISTINA GOETTSCH MITTERMEIER, HOTSPOTS: EARTH'S BIOLOGICALLY RICHEST AND MOST ENDANGERED TERRESTRIAL ECOREGIONS (1999). The importance of such hotspots is demonstrated by such statistics as the occurrence of 70% of all vascular plants within the 1.44% of the earth's surface occupied by the hotspots; and the presence of 35.5% of all bird, mammal, reptile and amphibian species only in those hotspots. *Id.* at 34, 37. For another list of hotspots, and a proposed Vital Ecosystems Preservation Act to protect them, *see* John Charles Kunich, *Preserving the Womb of Unknown Species with Hotspots Legislation*, 52 HASTINGS L.J. 1149, 1209–50 (2001).

B.   WHY WE CARE ABOUT BIODIVERSITY

Testimony of Dr. Michael Grever
Medicinal Uses of Plants; Protection for Plants Under the Endangered Species Act: Hearing Before the Subcommittee on Environment and Natural Resources of the House Committee on Merchant Marine and Fisheries

United States House of Representatives, 1993.
103d Cong., 1st Sess., pp. 20–32.

I am Dr. Michael Grever, Associate Director for the Developmental Therapeutics Program (DTP) of the National Cancer Institutes's (NCI) Division of Cancer Treatment, at the National Institutes of Health (NIH).... Thank you for the opportunity to appear before you today to discuss the NCI's efforts to locate and develop medicinal compounds from plants to combat cancer and HIV (human immunodeficiency virus) infection and its sequelae. NCI is exploring and supporting a broad spectrum of ways to combat these two diseases, from prevention and diagnosis through treatment to rehabilitation and psychosocial care of patients. Drug development from natural products is just one avenue of emphasis; but until we can completely prevent cancer or HIV infection from occurring, it is an extremely important part of our effort to develop and design more effective and less toxic treatments.

In 1993, over 1 million new cases of cancer will be diagnosed in the United States and about 526,000 people will die of the disease. Since cancer incidence increases with age, most cases occur in adults at mid-life or older. There has been a steady rise in the cancer mortality rate in the United

States in the last 50 years, with the major causes of this increase being lung, prostate, and breast cancers. The impact of cancer in general on minority and underserved populations is disproportionately great.

AIDS (acquired immune deficiency syndrome) was recognized as a distinctive syndrome over ten years ago and since then NCI has been involved in multiple disciplines of AIDS research. AIDS is not the primary mission of NCI; however, NCI leads the NIH's efforts in pediatric AIDS clinical studies, making advances in the identification and evaluation of potential therapies for HIV-infected children. Similarly, NCI heads efforts to develop therapies for HIV-associated malignancies, and has developed an extensive and comprehensive program to design and develop anti-HIV drug therapies. Following the discovery of the antiviral action of AZT (azidothy-midine), ddi (dideoxyinosine), and ddc (dideoxycytidine) in the intramural NCI in the mid–1980's, the cancer drug screen was adapted for anti-HIV drug screening.

For our natural products drug screening effort to be at all successful, we must have available to us a multitude of species to study, and preservation of the species is critical to this effort. Global consumption patterns, perverse policy incentives and population pressures threaten biodiversity worldwide. In the countryside, exploitative resource management practices deplete soil and contaminate water supplies; deforestation for farming, pasture and building material leads to erosion and heavy flooding. The resulting disappearance of natural habitats has profound economic, environmental and scientific consequences. Among the ultimate consequences will be a loss of raw materials from which medicinal products might be developed.

BACKGROUND

Throughout the ages humans have looked to nature as a source of medicines for the treatment of a wide variety of diseases. Plants have formed the basis for sophisticated systems of traditional medicine, which have been in existence for thousands of years throughout the globe. Microorganisms and marine organisms have, however, played lesser roles in such traditional systems.

Natural products also play an essential role in the health care systems of developed countries, in providing new types of biologically active substances that either cannot be made chemically or would not have been conceived by chemists. Well-known examples of plant-derived medicinal agents include the antimalarial drug quinine, the analgesics codeine and morphine, the tranquilizer reserpine, and the cardiac glycoside, digitalis. The role of microbial fermentations has been predominant in the development of antibiotics, with over 8,500 substances isolated from microbial sources, and close to 100,000 prepared by chemical modification of the native material. Well-known classes of antibiotics are the penicillins, cephalosporins, and tetracyclines, while other microbial products include immunosuppressive, antiparasitic, and antifungal agents. Until the development of SCUBA (Self–Contained Underwater Breathing Apparatus), the exploration of the marine environment was virtually impossible, with the result that few marine natural products of medicinal value have been developed to date. That, however, is changing rapidly, and more thorough investigation of this area is yielding an increasing number of novel active substances. . . .

RECENT NCI DISCOVERIES

A total of 54,000 extracts, derived from all natural product sources, have been submitted for anti-HIV screening since about 1986. To date, over 35,000 plant samples have been collected by the NCI contractors, and over 25,000 have been extracted to yield more than 50,000 plant extracts. Over 25,000 of these plant extracts have been tested in the anti-HIV screen, and about 2,700 have exhibited some intro activity; of these, close to 2,400 are aqueous extracts, and in the majority of cases the activity has been attributed to the presence of ubiquitous types of chemicals, such as polysaccharides and tannins. Such compounds are not a current NCI focus for drug development and typically are eliminated early in the discovery process. Therefore, the number of extracts undergoing active investigation is much smaller.

A number of novel in vitro active anti-HIV agents have been isolated and selected for preclinical development. The dimeric alkaloid, michellamine B, has been isolated from the leaves of a tropical vine collected in the rain forest regions of southwestern Cameroon. Michellamine B shows in vitro activity against both the HIV–1 and HIV–2, and is in advanced preclinical development. Preliminary surveys of the occurrence and abundance of the species, as well as cultivation experiments, have been carried out by Missouri Botanical Garden through its contract with the NCI. Surveys thus far indicate that its range and abundance are very limited, but fallen leaves collected from the forest floor have been shown to contain reasonable quantities of michellamine B; the collection of these leaves has obviated the large-scale harvest of fresh leaves, and avoided possible endangerment of the wild species. Fallen leaf collections will provide sufficient michellamine B to complete preclinical studies, but the NCI is proceeding with feasibility studies of the cultivation of the plant through contract mechanisms. The collections and cultivation experiments are being performed with the full participation of Cameroon authorities and scientists, as well as through close collaboration with the World Wide Fund for Nature, which is coordinating conservation projects in the Korup region of Cameroon. Thus far, no other related species have shown significant anti-HIV activity.

Calanolide A is a novel coumarin isolated from the leaves and twigs of a tree collected in the rain forest regions of Sarawak, Malaysia. Calanolide A shows potent intro activity against HIV–1 and several resistant strains of the virus, but not against HIV–2, and is in early preclinical development. Recollections of plant material of the original plant species from the same general location have shown a range of test results varying from reasonable activity to total lack of activity. It is apparent that the production of calanolide A is dependent on various factors, possibly including the immediate growth environment and the time of harvest. Thus far, calanolide A has not been detected in any of the recollections, and the original source tree cannot be located. Careful taxonomic and chemotaxonomic studies of this species are being performed by the UIC, under contract to and in collaboration with the NCI and scientists from Sarawak. A survey of related species has shown that the latex of another species collected in the same region yields the related compound, costatolide, which has significant in vitro anti-HIV activity, though being somewhat less active than calanolide A. Costatolide and a derivative have also been approved for preclinical development. The latex contains high yields of costatolide, and would be an excellent renewable source of the compound, should it advance to clinical development. In addition, the synthesis of calanolide A has recently been reported.

A novel chemical compound, conocurvone, has been isolated from a plant species endemic to Western Australia; this plant was originally collected for the NCI program by the USDA in 1981. Conocurvone exhibits potent in vitro activity against HIV–1 and is in early preclinical development. Conocurvone has been synthesized from a simpler chemical which can also be isolated from the plant, and in addition, other simpler analogs have been synthesized and shown to possess equivalent in vitro anti-HIV activity. The development of conocurvone or related compounds will be undertaken in close collaboration with Australian scientists, and surveys of the occurrence and abundance of the source plant and related species are being carried out by the Western Australian Department of Conservation and Land Management.

Another potential anti-HIV agent, prostratin, has been isolated from the stemwood of a Western Samoan tree. This tree is used in Western Samoa for the treatment of a variety of diseases, including yellow fever, and an extract of the stemwood was provided by Dr. Paul Cox of Brigham Young University. The prostratin belongs to the phorbol class of compounds which frequently exhibit significant tumor promoting properties, it does not appear to be associated with tumor promotion, and has been selected for early preclinical development.

Of the approximately 30,000 extracts tested so far in the intro human cancer cell line screen, which started after the anti-HIV screen, a very small percentage (< 1.0 percent) have shown some degree of selective cytotoxicity. Interesting, novel patterns of differential cytotoxicity have been observed, and while some have been associated with known classes of compounds, others appear to be new leads which are being investigated further. Two natural products currently approved for preclinical development are halomon, isolated from a red algae collected in the Philippines, and halichondrin B, isolated from a species of marine sponge found in the western pacific ocean. The procurement of materials for future development is underway. . . .

Notes & Questions

1. Scientists continue to examine numerous plants as they search for a cure for AIDS. A University of California, Irvine scientist has identified a

Bolivian plant that contains a chemical that prevents the AIDS virus from reproducing and infecting healthy cells, though a decade of testing may be needed to determine whether the chemical can actually be used against AIDS in humans. The Centers for Disease Prevention and Control report that "[h]erbs have been used extensively in hopes of improving immune response and reducing symptoms. Aloe vera, St. Johnswort, echinacea, licorice, and ginseng are just a few of the herbs used to treat HIV/AIDS." CDC, *Fact Sheet: HIV/AIDS and Alternative Therapies* <http://www.aegis.com/pubs/cdc_fact_sheets/1994/cdc94033.html>. To date, however, the cure for AIDS remains undiscovered.

Besides helping in the fight against AIDS, plants and animals are an important source of drugs and other medical treatments. Numerous exotic plants have yielded live-saving drugs. "Bark from the white willow gave us salicin, an ancient version of aspirin; the Grecian foxglove provided digoxin, a cardiac medication; bear bile is the origin of ursodiol, a gallstone dissolver; deadly nightshade led to atropine, an eye dilator and anti-inflammatory; the velvet bean produced L-dopa, a treatment for Parkinson's disease; and everyone knows the story of penicillin, the bacteria slayer discovered accidentally in a mold." CHARLES C. MANN & MARK L. PLUMMER, NOAH'S CHOICE: THE FUTURE OF ENDANGERED SPECIES 120–21 (1995). Animals, fish, amphibians and insects possess medical value as well. *See* THE ENDANGERED SPECIES COALITION, THE ENDANGERED SPECIES ACT: A COMMITMENT WORTH KEEPING 10 (1995) (citing the medical benefits of bats, crustaceans, mollusks, insects, and snakes); MYERS, *supra*, at 265 (describing how frogs can provide antitoxins and pain killers, an octopus produces a substance that relieves hypertension, and insects secrete substances similar to hormones). "Coral reef products include anti-inflammatory, antiviral, and anti-tumor compounds isolated from a variety of invertebrates." BEATTIE & EHRLICH, *supra*, at 202–204. Traditional medicines are especially dependent upon products derived from plants, animals, and other natural sources. Herbal medicines are becoming increasingly popular in the United States and other developed countries as well. And we are only beginning to study most species to learn if they possess any medical value. *See Endangered Species Act: Washington, DC–Part II: Oversight Hearing Before the Task Force on Endangered Species Act of the Comm. on Resources, House of Representatives*, 104th Cong. 35 (1995) (testimony of Dr. Kevin H. Browngoehl, Biodiversity Committee, National Physicians for the Environment) (stating that "[o]nly 5 percent of our flowering plant species have been studied chemically with any thoroughness to look for their medicinal value"); MANN & PLUMMER, *supra*, at 121–22 (observing that "biologists frequently liken the world's biodiversity to a library in which the vast majority of books has never been read.... Reading the books in the species library once will not be enough.... Each generation will profit from reading them over and over again.").

Biodiversity contributes to human health in other ways as well. A wide variety of animals serve as medical models: sharks rarely develop tumors, armadillos acquire leprosy, the large optic nerves of horseshoe crabs enable the study of human vision, and the failure of bears to lose bone mass might help address osteoporosis. Conversely, the absence of biodiversity can have lethal consequences for humans. A reduction in the predators of the deer mouse in the southwestern United States resulted in a population surge in the mice that caused a respiratory syndrome epidemic that killed 13 people in 1993. For additional information about such medical issues, see BIODIVERSITY AND HUMAN HEALTH (Francesa Grifo & Joshua Rosenthal eds. 1997); and Francesco T. Grifo & Eric Chivian, *The Implications of Biodiversity Loss for Human Health*, in THE LIVING PLANET IN CRISIS: BIODIVERSITY SCIENCE AND POLICY 197–208 (Joel Crancraft & Francesca T. Grifo eds. 1999).

2. Biodiversity offers many benefits beyond medicines and human health. Congress has explained that "species of fish, wildlife, and plants are of aesthetic, ecological, educational, historical, recreational, and scientific value to the Nation and its people." 16 U.S.C. § 1531(a)(3). These are all utilitarian justifications for protecting endangered species of wildlife and plants. The benefits are impressive:

- *Food*—People eat animals and plants, and not much else. This gives tasty or nutritious animals and plants real economic value. Today "the constant infusion of genes from wild plant species adds approximately $1 billion per year to U.S. agricultural production." *Washington ESA Hearing Part II, supra,* at 175 (statement of the National Wildlife Federation). Producing food provides jobs for millions of Americans. If an edible species disappears, so do the food and the jobs they provide. Thus some members of the Pacific Northwest's salmon industry are among the ESA's biggest supporters. Moreover, genetic diversity within species has helped increase the yields of many agricultural crops.

- *Aesthetics, tourism and recreation*—People enjoy visiting, photographing, painting, and just looking at wildlife. The aesthetic value of a beautiful animal or plant often produces a tangible economic value in the form of ecotourism. Grizzly bears attract millions of people to Yellowstone and Glacier National Parks annually. Whales bring visitors to California, Hawaii, and New England. Tourists travel to numerous areas to visit bald eagles. These visits produce substantial economic value. The FWS recently reported that the 76 million Americans who watched, photographed and fed birds and other wildlife in 1991 spent $18.1 billion on those activities. *See* James D. Caudill, U.S. Fish and Wildlife Service, 1991 Economic Impacts of Nonconsumptive Wildlife–Related Recreation 6–7 (1997). Another report calculated that birdwatching alone is a $15 billion dollar business annually. These general economic benefits also result from endangered species in particular. Whooping cranes and other wildlife generate $5 million annually to the economy of the area surrounding the Aransas National Wildlife Refuge in Texas. A rancher in the Texas Hill Country earned $14,000 from groups that came to see the endangered golden-cheeked warbler and black-capped vireo. Developing countries with a wealth of biodiversity benefit from increasing ecotourism, such as the $87 million in revenues that Costa Rica's protected areas account for annually. The aesthetic appeal of many ecosystems and endangered species extends to those who never see such an ecosystem or species in the wild, or who never will.

Biodiversity possesses other utilitarian values, too. Much clothing is made from natural fibers. Bioremediation employs bacteria and other organisms to fight oil spills and remove other forms of pollution. Copper production is facilitated by the use of bacteria to extract metals. The trees, animals and other life that form a natural landscape give many people a sense of place that contributes to their well being. Various plants and animals have served as a blueprint for human efforts to develop robotics, fire and smoke detectors, air conditioning, and energy. Biodiversity provides raw materials for the biotechnology industry. All of these benefits are cataloged in ANDREW BEATTIE & PAUL EHRLICH, WILD SOLUTIONS: HOW BIODIVERSITY IS MONEY IN THE BANK (2001); NATIONAL RESEARCH COUNCIL, PERSPECTIVES ON BIODIVERSITY 43–67 (1999); YVONNE BASKIN, THE WORK OF NATURE: HOW THE DIVERSITY OF LIFE SUSTAINS US (1997).

Yet biodiversity can be a two-edged sword. Is the unknown monkey in the Amazon rainforest the source of the cure for AIDS, or is it the bearer of the next AIDS virus? Species that stray outside of their natural habitats can have deadly consequences for people and biodiversity alike, as demonstrated by the effects of the smallpox virus on Native Americans and by the devastation wrought on native Hawaiian wildlife by goats, sheep, and mosquitos that were introduced (sometimes intentionally, sometimes not) by western settlers. National Geographic recently published the journal of a nature photographer whose assignment at a new Bolivian national park led him to encounter toxic moths, stinging black ants, burrowing maggots, and a sand fly whose bite spread flesh-eating parasites that created a wound requiring surgery and 21 days of intravenous treatment with an antimony compound. Joel Sartore, *Bugging Out*, NAT'L GEO., Mar. 2000, at 24–29. Moreover, not all species or ecosystems are equal in providing benefits to humanity. People depend upon about two dozen species for most of their food, but even considering that another 3,000 plant species may be a source of food, that still leaves many species with no nutritional value for humans. Efforts to identify plants and animals with medicinal uses have identified far more "useless" species than helpful ones. Beauty may be in the eye of the beholder, but if we are willing to designate certain areas as worthy of special protection because of their unique features, why not make such distinctions among plants and animals? Tourists make such choices daily: whatever abstract arguments can be made about their aesthetic appeal, the wildlife of the Everglades National Park attracts far more visitors than most local nature preserves. Charles Mann and Mark Plummer have thus concluded that "biodiversity as a whole has overwhelming utilitarian value, but most individual species do not." MANN & PLUMMER, *supra*, at 133. Indeed, they suggest that "the entire discussion of utilitarian value, though often invoked as a reason to conserve biodiversity, is a red herring." *Id.* at 134. Are they right?

Sandra Postel and Stephen Carpenter, *Freshwater Ecosystem Services*

NATURE'S SERVICES: SOCIETAL DEPENDENCE ON NATURAL ECOSYSTEMS (Gretchen Daily ed. 1997). Pages 195–207.

It is no coincidence that early human civilizations sprang from river valleys and floodplains. Sufficient quantities of freshwater have underpinned the advancement of human societies since their beginning. Today, we rely on the solar-powered hydrological cycle not only for water supplies, but also for a wide range of goods and life-support services, many of which are hidden and easy to take for granted.

Only a small portion of earth's water wealth consists of liquid water that is fresh enough to drink, grow crops, and satisfy other human needs. Of the total volume of water on the planet (an estimated 1,386,000,000 cubic kilometers, or km^3), only 2.5 percent is fresh-and two-thirds of that is locked in glaciers and ice caps. Merely 0.77 percent of all water is held in lakes, rivers, wetlands, underground aquifers, soil pores, plant life, and the atmosphere.

Of particular importance to the sustenance of earth's biological richness is precipitation on land, an estimated 110,000 km^3 per year. This water is made available year after year by the hydrological cycle and constitutes the total terrestrial renewable freshwater supply. Natural systems, such as forests, grasslands, and rivers, as well as many human-dominated landscapes, such as croplands and pasture, depend upon this rainfall and are finely tuned to natural precipitation patterns.

In some sense, this water is infinitely valuable, since without it land-based life as we know it would disappear. In this chapter, however, we focus not on the entire hydrological cycle, but on the benefits to the human enterprise provided by freshwater systems—primarily, rivers, lakes, aquifers, and wetlands. We attempt to estimate the total value of selected goods and services provided by these systems and, where data exist, offer some estimates of marginal values as well.

The benefits provided by freshwater systems fall into three broad categories: (1) the supply of water for drinking, irrigation, and other purposes; (2) the supply of goods other than water, such as fish and waterfowl; and (3) the supply of nonextractive or "instream" benefits, such as recreation, transportation, and flood control....

Water Supply Services

Once precipitation falls on land, it divides into two parts—evapotranspiration (representing the water supply for all nonirrigated vegetation) and runoff (overland flow back toward the sea via rivers, streams, and under ground aquifers). Through their role in the hydrological cycle, rivers, lakes, and underground aquifers provide a renewable source of freshwater for the human economy to tap. They are the principal source of freshwater for irrigation, households, industries, and other uses that require the removal of water from its natural channels.

Human demands for this water have increased rapidly in recent decades as a result of population growth, changes in diet, and higher levels of material consumption: withdrawals or extractions of water from the aquatic environment have more than tripled since 1950. Today, the volume of water removed from rivers, lakes, and aquifers for human activities worldwide totals some 4,430 km^3 per year. Because accessing this water typically requires the construction of dams, reservoirs, canals, groundwater wells, and other infrastructure, there is a direct and tangible economic cost associated with it; this water supply service is not totally free. However, the full value of the service comes to light by considering the cost of replacing natural sources of freshwater with the next best alternative.

Unlike oil, coal, or tin, for which substitutes exist, freshwater is largely nonsubstitutable. The next best alternative is water processed by technological desalination-the removal of salt from seawater, the function performed naturally by the hydrological cycle. Worldwide, desalination accounts for less than 0.1 percent of total water use. It is a highly energy-intensive process and therefore an expensive supply option. The cost of desalination is in the neighborhood of $1–2 per cubic meter ($m^3$)—four to eight times more than the average cost of urban water supplies today, and at least 10–20 times what most farmers currently pay. Not surprisingly, some 60 percent of the world's desalting capacity is in the Persian Gulf, where fossil energy sources are abundant and freshwater is scarce. Through desalination, countries in this region have essentially been turning oil into water to satisfy drinking and other household need. ...

Supply of Goods Other Than Water

In addition to supplying water, aquatic ecosystems provide many other goods of value to the human economy. Among the most important are fish waterfowl, shellfish, and pelts.

The global freshwater fishery harvest offers a lower-bound estimate of the commercial value of freshwater fish. The annual harvest in 1989–91 was about fourteen million tons, and was valued at some $8.2 billion. This figure does not include the values of the distribution economy or other components of the total economic impact of fishing.

Perhaps surprisingly, the value of sport fisheries often exceeds that of commercial fisheries-in some areas by one hundred-fold or more. Sport fishing is a substantial recreational pursuit in the United States. In 1991 thirty-one million anglers fished an average of fourteen days each in the United States. Expenditures—including equipment, travel costs, etc.—totaled about $16 billion. The full economic impacts of freshwater angling, however, are far larger than direct expenditures. These impacts include changes in income or employment resulting from angling, spending on intermediate goods and services by firms that benefit directly from angling and the economies supported by those firms. In the United States alone, the total economic output of freshwater fishing in 1991 was approximately $46 billion.

Waterfowl hunting in the United States in 1991 involved approx. 3 million hunters who, on average, spent about seven days each hunting migratory ducks and geese. Expenditures for these activities totaled $670 million. This figure underestimates the total economic value of waterfowl hunting, however, because it does not include secondary economic impacts.

Although the total global value of fish, waterfowl, and other goods extracted from freshwater systems cannot be estimated from available data, it certainly exceeds $100 billion per year and may be several times that amount. Moreover, the marginal value of these benefits is increasing in many places, as more people desire to spend time and money on these outdoor activities.

A wide variety of human activities threaten to diminish the benefits derived from living resources extracted from aquatic ecosystems. Overexploitation threatens to permanently diminish fish stocks. Toxic pollutants can render fish and other aquatic organisms unsafe to eat or reduce their productivity. Eutrophication, which can be caused by erosion, sewage inputs, or loss of riparian ecosystems, is correlated with undesirable shifts in fish communities. And to the extent that exotic species are introduced to develop sport fisheries, unexpected costs may result—such as collapse of native fish stocks and the spread of disease-that offset the benefits of the new fishery.

Nonextractive or Instream Benefits

Freshwater provides a host of services to humanity without ever leaving its natural channel or the aquatic system of which it is a part. These are the services most easily taken for granted, because they are provided with minimal or no investment or action on our part. They are also the services most rapidly being lost, since water and land management decisions frequently do not adequately value them or take them into account.

Most instream benefits have strong "public good" characteristics that make it difficult to capture their full value in the marketplace. For example, rivers, lakes, and reservoirs can provide environmental and recreational benefits to many people simultaneously (known in the economics lexicon as "nonrivalry in consumption"). It is also frequently difficult or impossible to exclude anyone from enjoying the benefits of public good resources, whether they pay for that enjoyment or not (known as "nonexcludability").

The value of at least some instream services provided by aquatic systems depends on cultural and societal factors, which makes it impossible to derive an estimate of their total global value. Recreational uses, for example, may be valued highly in wealthy countries but very little in poor countries, where people do not have as much free time or money to enjoy leisure activities. By contrast flood-recession farmers, fishers, and pastoralists may value certain instream services more than the rich, because they depend directly on them for their livelihoods. The value placed on protection of habitat for fish, birds, and other wildlife also may vary with the cultural and economic setting in which the aquatic habitat resides. What follows is a discussion of a few of the nonextractive or instream benefits provided by freshwater systems, along with some estimates of their value—either by way of rough global figures, or by regional or local examples.

Pollution Dilution

In late 1994 and early 1995, an estimated forty thousand migratory birds died at a reservoir in central Mexico. Scientists identified the cause to be an extremely high concentration of untreated human sewage in the water body, which allowed botulism bacteria to spread and poison the food eaten by ducks and other migratory waterfowl. During the months when most of the birds died, the reservoir reportedly consisted almost entirely of raw sewage. Given the vast quantities of sewage produced by the world's 5.7 billion people, such incidents might be commonplace were it not for a key environmental service performed by freshwater systems: the dilution of pollutants.

Freshwater remaining in its natural channels helps keep water quality parameters at levels safe for fish, other aquatic organisms, and people. Today, some 1.2 billion people—about one out of every three in the developing world—lack access to safe supplies of drinking water, and 1.7 billion lack adequate sanitation services. As a result, water-borne diseases are primary killers of the world's poorest. The number of deaths due to unsafe water and inadequate sanitation—which include at least 2 million children each year—would be far higher were it not for the dilution of pollution by freshwater systems.

The old adage "Dilution is the solution to pollution" described the basic approach to pollution control up until about 1970, when, in response to pollution episodes like the Cuyahoga River catching fire in the United States, laws began to be passed requiring that cities and industries treat their waste before releasing it into the environment. Large sums were spent to restore and protect water quality. Virtually all countries, however, still depend heavily upon the diluting capacity of natural waters. Even in the OECD countries, domestic wastewater treatment is estimated to cover only about 60 percent of the population. Information for developing countries is sparse, but treatment coverage is certainly far lower. Moreover, few regions control for farm runoff and other dispersed pollution sources that add substantial quantities of sediment, pesticides, and fertilizers to water bodies. Dilution alone is certainly not sufficient to protect water quality or human health where pollution is highly concentrated or toxic or where people lack access to safe drinking water supplies or adequate sanitation. But without the dilution function, things would be much worse.

One way of gauging the value of dilution as an instream service is to estimate what it would cost to remove all nutrients and contaminants from wastewater technologically.... The combined cost of $150 billion/year likely underestimates the total value of the dilution function, because a portion of agricultural drainage water would also require treatment to remove nitrates, pesticides, and other contaminants, a cost we do not attempt to estimate here.

Society already pays some of this price because pollution loads often exceed what nature can absorb, process, or dilute. But were the natural dilution service to be completely absent, the economic costs of keeping water pollution at harmless or tolerable levels would rise greatly. The risk today is that as increasing quantities of water are diverted from rivers and other water bodies to satisfy rising water demands, less water remains instream to provide this important ecosystem service. Decisions to divert water from its natural channels need to take into account the increased treatment costs that may be incurred as a result, as well as the potential costs to downstream water users of lower-quality water.

Transportation

In many parts of the world, inland waterways offer convenient and relatively expensive pathways for the transport of goods from one place to another. One way of valuing this instream service would be to estimate the cost of the next best alternative means of freight transportation in each area where navigation is used, and then to calculate the total cost-savings from navigation—an extremely difficult task since the next best alternative and its cost would vary from place to place. An easier approach is to examine the revenue derived from transportation by freshwater, averaged over all types of goods transported, exclusive of taxes. (Ideally, we should subtract from such figures the cost of maintaining navigation channels in order to arrive at a more accurate value of the ecosystem service, but we do not do that here.) In the United States, such revenues total $360 billion per year, and in Western Europe they total $169 billion per year.

Unfortunately, consistent or reliable figures for transportation revenues are not available for Asia, Africa, or South America. However, the major rivers of these continents are important arteries for commerce. In China, for example, waterways accounted for 9 percent of the cargo shipped in 1988.

Thus, the combined revenue derived from transportation by water in the United States and Western Europe—$529 billion per year—provides a lower-bound estimate of the value of this instream service. The additional value from water transport in other geographic areas, along with the benefit of waterways for human travel (which is not included in these revenue figures), would raise the total value of this important instream service considerably. These transportation benefits are placed at risk by river diversions that reduce flows to levels too low to support navigation, by land-use practices that result in siltation of waterways, and by other activities that impair the use of freshwater systems for shipping.

Recreation

Freshwater systems provide numerous and varied opportunities for recreation—including swimming, sports fishing, kayaking, canoeing, and rafting. Like most other instream benefits, these recreational services have "public good" characteristics that make it difficult to capture their full value in the marketplace. In countries such as the United States, where enjoyment of the outdoors is on the rise, a large group of people benefit from these recreational services, but the total value of their enjoyment is difficult to measure. There is no charge levied or donation made that fully captures their collective willingness to pay....

Instream recreational uses of water also generate substantial additional benefits to local economies in the form of recreation-related expenditures, such as boating, fishing, and camping equipment. One study, for example, found that boaters on a twenty-mile stretch of the Wisconsin River spurred more than $800,000 in sales by local businesses during the summer season. Such sales are a key source of livelihood for small towns and Native American reservations in the western United States. . . .

[A]t least during low-flow periods, the marginal value of water for instream recreational uses appears to be equal to or greater than the marginal value of water used in a substantial portion of irrigated agriculture in the western United States. The key policy message is similar to that for pollution dilution: Were these instream recreational values properly taken into account, fewer diversions for offstream uses would be economically justified. And a corollary: If water markets were able to operate more freely and purchases of water for instream recreational uses were more feasible, water would likely shift out of agriculture to the protection of instream recreational services.

Provision of Habitat

The supply of vital habitat by aquatic ecosystems depends greatly upon the dynamic connection between water and land, physical processes such as water and sediment flows, and a host of biophysical conditions such as water quality, temperature, and food web relationships. Freshwater ecosystems contain abundant life, including 41 percent of the world's known fish species and most of the world's endangered fish species. Decades of large-scale water engineering have disrupted many critical ecosystem functions and processes, with consequences that are just beginning to be recognized.

The provision of habitat in many large river systems, for example, depends critically on the annual flood. Floodplains are not only highly productive biologically, they offer a variety of aquatic habitats, including backwaters, marshes, and lakes. During a flood, many aquatic organisms leave the river channel to make use of these floodplain habitats as spawning, breeding, and nursery grounds. As floodwaters recede, young fish, water-fowl, and other organisms get funneled back into the main channel, along with nutrients and organic matter from the floodplain. In turn, the floodwaters deposit a new supply of sediment that enhances the floodplain's fertility. In this way, so called "flood pulses" provide critical habitat and increase the productivity of both the floodplain and the main river channel. Examples of large river-floodplain ecosystems that are world renowned for their wildlife and other habitat benefits include the Gran Pantanal of the Paraguay River in South America, which alone harbors 600 species of fish, 650 species of birds, and 80 species of mammals; the Sudd swamps on the White Nile in Sudan; and the Okavango River wetlands in Botswana.

In addition, the timing, volume, and quality of water flowing in its natural channel greatly affect the supply of habitat for fish and other aquatic organisms. Migrating fish species, for instance, may require certain minimum flow volumes at particular points in their life cycle. And many species have specific temperature, water quality, and other needs that must be met if they are to survive in a given river system.

The value of natural river, lake, and wetland systems as habitat for fish, waterfowl, and wildlife is even harder to estimate than recreational values, since the beneficiaries and benefits are much less clear and direct. In some cases, these values become visible only when they are lost or destroyed. In the Aral Sea basin in Central Asia, for instance, what was once the world's fourth largest inland lake has lost two-thirds of its volume because of excessive river diversions for irrigated agriculture. Some 20 of the 24 native fish species have disappeared, and the fish catch, which totaled approx. 40,000 tons a year in the 1950s and supported 60,000 jobs, has dropped to zero.

Wetlands have shrunk by 85 percent, which, combined with high levels of agricultural chemical pollution, has greatly reduced waterfowl populations. In the delta of the Syr Dar'ya River—one of the Aral Sea's two major sources of inflow—the number of nesting bird species has dropped from an estimated 173 to 38. This region illustrates vividly how economic and social decline may follow close on the heels of ecological destruction.

In the western United States, the emergence of active water markets combined with growing public interest in preserving fish species, bird populations, and wildlife generally has begun to attach some market values to the critical habitat supplied by aquatic ecosystems. During 1994, there were nineteen reported water transactions in the western United States that had the purpose of securing more water for aquatic habitats, especially rivers and wetlands. A sampling of such transactions during recent years [illustrates that] the value of water for habitat protection in the western United States, as with the value of instream water for recreation, appears to equal or exceed that for some offstream uses, particularly in agriculture.

Option, Bequest, and Existence Values

Because of freshwater's central role in maintaining uniquely beautiful natural areas, critical habitat, or highly valued recreational sites, "non-user" values of water can be substantial. Estimating people's willingness to pay to preserve the option of enjoying a site in the future (option value), to ensure that descendants will be able to enjoy a site (bequest values), or simply to know that a site will continue to exist (existence values) is not easy. These values are important, however, particularly when irreversible decisions are to be made, such as constructing a dam that will flood a beautiful mountain canyon, or channeling through a wetland that will permanently destroy wildlife habitat. According to Colby, "existence, bequest and option values ranging from $40–$80 per year per non-user household have been documented for stream systems in Wyoming, Colorado, and Alaska." It is estimated that the total (user and non-user) benefits of preserving Mono Lake levels amount to about $40 per California household, 80 percent of which is attributed to option, bequest, and existence values.

Notes & Questions

1. Gretchen Daily defines ecosystem services as "the conditions and processes through which natural ecosystems, and the species that make them up, sustain and fulfill human life.'' Gretchen C. Daily, *Introduction: What Are Ecosystem Services?*, in Nature's Services: Societal Dependence on Natural Ecosystems 3 (Gretchen C. Daily ed. 1997). In other words, ecosystems are valuable because they serve *us*—the people who live in those ecosystems. The contributions of freshwater ecosystems are complemented by the services provided by other kinds of ecosystems. Forests prevent erosion that causes flooding, sedimentation that interferes with hydroelectric dams and coral reefs, and the depletion of the ozone layer with its attendant exposure to harmful ultraviolet sun rays. Oceans distribute chemical and biological materials, detoxify or sequester wastes, and host abundant fisheries. Grasslands conserve soils, maintain the genetic library, and maintain the composition of the atmosphere. Soils moderate the climate, support plants, and dispose of dead organic matter. Specific insects pollinate specific plants, so if the diversity of insects declines, the diversity of plant life will decline, too. Natural predators, parasites, and pathogens operate to control pests that threaten agricultural crops. For more examples and analysis of the services that ecosystems provide, see the other chapters in Nature's Services, and the collection of articles contained in 20 Stan. Envtl. L.J. 309 (2001). We explore the topic further in Chapter 7.

2.   The value of ecosystem services is exceedingly difficult to measure. The most obvious measure is the cost of providing the services via alternative means. How expensive would it be to stop the flooding that forests prevent, to replicate the recreational opportunities afforded by wetlands, or to dispose of wastes that the soil naturally processes? Such questions have yielded dramatically different figures when economists have attempted to calculate their answers. The values of using an ecosystem, though, are actually the least difficult to determine. Ecosystems also afford many non-use values, including the intangible benefit that many people experience simply by knowing that the ecosystem exists. Economists have constructed numerous studies for measuring those values, including a contingent valuation approach that asks people how much they would be willing to spend to preserve a resource. Any means of measurement is controversial, yet all agree that the value of ecosystems extends beyond the types of commodities whose economic value the market quantifies. Such challenges to any effort to measure the value of ecosystem services are analyzed in Natural Research Council, *supra*, at 87–115; Dominic Moran & David Pearce, *The Ecomomic Consequences of Biodiversity Loss*, in Cracraft & Grifo, *supra* at 217–29; Lawrence H. Goulder & Donald Kennedy, *Valuing Ecosystem Services: Philosophical Bases and Empirical Methods*, in Daily, *supra*, at 23–47; and Robert Costanza & Carl Folke, *Valuing Ecosystem Services with Efficiency, Fairness, and Sustainability as Goals*, in Daily, *supra*, at 49–68.

3.   Suppose that we could obtain all of these services—plus all of our medicinal, food, clothing, recreational and other utilitarian needs—from a handful of species and ecosystems. Why would we need a *diversity* of species and ecosystems if we could satisfy all of our needs with just a few species and ecosystems? The law continues to struggle with the importance of diversity in a number of contexts, including affirmative action, broadcasting, political parties, language, and agriculture. *See* Jim Chen, *Diversity in a Different Dimension: Evolutionary Theory and Affirmative Action's Destiny*, 59 OHIO ST. L. J. 811, 834–62 (1998). Indeed, "diversity as an affirmative goal of civil society and of government, has gained broad acceptance in America only very recently and only after a very long struggle with the forces favoring ethnic homogeneity, forces that have prevailed in virtually all other societies at all other times." Peter H. Schuck, *The Perceived Values of Diversity, Then and Now*, 22 CARDOZO L. REV. 1915 (2001). The diversity of human cultures throughout the world has received growing attention as globalization and other developments threaten the existence of many native cultures. Several writers have even analogized biodiversity and cultural diversity. *See* NOSS & COOPERRIDER, *supra*, at 13–14; Kieran Suckling, *A House on Fire: Linking the Biological and Linguistic Diversity Crises*, 6 ANIMAL L. 193 (2000). Diversity is usually seen as a virtue in those settings, though the role of the law in achieving diversity is often controversial. There are times, though, where diversity is irrelevant or actually undesirable. Dan Perlman and Glenn Adelson have grouped the importance of the diversity of the features of their conservation biology students as follows:

- "Features in which we seek diversity"—academic background, nationality, home culture, gender, year in college, and leaves of absence.
- "Features in which we ignore diversity"—blood type, hair color, right or left handedness, and musical ability.
- "Features in which we actively avoid diversity"—general intelligence, writing ability, interest in subject, and commitment to class.

DAN L. PERLMAN & GLENN ADELSON, BIODIVERSITY: EXPLORING VALUES AND PRIORITIES IN CONSERVATION 53 (1997). To take another example, consider the infamous response that Senator Roman Hruska offered to those who questioned the abilities of Supreme Court nominee G. Harold Carswell: "Even if he was mediocre, there are a lot of mediocre judges and people and lawyers. They are entitled to a little representation, aren't they, and a little chance?" 116 CONG. REC. 7498 (1970). What makes a diversity of genes, species, and ecosystems more important than a diversity of intelligence among the students in the classroom or the Justices on the United States Supreme Court?

Or suppose that we could obtain all of these services through human efforts rather than through natural means. Would there be any reason for preferring the services provided by ecosystems instead of services that we would provide ourselves?

Bruce Babbitt
Between the Flood and the Rainbow: Our Covenant to Protect the Whole of Creation
2 ANIMAL LAW 1 (1996).

I began 1995 with one of the more memorable events of my lifetime. It took place in the heart of Yellowstone National Park during the first week of January, a time when a layer of deep, pure snow blanketed the first protected landscape in America. But for all its beauty, the last sixty years had rendered this landscape an incomplete ecosystem. By the 1930s, government-paid hunters had systematically eradicated the predator at the top of the food chain: the American grey wolf. I was there on that day, knee deep in the snow, because I had been given the honor of carrying the first wolves back into that landscape. Through the work of conservation laws, I was there to restore the natural cycle—to make Yellowstone complete.

The first wolf was an Alpha female. After I set her down in the transition area, where she would later mate and bear wild pups, I looked through the grate into the green eyes of this magnificent creature within a spectacular landscape. I was profoundly moved by the elevating nature of America's conservation laws—laws with the power to make creation whole. Upon returning to Washington, I witnessed a new Congress wielding a power of a different kind.

Attack on Water, Land, and Creatures

... [M]ore than any of our environmental laws, the act they have most aggressively singled out for elimination—one that made Yellowstone complete—is the Endangered Species Act (ESA). Never mind that this Act is working, having saved ninety-nine percent of all listed species. Never mind that it effectively protects hundreds of plants and animals, from grizzly bears to whooping cranes to greenback cutthroat trout. Never mind that it is doing so while costing each American sixteen cents per year. Although the new Congress may list some species as endangered, they can find absolutely no reason to protect all species in general. Who cares, they ask, if the spotted owl goes extinct? We won't miss it or, for that matter, the Texas blind salamander or the kangaroo rat. That goes double for the fairy shrimp, the burying beetle, the Delhi sands flower-loving fly, and the virgin spine dace. If they get in our way, and humans drive some creatures to extinction, that is just too bad. This is a fairly accurate summary of how the new majority in Congress has expressed its opinion of the ESA.

The Values of Children

Fortunately, there are other Americans who have expressed their opinion on this issue. I recently read an account of a Los Angeles "Eco–Expo," where children were invited to write down their answers to a basic question: "Why save endangered species?" One child, Gabriel, answered, "Because God gave us the animals." Travis and Gina wrote, "Because we love them." A third answered, "Because we'll be lonely without them." Still another wrote, "Because they're a part of our life. If we didn't have them, it would not be a complete world. The Lord put them on the earth to be enjoyed, not destroyed."

In my lifetime I have heard many political, agricultural, scientific, medical, and ecological reasons for saving endangered species. In fact, I have hired biologists and ecologists for just that purpose. All their reasons have to do with providing humans with potential cures for disease, yielding humans new strains of drought-resistant crops, offering humans bioremediation of oil spills, or thousands of other justifications for why species are useful to humans. However, none of their reasons moved me like those of the children. These children are using plain words to express a complex notion that has either been lost, forgotten, or never learned by some members of Congress and, indeed, by many of us. The children expressed the moral and spiritual imperative that there may be a higher purpose inherent in creation, one demanding our respect and our stewardship quite apart from whether a particular species is or ever will be of material use to mankind. They see in creation what our adult political leaders refuse to acknowledge. They express an answer that can be reduced to one word: values.

A Sacred Blue Mountain

I remember when I was their age as a child growing up in a small town in northern Arizona. I learned my religious values through a church that kept silent on our moral obligation to nature. By its silence, the church implicitly sanctioned the prevailing view of the earth as something to be used and disposed of however we saw fit, without any higher obligation. In all the years that I attended services, there was never any reference nor any link to our natural heritage or to the spiritual meaning of the land surrounding us. However, outside that church, I always had a nagging instinct that the vast landscape was somehow sacred and holy. It was connected to me in a sense that my religious training ignored.

At the edge of my home town, a great blue mountain called the San Francisco Peaks soars up out of the desert to a snowy summit, snagging clouds on its crest, changing color with the seasons. It was always a mystical, evocative presence in our daily lives. To me, that mountain, named by Spanish missionaries for Saint Francis, remains a manifestation of the presence of our Creator. That I was not alone in this view was something I had to discover through a very different religion, because the Hopi Indians lived on the opposite side of the blue mountain in small pueblos on the high mesas that stretch away toward the north. It was a young Hopi friend who taught me that the blue mountain was truly a sacred place.

One Sunday morning in June, my friend led me out to the mesa top villages where I watched as the Kachina filed into the plaza, arriving from the snowy heights of the mountain and bringing blessings from another world. Another time, he took me to the ceremonials where the priests of the snake clan chanted for rain and then released live rattlesnakes to carry their prayers to the spirits deep within the earth. Later, I went with my friend to a bubbling spring, deep in the Grand Canyon, lined with pahoes—the prayer feathers—where his ancestors had emerged from another world to populate this earth. By the end of that summer, I came to deeply and irrevocably believe that the land, the blue mountain, and all the plants and animals in the natural world together are a direct reflection of divinity.

Genesis and the Deluge

That awakening made me acutely aware of a poverty amidst my own rich religious tradition. I felt I had to either embrace a borrowed culture, or turn back and have a second look at my own. While priests then, as now, were not too fond of people rummaging about in the Bible to draw their own meanings, I chose to do so, asking, "Is there nothing in our Western, Judeo–Christian tradition that speaks to our natural heritage and the sacredness of that blue mountain? Is there nothing that can connect me to the surrounding Creation?" There are those who argue that there is not. There are those industrial apologists who, when asked about Judeo–Christian values relating to the environment, reply that the material world, including the environment, is just an incidental fact of no significance in the relation between us and our Creator. They cite the first verses of Genesis, concluding that God gave Adam and his descendants the absolute, unqualified right to "subdue" the earth and gave man "dominion over the fish of the sea, over the fowl of the air, and over every living thing that moveth upon the earth."[24] God, they assert, put the earth here for the disposal of man in whatever manner he sees fit. However, if they read a few verses further, they would discover in the account of the Deluge that the Bible conveys a far different message about our relation to God and to the earth. In Genesis, God commanded Noah to take into the ark two by two and seven by seven of every living thing in creation, the clean and the unclean.[25] God did not specify that Noah should limit the ark to two charismatic species, two good for hunting, two species that might provide some cure down the road, and two that draw crowds to the city zoo. He specified the whole of creation. When the waters receded and the dove flew off to dry land, God set all of the creatures free, commanding them to multiply upon the earth.[28] Then, in the words of the covenant with Noah, "when the rainbow appears in the clouds, I will see it and remember the everlasting covenant between me and all living things on earth."[29] We are thus instructed that this everlasting covenant was made to protect the whole of creation, not for the exclusive use and disposition of mankind, but for the purposes of the Creator.

Now, we all know that the commandment to protect creation in all its diversity does not come to us with detailed operating instructions. It is left to us to translate a moral imperative into a way of life and into public policy. Compelled by this ancient command, modern America turned to the national legislature which forged our collective moral imperative into one landmark law—the Endangered Species Act of 1973.

. . . Whenever I confront some of these bills that are routinely introduced—bills sometimes openly written by industrial lobbyists; bills that systematically attempt to eviscerate the Endangered Species Act–I take refuge and inspiration from the simple written answers of those children at the Los Angeles expo. However, I sometimes wonder if children are the

---

24. *Genesis* 1:24.

25. *Genesis* 6:20, 7:2–3.

28. *Genesis* 8:17.

29. *Genesis* 9:12–16.

only ones who express religious values when talking about endangered species. I wonder if anyone else in America is trying to restore an ounce of humility to mankind, reminding our political leaders that the earth is a sacred precinct designed by and for the purpose of the Creator.

I recently got my answer. I read letter after letter from five different religious orders, representing tens of millions of churchgoers, all opposing a House bill to weaken the Endangered Species Act. They opposed it not for technical or scientific or agricultural or medicinal reasons, but for spiritual reasons. I was moved not only by how such diverse faiths could reach so pure an agreement against this bill, but by the common language and terms with which they opposed it—language that echoed the voices of the children. Suddenly, I understood exactly why some members of Congress react with such unrestrained fear and loathing towards the Endangered Species Act. I understood why they tried to ban all those letters from the Congressional Record. I understand why they are so deeply disturbed by the prospect of religious values entering the national debate, because if they heard that command of our Creator—if they truly listened to His instructions to be responsible stewards—then their entire framework of human rationalizations for tearing apart the Act would unravel. Those religious values remain at the heart of the Endangered Species Act. They make themselves manifest through the green eyes of the grey wolf, through the call of the whooping crane, through the splash of the Pacific salmon, and through the voices of America's children. We are living between the flood and the rainbow—between the threats to creation on the one side and God's covenant to protect life on the other. Why should we save endangered species? Let us answer this question with one voice, the voice of the child at that expo, who scrawled her answer at the very bottom of the sheet: "Because we can."

Notes & Questions

1.   The limitations of the utilitarian arguments for preserving endangered species have led many to consider the moral, ethical, and religious arguments for protecting biodiversity. Besides Secretary Babbitt, who was raised Catholic, consider the position of the Evangelical Environmental Network (EEN), an evangelical Protestant group whose recent lobbying and nationwide advertising campaign regarding the ESA gained widespread attention. The EEN believes that "[a] species should be preserved because it has been created by God." Thus the EEN "oppose[s] any Congressional action that would weaken, hamper, reduce or end the protection, recovery and preservation of God's creatures, including their habitats, especially as accomplished under the Endangered Species Act." Likewise, Rabbi David Saperstein testified before a House ESA task force that "[e]very species is sacred."

The story of Noah features prominently in the positions of Secretary Babbitt, EEN and other Christian and Jewish defenses of the ESA. God "did not specify that Noah should limit the ark to two charismatic species, two good for hunting, two species that might provide some cure down the road, and, say, two that draw crowds to the city zoo." Noah invested much time, money and resources in building the ark and collecting all of the species. The resulting covenant between God and Noah "was made to protect the whole of creation, not for the exclusive use and disposition of mankind, but for the purposes of the Creator." Some have even suggested that the story implies that "God was more concerned about preserving animal species than sinful people." Moreover, the flood confronted by Noah has been compared to the flood of people and pollution that threatens biodiversity today. These and more lessons learned from Noah continue to inform wildlife protection thousands of years after the story was recorded.

Those seeking to avoid a duty to protect every endangered species are more likely to turn to other parts of the scriptures. Genesis records that God entrusted man with dominion over the earth, a notoriously difficult passage for environmentalists. *See Gen.* 1:26 ("Then God said ... 'let [man] have dominion over the fish of the sea, over the birds of the air, and over the cattle, over all the earth and over every creeping thing that creeps on the earth' "); *Gen.* 1:28 (commanding man to "fill the earth and subdue it; have dominion over the fish of the sea, over the birds of the air, and over every living thing that moves on the earth"). Thus former Interior Secretary Manuel Lujan cited the dominion command to justify actions that could result in the loss of a species. California Representative William Dannemeyer complained that the ESA "has reversed" God's grant of dominion so that "[a]nimals today are more important than people." Oregon Representative Wes Cooley expressed his belief that "we are going against God" by protecting endangered species, adding that "maybe we might be the higher creatures of God's creation." Likewise, the head of the National Association of Evangelicals has worried that "[t]here is a pantheistic element in all of this," contrary to the biblical mandate.

But many insist that the dominion of which Genesis speaks should not be understood as a license for humans to do whatever they want to the world for whatever purpose they happen to have in mind. The word "dominion" is used elsewhere in the scriptures to refer to a peaceful rule designed to serve those living subject to it. Conversely, those who exercise dominion in a way that serves only their own desires receive harsh criticism. Perhaps a better understanding of the implications of the dominion command of Genesis can be gleaned from three models presented in the biblical account. Dominion as kingship reflects the just, righteous rule that God expected of Israelite kings. Dominion as servanthood imitates the way in which God provides for creation and the kind of rule that Jesus described in the New Testament. Dominion as stewardship posits that God is the owner of creation who has asked us to serve as a trustee responsible for managing it on God's behalf.

Stewardship is the prevalent model. God's status as the owner of creation is confirmed throughout the scriptures and elsewhere: Genesis reports that God created all species, other biblical passages refer to the resulting creation (including wildlife) as belonging to God, and references to all endangered species as "God's creatures" occur frequently in the debate over the ESA. Like other trustees, people should not act to further their own best interests, but instead to best serve the owner. And God—the owner—values each part of creation. God also values the diversity of creation. Moreover, the biblical account anticipates a world that is to thrive in its own right and that will eventually be redeemed. This stewardship model is further supported by the instruction given by God to Adam in Genesis to tend the earth and to "keep" it. The stewards of creation, therefore, must treat the creation in a manner that reflects the value and purpose that God places on the creation.

Even under this understanding, though, the biblical commands conflict with ethical theories that treat people and other species in an identical fashion. Thus the claim of "speciesism"—a discriminatory preference for the human species over all other species—has been levelled against Christian environmentalists and others who acknowledge a higher ethical position for humanity. *See, e.g.*, PETER SINGER, ANIMAL LIBERATION (2d ed. 1990). The balance between human needs and the needs of animals will be set differently by someone seeking to adhere to the biblical account and by someone seeking to treat all species (including humans) alike.

Of course, the story of Noah does not instill an obligation to preserve all species for those who do not credit the authority of the scriptures. Nor is it obvious that the duty suggested by Noah's example (or any other moral, ethical or religious duty for protecting endangered species) must be translated into a legal duty. The books in which the story of Noah appears—the Bible in general and Genesis in particular—contain countless other commands that have never been adopted in statutory law. None of the proponents of an ESA based on Noah have explained why this is one of the commands demanding legal recognition.

Calls to imitate Noah have met with opposition. Several writers have questioned whether the analogy was appropriate, noting that "Noah did not seize anyone's property to build the ark, nor did he tax anyone to finance the operation." Marlo Lewis, Jr., *EPA Would Have Arrested Noah*, WASH. TIMES, Mar. 14, 1996, at A21. Idaho Representative Helen Chenoweth spoke at length against the incorporation of personal beliefs into laws that punish those who do not share those beliefs. *See* 142 CONG. REC. H1002–05 (daily ed. Jan. 31, 1996) (objecting to Secretary Babbitt's speech, characterizing environmentalism as a religion, and concluding that "this religious vision is not shared by every American and no American should be forced to promote a religious vision contrary to their own beliefs"); *see also* 134 CONG. REC. S9759 (daily ed. July 25, 1988) (statement of Sen. Symms) (criticizing the ESA as an effort to change moral attitudes by "the sheer brute force of Government"); The Thoreau Institute, *Fixing the Endangered Species Act*, Feb. 1996 (comparing the ESA to Lenin's efforts to force the evolution of ethics). The sponsors of one of the ESA reform bills admonished EEN to "keep the debate honest and don't use the pulpit to mislead people." In short, not everyone recognizes Noah's dictate to guide federal environmental law.

There are also a number of potential constitutional obstacles to basing a law on a biblical imperative. The establishment clause is sometimes interpreted to preclude statutes enacted for a religious purpose. Conversely, the free exercise clause may override the ESA for individuals whose religious beliefs demand the taking of a particular endangered species. Other laws that are viewed as being based on contested moral arguments have not fared well recently. *See* Romer v. Evans, 517 U.S. 620 (1996). Remember, too, that the government possesses limited ability to determine which living organisms are worthy of legal protection. *See* Akron v. Akron Center for Reproductive Health, Inc., 462 U.S. 416, 443 (1983) (concluding that "a State may not adopt one theory of when life begins" to justify its regulation of abortions); Roe v. Wade, 410 U.S. 113, 159 (1973) (invalidating state abortion law because the state cannot choose among contested views of "life").

2. Would different moral, ethical, or religious perspectives lead to different conclusions about protecting biodiversity? Secretary Babbitt referred to the imperative to protect rare wildlife that flows from the Native American spiritual tradition. As one writer has explained:

> The attitude of Indians toward the natural environment was basically what we would call spiritual or religious, although religion for them was not separated from the rest of life.... The Indians saw themselves as one with nature. All of their traditions agree on this. Nature is the larger whole of which mankind is only a part. People stand within the natural world, not separate from it; and are dependent on it, not dominant over it. All living things are one, and the people are joined with trees, predators and prey, rocks and rain in a vast, powerful, interrelationship.... Because of this deep kinship, Indians accorded to every form of life the right to live, perpetuate its species, and follow the way of its own being as a conscious fellow creature. Animals were treated with the same consideration and respect as human beings.

J. DONALD HUGHES, AMERICAN INDIAN ECOLOGY 14–17 (1983). A recent manifestation of those beliefs was expressed by four northwestern tribes responding to the disappearance of salmon from the Columbia River. The tribes explained that their unwritten laws "begin with the recognition of nature's bounty as a gift from the Creator, that everything in nature has a purpose, and that human society has a need to harmonize itself with the structures and rhythms of nature. When the first salmon comes up the river, the human world stops to honor the returning spirit of the salmon." COLUMBIA RIVER INTERTRIBAL FISH COMMISSION, THE TRIBAL VISION FOR THE FUTURE OF THE COLUMBIA RIVER BASIN & HOW TO ACHIEVE It 2 (1999), *available at* <http://www.critfc.org/legal/vision.pdf>. As with Christian teachings, though, the extent to which Native Americans have acted consistently with these principles has been questioned, as discussed below at pages 115–16.

Animal rights ideas offer another alternative. The premise of animal rights arguments is that each living creature is important not because of its importance to people (as with utilitarian arguments) or because of its importance to God (as with many religious arguments), but because of its intrinsic importance. The physical and mental characteristics that more developed animals share with humans are emphasized in advocating similar legal rights for animals. Such claims have met limited success in a legal system that has long regarded animals and plants as human property or at least unworthy of the legal protections afforded people. The case for extending rights to animals is made in STEVEN M. WISE, RATTLING THE CAGE: TOWARD LEGAL RIGHTS FOR ANIMALS (2000), while the current legal treatment of animals is described in PAMELA D. FRASCH ET AL., ANIMAL LAW (2000).

What other religious or moral arguments would support efforts to prevent species from going extinct? Do any of those arguments suggest that the preservation of a species may give way to other considerations in some circumstances? More generally, is it appropriate to rely upon any such contested theories? Is it possible *not* to rely on such theories?

C. THREATS TO BIODIVERSITY

Determination of Endangered Status for Five Freshwater Mussels and Threatened Status for Two Freshwater Mussels From the Eastern Gulf Slope Drainages of Alabama, Florida, and Georgia

United States Fish and Wildlife Service, 1998.
63 Federal Register 12664.

The fat threeridge, shinyrayed pocketbook, Gulf moccasinshell, Ochlockonee moccasinshell, oval pigtoe, Chipola slabshell, and purple bankclimber are freshwater mussels of the family Unionidae found only in eastern Gulf Slope streams draining the Apalachicolan Region, defined as streams from the Escambia to the Suwannee river systems, and occurring in southeast Alabama, southwest Georgia, and north Florida. The Apalachicolan Region is known for its high level of endemicity, harboring approximately 30 species of endemic (found only in the region) mussels. The Region drains primarily the Coastal Plain Physiographic Province. Only the headwaters of the Flint and Chattahoochee rivers, in the Apalachicola–Chattahoochee–Flint (ACF) River system, occur above the Fall Line in the Piedmont Physiographic Province in west-central Georgia.

The decline of some of the species included in this rule was evident decades ago. The fat threeridge, oval pigtoe, Chipola slabshell, and purple bankclimber were considered rare, but locally abundant, in the 1950's. The Gulf moccasinshell, oval pigtoe, and purple bankclimber were recognized in a list of rare species in 1970, and the fat threeridge was added to the list of regionally rare mussels a year later.

The purple bankclimber

... Freshwater mussel adults are filter-feeders, positioning themselves in substrates to facilitate siphoning of the water column for oxygen and food. Their food includes primarily detritus, plankton, and other microorganisms.

As a group, freshwater mussels are extremely long-lived, with life spans of up to 130 years for certain species. Life spans of these seven species are unknown. Based on the longevity of a congener of the fat threeridge (the threeridge [*Amblema plicata*]); the longevity of thick-shelled species, and the large size attained by the fat threeridge and purple bankclimber, the latter two species probably have long lifespans....

Summary of Factors Affecting the Species

... A species may be determined to be an endangered or threatened species due to one or more of the five factors described in section 4(a)(1) [of the Endangered Species Act]. These factors and their application to the fat threeridge (*Amblema neislerii*), shinyrayed pocketbook (*Lampsilis subangulata*), Gulf moccasinshell (*Medionidus penicillatus*), Ochlockonee moccasinshell (*Medionidus simpsonianus*), oval pigtoe (*Pleurobema pyriforme*), Chipola slabshell (*Elliptio chipolaensis*), and purple bankclimber (*Elliptoideus sloatianus*) are as follows.

A. *The Present or Threatened Destruction, Modification, or Curtailment of its Habitat or Range*

Historically, mussel faunas in the United States have declined extensively as an unintended consequence of human development. The mussel fauna in much of the Apalachicolan Region has been negatively impacted by impoundments, siltation, channelization, and by water pollution. The cumulative effect of these factors on the aquatic ecosystems of the ACF River basin has not been systematically evaluated; an ongoing USGS National Water Quality Assessment is currently addressing this task.

Impoundments have permanently altered a significant portion of the ACF River system, which has 16 mainstem impoundments. Impoundments affect mussels by altering current, substrate, and water chemistry, factors which are important to riverine mussels. Lack of mussel recruitment in impoundments may be due to loss of glochidia in the substrate, attacks on glochidia by microorganisms, or the juveniles' inability to survive in silt.

The Chattahoochee River has 13 dams, including three locks and dams along its lower half; the lower mainstem is inundated for approximately 400 km (248 mi). An additional 85 km (53 mi) of mainstem habitat are impounded upstream of Atlanta, making approximately 485 km (301 mi) of the mainstem's 700 km (434 mi) total length (69 percent) impounded. The lower portions of many tributaries were permanently flooded because of these reservoirs, including a known site for the shinyrayed pocketbook in Walter F. George Reservoir.

... Many regional streams have increased turbidity levels due to siltation. These seven mussels probably attract host fishes with visual cues. Such a reproductive strategy depends on clear water. Turbidity is a limiting factor impeding sight-feeding fishes, and may have contributed to the decline of these seven species.

Light to moderate levels of siltation are common in many Apalachico-lan Region streams with populations of these seven species, while heavy siltation has occurred in the Piedmont, which is well known for its highly erodible soils. Most of the topsoil in the Piedmont was eroded by 1935. Clench attributed the decline of the rich mussel fauna of the Chattahoochee River to erosion from intensive farming before the Civil War. The steep slopes characteristic of the Fall Line Hills and the Piedmont result in higher erosion rates than slopes on more level lands.

Couch et al. indicated that all parts of the ACF Basin have been subject to alteration of forest cover. They attributed severe historical erosion and sedimentation in the Blue Ridge Province to mining and logging. The Service believes that while deforestation historically represented a threat to these mussels, current silvicultural activities following best management practices are compatible with the continued existence of the species.

Because of their sedentary characteristics, mussels are extremely vul-nerable to toxic effluents. There are discharges from 137 municipal waste water treatment facilities in the ACF River basin. Although the quality of effluents has improved since the 1980's due to improved waste water treatment and a 1990 phosphate detergent ban in Georgia, two-thirds of the 938 stream miles in the Georgia portion of the ACF River basin do not meet the designated water use classifications under the requirements of the Clean Water Act.

Agricultural influences include nutrient enrichment from confined feeding of poultry and livestock (primarily in the Piedmont Province), and inputs of pesticides and fertilizers from row crop agriculture (primarily in the Coastal Plain).

An estimated 3.6 billion liters (0.95 billion gallons) of chemical-laden rinse, stripping, cleaning, and plating solutions were discharged through a short canal into the Flint River from 1955 to 1977 at a Department of Defense facility in Albany, Georgia. The Service believes the long-term release of this effluent likely had, and may continue to have, a chronic toxic effect on Flint River mussel populations. The canal and other portions of the facility are a Superfund site.

Abandoned battery salvage operations affect water quality in the Chipola River. Concentrations of heavy metals (e.g., chromium and cadmium) in Asian clams and sediments increased in samples taken downstream from two operations. Dead Lake, on the lower mainstem, was considered a contaminant sink. Chromium was found at levels known to be toxic to mussels in sediment samples from Dead Lake downstream. A large population of the fat threeridge has been extirpated in Dead Lake, possibly from such contamination.

Residential development in Georgia is resulting in the conversion of farmland to subdivisions in areas relatively distant from the cities of Albany, Atlanta, and Columbus. Development and land clearing increases siltation from erosion, runoff and transport of pollutants from stormwater, and municipal waste water facility effluents. Lenat and Crawford found that in Piedmont drainages, urban catchments had higher maximum average concentrations of heavy metals than agricultural or forested catchments. Urban waterways may harbor human-produced contaminants in concentrations sufficient to significantly affect fish health.

Additional water supply impoundments may be planned to satisfy expanding urban and suburban demand. Any impoundments on streams that support these species may have impacts on their long-term survival. Impoundments on streams that do not harbor these species could be designed in ways to minimize or eliminate potential impacts to these mussels and their habitat downstream. Future impoundments, particularly in the metropolitan Atlanta area, could impact stream habitats where small populations of the shinyrayed pocketbook, Gulf moccasinshell, and oval pigtoe exist.

In-stream and near-stream gravel mining has occurred in various portions of the Apalachicolan Region. Jenkinson recorded the shinyrayed pocketbook, oval pigtoe, Gulf moccasinshell, and ten other species in Little Uchee Creek, a tributary of the Chattahoochee River in Alabama. The creek had supported in-stream gravel mining; only a few shell fragments were found at Jenkinson's site in the status survey, although living shinyrayed pocketbooks were found at another site in Little Uchee Creek. Gravel mining operations in the Chattahoochee River do not pose a threat to these mussels since no populations exist there now. However, where in-stream gravel operations are conducted in the vicinity of populations of these species, mussels may be displaced, crushed, or covered by bottom materials.

Some artifact and fossil collectors have used suction dredges to scour benthic habitats in the ACF system. This can destroy mussel habitat at the collection site and resuspend silt, impacting downstream areas. In a study on the effects of suction dredging for gold on stream invertebrates, Harvey concluded that impacts from suction dredges can be expected to be more severe in streams with softer substrates (*e.g.*, sand, gravel), as is typical for most Apalachicolan Region streams.

Many of the impacts discussed above occurred in the past as unintended consequences of human development in the Apalachicolan Region. Improved understanding of these consequences has led to regulatory (*e.g.*, the Clean Water Act) and voluntary measures (*e.g.*, best management practices for agriculture and silviculture) and improved land use practices that are generally compatible with the continued existence of these mussels. Nonetheless, the seven mussel species currently are highly restricted in numbers and distribution and show little evidence of recovering from historic habitat losses.

B. *Overutilization for Commercial, Recreational, Scientific, or Educational Purposes*

The threeridge (a relative of the fat threeridge) and the washboard (Megalonaias nervosa), which is superficially similar to both the fat threeridge and purple bankclimber, are heavily utilized as sources of shell for nuclei in the cultured pearl industry. The Service has been informed by commercial shell buyers that shells from the ACF River system are of poor quality. However, shell material from this area may be used as "filler" for higher quality material from elsewhere. In the 1980's, the price of shell increased, resulting in increased competition for the harvesting of shell beds in the Apalachicolan Region.

Biological supply companies have used the Flint River and possibly the Ochlockonee River as sources for large mussel specimens, including the purple bankclimber and possibly the fat threeridge, to sell to academic institutions for use in laboratory studies. The practice of dissecting mussels in introductory laboratory courses is no longer widespread, and the threat posed to large species such as the fat threeridge and purple bankclimber is probably decreasing.

Nonetheless, harvest of the fat threeridge and purple bankclimber for these purposes could decimate their remaining populations. The increasing rarity of these mussels potentially makes them more appealing to shell collectors. Revealing specific stream reaches harboring these species could pose a threat from collectors....

C. *Disease or Predation*

Diseases of mussels are virtually unknown; this factor is not currently known to affect these seven species.

Juvenile and adult mussels may serve as prey for various animals, mostly fishes, turtles, birds, and mammals. The muskrat has been implicated in potentially jeopardizing recovery of federally listed mussels. Although muskrats are not common within the range of these species, Piedmont populations of the shinyrayed pocketbook, Gulf moccasinshell, and oval pigtoe in the upper Flint River system may be subject to some degree of muskrat predation. . . .

E. *Other Natural or Manmade Factors Affecting Its Continued Existence*

Because of slow growth and relative immobility, mussel recolonization of impacted river reaches is a lengthy process, achieved by dispersal of newly metamorphosed juveniles via infected host fish, passive adult movement downstream, and active migration or passive movement downstream of small individuals. Establishment of self-sustaining populations requires decades of immigration and recruitment, even for common species that may occur in high densities. A mussel species should be considered stable only when active population recruitment is demonstrated and a significant number of viable populations exists.

The exotic Asian clam (*Corbicula fluminea*) has invaded all of the rivers where these seven mussels occur. First reported from the Apalachicolan Region about 1960, this species may compete with native mussels for nutrients and space. Densities of Asian clams are sometimes high in Apalachicolan Region streams, with estimates ranging from approximately 100/m (9/ft) square to over 2,100/m (195/ft) square. In some streams, the substrate has changed from homogenous silty sand or sand to one with a gravel-like component comprised of huge numbers of live and dead Asian clams.

. . . Another introduced bivalve, the zebra mussel (*Dreissena polymorpha*), has caused the extirpation of numerous native mussel populations and may pose a threat to these mussels in the future. Introduced into the Great Lakes in the late 1980's, this exotic species has been rapidly expanding its range in the South, but has not been reported yet from Apalachicolan Region streams.

The complex life cycle of mussels increases the probability that weak links in their life history will preclude successful reproduction and recruitment. Egg formation and fertilization are critical phases in the life history, because many mussels fail to form eggs or fertilization is incomplete. Fertilization success has been shown to be strongly correlated with spatial aggregation; excessively dispersed populations may have poor success. The need for specific fish hosts and the difficulty in recolonizing areas where mussels have been decimated are other life history attributes which make them vulnerable.

These seven species have been rendered vulnerable to extinction due to significant habitat loss, range restriction, and population fragmentation and size reduction. Most of their populations have been extirpated from the Piedmont portion of their historical ranges, four of five species are extirpated from Alabama, and none of the species remain in the Chattahoochee

River. The restricted distribution of these seven species also makes localized populations susceptible to catastrophic events and collection.

Notes & Questions

1. Why are these mussels threatened with extinction? What actions would help prevent the mussels from becoming extinct?

2. Habitat destruction is the greatest threat to the survival of most endangered species. What is destroying the habitat of the mussels and so many other species? What can be done to preserve that habitat? What are the possible consequences of protecting the habitat of plants and wildlife?

3. The other threats faced by the seven Gulf Slope mussels confront numerous other species as well. Pollution raises a concern for many fish and aquatic species that depend upon clean water for their survival. Nor are the seven Gulf Slope mussels alone in having to compete with the zebra mussel, an exotic species that has wreaked havoc in the United States. As environmental historian J.R. McNeill describes:

> The zebra mussel is a striped mollusc, native to the Black and Caspian Seas. It hitched a ride to the Great Lakes in ballast water in 1985 or 1986. By 1996 it had colonized all the Great Lakes and the St. Lawrence, Illinois, Ohio, Tennessee, Arkansas, and most of the Mississippi Rivers. The zebra mussel filters water to feed and removes numerous pollutants and algae, leaving cleaner and clearer water wherever it goes. Preferring hard, smooth surfaces, it delighted in the industrial infrastructure of the Great Lakes region, building thick colonies that sank navigational buoys and clogged water intakes on factories, power plants, and municipal water filtration systems. By the early 1990s it had temporarily shut down a Ford Motor plant, a Michigan town's water power supply, and cost the United States about a billion dollars a year. In dollar terms it threatened to become the most costly invader in U.S. history, a distinction previously held by the boll weevil.

J.R. McNeill, Something New Under the Sun: An Environmental History of the Twentieth-Century World 259 (2000). The broader problem presented by invasive species is described in Invasive Species in a Changing World (Harold A. Mooney & Richard J. Hobbs eds. 2000).

Concern that the mussels will be sought because of their rarity demonstrates a much broader threat to many species. The Fish & Wildlife Service routinely refuses to indicate where an endangered plant species is located precisely because of the fear that collectors will take the plants at the site. Such purposeful exploitation has resulted in the demise of many kinds of species. Writing in 1913, conservationist William T. Hornaday (about whom more later at 864) insisted that "[i]t is high time for the whole civilized world to know that many of the most beautiful and remarkable birds of the world are now being *exterminated* to furnish millinery ornaments for women's wear." William T. Hornaday, Our Vanish-

ING WILD LIFE: ITS EXTERMINATION AND PRESERVATION 114 (1913). Yet over eighty years later, the Tibetan antelope faces a nearly identical threat. Scarves and shawls made from the wool from the antelope (known as "shatoosh") "sell for US$1,000 to $5,000 and more, and have become the rage among the rich, famous, and fashionable." Over $100,000 worth of shatoosh shawls were sold in one evening during a benefit for the Memorial Sloan–Kettering Cancer Center held at New York City's Mayfair Hotel. But shatoosh can only be obtained by killing the wild and shy antelope, often by poachers using automatic weapons against entire herds. Awareness of the rapid disappearance of the Tibetan antelope has yielded both legal sanctions against poachers and societal stigma against consumers, but the animal remains threatened by the underground trade in shatoosh. *See* Judy Mills, *Fashion Statement Spells Death for Tibetan Antelope*, *available at* <http://www.worldwildlife.org/news/pubs/shahtoosh/shahtoosh3.pdf>; Susan Saulny, *Shawls Sold at Charity Event: So Soft and So Illegal*, N.Y. TIMES, Jan. 3, 2001, § B, at 2 .

Other species face threats not encountered by the seven Gulf Slope mussels. Bighorn sheep, black-footed ferrets, and other animals are often the accidental victims of collisions with motor vehicles. And sometimes the threat results from natural causes rather than human activities. The Karner blue butterfly, for example, depends upon a specific kind of ecosystem that itself is constantly changing. Floods, droughts, hurricanes, and other natural disasters can cause species to disappear. There are also some instances where one species threatens the survival of another rare species. In southern California, the kit fox—a candidate for listing under the ESA—preys upon the endangered kangaroo rat, which itself eats certain endangered plants.

4. Countless species have disappeared from the earth in the past several centuries. *See* STEIN, KUTNER & ADAMS, *supra*, at 323–35 (listing the animals and plants that have gone extinct in the United States); NILES ELDREDGE, LIFE IN THE BALANCE: HUMANITY AND THE BIODIVERSITY CRISIS 195–207 (1998) (listing all of the animal species throughout the world that have gone extinct since 1600); WORLD CONSERVATION MONITORING CENTRE, *supra*, at 206–26 (listing animal species extinct since circa 1600 and extinct higher plant taxa). Most recently, the ivory billed woodpecker was declared extinct in 2000. Its demise was preceded by species such as the dodo, great auk, red gazelle of northern Africa, Steller's sea cow, Bavarian pine-vole, pink-headed duck of India, Caribbean monk seal, desert rat-kangaroo of central Australia, and Carolina parakeet. Those species are pictured and the reasons for their extinction are described in TIM FLANNERY & PETER SCHOUTEN, A GAP IN NATURE: DISCOVERING THE WORLD'S EXTINCT ANIMALS (2001).

The passenger pigeon provides the most dramatic extinction story in American history. The passenger pigeon flew about 60 miles per hour, was a prodigious breeder, and received its name because its migrations were so spectacular. It was the most common bird in North America as recently as two centuries ago—it represented as many as four of every ten birds on the continent—with flocks containing up to two billion birds extending from

Florida to Arkansas in the winter and New York to northern Canada in the summer. Then it was shot to death. Passenger pigeons were viewed as a pest by farmers precisely because of their massive numbers, they were sold for meat in city markets, and they were hunted simply for sport. In the 1860s, John James Audubon described how an approaching flock of passenger pigeons was so large that it blocked his view of the sun, but after the flock encountered the hunters, "[t]he pigeons were picked up and piled in heaps, until each had as many as he could dispose of, when the hogs were let loose to feed on the remainder." The advent of the breech-loading shotgun around 1870 led to even more frenzied hunting, including a competition that was won by a hunter who killed 30,000 pigeons. Within twenty years, the survival of the species was doubtful, and the last wild bird was killed in Ohio in 1900. Martha, the last passenger pigeon, died in her cage in the Cincinnati Zoo in 1914. Her stuffed remains are now on view in the Smithsonian's National Museum of Natural History in Washington. *See* TIM FLANNERY, THE ETERNAL FRONTIER: AN ECOLOGICAL HISTORY OF NORTH AMERICA AND ITS PEOPLES 312–15 (2001).

The Passenger Pigeon

Audubon's painting of a male and female passenger pigeon

Of course, extinction is not an exclusively modern phenomenon. *Jurassic Park* notwithstanding, dinosaurs have long since departed the earth. Indeed, mass extinctions occurred late in each of five geologic periods. That leads some observers to argue that there is no extinction crisis today because extinction is simply a fact of natural life. Most scientists, though, believe that the rate of extinction is much higher today than it has ever been. E.O. Wilson estimates that 27,000 species become extinct each year. The calculation of how many species have gone extinct is easier for species that are well known or live in places that are well known, whereas many obscure species in remote areas can go extinct without anyone knowing that they had existed.

It can even be difficult to know whether a particular species is extinct. Like other efforts to prove a negative, it is nearly impossible to know for certain whether a species no longer exists in the wild. For example, the Tasmanian tiger was last seen in the wild in the 1930s, yet there are some people today who insist that the tiger survives deep within the western Tasmanian forests. And the Borneo River shark was thought to be extinct for nearly a century until Malaysian fishermen caught one in 1997. The list of species protected by the ESA today includes numerous other species that were thought extinct for decades before they were rediscovered.

Coral Reef Conservation and Restoration Partnership Act of 2000

House Report 106–762.
106th Congress, 2d Session (2000).

BACKGROUND AND NEED FOR LEGISLATION

Coral reefs are among the world's most biologically diverse and productive ecosystems. Coral reefs provide habitat for one-third of all marine fish species, protect coastlines from waves and storms, provide opportunities for fishing, support tourism worth billions of dollars, and harbor organisms that are important for various pharmaceutical products. Over the past few decades, public awareness of coral reefs has risen in response to the escalating loss of coral ecosystems. Worldwide, it is estimated that 10 percent of coral reefs are degraded beyond recovery and another 30 percent are in such critical condition that they require immediate attention to prevent their loss. Several federal agencies have jurisdiction for managing different aspects of coral and coral reefs. Congressional action is needed to ensure the effective coordination of federal management activities and to provide assistance to carry out coral reef conservation projects.

Corals are found in all oceans of the world from the tropics to the polar regions, but not all coral species are reef builders. Reef-building (or stony) corals are of the order Scleractinia in the class Anthozoa of the phylum Cnidaria. Approximately 6,000 species of Anthozoans exist in the marine environment. Coral reefs are formed by tiny colonial animals (called coral polyps) that secrete a hard calcium carbonate skeleton as they grow. Living coral is found only on the surface of the skeletal structure, and actually consists of two distinct organisms, formed by a symbiotic relationship between an animal (coral polyp) and a plant (algae). These intertwined organisms require warm, nutrient poor water that is free of sediments to flourish. Coral reefs are widely distributed in relatively shallow tropical and equatorial waters around the globe, typically occurring between 30 degrees North and South latitude. Although corals can be found from the water's surface to depths of 19,700 feet, reef-building corals are generally found at depths of less than 150 feet where sunlight penetrates. The United States has jurisdiction over coral reef ecosystems covering over 17,000 square kilometers within the boundary of the 200–mile exclusive economic zone (EEZ). Approximately 90 percent of these reefs are located in the Western Pacific Ocean, including marine areas surrounding American Samoa, Guam, Hawaii, and the Commonwealth of the Northern Mariana Islands. Coral reefs are also well-developed in the coastal waters surrounding Florida, Puerto Rico and the U.S. Virgin Islands.

Coral reefs throughout the world suffer from anthropogenic and natural disturbances. Human activities including destructive fishing practices (such as the use of explosives and poisons), nutrient enrichment (eutrophication), increasing tourism pressure, and the physical destruction of reefs by vessel groundings contribute to the deterioration of coral reef ecosystems. Dramatic natural events such as hurricanes can also damage coral reefs. Coral reefs are biologically complex yet delicate ecosystems that are sensitive to changes in temperature, salinity, sediment deposition, and nutrients. Scientists believe that changes in natural conditions coupled with anthropogenic factors can place severe biological stress on the coral organisms, making them more susceptible to diseases such as white plague and black band. In 1998, coral reefs worldwide displayed wide spread "bleaching," which scientists believe was caused in part by warmer than normal sea surface temperature due to a strong El Nino event. Bleaching occurs when the algae living within the coral die or are ejected and is an indicator of biological stress. Bleaching, coral disease, and the physical destruction of coral reefs are all manifestations of declining coral health.

Notes & Questions

1. How do the threats faced by ecosystems compare to the threats faced by individual species?

2. Coral reefs are among the most spectacular ecosystems. Consider President Clinton's explanation for establishing the Northwestern Hawaiian Islands Coral Reef Ecosystem Reserve:

> This vast area supports a dynamic reef ecosystem that supports more than 7,000 marine species, of which approximately half are unique to the Hawaiian Island chain. This incredibly diverse ecosystem is home to many species of coral, fish, birds, marine mammals, and other flora and fauna including the endangered Hawaiian monk seal, the threatened green sea turtle, and the endangered leatherback and hawksbill

sea turtles. In addition, this area has great cultural significance to Native Hawaiians as well as linkages to early Polynesian culture-making it additionally worthy of protection and understanding. This is truly a unique and special place, a coral reef ecosystem like no place on earth, and a source of pride, inspiration, and satisfaction for all Americans, especially the people of Hawaii.

Exec. Order No. 13178, 65 Fed. Reg. 76903 (2000). But coral reefs are also among the most vulnerable ecosystems. Besides the threats identified in the House committee report, coral reefs suffer from pollution, overfishing, mining of reefs for construction materials, the destruction of nearby mangrove trees that filter harmful sediments, predation by the crown-of-thorns starfish, and global warming. The bill reported by the House committee would encourage coordination among federal agencies whose activities affect coral reefs and would fund conservation programs, but Congress has yet to act upon it or any similar measure. Recent studies of the condition of coral reefs include CLIVE WILKINSON, STATUS OF CORAL REEFS OF THE WORLD 2000, and D. BRYANT ET AL., REEFS AT RISK (World Resources Institute 1998), while current information can be obtained at the Global Coral Reef Monitoring Network's web site at <http://coral.aoml.noaa.gov/gcrmn/index.html>. We explore these and other threats to coral reef ecosystems in more detail in Chapter 12.

3. Other ecosystems have declined as well. A 1995 report prepared for the National Biological Service within the U.S. Department of the Interior attempted to collect the many studies of individual ecosystems that are threatened in the United States and around the world. According to that report, ecosystems that have completely disappeared include coastal rocky headland in New Hampshire, bluegrass savanna-woodland in Kentucky, and coastal strand in San Diego County. Another group of ecosystems have suffered greater than 99% losses, including oak savanna in the midwest, native grasslands in California, basin big sagebrush in Idaho's Snake River plain, European primary forests, and prairies in Illinois, Texas, and elsewhere. *See* Reed F. Noss, Edward T. LaRoe III & J. Michael Scott, *Endangered Ecosystems of the United States: A Preliminary Assessment of Loss and Degradation*, 28 BIOLOGICAL REP. 1, 37–49 (1995), *available at* <http://biology.usgs.gov/pubs/ecosys.htm>.

CHAPTER 3

HOW TO PROTECT BIODIVERSITY

Chapter Outline:
A. Ways of Protecting Biodiversity
B. Private Actions to Protect Biodiversity
C. State Law Protection of Biodiversity
D. Tribal Protection of Biodiversity

———

Biodiversity can be protected in many ways by many people. The threats to particular species or ecosystems often dictate the strategies that are best designed to protect biodiversity. The management of land to promote the natural life that is found there is a popular measure because it addresses the destruction of habitat that affects so many plants and animals. Different actions may be required to combat other threats, including the elimination of pollution, the removal of exotic species, and the regulation of hunting or collecting. If a species is nearing extinction, then more aggressive measures such as captive breeding and the reintroduction of the species into other habitats may become necessary.

Who should take those actions is often more contested than the determination of the appropriate action itself. The threshold issue is to decide whether the imperative to protect biodiversity in a given instance must be augmented by a legal duty. Private individuals and institutions play a significant role in the effort to save ecosystems and species alike, yet such efforts are typically complemented by government actions as well. Sometimes the government—and the laws it enforces—relies upon carrots to encourage the preservation of biodiversity, while at other times resort is made to sticks to compel such preservation, with predictable responses from those who are affected by the law's commands. And the decision to enlist the government in the preservation of biodiversity raises the question of which government should enact such laws: municipal or county governments, states, the United States, or a collection of concerned nations throughout the world.

Most of the attention on recent actions intended to protect biodiversity focuses on two kinds of responses: in the United States, federal regulation of activities that interfere with biodiversity; and internationally, the development of treaties that regulate conduct adverse to biodiversity. The role of the federal government and actions of the international community will be explored in much more detail in Parts III and IV of this book. But there are many other options. This chapter examines the role of private actors, states, and Native American tribes in preserving biodiversity. First, though, this chapter surveys the types of actions that can be taken to protect ecosystems, species, and other aspects of biodiversity. The two questions that arise throughout these materials are which actors are best suited to protect biodiversity in certain instances, and what kinds of actions are likely to be most effective in achieving that end.

A. WAYS OF PROTECTING BIODIVERSITY

National Audubon Society v. Hester

United States Court of Appeals for the District of Columbia Circuit, 1986.
801 F.2d 405.

■ EDWARDS, STARR AND SILBERMAN, Circuit Judges.

■ PER CURIAM

The California condor, the largest winged inhabitant of North America, has been decimated to the point where only twenty-six members of the species remain in existence. At the time this controversy began, all but six of the birds were kept in zoos in Los Angeles and San Diego as part of a breeding program designed to avert extinction of the species. This lawsuit arises from the U.S. Fish & Wildlife Service's decision to bring the remaining condors in from the wild. The district court granted plaintiff National Audubon Society's request for a preliminary injunction barring the Service from carrying out this decision. Because we believe that the agency's decision constituted a reasoned exercise of its discretion in fulfilling its statutory mandate, we reverse.

I.

In recent years, the Wildlife Service's energies have been engaged in inauspicious efforts to stem the condor flock's steady decline. In 1979, working in tandem with public and private groups (including the plaintiff), the Service developed a "Condor Recovery Plan." This plan had two principal elements: extensive tracking and study of wild birds, and the commencement of a captive propagation program. At the time, it was hoped that better information about the birds' lifestyle (and causes of death), together with enhanced breeding in capacity, could save the condor. The mortality rate among wild birds, however, proved to be alarming: in the winter of 1984–85, six of the then fifteen wild condors vanished. A common cause of death was believed to be lead poisoning following the birds' feeding on the carcasses of animals shot by hunters (the condor is a member of the vulture family).

After considering a wide range of scientific opinion, the Wildlife Service issued an Environmental Assessment in October, 1985 setting forth seven alternative courses of action for condor preservation. The option chosen by the agency combined capture of birds whose genes were poorly represented among the captive flock, maintenance of a small wild flock, and eventual release of young birds bred in captivity. This choice struck a balance between the competing considerations at stake (as well as the contending

The California condor

views of biologists and naturalists): on the one hand, bringing in the remaining wild condors would minimize mortality and increase the genetic diversity of the captive flock; on the other hand, preservation of a wild flock would provide "guide birds" available to lead captive-bred condors ultimately released, facilitate the improvement of techniques of protecting the birds, and prevent the erosion of public support for preserving the condors' habitat.

Shortly after this report was released, however, troubling news began reaching the Wildlife Service. One of the birds scheduled to remain in the wild appeared to be courting one of the birds slated for capture. Second, due to apparent zoo mismanagement, the young condors selected for release into the wilderness in the next year had grown too tame. And, most importantly, one condor inhabiting an area regarded as very safe, where "clean" carcasses were provided for the birds, nonetheless came down with lead poisoning (and has since died). In late December, the agency reversed its earlier decision and announced that all remaining wild birds would now be brought in. The federal Council on Environmental Quality certified that an emergency existed and that immediate documentation of the environmental effects of this decision was unnecessary. In any event, on December 23 the Service issued an "Addendum" to its October Environmental Assessment explaining the reasons why the agency now believed a different plan of action was called for.

This lawsuit followed. Audubon claimed that the Wildlife Service's action violated the Administrative Procedure Act (APA), 5 U.S.C. § 701 *et seq.* (1982), the Endangered Species Act (ESA), 16 U.S.C. § 1531 *et seq.* (1982), and the National Environmental Policy Act (NEPA), 42 U.S.C. § 4321 *et seq.* (1982), and moved for a preliminary injunction barring the capture of the wild condors. The district court granted Audubon's motion, finding that the plaintiff had demonstrated a likelihood of success on the merits and a balance of hardships in its favor. *See* 627 F. Supp. 1419 (D.D.C.1986). While acknowledging that a "reviewing court must be wary of substituting its own judgment for that of the agency," *id.* at 1422 (citing Citizens to Preserve Overton Park, Inc. v. Volpe, 401 U.S. 402, 416 (1971)), the district court nevertheless concluded that the agency's decision was fatally flawed. The court opined that the Wildlife Service had exhibited insufficient analysis and explanation of its departure from past policy. In the court's view, this change of policy amounted to arbitrary and capricious action in violation of the above-mentioned statutes and threatened irreparably to harm the plaintiff's interests.

II.

This court customarily reviews a district court's grant of preliminary equitable relief under the deferential abuse of discretion standard. A preliminary injunction premised upon an erroneous view of the law, however, is not insulated from appellate review. In this case, the district court's decision appears to have rested entirely on its view that the Wildlife Service had failed to justify its change of policy; the court relied upon this point not only in determining the plaintiff's likelihood of success on the merits, but also in concluding that the "balance of harms" favored Audubon. Since we believe, contrary to the district court, that the agency fully considered all appropriate courses of action and adequately explained its policy choice, we cannot uphold the district court's decision.

Although the district court relied upon the ESA and NEPA as well as the Administrative Procedure Act, it is clear that those statutes essentially place the same demands on agency decisionmakers as does the APA. Under the ESA, an agency's determination that its action will not threaten endangered species is to be set aside only if arbitrary and capricious. Under NEPA, agencies must prepare an Environmental Impact Statement (EIS) whenever proposed major federal action will significantly affect the quality of the human environment, *see* 42 U.S.C. § 4332(2)(C) (1982); an agency's decision *not* to prepare an EIS—because the proposed action will not significantly affect the environment—may be overturned, again, only if arbitrary and capricious. The question for reviewing courts is not whether an agency decision is "correct," but rather whether the decision reflects sufficient attention to environmental concerns and is adequately reasoned and explained. The ESA and NEPA were intended to ensure that agencies, in discharging their various functions, do not blithely disregard the environmental effects of their decisions; it is obvious, then, that judicial review of agency decisionmaking is at its most deferential where, as here, the agency action is based *solely* upon environmental considerations—where the challenge to agency action simply represents a quarrel with the agency's means of pursuing a universally desired end.

We believe the Wildlife Service's decision to capture the remaining wild condors was manifestly defensible. That decision represented a reevaluation by the responsible agency of the costs and benefits associated with the existence of captive and wild condor flocks. Contrary to the plaintiff's assertion, the decision was not markedly at odds with previous policy. In its October Environmental Assessment, while endorsing the maintenance of a small wild flock, the Service had recognized that there were weighty arguments to the contrary and that the question was close; it noted that "if the condor population appears to continue steadily downward after implementation of this option, we stand ready to reevaluate the taking into captivity of all, or a significant portion of, the remaining [wild] population." This is, in part, precisely what happened: the agency reconsidered its policy after learning of recent developments, including the lead poisoning suffered by a bird inhabiting what was thought to be one of the safest locations. The Wildlife Service simply exercised its discretion to "adapt [its] rules and policies to the demands of changing circumstances." Permian Basin Area Rate Cases, 390 U.S. 747, 784 (1968). The district court, however, concluded that the factual developments supporting the agency's change of course were not "new"—that they had been known at an earlier stage, when the agency reaffirmed its commitment to preserving a wild flock. The flaw in this reasoning is that in a case like this one there is no particular significance attached to the exact date that factual information reaches any official of an agency. It takes time for information to be disseminated from the lower echelons of agency staff to the agency's decisionmakers, and still more time for the decisionmakers to appraise the policy implications of that information.

More fundamentally, agencies are entitled to alter their policies " 'with *or without* a change in circumstances,' " so long as they satisfactorily explain why they have done so. Motor Vehicle Mfrs. Ass'n v. State Farm Mut. Auto. Ins. Co., 463 U.S. 29, 57 (1983) (quoting Greater Boston Television Corp. v. FCC, 444 F.2d 841, 852 (D.C.Cir.1970), *cert. denied*, 403 U.S. 923 (1971)) (emphasis added). We have little problem concluding that the Wildlife Service met its burden of justifying its change of course. The agency's October Environmental Assessment thoroughly surveyed the competing factors at stake and examined the desirability of seven alternative courses of action. Its December Addendum incorporated the earlier document's reasoning and additionally set forth the agency's reasons for preferring a different option than before. The Service's documentation may have been succinct, but nonetheless adequately discloses the concerns underlying the agency's decision and demonstrates that that decision rests on a rational basis. That much being so, our inquiry is at an end: NEPA's "mandate to the agencies is essentially procedural. . . . It is to insure a fully informed and well-considered decision, not necessarily a decision [federal judges] would have reached had they been members of the decisionmaking unit of the agency." Vermont Yankee Nuclear Power Co. v. NRDC, 435 U.S. 519, 558 (1978). The decision of the district court is therefore REVERSED.

Notes & Questions

1. The legal case for the removal of the last condors from the wild was aided by an associate solicitor at the Department of the Interior, Gail Norton. In April 2001, now Secretary of the Interior Norton released five captively-bred condors into the wilderness along California's Big Sur Coast. During the intervening fifteen years, the number of California condors grew from 24 to 184. Three facilities—the San Diego Wild Animal Park, the Los Angeles Zoo, and the World Center for Birds of Prey in Boise—have worked to breed the condors in captivity. The first condors were released back into the wild in 1992, and 33 condors were in California and 22 in Arizona by 2001. The goal of the recovery program is to establish two separate wild populations comprised by 150 condors (and 15 breeding pairs) each.

The captive breeding of the condors has been a success, but the birds still face numerous challenges in the wild. Numerous condors died from contact with electrical power lines, so young captive condors are taught to avoid perching on such lines. Condors have also been poisoned by feeding on carrion laced with lead shotgun pellets, a danger that can be avoided by non-toxic—but more expensive—bullets. Coyotes killed two condors, ravens and golden eagles threaten condor eggs, and two condors drowned. Still other condors have moved dangerously close to humans whose homes now occupy the ancestral nesting sites of the condors. Condors breed slowly, laying only one egg every year or two, and no condors have successfully reproduced in the wild since they were returned in 1992. And the people caring for the captive condors continue to place still-born cattle in remote wilderness areas to provide food for the wild condors.

The ongoing challenges facing the condors prompted one writer to ask, "What is the good of captive breeding if the birds are reintroduced to a habitat that still threatens their extinction? How long can human nursemaids guard against their missteps in the vast outdoors?" Michael K. Burns, *Rare Condors Return to Wild*, BALTIMORE SUN, July 18, 2001, at 2A. Holly Doremus responds that active human management will often remain part of the protection of biodiversity:

> [A] strict hands-off strategy is inconsistent with the protection of species, ecosystems, or natural processes. No place in the United States remains entirely unaffected by human actions. Ongoing management efforts are often necessary to compensate for the effect of past actions, or current actions outside the designated reserves. Competition with or predation by alien species, for example, is one of the leading threats to

domestic biodiversity. Once introduced, alien species often spread rapidly and are difficult, if not impossible to remove. Protecting native species from the threat of such exotics requires ongoing management. Intensive management may also be required to substitute for changes in historic fire regimes, predation levels, and other elements of the biophysical environment. Given the extensive changes in background conditions, ecologists tell us that most areas dedicated to the preservation of nature cannot simply be left to their own devices, but will require active human management.

Holly Doremus, *The Rhetoric and the Reality of Nature Protection: Toward a New Discourse*, 57 WASH. & LEE. L. REV. 11, 56–57 (2000). Note, too, that the effort to save the California condor has cost over $25 million. For current information about the status of the condors, visit the San Diego Wild Animal Park's site "Condor Ridge" at http://www.sandiego-zoo.org/special/condor/home.html, the Los Angeles Zoo's "California Condor Conservation" site at http://www.lazoo.org/cnews.htm.

2. Zoos play an active role in preserving biodiversity. The San Diego Wild Animal Park hosts a "Condor Ridge" exhibit that displays condors along with other native wildlife, such as black-footed ferrets and desert bighorn sheep. The exhibit has two purposes: to educate the public about the condor and endangered species generally, and to allow older condors to train younger condors so that they can be released in to the wild. *See* Jane Hendron, *Return to the Wild*, 25 ENDANGERED SPECIES BULLETIN 10 (May/June 2000). Other zoos feature species from more distant homes. The recovery of the Virgin Islands boa—the subject of an ESA dispute described below at page 189—is led by the Toledo Zoo. A variety of zoos have conducted captive breeding programs for such endangered species as black-footed ferrets, Hawaiian geese, Mongolian wild horses, and giant pandas. More generally, the American Zoo and Aquarium Association has established a program of Species Survival Plans (SSP). "Each SSP carefully manages the breeding of a species in order to maintain a healthy and self-sustaining captive population that is both genetically diverse and demographically stable. Beyond this, SSPs include a variety of other cooperative conservation activities, such as research, public education, reintroduction, and field projects. Currently, 90 SSPs covering 119 individual species are administered by the American Zoo and Aquarium Association, whose membership includes 185 accredited zoos and aquariums throughout North America." American Zoo and Aquarium Association, *AZA Species Survival Plan* <http://www.aza.org/Programs/SSP/about.htm>. The Center for Plant Conservation promotes a similar program to conserve and restore America's native plants, including a National Collection of Endangered Plants that houses more than 570 rare plants in institutions throughout the United States. For example, the San Antonio Botanical Gardens is working to save such plants as the Texas snowbell, the Navasota ladies-trusses, the black lace cactus, and the big red sage. *See* Plant Conservation: Plant Conservation at the San Antonio Botanical Gardens <http://www.sa-bot.org/CPC/>.

The relationship of zoos to the conservation of biodiversity can be problematic. David Hancocks, the director of the Open Range Zoo in Werribee, Australia, sees zoos as presenting "a perpetual dichotomy, which is the reverence that humans hold for Nature while simultaneously seeking to dominate it and smother its very wildness." He asserts that ":[k]eeping wild animals in captivity warrants stronger justification than the setting for a social gathering" for "a family day out." He sees zoos as having a tremendous opportunity for teaching about biodiversity, but he laments that such a role has been largely neglected. Hancocks also questions the AZA's work by observing "SSP is more accurately an acronym for a Self Supporting Program for zoos." And Hancocks calls for greater attention to the native biodiversity that exists near each zoo, citing the example of the San Diego zoos that are working to protect the condor and other species:

> Visitors to the San Diego Zoo can hear messages about the threat of tiger extinction and nod their head in concern, then drive north to the San Diego Wild Animal Park and hear messages about the depradation of elephants by poaching and comfort themselves by agreeing that they will never purchase ivory. The road they will have traveled between these sister zoos will have carried them through a region in which virtually every square inch of native chaparral habitat has been destroyed. It has been replaced by suburban sprawl and monoculture farms. Southern California contains one-fourth of all plant species known in the United States, half of them found nowhere else in the world, and it is one of the most endangered ecosystems on Earth. Yet there will be no hint from these Southern California zoos revealing this problem. It is too uncomfortably close to home.

David Hancocks, A Different Nature: The Paradoxical World of Zoos and Their Uncertain Future xvii, 6, 170–71 (2001). How can zoos that are interested in preserving biodiversity respond to Hancocks? What is the appropriate role for zoos in conserving biodiversity?

3. The condor case demonstrates that the captive breeding of endangered species can be especially controversial. Hancock complains that captive breeding "has nothing to do with conservation of biological diversity" because biodiversity is not "a thing, like a tiger, that can be saved and put on display." *Id.* at 159. The FWS and NMFS agree that "[c]ontrolled propagation is not a substitute for addressing factors responsible for an endangered or threatened species' decline," but they see a role for such breeding programs nonetheless. According to the official policy published by the agencies, "the controlled propagation of threatened and endangered species" ordinarily will be employed only if 14 criteria are satisfied: (1) measures to save the species in the wild are inadequate, (2) captive breeding is coordinated with efforts to provide secure and suitable habitat for the species, (3) an official recovery strategy recommends captive breeding, (4) the potential ecological and genetic effects of the removal of individuals from the wild are considered, (5) sound scientific principles to conserve genetic variation and species integrity support captive breeding, (6) a genetics management plan is developed, (7) all known precautions against the spread of diseases and parasites into the environment are taken, (8) the escape or accidental introduction of individuals outside their historic range is prevented, (9) more than one location is used, (10) other appropriate organizations and individuals are involved, (11) information needs and accepted protocols and standards are satisfied, (12) a commitment to funding is secured in advance, (13) captive breeding is tied to development of a reintroduction plan, and (14) regulations implementing the Endangered Species Act, Marine Mammal Protection Act, Animal Welfare Act, Lacey Act, Fish and Wildlife Act of 1956, and the Services' procedures relative to NEPA are followed. Policy Regarding Controlled Propagation of Species Listed Under the Endangered Species Act, 65 Fed. Reg. 59,616, 59,619–21 (2000). Is captive breeding appropriate in those circumstances? Is it appropriate even if those criteria are not satisfied?

Would satisfaction of those criteria also justify the cloning of an endangered species? In November 2000, the world's first cloned endangered species—an ox-like gaur that is native to southeast Asia—was born to an Iowa cow, but the baby gaur only lived two days. The scientists who cloned the gaur also described plans to clone the African bongo antelope, the Sumatran tiger, and the giant panda, and to reincarnate the recently extinct bucardo mountain goat of Spain. They defend cloning both as "a way to preserve and propagate endangered species that reproduce poorly in zoos until their habitats can be restored and they can be reintroduced into the wild," and as a means for enabling "researchers to introduce new genes back into the gene pool of a species that has few remaining animals." Robert P. Lanza, Betsy L. Dresser & Philip Damiani, *Cloning Noah's Ark: Biotechnology Might Offer the Best Way to Keep Some Endangered Species From Disappearing From the Planet*, SCIENTIFIC AMERICAN, Nov. 2000. It has even been suggested that long extinct animals such as the woolly mammoth could be recreated thanks to modern scientific breakthroughs. *See* Corey A. Salsberg, *Resurrecting the Woolly Mammoth: Science, Law, Ethics, Politics, and Religion*, 2000 STAN. TECH. L. REV. 1. At the same time, the possibility of cloning of humans has generated tremendous controversy, as evidenced by the approval of the Human Cloning Prohibition Act of 2001 by the U.S. House of Representatives in August 2001. Should the cloning of endangered species face the same objections as the cloning of humans?

4. Would biodiversity be satisfactorily protected if every species could be kept alive in zoos and every ecosystem could be contained in nature preserves? Holly Doremus argues that "none of the congressionally listed values of species can be fully protected in captivity." She explains:

> The esthetic benefits of wild and captive animals, for example, are quite different. Wild creatures, unconfined and uncontrolled by any human volition, inspire awe and wonder that captive animals cannot match. A bald eagle soaring above a river choked with spawning salmon offers a far richer esthetic experience than a caged eagle feasting on canned fish. A butterfly pinned to a display card, although beautiful and easy to view at leisure, cannot approach the beauty of

one glimpsed passing on the wing, or perched on a native flower. Captivity similarly diminishes the recreational value of species. The joy of wildlife-based recreation derives in large part from the quarry's lack of domestication. The thrill of the chase, the skill and knowledge it demands, are more important than the photograph or trophy it produces. Captive species, divorced from any natural ecosystem, have no ecological value. They provide only pale echoes of the educational and scientific value of species in their natural habitats. Even historic value is reduced by captivity. The bison that today roam free in Yellowstone National Park offer a far closer link to the continent's history than their semi-domesticated predecessors.... Wildness, understood as unpredictability or freedom from human control, imparts an aura that cannot be duplicated by captive species. That aura attracts and inspires us. It makes us care about wild places and wild creatures, and leads us to believe they merit special protection. Without wildness, the level of human concern for other species would be reduced.

Holly Doremus, *Restoring Endangered Species: The Importance of Being Wild*, 23 Harv. Envtl. L. Rev. 1, 12–13 (1999). Is there a contrary argument that would support the preservation of certain species only in zoos or elsewhere in captivity?

Bradley C. Karkkainen, *Biodiversity and Land*

83 Cornell Law Review 1, 10–14, 103–04 (1997).

A.   Large Reserves Where Possible, Small Reserves Where Necessary

There is a broad, though not universal, consensus within the scientific community that a biodiversity conservation strategy should be built on the foundation of a system of biological reserves containing relatively undisturbed habitats for diverse communities of species, linked where possible by a network of wildlife migration corridors. It is also widely agreed that, other things being equal, large reserves are preferable to small ones. This preference exists for several reasons. First, reserves large enough to protect naturally functioning ecosystems containing viable populations of resident species protect far more biodiversity—at the genetic, species, and ecosystem levels—at a far lower cost than do species-by-species management strategies. These reserves are also generally cheaper to acquire and maintain per unit of protected area than a series of smaller reserves of comparable total size. Second, some species, especially large mammals, have large home ranges, low natural growth rates, and low population densities, and therefore require large areas of protected habitat. Third, larger protected areas can generally support larger and more genetically diverse populations and "metapopulations," thus supporting genetic diversity within species and reducing the risk of extinction from human or natural disturbances, invasion by exotics, predation, disease, demographic events, or genetic depression. Fourth, other things being equal, larger protected areas are less likely to suffer from adverse "edge effects," including both human and natural disturbances from adjacent unprotected lands.

Nonetheless, large reserves are not always possible. In many parts of the country, habitats are already so fragmented that large blocks of relatively undisturbed land may be impossible to assemble. In such regions, species and ecosystems are most likely to be threatened or endangered precisely because habitat fragmentation is so advanced. In these regions, small reserves may be essential to protect the last valuable ecosystem fragments and habitat patches which, despite their shortcomings, may represent the best hope for survival of species on the brink of extinction.

There is also a high degree of consensus on reserve selection and management principles. R. Edward Grumbine, for example, identifies the following goals of conservation planning:

1. Maintain viable populations of all native species in situ.

2. Represent, within protected areas, all native ecosystem types across their natural range of variation.

3. Maintain evolutionary and ecological processes (e.g., disturbance regimes, hydrological processes, nutrient cycles, etc.).

4. Manage, over substantial periods of time, to maintain the evolutionary potential of species and ecosystems.

5. Accommodate human use and occupancy within the above constraints.

Similarly, Reed Noss proposes that the "fundamental objectives" of biodiversity conservation planning are to (1) represent, in a system of protected areas, all native ecosystem types and several stages across their natural range of variation; (2) maintain viable populations of all native species; (3) maintain ecological and evolutionary processes; and (4) design and manage the system to be responsive to environmental changes and to maintain evolutionary potential. Noss would select for immediate protection areas of high species richness, high endemism, high sensitivity to human pressure, and high levels of stress from human-caused disturbances. Michael Soule and Daniel Simberloff concur, suggesting that reserves should be selected so as to include optimal habitats for any species of special concern; areas where habitat and species diversity are greatest; areas of maximum endemicity; and, finally, sites that are particularly secure for long-term conservation.

B. Buffer Zones

Many commentators have pointed out that areas set aside as reserves are often relatively small "islands" representing only fragments of larger regional ecosystems. Consequently, they argue, the ability of reserve managers to achieve biodiversity conservation objectives is limited. One obvious solution is to create larger reserves. However, this is prohibitively costly not only in terms of land acquisition and management costs, but also in consideration of the opportunity costs of foregoing development on all the land necessary to protect the full array of representative ecosystems. Consequently, scientists and policy experts have recommended the establishment of "buffer zones" adjacent to protected reserves, thereby allowing some productive land uses, but restricting other uses to provide extended habitat for some species and limit adverse spillover effects on the core protected reserve.

As early as 1933, the visionary ecologist Victor Shelford recommended the core-and-buffer concept as a strategy for wildlife conservation. Its more recent incarnations include the UNESCO Man and the Biosphere program's biosphere reserve concept and the "multiple-use module" (or "MUM") concept developed by Reed Noss and Larry Harris, essentially an elaboration on the biosphere reserve concentric zoning model.

* * *

This Article proposes a system of federally owned core biological reserves—selected on the basis of such features as ecosystem and species representativeness, endemism, and species richness—surrounded by buffer zones in which the federal government would permit private ownership and compatible economic activities, but would regulate land use to limit and control adverse spillover effects in the protected core reserves. The proposal thus offers a way of giving legal and practical effect to UNESCO's Man and the Biosphere Program concept of "biosphere reserves"—core reserve areas selected for the importance of their biological resources, surrounded by "buffer" zones of limited economic activity compatible with conservation of the core.

Since its unveiling in 1968, the biosphere reserve concept has remained little more than a widely hailed idea. In this country, as throughout much of the world, "biosphere reserve" designation currently does not confer any special protections or binding legal obligations. Instead, the designation merely serves to signal interested parties that an area is of special concern for the value of its biological riches. Under my proposal, compatible (and therefore permissible) land uses permitted in the buffer areas certainly could include such activities as scientific research, education, and some recreational uses, even in areas immediately adjacent to the protected core. Farther away from the core, the government could permit (or encourage) increasingly intensive forms of economic activity on a sustainable basis, through application of traditional land-use planning techniques, participatory ecosystem-level planning by private stakeholders and affected units of state and local government, and creative use of innovative market-based approaches.

So long as the government permits viable economic uses on private lands in these outer concentric zones, the takings doctrine should not pose any problems to this scheme. And although this proposal does envision a strong federal role in land-use planning, that role would be much more confined, both in geographical scope and regulatory ambition, than in other, more expansive proposals for direct federal regulation of private lands for biodiversity conservation. With the federal role in land use regulation aimed primarily at protecting federally owned biodiversity reserves against negative spillovers from adjacent lands, the federalism and takings concerns ... may not be eliminated entirely, but would at least be cabined. This approach therefore represents a reasonable accommodation of the inherent tension between the global nature of the benefits of biodiversity conservation and the localized nature of its costs.

Finally, this Article suggests that it is time to move beyond our current thinking about the respective roles of "publicly owned lands" and "private lands" in biodiversity conservation—that some impenetrable and immovable barrier exists between these two categories. Although some lands that public land management agencies currently hold are extremely valuable for biodiversity conservation purposes, others are less so, and the same may be said of lands that private landowners currently hold. The challenge is to identify the most biologically valuable lands and devise workable strategies to protect them.

This Article proposes a core-and-buffer approach, with publicly owned lands (or, where appropriate, public ownership of less-than-fee interests) as the centerpiece. This would require a major overhaul of our public lands management strategy, placing biodiversity conservation at the pinnacle of public values to be served by federal land ownership and management. It also would necessitate a major reshuffling of the federal land portfolio, divesting lands of lesser biological value in favor of acquisitions of higher biological value. Admittedly, this approach necessitates a massive undertaking. Many will think the federal government is not up to the task. It is a daunting task, but the leading alternatives—continuing to muddle through on the road to extinction, or conducting a massive and unprecedented federal intervention in land-use regulation on a generalized and nationwide basis—are almost certainly worse.

Notes & Questions

1. The importance of land to the preservation of biodiversity is easily explained: habitat destruction is the greatest threat to ecosystems and individual species, but most habitat destruction occurs only with the approval of the landowner. A shopping center cannot displace a wetland, for example, if the landowner does not want to build a shopping center there. Thus a landowner can do much to protect biodiversity simply by managing the land with that goal in mind. The challenge, then, is to persuade landowners that the protection of biodiversity is a worthy goal.

There are, of course, millions of landowners. The largest landowner in the United States is the federal government, which owns about 650 million acres, or nearly one-third of the land in the nation. Federal ownership is especially prevalent in the western states, where the government owns 83% of the land in Nevada and more than half of the land in four other states. The management of federal land is divided among numerous agencies, each of which has different priorities in managing the land. Karkkainen highlights four federal agencies in his article:

- the Bureau of Land Management (BLM), which manages the most land (264 million acres), primarily in the west, that was never reserved for a special purpose but which is "relatively rich in biodiversity" (and whose statutory responsibilities with respect to biodiversity are described in more detail in chapter 9);

- the Forest Service, which oversees 191 million acres of national forests throughout the nation (and whose statutory responsibilities with respect to biodiversity are described in more detail in chapter 8);

- the FWS, which administers the National Wildlife Refuge system that includes 511 national wildlife refuges totaling 92 million acres, with the purposes of individual refuges ranging from protecting migratory bird habitat and elk refuges to "vast Alaskan holdings [that] may come closer than any other category of federal lands to constituting genuine biodiversity reserves" (and whose statutory responsibilities with respect to biodiversity are described in more detail in chapter 7); and

- the National Park Service, which contains 376 national parks, monuments, and other areas comprised of 81 million acres that "contribute less to biological diversity than might be expected," (and whose statutory responsibilities with respect to biodiversity are described in more detail in chapter 12).

Each agency must abide by different statutes that regulate what kinds of activities can occur on each type of land, with the importance of biodiversity varying significantly in each statute. The result is that "wildlife conservation is not a commanding force in federal land management. It is one of several management objectives on all the major federal land classifications. . . . It is not the exclusive, or even the dominant, goal on any lands but the national wildlife refuges." MICHAEL J. BEAN & MELANIE J. ROWLAND, THE EVOLUTION OF NATIONAL WILDLIFE LAW 278 (3d ed. 1997). The statutes governing each agency are complemented by statutes such as the National Environmental Policy Act (NEPA) and the ESA (about which much more in Part II of this book) that require federal agencies to account for environmental values, and biodiversity in particular, when making land use decisions.

2. Federal land ownership is not static. The federal government transfers some lands to private parties and state and local governments even as it acquires other lands. Federal acquisition of land holds special promise for the protection of biodiversity. As Professor Karkkainen observes, "in some areas—Hawaii, Florida, Texas, parts of the southeast, and high-population coastal areas generally—where combinations of rare, unique, or especially fragile ecosystems combine with intense development pressure to create a heightened threat of species and ecosystem extinction, targeted federal acquisitions of the last remaining habitat fragments may be the last best hope for conserving biodiversity." *Id.* at 51. But coastal property in rapidly developing areas is expensive, too. A number of federal programs empower the government to purchase lands of special biological value, including section 5 of the ESA and the Land and Water Conservation Fund (LWCF). Critics complain, however, that those programs are chronically underfunded. Congress is considering a substantial infusion of money for conservation purposes in the proposed Conservation and Reinvestment Act (CARA), which would automatically invest $3 billion in annual revenues from federal offshore oil and gas leases into conservation programs. The National Wildlife Federation explains that "[t]he logic behind CARA ... is simple and sensible—revenues derived from the exploitation of the nation's non-renewable oil and gas resources should be reinvested in the protection and restoration of renewable natural resources such as our wildlife, public lands, and coasts." National Wildlife Federation, *Conservation Funding: For Wildlife and Wild Places* <http:www.nwf.org/naturefunding/>. The contrary view, expressed by analysts with the Heritage Foundation, attacks CARA as "little more than a pork-filled land grab by federal and state land management and recreation agencies" that will divert money from other priorities, interfere with local land use decisions, and give the government an unfair advantage in buying land. Gregg VanHelmond & Angela Antonelli, *Why CARA is Fiscally Irresponsible and a Threat to Local Land Use Decisions*, The Heritage Foundation Backgrounder (May 9, 2000), *available at* <http:www.heritage.org/library/backgrounder/bg1370.html>.

3.   Besides the federal government, many other actors own land that contains abundant biodiversity. Increasingly, state and local governments are purchasing land in order to conserve the biodiversity it contains. Thus the State of Illinois spent $21 million to obtain property that contained great blue heron rookeries and other wildlife once the Supreme Court held in Solid Waste Agency of Northern Cook County v. U.S. Army Corps of Engineers, 531 U.S. 159 (2001), that federal Clean Water Act regulations could not prevent local governments from using the site to store its municipal wastes. Nature preserves also play a prominent role in the efforts of many other countries to protect biodiversity, as detailed in chapter 13. And many private landowners seek to protect biodiversity around the world. These actions range from the decisions of individual landowners to attract and sustain a diverse range of flora and fauna on their property to the purposeful efforts of international groups to acquire land that is especially noted for its biodiversity. (The work of the Nature Conservancy, the largest organization focused on the acquisition of lands for biodiversity, is described below).

4.   Karkkainen identifies two alternatives to his proposal: the status quo and more restrictive regulation of land use to protect biodiversity. Is his plan better than those alternatives? Are there any other alternatives to Karkkainen's proposal besides the ones he identifies?

B.   Private Actions to Protect Biodiversity

Martha Hodgkins Green *How Green Is My Valley?*

47 Nature Conservancy 18 (Sept./Oct. 1997).

... There is a valley in Idaho, the valley of the Silver Creek, caught between the mountains and the desert, between the pressure to change and a longing to stay the same.

This small valley—a five-mile-wide basin planed flat and arable between dry, stark hills—carries a high price on its head, valued variously for its world-famous creek, its ranches and farms, its wildlife, its proximity to posh resorts and its development potential.

Despite rising land prices, many landowners know the valley's worth cannot be measured in dollars per acre alone, but in something more—in open space, in silvery stands of aspen, in sandhill cranes picking their way among stubble fields, in a disappearing rural way of life. And it is their vision that is winning out in the Silver Creek Valley, as a handful of landowners have protected more than 9,600 acres of private land, permanently. Their protection tool has been the conservation easement, a legal agreement landowners voluntarily make with land trusts restricting the type and amount of development that can take place, regardless of who owns the property in the future. In this valley, the Conservancy, farmers and ranchers have relied on easements to conserve a watershed and to hold development at bay.

"This is where we're going to make our stand," says John Stevenson, one of the first Silver Creek landowners to donate an easement to the Conservancy, referring to the development schemes that continue to make inroads into this valley of some 1,000 residents. His voice is confident, though he eyes the Timmerman Hills behind his 4,000–acre Hillside Ranch with concern. Just beyond, a subdivision is planned. "It's got to be protected." Stevenson and many of his neighbors agree, there's simply too much at stake here.

Riffles and Waves

Tucked beneath clumps of willow and river birch, Silver Creek gurgles to life at dozens of springs scattered widely across the valley, from Gannett southeast to Picabo. The cold springs are rare in the arid high desert, and the water's steady surge outward and near-constant temperatures make Silver Creek ideal for trout. Big trout. Along the creek's wide, sweeping S-turns and in the deep reaches between riffles, you can see the shadowy outlines of two-foot-long rainbows. Some estimates gauge more than 5,100 fish per mile—one of the densest gatherings of wild trout in any stream in the country.

In fishing circles, Silver Creek is legendary. "You'll love it here," Ernest Hemingway wrote his 17–year-old son, Jack, in 1940. "Saw more big trout rising than have ever seen.... Just like English chalk streams." Eventually the creek worked its magic on the younger Hemingway, for it was he, as Idaho's Fish & Game Commissioner in the 1970s, who first contacted the Conservancy to help rescue the creek. Years of overfishing had made Silver Creek "a very sick stream," recalls Nif Sullivan, whose family owned land along the creek in the first half of the century. The trout were puny, "more like snakes than fish."

In 1976, the Conservancy purchased the 479–acre Sun Valley Ranch on Silver Creek—a ranch once owned by Averell Harriman and Union Pacific Railroad as a hunting and fishing accompaniment to the country's first destination ski resort they had developed in Sun Valley, some 30 miles to the north. A few years later the Conservancy expanded its Silver Creek Preserve by 400 acres when it rescued an adjacent property from development.

Although land acquisition has been a tried and true approach for the Conservancy over the years, it alone cannot do the job. "We're here for the watershed," says Paul Todd, the Conservancy's Silver Creek area manager. "But there's no way we could ever own, or want to own, all the land we think is necessary to protect the creek."

For one, land prices started going up in the 1970s, when "things got real crazy in Blaine County," says Nick Purdy, a fourth-generation rancher in the valley. As chairman of the county's fledgling planning and zoning committee then, Purdy had a firsthand look at the development pressures hitting the north end of the county—from Sun Valley to Hailey. He takes well-deserved pride in the ordinances they passed during his tenure, such as those limiting sprawl between towns and preventing it from creeping up the sere hills. But Purdy also had a too-close-for-comfort look at the pressures extending into the Silver Creek Valley at the county's southern end-into the agricultural land that his great-grandfather had laid claim to in the 1880s.

John Stevenson, too, was wary of the southward march of development. Emigrating from the East in the 1970s to farm in the valley, he was perhaps more attuned than his neighboring ranchers to developers' appetite for land that was devouring other rural areas around the country. Whatever his foresight, he says he decided to put an easement on part of his land because he and his wife, Elizabeth, found their new ranch, with its shallow wetlands scattered among barley fields and pastures, a "neat ecological area." They didn't savor the thought that it might ever be developed, even after their lifetimes. So in 1983, the Stevensons donated their first easement to the Conservancy: 200 acres along Stalker Creek.

Land ownership in the United States carries with it a bundle of rights—the right to construct buildings, to subdivide, to restrict access or to harvest timber, among others. When the Stevensons decided to give away the right to development on that 200–acre parcel, they "eased" it to the Conservancy. The legal agreement is binding, no matter who owns the land in the future. Because taxes are based on a property's "highest and best use," which usually means development, the Stevensons received a tax break, as the parcel's development potential no longer exists. Stevenson acknowledges that land valuation and tax issues have been a motivating factor over the years as he proceeded to donate another four easements covering more than 1,000 acres.

"I think there's some fear among landowners about losing control of land under easement," says Stevenson, but goes on to add that it's unfounded. His easements with the Conservancy are tailored, like all easements, to the unique conditions of the land and to his wishes. He continues to grow barley that he sells mainly to beer companies. To address the issue of potential agricultural runoff into streams, a buffer zone between his fields and the tributaries is established in the easement. The Conservancy shoulders some management responsibility, as evidenced in a small dam constructed 10 years ago to revive a marshy pond. On the pasture that Stevenson leases to ranchers, the Conservancy has constructed fencing to keep cattle-and soil—out of waterways.

Stevenson's easement surely made waves in the valley. "Some neighbors were pretty suspicious at first," he says, "but things have changed. I've seen even a few of the old-time ranchers come around." Indeed, some 8,700 acres of ranchland and farmland, and 25 miles of stream, are now under easement-more than 10 times what the Conservancy owns in the valley. The ripple-effect from those waves eventually reached many a stream bank.

At the general store in Picabo, there's a historical collection of sorts in the far corner, back behind the post office, the tinfoil and toothpaste, the homemade chili and burritos. It quietly boasts railroad ties and wagon-wheel rim blocks, a crumbling ranch ledger, newspaper and magazine clippings, and some yellowing photographs. One shows Gary Cooper and Ernest Hemingway, happy after a day of hunting, sharing a laugh with a young man. A nearby photo portrays this same man, now older, on horseback next to another rugged cowboy—the Marlboro man, the real one from billboard fame.

The man from the photos is Bud Purdy, now 80, owner of the Picabo Livestock Company and patriarch of a local ranching family whose presence in the valley stretches back several generations. The store is his, and it is the hub of Picabo and of 50,000 acres of ranchland that he manages with his son, Nick. Standing before the artifacts and photos, you get a strong sense of what this place means to the Purdys. Their identity is written on the land as it has been since an ancestor working for the railroad put down roots here along with rails.

The Purdys have shown a tenacious streak in keeping their land in the family for 112 years. They've had to diversify their cattle operation along the way, and so started an agricultural irrigation company in Boise. "Farming here is not fun," says Nick, whose sons moved on to Boise to run the business. Then in the 1990s, with land along Silver Creek valued at $10,000 an acre-a tenfold increase in 25 years-the Purdys faced a huge obstacle: how to hold on to their land when exorbitant inheritance taxes would come due once Bud died.

After watching 10 other valley landowners donate easements to the Conservancy, the Purdys donated one of their own in 1994—over 3,400 creekside acres. Without the easement, that same land might have been valued, in Bud's estate, based on the potential development of dozens of "ranchettes"—10–to 20–acre-plot subdivisions. Nick believes the taxes levied on Bud's estate, assessed at full development potential, would have forced them to sell off part of the land to keep the remainder.

"The easement also makes the kids manage the land the way we want," says Nick. And it allows them to develop three homesites in the future—keeping open options for their hoped-for homecoming. "It shows our commitment to the land. Dad and I think it's important for the preservation of the agricultural way of life." Not that their commitment needed much more underscoring. In 1995 Bud and his wife, Ruth, received the Environmental Stewardship Award from the National Cattlemen's Association in recognition of their grazing practices, willow plantings and reintroduction of beavers, whose dams have helped create meadows and reduce erosion.

"We treat the land as something permanent, for future generations," says Bud, former chairman of the National Cattlemen's Association. "Up in the north county, land is treated like a commodity, to be bought and sold for profit."

Both he and his son cringe at the idea of housing developments one day flanking their ranch. "The two don't mix," Nick says distastefully. "When the rural way of life is gone, you're out of there, you know?"

Here to Stay

Midway up a sage-covered hill, Paul Todd and Jerry Bashaw stand in full afternoon sun surveying Bashaw's farm. Amid sculpted brown fields below snakes a cord of green—the native sedges, shrubs and trees marking a stream, one of the many feeding Silver Creek. "When I moved here a few years ago," says Bashaw, "that creek was dry." Diversions had deprived the low spots of water. His interest in restoring the wetlands on his property led him to the Conservancy and to donating an easement on 1,000 acres. Today the verdant scene at their feet attests to their mutual success, underscored by Bashaw's story of a moose he'd recently seen along the stream, and the sound of drumming sage grouse. Todd is excited at this eye-witness report. Their conversation ranges easily from the moose to spring planting and the rows of barley that curve close to the waterway. As with all the easements the Conservancy holds in the Silver Creek Valley, Todd is responsible for monitoring them to make sure landowners comply with the easement terms.

"Easements aren't just technical legal tools," says Phil Tabas, director of conservation programs in the Conservancy's Eastern Regional Office. "They're partnerships, and partners must communicate openly and often. Also, when we take an easement, we've got to be prepared to stick with it for the long haul." He notes the hard work and financial responsibility the Conservancy assumes when accepting an easement—as well as a willingness to defend it legally in the event it is ever violated.

"The landowners who've given us easements are really dedicated to this place," says Todd, "and proud." Bud Purdy, for instance, has told him that when he was a boy he never saw cranes and geese like he does now. "I think they all see that conservation is what's ultimately going to save the valley."

Larry Schoen is clearly of this mind. A young urban refugee, Schoen dived into the Silver Creek Valley and its protection with a passion after he moved here six years ago. He bought parcels that already carried easements and then donated others. "Subdivision is chewing up this country," he says, echoing Nick Purdy's thoughts that suburban sprawl and agriculture are a clash of two worlds.

A businessman-turned-farmer, Schoen fortunately is not without a sense of humor about his idealism. "At first, I tried to plant an English hedge-row behind the house," he says wryly. "Do you know how many buckets of water I hauled trying to keep that thing alive?" Soon after a local farmer took Schoen under his wing—taught him to operate a combine, when to plant, how to judge manure loads.

These days Schoen is president of the Blaine County Ranchers Association, which has a conservation bent to its pro-agriculture agenda; Nick Purdy and John Stevenson are on the board. Schoen often talks to neighbors about donating easements to protect more land. "I tell them, look, I'm still farming and managing the way I want," he says. And what about the lower land valuation after an easement? "This place is much more valuable to me whole," he says. Yet, despite the easements, Schoen still worries about the strong growth pressure leaning on the valley.

"Some folks confuse newcomers like Schoen and Bashaw with that pressure," says Stevenson. "But they have the energy and resources to sink into the land to keep it whole."

Whatever their longevity, many farmers and ranchers in the Silver Creek Valley so far have managed to keep sprawl at a distance. With their land under easement and more likely to bear easements in the near future, there's a complete, solid feel to the valley. Like things just might stay the same.

"When I first moved here, I was on city time," says Schoen. Used to running at that pace, he admits the thought of making a quick turn-around on the farm he was piecing together flashed through his mind at some point. "But soon it hit me: This place moves in its own sense of time." He feels it when walking among the aspen, in the tall sage, in the fields at night where the calls of snipe and rails are sharp. With this new-found sense of time, he's grown deeply attached to the valley. "I can see living here for life now."

Notes & Questions

1.   What advantages are there to relying upon privately owned easements for protecting biodiversity? Are there any disadvantages? Note that some of the easements held by the Nature Conservancy in central Washington are being used to conserve the fifty remaining pygmy rabbits that were listed as endangered species on an emergency basis in November 2001. *See* Emergency Rule to List the Columbia Basin Distinct Population Segment of the Pygmy Rabbit (Brachylagus idahoensis) as Endangered, 66 Fed. Reg. 59724, 59740 (2001). For a more detailed analysis of the use of conservation easements to protect biodiversity, see Peter M. Morisette, *Conservation*

Easements and the Public Good: Preserving the Environment on Private Lands, 41 NAT. RESOURCES J. 373 (2001).

More generally, the Nature Conservancy operates the largest private system of nature sanctuaries in the world. The organization's mission is to "preserve plants, animals and natural communities that represent the diversity of life on Earth by protecting the lands and waters they need to survive." The Nature Conservancy relies upon private donations to purchase the land that it identifies as important wildlife and plant habitat in need of protection. In 1997, the organization received over $226 million in contributions, dues and grants. It also received gifts of land valued at nearly $64 million. Since 1953, the Nature Conservancy has helped to protect 10.5 million acres of habitat in the United States and sixty million acres throughout the rest of the world. It currently manages over one million acres of land that it owns or that on which it owns a conservation easement. It made its largest purchase in 2000 when it spent $35 million for Palmyra Atoll, a marine wilderness filled with coral reefs and emerald islets that had been targeted for a casino or a nuclear waste repository. The land obtained or protected by the Nature Conservancy is home to a wide variety of ecosystems and species. *See* About the Nature Conservancy, http://www.tnc.org/welcome/about/about.htm; Bennett A. Brown, *Landscape Protection and the Nature Conservancy*, in LANDSCAPE LINKAGES AND BIODIVERSITY 66–71 (Wendy E. Hudson ed. 1991).

Like any organization, though, the Nature Conservancy is not without its detractors. Property rights groups fear that the Conservancy's activities threaten property values in certain communities and interfere with the decisions of individual landowners. These fears were given credence by a letter in which a local Conservancy official threatened that the government would exercise its eminent domain power to purchase land for a new national wildlife refuge if the existing owner did not sell his property voluntarily, but the President of the Conservancy quickly disavowed such tactics. *See The Nature Conservancy's Condemnation Threat Letter* (reprinting correspondence between a local Illinois landowner, the Nature Conservancy, and Congressman Glenn Poshard), *available at* <http://members.aol.com/JWaugh7596/TNCthreat.html>. At the same time, some environmentalists suspect that the Conservancy is too willing to work with developers and private landowners. For example, a group of Arizona and New Mexico environmental groups accused the Conservancy of "allowing itself to be used as a shield against the growing public pressure to reform cattle grazing practices on Federal and State lands," and of supporting amendments that would undercut the effectiveness of the federal Endangered Species Act. *See* Memorandum from Southwest Center for Biological Diversity et al. to Andy Laurenzi, Nature Conservancy, Jan. 29, 1999, *available at* <http:www.sw-center.org/swcbd/activist/tnc.html>.

The Nature Conservancy is the largest group committed to acquiring land to protect biodiversity, but it is not the only such organization. And even without an organization, individual landowners can play a significant role with respect to particular places or species. Tom Aley, for example, is a "recognized cave specialist and expert karst hydrogeologist" who owns the cave in southwestern Missouri that is the only known home to the Tumbling Creek cavesnail, which was listed as endangered on an emergency basis in December 2001. *See* Listing the Tumbling Creek Cavesnail as Endangered, 66 Fed. Reg. 66803, 66804, 66806 (2001).

2.   Private groups have been particularly creative in soliciting funding for their conservation work. For example, Defenders of Wildlife has encouraged individuals to fund the ongoing work to protect wolves, the Arctic National Wildlife Refuge, and other biodiversity by donating the rebate that they received as a result of the 2001 federal income tax break legislation. Many organizations encourage private efforts to protect biodiversity by allowing individuals to "adopt" a particular rare animal. The National Zoo sponsors an "Adopt a Species" program that provides you with a photograph of "your" animal or plant (chosen from among dozens of species) and other benefits in exchange for a monetary contribution toward the preservation of that species. Defenders of Wildlife offers a choice of a bear, dolphin, polar bear, sea otter, snowy owl, whale, or wolf for adoption. The Adopt–A–Manatee program offers an adoption certificate, photo, biography of "your" manatee and other materials to those who pay an annual membership fee used to protect endangered Florida manatees. Overseas, the Chengdu Giant Panda Breeding and Research Center sponsors a more ambitious program for the preservation of pandas. *See Taiwan Pop Singers Adopt Two Chengdu Giant Pandas*, Xinhua News Agency, Oct. 12, 1998 (reporting that two entertainers paid $5,000 to adopt two pandas and that most of those who adopt pandas travel to visit them each year). The Nature Conservancy allows individuals to adopt an acre of wildlife habitat.

Private groups have also encouraged the preservation of biodiversity by compensating those who suffer losses caused by an endangered animal. The most notable program is sponsored by the Defenders of Wildlife, which pays ranchers whose livestock is killed by rare wolves. Since 1987, the organization has paid over $206,000 to 179 ranchers who lost a total of 249 cattle, 545 sheep and 28 other animals to attacks by wolves. *See* The Bailey Wolf Foundation Wolf Compensation Trust: Payments to Ranchers from Defenders' Wolf Compensation Program, *available at* <http://www.defenders.org/wildlife/wolf/wolfcomp.pdf>.

Besides giving money or land, private groups and individuals have worked in other ways to protect endangered species. For example, businesses have allowed their employees to take time off to remove invasive vegetation that threatens the habitat of rare wildlife. Ranchers allowed researchers working to save the California condor to use their property free of charge. Members of a Florida Boy Scout troop worked with National Wildlife Federation volunteers to collect melaleuca leaf weevils so that the insects could be released elsewhere in the Picayune State Forest to feed on Australian melaleuca trees that were destroying the habitat of Florida panthers. What other actions can private parties take to protect endangered species?

3. Notwithstanding all of these efforts, few believe that private actions alone will be sufficient to preserve biodiversity. Why?

Lujan v. Defenders of Wildlife

Supreme Court of the United States, 1992.
504 U.S. 555.

■ JUSTICE SCALIA delivered the opinion of the Court with respect to Parts I, II, III–A, and IV, and an opinion with respect to Part III–B, in which THE CHIEF JUSTICE, JUSTICE WHITE, and JUSTICE THOMAS join.

This case involves a challenge to a rule promulgated by the Secretary of the Interior interpreting § 7 of the Endangered Species Act of 1973 (ESA), 87 Stat. 892, as amended, 16 U. S. C. § 1536, in such fashion as to render it applicable only to actions within the United States or on the high seas. The preliminary issue, and the only one we reach, is whether respondents here, plaintiffs below, have standing to seek judicial review of the rule.

I.

The ESA, 87 Stat. 884, as amended, 16 U. S. C. § 1531 *et seq.*, seeks to protect species of animals against threats to their continuing existence caused by man. *See generally* TVA v. Hill, 437 U.S. 153 (1978). The ESA instructs the Secretary of the Interior to promulgate by regulation a list of those species which are either endangered or threatened under enumerated criteria, and to define the critical habitat of these species. 16 U.S.C. §§ 1533, 1536. Section 7(a)(2) of the Act then provides, in pertinent part:

"Each Federal agency shall, in consultation with and with the assistance of the Secretary [of the Interior], insure that any action authorized, funded, or carried out by such agency ... is not likely to jeopardize the continued existence of any endangered species or threatened species or result in the destruction or adverse modification of habitat of such species which is determined by the Secretary, after consultation as appropriate with affected States, to be critical." 16 U. S. C. § 1536(a)(2).

In 1978, the Fish and Wildlife Service (FWS) and the National Marine Fisheries Service (NMFS), on behalf of the Secretary of the Interior and the Secretary of Commerce respectively, promulgated a joint regulation stating that the obligations imposed by § 7(a)(2) extend to actions taken in foreign nations. The next year, however, the Interior Department began to reexamine its position. A revised joint regulation, reinterpreting § 7(a)(2) to require consultation only for actions taken in the United States or on the high seas, was proposed in 1983, and promulgated in 1986, 51 Fed. Reg. 19926; 50 CFR 402.01 (1991).

Shortly thereafter, respondents, organizations dedicated to wildlife conservation and other environmental causes, filed this action against the Secretary of the Interior, seeking a declaratory judgment that the new regulation is in error as to the geographic scope of § 7(a)(2) and an injunction requiring the Secretary to promulgate a new regulation restoring the initial interpretation. [The Eighth Circuit rejected the Secretary's motion to dismiss the case for lack of standing]. . . .

Over the years, our cases have established that the irreducible constitutional minimum of standing contains three elements. First, the plaintiff must have suffered an "injury in fact"—an invasion of a legally protected interest which is (a) concrete and particularized, and (b) "actual or imminent, not 'conjectural' or 'hypothetical.' " Second, there must be a causal connection between the injury and the conduct complained of—the injury has to be "fairly . . . trace[able] to the challenged action of the defendant, and not . . . the result [of] the independent action of some third party not before the court." Third, it must be "likely," as opposed to merely "speculative," that the injury will be "redressed by a favorable decision." . . . .

III.

We think the Court of Appeals failed to apply the foregoing principles in denying the Secretary's motion for summary judgment. Respondents had not made the requisite demonstration of (at least) injury and redressability.

A.

Respondents' claim to injury is that the lack of consultation with respect to certain funded activities abroad "increas[es] the rate of extinction of endangered and threatened species." Of course, the desire to use or observe an animal species, even for purely esthetic purposes, is undeniably a cognizable interest for purpose of standing. "But the 'injury in fact' test requires more than an injury to a cognizable interest. It requires that the party seeking review be himself among the injured." To survive the Secretary's summary judgment motion, respondents had to submit affidavits or other evidence showing, through specific facts, not only that listed species were in fact being threatened by funded activities abroad, but also that one or more of respondents' members would thereby be "directly" affected apart from their " 'special interest' in the subject."

With respect to this aspect of the case, the Court of Appeals focused on the affidavits of two Defenders' members—Joyce Kelly and Amy Skilbred. Ms. Kelly stated that she traveled to Egypt in 1986 and "observed the traditional habitat of the endangered nile crocodile there and intend[s] to do so again, and hope[s] to observe the crocodile directly," and that she "will suffer harm in fact as the result of [the] American . . . role . . . in overseeing the rehabilitation of the Aswan High Dam on the Nile . . . and [in] developing . . . Egypt's . . . Master Water Plan." Ms. Skilbred averred that she traveled to Sri Lanka in 1981 and "observed the habitat" of "endangered species such as the Asian elephant and the leopard" at what is now the site of the Mahaweli project funded by the Agency for International Development (AID), although she "was unable to see any of the endangered species"; "this development project," she continued, "will seriously reduce endangered, threatened, and endemic species habitat including areas that I visited ... [, which] may severely shorten the future of these species"; that threat, she concluded, harmed her because she "intend[s] to return to Sri Lanka in the future and hope[s] to be more fortunate in spotting at least the endangered elephant and leopard." When Ms. Skilbred was asked at a subsequent deposition if and when she had any plans to return to Sri Lanka, she reiterated that "I intend to go back to Sri Lanka," but confessed that she had no current plans: "I don't know [when]. There is a civil war going on right now. I don't know. Not next year, I will say. In the future."

We shall assume for the sake of argument that these affidavits contain facts showing that certain agency-funded projects threaten listed species—though that is questionable. They plainly contain no facts, however, showing how damage to the species will produce "imminent" injury to Mses. Kelly and Skilbred. That the women "had visited" the areas of the projects before the projects commenced proves nothing. As we have said in a related context, " 'Past exposure to illegal conduct does not in itself show a present case or controversy regarding injunctive relief ... if unaccompanied by any continuing, present adverse effects.' " And the affiants' profession of an "intent" to return to the places they had visited before—where they will presumably, this time, be deprived of the opportunity to observe animals of the endangered species—is simply not enough. Such "some day" intentions—without any description of concrete plans, or indeed even any specification of *when* the some day will be—do not support a finding of the "actual or imminent" injury that our cases require.

Besides relying upon the Kelly and Skilbred affidavits, respondents propose a series of novel standing theories. The first, inelegantly styled "ecosystem nexus," proposes that any person who uses *any part* of a "contiguous ecosystem" adversely affected by a funded activity has standing even if the activity is located a great distance away. This approach, as the Court of Appeals correctly observed, is inconsistent with our opinion in [Lujan v. National Wildlife Federation, 497 U.S. 871 (1990)], which held that a plaintiff claiming injury from environmental damage must use the area affected by the challenged activity and not an area roughly "in the vicinity" of it. It makes no difference that the general-purpose section of the ESA states that the Act was intended in part "to provide a means whereby the ecosystems upon which endangered species and threatened species depend may be conserved," 16 U. S. C. § 1531(b). To say that the Act protects ecosystems is not to say that the Act creates (if it were possible) rights of action in persons who have not been injured in fact, that is, persons who use portions of an ecosystem not perceptibly affected by the unlawful action in question.

Respondents' other theories are called, alas, the "animal nexus" approach, whereby anyone who has an interest in studying or seeing the endangered animals anywhere on the globe has standing; and the "vocational nexus" approach, under which anyone with a professional interest in such animals can sue. Under these theories, anyone who goes to see Asian elephants in the Bronx Zoo, and anyone who is a keeper of Asian elephants in the Bronx Zoo, has standing to sue because the Director of the Agency for International Development (AID) did not consult with the Secretary regarding the AID-funded project in Sri Lanka. This is beyond all reason. Standing is not "an ingenious academic exercise in the conceivable," but as we have said requires, at the summary judgment stage, a factual showing of perceptible harm. It is clear that the person who observes or works with a particular animal threatened by a federal decision is facing perceptible harm, since the very subject of his interest will no longer exist. It is even plausible—though it goes to the outermost limit of plausibility—to think that a person who observes or works with animals of a particular species in the very area of the world where that species is threatened by a federal decision is facing such harm, since some animals that might have been the subject of his interest will no longer exist, *see* Japan Whaling Assn. v. American Cetacean Society, 478 U.S. 221, 231, n. 4 (1986). It goes beyond the limit, however, and into pure speculation and fantasy, to say that anyone who observes or works with an endangered species, anywhere in the world, is appreciably harmed by a single project affecting some portion of that species with which he has no more specific connection.

B.

Besides failing to show injury, respondents failed to demonstrate redressability. Instead of attacking the separate decisions to fund particular projects allegedly causing them harm, respondents chose to challenge a more generalized level of Government action (rules regarding consultation), the invalidation of which would affect all overseas projects. This programmatic approach has obvious practical advantages, but also obvious difficulties insofar as proof of causation or redressability is concerned. As we have said in another context, "suits challenging, not specifically identifiable Government violations of law, but the particular programs agencies establish to carry out their legal obligations . . . [are], even when premised on allegations of several instances of violations of law, . . . rarely if ever appropriate for federal-court adjudication."

The most obvious problem in the present case is redressability. Since the agencies funding the projects were not parties to the case, the District Court could accord relief only against the Secretary: He could be ordered to revise his regulation to require consultation for foreign projects. But this would not remedy respondents' alleged injury unless the funding agencies were bound by the Secretary's regulation, which is very much an open question. Whereas in other contexts the ESA is quite explicit as to the Secretary's controlling authority, *see, e.g.,* 16 U. S. C. § 1533(a)(1) ("The Secretary shall" promulgate regulations determining endangered species); § 1535(d)(1) ("The Secretary is authorized to provide financial assistance to any State"), with respect to consultation the initiative, and hence arguably the initial responsibility for determining statutory necessity, lies with the agencies, *see* § 1536(a)(2) ("*Each Federal agency shall*, in consultation with and with the assistance of the Secretary, insure that any" funded action is not likely to jeopardize endangered or threatened species) (emphasis added). When the Secretary promulgated the regulation at issue here, he thought it was binding on the agencies, *see* 51 Fed. Reg. 19928 (1986). The Solicitor General, however, has repudiated that position here, and the agencies themselves apparently deny the Secretary's authority. (During the period when the Secretary took the view that § 7(a)(2) did apply abroad, AID and FWS engaged in a running controversy over whether consultation was required with respect to the Mahaweli project, AID insisting that consultation applied only to domestic actions.). . . .

A further impediment to redressability is the fact that the agencies generally supply only a fraction of the funding for a foreign project. AID, for example, has provided less than 10% of the funding for the Mahaweli project. Respondents have produced nothing to indicate that the projects they have named will either be suspended, or do less harm to listed species, if that fraction is eliminated. . . . [I]t is entirely conjectural whether the non-agency activity that affects respondents will be altered or affected by the agency activity they seek to achieve. There is no standing.

Notes & Questions

1.   Disputes concerning biodiversity feature prominently in the Supreme Court's controversial standing jurisprudence. *See, e.g.*, Bennett v. Spear, 520 U.S. 154 (1997) (granting standing to irrigation districts to challenge a biological opinion issued pursuant to the ESA); Lujan v. National Wildlife Federation, 497 U.S. 871 (1990) (denying standing to the National Wildlife Federation to challenge the Bureau of Land Management's proposal to reclassify certain federal lands); Sierra Club v. Morton, 405 U.S. 727 (1972) (denying standing to the Sierra Club to challenge a proposed Disney ski resort in a national game refuge). Why does the Constitution—and the Court—restrict who may bring suit to object to governmental decisions? Was Justice Blackmun right to imply that the Court's decision implied that plaintiffs raising environmental claims were being subjected to more rigorous standing requirements? *See Defenders of Wildlife*, 504 U.S. at 595 (Blackmun, J., dissenting). Or was Justice Scalia right to suggest in another part of his opinion that "[t]o permit Congress to convert the undifferentiated public interest in executive officers' compliance with the law into an 'individual right' vindicable in the courts is to permit Congress to transfer from the President to the courts the Chief Executive's most important constitutional duty, to 'take Care that the Laws be faithfully executed,' Art. II, § 3." *Id.* at 577.

2.   Private individuals and organizations can bring suit to enforce the terms of the ESA by virtue of the act's citizen suit provision. That provision, and the similar provisions in nearly every federal environmental law, gives private parties a role in implementing the law. Generally, the federal government's FWS (or the NMFS) plays the lead role in enforcing the ESA, but Congress created the citizen suit provision to provide an alternative means of seeking compliance with the statute's demands. Likewise, many state biodiversity statutes contain analogous provisions. *See*

Susan George, William J. Snape III & Rina Rodriguez, *The Public in Action: Using State Citizen Suit Statutes to Protect Biodiversity,* 6 U. Balt. J. Envtl. L. 1 (1997). Of course, a citizen suit provision can also be employed by developers, agricultural interests, and other organizations that are concerned about how the law is being applied. As you read through the cases in this book, note whether the plaintiff was a governmental official or agency, an environmental or trade organization, or a private individual, and consider why it was the government or a private party that brought a particular lawsuit.

3.   Another alternative is for the affected creatures to bring suit to protect themselves. The reported decisions include many lead plaintiffs such as the loggerhead turtle, the northern spotted owl, etc. Whether such animals are entitled to sue is questionable. An early article written by Christopher Stone argued that trees should have standing to sue in court, *see* Christopher Stone, *Should Trees Have Standing: Toward Legal Rights for Natural Objects,* 45 S. Cal. L. Rev. 450 (1972), a theory that Justice Douglas adopted in *Sierra Club v. Morton,* 405 U.S. at 742 (Douglas, J., dissenting), but which failed to attract the support of the rest of the Court. More recent commentators still insist that animals can, or at least should, be entitled to sue. *See* Cass R. Sunstein, *Standing for Animals (With Notes on Animal Rights),* 47 U.C.L.A. L. Rev. 1333, 1335 (2000) (concluding that "Congress can accord standing to animals if it chooses to do so"). The issue remains unresolved by the decided cases. As Judge Becker explained in the context of the federal ESA:

> In several cases, standing has been extended without significant analysis to members of protected species that have allegedly been injured. *See* Palila v. Hawaii Dep't of Land and Natural Resources, 852 F.2d 1106, 1107 (9th Cir.1988) (the Loxioides bailleui "has legal status and wings its way into federal court as a plaintiff in its own right"); *see also* Marbled Murrelet v. Babbitt, 83 F.3d 1068 (9th Cir.1996); Mt. Graham Red Squirrel v. Yeutter, 930 F.2d 703 (9th Cir.1991); Northern Spotted Owl v. Hodel, 716 F.Supp. 479 (W.D.Wash.1988); Northern Spotted Owl v. Lujan, 758 F.Supp. 621 (W.D.Wash.1991); Cabinet Mountains Wilderness/Scotchman's Peak Grizzly Bears v. Peterson, 222 U.S. App. D.C. 228, 685 F.2d 678 (D.C.Cir.1982). Additionally, in Marbled Murrelet v. Pacific Lumber Co., 880 F.Supp. 1343 (N.D.Cal. 1995), the district court determined, without resort to the authorizing provision of the ESA, that because of its protected status under the ESA, the Marbled Murrelet "had standing to sue in its own right." *Id.* at 1346 (citations omitted); *see also* Loggerhead Turtle v. County Council of Volusia County, Florida, 896 F.Supp. 1170, 1177 (M.D.Fla. 1995) (same).

> On the other hand, in two reported cases in which the naming of an animal as a party was explicitly challenged, the courts, in thoughtful opinions, concluded that a protected animal did not have standing to bring suit. *See* Citizens to End Animal Suffering & Exploitation, Inc. v. New England Aquarium, 836 F.Supp. 45, 49–50 (D.Ma.1993) (grant-

ing defendants' motion to remove dolphins name from caption of case because they lacked standing to sue under the Marine Mammal Protection Act); Hawaiian Crow v. Lujan, 906 F.Supp. 549, 551–52 (D.Haw. 1991) (holding that Hawaiian Crow was not a "person" with standing to sue under § 11 of ESA). In reaching this conclusion, these courts analyzed the language of section 11 of the ESA. The provision expressly authorizes citizen suits brought by "any person," 16 U.S.C. § 1540(g)(1), and the Act defines the term "person" to mean "an individual, corporation, partnership, trust, association, or any other private entity." 16 U.S.C. § 1532(13). Accordingly, the courts reasoned that Congress's use of the term "person" as defined in § 1523(13) does not include the non-"private," un-"associated" animal. Moreover, Judge Wolf observed that if Congress "intended to take the extraordinary step of authorizing animals ... to sue, they could, and should, have said so plainly." *Citizens to End Animal Suffering and Exploitation*, 836 F. Supp. at 49.

Hawksbill Sea Turtle v. Federal Emergency Mgmt. Agency, 126 F.3d 461, 466 n. 2 (3d Cir.1997). Who has the better argument under the ESA? *Should* animals be allowed to sue for themselves? If so, who should be entitled to represent them?

4. Filing a lawsuit is just one of the ways in which private parties can invoke the law to protect biodiversity. Individuals and environmental groups often lobby legislators to enact laws conducive to biodiversity. Agency officials solicit and consider public comments on such decisions as the listing of a species as endangered and the proper management plan for national forests. Moreover, many laws are designed to provide information about the environmental consequences of a governmental action to any interested parties. The National Environmental Policy Act itself cannot block a federal project even if an environmental impact statement (EIS) identifies threats to an ecosystem or a species, but the information gained from an EIS can be used by private organizations and individuals to apply political pressure against that project.

C. STATE LAW PROTECTION OF BIODIVERSITY

Barrett v. New York

Court of Appeals of New York, 1917.
220 N.Y. 423, 116 N.E. 99.

■ ANDREWS, J.

At one time beaver were very numerous in this state. So important were they commercially that they were represented upon the seal of the New Netherlands and upon that of the colony as well as upon the seals of New Amsterdam and of New York.

Because of their value they were relentlessly killed, and by the year 1900 they were practically exterminated. But some fifteen animals were left, scattered through the southern portion of Franklin county.

In that year the legislature undertook to afford them complete protection, and there has been no open season for beaver since the enactment of chapter 20 of the Laws of 1900.

In 1904 it was further provided that "No person shall molest or disturb any wild beaver or the dams, houses, homes or abiding places of same." (Laws 1904, ch. 674, section 1.) This is still the law, although in 1912 the forest, fish and game commission was authorized to permit protected animals which had become destructive to public or private property to be taken and disposed of. (Laws 1912, ch. 318.)

By the act of 1904 $500 was appropriated for the purchase of wild beaver to restock the Adirondacks, and in 1906 $1,000 more was appropriated for the same purpose. The commission, after purchasing the animals, was authorized to liberate them.

Under this authority twenty-one beaver have been purchased and freed by the commission. Of these four were placed upon Eagle creek, an inlet of the Fourth Lake of the Fulton Chain. There they seem to have remained and increased.

Beaver are naturally destructive to certain kinds of forest trees. During the fall and winter they live upon the bark of the twigs and smaller branches of poplar, birch and alder. To obtain a supply they fell even trees of large size, cut the smaller branches into suitable lengths and pull or float them to their houses. All this it must be assumed was known by the legislature as early as 1900.

The claimants own a valuable tract of woodland upon Fourth Lake bounded in the rear by Eagle creek. The land was held by them for building sites and was suitable for that purpose. Much of its attractiveness depended upon the forest grown upon it. In this forest were a number of poplar trees. In 1912 and during two or three years prior thereto 198 of these poplars were felled by beaver. Others were girdled and destroyed. The Board of Claims has found, upon evidence that fairly justifies the inference, that this destruction was caused by the four beaver liberated on Eagle creek and their descendants, and that the claimants have been damaged in the sum of $1,900. An award was made to them for that sum and this award has been affirmed by the Appellate Division.

To sustain it the respondents rely upon three propositions. It is said, *first*, that the state may not protect such an animal as the beaver which is known to be destructive; *second*, that the provision of the law of 1904 with regard to the molestation of beaver prohibits the claimants from protecting their property and is, therefore, an unreasonable exercise of the police power; and, *third*, that the state was in actual physical possession of the beaver placed on Eagle creek and that its act in freeing them, knowing their natural propensity to destroy trees, makes the state liable for the damage done by them.

We cannot agree with either of these propositions.

As to the first, the general right of the government to protect wild animals is too well established to be now called in question. Their ownership is in the state in its sovereign capacity, for the benefit of all the people. Their preservation is a matter of public interest. They are a species of natural wealth which without special protection would be destroyed. Everywhere and at all times governments have assumed the right to prescribe how and when they may be taken or killed. As early as 1705 New York passed such an act as to deer. (Colonial Laws, vol. 1, p. 585.) A series of statutes has followed protecting more or less completely game, birds and fish.

"The protection and preservation of game has been secured by law in all civilized countries, and may be justified on many grounds. * * * The measures best adapted to this end are for the legislature to determine, and courts cannot review its discretion. If the regulations operate, in any respect, unjustly or oppressively, the proper remedy must be applied by that body." (Phelps v. Racey, 60 N. Y. 10, 14.)

Wherever protection is accorded harm may be done to the individual. Deer or moose may browse on his crops; mink or skunks kill his chickens; robins eat his cherries. In certain cases the legislature may be mistaken in its belief that more good than harm is occasioned. But this is clearly a matter which is confided to its discretion. It exercises a governmental function for the benefit of the public at large and no one can complain of the incidental injuries that may result.

It is sought to draw a distinction between such animals and birds as have ordinarily received protection and beaver on the ground that the latter are unusually destructive and that to preserve them is an unreasonable exercise of the power of the state.

The state may exercise the police power "wherever the public interests demand it, and in this particular a large discretion is necessarily vested in the legislature to determine, not only what the interests of the public require, but what measures are necessary for the protection of such interests. To justify the state in thus interposing its authority in behalf of the public, it must appear, *first*, that the interests of the public generally, as distinguished from those of a particular class, require such interference; and, *second*, that the means are reasonably necessary for the accomplishment of the purpose, and not unduly oppressive upon individuals." (Lawton v. Steele, 152 U.S. 133, 136.)

The police power is not to be limited to guarding merely the physical or material interests of the citizen. His moral, intellectual and spiritual needs may also be considered. The eagle is preserved, not for its use but for its beauty.

The same thing may be said of the beaver. They are one of the most valuable of the fur-bearing animals of the state. They may be used for food. But apart from these considerations their habits and customs, their curious instincts and intelligence place them in a class by themselves. Observation of the animals at work or play is a source of never-failing interest and instruction. If they are to be preserved experience has taught us that protection is required. If they cause more damage than deer or moose, the degree of the mischief done by them is not so much greater or so different as to require the application of a special rule. If the preservation of the former does not unduly oppress individuals, neither does the latter.

In the determination of what is a reasonable exercise of the powers of the government, the acts of other governments under similar circumstances have some bearing. In Wyoming, Utah, North Dakota, Wisconsin, Maine, Colorado and Vermont beaver are absolutely protected. In Michigan they are protected except between November 1st and May 15th of each year. In South Dakota except between November 15th and April 2d. In Quebec for a number of years there was no open season. Lately there has been an open season for a short time in the autumn.

We, therefore, reach the conclusion that in protecting beaver the legislature did not exceed its powers. Nor did it so do in prohibiting their molestation. It is possible that were the interpretation given by the respondents to this section right a different result might follow. If the claimants, finding beaver destroying their property, might not drive them away, then possibly their rights would be infringed. In Aldrich v. Wright (53 N. H. 398) it was said in an elaborate opinion, although this question we do not decide, that a farmer might shoot mink even in the closed season should he find them threatening his geese.

But such an interpretation is too rigid and narrow. The claimants might have fenced their land without violation of the statute. They might have driven the beaver away, were they injuring their property. The prohibition against disturbing dams or houses built on or adjoining watercourses is no greater or different exercise of power from that assumed by the legislature when it prohibits the destruction of the nests and eggs of wild birds even when the latter are found upon private property.

The object is to protect the beaver. That object as we decide is within the power of the state. The destruction of dams and houses will result in driving away the beaver. The prohibition of such acts, being an apt means to the end desired, is not so unreasonable as to be beyond the legislative power.

We hold, therefore, that the acts referred to are constitutional. But had we reached a different conclusion the respondents would not be aided. We know of no principle of law under which the state becomes liable because of the adoption of an unconstitutional statute. Such a statute is no protection to officers assuming to proceed under its authority. The state itself, if it permits such a claim to be enforced against it, may become liable for what they do. But the statute itself is void. No one need obey it. If no affirmative act is done under its supposed authority neither the state nor its officers are liable, because the citizen chooses to obey where he need not have done so.

Somewhat different considerations apply to the act of the state in purchasing and liberating beaver.

The attempt to introduce life into a new environment does not always result happily. The rabbit in Australia—the mongoose in the West Indies have become pests. The English sparrow is no longer welcome. Certain of our most troublesome weeds are foreign flowers.

Yet governments have made such experiments in the belief that the public good would be promoted. Sometimes they have been mistaken. Again the attempt has succeeded. The English pheasant is a valuable addition to our stock of birds. But whether a success or failure the attempt is well within governmental powers.

If this is so with regard to foreign life, still more is it true with regard to animals native to the state, still existing here, when the intent is to increase the stock upon what the Constitution declares shall remain forever wild forest lands. If the state may provide for the increase of beaver by prohibiting their destruction it is difficult to see why it may not attain the same result by removing colonies to a more favorable locality or by replacing those destroyed by fresh importations.

Nor are the cases cited by the respondents controlling. It is true that one who keeps wild animals in captivity must see to it at his peril that they do no damage to others. But it is not true that whenever an individual is liable for a certain act the state is liable for the same act. In liberating these beaver the state was acting as a government. As a trustee for the people and as their representative it was doing what it thought best for the interests of the public at large. Under such circumstances we cannot hold that the rule of such cases as those cited is applicable.

We reach the conclusion that no recovery can be had under this claim. It is assumed, both by the respondents and by the appellant, that the Board of Claims had jurisdiction to determine the questions involved. That we do not discuss.

The judgment of the Appellate Division and the determination of the Board of Claims must be reversed and the claim dismissed, with costs in Appellate Division and in this court.

Notes & Questions

1. The saga of the beaver illustrates how states responded to the first indications of a crisis regarding biodiversity. Beaver were plentiful when the first European settlers arrived in America. Over the next several centuries, French, English and Native American trappers aggressively hunted beavers to the extent that the beavers began to disappear. Shepard Krech III picks up the story at that point:

> By the late nineteenth century, the beaver harvest was 10 percent of its level one century before, and beavers were scarce or locally extinct in North America. They disappeared from New Jersey by 1820 and New Hampshire by 1865. By 1890, they were rare or absent in

Pennsylvania, Wisconsin, Minnesota, most of New York, many parts of Quebec and Ontario, and elsewhere.

Concerned legislators passed laws designed to halt the destruction of beavers as they had with deer. Men and women active in the conservation movement that formed in the last three decades of the nineteenth century were appalled at the eradication of buffaloes, passenger pigeons, and other wildlife including beavers. New conservationists spoke of a "mad rush at the counter for fur and psuedo-fur" and the fashion for fur as a "craze." In the twentieth century, conservation sentiments and regulations had taken a stronger hold and for beavers, the tide turned.... In the first two decades of the twentieth century, restocking programs were instituted widely in the United States and Canada. Together with stringent laws restricting trapping, the programs succeeded—to the extent that within just a few years in the Adirondacks, where beavers had been extinct, New York's Conservation Commission called them "interesting but destructive," responsible for flooding highways and railroads. This success brought renewed trapping during fur booms in the 1920s and 1940s. Soon most states again allowed beaver trapping and the annual harvest in North America climbed to hundreds of thousands of pelts.

Like white-tailed deer, beavers survived to recover much of their former range.... Beavers recovered as a consequence of trapping restrictions, restocking, changes in fashion, and conservation. In the 1990s, antifur lobbies and changing fashions have cast trapping as a pariah profession, leaving beaver populations unchecked. Anthropomorphized, beavers are loved in the abstract—until like deer their unbridled populations explode into suburban cultural landscapes as pests, attracting headlines like "Busy Beavers Gnaw on Suburban Nerves" and "Besieged by Beavers in Rural New York." As the millennium approaches, these "annoying overachievers" once again are busily and eagerly altering every conceivable habitat in North America.

Shepard Krech III, The Ecological Indian: Myth and History 178–79 (1999).

2. The *Barrett* court held that the protection of wildlife is within the state's police power. Historically, states played the leading role in regulating wildlife within their borders. Geer v. Connecticut, 161 U.S. 519 (1896), is the leading case establishing the primacy of states with respect to wildlife. In *Geer*, the defendant was convicted for violating a state law prohibiting the possession of game birds with the intent to transfer them out of state. Geer charged that the exclusive congressional power to regulate interstate commerce rendered the Connecticut law unconstitutional. The Supreme Court disagreed, with Justice White explaining that states had the right "to control and regulate the common property in game ... as a trust for the benefit of the people." *Id.* at 528–29. *Geer* came to stand for the proposition that the states owned the wildlife within their borders. To some, it also indicated that *only* the states could regulate wildlife, so that the federal government lacked the constitutional power to do so. That claim was soon called into question when Congress enacted the Lacey Act in

1900, it further eroded when the Court upheld the Migratory Bird Act of 1913 as a valid exercise of the federal treaty power in Missouri v. Holland, 252 U.S. 416 (1920), and it has grown even more difficult to sustain in the aftermath of several Court decisions since the 1970s that have found that various federal wildlife statutes are authorized by either the congressional power to regulate interstate commerce or to regulate the property of the United States. *See generally* BEAN & ROWLAND, *supra*, at 15–27 (reviewing the federal constitutional authority to regulate wildlife).

Nonetheless, the scope of the federal power to protect wildlife remains contested. Lower court judges have reached different conclusions about the ability of the federal government to regulate an endangered species that lives in only one state and that is not an object of commerce. *See* Gibbs v. Babbitt, 214 F.3d 483 (4th Cir.2000) (holding 2–1 that the commerce clause permits federal regulation of the killing of endangered wolves on private land), *cert. denied*, 531 U.S. 1145 (2001); National Ass'n of Home Builders v. Babbitt, 130 F.3d 1041 (D.C.Cir.1997) (holding 2–1 that the commerce clause permits federal regulation of activities that would interfere with the habitat of the Delhi Sands Flower–Loving Fly), *cert. denied*, 524 U.S. 937 (1998); *see generally* John Copeland Nagle, *The Commerce Clause Meets the Delhi Sands Flower–Loving Fly*, 97 MICH. L. REV. 174 (1998) (analyzing the commerce clause issues raised by the ESA). The Supreme Court has declined to consider the issue, though it has recently questioned the extent to which Congress may rely upon the commerce clause to protect wetlands that are home to migratory birds. *See* Solid Waste Agency of Northern Cook County v. U.S. Army Corps of Engineers, 531 U.S. 159 (2001). *SWANCC* and the commerce clause issue are explored in more detail in chapter 10.

3. The defendant in *Barrett* also demanded that the state compensate him for the damage that the beavers caused to his property. Why did Judge Andrews reject that claim? Did the state government cause the injury suffered by the landowner? *Cf.* Palsgraf v. Long Island R. Co., 162 N.E. 99, 101 (1928) (Andrews, J., dissenting) (insisting that the railroad was the proximate cause of Mrs. Palsgraf's injuries). Damage done by wildlife protected—or even introduced—by the government continues to give rise to claims for compensation. *See* Christy v. Hodel, 857 F.2d 1324 (9th Cir.1988) (holding that the federal government did not take a Montana rancher's property when a grizzly bear killed his livestock), *cert. denied*, 490 U.S. 1114 (1989); State v. Sour Mountain Realty, Inc., 276 A.D.2d 8, 17–18 (N.Y.App.Div.2000) (holding that the state did not take a quarry owner's property when it prohibited the landowner from removing protected rattle-snakes). The need for the government to compensate property owners who cannot use their land as they like because of the presence of legally protected wildlife is discussed in more detail below at page 272.

In re Killington, Ltd.

Supreme Court of Vermont, 1992.
159 Vt. 206, 616 A.2d 241.

■ DOOLEY, J.

The Vermont Environmental Board denied an Act 250 permit to Killington, Ltd., for construction of a pond intended to enhance Killington's snowmaking capacity at its ski area. Killington appeals that ruling, which was based on the Board's conclusion that the proposed construction would not satisfy 10 V.S.A. § 6086(a)(8)(A) because it would imperil a habitat necessary to the survival of a population of black bears. We affirm.

Application for the permit was filed on February 18, 1986, with District Environmental Commission No. 1. The application sought approval to divert the waters of Madden Brook to build a snowmaking pond in an area known as Parker's Gore East in the Town of Mendon. After extensive hearings and procedural maneuvering, the District Commission denied the application on July 14, 1987. The Commission found that construction of the pond would threaten a population of black bears by making inaccessible a stand of beech trees used by the bears for food. The Commission also found that the construction would prevent the bears from using a tract of spruce and fir to travel into Parker's Gore East, where they build up body fat in the fall in preparation for winter hibernation.

Pursuant to 10 V.S.A. § 6089(a), Killington appealed the Commission's denial to the Environmental Board on August 13, 1987. The appeal challenged the Commission's decisions to grant party status to certain organizations and municipalities and to impose conditions on any permit that might be issued. Its fundamental challenge, however, was to the conclusion, under 10 V.S.A. § 6086(a)(8)(A) (hereinafter criterion 8(A)), that the project would significantly impair necessary wildlife habitat, would not utilize all feasible and reasonable means of lessening the alleged impairment, and would be a detriment to the general welfare of the public. Killington asserted that the Commission's findings did not support its conclusions and that the conclusions were based on a fundamental misconstruction of the term "necessary wildlife habitat" as used in the statute.

On May 11, 1989, the Board ruled that the term "necessary wildlife habitat" in criterion 8(A) covered habitat critical to the survival of the particular wildlife population dependent on it, and that Parker's Gore East constituted necessary wildlife habitat. It also concluded that construction and operation of the pond would destroy or significantly impair that habitat in a number of respects. On September 21, 1990, the Board issued its final findings. It concluded that Killington met none of the subcriteria of criterion 8(A),[1] which, if satisfied, allow the issuance of a permit regardless

---

1. These are set forth in 10 V.S.A. § 6086(a)(8)(A)(i)-(iii) as follows:

(A) Necessary wildlife habitat and endangered species. A permit will not be granted if it is demonstrated by any party opposing the applicant that a development or subdivision will destroy or significantly imperil necessary wildlife habitat or any endangered species, and

(i) the economic, social, cultural, recreational, or other benefit to the public from the development or subdivision will not outweigh the economic, environmental, or recreational loss to the public from the destruction or imperilment of the habitat or species, or

(ii) all feasible and reasonable means of preventing or lessening the destruction, diminution, or imperilment of the habitat or species have not been or will not continue to be applied, or

of the imperilment of necessary wildlife habitat. Therefore, the Board also denied the permit.

[Killington claims] that the Board misapprehended the meaning of 10 V.S.A. § 6086(8)(A), which requires the denial of a permit if a proposed project "will destroy or significantly imperil necessary wildlife habitat or any endangered species." Before the Board, Killington argued that the statute must be read to allow denial only when the existence of a particular wildlife species is threatened by the proposed project. Thus, Killington contended that the Board could not deny the permit under criterion 8(A) because only the population of black bears which lived in or utilized Parker's Gore East would be affected by the pond construction, and the existence of black bears elsewhere in the state would be unaffected. The Board ruled, as it had in previous decisions, that the destruction or significant imperilment of the habitat of a population of wildlife triggered criterion 8(A) review, irrespective of whether the species as a whole was threatened with extinction in the state. Since the Board's decision in the instant case, we have unanimously affirmed the Board's view in a case in which precisely this issue was presented. In re Southview Associates, 153 Vt. [171,] 175–76, 569 A.2d [501,] 503 [(1989)]. Killington asks that we overrule that decision. We see no reason to do so.

In *Southview*, we upheld the Board's denial of a permit for a development that would have threatened the habitat of a population of approximately twenty deer. We said that "a 'necessary wildlife habitat' under Act 250 is one that is decisive to the survival of the population of a particular species that depends upon the habitat." We are bound by *Southview* as a matter of stare decisis. Our adherence to precedent is reinforced by the fact that the Legislature has met twice since the *Southview* decision and has not amended the statute in response to that decision. . . .

Finally, Killington claims that the Board's findings do not support its conclusions that Parker's Gore East contains necessary black bear habitat and that such a habitat would be destroyed or significantly imperiled by the pond construction and operation. We note that Killington does not assert that the findings themselves are unsupported by substantial evidence, and so the only issue is whether they support the Board's conclusions.

Killington's attack on the Board's conclusion that the wetlands and the beech trees are necessary bear habitat is based on the absence of a finding that all the bears in Parker's Gore East would die without this habitat. A similar claim was made in *Southview*. Our response in that case is equally applicable here:

> Southview appears to argue that because the evidence did not prove that all the deer would perish if the project were completed, the project's opponents failed to prove that the deeryard was "necessary wildlife habitat." This argument again misconstrues the terms of the statute. A concentrated, identifiable deeryard is "necessary wildlife

(iii) a reasonably acceptable alternative site is owned or controlled by the applicant which would allow the development or subdivision to fulfill its intended purpose.

habitat'' if it is decisive to the survival of the white-tailed deer that use it during the winter—that is, if the deer require that sort of habitat to survive the winter. Of course, many of the individual animals might survive in another deeryard elsewhere in the state if the project were built; that fact does not render the area to be developed unnecessary to their survival in the sense contemplated by the Act.

Id. at 177 n.3, 569 A.2d at 504 n.3.

Based on this understanding of the statute, the findings are ample to support the Board's conclusion on necessary wildlife habitat. The Board detailed the importance of the wetlands as a source of food for the bears when they emerge from hibernation in the spring. It concluded that because the bears ''are dependent upon wetlands for their spring food supply'' and there are no other wetlands in Parker's Gore East, the wetland at the pond site was necessary bear habitat. Similar analysis was used with respect to the beech trees.

Killington makes a related argument that the findings are inadequate to support the conclusion that the construction and operation of the pond will destroy or significantly imperil necessary wildlife habitat. There is no question that the construction of the pond would eliminate or virtually eliminate the wetland. The findings support the conclusion.

Bolsa Chica Land Trust v. The Superior Court of San Diego County

Court of Appeal of California, Fourth Appellate District, Division One, 1999.
71 Cal.App.4th 493, 83 Cal.Rptr.2d 850.

■ Benke, J.

This case concerns development plans for a large tract of land in southern Orange County known as Bolsa Chica. Although the California Coastal Commission (Commission) approved a local coastal program (LCP) for Bolsa Chica, the trial court found defects in the program and remanded it to Commission for further proceedings. In this court both the opponents and proponents of the LCP contend that the trial court erred.

The opponents of the LCP contend the trial court erred in finding a planned relocation of a bird habitat was permissible under the Coastal Act. The proponents of the LCP contend the trial court erred in preventing residential development of a wetlands area and in requiring preservation of a pond that would have been eliminated under the LCP in order to make room for a street widening. The proponents also attack the trial court's award of attorney fees to the opponents of the LCP.

We find the trial court erred with respect to relocation of the bird habitat. The Coastal Act does not permit destruction of an environmentally sensitive habitat area (ESHA) simply because the destruction is mitigated offsite. At the very least, there must be some showing the destruction is needed to serve some other environmental or economic interest recognized by the act.

Factual Background

Bolsa Chica is a 1,588–acre area of undeveloped wetlands and coastal mesas. Urban development surrounds Bolsa Chica on three sides. On the fourth side is the Pacific Ocean, separated from Bolsa Chica by a narrow strip of beach, coastal dunes and coastal bluffs.

Approximately 1,300 acres of Bolsa Chica consist of lowlands ranging from fully submerged saltwater in Bolsa Bay to areas of freshwater and saltwater wetlands and islands of slightly raised dry lands used by local wildlife for nesting and foraging. However, a large part of the lowlands is devoted to an active oil field and at one time the area was farmed.

The lowlands are flanked by two mesas, the Bolsa Chica Mesa on the north and the Huntington Mesa on the south. The Bolsa Chica Mesa consists of 215 acres of uplands hosting a variety of habitat areas. Although much of Huntington Mesa is developed, a long narrow undeveloped strip of the mesa abutting the lowlands is the planned site of a public park.

In 1973 the State of California acquired 310 contiguous acres of the Bolsa Chica lowlands in settlement of a dispute over its ownership of several separate lowland parcels and the existence of a public trust easement over other lowland areas.

In 1985 the County of Orange and Commission approved a land use plan for Bolsa Chica which contemplated fairly intense development. The 1985 plan allowed development of 5,700 residential units, a 75–acre marina and a 600–foot-wide navigable ocean channel and breakwater.

By 1988 substantial concerns had been raised with respect to the environmental impacts of the proposed marina and navigable ocean channel. Accordingly, a developer which owned a large portion of Bolsa Chica, a group of concerned citizens, the State Lands Commission, the County of Orange and the City of Huntington Beach formed the Bolsa Chica Planning Coalition (coalition). The coalition in turn developed an LCP for Bolsa Chica which substantially reduced the intensity of development. The coalition's LCP was eventually adopted by the Orange County Board of Supervisors. Commission approved the LCP with suggested modifications which were adopted by the board of supervisors.

As approved by Commission, the LCP eliminated the planned marina and navigable ocean channel, eliminated 3 major roads, reduced residential development from a total of 5,700 homes to 2,500 homes on Bolsa Chica Mesa and 900 homes in the lowlands and expanded planned open space and wetlands restoration to 1,300 acres.

The material features of the LCP which are in dispute here are: the replacement of a degraded eucalyptus grove on Bolsa Chica Mesa with a new raptor habitat consisting of nesting poles, native trees and other native vegetation on Huntington Mesa at the sight of the planned public park; the residential development in the lowland area which the LCP permits as a means of financing restoration of substantially degraded wetlands; and the elimination of Warner Pond on Bolsa Chica Mesa in order to accommodate the widening of Warner Avenue.

Throughout the approval process several interested parties and public interest groups, including the Bolsa Chica Land Trust, Huntington Beach Tomorrow, Shoshone–Gabrieleno Nation, Sierra Club and Surfrider Foundation (collectively the trust) objected to these and other portions of the LCP. . . .

IV.   Eucalyptus Grove

A.   *History and Condition of the Grove*

The LCP would permit residential development over five acres of a six-and-one-half-acre eucalyptus grove on Bolsa Chica Mesa. The five acres where development would be permitted is owned by Koll; the remainder of the grove is owned by the state.

The eucalyptus grove is not native to the area and was planted almost 100 years ago by a hunting club which owned large portions of Bolsa Chica. Since the time of its planting, the original 20–acre grove has diminished considerably because of development in the area and the lack of any effort to preserve it. Indeed, although the eucalyptus grove was nine and two-tenths acres large as recently as 1989, it had shrunk to no more than six and one-half acres by 1994 and portions of it were under severe stress. According to expert testimony submitted to Commission, the grove is probably shrinking because of increased salinity in the soil.

Notwithstanding its current diminished and deteriorating condition, Commission identified the grove as an ESHA within the meaning of Public Resources Code section 30107.5. The ESHA identification was based on the fact the grove provided the only significant locally available roosting and nesting habitat for birds of prey (raptors) in the Bolsa Chica area. At least 11 species of raptors have been identified as utilizing the site, including the white-tailed kite, marsh hawk, sharp skinned hawk, Cooper's hawk and osprey. According to Commission, a number of the raptors are dependent upon the adjacent lowland wetlands for food and the eucalyptus grove provides an ideal nearby lookout location as well as a refuge and nesting site.

B.   *Section 30240*

Under the Coastal Act, Commission is required to protect the coastal zone's delicately balanced ecosystem. (§ 30001, subds. (a)-(c), 30001.5, subd. (a); City of San Diego v. California Coastal Com. (1981) 119 Cal. App. 3d 228, 233 [174 Cal. Rptr. 5]; Sierra Club v. California Coastal Com. (1993) 12 Cal. App. 4th 602, 611 [15 Cal. Rptr. 2d 779] .) Thus in reviewing all programs and projects governed by the Coastal Act, Commission must consider the effect of proposed development on the environment of the coast.

In terms of the general protection the Coastal Act provides for the coastal environment, we have analogized it to the California Environmental Quality Act (CEQA) (§ 21000–21174). (Coastal Southwest Dev. Corp. v. California Coastal Zone Conservation Com. (1976) 55 Cal. App. 3d 525, 537 [127 Cal. Rptr. 775].) We have found that under both the Coastal Act and

CEQA: " 'The courts are enjoined to construe the statute liberally in light of its beneficent purposes. The highest priority must be given to environmental consideration in interpreting the statute.' "(*Ibid.*)

In addition to the protection afforded by the requirement that Commission consider the environmental impact of all its decisions, the Coastal Act provides heightened protection to ESHA's. Section 30107.5 identifies an ESHA as "any area in which plant or animal life or their habitats are either rare or especially valuable because of their special nature or role in an ecosystem and which could be easily disturbed or degraded by human activities and developments." "The consequences of ESHA status are delineated in section 30240: '(a) Environmentally sensitive habitat areas shall be protected against any significant disruption of habitat values, and only uses dependent on those resources shall be allowed within those areas. [P] (b) Development in areas adjacent to environmentally sensitive habitat areas and parks and recreation areas shall be sited and designed to prevent impacts which would significantly degrade those areas, and shall be compatible with continuance of those habitat and recreation areas.' Thus development in ESHA areas themselves is limited to uses dependent on those resources, and development in adjacent areas must carefully safeguard their preservation."

Commission found that residential development in the eucalyptus grove was permissible under section 30240 because the LCP required that an alternate raptor habitat be developed on Huntington Mesa. Commission reasoned that section 30240 only requires that "habitat values" be protected and that given the deteriorating condition of the grove, creation of a new raptor habitat on Huntington Mesa was the best way to promote the "habitat values" of the eucalyptus grove.

The reasoning Commission employed is seductive but, in the end, unpersuasive. First, contrary to Koll's argument, we are not required to give great weight to the interpretation of section 30240 set forth by Commission in its findings approving the LCP.... Secondly, the language of section 30240 does not permit a process by which the habitat values of an ESHA can be isolated and then recreated in another location. Rather, a literal reading of the statute protects the area of an ESHA from uses which threaten the habitat values which exist in the ESHA. Importantly, while the obvious goal of section 30240 is to protect habitat values, the express terms of the statute do not provide that protection by treating those values as intangibles which can be moved from place to place to suit the needs of development. Rather, the terms of the statute protect habitat values by placing strict limits on the uses which may occur in an ESHA and by carefully controlling the manner uses in the area around the ESHA are developed.

Thirdly, contrary to Commission's reasoning, section 30240 does not permit its restrictions to be ignored based on the threatened or deteriorating condition of a particular ESHA. We do not doubt that in deciding whether a particular area is an ESHA within the meaning of section 30107.5, Commission may consider, among other matters, its viability.

However, where, as is the case here, Commission has decided that an area is an ESHA, section 30240 does not itself provide Commission power to alter its strict limitations. There is simply no reference in section 30240 which can be interpreted as diminishing the level of protection an ESHA receives based on its viability. Rather, under the statutory scheme, ESHA's, whether they are pristine and growing or fouled and threatened, receive uniform treatment and protection.

In this regard we agree with the trust that Commission's interpretation of section 30240 would pose a threat to ESHA's. As the trust points out, if, even though an ESHA meets the requirements of section 30107.5, application of section 30240's otherwise strict limitations also depends on the relative viability of an ESHA, developers will be encouraged to find threats and hazards to all ESHA's located in economically inconvenient locations. The pursuit of such hazards would in turn only promote the isolation and transfer of ESHA habitat values to more economically convenient locations. Such a system of isolation and transfer based on economic convenience would of course be completely contrary to the goal of the Coastal Act, which is to protect all coastal zone resources and provide heightened protection to ESHA's.

In short, while compromise and balancing in light of existing conditions is appropriate and indeed encouraged under other applicable portions of the Coastal Act, the power to balance and compromise conflicting interests cannot be found in section 30240.

C. Section 30007.5

Koll argues that even if transfer of habitat values was not permissible under section 30240, such a transfer was permissible under the provisions of section 30007.5 and our holding in [Sierra Club v. California Coastal Comm'n, 19 Cal.App.4th 547, 23 Cal.Rptr.2d 534 (1993) (*Batiquitos Lagoon*)]. Section 30007.5 states: "The Legislature further finds and recognizes that conflicts may occur between one or more policies of the [Coastal Act]. The Legislature therefore declares that in carrying out the provisions of this division such conflicts be resolved in a manner which on balance is the most protective of significant coastal resources. In this context, the Legislature declares that broader policies which, for example, serve to concentrate development in close proximity to urban and employment centers may be more protective, overall, than specific wildlife habitat and other similar resource policies."

In *Batiquitos Lagoon* we were confronted with "the conflicting interests of fish and fowl." Each interest was protected by a specific provision of the Coastal Act: The fish were protected by section 30230 which directed that marine resources be preserved and, where feasible, restored; the fowl were protected by the requirement of section 30233, subdivision (b), that the very substantial dredging needed to restore the fish habitat avoid significant disruption of the bird habitat. We found that under section 30007.5, Commission could resolve these conflicting policy interests by favoring long-term restoration of the fish habitat over the short-term, but significant, disruption of the bird habitat.

Here, in contrast to the situation in *Batiquitos Lagoon*, the record at this point will not support application of the balancing power provided by section 30007.5. Unlike the record in that case, here our review of the proceedings before Commission does not disclose any policy or interest which directly conflicts with application of section 30240 to the eucalyptus grove.

Although the Coastal Act itself recognizes the value and need for residential development, nothing in the record or the briefs of the parties suggests there is such an acute need for development of residential housing in and around the eucalyptus grove that it cannot be accommodated elsewhere. Rather, the only articulated interests which the proposed transfer of the "habitat values" serves is Commission's expressed desire to preserve the raptor habitat values over the long term and Commission's subsidiary interest in replacing nonnative eucalyptus with native vegetation. However, as the trust points out, there is no evidence in the record that destruction of the grove is a prerequisite to creation of the proposed Huntington Mesa habitat. In the absence of evidence as to why preservation of the raptor habitat at its current location is unworkable, we cannot reasonably conclude that any genuine conflict between long-term and short-term goals exists.

In sum then the trial court erred in sustaining that portion of the LCP which permitted development of the eucalyptus grove.

[The court also upheld the Commission's conclusion that neither residential development in the wetlands nor destruction of the pond is permissible].

Department Of Community Affairs v. Moorman

Supreme Court of Florida, 1995.
664 So.2d 930.

■ Kogan, J.

. . . This case involves the validity of a land-use ordinance enacted to protect an endangered species, the miniature Florida Key deer. The regulation affects Big Pine Key where the deer now are largely concentrated. Human development on the Key has put the deer perilously close to extinction, and their numbers are estimated to be only 350 to 400 animals. The minimum number needed to sustain a viable species is considered 100 to 250 animals. The animals are further endangered by human attempts to feed them, by pet dogs that may kill them, and by automobiles.

The ordinance in question prohibits the erection of fencing in portions of Big Pine Key, where the respondents own property. It was enacted because, in a natural environment, Key deer must roam freely over slash pinelands and wetlands in search of food and water. This necessarily means the deer also must roam over some privately owned lands. The blanket prohibition on fencing was meant as an interim restriction to be replaced within a year by a more comprehensive regulation that would better identify where fence restrictions would be proper and where they were unnecessary. Despite its interim nature, the ordinance had been on the books for more than five years before the times in question here.

The Florida key deer

Charles Moorman owns a lot located in the slash pinelands of Big Pine Key. He also is owner of Your Local Fence, a company that is a respondent in this review. Moorman filed for a permit to build a six-foot-high 400–foot-long fence, which Monroe County granted. This was done notwithstanding the "interim" county ordinance. The record contains evidence that Moorman's fence is in a location that will adversely affect the Key deer. The fact the land in question sits in an area of critical state concern is crucial to the result in this case, because it identifies an environmental concern unique to Big Pine Key.

The Department of Community Affairs ("DOCA") appealed the County's decision pursuant to DOCA's authority over areas of critical state concern. The Moorman lot sits in an area designated as a "critical state concern" in 1979.[2] The fact the land in question sits in an area of critical state concern is crucial to the result in this case, because it identifies an environmental concern unique to Big Pine Key.

The case was referred to a Division of Administrative Hearings ("DOAH") officer, who found the permits improper. The finding specifically noted that the permits were issued as a matter of right. The officer

---

2. The fact the land in question sits in an area of critical state concern is crucial to the result in this case, because it identifies an environmental concern unique to Big Pine Key.

recommended that the Cabinet, sitting as the Florida Land & Water Adjudicatory Commission ("Commission"), rescind the permits, and the Cabinet agreed. The Moormans then appealed to the Third District, which ruled the anti-fencing regulation facially unconstitutional.

The nub of the issue here is a simple failure to revisit an "interim" land-use regulation that was never intended to be a permanent feature of the county code book, as it seemed to have become by the times in question. While Monroe County chose not to apply the ordinance to respondents, DOCA now has taken the position that the ordinance must be enforced according to its strict letter. DOCA likewise contends that sufficient justification exists for such a blanket prohibition on fencing in the affected area.

We do not dispute that the State has a legitimate interest in protecting the natural habitat of the Keys and most especially of the Key deer. To this end, the State does in fact have a right to use its police power to establish land-use regulations addressing environmental concerns. The clear policy underlying Florida environmental regulation is that our society is to be the steward of the natural world, not its unreasoning overlord. As the Constitution itself states:

It shall be the policy of the state to conserve and protect its natural resources and scenic beauty. Adequate provision shall be made by law for abatement of air and water pollution and of excessive and unnecessary noise.

Art. II, § 7, Fla. Const. There is an obvious public interest in such a policy, given the fact that environmental degradation threatens not merely aesthetic concerns vital to the State's economy but also the health, welfare, and safety of substantial numbers of Floridians.

The right of equal protection embodied in article I, section 2 stands for many things, but it does not restrict the State's ability to establish or mandate reasonable environmental regulations, even those that may apply only in a certain area, where the State is addressing an environmental problem peculiar to the area. Even if equal protection is implicated here, the State would only need a rational basis for the zoning restriction. Here, the State has identified a sufficient interest in this instance to justify the classification in question. The interest is plainly stated in article II, section 7 of the Constitution and is only underscored by the unique problem of the Key deer.

The right of privacy set forth in article I, section 23 also does not apply here for a self-evident reason: The decision to use land in a manner contrary to lawful public environmental policy is simply not a private act. Art. I, § 23, Fla. Const. Landowners do not have an untrammeled right to use their property regardless of the legitimate environmental interests of the State.

The right of due process contained in article I, section 9 poses a somewhat different problem, because it does in fact guarantee the right to enjoy property. Within limits, that right can include decisions regarding the improvement of property. Nevertheless, this personal right does not necessarily supervene the rational concerns of public environmental policy. Due process, in other words, seeks to find a balance between public and private interests, not to make the landowner lord over the State, nor the State lord over the landowner.

The State is given wide range in exercising its lawful powers to regulate land use for environmental reasons, and any such land-use regulations thus are valid if supported by a rational basis consistent with overall policies of the State. Landowners simultaneously are protected by yet another feature of the law: Any resulting erosion of property value beyond reasonable limits is recompensable by means of inverse condemnation. Art. X, § 6, Fla. Const.; art. I, § 9, Fla. Const. In sum, the rights of property owners are limited by the lawful environmental policies of the State, and the State acting within its lawful power to regulate property is likewise limited by the depth of its purse.

However, we have repeatedly held that zoning restrictions must be upheld unless they bear no substantial relationship to legitimate societal policies. Here, the record contains competent substantial evidence that the unregulated erection of fencing in the affected area is contrary to Florida's overall policy of environmental stewardship. In this sense, we agree with the argument of DOCA and disagree with the district court below. Because the ordinance promotes a valid public policy, it should not have been stricken on its face.

Nevertheless, we are constrained to determine whether any valid basis existed for denying the Moormans a permit. We of course recognize that the enactment in question today was intended merely as an "interim" rule. This fact in turn means the blanket prohibition against fencing was never regarded as an essential feature of public policy. Rather, the underlying intent was to replace the ordinance with something less restrictive, although this intent regrettably was never carried forward into action.... However, the uncontroverted expert evidence clearly indicated that the Moormans' fence—the only one at issue here today—was harmful to Key deer habitat.

Based on the foregoing, we find that the application of the ordinance to this particular fence permit was constitutional because it was based on a rational basis consistent with state environmental policy....

■ Grimes, Chief Justice [joined by Wells, J.], concurring in part and dissenting in part.

I agree that the ordinance is constitutional on its face. I cannot agree that it was necessarily constitutional as applied. At the hearing, Charles Moorman testified that the purpose for his fence was to keep his neighbor's two children from falling into his hot tub and to keep the Key Deer from eating his plants. He also testified that his fence was only about 400 feet long.

Despite the fact that this interim ordinance had been on the books for more than five years, this case was tried on an all-or-nothing basis upon the assumption that the Moormans' permit must either stand or fall without adjustment. By virtue of its authority to attach conditions and restrictions to its decision, the Cabinet had the power to fashion a remedy that would soften the blanket prohibition against fencing while at the same time honor environmental stewardship.

I would remand the case for another hearing directed toward whether the regulation was unconstitutional as applied to the Moormans or, alternatively, whether the permit should be modified.

Notes & Questions

1. *Killington, Bolsa Chica Land Trust,* and *Moorman* are examples of state law being used to protect biodiversity. How are the statutes at issue in each case different? Which one seems best suited to protect biodiversity?

Many states have their own statutes protecting the rare wildlife and plants within their borders. Some of those laws offer protections similar to those provided by the federal laws such as the Endangered Species Act. The Defenders of Wildlife have published a comprehensive analysis of the effectiveness of state laws protecting endangered species. *See* Susan George, William J. Snape III & Michael Senatore, State Endangered Species Acts: Past, Present and Future, *available at* <http://www.defenders.org/pb-bst00.-html>. According to that study, 45 states have some kind of legislation that requires the protection of endangered species. (The exceptions are Alabama, Arkansas, Utah, West Virginia, and Wyoming). The report concludes, however, that "most acts easily fall far short of what is needed to adequately protect a state's imperiled species." Specifically, the report identifies shortcomings in the procedures by which species are listed, the protection of critical habitat, the conduct prohibited by the laws, recovery plans, and the penalty and enforcement provisions. The state listing procedures are often similar to the federal ESA (which is described below in chapter 4), with a handful of exceptions such as New Jersey's provision that listing is optional, the Virginia provision that plants and insects need not be listed if that is not "in the best interest of man," and the Maine and Montana provisions that final listing decisions are made by the legislature. The critical habitat provisions are more troubling: "Only six states have provisions requiring critical habitat designation and they are rarely used." States also differ from the federal ESA in failing to prohibit the destruction of the habitat of a listed species. While the U.S. Supreme Court has held that the federal prohibition on the "take" of an endangered species can apply to some habitat destruction (as discussed below at page 246), only in Massachusetts is the destruction of habitat a prohibited "take" of the species. Instead, most state laws prohibit hunting, capturing, or otherwise actually injuring a listed species. Recovery plans are similarly absent from most state acts, again unlike the federal ESA's requirement that the Fish & Wildlife Service prepare a plan detailing the steps that are needed to help each listed species recover from the brink of extinction. Finally, the report finds that "most states suffer from a lack of proper enforcement" of their endangered species laws, including the omission of a citizen suit provision that would allow concerned individuals to seek to enforce the provisions of the law themselves.

2.   There may be other ways of using state law to protect biodiversity besides state statutes. The public trust doctrine offers one approach. *See* Ralph W. Johnson & William C. Galloway, *Can the Public Trust Doctrine Prevent Extinctions?*, in BIODIVERSITY AND THE LAW 157–64 (William J. Snape III ed. 1996). Nuisance law might provide another alternative. Recall that a public nuisance is "an unreasonable interference with a right common to the general public." RESTATEMENT (SECOND) TORTS § 821B. An interference is "unreasonable" if it significantly interferes with public health, safety, comfort or convenience, if it is illegal, or if it is continuing or long-lasting and known to have a significant effect. *Id.* One author has argued that public nuisance actions may be used to protect biodiversity because "[a] healthy and viable population of native wildlife is arguably a public right." Siobhan O'Keeffe, *Using Public Nuisance Law to Protect Wildlife*, 6 BUFF. ENVT'L. L.J. 85 (1998). But Richard Epstein counters that "common law actions for trespass and nuisance are forlorn in the context of habitat preservation, because so much of the modification and destruction of habitat targeted by the ESA typically comes from ordinary husbandry of land: the clearing of trees, the construction of houses, the diversion of waters for drink and irrigation (subject to its own distinctive rules), and the like. It would take a stunning reversal of hundreds of years of legal history if these activities, generally productive, were now, for the first time, castigated by the common law as generally harmful." Richard A. Epstein, *Habitat Preservation: A Property Rights Perspective*, in WHO OWNS THE ENVIRONMENT? 229 (Peter J. Hill & Roger E. Meiners eds. 1998).

In *Moorman*, the court noted that the Florida state constitution contains a policy of conserving and protecting the natural resources of the state. Numerous state constitutions contain provisions guaranteeing certain environmental rights, and while those provisions have not typically been viewed as creating any judicially enforceable individual rights, one recent decision suggests a broader role for such provisions. *See* Montana Environmental Information Center v. Department of Envtl. Quality, 988 P.2d 1236 (Mt.1999) (holding that Montana's constitutional right to a clean and healthful environment mandates that a state statute permitting certain water pollution must satisfy strict scrutiny).

Could constitutional provisions play a greater role with respect to biodiversity? Rodger Schlickeisen, the president of Defenders of Wildlife, has argued that the federal constitution should be amended to include a provision protecting biodiversity. His proposed amendment would state:

> The living resources in the United States are the common property of all the people, including generations yet to come. All persons and their progeny have an inalienable, enforceable right to the benefits of those resources for themselves and their posterity. The United States and every state shall assure that use of those resources is sustainable and

that they are conserved and maintained for the benefit of all the people.

Rodger Schlickeisen, *The Argument for a Constitutional Amendment to Protect Living Nature*, in BIODIVERSITY AND THE LAW 234 (William J. Snape III ed. 1996). What is the advantage of including the protection of biodiversity within the constitution? Are there any disadvantages? What laws or actions would be unconstitutional under Schlickeisen's proposed amendment? For an argument against constitutional provisions addressing environmental rights, *see* J.B. Ruhl, *The Metrics of Constitutional Amendments: And Why Proposed Environmental Quality Amendments Don't Measure Up*, 74 NOTRE DAME L. REV. 245 (1999).

3. One of the factors that the federal Endangered Species Act considers when deciding to list a species is the adequacy of the existing regulatory mechanisms available for the protection of the species. *See* ESA § 4(a)(1)(C). This analysis considers both federal and state laws, and thus offers a helpful means of comparing the protection offered by the federal government versus individual states. For example, the listing notice for the seven southeastern mussels indicated that Alabama has commercial harvest guidelines for mussels, Florida imposed a moratorium on the commercial mussel harvest in 1996, and Georgia law places numerous restrictions on the collection of mussels, but that such restrictions are difficult to enforce. *See* Endangered and Threatened Wildlife and Plants; Determination of Endangered Status for Five Freshwater Mussels and Threatened Status for Two Freshwater Mussels from the Eastern Gulf Slope Drainages of Alabama, Florida, and Georgia, 63 Fed. Reg. 12,664, 12,682–12,683 (1998). The federal ESA, by contrast, prohibits a much larger range of conduct.

Consider also the Fish and Wildlife Service's analysis of the state laws available to protect the population of the Arkansas River shiner (ARS) that lives in three rivers crossing four states. The agency explained:

> Federal and state laws and regulations can protect the ARS and its habitat to some extent. The State of Kansas lists the ARS as a State endangered species. The [Kansas Department of Wildlife and Parks] has designated portions of the mainstem Cimarron, Arkansas, South Fork Ninnescah, and Ninnescah rivers as critical habitat for the shiner. A permit is also required by the State of Kansas for public actions that have the potential to destroy listed individuals or their critical habitat. Subject activities include any publicly funded or State or federally assisted action, or any action requiring a permit from any other State or Federal agency. Violation of the permit constitutes an unlawful taking, a Class A misdemeanor, and is punishable by a maximum fine of $2,500 and confinement for a period not to exceed 1 year. Kansas does not permit the commercial harvest of bait fish from rivers and streams.

> The State of New Mexico lists the ARS as a State endangered species. This listing prohibits the taking of the ARS without a valid scientific collecting permit but does not provide habitat protection. The State of Oklahoma lists the ARS as a State threatened species, but like

New Mexico, this listing does not provide habitat protection. The States of Arkansas and Texas provide no special protection for the species or its habitat.

While Kansas, New Mexico, and Oklahoma protect the ARS from take and/or possession, only Kansas addresses the problem of habitat destruction or modification. Only New Mexico provides significant protection from the potential introduction of non-native, competitive species. Licensed commercial bait dealers in New Mexico may sell bait minnows only within the drainage where they have been collected and cannot sell any State-listed fish species.

The Kansas legislature can identify a minimum desirable stream-flow for a stream as part of the Kansas Water Plan. The Chief Engineer is then required to withhold from appropriation the amount of water necessary to establish and maintain the minimum streamflow. New Mexico and Oklahoma water law does not include provisions for acquisition of instream water rights for protection of fish and wildlife and their habitats. However, Oklahoma indirectly provides some protection of instream uses, primarily by withholding appropriations for flows available less than 35 percent of the time. . . .

The status and threats to the ARS reflect, in part, the inability of these laws and regulations to adequately protect and provide for the conservation of the ARS. Even listing as threatened or endangered by the States of Kansas, New Mexico, and Oklahoma has not reversed the decline of this species.

Endangered and Threatened Wildlife and Plants; Final Rule to List the Arkansas River Basin Population of the Arkansas River Shiner (*Notropis girardi*) as Threatened, 63 Fed. Reg. 64,772, 64,795 (1998).

4. Many environmentalists insist that states are incapable of protecting biodiversity. One concern is that many species live in more than one state, thereby complicating the efforts of any particular state to preserve the species. Another contention posits that states have a disincentive to protect endangered species. As one court explained, "states are motivated to adopt lower standards of endangered species protection in order to attract development." National Association of Home Builders v. Babbitt, 130 F.3d 1041 (D.C.Cir.1997), *cert. denied*, 524 U.S. 937 (1998). Are those fears well-founded? Are there any ways in which states actually have a greater incentive than the federal government to protect biodiversity?

D. TRIBAL PROTECTION OF BIODIVERSITY

The Northern Arapahoe Tribe v. Hodel

United States Court of Appeals for the Tenth Circuit, 1987.
808 F.2d 741.

■ SEYMOUR, CIRCUIT JUDGE.

The Northern Arapahoe and Shoshone Tribes jointly inhabit the Wind River Indian Reservation in western Wyoming. At the request of the

Shoshone Tribe (the Shoshone), the Secretary of the Interior promulgated regulations establishing a game code regulating hunting on the reservation. The Arapahoe Tribe (the Arapahoe) sued the Secretary and other federal officials, seeking declaratory and injunctive relief to prevent enforcement of the regulations. The Shoshone intervened in the litigation as a defendant. The court held a two-day hearing on the request for a preliminary injunction. It thereafter entered an order denying the Arapahoe request for temporary relief and, at the same time and without prior notice, deciding the case on the merits and denying a permanent injunction.

On appeal, the Arapahoe contend that the district court's denial of a permanent injunction should be reversed because the Secretary has no authority to regulate hunting on the reservation and because the Secretary violated the Administrative Procedure Act. Alternatively, they assert that the trial court erred in consolidating the preliminary injunction hearing with a trial on the merits without prior notice, and ask that the case be remanded for trial. We affirm in part, reverse in part, and remand for further proceedings.

I.

BACKGROUND

The Wind River Indian Reservation was established in 1868 pursuant to the Treaty of Fort Bridger, which set aside territory "for the absolute and undisturbed use and occupation of the Shoshonee Indians . . ., and for such other friendly tribes or individual Indians as from time to time they may be willing, with the consent of the United States, to admit amongst them." Treaty between the United States of America and the Eastern Band of Shoshonees and the Bannack Tribe of Indians, July 3, 1868, 15 Stat. 673, 674 (the Treaty). Ten years after signing the Treaty, the United States broke this covenant when it brought a band of Northern Arapahoe onto the reservation under military escort. The Arapahoe had been allies of the Sioux, who were antagonistic toward the Shoshone. Despite the Shoshone's continual and vigorous efforts to have the Arapahoe removed, the United States failed to respond, dealing instead with the two tribes as lawful occupants and equals. The Shoshone ultimately were compensated for the taking of part of the reservation in an amount equal to one-half the value of the land, including the timber and mineral resources.

Today, both tribes inhabit the reservation. According to the superintendent of the Wind River Indian Agency, the combined adult population of the tribes is approximately 5,900. Each tribe governs itself separately by vote of the tribal membership at general council meetings or by vote of its elected business council. A joint business council of representatives from both tribes deals with certain matters of common interest.

The reservation itself encompasses nearly 1.9 million acres and ranges in altitude from 4,200 to over 13,000 feet. This topographical diversity provides habitat for a variety of wildlife, from waterfowl to big and small game. Only enrolled members of the tribes may hunt on the reservation. For the Shoshone and the Arapahoe, hunting is a traditional activity and a source of food.

Over the years, the tribes have submitted the issue of tribal game codes to their general memberships for decision. They have managed reservation wildlife both jointly and separately. In 1948 they enacted a joint game code, but abolished it five years later. Since that time, the only joint regulations have been prohibitions against waste, spotlighting, and the selling or trading of game meat.

In 1977 the tribes expressed concern for game management and called for a study of reservation wildlife. The joint business council passed Resolution No. 3923, which provided:

"WHEREAS: The Joint Shoshone and Arapahoe Business Council is aware of the potential of the wildlife habitat available and increasing game herds on the Wind River Indian Reservation, and

WHEREAS: A sound wildlife program is based on high-quality habitat and proper management of wildlife species, and

WHEREAS: Lack of management and protection of wildlife in the past and unrestricted harvest of wildlife species have occurred,

NOW, THEREFORE BE IT RESOLVED, that the US Fish and Wildlife Service be requested to establish a wildlife biologist position in Lander to assist in collecting data to protect habitat and wildlife, and to manage and insure the optimum potential of wildlife species on the Wind River Indian Reservation now, and for the future."

Pursuant to the joint resolution, the United States Fish and Wildlife Service (FWS) undertook a series of habitat and species studies. FWS reported its findings in various separate reports from 1980 to 1982 and in a comprehensive report in 1982 entitled "A Plan for the Management of Wildlife on the Wind River Reservation" (the 1982 Report). The 1982 Report concluded that the tribes' concern about dwindling herds was justified, and recommended management of all wildlife, particularly big game.[2] In December 1983 and February 1984 FWS conducted aerial big game surveys. Richard Baldes, project leader at the Lander FWS office, testified about the results of these surveys: "The information that we collected is the same. The herds are still going down, and we have serious problems. It hasn't changed. It just strengthened what we were saying before."

The tribes disagreed on the proper course of action in light of the conclusions of the FWS studies. In 1980, two years before the publication of the comprehensive 1982 Report, the Shoshone enacted a game code to govern the tribe's own members, which the Arapahoe General Council subsequently rejected as too restrictive. The Shoshone asked the Secretary

---

2. The regulations promulgated by the Secretary define big game as "any one of the following species of animals: elk, mule deer, whitetail deer, bighorn sheep, moose, antelope, black and grizzly bear, and mountain lion." 25 C.F.R. § 244.2 (1986).

to impose a moratorium on all hunting on the reservation until the two tribes could agree on a game code. The Associate Solicitor of Indian Affairs declined to intervene, opting instead to encourage the tribes to resolve the matter. In June 1983 the Arapahoe membership again considered enactment of a game code, but voted to table the issue.

At various times during the spring and fall of 1983, officials from the Bureau of Indian Affairs (BIA) met with the tribes both jointly and separately and expressed their concern about the need for a game code. The biggest meeting was held in September in Billings, attended by representatives from both tribes, their attorneys, and BIA officials from Washington and the regional office. At that meeting, BIA officials discussed the possibility that the federal government would have to issue regulations in order to fulfill its trust responsibility. The Arapahoe tribal council asked for more time, indicating that it would try to establish a code by January 1984.

The hard winter of 1983–1984 forced many of the big game herds to seek shelter and forage in the lower elevations of the reservation. This movement made big game vulnerable to hunters, particularly those with snowmobiles and four-wheel drive vehicles. Reports of a "massive elk kill" in December 1983 prompted wide publicity and a plea from the chairmen of the Shoshone and Arapahoe business councils for tribal members to exercise restraint in the hunting of big game.

The Arapahoe failed to enact a game code by January 1984, its preferred target date. Meetings among the tribes and BIA officials continued through the spring. The Arapahoe membership voted in May not to enact a game code. After meeting further with both tribes in July, BIA officials informed them it would accede to the Shoshone request to impose federal regulations.

On October 5 the BIA published an interim rule entitled "Wind River Reservation Game Code," which is the subject of this lawsuit. *See* 49 Fed. Reg. 39,308 (1984) (codified at 25 C.F.R. pt. 244 (1986)). The game code was designed to establish "a controlled wildlife hunting program on the Wind River Reservation in order to conserve, protect and increase the existing wildlife in the reservation area.".... By its terms, the game code will remain in effect only until the tribes jointly enact a game code. *See* 25 C.F.R. § § 244.1(a).

The game code provides that in 1984 the Wind River Agency Superintendent, an employee of the BIA, would establish the hunting seasons, define the hunting areas, set permit fees, and establish season limits for all wildlife. In subsequent years, the superintendent is to make those determinations before each June 1 after consulting with the tribes. In 1984, all enrolled members who wished to hunt were required to purchase a permit at a cost of $5.00, as well as big game tags at $1.00 per species. The only big game that could legally be harvested were elk, antlered deer, and buck antelope, and those only during specific hunting seasons. Hunting of bighorn sheep, moose, black bear, and mountain lion was prohibited. Other rules were established for hunting or trapping furbearing animals, upland game, and waterfowl. Substantial civil and criminal penalties and forfeitures may be imposed for violations of the regulations. After the game code was implemented, the Wind River Agency Superintendent obtained equipment and employed five federal enforcement agents.

On October 23, 1984, shortly after the regulations were imposed, the Arapahoe filed this action for declaratory and injunctive relief and moved for a temporary restraining order (TRO) to prevent enforcement of the game code. The Shoshone moved to intervene and opposed the issuance of temporary or preliminary relief. On October 24 and 25 the district court held hearings on the motion for a TRO. The Arapahoe filed fifty identical affidavits from members of the tribe, each averring that the affiant had hunted regularly on the reservation for a number of years, that he planned to hunt during the fall, that he hunted only for food, and that enforcement of the regulations would leave him and his family without an adequate food supply. The Arapahoe relied entirely on these affidavits at the hearing, presenting no other evidence.

The Government offered the testimony of two witnesses. Richard Baldes of FWS stated his opinion that the regulation was necessary to protect the reservation's game resources. He testified that, if hunting continued at its present rate, moose and bighorn sheep might become extinct or endangered on the reservation. He further stated that, while prong horn antelope and mule deer were in danger of being eliminated from the reservation if hunting continued at past rates, elimination probably would not occur for several years. Baldes also stated that management of furbearing animals was neither necessary nor a concern of the tribes. Lavern William Collier, Wind River Agency superintendent, testified that since implementation of the code, 103 permits had been sold, no violations had been found, and hunting had been somewhat light. In response to the allegations by Arapahoe tribal members that the regulation would leave them without sufficient food, Collier testified that each enrolled Arapahoe member received a $235 monthly allotment derived from the reservation's assets and that various food and assistance programs were available to needy families.

[The district court denied the Arapahoe motion for a temporary restraining order, and the Arapahoe appealed.]

II.

REGULATION OF HUNTING ON THE WIND RIVER RESERVATION

... The Arapahoe contend that the Secretary lacks any authority to regulate on-reservation hunting by Indians and that, even if he has such authority in exigent circumstances, the fact-finding below was inadequate to determine that the wildlife on the reservation have been hunted "to a point of endangerment or extinction." The Secretary and the Shoshone maintain that, in the absence of a jointly-adopted tribal game code, the Secretary has authority to regulate hunting on the reservation pursuant to his responsibility as trustee in accordance with the Treaty and his general authority under 25 U.S.C. § § 2 and 9 (1982).

A.   The Rights of Both Tribes to Hunt and Fish

As an initial matter, the Shoshone claim that they possess exclusive treaty rights to hunt and fish on the reservation and that the Arapahoe have no such special rights. Although the Treaty does not expressly mention hunting or fishing rights, these rights were included by implication in the setting aside of the reservation for the Shoshone's "absolute and undisturbed use and occupation," Treaty of July 3, 1868, 15 Stat. 673, 674. The Shoshone contend that their treaty rights to hunt and fish were neither lost nor diminished by the congressional and executive acts recognizing the settlement of the Arapahoe on the reservation, and that the Arapahoe cannot claim treaty rights to hunt and fish by virtue of their coexistence on the reservation. The Shoshone further submit that, although they were compensated for the Arapahoe presence on the reservation in an amount equal to one-half the value of the land including timber and mineral resources, they were never compensated for the loss of their exclusive treaty rights to hunt and fish.

We are not persuaded. The very principles of Indian law which dictate that the Shoshone have hunting and fishing rights notwithstanding the lack of an express treaty provision dictate that the Arapahoe have equivalent rights. The Arapahoe have rights to the reservation derived from their status as occupants of the land confirmed by congressional and executive acts. The rights to hunt and fish are part of the tribes' larger rights of possession. Whether by treaty or by congressional and executive acts, the Shoshone and the Arapahoe have equal rights to hunt on the reservation.

B.   Authority of the Secretary

Actions of the Secretary and those under his authority are subject to judicial review under principles of administrative law. Furthermore, in Indian matters, as in other areas, federal executive officials are limited to the authority conferred on them by Congress. "The rulemaking power granted to an administrative agency ... is not the power to make law. Rather, it is 'the power to adopt regulations to carry into effect the will of Congress....'" Santa Fe Industries, Inc. v. Green, 430 U.S. 462, 472 (1977) (quoting Ernst & Ernst v. Hochfelder, 425 U.S. 185, 213–14 (1976)). We thus must determine if the Secretary has been granted authority sufficient to support his enactment of the regulations.

The district court found that "historically, the Shoshone and Northern Arapahoe Tribes now occupying the Wind River Reservation have been free to self-regulate the hunting activities of tribal members on the reservation." The Government does not dispute that the primary authority to regulate hunting lies with the tribes, consistent with their sovereignty over the reservation land and resources. The narrower question presented in this appeal is whether authority exists for the Secretary, at the request of one of the tribes, to adopt interim hunting regulations as a necessary conservation measure to protect endangered wildlife and game on the reservation.

1.   25 U.S.C. § § 2 and 9.

Congress has delegated to the Secretary broad authority to manage Indian affairs, *see* 25 U.S.C. § 2, and to promulgate regulations relating to Indian affairs, *see id.* § 9. Sections 2 and 9, however, do not vest the Secretary with general regulatory authority. Section 2 delegates the general management of Indian affairs and relations to the Secretary of the Interior and Commissioner of Indian Affairs. The language of section 9 vests authority "for carrying into effect the various provisions of any act relating to Indian affairs." 25 U.S.C. § 9. Given the language of the statute and the fact that hunting on the reservation has historically been a matter of tribal self-regulation, we are reluctant to hold that sections 2 and 9 by themselves could support the regulations. For the reasons set out below, however, we conclude that Sections 2 and 9 together with the Treaty, construed in accordance with the special relationship between the United States and Indian tribes, provide the necessary authority for the Secretary to enact these regulations.

2.   The Relationship Between the Federal Government and Indian Tribes

The United States has a unique relationship with Indian tribes "derived from [their] separate constitutional status,"and their existence as quasi-sovereign governments. Drawing upon the concept of a protectorate or alliance relationship founded upon agreement by treaty, Chief Justice John Marshall described Indian tribes as "domestic dependent nations" which "look to our government for protection." Cherokee Nation v. Georgia, 30 U.S. (5 Pet.) 1, 17 (1831); *see* Worcester v. Georgia, 31 U.S. (6 Pet.) 515, 551–52 (1832). The Supreme Court has recognized "the undisputed existence of a general trust relationship between the United States and the Indian people." United States v. Mitchell, 463 U.S. 206 (1983). "This principle has long dominated the Government's dealings with Indians." *Id.*

3.   The Treaty and the Trust Responsibility

The Treaty of 1868 was intended to preserve for the Shoshone a reservation land base with sufficient resources to supply the needs of the Indian people who settled thereon. In the Treaty, the Government undertook the responsibility to protect the persons and property of the Shoshone from wrongdoers "among the whites, or among other people subject to the authority of the United States." Treaty of July 3, 1868, 15 Stat. 673. The right to hunt on the reservation is held in common by both tribes, and one tribe cannot claim that right to a point of endangering the resource in derogation of the other tribe's rights. Under the Treaty, the Government has the right upon request of the Shoshone to protect the resources guaranteed the Shoshone by treaty from misappropriation by third parties. The Government's right extends to preventing overuse by the Arapahoe of their shared right when that overuse endangers the resource and threatens to divest the Shoshone of their right. Because the right to the resource is shared, however, federal regulation of hunting on the reservation must accommodate the rights of both tribes.

When viewed in light of the trust responsibility and the Shoshone's specific request for regulation, the Treaty along with 2 U.S.C. § § 2 and 9 support the Secretary's authority to establish an interim game code on the reservation when there exists a risk of extinction or endangerment of the wildlife. If the facts support such a risk, the Secretary may implement reasonable interim measures so long as the tribes fail to enact their own game code.

[The court also rejected several procedural objections to the Game Code, and remanded for a trial on the merits.]

Notes & Questions

1. The interim game code was finalized with minor modifications soon after the Tenth Circuit's decision. *See* Wind River Reservation Game Code, 52 Fed. Reg. 23805 (1987).

2. Native Americans have an excellent reputation for their treatment of biodiversity. The romanticized version of the relationship of Native Americans to their natural surroundings asserts that "[w]hen Indians alone cared for the American earth, this continent was clothed in a green robe of forests, unbroken grasslands, and useful desert plants, filled with an abundance of wildlife." J. DONALD HUGHES, AMERICAN INDIAN ECOLOGY 1–2 (1983). The continuing concern that Native Americans have for biodiversity is evidenced by stories of the restoration of wetlands and other degraded habitats and by efforts to facilitate the recovery of disappearing species of trout and wolves. Native Americans often possess information about native wildlife and vegetation that conservation biologists have not yet uncovered. *See generally* BIODIVERSITY AND NATIVE AMERICA (Paul E. Minnis & Wayne J. Elisens eds. 2000) (collection of essays exploring how Native Americans have addressed biodiversity).

But not every observer sees a consistent pattern of regard for wildlife and plants. Consider two different accounts of the native treatment of buffalo in the nineteenth century. On one view, "[t]he Plains hunters expressed their thanks for the gifts of the buffalo by killing only as many as they needed and by using every part of the animals." *Id.* at 7. By contrast, other scholars describe "Indians who ate only the buffalo's tongue, only the fetus, or only the hump, or who abandoned bulls because they preferred cows." KRECH, *supra*, at 142. Many of the current disputes are the result of the tribal view of wildlife and plants as essential resources for their physical and cultural survival. That attitude has prompted conflicts between tribes and contemporary environmentalists on several issues. *See, e.g.*, Cook Inlet Beluga Whale v. Daley, 156 F.Supp.2d 16,18, 20 (D.D.C. 2001) (describing how Native American hunting of about 77 Cook Inlet Beluga whales annually is "the single most significant factor in the population decline" of the whale). Native American hunting practices have evoked complaints from those who want to prohibit hunting of certain species, or to prohibit hunting altogether. The desperate economic conditions that exist on many Indian reservations have prompted some tribes to authorize development activities opposed by many environmentalists. And, as discussed at page 237, Native American religious ceremonies sometimes employ parts of bald eagles, panthers, and other species that are now protected by federal and state law. In each instance, the tribal way of life has presented challenges to today's conceptions about the preservation of biodiversity.

3. Indian tribes enjoy a unique relationship to each state and to the United States as a whole. Early in the nineteenth century, Chief Justice John Marshall explained that Indian tribes are sovereign nations that are subject to the higher power of the federal government. *See* Worcester v. Georgia, 31 U.S. (6 Pet.) 515 (1832); Cherokee Nation v. Georgia, 30 U.S. (5 Pet.) 1 (1831). The sovereign status of Indian tribes led the federal government to enter into numerous treaties with individual tribes throughout the nineteenth century. Like the Treaty of Fort Bridger at issue in *The Northern Arapahoe Tribe,* many of those treaties provided that the tribes ceded most of their lands in exchange for rights to use the lands that the tribes retained. As Sandi Zellmer explains, "These retained lands are critical to tribal sovereignty and, indeed, the very survival of tribes as distinct cultural and political communities." Sandi B. Zellmer, *Indian Lands as Critical Habitat for Indian Nations and Endangered Species: Tribal Survival and Sovereignty Come First,* 43 S.D. L. Rev. 381 (1998).

The relative rights of tribes, the federal government, and the state government regarding wildlife continues to generate substantial litigation. *See, e.g.,* Minnesota v. Mille Lacs Band of Chippewa Indians, 526 U.S. 172 (1999) (holding 5–4 that the Chippewa Indians still retain hunting, fishing, and gathering rights guaranteed by a 1837 treaty with the United States); Bean & Rowland, *supra,* at 450–64 (analyzing state authority to regulate hunting and fishing by Native Americans on reservation lands and off reservation lands, federal authority to regulate Native American hunting and fishing, and tribal regulation of hunting and fishing by non-Indians on reservation lands). In particular, there is a tension between federal efforts to protect biodiversity the sovereignty enjoyed by tribes. Notwithstanding the *Billie* case reprinted below at page 231, it is still uncertain whether the ESA abrogates treaties affording tribal hunting and fishing rights when endangered species are at stake. Another controversy involves the designation of tribal lands as the critical habitat of a species pursuant to the ESA. Professor Zellmer contends that critical habitat "designation flies in the face of the United States' solemn promises to preserve tribal homelands for the undisturbed use of Indian nations and to protect tribal sovereignty from external incursions." Zellmer, *supra,* at 382. For the federal government's view of how the ESA affects tribes, visit U.S. Fish & Wildlife Service, *American Indian Tribal Rights, Federal–Tribal Trust Responsibilities, and the Endangered Species Act* <http://endangered.fws.gov/tribal/index.html>.

PART TWO

THE ENDANGERED SPECIES ACT

The Endangered Species Act is the most revered and reviled of federal environmental laws. Its champions praise it for saving the bald eagle from extinction, for blocking many misconceived development projects, and for providing a tool to protect ecosystems ranging from the southern California coast to the majestic forests of Pacific northwest. Its detractors accuse it of sacrificing timber jobs for obscure owls, nearly completed dams for tiny fish, and small farmers for unknown rodents. The basis for these claims lies in the unparalleled stringency of the ESA's provisions. Most other environmental statutes contain numerous opportunities for environmental interests to be balanced against other human needs. The ESA, by contrast, has long been viewed as requiring efforts "to halt and reverse the trend toward species extinction, *whatever the cost*." Tennessee Valley Authority v. Hill, 437 U.S. 153, 184 (1978) (emphasis added). There are some who question whether the ESA is really so intransigent, *see* Oliver A. Houck, *The Endangered Species Act and Its Implementation by the U.S. Departments of Interior and Commerce*, 64 U. COLO. L. REV. 277, 292 (1993) (asserting that the actual implementation of the ESA is much more relaxed), but the fact that environmentalists turn to the ESA to save whole ecosystems when other laws fail suggests that the common impression of the ESA is well founded.

The ESA was the culmination of nearly a century of legislation to protect rare wildlife. Most of the early statutes were enacted by states, consistent with Supreme Court's view that the states owned all of the wildlife within their borders. *See* Geer v. Connecticut, 161 U.S. 519 (1896). But the problem of extinctions soon gained national attention. John Lacey, a member of Congress from Iowa, worried that the passenger pigeon, "formerly in this country in flocks of millions, has entirely disappeared from the face of the earth." So in 1900, Congress passed the Lacey Act— the first federal wildlife statute—to aid state wildlife preservation efforts by making it a federal crime to transport across state lines any wildlife killed in violation of state law. But the law did not prevent the passenger pigeon from going extinct when the last bird died in the Cincinnati Zoo in 1914, nor did it save the Carolina parakeet, the heath hen, and a growing number of other species once found in the United States.

The whooping crane is often credited with motivating federal efforts to devise a program to save rare wildlife from extinction. At one time, the whooping crane lived throughout the middle of the United States and in Canada, but fewer than thirty survived by the 1940's. The seemingly imminent demise of the whooping crane prompted federal officials, the Canadian government, and the National Audubon Society to begin the first concerted attempt to save a particular species from extinction. Still, it was not until 1966 that Congress enacted the first federal statute aimed at saving vanishing wildlife and plants. The Endangered Species Preservation Act, Pub. L. No. 89–669, 80 Stat. 926 (1966), directed the Secretary of the Interior to use existing land acquisition authorities to purchase the habitat of native fish and wildlife that were threatened with extinction, and it instructed the Secretary to "encourage other Federal agencies to utilize, where practicable, their authorities" to further the preservation effort. Congress expanded the effort three years later with the Endangered Species Conservation Act of 1969, Pub. L. No. 91–135, 83 Stat. 275 (1969), which authorized the creation of a list of species "threatened with world-wide extinction" and prohibited the importation of most such species into the United States. Yet even then the concept of an "endangered species" was so new that the term itself did not appear in a federal court decision until 1973—and then it was used to describe not an animal on the brink of extinction, but the Internal Revenue Code. *See* Dennis v. Commissioner, 473 F.2d 274, 286 (5th Cir.1973).

Almost immediately, the 1969 law was criticized for not going far enough. The law did not prohibit the hunting or collecting or killing of a listed species, it did not regulate conduct that destroyed the habitat of a species, and it did not offer any protection at all to plants. Throughout 1972 and 1973, Congress considered a range of proposed bills that would provide much more powerful protections for any wildlife or plant species that was facing extinction. But the debate in Congress always referred to bald eagles, grizzly bears, whooping cranes, alligators, whales, and other prominent species now described as "charismatic megafauna." Few members of Congress wanted to be seen as opposed to such popular animals, and few were. The Senate approved its bill 92–0, and after several minor changes, the final bill passed the House 355–4. So on December 28, 1973, President Nixon signed the Endangered Species Act, Pub. L. No. 93–205, 87

Stat. 884 (1973) (codified at 16 U.S.C. §§ 1531–1544)—what we have known since as the ESA.

The first reported case under the new law involved a long since forgotten water dispute between cattle ranchers in Nevada and the endangered pupfish. *See* United States v. Cappaert, 508 F.2d 313 (9th Cir.1974). Shortly thereafter, though, another case emerged that has colored the perception of the ESA ever since. Much to the chagrin of the United States Fish & Wildlife Service (FWS) and many of the members of Congress who had just voted for the ESA, the listing of the snail darter as endangered resulted in a Supreme Court decision holding that the nearly completed Tellico Dam could not be finished because the resulting reservoir would wipe out the fish. (The case—*Tennessee Valley Authority v. Hill*—is reprinted below at page 198). That decision caused Congress to amend the statute, albeit in a relatively modest fashion. Several other amendments occurred in 1982, but since then the law has remained virtually unchanged.

The congressional failure to amend the law is not for want of trying. During the early 1990s, environmentalists pressed to expand the coverage of the ESA to include whole ecosystems that were imperilled by human developments or other causes. Conversely, the ESA was blamed for causing economic dislocation throughout the Pacific northwest as a result of the listing of the northern spotted owl. The most sustained effort to reform—or gut, depending on your perspective—the ESA occurred in 1994. Speaker of the House Newt Gingrich established an ESA task force that held hearings across the country in areas that had chafed under the restrictions of the law. Landowners and developers told horror stories of widows losing their life's savings when the presence of an endangered songbird prevented them from building on their land and of farmers facing federal prosecutions for attempting to prevent fires in a manner that harmed endangered kangaroo rats. Several bills were introduced to amend the ESA by requiring more rigorous scientific evidence before a species could be listed, helping private landowners who confront a listed species on their property, and speeding recovery efforts so that a species could be delisted. The bills stalled in the face of a certain presidential veto and pressure from environmentalists, religious leaders, moderate eastern politicians, and others who were intent on saving rare wildlife.[1] Despite renewed efforts in recent years, it appears unlikely that any significant changes to the ESA are forthcoming.

As amended, the ESA has ten sections. The statute's principal provisions are as follows:

§ 2—congressional findings and purposes of the law

§ 3—definitions of key statutory terms

§ 4—procedures for listing a species as "endangered" or "threatened"

---

1. For an account describing perhaps the unlikeliest foe of the proposed changes to the ESA, *see* Michael J. Bean, *The Gingrich* *That Saved the ESA*, ENVTL. FORUM, at 26–32 (1999).

§ 5—authority to purchase land to conserve wildlife and plants, not just listed species

§ 6—means of encouraging federal cooperation with state preservation efforts

§ 7—duties imposed on federal agencies to conserve listed species and not to jeopardize the continued existence of a species or its critical habitat

§ 8—means of encouraging cooperation with the efforts of foreign governments to preserve listed species

§ 9—prohibition on killing, harming, smuggling, or any other "taking" of any endangered species

§ 10—permits and exceptions from the prohibitions in section 9

§ 11—enforcement mechanisms and penalties for violating the law

§ 18—Secretary of the Interior's duty to provide an annual report to Congress on the cost of measures to preserve each listed species

The ESA charges the Secretary of the Interior with the primary responsibility for implementing the act with respect to most kinds of species. The Secretary, in turn, has delegated that authority to the United States Fish & Wildlife Service (FWS), an agency within the Department of the Interior. An exception to that structure exists for marine species, which are under the jurisdiction of the Secretary of Commerce, who has delegated that statutory authority to the National Marine Fisheries Service (NMFS). Thus, the FWS and the NMFS are the federal agencies with the most responsibility for enforcing the ESA, though the law also calls upon all federal agencies to support that effort (as described in chapter 5).

The following three chapters examine the three major issues raised by the ESA. Chapter 4 explains and analyzes the procedure for identifying which species are "endangered" or "threatened" and thus entitled to the statute's protections. Chapter 5 addresses the duties that the ESA imposes on the FWS and other federal agencies. Chapter 6 details the increasingly controversial provisions that restrict the actions of private individuals, corporate developers, and state and local governments alike. The concluding section of this Part evaluates the success of the ESA and considers the statute's future.

There are, of course, many extremely helpful sources of information about the ESA. The provisions of the law are detailed in MICHAEL J. BEAN & MELANIE J. ROWLAND, THE EVOLUTION OF NATIONAL WILDLIFE LAW 198–276 (3d ed. 1997); RICHARD LITTELL, ENDANGERED SPECIES AND OTHER PROTECTED SPECIES: FEDERAL LAW AND REGULATION (1992); and STANFORD ENVIRONMENTAL LAW SOCIETY, THE ENDANGERED SPECIES ACT (2001). Current information about the law, lists of protected species, and succinct explanations of the ESA's provisions are available at the FWS's web site at <http://endangered.fws.gov/endspp.html>. For an eloquent, influential, but controversial account of the ESA and its implementation, *see* CHARLES C. MANN & MARK L. PLUMMER, NOAH'S CHOICE: THE FUTURE OF ENDANGERED SPECIES (1995).

CHAPTER 4

IDENTIFYING WHICH SPECIES TO PROTECT

Chapter Outline:
A. Identifying Species
B. Identifying Threats to Species
C. The Listing Decision Process
D. Case Study: The Black–Tailed Prairie Dog

The title of the ESA provides an accurate description of the scope of the law. The law only applies to a species that is "endangered" or "threatened." The ESA offers no help to deer and cardinals and dolphins because they are not endangered or threatened with extinction, so a landowner can "take" a deer without threat of federal sanction. More importantly, the ESA protects only those species that have been formally listed as endangered or threatened. The ESA operates to protect species only once they are formally listed, no matter how endangered they may be in fact.

The ESA's other limitation is that it applies to species, not to individual animals or entire ecosystems. The demise of a popular animal such as Ling–Ling, the panda who lived in Washington's National Zoo, can be cause for sadness, but the ESA's sole concern is about the survival of the panda as a species. The only exceptions to the focus on a whole species involve the ESA's protection of "subspecies" and of separate populations of a species. Similarly, while increased understanding of the importance of ecosystems has prompted numerous calls for an endangered ecosystems act, no such statute has yet emerged from Congress.

Section 4 of the ESA describes the procedure by which a species is designated as "endangered" or "threatened." The procedure has been criticized in recent years both by landowners and developers who believe that many species are listed without adequate scientific studies, and by environmentalists who complain that the federal government has moved much too slowly to add disappearing species to the list. The controversy about the listing process even resulted in a temporary moratorium on the listing of any new species during most of 1995 and the beginning of 1996. The moratorium was supposed to have enabled Congress to address the problems with the listing process, but the absence of a sufficient consensus about which changes were appropriate has left the listing process unchanged since the last congressional amendments to section 4 in 1982.

A.   IDENTIFYING SPECIES

United States v. Guthrie

United States Court of Appeals for the Eleventh Circuit, 1995.
50 F.3d 936.

■ CARNES, CIRCUIT JUDGE:

[In 1990, Robert Waites Guthrie asked an undercover agent working for the U.S. Department of the Interior whether the agent could obtain any Alabama red-bellied turtles, a species that had been listed as endangered in 1987. Guthrie explained that he planned to buy up the remaining wild of Alabama red-bellied turtles that lived in the wild, and then apply for a government grant to reintroduce the turtles into the wild from his own private stock. Guthrie was charged with illegally taking, possessing, selling and transporting the turtles in violation of the ESA. He moved to dismiss the charge alleging that the Alabama red-bellied turtle is a hybrid, and thus it was improperly listed as a "species" under the ESA. Guthrie asked the court to authorize a DNA study to determine whether the turtle was a hybrid or a pure species. When the district court refused, Guthrie entered a conditional guilty plea that preserved his right to appeal the court's ruling on the species question. The district court sentenced Guthrie to thirteen months in prison with three years supervised release, and it ordered him to pay $150 in special assessments and $5,000 as a fine or as a donation to the state or federal government efforts to preserve the Alabama red-bellied turtle.]

The Alabama red-bellied turtle

The Secretary of the Interior ("the Secretary") has authority under the ESA to promulgate a list of endangered and threatened species to be protected under the Act. *See* 16 U.S.C.A. § 1533 (1985).[4] The ESA defines "endangered species" as "any species which is in danger of extinction throughout all or a significant portion of its range," 16 U.S.C.A. § 1532(6) (1985), and "threatened species" as "any species which is likely to become an endangered species within the foreseeable future throughout all or a significant portion of its range." 16 U.S.C.A. § 1532(20) (1985). In July 1986, the Secretary proposed listing the Alabama red-bellied turtle as a threatened species. 51 Fed. Reg. 24,727 (1986). Eleven months later, he adopted a regulation listing the turtle as an endangered species. 52 Fed. Reg. 22,939 (1987). During the process, the Secretary changed the status of the turtle from threatened to endangered, and he did so in response to one of the public comments received in support of the turtle's protection. Neither Guthrie nor anyone else submitted any public comments opposed to listing the turtle.

The ESA provides a petition process for agency review, *see* 16 U.S.C.A. § 1533(b)(3)(A) (1985 & Supp.1994), and it also authorizes citizen suits to challenge whether the Secretary has met his duties under the act. 16 U.S.C.A. § 1540(g) (1985). In the seven years since the regulation listing the Alabama red-bellied turtle as an endangered species was adopted, neither Guthrie nor anyone else has attempted to have the regulation reviewed by either of these authorized means. Instead, having willfully violated the regulation, Guthrie now seeks to collaterally attack its validity in this criminal proceeding. Guthrie's sole defense to the charge that his conduct violated the ESA is that the Alabama red-bellied turtle is not a species, that the Secretary therefore lacked the authority to list it as an endangered species, and that as a result, the ESA does not entitle the Alabama red-bellied turtle to protection.

. . . Guthrie did not seek to change the agency regulation. He chose to violate the law. We will not reward that choice by allowing him to bypass the agency and receive judicial review of the regulation in light of the new DNA study. Instead, Guthrie at most is entitled to the same review he would have received had he sought direct review of the agency regulation at the time it was promulgated. Such review is limited to the evidence before the agency at that time. . . .

---

4. The Secretary of the Interior shares authority with the Secretary of Commerce, with each being authorized to list different animals. *See* 16 U.S.C.A. § 1532 (1985). With the general exception of marine animals, most species (including the Alabama red-bellied turtle) fall within the authority of the Secretary of the Interior, who has delegated that authority to the Fish and Wildlife Service. *See* 48 Fed. Reg. 29,990, 29,997 (1983); *see also* 50 C.F.R. § 402.01(b) (1993); Reorganization Plan No. 4 of 1970, *reprinted in* 1970 U.S.C.C.A.N. at 6326. Because the Fish and Wildlife Service is part of the Department of the Interior and thus under the Secretary's control, for simplicity's sake we refer to the Secretary, even though the Fish and Wildlife Service actually took the administrative actions in question.

On direct review of the Secretary's decision to list the Alabama red-bellied turtle as an endangered species, we would hold that decision unlawful only if we found it to be "arbitrary, capricious, an abuse of discretion, or otherwise not in accordance with law." This arbitrary and capricious standard applies to the Secretary's finding that "the Alabama red-bellied turtle is considered to be a valid species." Having reviewed the sources and studies cited by the Secretary in support of this finding, we conclude that the Secretary did not act arbitrarily or capriciously when he found the Alabama red-bellied turtle to be a valid species.

In its proposed regulation, the Secretary acknowledged that "the taxonomic status of this turtle has been questioned ... and questions still remain regarding its relationships with other members of the *Pseudemys rubriventris* group, specifically the Florida red-bellied turtle...." The Secretary then cited two texts by the same herpetologist questioning the taxonomic status of the species. In the final regulation, the Secretary listed six other texts in support of his finding that the Alabama red-bellied turtle is nevertheless a valid species. The eight texts cited by the Secretary reflect the scientific history behind the Secretary's conclusion that the Alabama red-bellied turtle is a separate species.

The Alabama red-bellied turtle was first classified as a species with the name "Pseudemys alabamensis" by a herpetologist named Baur in 1893. In the 1930s, another herpetologist named Archie Carr authored an article and a book questioning whether the Alabama red-bellied turtle was a species. In the article, Carr concluded that Pseudemys alabamensis was not a species but likely "a mutant occurring in [Pseudemys] mobiliensis and suwanniensis." In the book, Carr catalogued the morphological traits of several species from the floridana group of turtles before stating that "it seems nearly certain that another Gulf coast turtle known to herpetologists for the last fifty years as P. alabamensis, and characterized by having a deep notch and toothlike cusps at the tip of the upper jaw, is really just a variant form that occurs with apparently equal frequency among [two different subspecies of turtle in the family Emydidae]."

Five years after his book was published, Carr changed his position in an article he co-authored with John W. Crenshaw, Jr. Carr and Crenshaw explained that in reaching his earlier conclusion that the Alabama red-bellied turtle was not a species, Carr had examined specimens of the turtle from New Orleans, Mobile, and Crystal River in Florida, and had observed that they exhibited characteristics of other turtles in those areas. Carr and Crenshaw then distinguished the Alabama red-bellied turtle from Pseudemys floridana based on different morphological characteristics from the ones Carr had noted previously, especially the presence of head markings in the shape of a pre-frontal arrow. The herpetologists also suggested that at least one of Carr's Crystal River turtles may have been misidentified. They concluded that the Alabama red-bellied turtle was indeed a separate species, despite its apparent intergradation with two different groups of emydid turtles.

Since Carr and Crenshaw's publication, the authors of five other books and studies cited by the Secretary have agreed that the Alabama red-bellied turtle, Pseudemys alabamensis, deserves separate species status. Some of these scientists have concurred in this status even while noting that unresolved taxonomic questions remain.

The final study relied upon by the Secretary was published just two years before the Secretary's decision. It reclassifies some of the turtle specimens labeled Pseudemys alabamensis by Carr and Crenshaw, and finds that the Alabama red-bellied turtle is a "valid species" that "is endemic to Alabama." After trapping twelve rivers and the lakes on one island, Dobie found Alabama red-bellied turtles only along the Mobile River system, suggesting that earlier studies which found the Alabama red-bellied turtle as widely spread as Texas and Florida had misidentified the examined turtles. Dobie and the later sources thus re-affirmed the resurrected species status of Pseudemys alabamensis, the Alabama red-bellied turtle.

Given scientific support from numerous herpetologists, the Secretary did not act arbitrarily or capriciously when he listed the Alabama red-bellied turtle as an endangered species. This circuit is "highly deferential" to an agency's consideration of the factors relevant to its decision. The Secretary noted that the Alabama red-bellied turtle's status had indeed been questioned in the past, and thus he did not "entirely fail[ ] to consider an important aspect of the problem." Nor was the Secretary's finding "so implausible that it could not be ascribed to a difference in view or the product of agency expertise." Having examined the articles, studies, and books relied upon by the Secretary when he concluded that the Alabama red-bellied turtle is a separate taxonomic species, we are satisfied that, despite the absence of total agreement within the scientific community, his finding is entirely reasonable. It certainly is not arbitrary or capricious.

Notes and Questions

1. How is the Alabama red-bellied turtle different from other turtles? Do those differences support the conclusion that the turtle is a separate species? Are those differences sufficient to support efforts to keep the turtle from going extinct?

2. The Alabama red-bellied turtle is the official reptile of Alabama. *See* Ala. Code § 1–2–25 (1999). Should that suffice to justify protection of the turtle, whether or not it is faced with the threat of extinction? *Cf.* 42 U.S.C. § 9605(a)(8)(B) (authorizing each state to select one hazardous waste site for inclusion on a list of the highest priority cleanup sites throughout the United States, regardless of whether the state's site is actually as contaminated as other sites that are placed on the list based on a scientific assessment of their hazardousness).

3. *Guthrie* is not the only instance in which the question of what constitutes a species has been contested. The Arizona leatherflower, for example, was once regarded as a separate species because its leaflet lobes were wider and its leaves were more spread out than its nearest relative, but the FWS removed the flower from consideration for listing under the ESA once a new scientific study showed that there were no clear differences between the two flowers. *See* Notice of Reclassification of a Candidate Taxon: *Clematis Hirsutisima var. Arizonica* (Arizona Leatherflower), 63 Fed. Reg. 1418 (1998). It is even possible that a plant or animal can be removed from the list of species protected by the ESA if the FWS determines that it is no longer a distinct species.

The FWS must make such determinations without any statutory definition of what constitutes a "species." The standard scientific taxonomy dates from eighteenth century Swedish botanist Carl Linnaeus, who developed a hierarchy that now progresses from kingdom to phylum to class to order to family to genus to species. The Alabama red-bellied turtle, for example, is a member of the animal kingdom, the reptilia class, the Testudines order, the Emydidae family, the Pseudemys genus, and the alabamensis species—with the latter two designations comprising the Latin name of the species, *Pseudemys alabamensis*. What that neat formula omits is any indication of what features support the characterization of the Alabama red-bellied turtle as a distinct species from, say, the Florida red-bellied turtle or the Alabama map turtle. As explained in chapter 2, the prevailing explanation adheres to the definition of a species offered by Harvard evolutionary biologist Ernst Mayr: "A species is a reproductive community of populations, reproductively isolated from other populations, that occupies a specific niche in nature." ERNST MAYR, THE GROWTH OF BIOLOGICAL THOUGHT 273 (1982). But that approach still leaves ample room for debate about the nature of a "community of populations," the meaning of reproductive isolation, and the scope of the relevant niche in nature. As Professor Kevin Hill has written, "Ambiguity is inherent in the taxonomic classification of endangered species." Kevin D. Hill, *What Do We Mean By Species?*, 20 B.C. ENVTL. AFF. L. REV. 239, 257 (1993). Even accounting for that ambiguity, Hill faults "the drafters of the ESA and frequently the Fish and Wildlife Service" because they "seem to have a very simplistic view of what constitutes a species. Quite often, under the Act, species are treated as discrete entities under a traditional typological approach emphasizing physical characteristics. Thus, a species is defined if it has a particular kind of shape, size, color, or other attribute. The purpose of many endangered species programs was to preserve this particular snapshot of present day characteristics, ignoring the changes caused by evolutionary adaptation." *Id.* at 253.

Hybrids present an especially acute definitional problem. A hybrid occurs when two different species combine to reproduce, notwithstanding the reproductive isolation inherent in the very understanding of a species. Robert Guthrie argued that the Alabama red-bellied turtle was actually a hybrid, but the FWS disagreed. The FWS has deleted a plant from the list of endangered species when new scientific information indicated that the plant is a hybrid rather than a distinct species. *See, e.g.,* Final Rule to Remove the Plant *"Echinocereus lloydii"* (Lloyd's Hedgehog Cactus) from the Federal List of Endangered and Threatened Plant, 64 Fed. Reg. 33796 (1999) (concluding that a cactus that had been listed as endangered in 1979 was actually a hybrid rather than a distinct species, and thus removing the cactus from the ESA's list of protected species). But it remains uncertain whether the ESA protects another kind of hybrid: the offspring of an endangered species and another species, such as Florida panthers that have mated with captive panthers from South America. *See* LITTELL, *supra*, at 21.

4. The ESA protects nearly every kind of species that biologists recognize. Section 3(16) indicates that the Act applies to "fish or wildlife or plants." "Fish or wildlife," in turn, "means any member of the animal kingdom, including without limitation any mammal, fish, bird . . . amphibian, reptile, mollusk, crustacean, arthropod or other invertebrate." ESA § 3(8). And "[t]he term plant means any member of the plant kingdom." ESA § 14. The only exception to the universal coverage of the ESA concerns "a species of the Class Insecta determined by the Secretary to constitute a pest whose protection under the provisions of this chapter would present an overwhelming and overriding risk to man." ESA § 3(6); 50 C.F.R. § 424.02(k).

As of June 30, 2002, a total of 1,817 species had been listed under the ESA. There are 1,500 endangered species and 317 threatened species. The list of endangered or threatened species includes 713 flowering plants, 342 mammals, 273 birds, 126 fish, 115 reptiles, 72 clams, 48 insects, 33 snails, 26 ferns and lillies, 21 crustaceans, 12 arachnids, five conifers and cycads, and two lichens. 1,259 of those species live in the United States; the balance live in other countries throughout the world. Within the United States, Hawaii has the most species with 317, followed by California (294) Alabama (125), Florida (111), and Tennessee (101). Only eight listed species are found in Vermont, with North Dakota having just nine listed species. (For the latest totals, *see* the FWS web site at <http://ecos.fws.gov/tess/html/boxscore.html>. A map indicating how many listed species live in each state is available at <http://ecos.fws.gov/web-page/usmap.html?&status=listed>).

5. The ESA defines "species" to include "any subspecies of fish or wildlife or plants." 16 U.S.C. § 1532(16). "Subspecies" is undefined in the statute and imprecisely defined elsewhere. *See, e.g.,* EDWARD O. WILSON, THE DIVERSI-TY OF LIFE 406 (1992) (definition of a "subspecies" is "[s]ubdivision of a species. Usually defined narrowly as a geographical race: a population or series of populations occupying a discrete range and differing genetically from other geographical races of the same species."). The protection of subspecies has potentially sweeping consequences. The Delhi Sands Flow-er–Loving Fly is a subspecies; so is the northern spotted owl whose listing resulted in tremendous controversies in the Pacific northwest.

Why protect subspecies? That question elicited contrasting responses during the 1995 congressional hearings on the ESA. Opponents of the listing of subspecies argued that they need not be protected unless their survival was essential to the survival of the species as a whole. At a minimum, they argued, "[t]he sub-species or sub-sub-species of kangaroo rat should not receive the same treatment as the California condor." *Endangered Species Act—Vancouver, Washington: Hearing Before the Task*

Force on Endangered Species of the House Resources Comm., 104th Cong., 1st Sess. 65 (1995) (testimony of Barbara Tilly, Chairman, Chelan County Public Utility District Board of Commissioners). Secretary of the Interior Bruce Babbitt answered that "[s]hould a subspecies begin to decline, this may be a warning that the species as a whole may be in danger," and that an early response to that trend can improve the likelihood and decrease the cost of protection efforts. He added that the existence of genetically distinct subspecies "improve the ability of the species as a whole to survive." *Endangered Species Act: Washington, DC—Part III: Hearing Before the Task Force on Endangered Species of the House Resources Comm.*, 104th Cong., 1st Sess. 261–62 (1995).

6. The ESA also protects populations of a vertebrate species—but not an invertebrate or a plant—that is endangered in one place but not in another. *See* ESA § 3(16) (defining "species" to include "any distinct population segment of any species of vertebrate fish or wildlife that interbreeds when mature"). The FWS has listed so-called "distinct population segments" of a species if the population (1) is isolated from other members of the species, (2) occupies an ecosystem that is danger of destruction throughout much of its historical distribution, (3) is the only occurrence of the species in the United States, or (4) lives within different political jurisdictions that provide varying degrees of protection. Thus the FWS listed the population of the Canada lynx that lives in the contiguous United States from New Hampshire to Montana despite the presence of many lynx in Alaska and Canada. The agency explained that the lynx population "in the contiguous United States may be considered biologically and ecologically significant simply because of the climatic, vegetational, and ecological differences between lynx habitat in the contiguous United States and that in northern latitudes in Canada and Alaska." Determination of Threatened Status for the Contiguous U.S. Distinct Population Segment of the Canada Lynx and Related Rule, 65 Fed. Reg. 16052, 16060 (2000). For another example of a distinct population segment involving a species that lives both in the United States and another country, *see* Threatened Status for the Alaska Breeding Population of the Steller's Eider, 62 Fed. Reg. 31748 (1997) (listing the population of the Steller's eider—a sea duck—that breeds in Alaska, but declining to list the balance of the species that lives in Russia).

Or consider the FWS's 2000 decision to list anadromous Atlantic salmon in the Gulf of Maine as a distinct population segment. "Atlantic salmon from fish farms fill seafood coolers in supermarkets, but wild strains have disappeared from old American haunts," including several rivers in Maine. Andrew C. Revkin, *Maine Salmon Are a Breed Apart, Panel Reports*, N.Y. TIMES, Jan. 8, 2002, at A17. The status of the salmon in Maine is further complicated by the mixing of the original wild populations with salmon stocks that were added to the river beginning in the 1870's and by salmon that escaped from fish farms. The FWS concluded that "[t]he conservation of the populations of the Gulf of Maine population segment is essential because these Atlantic salmon represent the remaining genetic legacy of ancestral populations that were locally adapted to the rivers and streams of the region that formerly extended from the Housatonic River in Connecticut to the headwaters of the Aroostook River in Maine." Final Endangered Status for a Distinct Population Segment of Anadromous Atlantic Salmon (*Salmo salar*) in the Gulf of Maine, 65 Fed. Reg. 69459, 69460 (2000). An interim report released by the National Academy of Sciences in January 2002 confirmed the genetic distinctiveness of the salmon in Maine, concluding that "the natural salmon spawning in Maine's DPS-designated rivers are 'Maine salmon,' not just 'salmon in Maine.' " NATIONAL ACADEMY OF SCIENCES, GENETIC STATUS OF ATLANTIC SALMON IN MAINE: INTERIM REPORT 35 (2002), *available at* <http://books.nap.edu/books/0309083117/html/index.html>.

The listing of distinct population segments elicits many of the same criticisms and defenses as the listing of subspecies. Or, to put it another way, consider why the law allows the bald eagle to be listed as endangered in the lower 48 states even though there have always been many bald eagles living in Alaska. Michael Bean offers one answer when he states that "[a]voiding local extirpation of a species is desirable not only because a series of local extirpations frequently leads to endangerment of the species as a whole, but also because of the ecological, recreational, aesthetic, and other values populations provide in their localities." BEAN, *supra*, at 200. By contrast, when federal officials threatened to remove a beaver that had been eating the cherry trees surrounding Washington's Tidal Basin, Representative Helen Chenoweth mockingly suggested that "[t]his distinct population segment of the Rodentia family must be saved." 145 CONG. REC. H1840 (daily ed. Apr. 12, 1999).

B. IDENTIFYING THREATS TO SPECIES

Final Rule to List the Illinois Cave Amphipod as Endangered

United States Fish and Wildlife Service, 1998.
63 Fed. Reg. 46900.

[The Illinois cave amphipod, *Gammarus acherondytes*, lives in the dark zone of streams in six cave systems in central Illinois. An amphipod is a scavenger that feeds on dead animal and plant matter or the thin bacterial film covering most of the submerged surfaces in their aquatic habitat. The Illinois cave amphipod is distinguished from other amphipods by its light gray-blue color, its small kidney-shaped eyes, and the size of one of its two antennas that stretches nearly half the length of its less than an inch long body.]

. . . The cave streams from which this species is historically known are each fed by a distinct watershed or recharge area; and there are no known interconnections between them, or with other cave systems. Two of the six caves may become hydrologically connected during extremely high rainfall over short periods of time. Thus, it is believed that there is virtually no opportunity for this species to become distributed to other cave systems via natural pathways.

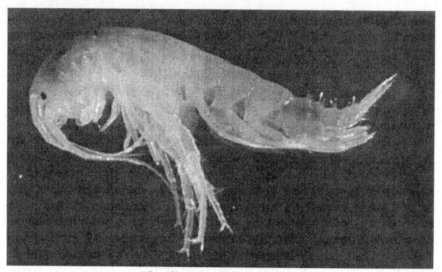

The Illinois cave amphipod

There are few data or adequate survey techniques on which to base population, productivity, or trend estimates for this species. Sampling for cave fauna is difficult at best, and the challenges of surveying are compounded by the relatively small size of this species and the difficulty of researchers to distinguish it from other similar amphipods in the field. Thus, survey data are not sufficient to accurately record numbers of this small subterranean invertebrate; however, they do demonstrate a reduction in its range and the number of extant populations. . . .

The most recent and extensive sampling effort was in 1995 in which the Illinois Natural History Survey (INHS) investigated 25 caves in the Illinois Sinkhole Plain and confirmed the presence of the species in only 3 of the original 6 cave systems, all in Monroe County. The species was not found in any additional caves. In 1995, 56 specimens were taken from Illinois Caverns, 19 specimens from Fogelpole Cave, and 2 specimens from a third, privately owned cave. The species appears to be extirpated from the two caves where no specimens were collected in 1965 or 1986. Its status in a sixth cave is currently unknown because the cave entrance has been closed by the landowner, thus the cave has not been re-surveyed since 1965. Due to the extensive searches by INHS, it is possible, but unlikely, that there are populations in other caves in the Illinois Sinkhole Plain. The INHS made an intensive effort to collect in all small side rivulets and drip pools in the 25 caves it sampled and believes that the collection results reasonably reflect the relative abundance of the species in cave streams of the Sinkhole Plain.

. . . The Service received comments [on its proposal to list the Illinois cave amphipod as an endangered species] from 27 individuals and organizations during the comment periods; some parties provided more than one comment letter. Eight commenters supported the proposal. Twelve parties expressed concern over the possible effect the listing may have on their area of interest (agriculture or cave visitation), and several offered rebuttals to the Service's rationale but did not directly oppose the proposal. Four commenters expressed opposition to the proposal.

... Issue 3: The Service lacks the scientific data to justify listing this species since there has been inadequate sampling conducted: one cave in which the species historically occurred could not even be surveyed.

Service Response: The Service believes that the sampling efforts conducted in 1993 and 1995 were by far the most intensive and extensive to date, and were appropriate to demonstrate the decline in the species' range with a high degree of certainty. In 1995 the [Illinois Natural History Survey] sampled 25 caves in the Illinois Sinkhole Plain and found *Gammarus acherondytes* in only 3 caves. In 1 cave that historically contained *G. acherondytes*, for example, a total of 561 amphipods from other species were collected without collecting any *G. acherondytes*. In a second cave that historically contained the species, 673 amphipods were collected without taking any *G. acherondytes*. If it is present in either of these caves, it would have to be extremely rare, constituting less than 2 individuals per 1000 amphipods sampled. By comparison, *G. acherondytes* appeared in higher numbers in much smaller amphipod samples in Fogelpole Cave (at a rate of more than 50 individuals per 1000 sampled) and Illinois Caverns (at a rate of about 250 individuals per 1000 sampled). If the species is present in significant numbers in the other 2 caves, it should have been readily collected in mainstream samples at the level of sampling intensity that was carried out in the 1993 and 1995 surveys. More intensive collecting, in which thousands of amphipod specimens are taken from each cave for later identification, might be inappropriate and probably unhealthy for the cave community. Such intensive collecting might decimate or extirpate an amphipod species whose numbers already are extremely low. Although survey data cannot unequivocally prove that the species is extirpated from any cave, they demonstrate that the most optimistic scenario is that the species is extremely rare, and its numbers have decreased since the surveys done prior to 1993.

The Service recognizes that the species may still occur in the one cave whose entrance has been closed by the landowner, and we have not made the assumption that it has been extirpated from that location. However, even if it does still occur there, the data indicate that the species' range has decreased from six caves to three or four.

Issue 4: Recent sampling efforts have yielded more specimens than previous efforts, indicating that species numbers may actually be increasing.

Service Response: The Service acknowledges a remote possibility that the species may be found in other cave streams in the sinkhole plain. There is also a chance that it may be found in other locations within Fogelpole Cave and Illinois Caverns. However, the Service believes the sampling effort that was expended looking for this species is more than adequate and reasonably reflects the relative abundance and diminishing distribution of the species in cave streams of the sinkhole plain. The Service does intend to keep looking for this species in other locations, however.

With regard to estimating the actual population of this species, the Service acknowledges that it is not likely to ever achieve that goal, regardless of the amount of effort put into surveys. The nature of this species and its habitat make it difficult, at best, to survey for it. Furthermore, the current identification technique for the species requires that it be sacrificed. It would be counter productive to sacrifice substantial numbers of an extremely rare species in order to obtain a more precise population estimate.

However, obtaining an accurate estimate of species numbers is not necessary for the Service to determine that the species warrants protection under the Act. What must be demonstrated is that its range has been significantly reduced and the threats to the species continue and can reasonably be expected to result in a further decline. An accurate population estimate also is not necessary to establish and achieve recovery goals for the species. Recovery can be achieved by protecting the quality of its habitat and by restoring stable and viable populations to the caves from which it has been extirpated. Once listed, the amphipod's relative abundance and population trend will be monitored safely using standard scientific methods.

Peer Review

In accordance with policy promulgated July 1, 1994, the Service solicited the expert opinions of independent specialists regarding pertinent scientific or commercial data relating to the supportive biological and ecological information for species under consideration for listing. The purpose of such review is to ensure listing decisions are based on scientifically sound data, assumptions, and analyses, including input of appropriate experts and specialists.

Following the publication of the listing proposal, the Service solicited the comments of two biologists having recognized expertise in invertebrate zoology and one individual having recognized expertise in karst hydrology and underground environments and requested their review of the available data concerning the Illinois cave amphipod. In order to ensure an unbiased examination of the data, the Service selected individuals who had only minor or no involvement in previous discussions on the possible listing of the species.

Comments were received from all three peer reviewers within the comment period. The two biological reviewers concurred with the Service on factors relating to the taxonomic, biological, and ecological information and concurred with the proposal to list the Illinois cave amphipod as an endangered species. The karst hydrologist provided additional clarification of the importance of oxygen depletion as the primary mechanism by which the species is being harmed. That reviewer also concurred that the Illinois cave amphipod is in danger of extinction in the foreseeable future.

Summary of Factors Affecting the Species

Section 4 of the Act and regulations promulgated to implement the listing provisions of the Act set forth the procedures for adding species to the Federal lists. A species may be determined to be threatened or endangered due to one or more of the five factors described in section 4(a)(1). These factors and their application to the Illinois cave amphipod (*Gammarus acherondytes*) of are as follows:

A. *The Present or Threatened Destruction, Modification, or Curtailment of Its Habitat or Range*

The degradation of habitat through the contamination of groundwater is believed to be the primary threat to the Illinois cave amphipod. . . . There are several sources of groundwater contamination affecting the amphipod's habitat: (1) the application of agricultural chemicals, evidence of which has been found in spring and well water samples in Monroe County; (2) bacterial contamination from human and animal wastes, which finds its way to subsurface water via septic systems, the direct discharge of sewage waste into sinkholes, or from livestock feedlots; (3) the application of residential pesticides and fertilizers; and (4) the accidental or intentional dumping of a toxic substance into a sinkhole. . . .

B. *Overutilization for Commercial, Recreational, Scientific, or Educational Purposes*

Overexploitation or scientific collecting are not believed to be factors affecting the species' continued existence at this time, but the Federal listing will prohibit unauthorized collection of individuals of the species. Exact numbers are unknown, but at a minimum 139 specimens have been collected from 6 caves over a 55–year period. Protection from collection may become important because collectors may seek the species once it becomes listed.

C. *Disease or Predation*

The importance of these factors is presently unknown.

D. *The Inadequacy of Existing Regulatory Mechanisms*

This species currently has no protection under Federal law. The Federal Cave Resources Protection Act of 1988 seeks to secure, protect, and preserve significant caves on Federal lands for the perpetual use, enjoyment, and benefit of all people. However, at this time, the Cave Resources Protection Act provides no protection to any caves containing, or potentially containing, Illinois cave amphipods, because none of the caves are on or under Federal land or are located in the immediate vicinity of Federal ownership. Therefore, these caves are ineligible for Federal protection under the Cave Resources Protection Act.

The Illinois cave amphipod is listed as an endangered species under the Illinois Endangered Species Protection Act. As such, it is protected from direct taking (*i.e.*, injury or mortality) regardless of whether it is on public or private land. However, "take" under State law does not include indirect harm through such mechanisms as habitat alteration. As long as the actions of private landowners are otherwise in compliance with the law, actions which destroy or degrade habitat for this species are allowed under Illinois law....

As mentioned under Factor A of this section, several of the entrances to caves containing the species are dedicated as Illinois Nature Preserves which is the strongest land protection mechanism in Illinois. Such dedication restricts future uses of the land, in perpetuity, for the purpose of preserving the site in its natural state. The removal of biota from the site is prohibited except by permit and for scientific purposes only. Allowable uses of the site are limited to nonconsumptive, nondestructive activities. The landowner may decide whether to allow public access to the site, and management is accomplished in accordance with a master management plan prepared jointly by the Illinois Nature Preserves Commission and the landowner. Dedicated properties cannot be subdivided, and the dedication instrument is attached to the deed and recorded.

Ownership or protection of cave entrances does not necessarily ensure protection of the caves' environment, particularly water quality. Water quality is largely a function of land use in the cave stream recharge areas on the land surface, and the vast majority of the watersheds of all caves containing the amphipod is in private ownership, and land use is primarily agriculture. Recharge areas may be several square miles in size, and runoff and seepage from thousands of acres of agricultural land may be funneled into one cave system, thus increasing the magnitude of any toxic hazard posed by the use of agricultural chemicals.... Current State and local regulations are inadequate for protecting water quality in a sensitive geological formation like karst. St. Clair and Monroe counties are rapidly developing as residential communities for the St. Louis, Missouri, Metropolitan Area with most home sites being served by individual wells and septic systems. Septic systems may not perform as designed, and, in some cases, septic effluent drains directly into sinkholes. Studies have shown that there is no general housing density zoning in karst terrain that assures that groundwater quality will be protected when septic systems are used. The more houses there are in a spring or cave stream recharge area, the greater the chance that some of them will introduce contaminants into the groundwater system, and the greater the chance that one or more of the septic field systems will constitute a major source of groundwater contamination.

E. *Other Natural or Manmade Factors Affecting Its Continued Existence*

Because of the low numbers of the Illinois cave amphipod and a highly restricted range, even the loss of a few individuals to natural events may be significant to the species' survival. As a group, aquatic amphipods have adapted to the extremes of natural events such as spring floods or high water discharges following rainstorms and, no doubt, some individuals are washed out of the cave environment during such events. Because the species is extant in only three or four cave systems within a relatively small geographic area, it is conceivable that a heavy spring snowmelt or rainstorm could cause a flushing of all systems at one time significantly affecting each population.

The risk of extinction due to the threats to the Illinois cave amphipod (*Gammarus acherondytes*) posed by the above factors is exacerbated by the small number of low density populations that remain. Although *Gammarus acherondytes* was always rare, the current population densities are likely much lower (due to the previously identified threats) than historical levels. Despite any adaptations to conditions which result in rarity, habitat loss and degradation increase a species' vulnerability to extinction. Environmental variation, whether random or predictable, naturally causes fluctuations in populations. However, populations with small numbers are more likely to fluctuate below the minimum viable population (*i.e.*, the minimum number of individuals needed for a population to survive). If population levels stay below this minimum size, an inevitable, and often irreversible, slide toward extinction will occur. Small populations are also more susceptible to inbreeding depression and genetic drift. Populations subjected to either of these problems usually have low genetic diversity, which reduces fertility and survivorship. Lastly, chance variation in age and sex ratios can affect birth and death rates. Changes to demographics may lead to death rates exceeding the birth rates, and when this occurs in small populations there is a higher risk of extinction.

The Service has carefully assessed the best scientific and commercial information available regarding the past, present, and future threats faced by this species in determining to make this rule final. Based on this evaluation, the preferred action is to list the Illinois cave amphipod as endangered....

Notes and Questions

1. Is the Illinois cave amphipod in danger of becoming extinct? How do we know? The agency admits that it will probably never be able to estimate the actual population of the species. Moreover, the agency agrees that natural causes play some role in the disappearance of the Illinois cave amphipod, and that the precise effect of various human activities is not entirely understood. Note, too, that the agency acknowledges that the amphipod was always *rare*. How can we distinguish between a species that is rare and a species that is endangered?

2. Why did the FWS list the Illinois cave amphipod as endangered instead of as threatened? A species is endangered if it "is in danger of extinction," whereas a species is threatened if it "is likely to become an endangered species in the foreseeable future." ESA § 3(6), 3(20). Congress included the "threatened" category "not only as a means of giving some protection to species before they become endangered, but also as a means of gradually reducing the level of protection for previously endangered species that had been successfully 'restored' to the point at which the strong protective measures for that category were no longer necessary." BEAN, *supra*, at 201. There are only a few practical consequences between the two categories. The restrictions on "taking" a species (discussed in chapter 6) automatically apply to endangered species, but their application to threatened species is within the discretion of the FWS. Most of the time the agency simply extends the takings prohibitions to threatened species, but in an increasing number of controversial cases the agency has developed somewhat less stringent prohibitions for threatened species. Another difference between endangered and threatened species lies in the greater penalties that the ESA imposes on those who violate the law with respect to an endangered species.

3. Of the five factors specified in ESA section 4(a)(1) that the FWS must consider when determining if a species is endangered or threatened, four require consideration of the actual threats to the species. The fifth factor— "the inadequacy of existing regulatory mechanisms"—examines the protection that the species already enjoys under federal, state or local law. This factor presumes that a species need not be listed under the ESA if alternative legal measures are sufficient to prevent the species from becoming extinct. The FWS determined that the Federal Cave Resources Protection Act of 1988 and the Illinois Endangered Species Protection Act failed to provide that assurance for the Illinois cave amphipod. By contrast, the FWS withdrew a proposal to list the pecos pupfish once the federal Bureau of Land Management (BLM) restricted oil and gas operations, excluded ORVs, and retired a grazing lease near the Pecos River, and the states of New Mexico and Texas modified their fishing regulations, controlled harmful exotic fish, and promised to take several additional steps to secure the pupfish's river habitat. Withdrawal of Proposed Rule to List the Pecos Pupfish (Cyprinodon peconsensis) as Endangered, 65 Fed. Reg. 14513, 14514 (2000). The FWS and the NMFS have indicated that whether such actions can make an ESA listing unnecessary depends upon the certainty that the conservation effort will be implemented and the certainty that the effort will be effective. The courts have precluded reliance upon conservation measures that were speculative or unavailable for public scrutiny, *see, e.g.*, Friends of the Wild Swan, Inc. v. United States Fish & Wildlife Serv., 945 F.Supp. 1388 (D.Or.1996) (reprinted below at page 134), but they have accepted agreements that were included in the formal listing review process. *See* Defenders of Wildlife v. Babbitt, No. 97–CV–2330 TW(LSP), 1999 U.S. Dist. LEXIS 10366, at *23 (S.D.Cal. June 14, 1999) (concluding that "[t]o require FWS to not consider state conservation strategies and agreements simply because the strategy is 'newly implemented' would discourage states from engaging in any conservation efforts at all").

4. The ESA provides that science should be the determinative factor in deciding whether to list a species as endangered or threatened. *See* ESA § 4(b)(1)(A) (stating that the Secretary of the Interior's decision whether to list a species must be made "solely on the basis of the best scientific and commercial data available at the time the decision is made"). The problem posed in this circumstance—and in many others—is how to respond to scientific uncertainty regarding the status of a species. The ESA does not require the government or the party requesting the listing of a species to conduct any research concerning the status of a species. Instead, the listing decision must be made with whatever scientific information is available. Thus the state studies of the Illinois cave amphipod provided the primary source of data on which the FWS relied in concluding that the amphipod was endangered.

The quality of the scientific evidence supporting ESA listing decisions has been controversial in recent years. The complaints about the reliability and sufficiency of the evidence regarding the Illinois cave amphipod are similar to those voiced in many other listing decisions. For example, a proposal to list four California plants provoked objections that the scientific data supporting the listing "was either inaccurate, insufficient, inconsistent, erroneous, unsubstantiated, unverified, unjustified, based only on biased opinions in favor of listing the species, not peer-reviewed," collected only during drought years when the plants would not be present, and collected by illegal trespassing. *See* Determination of Threatened Status for Four Plants from the Foothills of the Sierra Nevada Mountains in California, 63 Fed. Reg. 49022, 49025 (1998). The FWS has responded that the statute "does not require us to possess detailed or extensive information about the general biology of a species or to make an actual determination of the causes for the species' status to make a listing determination." Final Rule to List the Flatwoods Salamander as a Threatened Species, 64 Fed. Reg. 15691, 15695 (1999).

The courts have rejected all attempts to require additional scientific research prior to the determination of whether a species should be listed. When a federal district court ordered the FWS to conduct an actual count of the number of Queen Charlotte goshawks to determine whether the bird should be listed, the court of appeals reversed because the ESA provides that the listing decision must "be made solely on the basis of the best scientific and commercial data available," so "the district court was without authority to order the Secretary to conduct an independent population count of the birds." Southwest Center for Biological Diversity v. Babbitt, 215 F.3d 58 (D.C.Cir.2000). Nor is the inadequacy of the existing data a sufficient cause for delay. In Defenders of Wildlife v. Babbitt, 958 F.Supp. 670 (D.D.C.1997), the FWS had concluded that listing the Canada lynx was unwarranted because there was not "any conclusive evidence of the biological vulnerability or real threats to the species in the 48 contiguous states." The court overturned that decision because the ESA "contains no requirement that the evidence be conclusive in order for a species to be listed." *Id.* at 679. Indeed, the FWS must proceed with its listing determination even if the best available scientific evidence is inherently flawed. *See* Defenders of Wildlife v. Babbitt, No.–97–CV–2330 TW (LSP) 1999 U.S. Dist. LEXIS 10366, at *14 (S.D.Cal. June 14, 1999) (upholding the FWS's use of questionable scat count data when listing the flat-tailed horned lizard).

5. In an effort to dispel the concerns about the adequacy of the scientific evidence used to list species, the FWS and the NMFS have adopted a scientific peer review policy that promises to "[s]olicit the expert opinions of three appropriate and independent specialists regarding pertinent scientific or commercial data and assumptions relating to the taxonomy, population models, and supportive biological and ecological information for species under consideration for listing." Notice of Interagency Cooperative Policy for Peer Review in Endangered Species Act Activities, 59 Fed. Reg. 34270 (1994). The Illinois cave amphipod listing briefly describes the results of such peer review. Nonetheless, some members of Congress continue to insist that the listing process must be informed by better science. One proposed bill would require the FWS to determine that any petition to list a species to contain the following information:

- scientific documentation from a published scientific source that the fish or wildlife or plant that is the subject of the petition is a species;

- a description of all available data on the historical and current range, population, and distribution of the species, an explanation of the methodology used to collect the data, and an identification of the location where the data can be reviewed;

- scientific evidence that the population of the species is declining or has declined from historic population levels and beyond normal population fluctuations for the species;

- an appraisal of the available data on the threats to the species or the cause of its decline;

- an identification of the information contained or referred to in the petition that has been peer-reviewed or field-tested;

- a bibliography of scientific literature on the species, if any, in support of the petition;

- the qualifications of any person cited in the petition as an expert on the species or the status of the species; and

- at least one study or credible expert opinion, by a person who is not affiliated with the petitioner, to support the action requested in the petition.

H.R. 3160, 106th Cong., 2d Sess. § 101 (2000). What effect might those standards have on listing decisions? Is it correct to suggest, as one opponent charges, that such requirements are "simply an attempt to use science to make the listing process more difficult?"

6. There is one way in which a species can be protected by the ESA even though it is neither endangered nor threatened. If a species is so similar in appearance to another listed species that protection of the listed species requires protection of the similar species, too, then the similar species can be listed even though it is not endangered or threatened itself. *See* ESA § 4(e). For example, when the Fish and Wildlife Service listed the northern population of the bog turtle as threatened, it also listed the southern population because of the two turtles were virtually indistinguishable. The agency explained that "[t]he listing of the southern population as threatened due to similarity of appearance eliminates the ability of commercial collectors to commingle northern bog turtles with southern ones or to misrepresent them as southern bog turtles for commercial purposes." *Endangered and Threatened Wildlife and Plants; Final Rule to List the Northern Population of the Bog Turtle as Threatened and the Southern Population as Threatened Due to Similarity of Appearance*, 62 Fed. Reg. 59,605, 59,622 (1997). Thus far, relatively few species have been listed because they are similar in appearance to another protected species.

C. THE LISTING DECISION PROCESS

Northern Spotted Owl v. Hodel

United States District Court for the Western District of Washington, 1988.
716 F.Supp. 479.

■ ZILLY, DISTRICT JUDGE

A number of environmental organizations bring this action against the United States Fish & Wildlife Service ("Service") and others, alleging that the Service's decision not to list the northern spotted owl as endangered or threatened under the Endangered Species Act of 1973, as amended, 16 U.S.C. § 1531 *et seq.* ("ESA" or "the Act"), was arbitrary and capricious or contrary to law.

Since the 1970s the northern spotted owl has received much scientific attention, beginning with comprehensive studies of its natural history by Dr. Eric Forsman, whose most significant discovery was the close association between spotted owls and old-growth forests. This discovery raised concerns because the majority of remaining old-growth owl habitat is on public land available for harvest.

In January 1987, plaintiff Greenworld, pursuant to Sec. 4(b)(3) of the ESA, petitioned the Service to list the northern spotted owl as endangered. In August 1987, 29 conservation organizations filed a second petition to list the owl as endangered both in the Olympic Peninsula in Washington and in the Oregon Coast Range, and as threatened throughout the rest of its range.

The ESA directs the Secretary of the Interior to determine whether any species have become endangered or threatened[1] due to habitat destruction, overutilization, disease or predation, or other natural or manmade factors. 16 U.S.C. § 1533(a)(1).[2] The Act was amended in 1982 to ensure

---

1. The ESA defines an "endangered species" as "any species which is in danger of extinction throughout all or a significant portion of its range...." 16 U.S.C. § 1532(6). A "threatened species" is "any species which is likely to become an endangered species with-

in the foreseeable future throughout all or a significant portion of its range." 16 U.S.C. § 1532(20).

2. Section 4(a)(1) ... provides that:

The Secretary [of Interior in the case of terrestrial species] shall ... determine that the decision whether to list a species as endangered or threatened was based solely on an evaluation of the biological risks faced by the species, to the exclusion of all other factors. *See* Conf. Report 97–835, 97th Cong. 2d Sess. (Sept. 17, 1982) at 19, *reprinted in* 1982 U.S. Code Cong. & Admin. News 2860.

The Service's role in deciding whether to list the northern spotted owl as endangered or threatened is to assess the technical and scientific data in the administrative record against the relevant listing criteria in section 4(a)(1) and then to exercise its own expert discretion in reaching its decision.

In July 1987, the Service announced that it would initiate a status review of the spotted owl and requested public comment. The Service assembled a group of Service biologists, including Dr. Mark Shaffer, its staff expert on population viability, to conduct the review. The Service charged Dr. Shaffer with analyzing current scientific information on the owl. Dr. Shaffer concluded that:

> the most reasonable interpretation of current data and knowledge indicate continued old growth harvesting is likely to lead to the extinction of the subspecies in the foreseeable future which argues strongly for listing the subspecies as threatened or endangered at this time.

The Service invited a peer review of Dr. Shaffer's analysis by a number of U.S. experts on population viability, all of whom agreed with Dr. Shaffer's prognosis for the owl, although each had some criticisms of his work.

The Service's decision is contained in its 1987 Status Review of the owl ("Status Review") and summarized in its Finding on Greenworld's petition ("Finding"). The Status Review was completed on December 14, 1987, and on December 17 the Service announced that listing the owl as endangered under the Act was not warranted at that time.[5] 52 Fed. Reg. 48552, 48554 (Dec. 23, 1987). This suit followed. Both sides now move for summary judgment on the administrative record before the Court. . . .

This Court reviews the Service's action under the "arbitrary and capricious" standard of the Administrative Procedure Act ("APA"). This

---

whether any species is an endangered species or a threatened species because of any of the following factors:

    (A) the present or threatened destruction, modification, or curtailment of its habitat or range;

    (B) overutilization for commercial, recreational, scientific, or educational purposes;

    (C) disease or predation;

    (D) the inadequacy of existing regulatory mechanisms; or

    (E) other natural or manmade factors affecting its continued existence.

5. The Service's Finding provides as follows:

A finding is made that a proposed listing of the northern spotted owl is not warranted at this time. Due to the need for population trend information and other biological data, priority given by the Service to this species for further research and monitoring will continue to be high. Interagency agreements and Service initiatives support continued conservation efforts.

standard is narrow and presumes the agency action is valid, but it does not shield agency action from a "thorough, probing, in-depth review." Courts must not "rubber-stamp the agency decision as correct."

> Rather, the reviewing court must assure itself that the agency decision was "based on a consideration of the relevant factors...." Moreover, it must engage in a "substantial inquiry" into the facts, one that is "searching and careful." This is particularly true in highly technical cases....

> Agency action is arbitrary and capricious where the agency has failed to "articulate a satisfactory explanation for its action including a 'rational connection between the facts found and the choice made.'"

The Status Review and the Finding to the listing petition offer little insight into how the Service found that the owl currently has a viable population. Although the Status Review cites extensive empirical data and lists various conclusions, it fails to provide any analysis. The Service asserts that it is entitled to make its own decision, yet it provides no explanation for its findings. An agency must set forth clearly the grounds on which it acted. Judicial deference to agency expertise is proper, but the Court will not do so blindly. The Court finds that the Service has not set forth the grounds for its decision against listing the owl.

The Service's documents also lack any expert analysis supporting its conclusion. Rather, the expert opinion is entirely to the contrary. The only reference in the Status Review to an actual opinion that the owl does not face a significant likelihood of extinction is a mischaracterization of a conclusion of Dr. Mark Boyce:

> Boyce in his analysis of the draft preferred alternative conclusions that there is a low probability that the spotted owls will go extinct. He does point out that population fragmentation appears to impose the greatest risks to extinction....

Dr. Boyce responded to the Service:

> I did not conclude that the Spotted Owl enjoys a low probability of extinction, and I would be very disappointed if efforts to preserve the Spotted Owl were in any way thwarted by a misinterpretation of something I wrote.

Numerous other experts on population viability contributed to or reviewed drafts of the Status Review, or otherwise assessed spotted owl viability. Some were employed by the Service; others were independent. None concluded that the northern spotted owl is not at risk of extinction. For example, as noted above, Dr. Shaffer evaluated the current data and knowledge and determined that continued logging of old growth likely would lead to the extinction of the owl in the foreseeable future. This risk, he concluded, argued strongly for immediate listing of the subspecies as threatened or endangered.

The Service invited a peer review of Dr. Shaffer's analysis. Drs. Michael Soule, Bruce Wilcox, and Daniel Goodman, three leading U.S.

experts on population viability, reviewed and agreed completely with Dr. Shaffer's prognosis for the owl.

For example, Dr. Soule, the acknowledged founder of the discipline of "conservation biology" (the study of species extinction), concluded:

> I completely concur with your conclusions, and the methods by which you reached them. The more one hears about *Strix occidentalis caurina*, the more concern one feels. Problems with the data base and in the models notwithstanding, and politics notwithstanding, I just can't see how a responsible biologist could reach any other conclusion than yours.

The Court will reject conclusory assertions of agency "expertise" where the agency spurns unrebutted expert opinions without itself offering a credible alternative explanation. Here, the Service disregarded all the expert opinion on population viability, including that of its own expert, that the owl is facing extinction, and instead merely asserted its expertise in support of its conclusions.

The Service has failed to provide its own or other expert analysis supporting its conclusions. Such analysis is necessary to establish a rational connection between the evidence presented and the Service's decision. Accordingly, the United States Fish and Wildlife Service's decision not to list at this time the northern spotted owl as endangered or threatened under the Endangered Species Act was arbitrary and capricious and contrary to law.

The Court further finds that it is not possible from the record to determine that the Service considered the related issue of whether the northern spotted owl is a threatened species. This failure of the Service to review and make an express finding on the issue of threatened status is also arbitrary and capricious and contrary to law.

In deference to the Service's expertise and its role under the Endangered Species Act, the Court remands this matter to the Service, which has 90 days from the date of this order to provide an analysis for its decision that listing the northern spotted owl as threatened or endangered is not currently warranted. Further, the Service is ordered to supplement its Status Review and petition Finding consistent with this Court's ruling.

Notes and Questions

1. Why was the FWS so reluctant to list the northern spotted owl under the ESA? Why were many environmental groups so eager to list the owl? Note that the FWS listed the northern spotted owl as a threatened species in 1990.

2. The ESA contains two procedures for the consideration of whether a species should be listed. The FWS or the NMFS can initiate the process itself based on information collected by agency scientists, as the FWS did for the Illinois cave amphipod and many other species. Alternately, any interested person may petition the appropriate agency to list a species, as illustrated by the thirty environmental groups that petitioned the FWS to list the northern spotted owl in 1987. The same substantive standards apply to the agency's subsequent determination of whether the species is endangered, threatened, or neither.

In 1982, Congress amended section 4 to establish a specific schedule for the evaluation of petitions to list a species. To the maximum extent practical, the agency must determine whether the petition presents substantial evidence that listing is warranted within 90 days after the petition is filed. The agency then has 12 months to decide whether listing is warranted, unwarranted, or warranted but precluded by higher agency priorities. Once the agency proposes to list the species, it must reach a final decision within 12 months (with a possible six month extension). During that time, the public has 60 days after the proposal to submit comments on the proposed listing, and a public hearing will be held if requested within 45 days of the proposal. At the end of the 12 months, the agency will either list the species or decide to withdraw the proposal to list the species.

Judicial review of any final decision regarding a species is available to any party who can establish standing to sue. Judicial review is also available if the agency fails to make a decision within the statutory period. The courts have generally been sympathetic to the FWS when it fails to act within the statutory deadlines, but they have ordered the agency to comply nonetheless. *See, e.g.,* Biodiversity Legal Foundation v. Badgley, 1999 WL 1042567 (D.Or.1999) (establishing a tight compliance schedule for findings that were over three years overdue); Biodiversity Legal Foundation v. Babbitt, 63 F.Supp. 2d 31 (D.D.C.1999) (ordering the FWS to act on a 1997 petition to list the Baird's sparrow). Conversely, the courts have been unwilling to void a decision to list a species when the FWS acted beyond the statutory period. *See* Idaho Farm Bureau Fed'n v. Babbitt, 58 F.3d 1392 (9th Cir.1995); Endangered Species Comm. of the Bldg. Ind. Ass'n of Southern Cal. v. Babbitt, 852 F.Supp. 32 (D.D.C.1994).

3.   The listings of the northern spotted owl, the Illinois cave amphipod, and numerous other species were opposed by local residents and developers who feared listing the species under the ESA would adversely effect commercial, recreational, and other activities. The ESA, however, precludes consideration of such factors when deciding whether to list a species. Instead, the ESA insists that the determination of endangered or threatened status is a purely scientific question, and that the effects of a listing can be considered at other points once a species is listed.

Why should scientific data alone determine whether a species should be listed? Professor Holly Doremus has argued that "the closed, technocratic decisionmaking process in the scientific community ... is inappropriate in the endangered species context because the relevant scientific questions are both intractable and closely intertwined with controversial value choices." Instead, she suggests, Congress "should separate the scientific aspects of listing determinations from the value judgments, including which groups should be considered for protection, what level of extinction risk is tolerable, and what the time line for evaluating extinction risks should be." Holly

Doremus, *Listing Decisions Under the Endangered Species Act: Why Better Science Isn't Always Better Policy*, 75 WASH. U. L. Q. 1029, 1036, 1153 (1997).

Or consider another alternative. When the FWS publishes a final rule listing a species as endangered or threatened, it indicates how many public comments supported the proposed listing and how many opposed it. The votes recorded on several recent listing proposals are as follows:

	Yes	No
Blackburn's sphinx moth	5	0
Flatwoods salamander	39	136
Kauai cave wolf spider & Kauai cave amphipod	3	2
Lake Erie water snakes	89	7
Pecos sunflower	7	7
Preble's meadow jumping mouse	35	86
Rough popcornflower	3	3
San Bernadino kangaroo rat	29	14
Sierra Nevada population of the California bighorn sheep	37	2
St. Andrew beach mouse	2	3
Topeka shiner	92	80

Do these results suggest that the Flatwoods salamander, Preble's meadow jumping mouse, and St. Andrew beach mouse should not be listed? How about the Pecos sunflower and the rough popcornflower? Should the listing of a species under the ESA depend upon a popular vote? If not, why solicit comments from the general public on a listing proposal?

Friends of the Wild Swan, Inc. v. United States Fish and Wildlife Service

United States District Court for the District of Oregon, 1996.
945 F.Supp. 1388.

■ ROBERT E. JONES, DISTRICT JUDGE:

[The bull trout is a freshwater fish found in the western United States, Canada, and Alaska. Its numbers have dropped significantly because of difficulties in migration in the continental United States and because of land and water management practices. In 1993, two environmental groups petitioned the Fish and Wildlife Service to list the bull trout under the ESA. The agency responded by concluding—in separate findings in 1994 and 1995—that the listing of the species was warranted but precluded by other agency priorities. The agency relied upon the Clinton Administration's Forest Plan for addressing environmental issues in the Pacific Northwest as one of the reasons why listing the bull trout was unnecessary at that time. The environmental groups sued.]

[When a party petitions to list a species], FWS must make a finding with 90 days of the petition's submission "as to whether the petition presents substantial scientific or commercial information indicating that the petitioned action may be warranted." If the action is warranted, FWS commences "a review of the status of the species concerned," and within 12 months must find either that: (1) the petitioned action is not warranted; (2) the petitioned action is warranted; or (3) "[t]he petitioned action is warranted, but"

> (I) the immediate proposal and timely promulgation of a final regulation implementing the petitioned action ... is precluded by pending proposals to determine whether any species is an endangered species or a threatened species, and

> (II) expeditious progress is being made to add qualified species to either of the lists ... and to remove from such lists species for which the protections of this chapter are no longer necessary....

16 U.S.C. § 1533(b)(3)(B).

The ESA requires FWS to establish as published agency guidelines "a ranking system to assist in the identification of species that should receive priority review under subsection (a) (1) of this section...." 16 U.S.C. § 1533(g). In compliance, FWS has published a 12–level ranking system based on three criteria. 48 Fed. Reg. 43098, 43102 (Sept. 21, 1983). First, FWS determines whether the magnitude of the threat to the species is "high," on the one hand, or "medium to low," on the other. Second, it determines whether the immediacy of the threat is "imminent" or "non-imminent." Finally, it looks to the taxonomic level of the species at issue— monotypic genus,[2] species, or subspecies. A monotypic genus facing a high magnitude and imminent threat receives a rank of "1"; a subspecies facing a moderate-to-low magnitude and non-imminent threat receives a rank of "12." In effect, the lower the ranking number, the more important it is for the species to be listed and receive the Act's protections.

FWS's guidelines do not specify factors for FWS to rely upon when it classifies a proposed species into each of the three criteria, nor has FWS delineated general factors besides its ranking scale that it will consider. Finally, FWS has emphasized that it sets only relative, not absolute, priorities.

Despite this obscurity and indeterminacy in the ranking process, a species' priority level effectively determines whether or not it is listed under the ESA: in 1994, for example, "because of limited staff and funding, the Service must give priority to the listing of species with a listing priority of 1 through 6." Because of the way FWS's ranking system is structured, this policy meant that in 1994 only species facing a "high" magnitude of threat receive the Act's protections....

In its 1994 Finding, FWS found that listing of the bull trout was warranted but precluded based on its ranking of the bull trout as a 9. This

---

2. A monotypic genus is a genus made up of only one species. 48 Fed. Reg. at 43103. FWS gives a higher priority to such species because they represent "highly distinctive or isolated gene pools...."

ranking, in turn, depends upon FWS's determination that the magnitude of the threat to the bull trout is "moderate," not "high." Therefore, this court's review will focus on whether FWS's determination that the bull trout face a "moderate" threat is arbitrary and capricious.

... FWS explicitly listed the factors that it relied upon in determining that the bull trout faced a "moderate" threat. "The Service considers that the threat to the bull trout's continued existence is moderate because of its widespread range, the existence of populations in protected areas, and ongoing management changes (e.g. Forest Plan) that are expected to benefit some populations." This court's review of whether FWS's determination is arbitrary and capricious is thus limited to the propriety of its reliance on these factors.

Several aspects of the record make FWS's reliance on these factors, without further explanation, so questionable and internally inconsistent as to render that reliance arbitrary and capricious. The record supports FWS's conclusions that the bull trout has a widespread range and that it exists in protected areas. However, FWS's own findings contradict any assertion that these factors reduce the threat of extinction to the bull trout. As FWS itself noted, "persistence of migratory life history forms and maintenance or re-establishment of stream migration corridors is crucial to the viability of bull trout populations...." The Service discussed in detail how hydropower dams have isolated many bull trout populations, concluding that "natural recolonization of historically occupied sites has become impossible." Moreover, the threat of hybridization with brook trout "is exacerbated when larger, migratory forms of bull trout have been eliminated and gene flow is prevented by the isolation of remnant bull trout populations." In short, "once isolated, bull trout populations face relatively high probabilities of extinction due to loss of gene flow and relatively low population size...."

Thus, in concluding that bull trout are sufficiently threatened with extinction to warrant listing under the ESA, FWS repeatedly emphasized the loss of the migratory life form as an important factor creating that risk of extinction. Therefore, its reliance on the species' widespread range—a range consisting largely and increasingly of isolated subpopulations—as a reason for viewing the threat as "moderate" is internally inconsistent and "entirely fail[s] to consider an important aspect of the problem." As such, FWS's reliance on this factor, without further explanation, supports a finding that the 1994 FINDING is arbitrary and capricious.

The record provides even less support for FWS's reliance on the presence of bull trout populations in protected areas. In its briefing statement for the Regional Director, prepared less than four months before the 1994 FINDING, the Washington State Supervisor concluded that "threats exist even in ecosystems not impacted by traditional physical habitat perturbations, such as Wilderness Areas and National Parks, where nonnative brook and lake trout pose a serious threat to the persistence of many bull trout populations." A briefing statement written three months later reached a similar conclusion:

Even in wilderness areas and national parks, the threat of extirpation due to introduced species is great. Extirpated populations are unlikely to be reestablished due to the loss of migratory life history forms. Many isolated populations face serious risks of extinction even with no further habitat loss.

FWS cites to nothing in the record, nor could this court find anything in that record, that supports its conclusion that the presence of bull trout in protected areas reduces the magnitude of the threat to the species. Therefore, FWS's reliance on the presence of bull trout in protected areas constitutes "an explanation for its decision that runs counter to the evidence before the agency" and necessitates a finding that the agency's decision is arbitrary and capricious.

Finally, FWS's reliance on land management plans of other federal agencies with future effect is both arbitrary and capricious and contrary to law, and for four reasons. First, FWS made clear in oral argument that the then-newly-promulgated Clinton Forest Plan was a key factor in its decision that the threat to the bull trout was moderate. The Clinton Forest Plan, however, is not part of the administrative record for the 1994 Finding. Therefore, FWS should not have relied upon it in determining the magnitude of the threat to the bull trout, and this fact alone is sufficient to render its decision arbitrary and capricious.

Second, FWS more generally relied on "ongoing management changes (e.g. Forest Plan) that *are expected* to benefit some populations." These management changes were for the future, with uncertain effect on the bull trout species. However, as FWS has acknowledged, it is required "to base listing decisions upon an analysis of existing threats." Thus, FWS determines for listing decisions whether a species "is an endangered species or threatened species," 16 U.S.C. § 1533(a)(1); 50 C.F.R. § 424.11(c), based on the current status of the species. *See* 16 U.S.C. § 1533(b)(1)(A). Moreover, it must make its listing determination "solely on the basis of the best scientific and commercial available" to it. It cannot rely upon its own speculations as to the future effects of another agency's management plans to put off listing a species, and such reliance here again requires a finding that the decision was arbitrary and capricious.

Third, reliance on management plans is at odds with the factors FWS must use to determine whether listing of the bull trout is warranted at all. The ESA establishes that "the inadequacy of existing regulatory mechanisms" is one basis for determining that a species is endangered or threatened. 16 U.S.C. § 1533(a)(1)(D). As such, FWS's finding that listing of the bull trout was warranted strongly implies that existing regulatory mechanisms are in fact inadequate to protect the trout. Specifically, in assessing this criterion, FWS found that:

> Federal and State laws designed to conserve fish resources or maintain water quality have not been sufficient to prevent past and ongoing habitat degradation and population fragmentation. Conservation measures provided for in many Federal regulations are merely advisory.

This court finds it difficult to understand, without further analysis and explanation by the agency, how FWS can find past and present protective measures so insufficient and yet rely on its future expectations of additional such measures to avoid listing the bull trout.[16]

Finally, FWS's reliance on the plans of other federal agencies is contrary to the provisions and purposes of the ESA. Since 1982, FWS may take "into account those efforts, if any, being made by any *State or foreign nation*, or any political subdivision of a *State or foreign nation*, to protect such species. . . ." 16 U.S.C. § 1533(b)(1)(A), as amended by Pub. L. 97–304 § 2(b)(1)(A), 96 Stat. 1411 (Oct. 13, 1982). Similarly, FWS "shall give consideration to species which have been— . . . (ii) identified as in danger of extinction, or likely to become so within the foreseeable future, by any *State agency or by any agency of a foreign nation* that is responsible for the conservation of fish or wildlife or plants." 16 U.S.C. § 1533(b)(1)(B)(ii), as amended by Pub. L. 97–304 § 2(b)(1)(B), 96 Stat. 1411–12 (Oct. 13, 1982) (emphasis added). No provisions explicitly authorize FWS to consider the actions of other federal agencies. . . .

The omission of other federal agencies furthers the purposes of the Act because the ESA imposes conservation duties on all federal agencies only *after* FWS has taken the initial step of listing the species as endangered or threatened. *See* 16 U.S.C. § 1531(c)(1) (announcing the policy of Congress "that all Federal departments and agencies shall seek to conserve endangered species and threatened species and shall utilize their authorities in furtherance of the purposes of this chapter"); 16 U.S.C. § 1536(a)(1) (requiring federal agencies to carry out programs "for the conservation of endangered and threatened species listed pursuant to section 1533"); 16 U.S.C. § 1536(a)(2) (requiring that federal agencies not jeopardize listed species). . . . Only after listing do other federal agencies' comprehensive duties to protect endangered and threatened species arise. The Act thus clearly indicates that federal protection of endangered and threatened species *begins* with the listing process, and it would be contrary to the purposes of the Act for FWS to allow the actions of one or two federal agencies, not required by the ESA, to deprive an otherwise warranted species of the broad protections the Act would require of all federal agencies if FWS completed the listing process.

FWS has further emphasized this limitation in its own regulations, establishing that when it makes listing determinations, it shall consider the efforts being made by states and foreign nations. 50 C.F.R. § 424.11(f). In contrast, the only provision it has made for other federal agencies is that

---

16. This is not to say that the factors for prioritizing species need be identical to the factors for determining whether to list a species in the first place; indeed, the act of prioritizing requires additional analysis. However, it is axiomatic that FWS cannot contradict the ESA in its prioritizing decisions. Moreover, FWS has given this court no indication of how it differentiates its review of "the inadequacy of existing regulatory mechanisms" for purposes of listing a species pursuant to 16 U.S.C. § 1533(a)(1) and its review of regulatory mechanisms for purposes of prioritizing species. In the absence of a clear explanation of the agency's policy and decisionmaking process, its asserted conclusion cannot stand.

FWS shall consult "as appropriate" with "other affected Federal agencies" "when considering any revision of the lists." 50 C.F.R. § 424.13.

"The plain intent of Congress in enacting [the ESA] was to halt and reverse the trend toward species extinction, whatever the cost." Tennessee Valley Auth. v. Hill, 437 U.S. 153, 184 (1978). When it gave FWS the authority to make "warranted but precluded" findings, Congress simultaneously warned courts that they "will, in essence, be called on to separate justifications grounded in the purposes of the Act from the foot-dragging efforts of a delinquent agency." Under the APA, FWS's decision is arbitrary and capricious if "the agency has relied on factors which Congress has not intended it to consider." Similarly, even if FWS's apparently ad hoc delineation of factors supporting its determination that the bull trout faced a "moderate" threat were entitled to *Chevron* deference, "this court must reject constructions that are contrary to clear congressional intent or that frustrate the policy Congress sought to implement." Because this court finds that (1) FWS relied upon the Clinton Forest Plan, which was never part of the administrative record before it; (2) FWS relied on one factor that Congress did not intend it to consider, contradicting congressional purposes; (3) FWS relied on a second factor that is contradicted by the record with no corresponding support; and (4) FWS relied on a third factor that appears to be inconsistent with its own biological findings, it concludes that FWS's determination that the bull trout faces a "moderate" threat is arbitrary and capricious. As such, plaintiffs are entitled to judgment as a matter of law on this issue.

[In 1995, FWS renewed its "warranted but precluded" determination for the bull trout, again giving the species a ranking of 9.]

SCOPE OF RELIEF

. . . Two aspects of the bull trout listing procedure argue against FWS's simplification of the importance of our determination that the 1994 FINDING was arbitrary and capricious. First, if FWS concludes that it should have ranked the bull trout as facing a "high" magnitude of threat, listing of the bull trout should proceed in accordance with the Act; as such, the 1995 FINDING will be without legal effect because the Act never would have required it.

Second, even if FWS affirms its "warranted but precluded" finding on legally defensible grounds, it will not be finished with the bull trout. As the Ninth Circuit noted, "FWS is required by the ESA to make a new determination every twelve months if it finds listing the bull trout warranted but precluded." Under this requirement, FWS should have again revisited the bull trout's status by June 1996, twelve months after its 1995 FINDING. To the best of this court's knowledge, FWS has not published the required 12–month finding for 1996. FWS's reconsideration of its 1994 FINDING in light of this opinion will, this court hopes, inform any future decisions FWS must make regarding listing of the bull trout, including emergency listing decisions.

FWS also cites lack of funding as a reason not to remand the bull trout findings to it. Funding considerations do not repeal or modify FWS's duties under the ESA. Environmental Defense Ctr. v. Babbitt, 73 F.3d 867, 871 (9th Cir.1995). Nevertheless, lack of funds, when Congress expressly prohibits expenditures for listing species, can excusably delay mandatory listing determinations. FWS does not argue that it is currently prohibited from expending funds—only that it does not think that the bull trout will fit into its current funding priorities. Under FWS's current guidelines, however, emergency listings receive the highest priority. Reconsideration of its earlier determinations could well mean that FWS will find that certain listing actions for the bull trout should be higher in FWS's current funding priorities. Viewing FWS's current situation as a whole in light of our determination that FWS's decisions regarding the bull trout—including its emergency listing decisions—are arbitrary and capricious, this court cannot conclude that a remand to the agency is irrational, as FWS argues.

Therefore, this court concludes that FWS's failure to address emergency listings of the bull trout and its determination that the magnitude of the threat facing that species is "moderate" rather than "high" have been arbitrary and capricious/contrary to law. It remands the 1994 FINDING to the agency for further consideration in accordance with this opinion, with the additional instructions that (1) FWS must limit its review to the record before it in 1994, and (2) if FWS concludes that it should have issued emergency listings and/or should have determined that the bull trout faced a "high" magnitude of threat in its 1994 FINDING, it will incorporate those new determinations into its current listing priorities. If FWS's reconsideration does not alter its ultimate conclusion that listing of the bull trout is "warranted but precluded," this court respectfully reminds the agency of its duties pursuant to 16 U.S.C. § 1533(b)(3)(C)(i) and (b)(3)(C)(iii) to treat plaintiffs' petition as resubmitted and to use its authority to issue emergency listings "to prevent a significant risk to the well-being of any such species."

Notes and Questions

1. The question of priorities exists because Congress has never appropriated enough funds to perform all of the tasks imposed by the ESA and it is unlikely to do so in the future. How should the FWS determine which listing actions have the greatest priority? Should the agency treat all species equally? Is saving some species more important than others? If the agency is forced to choose among species, what criteria should it use to do so?

2. There were 259 species that were formal candidates for listing under the ESA as of July 2002. The FWS defines a candidate species as "plants and animals for which the [FWS] has sufficient information on their biological status and threats to propose them as endangered or threatened under the [ESA], but for which development of a proposed listing regulation is precluded by other higher priority listing activities," while the

NMFS uses a slightly broader definition that includes species whose status is of concern but for which more information is needed. A species remains a candidate until the agency has the time and resources to conduct the determination of whether the species should be listed under the ESA. The status of each species is reevaluated annually. The bull trout remained a candidate species until the FWS proposed its listing as a threatened species in June 1997, which was followed by the final listing in November 1999. *See* Determination of Threatened Status for Bull Trout in the Coterminous United States, 64 Fed. Reg. 58910 (1999).

The protections of the ESA do not apply to candidate species. The combination of the threat of legal regulation resulting from a listing and the absence of any regulation prior to listing has caused some individuals to attempt to eliminate a species before it is protected. For example, a developer bulldozed one of the three remaining populations of the San Diego mesa mint just days before the plant was listed, and some western landowners expedited their extermination of black-tailed prairie dogs once that species was identified as a potential addition to the list of species covered by the ESA.

The FWS can avoid such strategies by listing a species on an emergency basis. Section 4(b)(7) of the ESA allows the FWS to immediately list any fish, wildlife or plant species for 240 days when there is an "emergency posing a significant risk to the well-being of" the species. These emergencies have included unauthorized county road construction that threatened the survival of the bull trout in Nevada's Jarbridge River, rapid development amidst the breeding sites of the California tiger salamander, and hungry mountain lions feeding on the California bighorn sheep in the Sierra Nevada mountains. The leading case involving an emergency listing—City of Las Vegas v. Lujan, 891 F.2d 927 (D.C.Cir.1989)—upheld the FWS's protection of the Mojave Desert population of the desert tortoise pursuant to a more deferential standard of judicial review appropriate given the exigencies of the situation. The FWS must then decide whether to list a species on a permanent basis by the end of the 240 days following an emergency listing.

Emergency listings are uncommon. According to Professor Francesca Ortiz, there were only sixteen such listings between 1980 and 1999. She finds this surprising for three reasons:

> First, Congress specifically indicated that emergency listings should be used "less cautiously," in a "shoot first and ask ... questions later" fashion, to prevent a significant risk to the well-being of candidate species.... Second, the evidentiary standard for emergency listings is less exacting than that for normal listings, suggesting that in appropriate cases, implementation of an emergency rule would be easily justified. Finally, the FWS is not financially incapacitated with regard to emergency listings, which have been given the highest priority under each listing priority guidance issued since the moratorium.

Francesca Ortiz, *Candidate Conservation Agreements as a Devolutionary Response to Extinction*, 33 Ga. L. Rev. 413, 459 (1999). Perhaps the FWS heeded Professor Ortiz's advice, for the agency listed five species on an emergency basis in the first two years after her article was published. Even so, why does the FWS exercise its emergency listing power so infrequently?

3. *Friends of the Wild Swan, Inc.* illustrates the application of the FWS ranking system for determining which species should be considered first. Additionally, Congress has authorized the FWS to establish priorities among the agency's various responsibilities under the ESA. The agency's guidelines give emergency listings highest priority, the processing of final decisions on proposed listings second priority, resolving the status of candidate species third priority, and processing petitions fourth priority. *See* Final Listing Priority Guidance for Fiscal Year 2000, 64 Fed. Reg. 57114 (1999). The Tenth Circuit rejected a challenge to an earlier version of those guidelines, reasoning that "the Service must retain the ability to order and prioritize its work, particularly when provided with limited resources, in order to adequately fulfill its mission." Biodiversity Legal Foundation v. Babbitt, 146 F.3d 1249, 1255 (10th Cir.1998).

4. In 1995, Congress enacted—and President Clinton reluctantly signed— a moratorium on the listing of any additional species as endangered or threatened. The proponents of the moratorium wanted to fix the problems that they perceived with the ESA before additional species became subject to its coverage, and they hoped that the contested issues regarding the reform of the ESA could be resolved while the moratorium was in place. Opponents of the moratorium complained that it would simply require additional efforts to save species in the future once they had become even more endangered. Indeed, one environmentalist characterized the moratorium as "one of the most shortsighted, mean-spirited, and counterproductive legislative actions in recent years." *Lifting of Moratorium on ESA Listings: Hearing Before the House Comm. On Resources*, 105th Cong., 2d Sess. 129 (1996) (statement of Eric R. Glitzenstein, Fund for Animals). The courts refused to intervene. In the *Environmental Defense Center v. Babbitt* decision cited in the bull trout case, the Ninth Circuit held that the listing duties imposed by the ESA remained in effect, but lack of funding to perform that duty precluded the Fish and Wildlife Service from taking any action on previous proposal to list the California red-legged frog. By contrast, when the agency was subject to a prior court order to designate critical habitat for the marbled murrelet, the lack of funds did not excuse the agency from proceeding with such a designation. *See* Marbled Murrelet v. Babbitt, 918 F.Supp. 318 (W.D.Wash.1996). Congress allowed President Clinton to lift the moratorium in the spring of 1996, and he did so immediately.

But funding difficulties continued to plague the listing process after the moratorium was lifted. In November 2000, the FWS announced that it would not be able to work on any listing proposals for the remaining ten months of its fiscal year because of court orders directing the agency to process critical habitat designation requests. "We just don't have the staff or the funding necessary to do anything that isn't ordered by a court," the FWS explained, but the Center for Biological Diversity accused the FWS of "playing serious politics, and the loser is America's endangered wildlife." *Wildlife Service Says Lawsuits Delaying Additions to Endangered List*, WASH. POST, Nov. 24, 2000, at A5. By August 2001, though, the FWS, the Center for Biological Diversity, and other interested groups reached an agreement pursuant to which the agency moved to list 29 species while the environmental groups refrained from litigating to obtain critical habitat designations. The compromise gained widespread acclaim, though the *New York Times* editorialized that "it would have never been necessary if, over the years, Congress had provided Interior with the resources it needed to enforce the act in a systematic, timely way." *A Victory for Endangered Species*, N.Y. TIMES, Sept. 3, 2001, at A14. The suggestion that more money would cure the problem is disputed by Utah Representative James Hansen, who has complained that "we continue to throw money at the ESA in the hope that somehow funding might recover species. This approach will not work." 145 CONG. REC. H3061 (daily ed. May 12, 1999).

5. Listing is not supposed to be forever. The goal of the ESA is to help a species recover so that it is no longer in danger of extinction and no longer in need of the law's protections. Toward that end, the FWS can delist a species when it is no longer endangered or threatened according to the same five criteria set forth in section 4(a)(1) that are used to determine whether a species should be listed in the first place. Ideally, a species will be removed from the list when it has recovered. A species may also be delisted when new scientific information indicates that it is more abundant than previously realized or that it is not really a distinct species at all. Less happily, a species can be delisted if it is extinct. As of July 2000, 31 species have been delisted: 11 had recovered, six because of new information and six because of taxonomical changes, and seven were believed to be extinct. Several other high profile species, including the bald eagle and the grey wolf, are the subject of pending delisting proposals. The list of delisted species is available at <http://ecos.fws.gov/webpage/webpage_delisted.html?module=521&listings=0>.

D. CASE STUDY: THE BLACK-TAILED PRAIRIE DOG

Imagine that you are the Director of the FWS. On July 4, 1998, the National Wildlife Federation (NWF) petitioned the FWS to list the black-tailed prairie dog under the ESA, both on a permanent and on an emergency basis. The Secretary of the Interior—your boss—has asked you to prepare a memorandum responding to the NWF's petition to list the black-tailed prairie dog under the ESA. In particular, he wants your opinion on the following matters: (1) whether the black-tailed prairie dog satisfies the statutory and regulatory criteria for listing under the ESA, (2) the legal obligations that would be imposed on the FWS if the black-tailed prairie dog is listed, (3) the legal obligations that would be imposed on other federal agencies if the black-tailed prairie dog is listed, (4) the legal obligations that would be imposed on private parties and on state and local governments if the black-tailed prairie dog is listed, and (5) how to respond to the policy arguments raised by the supporters and the opponents of the petition, wholly apart from the legal requirements of the ESA. You have gathered the following information.

The black-tailed prairie dog is shaped like a football, is colored brown except for its black-tailed tail, and is actually a burrowing ground squirrel that is labeled a dog because of its barking call. It is one of five species of prairie dogs in North America: the Mexican prairie dog is listed as endangered under the ESA and the Utah prairie dog is listed as threatened, while the white-tailed prairie dog and the Gunnison prairie dog are both abundant and are not listed. The black-tailed prairie dog can be found on plains and plateaus from southern Saskatchewan to northern Mexico. Of the nearly ten million black-tailed prairie dogs in the United States, half live in South Dakota, the next largest populations live in Montana and Wyoming, and the rest of the animals live in Colorado, Kansas, Nebraska, New Mexico, North Dakota, Oklahoma, and Texas.

Black-tailed prairie dogs live in towns comprised of thousands of individual animals. Those towns (also known as colonies) are a boon to nearby wildlife and to the entire short-grass prairie ecosystem. More kinds of birds and other mammals live in areas inhabited by prairie dogs than in adjacent parts of the prairie. The tunnels that the prairie dogs dig are also used by snakes, birds and the other mammals that live in the prairie. And the black-tailed prairie dog is prey for black-footed ferrets (which are listed as endangered under the ESA), mountain plovers (which the FWS has proposed to list as threatened), swift foxes (a candidate for ESA listing), burrowing owls, bobcats, badgers, and hawks. The black-footed ferret nearly became extinct because of the decline of the prairie dogs, and efforts to reintroduce the ferret into the wild are hampered by the lack of suitable sites containing prairie dog colonies. Similarly, mountain plovers depend upon large colonies of prairie dogs to reduce vegetation and thus create nesting sites for the birds.

At one time, about 1.5 billion black-tailed prairie dogs lived on about 100 million acres of habitat in North America. The animals thrived in huge colonies that extended more than one hundred miles and contained more than one million individual animals. Today, the number of black-tailed prairie dogs has dropped to about ten million, and the FWS estimates that only about 768,000 acres of habitat remain. Of that remaining land, 45% is privately owned, 30% is on Native American tribal reservations, 15% is federally owned, and 10% is state owned. The black-tailed prairie dog colonies that survive are often small, fragmented, and surrounded by lands that are unsuitable for expansion. The two largest colonies, numbering about one million animals each, are found in the Badlands National Park in South Dakota and in Custer National Forest in rural southeastern Montana. The three smallest colonies contain less than five hundred animals each: one survives on federal land in central Wyoming that is used by private ranchers for cattle grazing but which is managed by the Bureau of Land Management (BLM), another exists on BLM land just outside of Yellowstone National Park, and the third lives on the fringes of Denver on privately owned lands that are in the path of that city's recent housing boom.

The causes of the decline in the population of the black-tailed prairie dog are:

- Habitat destruction—Less than one percent of the original prairie dog habitat remains intact. Most of what was once the prairie dog's habitat is now used to grow crops. Today colonies of prairie dogs are often too small and widespread to support viable populations or to allow for critical movement between populations.

- Poison—Ranchers regard prairie dogs as pests akin to hazardous substances. Millions of prairie dogs have been killed by the application of poisons to the habitat and food of the animals. Poisoning wiped out the entire population of black-tailed prairie dogs that once lived in Arizona.

- Recreational shooting—Prairie dogs are used as targets by hunters throughout the west. Some rural communities host prairie dog shooting contests with cash prizes for those who kill the most animals in a day.

- Disease—Prairie dogs have no known immunities to Sylvatic plague, a disease spread among squirrels and other rodents by fleas. Whole colonies of prairie dogs have been killed by the disease. Only South Dakota has remained untouched by the disease thus far. On the other hand, prairie dogs are reported to transmit the plague to other animals and to humans.

All of the states in which the black-tailed prairie dog lives mandate or encourage the extermination of prairie dogs because they regard the animal as a pest. For example, a 1927 Colorado statute provides that "prairie dogs are such a grave menace to the agricultural and livestock industries that the state is directed to take immediate action to control, suppress and eradicate such rodents in the areas infested by them." South Dakota requires private landowners to keep prairie dogs at least one hundred yards from neighboring property or otherwise pay for their control. The federal government has long had the same view of prairie dogs. The United States Department of Agriculture (USDA) issues permits for the poisoning of 200,000 acres of prairie dog colonies annually. Nonetheless, attitudes toward prairie dogs are beginning to change. In 1998, the Forest Service banned the shooting of prairie dogs on 9,000 acres in the Buffalo Gap National Grassland in southwestern South Dakota, an area where hunters have killed hundred of thousands of prairie dogs and where only 200,000 remain. Earlier this year, the BLM called for a voluntary moratorium on prairie dog shooting in the area immediately adjacent to Yellowstone National Park. Since the NWF petitioned for the listing of the black-tailed prairie dog, several state agencies, trade associations, tribal governments and private landowners have begun to meet with the NWF to discuss ways of protecting the prairie dog and prairie ecosystems while respecting the concerns of local residents.

The NWF says that these facts support the listing of the black-tailed prairie dog under the ESA. Further, the NWF has petitioned for an emergency listing as well. Besides the facts already recounted, the NWF cites three additional reasons why an emergency listing is appropriate: (1) the number of events organized for the shooting of prairie dogs has risen from 200 to 600 in just the past three years, (2) the Denver metropolitan area is experiencing a housing construction boom that is expected to replace 100,000 acres of prairie dog habitat with residential subdivisions this year, and (3) the very act of asking that the black-tailed prairie dog be listed under the ESA is likely to encourage ranchers to wipe out as many of the animals as they can before it becomes illegal to do so.

Not surprisingly, the effort to extend the protections of the ESA to black-tailed prairie dogs has been extremely controversial. The FWS has received 3,300 comments on the petition, with 60% favoring the listing and 40% opposing it. Other environmentalists have joined the NWF in extolling the virtues of the prairie dog as a crucial component of the rapidly disappearing prairie ecosystems. But the proposed listing of the black-tailed prairie dog has many opponents.

Ranchers see prairie dogs as pests of the worst kind. Cattle, horses and ranchers themselves break their legs in the holes dug by prairie dogs. Also, ranchers have long complained that prairie dogs compete with cattle for grass to eat, though environmentalists insist that there is little overlap between the plants eaten by cattle and the plants eaten by prairie dogs. It is difficult for ranchers to believe that prairie dogs are becoming rare when they spend millions of dollars annually to get rid of them, and when they see the animals reproduce so quickly. The additional costs that prairie dogs impose on ranchers come at a time when many western ranchers are struggling to survive amidst falling livestock prices and when many rural western communities are economically depressed.

Hunters love prairie dogs. Forty thousand people belong to the Varmint Hunters Association, which specializes in hunting prairie dogs. Another group, the Varmint Militia, prides itself in "defending farmers from the true invaders: the prairie dogs." Some outfitters charge $350 per day to conduct prairie dog hunts. Other hunters participate in prairie dog killing contests. For example, the tiny town of Lucla, Colorado sponsors a contest that draws over one hundred hunters who pay $100 each for the privilege of taking shoot at prairie dogs. One Lucla contest produced 3,000 dead prairie dogs, $7,000 in prizes, and $3,000 for the town's budget. Alas, in 1997 the Colorado Wildlife Commission limited the number of prairie dogs that could be killed to five per day. Even so, hunters are opposed by numerous animal rights groups that object to any killing of prairie dogs.

Developers object to prairie dogs. The dangers presented by having so many animals and their tunnels near human homes have forced many developers to use vacuums and hoses to force prairie dogs from their tunnels so the animals can be moved elsewhere. But developers are most concerned about the effects of federal protection of prairie dogs. The Colorado Home Builders Association (CAHB) predicts that housing construction in the Denver area will grind to a halt if black-tailed prairie dogs are protected under the ESA. The CAHB is unmoved by assurances that many construction activities may be allowed if a special ESA "take" rule is issued if the species is listed as threatened instead of as endangered. Moreover, the CAHB notes that many residential developments must include golf courses to be economically viable, but prairie dogs can destroy a golf course by their tunneling. Not only could the ESA prohibit a golf course from taking steps designed to control prairie dogs, but the animals could inadvertently suffer from the toxic effects of pesticides and fertilizers used to maintain a course.

The military is suspicious of the consequences of listing the black-tailed prairie dog under the ESA. Prairie dog colonies keep appearing on Army bases in areas that have been used as a target range for the testing of new weapons. Air Force bases have struggled to maintain dirt landing strips on land where prairie dogs continue to build new holes and tunnels. The military also has a general concern about the disruptive effect that protecting wildlife can have on the military's primary defense mission.

State and local officials perceive the proposed listing of the prairie dog as yet another effort by Washington bureaucrats to wrest land use control from westerners. Colorado's Representative Schaffer reports that when a member of his staff called the FWS for information on the black-tailed prairie dog, an agency official asked "is that some kind of hunting dog or something?" Officials in Baca County recently persuaded the Colorado state legislature to ban the importation of prairie dogs into any county where they are not welcomed by local residents. Throughout the west, most local officials contend that prairie dogs are not worthy of any protection, but if such protection is necessary, then it should be provided on a local basis because county and town officials are most familiar with the habits of prairie dogs and the steps that would be needed to save them. Native American officials are more sympathetic to the plight of the prairie dog, but they, too, worry about the intrusive effects of federal regulation of tribal reservations and the effects of the ESA in particular.

Finally, and perhaps most surprisingly, Japanese pet owners object to any federal regulation of prairie dogs. Every year thousands of prairie dogs are exported from the United States to Japan, where the animals are sold in pet stores for several hundred dollars each. Pat Storer, the author of "Prairie Dog Pets," explains that "the Japanese are crazy about small animals because they are good luck to them." Some members of the Japanese Prairie Dog Association even place dirt on the top of their buildings so that prairie dogs can establish colonies there. But prairie dogs cannot be kept as pets in most U.S. states because the animals are considered wildlife.

CHAPTER 5

FEDERAL GOVERNMENT RESPONSIBILITIES

Chapter Outline:
A. Listing Agency Responsibilities Under the ESA
   1. Designation of Critical Habitat
   2. Development of Recovery Plans
B. Responsibilities of All Federal Agencies Under the ESA
   1. The Conservation Duty
   2. The Duty to Avoid Causing Jeopardy

―――――

Once a species has been listed as endangered or threatened, the substantive obligations of the Endangered Species Act (ESA) come into play. The ESA imposes a variety of duties and prohibitions on government and private parties. The responsibilities of the federal government will be detailed in this chapter; the responsibilities applicable to others—including private parties and state and local governments—will be described in chapter 6.

The substantive commands of the ESA for the federal government can be divided into two groups. The first set of commands apply to the agency that is responsible for the species once it is listed. This is usually the Fish & Wildlife Service (FWS), though the National Marine Fisheries Service (NMFS) will be the responsible agency for marine mammals and fish listed under the ESA. The second set of commands apply to *all* federal agencies. Those commands are often difficult to enforce because they require an agency to protect biodiversity even when such protection conflicts with the agency's primary statutory responsibility to, for example, promote military readiness or respond to natural disasters.

―――――

A. LISTING AGENCY RESPONSIBILITIES UNDER THE ESA

The ESA's substantive provisions are designed to eliminate the danger of extinction which led to the listing of the species. The listing agency—the FWS or the NMFS—is charged with the responsibility for finding a way to remove the species from the brink of extinction, and thus, from the ESA's list. The statute directs the agency to take two steps toward that goal upon the listing of the species. First, the ESA provides for the listing of the "critical habitat" of the species at the same time that the species itself is listed as endangered or threatened. Second, the ESA requires the agency to prepare a "recovery plan" that explains the measures that will be necessary to bring the species back from the brink of extinction. Neither provision, however, has performed the kind of invaluable role that the original supporters of the ESA had hoped, and the appropriate roles of the designation of critical habitat and recovery plans continue to be debated as changes to the ESA are being considered.

1.   DESIGNATION OF CRITICAL HABITAT

Natural Resources Defense Council v. United States Department of the Interior

United States Court of Appeals for the Ninth Circuit, 1997.
113 F.3d 1121.

■ PREGERSON, CIRCUIT JUDGE:

This case presents the question whether the defendants violated the Endangered Species Act by failing to designate critical habitat for the coastal California gnatcatcher.... The coastal California gnatcatcher is a songbird unique to coastal southern California and northern Baja California. The gnatcatcher's survival depends upon certain subassociations of coastal sage scrub, a type of habitat that has been severely depleted by agricultural and urban development. Approximately 2500 pairs of gnatcatchers survive in southern California today.

The California gnatcatcher

On March 30, 1993, the U.S. Fish and Wildlife Service listed the gnatcatcher under the Endangered Species Act as a "threatened species." Under section 4 of the Act, the listing of a threatened species must be accompanied by the concurrent designation of critical habitat for that species "to the maximum extent prudent and determinable." The designation of critical habitat in turn triggers the protections of section 7 of the Act. Section 7 requires that federal agencies consult with the Secretary of the Interior (the "Secretary") to ensure that actions authorized, funded, or carried out by federal agencies do not harm critical habitat.

At the time of the gnatcatcher's listing as a threatened species, the Service found that coastal sage scrub habitat loss posed "a significant threat to the continued existence of the coastal California gnatcatcher." Nevertheless, the Service concluded that critical habitat designation would not be "prudent" within the meaning of section 4 for two reasons. First, the Service claimed that the public identification of critical habitat would increase the risk that landowners might deliberately destroy gnatcatcher habitat. Second, the Service claimed that critical habitat designation "would not appreciably benefit" the gnatcatcher because most gnatcatcher habitat is found on private lands to which section 7's consultation requirement does not apply. . . .

THE SERVICE'S FAILURE TO DESIGNATE CRITICAL HABITAT

Section 4 of the Act requires that the gnatcatcher's listing as a threatened species be accompanied by concurrent designation of critical habitat "*to the maximum extent prudent* and determinable":

> The Secretary, by regulation promulgated in accordance with subsection (b) of this section and *to the maximum extent prudent* and determinable—
>
> (A) shall, concurrently with making a determination under paragraph (1) that a species is an endangered species or a threatened species, designate any habitat of such species which is then considered to be critical habitat; and
>
> (B) may, from time-to-time thereafter as appropriate, revise such designation.

16 U.S.C. § 1533(a)(3) (emphasis added).

The Act itself does not define the term "prudent." The Service has defined what would *not* be prudent, however, in the regulations promulgated under the Act. According to the regulations, critical habitat designation is not prudent "when one or both of the following situations exist":

> (i) The species is threatened by taking or other human activity, and identification of critical habitat can be expected to *increase the degree of such threat to the species*, or
>
> (ii) Such designation of critical habitat *would not be beneficial to the species*.

50 C.F.R. § 424.12(a)(1)(I)-(ii) (1995) (emphasis added).

When the Service published the gnatcatcher's final listing as a threatened species, the Service stated that critical habitat designation would not be prudent under either prong of the regulatory definition. The final listing fails to show, however, that the Service adequately "considered the relevant factors and articulated a rational connection between the facts found and the choice made"....

A. *Increased Threat to the Species*

The Service's first reason for declining to designate critical habitat was that designation would increase the degree of threat to the gnatcatcher. The final listing referred to eleven cases in which landowners or developers had destroyed gnatcatcher sites; in two of these cases, habitat was destroyed after the Service notified local authorities that gnatcatchers were present at a proposed development site. On the basis of this history, the Service concluded that because the publication of critical habitat descriptions and maps would enable more landowners to identify gnatcatcher sites, designating critical habitat "would likely make the species more vulnerable to [prohibited takings] activities."

This "increased threat" rationale fails to balance the pros and cons of designation as Congress expressly required under section 4 of the Act. Section 4(b)(2) states that the Secretary may only exclude portions of habitat from critical habitat designation "if he determines that the benefits of such exclusion *outweigh the benefits* of specifying such area as part of the critical habitat." 16 U.S.C. § 1533(b)(2) (emphasis added). In addition, the Service itself has said that it will forgo habitat designation as a matter of prudence only "in those cases in which the possible adverse consequences would *outweigh the benefits* of designation." 49 Fed. Reg. 38900, 38903 (1984) (emphasis added).

In this case, the Service never weighed the benefits of designation against the risks of designation. The final listing decision cited only eleven cases of habitat destruction, out of 400,000 acres of gnatcatcher habitat. The listing did not explain how such evidence shows that designation would cause more landowners to destroy, rather than protect, gnatcatcher sites. The absence of such an explanation is particularly troubling given that the record shows these areas had already been surveyed extensively in other gnatcatcher or coastal sage scrub studies published prior to the date of final listing.

By failing to balance the relative threat of coastal sage scrub takings both with and without critical habitat designation, the Service failed to consider all relevant factors.... The Service's reliance on the "increased threat" exception to section 4 designation was therefore improper.

B. *No Benefit to the Species*

The Service's second reason for declining to designate habitat was that designation "would not appreciably benefit the species." According to the Service's final listing decision, most populations of gnatcatchers are found on private lands to which section 7's consultation requirement would not apply. The final listing decision suggests that designation may only be deemed "beneficial to the species" and therefore "prudent" if it would result in the application of section 7 to "the *majority* of land-use activities occurring within critical habitat."

By rewriting its "beneficial to the species" test for prudence into a "beneficial to *most* of the species" requirement, the Service expands the narrow statutory exception for imprudent designations into a broad exemption for imperfect designations. This expansive construction of the "no benefit" prong to the imprudence exception is inconsistent with clear congressional intent.

The fact that Congress intended the imprudence exception to be a narrow one is clear from the legislative history, which reads in part:

> The committee intends that in most situations the Secretary will ... designate critical habitat at the same time that a species is listed as either endangered or threatened. *It is only in rare circumstances where the specification of critical habitat concurrently with the listing would not be beneficial to the species.*

H.R. Rep. No. 95–1625 at 17 (1978), *reprinted in* 1978 U.S.C.C.A.N. 9453, 9467 (emphasis added). *See also* Enos v. Marsh, 769 F.2d 1363, 1371 (9th Cir.1985) (holding that the Secretary "may only fail to designate critical habitat under *rare* circumstances") (emphasis added); Northern Spotted Owl v. Lujan, 758 F. Supp. 621, 626 (W.D.Wash.1991) ("This legislative history leaves little room for doubt regarding the intent of Congress: The designation of critical habitat is to coincide with the final listing decision absent *extraordinary* circumstances") (emphasis added).

By expanding the imprudence exception to encompass all cases in which designation would fail to control "*the majority* of land-use activities occurring within critical habitat," the Service contravenes the clear congressional intent that the imprudence exception be a rare exception. Since "the court, as well as the agency, must give effect to the unambiguously expressed intent of Congress," we reject the Service's suggestion that designation is only necessary where it would protect the majority of species habitat.

In the present case, the Service found that of approximately 400,000 acres of gnatcatcher habitat, over 80,000 acres were publicly-owned and therefore subject to section 7 requirements. Other privately-owned lands would also be subject to section 7 requirements if their use involved any form of federal agency authorization or action.

The Service does not explain why a designation that would benefit such a large portion of critical habitat is not "beneficial to the species" within the plain meaning of the regulations and "prudent" within the clear meaning of the statute. Accordingly, we conclude that the Service's "no benefit" argument fails to "articulate[ ] a rational connection between the facts found and the choice made".... The Service's reliance on the "no benefit" exception to section 4 designation was therefore improper.

C. Less Benefit to the Species

In addition to the above two rationales which were stated in the final listing, the defendants now offer a third argument in defense of the

Service's failure to designate critical habitat. The defendants contend that a "far superior" means of protecting gnatcatcher habitat is provided by the state-run "comprehensive habitat management program" created under California's Natural Communities Conservation Program ("NCCP"). The Service has endorsed the NCCP as a "special rule" for gnatcatcher protection under section 4(d) of the Act, 16 U.S.C. § 1533(d).

Regulations under the Act provide that "the reasons for not designating critical habitat will be stated in the publication of proposed and final rules listing a species." 50 C.F.R. § 424.12(a). The NCCP alternative was not identified in the Service's proposed or final listings as a reason not to designate critical habitat. Therefore, this argument is not properly before us for consideration.

Even if we were to consider the NCCP alternative, however, the existence of such an alternative would not justify the Service's failure to designate critical habitat. The Act provides that designation of critical habitat is necessary except when designation would not be "prudent" or "determinable." The Service's regulations define "*not* prudent" as "increasing the degree of [takings] threat to the species" or "*not* . . . beneficial to the species." 50 C.F.R. § 424.12(a)(1)(I)-(ii) (emphasis added). Neither the Act nor the implementing regulations sanctions nondesignation of habitat when designation would be merely less beneficial to the species than another type of protection.

In any event, the NCCP alternative cannot be viewed as a functional substitute for critical habitat designation. Critical habitat designation triggers mandatory consultation requirements for federal agency actions involving critical habitat. The NCCP alternative, in contrast, is a purely voluntary program that applies only to non-federal land-use activities. The Service itself recognized at the time of its final listing decision that "no substantive protection of the coastal California gnatcatcher is currently provided by city/county enrollments [in the NCCP]." Accordingly, we reject the defendants' post hoc invocation of the NCCP to justify the Service's failure to designate critical habitat. . . .

■ O'Scannlain, Circuit Judge, dissenting:

. . . The majority states that a determination of whether a designation would be prudent must include weighing the benefits of designation against its risks. Our cases do not support this conclusion, however; they generally require only that the agency follow a rational decision-making process. . . . That point aside, however, I believe that a fair reading of the Service's decision reveals that it did in fact conduct precisely the balancing test called for by the majority when it concluded that designation may cause the intentional destruction of habitat by private landowners, but would produce little benefit since most of the habitat is not on publicly-owned land. It is worth quoting the Service at length:

> [S]ome landowners or project developers have brushed or graded sites occupied by gnatcatchers prior to regulatory agency review or the issuance of a grading permit. In some instances, gnatcatcher habitat

was destroyed shortly after the Service notified a local regulatory agency that a draft environmental review document for a proposed housing development failed to disclose the presence of gnatcatchers on-site. On the basis of these kinds of activities, *the Service finds that publication of critical habitat descriptions and maps would likely make the species more vulnerable* to [prohibited] activities. . . .

Most populations of the coastal California gnatcatcher in the United States are found on private lands where Federal involvement in land-use activities does not generally occur. Additional protection resulting from critical habitat designation is achieved through the section 7 consultation process. *Since section 7 would not apply to the majority of land-use activities occurring within critical habitat, its designation would not appreciably benefit the species.*

58 Fed. Reg. 16742, 16756 (1993) (emphasis added). In my view, applying the majority's balancing requirement, the Service indeed weighed the benefits and risks of designation and came to a rational, defensible conclusion that designation was not prudent. . . .

The second situation in which designation would not be prudent exists when "such designation of critical habitat would not be beneficial to the species." 50 C.F.R. § 424.12(a)(1)(ii) (1996). . . . In my view, the majority takes too narrow a view of the phrase "beneficial to the species." The question should not be whether any member of the species would be better off by a slender margin, but whether the species as a whole would benefit from the designation. Even though the gnatcatchers in most of the habitat would not benefit, reasons the majority, some of the gnatcatchers would benefit, and hence designation would be beneficial for the species. The problem with this argument is that it overlooks the Service's expert opinion, to which we are required to defer, that designation may harm the gnatcatchers when landowners intentionally destroy the habitat. Even though individual pockets of gnatcatchers may benefit, the species as a whole may not.

Notes and Questions

1. One of the congressional purposes in enacting the ESA was "to provide a means whereby the ecosystems upon which endangered species and threatened species depend may be conserved." ESA § 2(b), 16 U.S.C. § 1531(b). How can that goal be accomplished? How could the designation of critical habitat help accomplish that goal? For an illustration of how the ESA's various provisions, including the designation of critical habitat, "can protect imperilled ecosystems far more effectively than is commonly thought," *see* William Snape III et al., *Protecting Ecosystems Under the Endangered Species Act: The Sonoran Desert Example*, 41 WASHBURN L. REV. 14, 15 (2001).

2. Most of the substantive provisions of the ESA focus on the species itself, rather than the habitat of a species. There are only a few provisions that specifically aim to protect habitat. Section 5 of the ESA authorizes the government to purchase land that serves as the habitat of an endangered or threatened species. Section 9 of the ESA has been interpreted to prohibit certain private or governmental actions that adversely affect the habitat of a species. *See* Babbitt v. Sweet Home Chapter of Communities for a Great Oregon, 515 U.S. 687 (1995). The gnatcatcher case illustrates the means by which the ESA designates the "critical habitat" of a species pursuant to ESA section 4.

The ESA defines "critical habitat" as "the specific areas within the geographical area occupied by the species, at the time it is listed .. on which are found those physical and biological features (I) essential to the conservation of the species and (II) which may require special management considerations or protections." ESA § 3(5)(i); *see also* ESA § 3(5)(ii) (adding that critical habitat can also include areas outside the current range of a listed species if the Secretary determines that "such areas are essential for the conservation of the species"). If critical habitat is designated, the only regulatory consequence is that ESA section 7 requires all federal agencies to insure that none of their actions "result in the destruction or adverse modification" of the designated critical habitat of a species. The most famous example of that prohibition is Tennessee Valley Authority v. Hill, 437 U.S. 153 (1978), which blocked the completion of the Tellico Dam because the resulting impoundment would have destroyed the endangered snail darter's critical habitat. The designation of critical habitat has no effect on the activities of private parties or state governments unless they are seeking federal funding or a federal permit that triggers the section 7 scrutiny. Moreover, the prohibition on federal actions that destroy or modify critical habitat usually overlaps the restrictions imposed by section 9's ban on the "take" of a protected species. Yet the official designation of critical habitat has been controversial. Why?

3.   The designation of critical habitat is to occur at the same time that the listing agency determines that a species is endangered or threatened. The court in the gnatcatcher case expected that the agency would decline to designate critical habitat only in "rare" circumstances. In fact, it is much more common for the agency *not* to designate the critical habitat of a species. As of July 2002, critical habitat had been designated for 155 species, or about twelve percent of the listed species in the United States. Fish are most likely to benefit from critical habitat designations, while birds and plants are much less likely. *See* James Saltzman, *Evolution and the Application of Critical Habitat Under the Endangered Species Act*, 14 Harv. Envtl. L. Rev. 311, 332 (1990) (table showing that 47% of fish, 39% of reptiles, 33% of amphibians, 29% of crustaceans, 27% of mammals, 14% of plants, 12% of birds, and no snails or clams listed as endangered or threatened had designated critical habitat as of 1988).

Plants pose a particular problem for the designation of critical habitat. Rare plants are attractive to collectors, and plants are stationary, so identifying where rare plants can be found is often counterproductive. For example, when declining to designate critical habitat for the San Diego thornmint and three other southern Californian plants, the Fish and Wildlife Service explained:

> Landowners may mistakenly believe that critical habitat designation will be an obstacle to development and impose restrictions on their use of their property. Unfortunately, inaccurate and misleading statements reported through widely popular medium available worldwide, are the types of misinformation that can and have led private landowners to believe that critical habitat designations prohibit them from making use of their private land when, in fact, they face potential constraints only if they need a Federal permit or receive Federal funding to conduct specific activities on their lands.... A designation of critical habitat on private lands could actually encourage habitat destruction by private landowners to rid themselves of the perceived endangered species problem. Listed plants have limited protection under the Act, particularly on private lands.... Thus, a private landowner concerned about perceived land management conflicts resulting from a critical habitat designation covering his property would likely face no legal consequences if the landowner removed the listed species or destroyed its habitat. For example, in the spring of 1998, a Los Angeles area developer buried one of the only three populations of the endangered *Astragalus brautonii* in defiance of efforts under the [California Environmental Quality Act] to negotiate mitigation for the species. The designation of critical habitat involves the publication of habitat descriptions and mapped locations of the species in the Federal Register, increasing the likelihood of potential search and removal activities at specific sites.

Determination of Endangered or Threatened Status for Four Plants from Southwestern California and Baja California, Mexico, 63 Fed. Reg. 54938, 54951–52 (1998). Are these explanations for resisting the designation of the critical habitat of endangered plants persuasive?

4. As the gnatcatcher case illustrates, the ESA requires the designation of critical habitat "to the maximum extent prudent and determinable." ESA § 4(a)(3). There have not been too many instances when the listed agency has refused to designate critical habitat because it was indeterminable. Critical habitat is not determinable if there is inadequate information about biological needs of the species or the impacts of a designation. 50 C.F.R § 424.12(a)(2). The courts have been wary of the instances in which the FWS has claimed that critical habitat is not determinable. *See* Northern Spotted Owl v. Lujan, 758 F.Supp. 621 (W.D.Wash.1991) (holding that the failure to provide any reason for concluding that the critical habitat of the northern spotted owl was not determinable was arbitrary and capricious); Colorado Wildlife Federation v. FWS, 36 E.R.C. (BNA) 1409 (D.Colo. 1992) (holding that the agency's fear of designating too large an area as critical habitat because of the inadequacy of current information violated the statutory directive to make the determination based on whatever information was presently available).

The most common reason for failing to designate critical habitat is that such a designation is not "prudent." Since April 1996, the FWS has determined that it would not be prudent to designate critical habitat in 228 out of 256 cases. But the Ninth Circuit's decision rejecting the agency's broad understanding of when a critical habitat designation is not prudent is already producing different results. On remand, the FWS designated over 500,000 acres in five southern California counties as the critical habitat of the California gnatcatcher. Final Determination of Critical Habitat for the Coastal California Gnatcatcher, 65 Fed. Reg. 63680 (2000). The FWS designated public and private land alike, omitting only those areas that were already protected by a habitat conservation plan. The Ninth Circuit's decision also prompted the FWS to reverse its original determination and conclude that it would be prudent to designate critical habitat for ten newly listed plants on Maui, Molokai, Lanai and Kahoolawe. *See* Final Endangered Status for 10 Plant Taxa from Maui Nui, HA, 64 Fed. Reg. 48307, 48319 (1999).

5. The designation of critical habitat is limited by the resources available to the Fish and Wildlife Service. The designation of critical habitat can cost the agency as much as $500,000 per species, whereas the agency's 1998 budget for all listing activities—not just determinations of critical habitat— was $5.19 million. Thus, when sued for not designating critical habitat for 245 recently listed Hawaiian plants, the agency responded that "if the FWS must complete prudency determinations for all 245 plants by 2000, it will need to suspend all other listing activity" in that region. Conservation Council for Hawai'i v. Babbitt, 24 F.Supp.2d 1074, 1078 (D.Hawai'i 1998). The environmental plaintiffs pointed out, though, that the agency had received all of the funds that it had requested from Congress. The court fashioned a compromise that required the agency to publish proposed rules regarding the designation of critical habitat for 100 of the plants by November 30, 2000, with the remaining plants to be addressed by April 30, 2002. Also, the court deferred to the agency's decision to consider plants in the same Hawaiian ecosystem first, rather than accepting the environmentalists's suggestion to first consider those plants most directly affected by federal actions. For another example of a court's response to the Fish and Wildlife Service's resource limitations with respect to designating critical habitat, *see* Forest Guardians v. Babbitt, 164 F.3d 1261 (10th Cir.1998) (holding that the Secretary of the Interior failed to perform his statutory duty to designate the critical habitat of the Rio Grande silvery minnow, and that claims of impossibility because of resource limitations were premature).

The FWS contends that its limited resources should not be devoted to critical habitat designations. In June 1999, the agency published a proposed rule to clarify the role of habitat in endangered species conservation. The agency expressed its belief that "the present system for determining and designating critical habitat is not working." The FWS sees itself caught between a statutory requirement of little value and a need to commit limited resources to more pressing needs. "[I]n most circumstances," the agency explained, "the designation of 'official' critical habitat is of little additional value for most listed species, yet it consumes large amounts of conservation resources. . . . Moreover, we have long believed that separate protection of critical habitat is duplicative for most species.'' But lawsuits like the gnatcatcher case and Hawaiian plant case that seek to compel the FWS to designate critical habitat threaten to overwhelm the agency. The agency insisted that ''[t]here are other species in Hawaii that are literally facing extinction while precious resources are being depleted on critical habitat litigation support and reexaminations of critical habitat prudency determinations for species already listed.''

With these concerns in mind, the FWS solicited public input in to the future of critical habitat under the ESA. What advice would you give to the agency? One possibility is reflected by S. 1100, a bill introduced by Senator Domenici in 1999 that would amend the ESA to require the completion of a recovery plan for a species before critical habitat is designated. That idea was endorsed by Interior Secretary Babbitt, who testified before Congress in the spring of 1999 that the designation of critical habitat under the ESA ''does not work. It does not produce good results. It should be modified, because the Courts are driving us to front-end determinations which, more properly, should be incorporated in recovery plans at the back end when we, in fact, have the information.'' 145 CONG. REC. S4424 (daily ed. Apr. 29, 1999). For other suggestions, *compare* Ben Neary, *Habitat Moratorium Tacked Onto War Bill*, SANTA FE NEW MEXICAN, May 12, 1999, at B–1 (describing a provision to be added to a bill funding the war in Kosovo that would impose a moratorium on the designation of any new critical habitat) *with* S. 1100, 106th Cong., 1st Sess. (1999) (proposed) *and* Oliver A. Houck, *The Endangered Species Act and Its Implementation by the U.S. Departments of Interior and Commerce*, 64 U. COLO. L. REV. 277, 296–315 (1993) (criticizing the FWS for failing to designate critical habitat and defending the need for such designations).

6. An area can be excluded from the designated critical habitat if the benefits of exclusion outweigh the benefits of inclusion, unless the failure to designate the area as critical habitat will result in the extinction of the species. In conducting that balance, the ESA directs the listing agency to consider ''the economic impact, and any other relevant impact, of specifying any particular area as critical habitat.'' 16 U.S.C § 1533(b)(2). What result should such balancing yield in the gnatcatcher case? The California gnatcatcher has sparked a serious conflict in land use in southern California. The gnatcatcher's habitat is on land along the Pacific coast around Los Angeles and San Diego—precisely the land that developers, companies, and homeowners desire for their own uses. Some of that land costs as much as $200,000 per lot. Does that mean that the economic impact of designating the gnatcatcher's critical habitat should outweigh any benefit from a designation?

And how should one calculate the economic impact attributable to the designation of critical habitat? The FWS has insisted that any economic impact of the listing of the species should not count toward the determination of the economic impact of designating critical habitat. Thus, for example, any economic losses that landowners suffer because the listing of the California gnatcatcher prevents them from clearing sage scrub from their property would not be relevant in calculating the economic impact of designating that land as the gnatcatcher's critical habitat. That baseline approach greatly reduced the likelihood that the costs of designating critical habitat would outweigh the benefits. But in New Mexico Cattle Growers Association v. FWS, 248 F.3d 1277 (10th Cir.2001), the court held that the statutory language of the ESA requires FWS to consider "all of the economic impact of the [critical habitat designation], regardless of whether those impacts are caused co-extensively by any other agency action (such as listing) and even if those impacts would remain in the absence of the" designation. *Id.* at 1283. Applying that approach, what would the FWS need to investigate to decide whether the economic impacts of designating certain land as the critical habitat of the California gnatcatcher outweigh the benefits?

7. The efforts of all of the interested parties to reach an agreement about how to comply with the ESA's protection of the gnatcatcher while also allowing development in the area are detailed in Charles C. Mann & Mark L. Plummer, *California v. Gnatcatcher*, AUDUBON, Jan.-Feb. 1995, at 38; and Dwight Holing, *The Coastal Scrub Solution*, NATURE CONSERVANCY, July/August 1997, at 16. More recently, the gnatcatcher has been the subject of the California Natural Communities Conservation Program, a regional Habitat Conservation Plan (HCP) of the type discussed in chapter 6.

2. DEVELOPMENT OF RECOVERY PLANS

Strahan v. Linnon

United States District Court for the District of Massachusetts, 1997.
967 F.Supp. 581.

■ WOODLOCK, DISTRICT JUDGE

[The Northern Right whale, the Humpback whale, and other endangered marine mammals live in the Atlantic Ocean off the coast of New England. The United States Coast Guard operates in the same area, and on several occasions its vessels struck and killed a whale. Therefore, the Coast Guard initiated formal consultations with the National Marine Fisheries Service (NMFS) regarding the effect of the Coast Guard's operations on the endangered whales and other mammals. Not satisfied with the results of those consultations, Max Strahan filed a citizen suit alleging, *inter alia*, that the NMFS had failed to prepare a recovery plan pursuant to ESA § 4 for several of the endangered species, and that the recovery plans that did exist for other species did not comply with the ESA's requirements.]

The plaintiff ... alleges that NMFS has violated § 4(f) of the ESA because it has not "developed and implemented plans for the conservation and survival of endangered species and threatened species...." 16 U.S.C. § 1533(f). Plaintiff makes two separate arguments with respect to NMFS's recovery plans. First, he contends that NMFS has violated the ESA because it has not developed any recovery plans for the Blue, Sei, Fin, or Minke whales, but only the Right and Humpback whales. The plaintiff then alleges that the existing Right whale recovery plan is insufficient because it does not "incorporate implementable site-specific management actions necessary to achieve the plan's goal" and because it does not "establish a realistic recovery goal." The Amended Complaint further alleges that the existing plan has not been revised.

With respect to the claim that NMFS has violated § 4(f) because it has not developed recovery plans for federally protected whales other than Right and Humpback whales, the defendants respond that the there are no time limits in § 4(f) within which the Secretary must develop, implement, or revise a recovery plan. I am persuaded by the defendants' argument. *See* Oregon Natural Resource Council v. Turner, 863 F. Supp. 1277, 1282–83 (D.Or.1994). The court observed:

> Congress recognized that the development of recovery plans for listed species would take significant time and resources. It therefore provided in the ESA that the Secretary could establish a priority system for developing and implementing such plans. This priority system allows the Secretary broad discretion to allocate scare resources to those species that he or she determines would most likely benefit from development of a recovery plan. Unlike other requirements under the ESA, such as the designation of critical habitat, the statute places no time constraints on the development of recovery plans. *See* 16 U.S.C. § 1533(f).[18]

Accordingly, the Secretary has developed a priority system for developing such recovery plans. *See* 55 Fed. Reg. 24296. I find, therefore, that the fact that NMFS has not issued recovery plans for Sei, Blue, and Fin whales[19] does not constitute a violation of § 4(f).

The plaintiff also asserts that the recovery plans that do exist, are deficient and thus violative of the statute. He claims that "as a general matter, [the recovery plan] does not contain objective, scientific, measurable criteria." More specifically, the plaintiff contends that the plan "fails to include . . . an annual census, a population viability analysis, modeling of ship-whale interactions, risk analysis, and interim numerical goals." The defendants assert that the discretionary nature of a recovery plan also

---

18. I am also persuaded, therefore, that there are no stringent time requirements for revising a recovery plan. The plaintiff argues, nevertheless, that NMFS itself determined that the Recovery Plan should be revised every three years for the first fifteen years and every five years thereafter. At oral argument, plaintiff argued that such revision was, therefore, nondiscretionary. The relevant language in the Recovery Plan states that "three-year intervals [for updating the Plan] are recommended." This is clearly the language of discretion. Moreover, inasmuch as 16 U.S.C. § 1533(f) requires NMFS to report to Congress "on the status of efforts to develop and implement recovery plans," the defendants represented at oral argument that the 1996 report was forthcoming. Considering these factors, I find the plaintiff's allegation with respect to revision to be meritless.

19. The plaintiff includes Minke whales in his list. Minke whales, however, are not listed as an endangered or threatened species under the ESA.

applies to the plan's content and that "it is not necessary for a recovery plan to be an exhaustively detailed document."

Case law instructs that the defendants are correct in their assertion that the content of recovery plans is discretionary. For example, in Fund for Animals, Inc. v. Rice, 85 F.3d 535 (11th Cir.1996), the plaintiffs' argument relied on the assumption that "Recovery Plan[s] [are] document[s] with the force of law." *Id.* at 547. The court rejected that characterization stating that "section 4(f) makes it plain that recovery plans are for guidance purposes only." *Id.* Similarly, the court in Morrill v. Lujan, 802 F. Supp. 424, 433 (S.D.Ala.1992), found that "the contents of [recovery] plans are discretionary." While it is true that § 4(f) "does not permit an agency unbridled discretion," and "imposes a clear duty on the agency to fulfill the statutory command to the extent that it is feasible or possible," Fund for Animals v. Babbitt, 903 F. Supp. 96, 107 (D.D.C.1995), the requirement does not mean that the agency can be forced to include specific measures in its recovery plan. In fact, all that is required in a recovery plan is "the identification of management actions necessary to achieve the Plan's goals for the conservation and survival of the species." *Id.* at 108.

In any event, the evidence does not support that the measures suggested by the plaintiff are "necessary to achieve the plan's goal for the conservation and survival of the species." 16 U.S.C. § 1533(f)(B)(ii). And in fact, some of the measures advocated by the plaintiff are currently being implemented by NMFS [such as a population study]. Experts in the field plainly have different opinions as to what measures should be taken most effectively to promote conservation efforts for Right whales. It is also plain, however ... that NMFS has considered the alternatives suggested by the plaintiff. The fact that NMFS did not adopt precisely the recommended measures in its recovery plan, does not make that plan deficient. Indeed, especially when expert, scientific judgments are involved, the court must afford the agency's decision a great deal of deference.

Last, I find that the recovery plan does contain "objective, measurable criteria," § 4(f)(1)(B)(ii), and "a description of site-specific management actions," § 4(f)(1)(B)(i). In terms of "objective, measurable criteria," the recovery plan states that the recovery goal is 7000 animals. The plaintiff argues that this goal is unrealistic and meaningless without a provision for interim goals. I find nothing in § 4(f) that mandates such interim goals. I also find that the Recovery Plan satisfies the "site-specific" requirement. The term "site-specific" has been interpreted to refer to geographical areas, requiring that the agency "in designing management actions, consider the distinct needs of separate ecosystems or recovery zones occupied by a threatened or endangered species." *Fund for Animals,* 903 F. Supp. at 106. The Recovery Plan meets this requirement because it considers the separate needs of the northern Atlantic population and the northern Pacific population. Additionally, the plan also addresses the different habitats of Northern Right whales at different times of year and contains measures specifically directed at each habitat. I find that the Recovery Plan is not arbitrary and capricious.

The plaintiff argues that even if the Recovery Plan is not arbitrary and capricious, NMFS has still violated the ESA because it has not implemented the plan. To support this contention, the plaintiff lists certain goals limned in the Recovery Plan that are not yet in effect. While it appears that some of the Recovery Plan's goals have not been implemented, *e.g.,* "appropriate seasonal or geographic regulations for the use of certain fishing gear in" the Bay of Fundy and the Southern Nova Scotia Shelf, I find plaintiff's allegations to be largely unfounded and needlessly technical. For example, the plaintiff states that no regulations on whale-watch vessels exist today. In February, 1997, however, NMFS issued a rule restricting all vessels and aircraft from approaching Right whales at a distance closer than 500 yards. Moreover, while the plaintiff asserts that "NMFS still has not located the unknown wintering area it alleges exists," [a government expert] avers that the research enabling NMFS to "find the unknown summer nursery and wintering grounds" is ongoing. After considering these efforts, I find that NMFS is taking steps to implement its Recovery Plan and that no ESA violation exists. . . .

[On appeal, the First Circuit affirmed in an unpublished opinion and rejected the challenges to the recovery plan "essentially for the reasons stated by the district court." *See* Strahan v. Linnon, 187 F.3d 623 (1st Cir.1998)].

Notes and Questions

1. Does the recovery plan for the whales satisfy the goals of the ESA? What actions would be necessary to be absolutely certain that the whales will not become extinct? Obviously, Max Strahan believes that far more needs to be done to save the whales, but his approach has alienated many environmentalists. *See* Carey Goldberg, *A Boston Firebrand Alienates His Allies Even as He Saves Whales*, N.Y. Times, Jan. 23, 1999, at A9 (reporting that Strahan has been barred from the New England Aquarium after ranting at visitors about whale-watching cruises and from government offices where he has made workers cry, and that he has been arrested 130 times).

2. Section 4 of the ESA directs the Secretary of the Interior to "develop and implement [recovery plans] for the conservation and survival of endangered and threatened species . . . unless he finds that such a plan will not promote the conservation of the species." ESA § 4(f), 16 U.S.C. § 1533(f). The FWS does not develop a recovery plan for a species if (1) the species is thought to be extinct, (2) state management plans serve as an adequate substitute, or (3) ecosystem initiatives addressing the recovery of multiple species exist. The law further describes how to establish priorities among species for recovery plans and what must be included in a recovery plan. To the maximum extent practicable, priority is to be given to listed species "that are most likely to benefit from such plans, particularly those species that are, or may be, in conflict with construction or other development projects or other forms of economic activity." ESA § 4(f)(1)(A), 16 U.S.C. § 1533(f)(1)(A). Taxonomic classification is not to be considered in establishing such priorities. The contents of a recovery plan—again, to the maximum extent practicable—must include (1) "a description of such site-specific management actions as may be necessary to achieve the plan's goal for the conservation and survival of the species," (2) "objective, measurable criteria which, when met, would result in a determination ... that the species be removed from the list" of endangered or threatened species; and (3) "estimates of the time required and cost to carry out those measures needed to achieve the plan's goal and to achieve intermediate steps toward that goal." ESA § 4(f)(1)(B), 16 U.S.C. § 1533(f)(1)(B). Public comments on a proposed recovery plan must be considered, and the Secretary must report to Congress every two years regarding the status of efforts to develop recovery plans and the status of species for which plans have been developed. ESA § 4(f)(1)(C), (D), 16 U.S.C. § 1533(f)(1)(C), (D).

3. Recovery plans were in place for 981 species as of June 2002. Some kinds of species—such as snails, clams, and ferns and other plants—are much more likely to have a recovery plan than other kinds of species, such as mammals. *See* John Copeland Nagle, *Playing Noah*, 82 MINN. L. REV. 1171, 1200 n.114 (1998) (indicating that recovery plans existed for 80% of arachnids, 78% of snails, 71% of ferns and certain other plants, 69% of clams, 62% of fishes, 58% of flowering plants, 57% of insects, 44% of amphibians, 33% of crustaceans, 27% of reptiles, 26% of birds, 25% of conifers, and 12% of mammals as of July 1991). Why are endangered and threatened snails so likely to have a recovery plan, but listed mammals are not?

4. Judicial review of the content and implementation of recovery plans has been deferential to the agency. One notable exception is The Fund for Animals v. Babbitt, 903 F.Supp. 96 (D.D.C.1995), where the court rejected many objections to the recovery plan for the grizzly bear, but the court ordered the agency to reconsider the plan because it failed to adequately address some of the threats to the grizzlies and because it failed to justify a population target that was lower than what the environmentalists sought. More typical is Morrill v. Lujan, 802 F.Supp. 424, 433 (S.D.Ala.1992), where the court stated that the contents of a recovery plan are discretionary, and therefore refused to order the Fish and Wildlife Service to protect an area occupied by the endangered Perdido Key beach mouse but where a lounge was being built. Should the courts take a more active role in managing the recovery process?

Recovery Plan for Koolau Mountain Plant Cluster

U.S. Fish & Wildlife Service, Pacific Region, 1996.
Pages 1–3, 47–52.

INTRODUCTION

Brief Overview

This recovery plan covers the following 11 plant taxa that were listed as endangered on the federal list of endangered and threatened species on

March 28, 1994: *Chamaesyce deppeana* (akoko), *Cyanea crispa* ([no common name] (NCN)), *Cyanea truncata* (haha), *Cyrtandra crenata* (haiwale), *Cyrtandra polyantha* (haiwale), *Eugenia koolauensis* (nioi), *Hesperomannia arborescens* (NCN), *Lobelia oahuensis* (NCN), *Melicope lydgatei* (alani), *Phlegmariurus nutans* (wawaeiole), and *Tetraplasandra gymnocarpa* (oheohe).

Ten of the 11 taxa are known to exist only on the island of Oahu, Hawaii; one species, *Hesperomannia arborescens,* also is found on the islands of Molokai and Maui. All 11 of these taxa are known from the Koolau Mountain Range on Oahu. Historically, three of the taxa were also found on Molokai *(Eugenia koolauensis* and *Hesperomannia arborescens),* Lanai *(Hesperomannia arborescens),* and Kauai *(Phlegmariurus nutans).* The 11 plant taxa and their habitats have been variously affected and are currently threatened by one or more of the following: habitat degradation and/or predation by feral or domestic animals (goats and pigs); competition for space, light, water, and nutrients by naturalized, alien vegetation; habitat loss from fires; predation by rats, human recreational activities; and military training exercises. Because of the low numbers of individuals and their severely restricted distributions, populations of these taxa face an increased likelihood of extinction from stochastic events....

The plant taxa addressed in this plan are all endemic to the eight "main Hawaiian Islands", which include: Niihau, Kanai, Oahu, Maui, Molokai, Lanai, Kahoolawe, and Hawaii (also known as "the Big Island"). The Hawaiian Islands are located over 2,000 miles (3,200 kilometers) from the nearest continent. This isolation has allowed the few plants and animals that arrived here to evolve into many varied and highly endemic species. In many cases, these species lack defenses against threats such as mammalian predation and competition with aggressive, weedy plant species that are typical of mainland environments.

General Description of Habitat

The island of Oahu was formed from the remnants of two large shield volcanoes, the younger Koolau volcano on the east and the older Waianae volcano to the west. Their original shield volcano shape has been lost as a result of extensive erosion, and today these volcanoes are called mountains or ranges, and consist of long, narrow ridges. The Koolau Mountains were built by eruptions that took place primarily along a northwest-trending rift zone and formed a range now approximately 37 miles (60 kilometers) long. Median annual rainfall for the Koolau Mountains varies from 40 to 280 inches (100 to 700 centimeters), most of which is received at higher elevations along the entire length of the windward (northeastern) side....

The vegetation communities of the Koolau Mountains, especially in the upper elevations to which many of the 11 plant taxa are restricted, are primarily lowland mesic and wet forests dominated by *Metrosideros polymorpha* (ohia) and/or other tree or fern taxa. Much of the Koolau Mountain Range is vegetated with of alien plant taxa. Most of the remaining native vegetation is restricted to steep valley headwalls and inaccessible summit ridges. The windswept ridges are very steep and are characterized by grasses, ferns, and low-growing, stunted shrubs.

The land that supports these 11 plant taxa is owned by the State of Hawaii (including land classified as natural area reserve and forest reserve), the Federal government, and various private parties. Plants on Federal land are located on the boundary of Schofield Barracks Military Reservation, under the jurisdiction of the U.S. Army. Populations of five taxa grow on land leased by the U.S. Army from private parties and the State. [There are no known populations of two of the listed plants, the haha and the haiwale.]

Overall Recovery Strategy

The 11 species covered in this plan are all dangerously close to extinction due to their extremely low numbers (most number fewer than 100 individuals in the wild) and their limited distributions. Immediate action must be taken to stabilize the few remaining wild populations. These actions include propagation and maintenance of genetic stock *ex situ*, and protection of remaining wild individuals from threats. Current threats to those species should be managed through fencing and/or hunting to control ungulates; control of alien plants; protection from fire; control of rodents; protection from human disturbance; a comprehensive monitoring program; and, if deemed necessary, protection from insects and disease. Simultaneously, surveys should be planned to determine the status of the two species with no known extant individuals *(Cyanea truncata* and *Cyrtandra crenata),* and those populations of other species that have not been observed in recent years. Individuals of these species may exist in former habitats, or may be present in areas that have not been surveyed recently.

Secondly, management units should be delineated to conserve not only these taxa, but their habitats as well. These units should be managed to preserve as many native species (flora and fauna) as possible, through threat-control and forest-restoration programs.

The next step in the recovery of these species is augmentation of small populations and re-establishment of new populations within the historical range of the species, when necessary to meet down/delisting objectives. This step includes selection of areas for augmentation and re-establishment, determination of the best methods for *ex situ* propagation and transplanting, selection of the best genetic stock for each area, propagation of suitable stock, preparation of sites for seeding and/or transplanting, and monitoring and maintenance of new individuals and populations as they are established.

A research program is also recommended to study the growth and reproductive viability of each taxon, determine the parameters of viable populations of each taxon, study the reproductive strategy and pollinators of each taxon, study possible pests and diseases, and use the results of such research to improve management practices.

Finally, the recovery criteria should be refined and revised as new information becomes available. . . .

RECOVERY

Objectives and Criteria

The long-range objective of this plan is recovery and delisting of all Koolau plant cluster taxa. Criteria are also provided for the interim objectives of stabilizing and downlisting these plants. [Before each plant can be delisted, a] total of 8 to 10 populations of each taxon should be documented on Oahu and at least one other island where they now occur or occurred historically. . . . Each of these populations must be naturally reproducing, stable or increasing in number, and secure from threats, with a minimum of 100 mature individuals per population for long-lived perennials and a minimum of 300 mature individuals per population for short-lived perennials. Each population should persist at this level for a minimum of 5 consecutive years.

Step-down Outline

1. Protect habitat and control threats

    11. Identify and map all extant wild populations.

    12. Ensure long-term protection for habitat.

    13. Identify and control threats.

        131. Control feral ungulates.

        1311. Construct and maintain fencing.

        1312. Evaluate the potential for controlling ungulates through eradication programs or establishment of game preserves.

        132. Conduct alien plant control.

        133. Provide necessary fire protection.

        134. Control rodents, if necessary.

135. Propagate and maintain genetic stock *ex situ*.

136. Ensure availability of pollination vectors.

137. Protect areas from human disturbance.

138. Control insects and disease, if necessary.

139. Control all other identified threats.

14. Delineate management units.

2. Expand existing wild populations.

21. Propagate and maintain genetic stock *ex situ*

22. Prepare sites and plant.

3. Conduct essential research.

32. Collect diagnostic data on crucial associated ecosystem components.

33. Study various aspects of growth.

34. Study reproductive viability.

35. Determine parameters of viable populations.

36. Determine the kind and degree of threat posed by introduced plants, animals, and diseases.

37. Determine effective control methods to combat threats found by completing task #36.

38. Evaluate results and use in future management.

4. Develop and maintain detailed monitoring plans for all species.

5. Reestablish wild populations within the historic range.

51. Investigate feasibility and desirability of reintroduction.

52. Develop and implement specific plans for re-establishment.

6. Validate recovery criteria.

61. Determine number of populations needed for long term survival.

62. Determine the number of individuals for long term survival.

63. Refine/revise downlisting and delisting criteria.

7. Design and implement a public education program.

[The recovery plan then describes each of these steps in somewhat more detail. The plan also notes "the order of tasks listed in the stepdown outline and narrative does not necessarily designate the order in which these tasks should be implemented."]

Notes and Questions

1. Hawaii has more endangered species than any other state. This is the understandable consequence of the unique forms of life that have lived only in Hawaii, and the threats to such species as the islands have been developed in recent years. For more information on the unique challenges to preserving biodiversity in Hawaii, *see* Elizabeth Royte, *Hawaii's Vanishing Species*, NATIONAL GEOGRAPHIC, Sept. 1995, at 2; and a list of websites relating to Hawaii's biodiversity at <http://www.hisurf.com/?enchanted/hawaii'shotlinks1.html>

2. Is the recovery plan likely to result in the preservation of the Koolau Mountain plants? What other steps could be taken to preserve the plants?

3. How should the plan be implemented? Consider item number 131 on the step-down outline, "control feral ungulates." Goats, pigs, cattle and other ungulates were introduced to Hawaii by early natives and by more recent settlers to provide opportunities for farming and ranching. Many of those animals escaped from domestication and now support a recreational hunting industry, but they also pose a threat to countless native species of plants and birds. Thus the recovery plan explains, "Controlling these ungulates to the point where they are no longer impacting native vegetation is absolutely imperative.... However, public support of hunting is fervent and the likelihood of an ungulate eradication program is remote." *Recovery Plan for Koolau Mountain Plant Cluster, supra,* at 55, 57. What, therefore, should be done to preserve the endangered plants from the goats, pigs and cattle? For other illustrations of the same problem, *see* Palila v. Hawaii Department of Land and Natural Resources, 852 F.2d 1106 (9th Cir.1988) (directing the state to take steps to protect an endangered bird from sheep and goats); Final Endangered Status for 10 Plant Taxa from Maui Nui, HA, 64 Fed. Reg. 48307, 48315 (1999) (detailing the history of the introduced animals in Hawaii and the threat that they pose to native plants).

4. Does the recovery plan satisfy the requirements of ESA § 4? More generally, do the requirements of section 4 assure that a species will recover from the danger of extinction? The role of recovery planning has been the subject of numerous recent reform proposals. H.R. 960, introduced by Representative Miller in 1999, would (1) expand recovery plans to include species that have not yet been listed as endangered or threatened but which are under consideration for listing, (2) require each affected federal agency to prepare specific implementation plans involving their relationship with a protected species, and (3) include in each recovery plan a list of activities that would be prohibited by the substantive commands of the ESA. Another bill introduced by Senator Kempthorne in 1997 would have established fixed deadlines for the preparation of recovery plans, required the plans to include specific numerical targets, and given states a greater role in the preparation of the plans. Which of these ideas would improve the role that recovery plans play in the ESA?

5. The Koolau Mountain plant cluster recovery plan estimates that it will cost almost $25 million to implement through the year 2016. That cost, remember, is for only ten of the species protected by the ESA. A 1994 study indicated that it would cost nearly $900 million to implement the recovery plans existing at that time, while the Fish and Wildlife Service had requested only $84 million to implement them. *See* NATIONAL WILDERNESS INSTITUTE, GOING BROKE? COSTS OF THE ENDANGERED SPECIES ACT AS REVEALED IN ENDANGERED SPECIES RECOVERY PLANS 1 (1994). The Fish and Wildlife Service cautions, however, that the cost estimates contained in recovery plans overstate the true cost of achieving the recovery goal for a species. *See* FWS Division of Endangered Species, *Endangered Species Recovery* 2, http://www.fws.gov/r9endspp/faqrecov.html ("The recovery plan is best thought of as a menu. To have a healthy meal in a restaurant, one would not total an entire menu to arrive at the cost of one dinner. Not all the tasks in a recovery plan need to be implemented to reach the recovery goal.").

Establishment of a Nonessential Experimental Population of the Mexican Gray Wolf in Arizona and New Mexico

United States Fish and Wildlife Service, 1998.
63 Fed. Reg. 1752.

BACKGROUND

Legislative

The Endangered Species Act Amendments of 1982, created section 10(j), providing for the designation of specific populations of listed species as "experimental populations." Under previous authorities of the Act, the Service was permitted to re-establish (reintroduce) populations of a listed species into unoccupied portions of its historic range for conservation and recovery purposes. However, local opposition to reintroduction efforts, stemming from concerns by some about potential restrictions, and prohibitions on Federal and private activities contained in sections 7 and 9 of the Act, reduced the effectiveness of reintroduction as a conservation and recovery tool.

Under section 10(j), a population of a listed species re-established outside its current range but within its probable historic range may be designated as "experimental" at the discretion of the Secretary of the Interior (Secretary). Reintroduction of the experimental population must further the conservation of the listed species. An experimental population must be separate geographically from nonexperimental populations of the same species. Designation of a population as experimental increases the Service's management flexibility.

Additional management flexibility exists if the Secretary finds the experimental population to be "nonessential" to the continued existence of the species. For purposes of section 7 [except section 7(a)(1), which requires Federal agencies to use their authorities to conserve listed species], nonessential experimental populations located outside national wildlife refuge or national park lands are treated as if they are proposed for listing. This means that Federal agencies are under an obligation to confer, as opposed to consult (required for a listed species), on any actions authorized, funded, or carried out by them that are likely to jeopardize the continued existence of the species. Nonessential experimental populations located on national wildlife refuge or national park lands are treated as threatened, and formal consultation may be required. Activities undertaken on private or tribal lands are not affected by section 7 of the Act unless they are authorized, funded, or carried out by a Federal agency.

Individual animals used in establishing an experimental population can be removed from a source population if their removal is not likely to jeopardize the continued existence of the species, and a permit has been issued in accordance with 50 C.F.R. part 17.22.

The Mexican gray wolf was listed as an endangered subspecies on April 28, 1976. The gray wolf species in North America south of Canada was listed as endangered on March 9, 1978, except in Minnesota where it was listed as threatened. This listing of the species as a whole continued to recognize valid biological subspecies for purposes of research and conservation.

Biological

This final experimental population rule addresses the Mexican gray wolf (*Canis lupus baileyi*), an endangered subspecies of gray wolf that was extirpated from the southwestern United States by 1970. The gray wolf species (*C. lupus*) is native to most of North America north of Mexico City. An exception is in the southeastern United States, which was occupied by the red wolf species (*C. rufus*). The gray wolf occupied areas that supported populations of hoofed mammals (ungulates), its major food source.

The Mexican gray wolf historically occurred over much of New Mexico, Arizona, Texas, and northern Mexico, mostly in or near forested, mountainous terrain. Numbering in the thousands before European settlement, the "lobo" declined rapidly when its reputation as a livestock killer led to concerted eradication efforts. Other factors contributing to its decline were commercial and recreational hunting and trapping, killing of wolves by game managers on the theory that more game animals would be available for hunters, habitat alteration, and human safety concerns (although no documentation exists of Mexican wolf attacks on humans).

The subspecies is now considered extirpated from its historic range in the southwestern United States because no wild wolf has been confirmed since 1970. Occasional sightings of "wolves" continue to be reported from U.S. locations, but none have been confirmed. Ongoing field research has not confirmed that wolves remain in Mexico. . . .

Recovery Efforts

The *Mexican Wolf Recovery Plan* was adopted by the Directors of the Service and the Mexican Direccion General de la Fauna Silvestre in 1982. Its objective is to conserve and ensure survival of the subspecies by maintaining a captive breeding program and re-establishing a viable, self-sustaining population of at least 100 Mexican wolves in a 5,000 square mile area within the subspecies' historic range. The plan guides recovery efforts for the subspecies, laying out a series of recommended actions. The recovery plan is currently being revised; the Service expects to release a draft for public review in 1998. The revised plan will more precisely define population levels at which the Mexican wolf can be downlisted to "threatened" status and removed from protection under the Act (i.e., delisted).

A captive breeding program was initiated with the capture of five wild Mexican wolves between 1977 and 1980, from Durango and Chihuahua, Mexico. Three of these animals (two males and a female that was pregnant when captured) produced offspring, founding the "certified" captive lineage. Two additional captive populations were determined in July 1995 to be pure Mexican wolves—each has two founders. The captive population included 148 animals as of January 1997–119 are held at 25 facilities in the United States and 29 at five facilities in Mexico.

[In 1992, the Service began planning for the reintroduction of the Mexican wolves. The Service held 14 public meetings and three public hearings, prepared an environmental impact statement, and received comments on the EIS from about 18,000 people. After consulting with state and local governments, other federal agencies, private landowners, native American tribes, and technical experts, the Service issued a proposed rule to reintroduce the Mexican wolves in 1996. Four additional public meetings were held in the affected area. Then in April 1997, the Service announced its intention to reintroduce captive-raised Mexican wolves as a designated nonessential experimental population in eastern Arizona within the designated Blue Range Wolf Recovery Area. The Service added that it would reintroduce wolves into the White Sands Wolf Recovery Area, the designated back-up area, if that becomes necessary for the recovery of the wolves and feasible.]

Mexican Wolf Recovery Areas

. . . The two wolf recovery areas are within the Mexican wolf's probable historic range. The Mexican wolf is considered extinct in the wild in the United States. Thus, both areas are geographically separate from any known, naturally-occurring, nonexperimental populations of wild wolves.

[T]his rule establishes a larger Mexican Wolf Experimental Population Area, which also is geographically separate from any known, naturally-occurring nonexperimental populations of wild wolves. The Service is not proposing to re-establish Mexican wolves throughout this larger area. The purpose of designating an experimental population area is to establish that any member of the re-established Mexican wolf population found in this larger area is a member of the nonessential experimental population, and subject to the provisions of this rule including, but not limited to, its capture and return to the designated recovery area(s).

Reintroduction Procedures

Captive Mexican wolves are selected for release based on genetics, reproductive performance, behavioral compatibility, response to the adaptation process, and other factors. Selected wolves have been moved to the Service's captive wolf management facility on the Sevilleta National Wildlife Refuge in central New Mexico where they have been paired based on genetic and behavioral compatibility and measures are being taken to adapt them to life in the wild. As wolves are moved to release pens, more will be moved to the Sevilleta facility. Additional wolves for reintroduction may be obtained from selected cooperating facilities that provide an appropriate captive environment.

Initially, wolves will be reintroduced by a "soft release" approach designed to reduce the likelihood of quick dispersal away from the release areas. This involves holding the animals in pens at the release site for several weeks in order to acclimate them and to increase their affinity for the area.... The releases will begin in 1998. Procedures for releases could be modified if new information warrants such changes.

In the Blue Range Wolf Recovery Area, approximately 14 family groups will be released over a period of 5 years, with the goal of reaching a population of 100 wild wolves. Approximately five family groups of captive raised Mexican wolves will be released over a period of 3 years into the White Sands Wolf Recovery Area, if this back-up area is used, with the goal of reaching a population of 20 wolves.

Management of the Reintroduced Population

The nonessential experimental designation enables the Service to develop measures for management of the population that are less restrictive than the mandatory prohibitions that protect species with "endangered" status. This includes allowing limited "take' "of individual wolves under narrowly defined circumstances. Management flexibility is needed to make reintroduction compatible with current and planned human activities, such as livestock grazing and hunting. It is also critical to obtaining needed State, Tribal, local, and private cooperation. The Service believes this flexibility will improve the likelihood of success.

Reintroduction will occur under management plans that allow dispersal by the new wolf populations beyond the primary recovery zones where they will be released into the secondary recovery zones of the designated wolf recovery area(s). The Service and cooperating agencies will not allow the wolves to establish territories on public lands wholly outside these wolf recovery area boundaries. With landowner consent, the Service also would prevent wolf colonization of private or tribal lands outside the designated recovery area(s).

No measures are expected to be needed to isolate the experimental population from naturally occurring populations because no Mexican wolves are known to occur anywhere in the wild. The Service has ensured that no population of naturally-occurring wild wolves exists within the recovery areas. Surveys for wolf signs in these areas have been conducted, and no naturally occurring population has been documented. No naturally occurring population of Mexican wolves has been documented in Mexico following four years of survey efforts there. Therefore, based on the best available information, the Service concludes that future natural migration of wild wolves into the experimental population area is not possible.

Findings Regarding Reintroduction

The Service finds that . . . the reintroduced experimental population is likely to become established and survive in the wild within the Mexican gray wolf's probable historic range. The Service projects that this reintroduction will achieve the recovery goal of at least 100 wolves occupying 5,000 square miles. The Blue Range Wolf Recovery Area comprises 6,854 square miles of which about 95% is National Forest.

Some members of the experimental population are expected to die during the reintroduction efforts after removal from the captive population. The Service finds that even if the entire experimental population died, this would not appreciably reduce the prospects for future survival of the subspecies in the wild. That is, the captive population could produce more surplus wolves and future reintroductions still would be feasible if the reasons for the initial failure are understood. The individual Mexican wolves selected for release will be as genetically redundant with other members of the captive population as possible, thus minimizing any adverse effects on the genetic integrity of the remaining captive population. . . . The United States captive population of Mexican wolves has approximately doubled in the last 3 years, demonstrating the captive population's reproductive potential to replace reintroduced wolves that die. In view of all these safeguards the Service finds that the reintroduced population would not be "essential" under 50 C.F.R. § 17.81(c)(2).

The Service finds that release of the experimental population will further the conservation of the subspecies and of the gray wolf species as a whole. Currently, no populations or individuals of the Mexican gray wolf subspecies are known to exist anywhere in the wild. No wild populations of the gray wolf species are known to exist in the United States south of Washington, Idaho, Wyoming, North Dakota, Minnesota, Wisconsin, and Michigan. Therefore, based on the best available information, the Service finds that the re-established population would be completely geographically separate from any extant wild populations or individual gray wolves and that future migration of wild Mexican wolves into the experimental population area is not possible. The Mexican wolf is the most southerly and the most genetically distinct of the North American gray wolf subspecies. It is the rarest gray wolf subspecies and has been given the highest recovery priority for gray wolves worldwide by the Wolf Specialist Group of the World Conservation Union (IUCN).

. . . Designation of the released wolves as nonessential experimental is considered necessary to obtain needed State, Tribal, local, and private cooperation. This designation also allows for management flexibility to mitigate negative impacts, such as livestock depredation. Without such flexibility, intentional illegal killing of wolves likely would harm the prospects for success.

Potential for Conflict With Federal and Other Activities

As indicated, considerable management flexibility has been incorporated into the final experimental population rule to reduce potential conflicts between wolves and the activities of governmental agencies, livestock operators, hunters, and others. No major conflicts with current management of Federal, State, private, or Tribal lands are anticipated. Mexican wolves are not expected to be adversely affected by most of the current land uses in the designated wolf recovery areas. However, temporary restrictions on human activities may be imposed around release sites, active dens, and rendezvous sites.

Also, the U.S. Department of Agriculture, Animal and Plant Health Inspection Service, Wildlife Services (WS) division will discontinue use of M–44's and choking-type snares in "occupied Mexican wolf range." Other predator control activities may be restricted or modified pursuant to a cooperative management agreement or a conference between the WS and the Service.

The Service and other authorized agencies may harass, take, remove, or translocate Mexican wolves under certain circumstances described in detail in this rule. Private citizens also are given broad authority to harass Mexican wolves for purposes of scaring them away from people, buildings, facilities, pets, and livestock. They may kill or injure them in defense of human life or when wolves are in the act of attacking their livestock (if certain conditions are met). In addition, ranchers can seek compensation from a private fund if depredation on their livestock occurs.

No formal consultation under section 7 of the Act would be required regarding potential impacts of land uses on nonessential experimental Mexican wolves. Any harm to wolves resulting solely from habitat modification caused by authorized uses of public lands that are not in violation of the temporary restriction provisions or other provisions regarding take or harassment would be a legal take under this rule. Any habitat modification occurring on private or tribal lands would not constitute illegal take. Based on evidence from other areas, the Service does not believe that wolf recovery requires major changes to currently authorized land uses. The main management goals are to protect wolves from disturbance during vulnerable periods, minimize illegal take, and remove individuals from the wild population that depredate livestock or otherwise cause significant problems.

The Service does not intend to change the "nonessential experimental" designation to "essential experimental," "threatened," or "endangered" and the Service does not intend to designate critical habitat for the Mexican wolf. Critical habitat cannot be designated under the nonessential experimental classification, 16 U.S.C. 1539(j)(2)(C)(ii). The Service foresees no likely situation which would result in such changes in the future. . . .

Summary of Public Participation

In June 1996, public open house meetings and formal public hearings were held in El Paso, Texas; Alamogordo and Silver City, New Mexico; and Springerville, Arizona. About 166 people attended these meetings and had an opportunity to speak with agency representatives and submit oral and written comments. Oral testimony was presented by 49 people at the hearings, and 150 people submitted written comments on the proposed rule. We received a petition supporting full endangered status for reintroduced Mexican wolves signed by 32 people; and a petition opposing the reintroduction of Mexican wolves signed by 91 people. In addition, many comments on the [draft] EIS were specific to the draft proposed rule or related management considerations. These comments also were considered in this revision of the proposed rule. . . .

Notes and Questions

1. On March 30, 1998, the first Mexican wolves were released in the Apache National Forest in eastern Arizona. By the end of the year, five wolves had been shot and killed, one was missing and presumed dead, and three were recaptured for their own protection, leaving only two wolves in the wild. The first wolf pup born in the wild in half a century was missing and presumed dead, too. One of the wolves was shot by a camper who believed his family was in danger when the wolf attacked his dog. Environmentalists criticized the camper for taking his dogs into wolf country and for not leaving with his family after their first encounter with the wolf. The government decided not to prosecute the camper. A reward of $50,000 was offered by the government, environmental groups, and Michael Blake (the author of "Dances With Wolves") for information regarding those who killed the other wolves, but even the offer of a reward had been controversial. *See* E.J. Montini, *Of Wolves and Men and Money*, ARIZ. REP., Nov. 2, 1998, at B1 (interviewing a man who noted that only $12,000 had been raised as a reward for his son's unsolved murder).

Secretary Babbitt vowed to continue the reintroduction effort when he returned to the area in January 1999. More than twenty wolves were released into the recovery zone in four different packs in 1999. This time the wolves wore bright orange collars. Several more pups were born in the wild, but they appear to have died from natural causes. The first confirmed killing of livestock by wolves occurred in July, with three more cows succumbing in the following months. The FWS set out elk and deer carcasses in an effort to steer the wolves away; the rancher was compensated for his loss by the Defenders of Wildlife. The agency plans to release additional wolves through 2002 and still hopes to achieve its goal of 100 wolves in the wild by 2008. For updates on the progress of the reintroduction effort, *see* the FWS website at <http://ifw2es.fws.gov/MexicanWolf>.

2. Wolves have long been controversial in the United States. As one environmental leader observed, "When Europeans reached this continent, they brought with them thousand of years of accumulated prejudice against the wolf. Widely expressed in mythology and in folk takes from Aesop's fables to The Three Little Pigs and Little Red Riding Hood, it easily overpowered the Native American view of the wolf as a respected fellow creature in the web of life." Rodger Schlickeisen, *A Positive Turn for Man and Nature; Wolf Reintroduction Marks True Progress*, ARIZ. REP., Jan. 26, 1998, at B5. Wolves were killed by the tens of thousands—with government support or by government agents—from the mid-nineteenth century until about 1920, by which time wolves were nearly eradicated from the entire country. The attitude toward wolves today is much more divided. Many ranchers and westerners retain the traditional fear of wolves, while others are fascinated by wolves, including tourists who flock to hear wolves howl in areas where wolves are more common. *See generally* WOLVES AND HUMAN COMMUNITIES: BIOLOGY, POLITICS, AND ETHICS (Virginia A. Sharpe, Bryan G. Norton & Strachan Donnelly eds. 2001) (collecting essays on the relationship of wolves to their ecosystems and to people).

3. Almost by definition, a species that has become endangered no longer lives in many areas where it once could be found. Efforts to save the species often seek to reintroduce it to areas from which it has disappeared, or even to areas where it never lived but which offer suitable habitat. The list of species that have been reintroduced into former or new habitat since the enactment of the ESA includes California condors, grizzly bears, black-footed ferrets, peregrine falcons, and many others. Many other recovery plans—including the plan for the Koolau Mountain plant cluster—list reintroduction as a possible step toward the preservation of a species. But a 1994 study of 145 reintroduction efforts involving 115 species concluded that only sixteen had produced populations that were sustaining themselves in the wild, and that only half of those species had been endangered. *See* Mark Derr, *As Rescue Plan for Threatened Species, Breeding Programs Falter*, N.Y. TIMES, Jan. 19, 1999, at F1. The reintroduction of the red wolf into the Great Smoky Mountains National Park was abandoned in 1998 when government officials removed the remaining wolves because the wolves could not find enough prey to survive. Reintroduction efforts face other criticisms as well. They are costly: reintroduction of the condor, wolf and black-footed ferret alone have cost a total of more than $50 million. They are dependent upon adequate habitat, and so face the same challenges as existing populations of wild species in the face of human development. And they can be controversial among local residents, as illustrated by the reaction of Arizona ranchers to the reintroduction of the Mexican wolf and the reaction of many Idaho residents to proposed grizzly bear reintroductions.

4. In 1982, Congress added section 10(j) to the ESA. That provision authorizes the Secretary to release an experimental population of an endangered or threatened species if the Secretary determines that such a release will further the conservation of the species. The purpose of section 10(j) is to promote the recovery of a species in a manner that encourages the cooperation of other federal agencies and private landowners. Any experimental population must be kept wholly separate from nonexperimental populations of the same species, *see* ESA § 10(j)(1), 16 U.S.C. § 1539(j)(1), and an experimental population is automatically listed as a separate threatened species. ESA § 10(j)(2)(C), 16 U.S.C § 1539(j)(2)(C). Because it is a threatened species, the Secretary may decide to issue a special rule imposing somewhat more relaxed restrictions on private landowners than those ordinarily provided by section 9 of the ESA. Additionally, section 10(j) requires the Secretary to determine whether the experimental population "is essential to the continued existence" of the species. ESA § 10(j)(2)(B), 16 U.S.C. § 1539(j)(2)(B). If the population is not essential to the survival of the species, then section 7 of the ESA (concerning the duties of federal agencies) does not apply unless the species is in a National Park or a National Wildlife Refuge. ESA § 10(j)(2)(C), 16 U.S.C. § 1539(j)(2)(C).

The FWS prefers to reintroduce species as experimental and not essential pursuant to section 10(j). In response to a public comment regarding the Mexican wolf reintroduction, the agency explained:

> The "experimental nonessential" terminology in section 10(j) of the Act is confusing. It does not mean that the animal is not near extinction and it does not mean the reintroduction is just an experiment. It is a classification designed to make the reintroduction and management of endangered species more flexible and responsive to public concerns to improve the likelihood of successfully recovering the species.

63 Fed. Reg. at 1757. That flexibility is important to the agency because it allows the crafting of special rules favored by local residents and by other federal agencies and because it permits the agency to more actively manage the reintroduced population. But that flexibility has its costs. Professor Holly Doremus has criticized the FWS for effectively eliminating the statutory distinction between essential and nonessential experimental populations by reading the law so that no population is ever deemed essential to the survival of the species. More broadly, she argues that "adoption of special rules permitting incidental and deliberate take of introduced populations is a choice to restrict the wildness of those populations." She explains that "species must be protected as wild creatures, rather than merely as biological entities," and that "none of the congressionally listed values of species can be fully protected in captivity." She adds that animals are wild only "if they enjoy natural autonomy, that is if their natural instincts determine such basic choices as where they sleep, what they eat, and how they select a mate." Holly Doremus, *Restoring Endangered Species: The Importance of Being Wild*, 23 HARV. ENVTL. L. REV. 1, 10–16,

38–47 (1999). Is Professor Doremus right about the importance of preserving biodiversity in its wild, natural state? Why isn't the ESA satisfied if species are preserved in zoos? And how can purposeful human efforts to reintroduce a species be reconciled with the wildness of a species? Should the FWS impose the same restrictions on private and government activity that the ESA imposes on naturally occurring populations of a listed species? Are the Mexican wolves truly wild if government employees use food to steer the pack in another direction?

Section 10(j)'s requirement that an experimental population be wholly separate from any existing population of the species has generated challenges to high profile reintroduction efforts. Several local ranchers and trade associations have sued the FWS alleging that the reintroduction of the Mexican wolves is illegal because some wolves still occur there naturally. The same argument succeeded in a hotly contested decision invalidating the reintroduction of northern Rocky Mountain gray wolves in Yellowstone National Park. In Wyoming Farm Bureau Federation v. Babbitt, 987 F.Supp. 1349 (D.Wyo.1997), *rev'd*, 199 F.3d 1224 (10th Cir.2000), the district court held that the range of the wolves reintroduced into Yellowstone overlapped the existing range of naturally occurring wolves, and thus the reintroduction was impermissible under section 10(j). That decision created an uproar among environmentalists until the Tenth Circuit reversed it over two years later. The Ninth Circuit upheld the Yellowstone wolf reintroduction program in United States v. McKittrick, 142 F.3d 1170 (9th Cir.1998). In *McKittrick*, a hunter shot and skinned one of the reintroduced wolves that had migrated out of Yellowstone to nearby Red Lodge, Montana (hence the Ninth Circuit's jurisdiction). The court held that the occasional appearance of a few native wolves did not constitute a "population" that overlapped with the reintroduction population. *Id.* at 1175: *accord id.* at 1179 (O'Scannlain, J., concurring) ("A single straggler does not a population make."). But suppose that the Wyoming district court was right. Should the reintroduced wolves be removed from Yellowstone—as the court suggested—or is some other remedy possible?

B.   RESPONSIBILITIES OF ALL FEDERAL AGENCIES UNDER THE ESA

Congress has charged the FWS (and in some circumstances, the NMFS) with the responsibility for taking the lead on federal efforts to protect species listed as endangered or threatened under the ESA. While the FWS leads, other federal agencies are supposed to follow. The ESA requires all federal agencies to consider the effects of their activities on listed species. That means that a Federal Highway Administration interstate project, an Air Force missile range, and a Corps of Engineers river dredging project must all account for any listed species that they might affect. In particular, the ESA imposes an affirmative duty and a negative duty on federal agencies. The affirmative duty directs all agencies to conserve listed species. The negative duty prohibits federal agencies from jeopardizing the continued existence of a listed species or adversely affecting the designated critical habitat of a species. (Federal agencies must also comply with the ESA's prohibition against taking a listed species, but that provision is discussed in chapter 6 because it applies to governmental and private parties alike). The jeopardy prohibition has received much more attention than the conservation duty to date, but there are indications that the conservation duty could begin to become a more prominent feature of the act.

1. The Conservation Duty

The Hawksbill Sea Turtle v. Federal Emergency Management Agency

United States District Court for the District of the Virgin Islands, 1998.
11 F.Supp.2d 529.

■ Brotman, District Judge, Sitting by Designation.

[In September 1995, Hurricane Marilyn swept through St. Thomas in the U.S. Virgin Islands, killing eleven people and displacing hundreds of others from their homes. Several months afterward, the Federal Emergency Management Agency (FEMA) provided funds to the Virgin Island Housing Authority for the construction of a temporary housing shelter project to house the low-income residents who were still homeless as a result of the hurricane. The project was to consist of prefabricated buildings that could house up to 800 people and that would be dismantled after the emergency abated. The site of the project, though, was near the habitat of three listed species: the endangered Virgin Island Tree Boa, a secretive snake that lives only on the eastern portion of St. Thomas; the endangered Hawksbill Sea Turtle, which lives in coral reefs off of St. Thomas and other islands; and the threatened Green Sea Turtle. FEMA worked with local officials in the Virgin Islands to prepare a "Tree Boa Mitigation Plan" that directed construction personnel to hand clear the area before heavy machinery was operated at the site. FEMA also took steps to assure that the construction activities would not degrade water quality to the detriment of the turtles. Finding these steps unsatisfactory, a group of residents who lived in the area of the proposed emergency housing project sued to block construction alleging, *inter alia*, that FEMA failed to comply with the conservation duties of the ESA.]

Plaintiffs allege that Defendants violated their Section 7(a)(1) duties to conserve the Tree Boa and the Sea Turtles. Section 7(a)(1) provides in pertinent part that:

> The Secretary [of the Interior] shall review other programs administered by him and utilize such programs in furtherance of the purposes of this chapter. All other Federal agencies shall, in consultation with and with the assistance of the Secretary, utilize their authorities in

furtherance of the purposes of this chapter by carrying out programs for the conservation[16] of endangered species and threatened species listed pursuant to section 1533 of this title.

16 U.S.C. § 1536(a)(1) (footnote added). Plaintiffs claim that Defendants' violated their Section 7(a)(1) conservation duties by constructing a housing project that would have a harmful effect on the Tree Boa and the Sea Turtles. They also aver that Defendants failed "to take the affirmative steps required by the ESA to conserve" the species.

This Court finds Plaintiffs' Section 7(a)(1) claims without merit. "Conservation plans under [Section] 7(a)(1) are 'voluntary measures that the federal agency has the discretion to undertake' and 'the [ESA] does not mandate particular actions be taken by Federal Agencies to implement [Section] 7(a)(1).'" Strahan v. Linnon, 967 F. Supp. 581, 596 (D.Mass. 1997) (quoting 51 Fed. Reg. 19926, 19931, 19934). "Agencies have 'some discretion in ascertaining how best to fulfill the mandate to conserve under section 7(a)(1).'" J. B. Ruhl, *Section 7(a)(1) of the "New" Endangered Species Act: Rediscovering and Redefining the Untapped Power of Federal Agencies' Duty to Conserve Species*, 25 ENVTL. L. 1107, 1132 (1995) (quoting Pyramid Lake Paiute Tribe of Indians v. United States Dep't of the Navy, 898 F.2d 1410 (9th Cir.1990)). "Reasonable people could disagree as to the proper level of activism required by an agency under the ESA. The court will not substitute its judgment for the agency's in deciding as a general matter that the totality of defendant's actions taken to protect threatened and endangered species were insufficient." Defenders of Wildlife v. Administrator, Environmental Protection Agency, 688 F. Supp. 1334, 1352 (D.Minn.1988), *aff'd in part and rev'd in part on other grounds*, 882 F.2d 1294 (8th Cir.1989). Defendants have provided evidence that they have adopted various, specifically targeted mitigation measures in an attempt to conserve the Tree Boa and the marine environment in Vessup Bay. On the other hand, Plaintiffs have not specified any conservation measures that Defendants have failed to take. Given Plaintiffs failure to specify alternative conservation measures, this Court must conclude that the Defendants' actions as a whole did not constitute an arbitrary and capricious failure to conserve threatened and endangered species. *See Strahan*, 967 F. Supp. at 595–96 (finding Section 7(a)(1) argument unpersuasive because plaintiff "has not demonstrated . . . specific measures that are necessary to prevent the loss of any 'endangered species' in addition to those adopted by the agency)."

---

16. The term "conservation" means "the use of all methods and procedures which are necessary to bring any endangered species or threatened species to the point at which the measures provided pursuant to this chapter are no longer necessary. Such methods and procedures include, but are not limited to, all activities associated with scientific resources management such as research, census, law enforcement, habitat acquisition and maintenance, propagation, live trapping, and transplantation, and, in the extraordinary case where population pressures within a given ecosystem cannot be otherwise relieved, may include regulated taking." 16 U.S.C. § 1532(3).

Notes and Questions

1. The "temporary" housing project at issue in *Hawksbill Turtle* was two and a half years old by the time the district court decided the case. In April 1998, the governor of the Virgin Islands ordered a stay of eviction of the families remaining in the project until alternative housing was arranged. FEMA advised the FWS of the delay, but FWS did not find any additional threat to the listed species so long as the plans to relocate the remaining families proceeded.

2. The conservation duty has been interpreted narrowly by most other courts. *See, e.g.*, Pyramid Lake Paiute Tribe of Indians v. U.S. Department of the Navy, 898 F.2d 1410 (9th Cir.1990) (holding that the Navy did not violate section 7(a)(1) when it took water from the Truckee River to suppress dust that interfered with flight training despite the possible adverse effect on endangered fish living in Pyramid Lake resulting from the reduction in the lake's water level); Strahan v. Linnon, 966 F.Supp. 111 (D.Mass.1997) (rejecting the claim that the Coast Guard violated section 7(a)(1) by refusing to impose speed limits and other constraints on vessels operating in waters containing the Northern Right whale and other endangered marine species); Center for Marine Conservation v. Brown, 917 F.Supp. 1128, 1149–50 (S.D.Tex.1996) (concluding that the federal agencies responsible for regulating fishery resources did not violate section 7(a)(1) when they refused to take additional steps to protect endangered sea turtles from commercial shrimping operations); National Wildlife Fed'n v. Hodel, 1985 WL 186671 (E.D.Cal.1985) (holding that FWS did not violate section 7(a)(1) when it permitted the use of lead shot in hunting migratory birds); *but see* Florida Key Deer v. Stickney, 864 F.Supp. 1222 (S.D.Fla. 1994) (finding that FEMA violated section 7(a)(1) when it failed to undertake *any* actions to protect the endangered Florida Key Deer); Defenders of Wildlife v. Andrus, 428 F.Supp. 167 (D.D.C.1977) (holding that the FWS violated section 7(a)(1) when it failed to consider whether permitting hunting during twilight would result in protected birds being shot by mistake).

The Fifth Circuit, has adopted a much more sweeping understanding of the conservation duty. In Sierra Club v. Glickman, 156 F.3d 606 (5th Cir.1998), the court held that the Department of Agriculture failed to comply with section 7(a)(1) when it declined to take actions necessary to prevent the contamination of the Edwards Aquifer in central Texas. The Edwards Aquifer provides water to irrigate millions of dollars worth of crops, but the aquifer is also home to the Texas blind salamander and four other endangered species. The court held that the Sierra Club had standing to challenge the effects of the USDA's actions on the listed species because the court interpreted section 7(a)(1) as follows:

> At first blush, this section appears to suggest that federal agencies have only a generalized duty to confer and develop programs for the benefit of endangered and threatened species—*i.e.*, not with respect to any particular species.... When read in the context of the ESA as a whole, however, we find that the agencies' duties under § 7(a)(1) are

much more specific and particular.... By imposing a duty on all federal agencies to use "all methods and procedures which are necessary to bring *any* endangered species or threatened species to the point at which the measures provided pursuant to this chapter are no longer necessary," 16 U.S.C. § 1532(2) (emphasis added), Congress was clearly concerned with the conservation of each endangered and threatened species. To read the command of § 7(a)(1) to mean that the agencies have only a generalized duty would ignore the plain meaning of the statute.

The court rejected the USDA's argument that it had complied with the conservation duty because the endangered species in the Edwards Aquifer had enjoyed incidental benefits from USDA programs carried out for other purposes. USDA did not work with FWS, as required by section 7(a)(1), and USDA's efforts were inadequate in any event.

3. Suppose that a recovery plan for a species encourages the removal of exotic plants from the area inhabited by the species. The land is owned by the Army, which uses it for infantry training exercises that seek to simulate the lush, tropical vegetation in other parts of the world where it might one day need to deploy. Does the conservation duty require the Army to follow the guidance of the recovery plan?

4. In an article on section 7(a)(1) that is cited in *Hawksbill Turtle*, one of us has observed that federal agencies "may possibly cultivate, channel, and control the potentially broad application of that duty to conserve in ways not possible under the core programs, thereby expanding the effectiveness of section 7(a)(1) without attracting the same backlash the core programs have suffered." Three features of section 7(a)(1) illustrate "this dual virtue of breadth and flexibility." First, "unlike jeopardy consultations under section 7(a)(2), the duty to conserve species under section 7(a)(1) applies to federal programs and not merely to federal actions." The difference is that while the jeopardy provision only applies on a case-by-case basis as individual problems arise, the conservation duty can be applied to entire federal agency "authorities" that can be used to implement species conservation "programs." Second, "unlike conservation plans under section 10(a)(1) and jeopardy consultations under section 7(a)(2), the duty to conserve species under section 7(a)(1) applies independent of take and jeopardy findings." Much of the controversy surrounding the take and jeopardy provisions results from the sometimes dramatic consequences for a proposed project of a finding that the project takes an endangered species or jeopardizes the continued existence of the species as a whole. Section 7(a)(1) avoids such dire results by encouraging efforts to protect listed species that operate independently of an agency or private party's desire to pursue a particular project. Third, "the duty to conserve species under section 7(a)(1), like jeopardy consultations under section 7(a)(2), applies to endangered and threatened animals and plants." Plants and threatened species receive limited protection under the ESA's take and jeopardy provisions, despite the significant values associated with many plants and despite the law's objective of preventing a threatened species from actually becoming endangered. The conservation duty offers a means by which plants and threatened species can be protected by all federal agencies. If the conservation duty is this appealing, then why has it not received more attention?

2. The Duty to Avoid Causing Jeopardy

Natural Resources Defense Council v. Houston

United States Court of Appeals for the Ninth Circuit, 1998.
146 F.3d 1118.

■ Tashima, Circuit Judge:

Various irrigation and water districts (Non-federal defendants) that rely on water from the Friant dam, appeal the district court's summary judgment decision that the Bureau of Reclamation (Bureau or Federal Defendant), violated the Endangered Species Act (ESA) by renewing water contracts prior to completing required endangered species consultations.... We affirm the district court's holding that the ESA was violated and its decision to rescind the contracts at issue.

Background

The Central Valley Project (CVP) is a multi-unit reclamation project administered by the Bureau. The Friant dam unit of the CVP was built on the San Joaquin River by the Bureau in the 1940s. Prior to construction of the dam, the San Joaquin River met the Sacramento River at the Sacramento—San Joaquin Delta, where they then flowed out to the Pacific Ocean. Since the time that the dam was completed, the Friant unit has impounded the San Joaquin River water behind the Friant dam and diverted the water to surrounding irrigation districts. This impoundment and diversion leaves a dry stretch of San Joaquin riverbed.

In the late 1940s, the Non-federal Defendants entered into 40–year Friant water service contracts with the government, pursuant to Section 9(e) of the Reclamation Act of 1939, 43 U.S.C. § 485h(e). The contracts typically provided that they would be renewed no later than one year prior to expiration on terms that "shall be agreed upon." In 1956, Congress mandated that contract holders had a right to renewal "under stated terms and conditions mutually agreeable to the parties." Contract holders had "a first right ... to a stated share or quantity of the project's available water supply...."

The first of these contracts, the contract with the Orange Cove Irrigation District (Orange Cove), expired in February of 1989. The Bureau began contract renewal negotiations with Orange Cove in June, 1988, and executed a renewal contract in May, 1989. By 1992, the Bureau had executed 13 additional water contracts. All 14 contracts provided for water delivery for a 40–year period under terms substantially similar to those in the previous contracts.

In 1992, Congress enacted the Central Valley Project Improvement Act (CVPIA), Pub. L. No. 102–575, § 3401 *et seq.*, which required the government to perform an environmental impact statement (EIS) on the Friant unit before it could execute the remaining renewal contracts. The CVPIA also limited the length of subsequently renewed contracts to 25 years. Therefore, of the 28 Friant water service contracts that were up for renewal, only the first 14 contracts are at issue.

Prior to construction of the Friant dam, the San Joaquin River supported a variety of fish species, including the chinook salmon. The annual spring floods also fed the surrounding wetlands with fresh water. After the Friant dam was built, the San Joaquin River terminated at the dam, and water from the Sacramento—San Joaquin Delta is exported upstream to water users below the dam through a process of pumping and reverse flows. This situation has adversely affected both wetlands and river fish, including the winter-run chinook salmon. The salmon, which was listed as threatened in August, 1989, and is now endangered, is under the protective jurisdiction of the National Marine Fisheries Service (NMFS). Other listed species under the jurisdiction of the Fish and Wildlife Service (FWS) are also located in the Friant Service Area. . . .

DISCUSSION

I. *Endangered Species Act*

A. *Overview*

Section 7(a)(2) of the ESA requires all federal agencies "to insure that any action authorized, funded, or carried out by such agency is not likely to jeopardize the continued existence" of any endangered or threatened species or result in the destruction of critical habitats. If an agency determines that its proposed action "may affect" an endangered or threatened species, the agency must formally consult with the relevant Service, the FWS and/or the NMFS, depending on the species that are protected in the area of the proposed action. After the formal consultation is completed, the relevant Service will issue a Biological Opinion evaluating the nature and extent of effect on the threatened or endangered species. If the Biological Opinion concludes that the proposed action is likely to jeopardize a protected species, the agency must modify its proposal. Section 7(d) of the ESA prohibits the "irreversible or irretrievable commitment of resources" during the consultation process.

As the district court observed, the ESA has "explicit substantive goals which [are] served by its procedural requirements." The district court concluded that the contracts amounted to an "irreversible and irretrievable commitment of resources" and all contracts executed prior to completion of the required consultations with the FWS and the NMFS violated § 7(d). The district court invalidated all of the contracts which had been executed prior to the completion of the required consultations. . . .

B. *Applicability of the ESA*

As a threshold question, the Non-federal Defendants argue that the ESA did not apply to the contract renewals because the renewals were not "agency action." *See* 16 U.S.C. § 1536(a)(2). This argument must fail. The term "agency action" has been defined broadly. In TVA v. Hill, 437 U.S. 153 (1978), the Court stated:

> One would be hard pressed to find a statutory provision whose terms were any plainer than those in § 7 of the Endangered Species Act. Its very words affirmatively command all federal agencies "to insure that actions authorized, funded, or carried out by them do not jeopardize the continued existence" of an endangered species or "result in the destruction or modification of habitat of such species...." This language admits of no exception.

Id. at 173. The regulation defining agency action states:

> Action means all activities or programs of any kind authorized, funded, or carried out, in whole or in part, by Federal agencies.... Examples include, but are not limited to: ... the granting of licenses, contracts, leases, easements, rights-of-way, permits or grants in aid.

50 C.F.R. § 402.02. Clearly, negotiating and executing contracts is "agency action."

Orange Cove, the Madera Irrigation District (Madera) and Chowchilla Water District (Chowchilla) contend that the Bureau had no discretion to alter the terms of the renewal contracts, particularly the quantity of water delivered. Where there is no agency discretion to act, the ESA does not apply. The federal reclamation laws, which provided the right to renewal, state that the government is to renew the contracts on "mutually agreeable" terms, that water rights are based on the amount of available project water,and that the Secretary of the Interior (Secretary) has the discretion to set rates to cover an appropriate share of the operation and maintenance costs. Clearly, there was some discretion available to the Bureau during the negotiation process....

C. Procedural Violations of the ESA

1. Failure to Consult with the NMFS

Before initiating any agency action in an area that contains threatened or endangered species or a critical habitat, the agency must (1) make an independent determination of whether its action "may affect" a protected species or habitat, or (2) initiate a formal consultation with the agency that has jurisdiction over the species. If an agency determines that an action "may affect" critical species or habitats, formal consultation is mandated. Formal consultation is excused only where (1) an agency determines that its action is unlikely to adversely affect the protected species or habitat, and (2) the relevant Service (FWS or NMFS) concurs with that determination.

The NMFS has jurisdiction over the winter-run chinook salmon, which was listed as a threatened species prior to execution of all but one of the water contracts. The Bureau independently determined that the renewal contracts and recommitment of all the Friant dam's water were not likely adversely to affect the salmon. The Bureau then sought the NMFS' concurrence with that assessment. On November 1, 1991, the Director of the NMFS refused to concur in the Bureau's opinion that the salmon would not be adversely affected. However, the NMFS also stated that formal consultation was not required.... The Bureau then proceeded to execute the water contracts without requesting a formal consultation with the NMFS. The Bureau argued that it reasonably relied on the NMFS' determination that a formal consultation was unnecessary....

The Bureau had an affirmative duty to ensure that its actions did not jeopardize endangered species, and the NMFS letter clearly disagreed with the agency's determination of no adverse impact. *See* 16 U.S.C. § 1536(a)(2); 50 C.F.R. § 402.14. Under those circumstances, regardless of the NMFS position that a formal consultation was "unnecessary," the Bureau had a clear legal obligation to at least request a formal consultation. *See* 50 C.F.R. §§ 402.13, 402.14. The reason that the NMFS gave for stating that a consultation was unnecessary was not supported by statute or regulation and had no rational relationship to the Bureau's independent obligations to ensure that its proposed actions were not likely adversely to affect the salmon. The district court did not err in concluding that it was arbitrary and capricious for the Bureau to forgo a formal consultation with the NMFS where the NMFS specifically refused to provide the required concurrence of "no adverse impact." Where the Bureau executed these 40–year contracts without first obtaining either the required concurrence from NMFS that the proposed action was not likely to affect a threatened species or a properly issued NMFS "no jeopardy" Biological Opinion, the Bureau acted arbitrarily and capriciously and not in accordance with the law. Therefore, all of these contracts were subject to rescission.

2. *Untimely Consultation with the FWS*

In addition to failing to request and follow through with a required consultation with NMFS, the Bureau also failed to follow its obligations under law with respect to its consultation with the FWS. The FWS has jurisdiction over several protected species in the Friant area, and the Bureau informally consulted with the FWS during 1990 and 1991. Formal consultation was requested on May 22, 1991. The FWS issued a "no jeopardy" Biological Opinion on October 15, 1991. Ten of the Friant contracts had already been executed by that time. The contracts contained Article 14, which allowed some contract modification pursuant to environmental review, and all but one of the contracts contained a provision modifying the terms dependent on the outcome of this litigation.

Section 7(d) of the ESA provides:

After initiation of consultation required under subsection (a)(2) of this section, the Federal agency and the permit or license applicant shall not make any irreversible or irretrievable commitment of resources with respect to the agency action which has the effect of foreclosing the formulation or implementation of any reasonable and prudent alternative measures which would not violate subsection (a)(2) of this section.

16 U.S.C. § 1536(d); *see* 50 C.F.R. § 402.09. The district court concluded that the 40–year water contracts constituted an irreversible and irretrievable commitment of resources and that the Bureau was not permitted to proceed until FWS found that the contracts were not likely to affect a protected species.

The Non-federal Defendants insist that even if the water contracts are an irreversible and irretrievable commitment of resources, Article 14 prevented the foreclosure of reasonable and prudent alternatives and, therefore, § 7(d) was not violated. We do not think that an agency should be permitted to skirt the procedural requirements of § 7(d) by including such a catchall savings clause in illegally executed contracts. However, even if such a clause could preserve the contracts, Article 14 is inadequate to serve that purpose here because it limits conservation-based modifications to minor adjustments and prohibits an adjustment in the amount of water delivered. Because Article 14 does not permit a reduction in the quantity of water delivered, the reasonable and prudent alternative of reallocating contracted water from irrigation to conservation is foreclosed. The district court did not err in concluding that the Bureau violated § 7(d) when it executed the contracts prior to completing the formal consultation process with the FWS, and the contracts executed prior to the issuance of the FWS Biological Opinion are subject to rescission.

[The court concluded that the district court's decision to rescind the contracts was not an abuse of discretion.]

Notes and Questions

1. When should consultation take place to determine whether there is any impact on endangered species in these cases?

- Permission to engage in oil and gas production on federal lands proceeds in several stages, including the issuance of a lease, pre-exploration activities, exploration, and production. *See* Conner v. Burford, 848 F.2d 1441 (9th Cir.1988), *cert. denied*, 489 U.S. 1012 (1989); North Slope Borough v. Andrus, 642 F.2d 589 (D.C.Cir.1980); Conservation Law Foundation v. Andrus, 623 F.2d 712 (1st Cir.1979); Swan View Coalition, Inc. v. Turner, 824 F.Supp. 923 (D.Mont.1992).

- The Bureau of Land Management issues general guidelines for allowing timber sales consistent with the conservation of the northern spotted owl. The guidelines do not authorize any particular timber sale, but each proposed sale is judged by reference to the strategy. *See* Lane County Audubon Society v. Jamison, 958 F.2d 290 (9th Cir.1992).

2. Suppose that a federal highway project is proposed for Las Vegas, the fastest growing city in the United States. Scores of other commercial, residential, and road projects are under consideration, about to begin construction, or already being built. Some of these projects are supported or authorized by federal agencies; others are state, local or private projects with no federal nexus. The area is also home to the endangered desert tortoise and a number of other listed species and species under consideration for listing. In determining the effect of the federal highway project, must the cumulative impact of all of the other projects be considered, or only some of them? Must the impact on species that are not listed be considered?

3. What happens if the federal agency action changes or additional information is learned about a species—or a new species is listed—after the consultation is completed? The FWS regulations provide that an agency must reopen consultation when (1) the amount or extent of a taking specified in a incidental take statement is exceeded, (2) new information reveals effects of the action that may affect listed species or critical habitat in a manner or to an extent not previously considered, (3) the identified action is subsequently modified in a manner that causes an effect to the listed species or critical habitat that was not considered in the biological opinion, or (4) a new species is listed or critical habitat designated that may be affected by the identified action. 50 C.F.R. § 402.16. The application of the regulations is illustrated by Sierra Club v. Marsh, 816 F.2d 1376 (9th Cir.1987), where the Army Corps of Engineers agreed as a result of a consultation with the FWS to purchase 200 acres of land to protect two endangered birds that could be affected by a Corps highway and flood control project. When the land proved to be difficult to obtain, the Corps continued work on the project but refused to reinitiate consultation with the FWS. The court held that the Corps could not proceed with the project until it further consulted with the FWS or it acquired the land because of the "institutionalized caution mandated by section 7 of the ESA."

Tennessee Valley Authority v. Hill

Supreme Court of the United States, 1978.
437 U.S. 153.

■ MR. CHIEF JUSTICE BURGER delivered the opinion of the Court.

The questions presented in this case are (a) whether the Endangered Species Act of 1973 requires a court to enjoin the operation of a virtually completed federal dam—which had been authorized prior to 1973—when, pursuant to authority vested in him by Congress, the Secretary of the Interior has determined that operation of the dam would eradicate an endangered species; and (b) whether continued congressional appropriations for the dam after 1973 constituted an implied repeal of the Endangered Species Act, at least as to the particular dam.

I

The Little Tennessee River originates in the mountains of northern Georgia and flows through the national forest lands of North Carolina into Tennessee, where it converges with the Big Tennessee River near Knoxville. The lower 33 miles of the Little Tennessee takes the river's clear, free-flowing waters through an area of great natural beauty. Among other environmental amenities, this stretch of river is said to contain abundant trout. Considerable historical importance attaches to the areas immediately adjacent to this portion of the Little Tennessee's banks. To the south of the river's edge lies Fort Loudon, established in 1756 as England's southwestern outpost in the French and Indian War. Nearby are also the ancient sites of several native American villages, the archeological stores of which are to a large extent unexplored. These include the Cherokee towns of Echota and Tennase, the former being the sacred capital of the Cherokee Nation as early as the 16th century and the latter providing the linguistic basis from which the State of Tennessee derives its name.

In this area of the Little Tennessee River the Tennessee Valley Authority, a wholly owned public corporation of the United States, began constructing the Tellico Dam and Reservoir Project in 1967, shortly after Congress appropriated initial funds for its development. Tellico is a multi-purpose regional development project designed principally to stimulate shoreline development, generate sufficient electric current to heat 20,000 homes, and provide flatwater recreation and flood control, as well as improve economic conditions in "an area characterized by underutilization of human resources and outmigration of young people." Of particular relevance to this case is one aspect of the project, a dam which TVA determined to place on the Little Tennessee, a short distance from where the river's waters meet with the Big Tennessee. When fully operational, the dam would impound water covering some 16,500 acres—much of which represents valuable and productive farmland—thereby converting the river's shallow, fast-flowing waters into a deep reservoir over 30 miles in length.

The Tellico Dam has never opened, however, despite the fact that construction has been virtually completed and the dam is essentially ready for operation. Although Congress has appropriated monies for Tellico every year since 1967, progress was delayed, and ultimately stopped, by a tangle of lawsuits and administrative proceedings. After unsuccessfully urging TVA to consider alternatives to damming the Little Tennessee, local citizens and national conservation groups brought suit in the District Court, claiming that the project did not conform to the requirements of the National Environmental Policy Act of 1969 (NEPA). After finding TVA to be in violation of NEPA, the District Court enjoined the dam's completion pending the filing of an appropriate environmental impact statement. The injunction remained in effect until late 1973, when the District Court concluded that TVA's final environmental impact statement for Tellico was in compliance with the law.

A few months prior to the District Court's decision dissolving the NEPA injunction, a discovery was made in the waters of the Little Tennessee which would profoundly affect the Tellico Project. Exploring the area around Coytee Springs, which is about seven miles from the mouth of the river, a University of Tennessee ichthyologist, Dr. David A. Etnier, found a previously unknown species of perch, the snail darter, or *Percina (Imostoma) tanasi.* This three-inch, tannish-colored fish, whose numbers are estimated to be in the range of 10,000 to 15,000, would soon engage the attention of environmentalists, the TVA, the Department of the Interior, the Congress of the United States, and ultimately the federal courts, as a new and additional basis to halt construction of the dam.

Until recently the finding of a new species of animal life would hardly generate a cause celebre. This is particularly so in the case of darters, of which there are approximately 130 known species, 8 to 10 of these having been identified only in the last five years.[7] The moving force behind the snail darter's sudden fame came some four months after its discovery, when the Congress passed the Endangered Species Act of 1973 (Act). This legislation, among other things, authorizes the Secretary of the Interior to declare species of animal life "endangered" and to identify the "critical habitat" of these creatures. When a species or its habitat is so listed, the following portion of the Act—relevant here—becomes effective:

> "The Secretary [of the Interior] shall review other programs administered by him and utilize such programs in furtherance of the purposes of this chapter. All other Federal departments and agencies shall, in consultation with and with the assistance of the Secretary, utilize their authorities in furtherance of the purposes of this chapter by carrying out programs for the conservation of endangered species and threatened species listed pursuant to section 1533 of this title and *by taking such action necessary to insure that actions authorized, funded, or carried out by them do not jeopardize the continued existence of such endangered species and threatened species or result in the destruction or modification of habitat of such species* which is determined by the Secretary, after consultation as appropriate with the affected States, to be critical." 16 U. S. C § 1536 (1976 ed.) (emphasis added).

In January 1975, the respondents in this case and others petitioned the Secretary of the Interior to list the snail darter as an endangered species. After receiving comments from various interested parties, including TVA and the State of Tennessee, the Secretary formally listed the snail darter as an endangered species on October 8, 1975. In so acting, it was noted that "the snail darter is a living entity which is genetically distinct and reproductively isolated from other fishes." More important for the purposes of this case, the Secretary determined that the snail darter apparently lives only in that portion of the Little Tennessee River which would be completely inundated by the reservoir created as a consequence of the Tellico Dam's completion.[12] The Secretary went on to explain the significance of the dam to the habitat of the snail darter:

---

7. In Tennessee alone there are 85 to 90 species of darters, of which upward to 45 live in the Tennessee River system. New species of darters are being constantly discovered and classified—at the rate of about one per year. This is a difficult task for even trained ichthyologists since species of darters are often hard to differentiate from one another.

12. Searches by TVA in more than 60 watercourses have failed to find other populations of snail darters. The Secretary has noted that "more than 1,000 collections in recent years and additional earlier collections from central and east Tennessee have not

"[The] snail darter occurs only in the swifter portions of shoals over clean gravel substrate in cool, low-turbidity water. Food of the snail darter is almost exclusively snails which require a clean gravel substrate for their survival. *The proposed impoundment of water behind the proposed Tellico Dam would result in total destruction of the snail darter's habitat.*" (emphasis added).

Subsequent to this determination, the Secretary declared the area of the Little Tennessee which would be affected by the Tellico Dam to be the "critical habitat" of the snail darter. [TVA tried to find an alternative river to which it could relocate the snail darters, but those efforts failed. TVA continued to seek congressional funding for the dam, which Congress and President Carter approved in a December 1975 appropriations bill containing funds for the continued construction of the dam. In February 1976, a University of Tennessee law student named Hiram Hill sued TVA to enjoin the completion of the dam based on section 7 of the ESA. The district court refused to issue an injunction, but the Sixth Circuit reversed an ordered a permanent injunction against the construction of the Tellico Dam. TVA officials testified before congressional committees at several times during the course of the litigation, and each time the committee stated its understanding that the dam should be completed notwithstanding the ESA.]

II

We begin with the premise that operation of the Tellico Dam will either eradicate the known population of snail darters or destroy their critical habitat. Petitioner does not now seriously dispute this fact. In any event, under § 4 (a)(1) of the Act, the Secretary of the Interior is vested with exclusive authority to determine whether a species such as the snail darter is "endangered" or "threatened" and to ascertain the factors which have led to such a precarious existence. By § 4(d) Congress has authorized—indeed commanded—the Secretary to "issue such regulations as he deems necessary and advisable to provide for the conservation of such species." As we have seen, the Secretary promulgated regulations which declared the snail darter an endangered species whose critical habitat would be destroyed by creation of the Tellico Reservoir. Doubtless petitioner would prefer not to have these regulations on the books, but there is no suggestion that the Secretary exceeded his authority or abused his discretion in issuing the regulations. Indeed, no judicial review of the Secretary's determinations has ever been sought and hence the validity of his actions are not open to review in this Court.

(A)

It may seem curious to some that the survival of a relatively small number of three-inch fish among all the countless millions of species extant

---

revealed the presence of the snail darter outside the Little Tennessee River." It is estimated, however, that the snail darter's range once extended throughout the upper main Tennessee River and the lower portions of its major tributaries above Chattanooga—all of which are now the sites of dam impoundments.

would require the permanent halting of a virtually completed dam for which Congress has expended more than $100 million. The paradox is not minimized by the fact that Congress continued to appropriate large sums of public money for the project, even after congressional Appropriations Committees were apprised of its apparent impact upon the survival of the snail darter. We conclude, however, that the explicit provisions of the Endangered Species Act require precisely that result.

One would be hard pressed to find a statutory provision whose terms were any plainer than those in § 7 of the Endangered Species Act. Its very words affirmatively command all federal agencies "to *insure* that actions *authorized, funded, or carried out* by them do not *jeopardize* the continued existence" of an endangered species or *"result* in the destruction or modification of habitat of such species. . . ." 16 U. S. C. § 1536 (1976 ed.). (Emphasis added.) This language admits of no exception. Nonetheless, petitioner urges, as do the dissenters, that the Act cannot reasonably be interpreted as applying to a federal project which was well under way when Congress passed the Endangered Species Act of 1973. To sustain that position, however, we would be forced to ignore the ordinary meaning of plain language. It has not been shown, for example, how TVA can close the gates of the Tellico Dam without "carrying out" an action that has been "authorized" and "funded" by a federal agency. Nor can we understand how such action will *"insure"* that the snail darter's habitat is not disrupted. Accepting the Secretary's determinations, as we must, it is clear that TVA's proposed operation of the dam will have precisely the opposite effect, namely the eradication of an endangered species.

Concededly, this view of the Act will produce results requiring the sacrifice of the anticipated benefits of the project and of many millions of dollars in public funds. But examination of the language, history, and structure of the legislation under review here indicates beyond doubt that Congress intended endangered species to be afforded the highest of priorities. . . .

The legislative proceedings in 1973 are, in fact, replete with expressions of concern over the risk that might lie in the loss of any endangered species. . . . Congress was concerned about the unknown uses that endangered species might have and about the unforeseeable place such creatures may have in the chain of life on this planet.

In shaping legislation to deal with the problem thus presented, Congress started from the finding that "[the] two major causes of extinction are hunting and destruction of natural habitat." Of these twin threats, Congress was informed that the greatest was destruction of natural habitats. Witnesses recommended, among other things, that Congress require all land-managing agencies "to avoid damaging critical habitat for endangered species and to take positive steps to improve such habitat." Virtually every bill introduced in Congress during the 1973 session responded to this concern by incorporating language similar, if not identical, to that found in the present § 7 of the Act. These provisions were designed, in the words of an administration witness, "for the first time [to] *prohibit* [a] federal agency from taking action which does jeopardize the status of endangered species," furthermore, the proposed bills would "[direct] all ... Federal agencies to utilize their authorities for carrying out programs for the protection of endangered animals." (Emphasis added.)

As it was finally passed, the Endangered Species Act of 1973 represented the most comprehensive legislation for the preservation of endangered species ever enacted by any nation. Its stated purposes were "to provide a means whereby the ecosystems upon which endangered species and threatened species depend may be conserved," and "to provide a program for the conservation of such ... species...." In furtherance of these goals, Congress expressly stated in § 2 (c) that "all Federal departments and agencies *shall* seek *to conserve endangered species* and threatened species...." (Emphasis added.) Lest there be any ambiguity as to the meaning of this statutory directive, the Act specifically defined "conserve" as meaning "to use and the use of *all methods and procedures which are necessary* to bring *any endangered species* or threatened species to the point at which the measures provided pursuant to this chapter are no longer necessary." (Emphasis added.) Aside from § 7, other provisions indicated the seriousness with which Congress viewed this issue: Virtually all dealings with endangered species, including taking, possession, transportation, and sale, were prohibited, except in extremely narrow circumstances. The Secretary was also given extensive power to develop regulations and programs for the preservation of endangered and threatened species. Citizen involvement was encouraged by the Act, with provisions allowing interested persons to petition the Secretary to list a species as endangered or threatened, and bring civil suits in United States district courts to force compliance with any provision of the Act.

Section 7 of the Act, which of course is relied upon by respondents in this case, provides a particularly good gauge of congressional intent. As we have seen, this provision had its genesis in the Endangered Species Act of 1966, but that legislation qualified the obligation of federal agencies by stating that they should seek to preserve endangered species only *"insofar as is practicable and consistent with [their] primary purposes...."* Likewise, every bill introduced in 1973 contained a qualification similar to that found in the earlier statutes. Exemplary of these was the administration bill, H. R. 4758, which in § 2 (b) would direct federal agencies to use their authorities to further the ends of the Act *"insofar as is practicable and consistent with [their] primary purposes...."* (Emphasis added.) Explaining the idea behind this language, an administration spokesman told Congress that it "would further signal to all ... agencies of the Government that this is the *first priority, consistent with their primary objectives."* (Emphasis added.) This type of language did not go unnoticed by those advocating strong endangered species legislation. A representative of the Sierra Club, for example, attacked the use of the phrase "consistent with the primary purpose" in proposed H. R. 4758, cautioning that the qualification "could be construed to be a declaration of congressional policy that other agency purposes are necessarily more important than protection of endangered species and would always prevail if conflict were to occur."

What is very significant in this sequence is that the final version of the 1973 Act carefully omitted all of the reservations described above.... It is against this legislative background that we must measure TVA's claim that the Act was not intended to stop operation of a project which, like Tellico Dam, was near completion when an endangered species was discovered in its path. While there is no discussion in the legislative history of precisely this problem, the totality of congressional action makes it abundantly clear that the result we reach today is wholly in accord with both the words of the statute and the intent of Congress. The plain intent of Congress in enacting this statute was to halt and reverse the trend toward species extinction, whatever the cost. This is reflected not only in the stated policies of the Act, but in literally every section of the statute. All persons, including federal agencies, are specifically instructed not to "take" endangered species, meaning that no one is "to harass, harm, pursue, hunt, shoot, wound, kill, trap, capture, or collect" such life forms. Agencies in particular are directed by §§ 2 (c) and 3 (2) of the Act to "use ... *all methods* and procedures which are necessary" to preserve endangered species. (emphasis added). In addition, the legislative history undergirding § 7 reveals an explicit congressional decision to require agencies to afford first priority to the declared national policy of saving endangered species. The pointed omission of the type of qualifying language previously included in endangered species legislation reveals a conscious decision by Congress to give endangered species priority over the "primary missions" of federal agencies.

It is not for us to speculate, much less act, on whether Congress would have altered its stance had the specific events of this case been anticipated. In any event, we discern no hint in the deliberations of Congress relating to the 1973 Act that would compel a different result than we reach here. Indeed, the repeated expressions of congressional concern over what it saw as the potentially enormous danger presented by the eradication of any endangered species suggest how the balance would have been struck had the issue been presented to Congress in 1973.

Furthermore, it is clear Congress foresaw that § 7 would, on occasion, require agencies to alter ongoing projects in order to fulfill the goals of the Act. Congressman Dingell's discussion of Air Force practice bombing, for instance, obviously pinpoints a particular activity—intimately related to the national defense—which a major federal department would be obliged to alter in deference to the strictures of § 7.... One might dispute the applicability of these examples to the Tellico Dam by saying that in this case the burden on the public through the loss of millions of unrecoverable dollars would greatly outweigh the loss of the snail darter. But neither the Endangered Species Act nor Art. III of the Constitution provides federal courts with authority to make such fine utilitarian calculations. On the contrary, the plain language of the Act, buttressed by its legislative history, shows clearly that Congress viewed the value of endangered species as "incalculable." Quite obviously, it would be difficult for a court to balance the loss of a sum certain—even $100 million—against a congressionally declared "incalculable" value, even assuming we had the power to engage in such a weighing process, which we emphatically do not. . . .

Notwithstanding Congress' expression of intent in 1973, we are urged to find that the continuing appropriations for Tellico Dam constitute an implied repeal of the 1973 Act, at least insofar as it applies to the Tellico Project. In support of this view, TVA points to the statements found in various House and Senate Appropriations Committees' Reports; as described in Part I, *supra*, those Reports generally reflected the attitude of the *Committees* either that the Act did not apply to Tellico or that the dam should be completed regardless of the provisions of the Act. Since we are unwilling to assume that these latter Committee statements constituted advice to ignore the provisions of a duly enacted law, we assume that these Committees believed that the Act simply was not applicable in this situation. But even under this interpretation of the Committees' actions, we are unable to conclude that the Act has been in any respect amended or repealed. There is nothing in the appropriations measures, as passed, which states that the Tellico Project was to be completed irrespective of the requirements of the Endangered Species Act. These appropriations, in fact, represented relatively minor components of the lump-sum amounts for the entire TVA budget. To find a repeal of the Endangered Species Act under these circumstances would surely do violence to the " 'cardinal rule . . . that repeals by implication are not favored.' ". . . . The doctrine disfavoring repeals by implication "applies with full vigor when . . . the subsequent legislation is an appropriations measure.". . . . Perhaps mindful of the fact that it is "swimming upstream" against a strong current of well-established precedent, TVA argues for an exception to the rule against implied repealers in a circumstance where, as here, Appropriations Committees have expressly stated their "understanding" that the earlier legislation would not prohibit the proposed expenditure. We cannot accept such a proposition. Expressions of committees dealing with requests for appropriations cannot be equated with statutes enacted by Congress, particularly not in the circumstances presented by this case. First, the Appropriations Committees had no jurisdiction over the subject of endangered species, much less did they conduct the type of extensive hearings which preceded passage of the earlier Endangered Species Acts, especially the 1973 Act. . . . Second, there is no indication that Congress as a whole was aware of TVA's position, although the Appropriations Committees apparently agreed with petitioner's views.

(B)

Having determined that there is an irreconcilable conflict between operation of the Tellico Dam and the explicit provisions of § 7 of the Endangered Species Act, we must now consider what remedy, if any, is appropriate. It is correct, of course, that a federal judge sitting as a chancellor is not mechanically obligated to grant an injunction for every violation of law. . . . As a general matter it may be said that "[since] all or almost all equitable remedies are discretionary, the balancing of equities and hardships is appropriate in almost any case as a guide to the chancellor's discretion." ... But these principles take a court only so far. Our system of government is, after all, a tripartite one, with each branch having certain defined functions delegated to it by the Constitution. While "[it] is emphatically the province and duty of the judicial department to say what the law is," Marbury v. Madison, 1 Cranch 137, 177 (1803), it is equally—and emphatically—the exclusive province of the Congress not only to formulate legislative policies and mandate programs and projects, but also to establish their relative priority for the Nation. Once Congress, exercising its delegated powers, has decided the order of priorities in a given area, it is for the Executive to administer the laws and for the courts to enforce them when enforcement is sought. Here we are urged to view the Endangered Species Act "reasonably,"and hence shape a remedy "that accords with some modicum of common sense and the public weal." But is that our function? We have no expert knowledge on the subject of endangered species, much less do we have a mandate from the people to strike a balance of equities on the side of the Tellico Dam. Congress has spoken in the plainest of words, making it abundantly clear that the balance has been struck in favor of affording endangered species the highest of priorities, thereby adopting a policy which it described as "institutionalized caution." Our individual appraisal of the wisdom or unwisdom of a particular course consciously selected by the Congress is to be put aside in the process of interpreting a statute. Once the meaning of an enactment is discerned and its constitutionality determined, the judicial process comes to an end. We do not sit as a committee of review, nor are we vested with the power of veto. The lines ascribed to Sir Thomas More by Robert Bolt are not without relevance here:

> "The law, Roper, the law. I know what's legal, not what's right. And I'll stick to what's legal.... I'm not God. The currents and eddies of right and wrong, which you find such plain-sailing, I can't navigate, I'm no voyager. But in the thickets of the law, oh there I'm a forester.... What would you do? Cut a great road through the law to get after the Devil? ... And when the last law was down, and the Devil turned round on you—where would you hide, Roper, the laws all being flat? ... This country's planted thick with laws from coast to coast—Man's laws, not God's—and if you cut them down ... d'you really think you could stand upright in the winds that would below then? ... Yes, I'd give the Devil benefit of law, for my own safety's sake." R. BOLT, A MAN FOR ALL SEASONS, Act I, p. 147 (Three Plays, Heinemann ed. 1967).

We agree with the Court of Appeals that in our constitutional system the commitment to the separation of powers is too fundamental for us to pre-empt congressional action by judicially decreeing what accords with "common sense and the public weal." Our Constitution vests such responsibilities in the political branches.

■ MR. JUSTICE POWELL, with whom MR. JUSTICE BLACKMUN joins, dissenting.

The Court today holds that § 7 of the Endangered Species Act requires a federal court, for the purpose of protecting an endangered species or its habitat, to enjoin permanently the operation of any federal project, whether completed or substantially completed. This decision casts a long shadow over the operation of even the most important projects, serving vital needs of society and national defense, whenever it is determined that continued operation would threaten extinction of an endangered species or its habitat. This result is said to be required by the "plain intent of Congress" as well as by the language of the statute.

In my view § 7 cannot reasonably be interpreted as applying to a project that is completed or substantially completed when its threat to an endangered species is discovered. Nor can I believe that Congress could have intended this Act to produce the "absurd result"—in the words of the District Court—of this case. If it were clear from the language of the Act and its legislative history that Congress intended to authorize this result, this Court would be compelled to enforce it. It is not our province to rectify policy or political judgments by the Legislative Branch, however egregiously they may disserve the public interest. But where the statutory language and legislative history, as in this case, need not be construed to reach such a result, I view it as the duty of this Court to adopt a permissible construction that accords with some modicum of common sense and the public weal.

... I have little doubt that Congress will amend the Endangered Species Act to prevent the grave consequences made possible by today's decision. Few, if any, Members of that body will wish to defend an interpretation of the Act that requires the waste of at least $53 million, and denies the people of the Tennessee Valley area the benefits of the reservoir that Congress intended to confer. There will be little sentiment to leave this dam standing before an empty reservoir, serving no purpose other than a conversation piece for incredulous tourists.

But more far reaching than the adverse effect on the people of this economically depressed area is the continuing threat to the operation of every federal project, no matter how important to the Nation. If Congress acts expeditiously, as may be anticipated, the Court's decision probably will have no lasting adverse consequences. But I had not thought it to be the province of this Court to force Congress into otherwise unnecessary action by interpreting a statute to produce a result no one intended.

■ [JUSTICE REHNQUIST dissented because he concluded that the district court's refusal to issue an injunction was not an abuse of discretion]

Notes and Questions

1. In 1979, six years after voting for the ESA, Tennessee Senator Howard Baker described the snail darter as "the bold perverter of the Endangered Species Act." 125 CONG. REC. 23867 (1979). Was he right?

2. What would happen if the snail darter was discovered just *after* the dam was completed, but before the darter's habitat was destroyed? Would TVA have to tear down the dam? At oral argument, when Justice Powell asked Zygmunt Plater—Hiram Hill's professor at the University of Tennessee and counsel for those opposing the dam—whether the ESA would require the removal of Arizona's Grand Coulee Dam if an endangered species was found there, Plater answered "yes." In fact, the ESA has never been applied to require the destruction or removal of an existing structure. Recently, though, environmentalists have been pressing to remove some of the dams on the Columbia River system in the Pacific Northwest in order to protect several endangered species of salmon. *See* Michael C. Blumm et al., *Saving Snake River Water and Salmon Simultaneously: The Biological, Economic, and Legal Case for Breaching the Lower Snake River Dams, Lowering the John Day Reservoir, and Restoring Natural River Flows*, 28 ENVTL. L. 997 (1998).

If the TVA had succeeded in its efforts to relocate the snail darter to any river, would that have been an acceptable solution?

3.   Section 7's jeopardy provision has been employed to block other notable projects. For example, one of the first ESA decisions, National Wildlife Federation v. Coleman, 529 F.2d 359 (5th Cir.), *cert. denied*, 429 U.S. 979 (1976), enjoined construction of a segment of Interstate 10 in Mississippi because of the highway's impact on the endangered Mississippi sandhill crane. Also, in Roosevelt Campobello International Park Commission v. EPA, 684 F.2d 1041 (1st Cir.1982), the court overturned EPA's issuance of a permit to build an oil refinery along the coast of Maine because EPA failed to adequately study the likelihood of an oil spill that would jeopardize the continued existence of several endangered whales.

Perhaps the most dramatic jeopardy finding occurred in April 2001 when the FWS and NMFS advised the Federal Bureau of Reclamation that its annual operating plan for the Klamath Reclamation Project would jeopardize the continued existence of the endangered Lost River sucker, the shortnose sucker, and the Southern Oregon/Northern California Coast coho salmon. Congress authorized the irrigation project as part of its effort to encourage the settlement of previously arid parts of Oregon. By 2001, though, farmers relying upon irrigation water competed with the suckers and salmon living in the Klamath River, bald eagles preying upon waterfowl in the Lower Klamath National Wildlife Refuge, and members of the Klamath and Yurok Tribes who revered the suckers and who possessed treaty rights to the rivers. Then a drought was predicted for 2001. An ESA section 7 consultation between FWS, NMFS, and the Bureau of Reclamation resulted in a biological opinion that the Bureau's plans to provide irrigation water from the project would jeopardize the suckers and the salmon, so the Bureau issued a revised plan that denied water to most of the farmers who had historically relied upon it. The farmers and their irrigation district then sought an injunction against that revised plan for the Klamath Reclamation Project. The district court denied the injunction because the balance of hardships did not tip sufficiently in favor of the plaintiffs: while there was "no question that farmers who rely upon irrigation water and their communities will suffer severe economic hardship" from the plan, the court cited *TVA v. Hill* as evidence that "[t]hreats to the continued existence of endangered and threatened species constitute ultimate harm." Kandra v. United States, 145 F.Supp.2d 1192, 1200–01 (D.Or.2001).

The jeopardy finding and the court's refusal to enjoin the revised plan triggered civil unrest in Oregon throughout the summer of 2001. Farmers watched as their crops died during the record drought. In early July, "100 to 150 people formed a human chain and shielded men who cut off the headgate's lock using a diamond-bladed chainsaw and a cutting torch, sending water from the Upper Klamath Lake into the canal." *Farmers Force Open Canal in Fight with U.S. Over Water*, N.Y. TIMES, July 6, 2001, at A10. The irrigation canal's headgates were opened three more times in a symbolic protest against the lack of irrigation water. The protestors withdrew after the September 11 terrorist attacks, but the controversy gained a racial edge when three men were arrested in December for driving through the town of Chiloquin, Oregon while firing shotguns and deriding Native Americans as "sucker lovers." And then the premise for the dispute was called into question when the National Academy of Sciences reported in February 2002 that the provision of water to the endangered suckers and salmon instead of the farmers did not yield any appreciable benefit to the fish. *See* NATIONAL ACADEMY OF SCIENCES, SCIENTIFIC EVALUATION OF THE BIOLOGICAL OPINIONS ON ENDANGERED AND THREATENED FISHES IN THE KLAMATH RIVER BASIN: INTERIM REPORT (2002).

4.   Professor Oliver Houck counters that the vast majority of section 7 consultations result in no jeopardy findings. His 1993 study determined that fewer than .02% of all consultations have resulted in the termination of the project in question. "No major public activity, nor any major federally-permitted activity was blocked" in any of the 99 FWS jeopardy opinions issued between 1987 and 1992 that Houck examined. Indeed, he argues that FWS has wrongly avoided the invocation of the jeopardy provision by narrowly viewing the agency action at issue, by failing to apply section 7 to federal projects overseas, by conflating jeopardy and critical habitat, and by improperly declining to list species in the first place. *See* Oliver A. Houck, *The Endangered Species Act and Its Implementation By the U.S. Departments of Interior and Commerce*, 64 U. COLO. L. REV. 277, 317–26 (1993). Daniel Rohlf reached a similar conclusion when he asserted that "the concept of jeopardy often amounts to little more than a vague threat employed by FWS and NMFS to negotiate relatively minor modifications to federal and non-federal actions." Daniel J. Rohlf, *Jeopardy Under the Endangered Species Act: Playing a Game Protected Species Can't Win*, 41 WASHBURN L. REV. 114, 115 (2001).

5.   The Supreme Court's decision in *TVA v. Hill* was not the end of the fight between the Tellico Dam and the snail darter. Later in 1978, Congress responded by adding a provision to the ESA that established an Endangered Species Committee empowered to waive the jeopardy prohibition in appropriate circumstances. This committee—commonly known as the "God Squad" because of its power to determine the fate of a species on the brink of extinction—is comprised of seven federal officials and one individual selected by the President from each of the states involved. The exemption process may be triggered by a request from the federal agency proposing the action, the governor of the state in which the action will occur, or a private applicant for federal permit or license that results in a jeopardy finding. The committee is empowered to waive the prohibitions of section 7 (and also section 9, whose provisions are discussed in chapter 6) only if at least five of its members determine that (1) there are no reasonable and prudent alternatives to the federal agency action, (2) the benefits of the agency action clearly outweigh alternatives that would protect the species, (3) the agency action is of national or regional significance, and (4) no irreversible or irretrievable commitment of resources are made prior to the committee's ruling. *See* ESA § 7(e).

Not surprisingly, the God Squad made its debut when TVA requested a waiver for the Tellico Dam. More surprisingly, in January 1979 the committee unanimously rejected the request. As Secretary of the Interior Cecil Andrus stated, "Frankly, I hate to see the snail darter get the credit for stopping a project that was ill-conceived and uneconomical in the first place."

The subtle approach having failed, congressional supporters of the Tellico Dam responded by adding a provision to the 1980 Interior appropriations bill that exempted the dam from *all* federal laws. The House approved the bill without debate, then it passed the Senate by a narrow 48–44 margin. President Carter opposed the exemption for the dam, but he signed the bill anyway because of the other things that it contained.

The fight against the dam had one last gasp. The congressional waiver protected the dam from any federal statutory requirements, but it did not—and probably could not—bar constitutional objections to the dam. A group of Native Americans claimed that the dam would result in flooding of their sacred sites and would thus violate their first amendment right to the free exercise of their religion. The courts said no. *See* Sequoyah v. Tennessee Valley Auth., 480 F.Supp. 608 (E.D.Tenn.1979), *cert. denied*, 449 U.S. 953 (1980).

Finally, on November 29, 1979, the Tellico Dam was completed and the Little Tennessee River was impounded. Again, the consequences were not at all expected. The economic development forecast by the supporters of the dam never quite materialized. An early proposal would have located a hazardous waste facility near the newly created reservoir, but instead a retirement community known as Tellico Village developed at the site. On the other hand, the snail darter did not become extinct. Nearly one year after the dam was completed, the same scientist who had discovered the snail darter on the Little Tennessee River in 1973 found other populations of snail darters in four other rivers in Tennessee whose conditions were not supposed to have been acceptable for the fish. Apparently the snail darters did not know that, though, and they survive in such abundance that the species was downlisted from endangered to threatened in 1984.

The saga of the snail darter and the Tellico Dam is told in CHARLES C. MANN & MARK L. PLUMMER, NOAH'S CHOICE: THE FUTURE OF ENDANGERED SPECIES

147–175 (1995); WILLIAM B. WHEELER & MICHAEL J. McDONALD, TVA AND THE TELLICO DAM, 1936–1979 (1986); Zygmunt Plater, *In the Wake of the Snail Darter: An Environmental Law Paradigm and Its Consequences*, 19 U. MICH. J. L. REF. 19 (1986).

5. Congress designed the Endangered Species Committee with the snail darter in mind, but the committee remains a permanent fixture in the ESA. It has been employed infrequently. Only a handful of applications have been made to invoke the committee to waive a jeopardy finding, and only twice has the committee agreed to do so. The first instance occurred in 1979, when the committee approved a settlement that allowed the operation of Nebraska's Greyrocks Dam and Reservoir consistent with the needs of the endangered whooping crane. The committee granted its second waiver in 1992 when it exempted 13 out of a requested 44 proposed Oregon timber sales despite the presence of the northern spotted owl, but a court decision questioning the committee's procedures and the advent of the Clinton Administration resulted in the timber sales never actually occurring. The exemption process has not been invoked since then, though there has been speculation that the listing of several species of salmon in the Pacific Northwest could trigger a request for an exemption if a federal water project is found to jeopardize the salmon. Most recently, the irrigation districts denied water pursuant to the jeopardy finding described above in note 3 sought to obtain an exemption in the summer of 2001, but Interior Secretary Gale Norton concluded that the districts were not among the parties empowered by the statute to invoke the exemption process.

Arizona Cattle Growers' Association v. United States Fish and Wildlife Service

United States Court of Appeals for the Ninth Circuit, 2001.
273 F.3d 1229.

■ WARDLAW, CIRCUIT JUDGE:

At issue in these consolidated cross-appeals is whether the United States Fish and Wildlife Service's provision of Incidental Take Statements pursuant to the Endangered Species Act was arbitrary and capricious under Section 706 of the Administrative Procedure Act. In separate actions, the Arizona Cattle Growers' Association ("ACGA") challenged the Incidental Take Statements set forth in the Biological Opinions issued by the Fish and Wildlife Service in consultation with the Bureau of Land Management (ACGA I) and the United States Forest Service (ACGA II) in response to ACGA's application for cattle grazing permits in Southeastern Arizona. In the district courts, each of the Incidental Take Statements was set aside, with one exception, as arbitrary and capricious actions by the Fish and Wildlife Service, due to insufficient evidence of a take.

We hold, based on the legislative history, case law, prior agency representations, and the plain language of the Endangered Species Act, that an Incidental Take Statement must be predicated on a finding of an incidental take. Further, the Fish and Wildlife Service acted in an arbitrary and capricious manner by issuing Incidental Take Statements imposing terms and conditions on land use permits, where there either was no evidence that the endangered species existed on the land or no evidence that a take would occur if the permit were issued. We also find that it was arbitrary and capricious for the Fish and Wildlife Service to issue terms and conditions so vague as to preclude compliance therewith.

I. Background

A. ACGA I

Arizona Cattle Growers' Association and Jeff Menges, a rancher seeking a grazing permit on the lands at issue (collectively "ACGA"), sued the Fish and Wildlife Service and the Bureau of Land Management to challenge Incidental Take Statements issued by the Fish and Wildlife Service in a Biological Opinion for certain grazing lands. Ariz. Cattle Growers' Ass'n v. U.S. Fish & Wildlife Serv., 63 F. Supp. 2d 1034 (D.Ariz.1998) (Ezra, C. J., presiding) ("ACGA I"). Menges sought livestock grazing permits for land within the area supervised by the Bureau of Land Management's Saffold and Tucson, Arizona field offices, and the Association represented members who claimed to be harmed by the government action. The Bureau of Land Management's livestock grazing program for this area affects 288 separate grazing allotments that in total comprise nearly 1.6 million acres of land. The Fish and Wildlife Service's Biological Opinion, issued on September 26, 1997, analyzes twenty species of plants and animals and concludes that the livestock grazing program was not likely to jeopardize the continued existence of the species affected nor was likely to result in destruction or adverse modification of the designated or proposed critical habitat. The Fish and Wildlife Service did, however, issue Incidental Take Statements for various species of fish and wildlife listed or proposed as endangered. . . .

B. ACGA II

In *ACGA II*, ACGA challenged Incidental Take Statements set forth in a second Biological Opinion issued by the Fish and Wildlife Service that concerns livestock grazing on public lands administered by the United States Forest Service. The Fish and Wildlife Service examined 962 allotments, determining that grazing would have no effect on listed species for 619 of those allotments and cause no adverse effects for 321 of the remaining allotments, leaving 22 allotments. These allotments were each roughly 30,000 acres, but several of the allotments were significantly larger. In its Biological Opinion, the Fish and Wildlife Service concluded that ongoing grazing activities on 21 out of the 22 allotments at issue would not jeopardize the continued existence of any protected species or result in the destruction or adverse modification of any critical habitat. It determined, however, that ongoing grazing activities would incidentally take members of one or more protected species in each of the 22 allotments, and it issued Incidental Take Statements for each of those allotments. ACGA contested the issuance of Incidental Take Statements for six of the allotments: Cow Flat, East Eagle, Montana, Sears–Club/Chalk Mountain, Sheep Springs, and Wildbunch.

[In both cases, ACGA then challenged the need for, and contents of, the Incidental Take Statements. The district court held that the FWS acted arbitrarily and capriciously by issuing Incidental Take Statements where it was not reasonably certain that the take of a listed species would occur. The court found that such a take would occur only at the Cow Flat Allotment, and it upheld only that Incidental Take Statement.]

. . . Section 7 of the Act imposes an affirmative duty to prevent violations of Section 9 upon federal agencies, such as the Bureau of Land Management and the U.S. Forest Service. 16 U.S.C. § 1536(a)(2). This affirmative duty extends to "any action authorized, funded, or carried out by such agency," including authorizing grazing permits on land owned by the federal government. *Id.*

To determine whether an "action may affect listed species or critical habitat," the agency may be required to create a Biological Assessment that "evaluates the potential effects of the action on listed and proposed species and . . . critical habitat and determines whether any such species or habitat are likely to be adversely affected by the action." 50 C.F.R. § 402.12. If the agency finds evidence of an adverse impact on any issued species, it must initiate formal consultation with the Fish and Wildlife Service. 50 C.F.R. § 402.14.

If formal consultation is necessary, the Fish and Wildlife Service will issue a Biological Opinion, summarizing the relevant findings and determining whether the proposed action is likely to jeopardize the continued existence of the species. 16 U.S.C. § 1536(b). If so, the Biological Opinion must list any "reasonable and prudent alternatives" that, if followed, would not jeopardize the continued existence of the species. 16 U.S.C. § 1536(b)(3)(A); 50 C.F. R. § 402.14.

Additionally, the Fish and Wildlife Service must specify whether any "incidental taking" of protected species will occur, specifically "any taking otherwise prohibited, if such taking is incidental to, and not the purpose of, the carrying out of an otherwise lawful activity." 16 U.S.C. § 1536(b)(4); 50 C.F.R. § 17.3. Its determination that an incidental taking will result leads to the publication of the "Incidental Take Statement," identifying areas where members of the particular species are at risk. Contained in the Incidental Take Statement is an advisory opinion which:

(i) specifies the impact of such incidental taking on the species,

(ii) specifies those reasonable and prudent measures that the Secretary considers necessary or appropriate to minimize such impact [and] . . .

(iv) sets forth the terms and conditions . . . that must be complied with by the Federal agency or applicant . . . or both, to implement the measures specified under clause (ii).

16 U.S.C. § 1536(b)(4) (subsection (iii) omitted).

Significantly, the Incidental Take Statement functions as a safe harbor provision immunizing persons from Section 9 liability and penalties for takings committed during activities that are otherwise lawful and in compliance with its terms and conditions. 16 U.S.C. § 1536(o). Any such incidental taking "shall not be considered to be a prohibited taking of the species concerned." *Id.* Although the action agency is "technically free to disregard the Biological Opinion and proceed with its proposed action . . . it does so at its own peril." [Bennett v. Spear, 520 U.S. 154, 170 (1997)]. Consequently, if the terms and conditions of the Incidental Take Statement are disregarded and a taking does occur, the action agency or the applicant may be subject to potentially severe civil and criminal penalties under Section 9.

V.   Determining When the Fish and Wildlife Service Must Issue an Incidental Take Statement

The Fish and Wildlife Service contends that the district courts erred in scrutinizing its decision to issue Incidental Take Statements because it is statutorily required pursuant to the ESA to "issue an ITS in all no-jeopardy determinations." In particular, it contests the *ACGA I* court's requirement that it provide evidence of a listed species' existence on the land and the *ACGA II* court's holding that issuing an Incidental Take Statement is "appropriate only when a take has occurred or is reasonably certain to occur." The Fish and Wildlife Service argues that both standards establish "an inappropriate and high burden of proof" and that it should be permitted to issue an Incidental Take Statement whenever there is any possibility, no matter how small, that a listed species will be taken. As we believe that Congress has spoken to the precise question at issue, we must reject the agency's interpretation of the ESA as contrary to clear congressional intent. . . .

The Fish and Wildlife Service argues that the plain language of the statute and implementing regulations "expressly direct" it to issue an Incidental Take Statement in every case. Section 7(b)(4) of the ESA provides:

> If after consultation under subsection (a)(2) of this section, the Secretary concludes that—
>
> (A) the agency action will not violate such subsection, or offers reasonable and prudent alternatives which the Secretary believes would not violate such subsection;
>
> (B) the taking of an endangered species or a threatened species incidental to the agency action will not violate such subsection; and
>
> (C) if an endangered species or threatened species of a marine mammal is involved, the taking is authorized pursuant to section 1371(a)(5) of this title;
>
> the Secretary shall provide the Federal agency and the applicant concerned, if any, with a written statement that—
>
> (i) specifies the impact of such incidental taking on
>
> the species, ...

16 U.S.C. § 1536(b)(4). The Fish and Wildlife Service relies on the statutory provision directing the Secretary to provide "a written statement that . . . specifies the impact of such incidental taking on the species." *Id.* . . .

When read in context, it is clear that the issuance of the Incidental Take Statement is subject to the finding of the factors enumerated in the ESA. The statute explicitly provides that the written statement is subject to the consultation and the Secretary's conclusions. A contrary interpretation would render meaningless the clause stating that the Incidental Take Statement will specify "the impact of *such* incidental taking." 16 U.S.C. § 1536(b)(4)(C)(i) (emphasis added). We therefore agree with ACGA that the plain language of the ESA does not dictate that the Fish and Wildlife Service must issue an Incidental Take Statement irrespective of whether any incidental takings will occur. *See* Nat'l Wildlife Fed'n v. Nat'l Park Serv., 669 F. Supp. 384, 389–90 (D.Wyo.1987) (holding that a careful reading of § 1536(b) supports the defendants' contention that an Incidental Take Statement is not required if no incidental takings are foreseen).

The plain language of the implementing regulations also supports ACGA's argument. One regulation specifically instructs the Fish and Wildlife Service that its "responsibilities during formal consultation are . . . to formulate a statement concerning incidental take, if such take *may* occur." 50 C.F.R. § 402.14(g)(7) (emphasis added). Moreover, the same regulation also instructs:

> (1) In those cases where the Service concludes that an action (or the implementation of any reasonable and prudent alternatives) and the resultant incidental take of listed species will not violate section 7(a)(2), . . . the Service will provide with the biological opinion a statement concerning incidental take that:
>
> (i) Specifies the impact, i.e., the amount or extent, of such incidental taking on the species;

50 C.F.R. § 402.14(i)(1) (2001). Thus, consistent with the language of the statute, the regulations only require the issuance of an Incidental Take Statement when the "*resultant* incidental take of listed species will not violate section 7(a)(2)." *Id.* (emphasis added).

Likewise, the legislative history supports this interpretation of the statute. If the sole purpose of the Incidental Take Statement is to provide shelter from Section 9 penalties, as previously noted, it would be nonsensical to require the issuance of a Incidental Take Statement when no takings cognizable under Section 9 are to occur. *See* H.R. Rep. No. 97–567, at 26 (1982).

The Fish and Wildlife Service's internal handbook does not alter our conclusion. The 1998 version of the agency's Section 7 Consultation Handbook provides that "when no take is anticipated" the agency should include in an Incidental Take Statement the following language: "The Service does not anticipate the proposed action will incidentally take any (species)." Indeed, one of Incidental Take Statements in the *ACGA II* consultation that the Fish and Wildlife Service issued contains this very language. That

Incidental Take Statement, however, is not before us in this appeal. It is true that "an agency's construction of the laws it administers is accorded considerable weight." We reiterate, however, that "the judiciary is the final authority on issues of statutory construction and must reject administrative constructions which are contrary to clear congressional intent." The Fish and Wildlife Service's handbook instruction to issue an Incidental Take Statement when no take will occur as a result of permitted activity is contrary to the plain meaning of the statute as well as the agency's own regulations. Accordingly, we hold that absent rare circumstances such as those involving migratory species, it is arbitrary and capricious to issue an Incidental Take Statement when the Fish and Wildlife Service has no rational basis to conclude that a take will occur incident to the otherwise lawful activity.

VI.   Review of the Incidental Take Statements under the Arbitrary and Capricious Standard Pursuant to the APA

Because we reject the Fish and Wildlife Service's interpretation of the ESA and hold that it is not required to provide an Incidental Take Statement whenever it issues a Biological Opinion, we must now examine each Incidental Take Statement at issue under Section 706. 5 U.S.C. § 706. . . .

A.   ACGA I

1. The Razorback Sucker

In the Biological Opinion issued in response to ACGA's first request for land use permits, the Fish and Wildlife Service concluded that the direct effects of cattle grazing are infrequent to the razorback sucker, a moderately sized fish listed as endangered in November 1991. Although once abundant in the project area, the Fish and Wildlife Service admitted that there have been no reported sightings of the razorback sucker in the area since 1991 and that "effects of the livestock grazing program on individual fish or fish populations probably occur infrequently." Nevertheless, the Fish and Wildlife Service issued an Incidental Take Statement for the fish, anticipating take as a result of the direct effects of grazing in the project area, the construction of fences, the construction and existence of stock tanks for non-native fish, as well as other "activities in the watershed." Because the Fish and Wildlife Service could not directly quantify the level of incidental take, it determined that authorized take would be exceeded if range conditions in the allotment deteriorated and cattle grazing could not be ruled out as a cause of the deterioration.

Despite the lack of evidence that the razorback sucker exists on the allotment in question, the Fish and Wildlife Service argues that it should be able to issue an Incidental Take Statement based upon prospective harm. While we recognize the importance of a prospective orientation, the regulations mandate a separate procedure for reinitiating consultation if different evidence is later developed:

Reinitiation of formal consultation is required and shall be requested by the Federal agency or by the Service, where discretionary Federal involvement or control over the action has been retained or is authorized by law and:

(a) If the amount or extent of taking specified in the incidental take statement is exceeded;

(b) If new information reveals effects of the action that may affect listed species or critical habitat in a manner or to an extent not previously considered;

(c) If the identified action is subsequently modified in a manner that causes an effect to the listed species or critical habitat that was not considered in the biological opinion; or

(d) If a new species is listed or critical habitat designated that may be affected by the identified action.

50 C.F.R. § 402.16. Additionally, the ESA provides for the designation of critical habitat outside the geographic area currently occupied by the species when "such areas are essential for the conservation of the species." 16 U.S.C. § 1532(5)(A)(ii). Absent this procedure, however, there is no evidence that Congress intended to allow the Fish and Wildlife Service to regulate any parcel of land that is merely capable of supporting a protected species.

The only additional evidence that the Fish and Wildlife Service offers to justify its decision is that "small numbers of the juvenile fish ... likely survived" in an unsuccessful attempt to repopulate the project area between 1981–1987. This speculative evidence, without more, is woefully insufficient to meet the standards imposed by the governing statute. *See* 50 C.F.R. § 402.14(g)(8) ("In formulating its biological opinion ... the Service will use the best scientific and commercial data available. . . ."). Likewise, the Fish and Wildlife Service failed to present evidence that an indirect taking would occur absent the existence of the species on the property. Although habitat modification resulting in actual killing or injury may constitute a taking, the Fish and Wildlife Service has presented only speculative evidence that habitat modification, brought about by livestock grazing, may impact the razorback sucker. The agency has a very low bar to meet, but it must at least attain it. It would be improper to force ACGA to prove that the species does not exist on the permitted area, as the Fish and Wildlife Service urges, both because it would require ACGA to meet the burden statutorily imposed on the agency, and because it would be requiring it to prove a negative.

Based on a careful review of the record, we find that it is arbitrary and capricious to issue an Incidental Take Statement for the razorback sucker when the Fish and Wildlife Service's speculation that the species exists on the property is not supported by the record. We agree with the district court's ruling that the Fish and Wildlife Service failed to establish an incidental taking because it did not have evidence that the razorback sucker even exists anywhere in the area. Where the agency purports to impose conditions on the lawful use of that land without showing that the species exists on it, it acts beyond its authority in violation of 5 U.S.C. § 706.

2.   The Cactus Ferruginous Pygmy-owl

As with the razorback sucker, the record does not support a claim that the cactus ferruginous pygmy-owl exists in the area of the allotment in question, and the Fish and Wildlife Service thus acted in an arbitrary and capricious manner in issuing an Incidental Take Statement for that species. . . .

B.   The ACGA II Consultation

1.   The Issuance of Incidental Take Statements

[For similar reasons, the court held that the Incidental Take Statements issued for four other allotments were arbitrary and capricious as well.]

E.   The Cow Flat Allotment (loach minnow and spikedace)

According to the Biological Opinion, the Blue River passes through or adjacent to approximately 3.5 miles of the Cow Flat Allotment, made up of 22,592 acres in the Apache–Sitgreaves National Forest. Surveys conducted in 1994, 1995, and 1996 found loach minnow throughout the Blue River. The Fish and Wildlife Service concluded that the segment of the Blue River that passes through or adjacent to the Cow Flat Allotment is considered occupied loach minnow habitat.

The loach minnow

Having determined that loach minnow exist on the allotment, Fish and Wildlife Service determined that the loach minnow are vulnerable to direct harms resulting from cattle crossings, such as trampling. Moreover, because the fish use the spaces between large substrates for resting and spawning, sedimentation resulting from grazing in pastures that settles in these spaces can adversely affect loach minnow habitat. The Biological

Opinion determines that this indirect effect, along with the direct crushing of loach minnow eggs and the reduction in food availability, will result in take of the loach minnow. The Incidental Take Statement, however, does not directly quantify the incidental takings of loach minnow and determines that such takings "will be difficult to detect." Defining the incidental take in terms of habitat characteristics, the Fish and Wildlife Service found that take will be exceeded if several conditions are not met. One such condition was if "ecological conditions do not improve under the proposed livestock management" plan.

We agree with the district court that the issuance of the Cow Flat Incidental Take Statement was not arbitrary and capricious. Unlike the other allotments in question, the Fish and Wildlife Service provided evidence that the listed species exist on the land in question and that the cattle have access to the endangered species' habitat. Accordingly, the Fish and Wildlife Service could reasonably conclude that the loach minnow could be harmed when the livestock entered the river. Additionally, the Fish and Wildlife Service provided extensive site-specific information that discussed not only the topography of the relevant allotment, but the indirect effects of grazing on the species due to the topography. The specificity of the Service's data, as well as the articulated causal connections between the activity and the "actual killing or injury" of a protected species distinguishes the Fish and Wildlife Service's treatment of this allotment from the other allotments at issue in the two consultations. Thus, we hold that because the Fish and Wildlife Service articulated a rational connection between harm to the species and the land grazing activities at issue, the issuance of the Incidental Take Statements for the Cow Flat Allotment was not arbitrary and capricious.

2. The Anticipated Take Provisions

We now turn to the question whether the Service acted arbitrarily and capriciously by failing to properly specify the amount of anticipated take in the Incidental Take Statement for the Cow Flat Allotment and by failing to provide a clear standard for determining when the authorized level of take has been exceeded. The district court upheld the Cow Flat take provision, including its conditions on the land use, issued by the Fish and Wildlife Service, finding that it was rationally connected to the proposed action of cattle grazing and thus did not violate the arbitrary and capricious standard.

In general, Incidental Take Statements set forth a "trigger" that, when reached, results in an unacceptable level of incidental take, invalidating the safe harbor provision, and requiring the parties to reinitiate consultation. Ideally, this "trigger" should be a specific number. *See, e.g.,* Mausolf v. Babbitt, 125 F.3d 661 (8th Cir.1997) (snowmobiling activity may take no more than two wolves); Fund for Animals v. Rice, 85 F.3d 535 (11th Cir.1996) (municipal landfill may take fifty-two snakes during construction and an additional two snakes per year thereafter); Mt. Graham Red Squirrel v. Madigan, 954 F.2d 1441 (9th Cir.1992) (telescope construction may take six red squirrels per year); Ctr. for Marine Conservation v.

Brown, 917 F. Supp. 1128 (S.D.Tex.1996) (shrimping operation may take four hawksbill turtles, four leatherback turtles, ten Kemp's ridley turtles, ten green turtles, or 370 loggerhead turtles). Here, however, the "trigger" took the form of several conditions. We must therefore determine whether the linking of the level of permissible take to the conditions set forth in the various Incidental Take Statements was arbitrary and capricious.

ACGA argues that the Incidental Take Statements fail to specify the amount or extent of authorized take with the required degree of exactness. Specifically, ACGA objected to the first condition:

> The service concludes that incidental take of loach minnow from the proposed action will be considered to be exceeded if any of the following conditions are met:

> [Condition 1] Ecological conditions do not improve under the proposed livestock management. Improving conditions can be defined through improvements in watershed, soil condition, trend and condition of rangelands (e.g., vegetative litter, plant vigor, and native species diversity), riparian conditions (e.g., vegetative and geomorphologic: bank, terrace, and flood plain conditions), and stream channel conditions (e.g., channel profile, embeddedness, water temperature, and base flow) within the natural capabilities of the landscape in all pastures on the allotment within the Blue River watershed.

We have never held that a numerical limit is required. Indeed, we have upheld Incidental Take Statements that used a combination of numbers and estimates. *See* Ramsey v. Kantor, 96 F.3d 434, 441 n. 12 (9th Cir.1996) (utilizing both harvesting rates and estimated numbers of fish to reach a permitted take); Southwest Ctr. for Biological Diversity v. U.S. Bureau of Reclamation, 6 F. Supp. 2d 1119 (D.Ariz.1997) (concluding that an Incidental Take Statement that indexes the permissible take to successful completion of the reasonable and prudent measures as well as the terms and conditions is valid); Pac. Northwest Generating Coop. v. Brown, 822 F. Supp. 1479, 1510 (D.Or.1993) (ruling that an Incidental Take Statement that defines the allotted take in percentage terms is valid).

Moreover, while Congress indicated its preference for a numerical value, it anticipated situations in which impact could not be contemplated in terms of a precise number. *See* H.R. Rep. No. 97–567, at 27 (1982) ("The Committee does not intend that the Secretary will, in every instance, interpret the word impact to be a precise number. Where possible, the impact should be specified in terms of a numerical limitation."); *see also* 50 C.F.R. § 402.14 (defining impact as "the amount or extent, of such incidental taking on the species."). In the absence of a specific numerical value, however, the Fish and Wildlife Service must establish that no such numerical value could be practically obtained.

We agree with the *ACGA II* court's conclusion that, "the use of ecological conditions as a surrogate for defining the amount or extent of incidental take is reasonable so long as these conditions are linked to the take of the protected species." Indeed, this finding is consistent with the Fish and Wildlife Service's Section 7 Consultation Handbook:

> When preparing an incidental take statement, a specific number (for some species, expressed as an amount or extent, e.g., all turtle nests not found and moved by the approved relocation technique) or level of disturbance to habitat must be described. Take can be expressed also as a change in habitat characteristics affecting the species (e.g., for an aquatic species, changes in water temperature or chemistry, flows, or sediment loads) where data or information exists which links such changes to the take of the listed species. In some situations, the species itself or the effect on the species may be difficult to detect. However, some detectable measure of effect should be provided ... If a sufficient causal link is demonstrated (i.e., the number of burrows affected or a quantitative loss of cover, food, water quality, or symbionts), then this can establish a measure of the impact on the species or its habitat and provide the yardstick for reinitiation.

Final ESA Section 7 Consultation Handbook, March 1998 at 4–47 to 4–48. By "causal link" we do not mean that the Fish and Wildlife Service must demonstrate a specific number of takings; only that it must establish a link between the activity and the taking of species before setting forth specific conditions.

ACGA argues that it is entitled to more certainty than "vague and undetectable criteria such as changes in a 22,000 acre allotment's 'ecological condition.' " In response, the Fish and Wildlife Service argues that "the [Incidental Take Statement] provides for those studies necessary to provide the quantification of impacts which the Cattle Growers claim is lacking."

We disagree with the government's position. The Incidental Take Statements at issue here do not sufficiently discuss the causal connection between Condition 1 and the taking of the species at issue. Based on the Incidental Take Statement, if "ecological conditions do not improve," takings will occur. This vague analysis, however, cannot be what Congress contemplated when it anticipated that surrogate indices might be used in place of specific numbers. Moreover, whether there has been compliance with this vague directive is within the unfettered discretion of the Fish and Wildlife Service, leaving no method by which the applicant or the action agency can gauge their performance. Finally, Condition 1 leaves ACGA and the United States Forest Service responsible for the general ecological improvement of the approximately 22,000 acres that comprise the Cow Flat Allotment.

Based upon the lack of an articulated, rational connection between Condition 1 and the taking of species, as well as the vagueness of the condition itself, we hold that its implementation was arbitrary and capricious. The terms of an Incidental Take Statement do not operate in a vacuum. To the contrary, they are integral parts of the statutory scheme, determining, among other things, when consultation must be reinitiated.

Thus, even though the Fish and Wildlife Service was not arbitrary and capricious in issuing Incidental Take Statements for the Cow Flat Allotment, its failure to properly specify the amount of anticipated take and to provide a clear standard for determining when the authorized level of take has been exceeded is arbitrary and capricious. As with the Incidental Take Statements for the other allotments, we therefore conclude that the issuance of the Cow Flat Allotment Incidental Take Statement was arbitrary and capricious.

VII. Conclusion

For the foregoing reasons, the decision of the ACGA I district court is AFFIRMED, and the decision of the ACGA II district court is AFFIRMED in part, and REVERSED in part.

Notes and Questions

1. Why did the FWS think that it could use the ESA to condition the grazing of cattle on federal land even though no species would be jeopardized by the grazing? Should it have that power? For a review of the effects of grazing on grassland ecosystems and the law governing the use of federal lands, see chapter 9.

2. Both section 7 and section 9 of the ESA limit the "take" of endangered species. The FWS argued in *Arizona Cattle Growers Association* that "take" should have a broader meaning in section 7 because of the protective nature of that provision. The court disagreed because "[i]nterpreting the statutes in the manner urged by the Fish and Wildlife Service could effectively stop the proposed cattle grazing entirely. Such a broad interpretation would allow the Fish and Wildlife Service to engage in widespread land regulation even where no Section 9 liability could be imposed." *Id.* at 1240. Should the FWS be empowered to conduct such regulation? That is precisely what many landowners dislike about section 9 of the ESA, as detailed in chapter 6. Section 9, moreover, states a firm prohibition, whereas the agencies or ranchers here could ignore the conditions contained in the incidental take statements. But the court dismissed that possibility as merely theoretical because "Biological Opinions exert a 'powerful coercive effect' in shaping the policies of the federal agencies whose actions are at issue." *Id.* (citing *Bennett*, 520 U.S. at 169).

3. What changes in the incidental take statement for the Cow Flat allotment would be sufficiently precise to satisfy the court's concerns?

4. The court cited the agency consultation handbook that FWS prepared with the NMFS to explain the details of how section 7 consultations should proceed. *See* FWS & NMFS, Endangered Species Consultation Handbook: Procedures for Consultation and Conference Activities Under Section 7 of the Endangered Species Act (March 1998), available at <http://endangered.fws.gov/consultations/s7hndbk/toc-glos.pdf>. The handbook was prepared for the scientists and officials working for both agencies, but it is an excellent source of information about the consultation process for all concerned parties (and students!). The handbook highlights four aspects of the consultation process: (1) the use of *sound science* as an "overriding factor" in making all determinations under section 7; (2) the need for *flexibility and innovation*, with biologists encouraged to "be creative in problem solving and look for ways to conserve listed species while still accommodating project goals;" (3) coordination with state agencies and affected tribal governments; and (4) efforts to "streamline consultation processes" and thus *shorten timeframes* "without giving up any protection for listed species/designated critical habitats or the use and review of the best available information." *Id.* at 25–26.

The handbook also offers the following "thoughts . . . as an expression of the philosophy guiding section 7 work":

- The biology comes first. Know the facts; state the case; and provide supporting documentation. Keep in mind the FWS's ecosystem approach to conservation of endangered and threatened species.

- Base the determination of *jeopardy/no jeopardy* on a careful analysis of the best available scientific and commercial data. Never determine the conclusion of a biological opinion before completing the analysis of the best available data.

- Clarity and conciseness are extremely important. They make consultation documents more understandable to everyone. A biological opinion should clearly explain the proposed project, its impacts on the affected species, and the Services' recommendations. It should be written so the general public could trace the path of logic to the biological conclusion and complete enough to withstand the rigors of a legal review.

- Strong interpersonal skills serve section 7 biologists well. Establishing a positive working relationship with action agencies enhances the Services' ability to do the job successfully. Remember, you are trying to assist the agency in meeting their section 7 responsibilities under the Act.

- Present a positive image as a representative of your Service.

- Section 7 consultation is a cooperative process. The Services do not have all the answers. Actively seek the views of the action agency and its designated representatives, and involve them in your opinion preparation, especially in the development of reasonable and prudent alternatives, reasonable and prudent measures, terms and conditions to minimize the impacts of incidental take, and conservation recommendations.

- Use all aspects of section 7, especially opportunities for informal consultation where solutions can be worked out prior to the structured process mandated by formal consultation. Be creative, and make the process work to the species' advantage.

- It is important to be consistent throughout a species' range when implementing section 7. Be flexible but not inconsistent. Study the law, the regulations and this handbook. Know the authorities and be flexible when it is prudent, but always stand firm for maintaining the substantive standards of section 7.

- Take advantage of professional support within and outside the Services. For example, the FWS Division of Engineering can provide valuable technical review of development proposals. Attorneys in the Regional and field offices of the FWS Solicitor/NMFS General Counsel can offer advice on section 7 regulations and the latitude within which to conduct consultation. Similarly, the Services' law enforcement personnel may be able to answer questions about direct or incidental take.

- Strive to solve problems locally.

- An effective section 7 biologist is a good teacher and a good student. Seek every opportunity to teach the section 7 process within and outside the Services in an informative and non-threatening way. Learn all you can about other Services' programs, Federal action agency's mandates and procedures, and State/tribal/private agency's/client's needs and expectations.

Id. at 1–2 to 1–3. Were those suggestions heeded in *Arizona Cattle Growers' Association*? What other suggestions would you offer to the scientists who are responsible for preparing biological opinions and enforcing section 7?

5. For an illustration of the contents of an actual biological opinion, consider the recent opinion that FWS provided to the federal Bureau of Prisons (BOP). In August 2001, BOP initiated consultation with FWS concerning a proposed new federal penitentiary to be built in Tucson. BOP's biological assessment indicated that the prison may affect the endangered Pima pineapple cactus but not the cactus ferruginous pygmy-owl or the lesser long-nosed bat. After consultation between the agencies, FWS provided its biological opinion in March 2002. The opinion contained six sections. First, it described the proposed action to build the minimum security prison on 631 acres in southern Tucson and the conservation measures proposed by BOP to minimize any adverse effects on the Pima pineapple cactus and its habitat. Second, it described the status of the cactus, including its listing as endangered in 1993, a scientific account of the life of the cactus, a positive report on the population of the cactus, the habitat and distribution of the cactus in the Sonoran desert scrub and semidesert grasslands of Arizona, and the effects of urbanization, farming, and exotic species on the survival of the cactus. ("Very little is known," the opinion explains, "regarding the effects of low to moderate levels of livestock grazing on Pima pineapple cactus distribution.") Third, the opinion describes the "environmental baseline": the condition of the land at the site of the proposed prison, the presence of the cactus on certain parts of the site, and the general suitability of the land for the cactus. Fourth, the opinion details the effects of the proposed action, notably the development of 200 acres of cactus habitat, and the BOP's proposal to set aside another 200 acres of land south of the construction zone where nearly 80% of the cactus plants had been found. Fifth, the opinion explained that the cumula-

tive effects of other actions in the area included continuing private development with little or no federal nexus. Based on all of these factors, the
opinion's conclusion stated that "the proposed action is not likely to
jeopardize the continued existence of Pima pineapple cactus." FWS based
that conclusion on the 200 acres that BOP had agreed to set aside as a
permanent conservation area for the cactus, plus BOP's ongoing monitoring of the cactus.

The letter that FWS sent to BOP contained three additional sections.
It said that no incidental take statement is needed. It recommended two
actions to fulfill the conservation duty imposed by ESA section 7(a)(1):
working with the Arizona–Sonora Desert Museum when relocating cactus
to the conservation area, and expanding the conservation area to include an
extra 231 acres that connect to other parcels of cactus habitat. The letter
finished by noting the circumstances in which the BOP would need to
reinitiate consultation under section 7. The entire document is contained in
the letter from David L. Harlow, Field Supervisor, FWS to David J.
Dorworth, Chief, Site Selection and Environmental Review Branch, BOP
(Mar. 18, 2002), *available at* <http://arizonaes.fws.gov/Documents/Biologi-
calOpinions/01101TucsonFederalPrison.pdf>. Other biological opinions can
be found at the web sites for the local offices of the FWS and the national
office of the NMFS.

CHAPTER 6

THE TAKE PROHIBITION

Chapter Outline:
A. Poaching, Smuggling, and Direct Takings of a Listed Species
B. Habitat Destruction and Indirect Taking of Endangered Species
   1. Defining Prohibited Take of Species
   2. The Scope of Liability
C. Limits on the Take Prohibition
   1. The Scope of Federal Jurisdiction
   2. Takes of Species and Takings of Property
D. Authorizing Incidental Take of Protected Species

Despite all of the land that it owns and all of its responsibilities, the efforts of the federal government alone cannot save most endangered species. Most listed species live on at least some privately owned land, and many of the threats to particular species are attributable to the actions of state and local governments and private individuals. The ESA addresses the conduct of all public and private parties in section 9. That provision makes it illegal to "import," "export," "possess," "transport," or "take" a listed species, just to list a few of the verbs contained in the law. ESA § 9(a)(1). In other words, the ESA prohibits the smuggling of a listed fish or wildlife species—dead or alive—into the United States, and it prohibits poaching or killing of such a species. (By contrast, these provisions of section 9 do not apply to plants.)

But section 9 applies to other conduct, too. "Take" is defined to include "to harass, harm, pursue, hunt, shoot, wound, kill, trap, capture, or collect, or to attempt to engage in any such conduct." ESA § 3(18). Those terms, in turn, open the possibility that the unintentional destruction of the habitat of a listed species runs afoul of the law. The extension of the ESA to private developers and landowners whose actions adversely impact the habitat of a species has resulted in most of the controversy surrounding the law in recent years. Private landowners object that while the discovery of gold or oil makes a property owner rich, the discovery of a rare animal or bird on one's property threatens bankruptcy if that blocks the landowner from using the property. Yet the preservation of habitat is essential for the survival of biodiversity, especially given that habitat destruction is the primary cause of the decline of most endangered species.

The ongoing debate about the application of the ESA to these and other scenarios has produced a rich body of legal and scientific literature. The controversies that have developed in the course of the application of section 9 have also led to regulatory changes, proposed legislation, and other ideas regarding the best way to prevent species from going extinct.

The response to these developments says much about the seriousness with which we take the ESA's stated purpose of "provid[ing] a means whereby the ecosystems upon which endangered species and threatened species depend may be conserved." ESA § 2(c)(1).

A.   Poaching, Smuggling and Direct Takings of a Listed Species

You may not shoot an endangered species. Nor may you "harass, harm, pursue, hunt ... wound, kill, trap, capture, or collect" an endangered species. ESA § 3(19). Most of these verbs connote some kind of direct action that targets a particular animal, bird, fish, or other protected species. Whether that is all that those terms connote has been a source of great controversy—answered in the negative by the FWS and the Supreme Court as discussed below at page 246—but there has never been any doubt about the illegality of shooting, hunting, killing or otherwise directly harming an endangered species. The statute also makes it plain that one cannot "possess, sell, deliver, carry, transport, or ship" any illegally taken species, ESA § 9(a)(1)(D); or "deliver, receive, carry, transport, or ship" any species in interstate or foreign commerce, ESA § 9(1)(E); or import, export, or sell any species, ESA § 9(1)(A), (F).

But people do it anyway. Bald eagles are shot, butterflies are collected, and rhinoceros horns are imported in violation of the law. The greatest challenge in most of those instances is to catch the perpetrators. Once caught, the appropriate sanction quickly becomes the crucial issue. Sometimes, though, a defendant accused of illegally taking an endangered species asserts a defense or other privilege that would allow his or her action. Those cases say much about the nature of the prohibition on taking an endangered species and the zeal with which it is applied.

United States v. Clavette

United States Court of Appeals for the Ninth Circuit, 1998.
135 F.3d 1308.

■ Reavley, Circuit Judge:

This is an appeal from the conviction of Paul Clavette for killing a grizzly bear in violation of the Endangered Species Act, 16 U.S.C. §§ 1538(a)(1)(G) and 1540(b)(1). We affirm.

I.   Background

On September 20, 1995, U.S. Fish and Wildlife Service Special Agent Tim Eicher began investigating the killing of a grizzly bear at a campsite southwest of Big Sky, Montana. At the campsite, Eicher discovered two pine trees with a pole suspended by rope between them. This was a "meat pole," used for stringing up and skinning large game animals. Underneath it, Eicher found traces of moose blood and hair, indicating that a moose had recently been dressed there. Eicher found the dead grizzly bear approximately 170 yards away, lying in a large pool of blood. The bear had been shot at least four times. Looking for bullets or spent shell casings, Eicher searched a conical area extending about 25 yards beyond the bear toward the campsite; he found one .7 mm casing by the meat pole and two bullets, one buried about two inches in the dirt at the base of a tree near the bear, and one on the surface of the ground next to the pool of the bear's blood.

Eicher located two bowhunters who had stopped at the campsite on September 17, 1995, to visit with an Oregon man skinning a freshly killed moose. The man seemed to be in a hurry and did not say anything about confronting or killing a grizzly. He did ask the bowhunters what would happen to someone who shot a grizzly bear. The bowhunters told him he had better be prepared to prove it was in self-defense.

Through these bowhunters and Montana hunting license records, Eicher identified the defendant, Paul Clavette, as the man at the campsite on September 17, 1995. Agents of the U.S. Fish and Wildlife Service in Portland, Oregon, obtained and executed a search warrant in defendant's home on November 2, 1995. During the course of that search, and after full *Miranda* warnings, Clavette admitted to killing the grizzly, claiming that it was in self-defense.

After a bench trial, the district court found Clavette guilty of illegally killing a grizzly bear. Clavette was sentenced to three years' probation. Additionally, Clavette was ordered to pay a fine of $2,000 and restitution of $6,250 to the United States Fish & Wildlife Service. . . .

III.   SUFFICIENCY OF THE EVIDENCE

Because Clavette moved to dismiss at the close of the Government's case-in-chief, the sufficiency of the evidence is reviewed *de novo*. If any reasonable person could have found each of the essential elements of the offense charged beyond a reasonable doubt, the evidence is sufficient to convict.

To find Clavette guilty of knowingly taking an endangered species, the Government must prove, beyond a reasonable doubt, that: (1) Clavette knowingly killed a bear; (2) the bear was a grizzly; (3) Clavette had no permit from the United States Fish & Wildlife Service to kill a grizzly bear; and (4) Clavette did not act in self-defense or in the defense of others. Pursuant to the regulations, a grizzly bear may be taken in self-defense or defense of others, but any such taking must be reported within five days to the U.S. Fish and Wildlife Service.[24]

There is no dispute that Clavette knowingly killed a grizzly bear without first obtaining a permit from the Fish & Wildlife Service. The only issue at trial was whether he acted in self-defense or in defense of his wife. Because Clavette presented evidence that he acted in self-defense, the Government must disprove self-defense beyond a reasonable doubt.

---

24.   50 C.F.R. § 17.40(b)(i)(B) (1996).

Clavette and his wife changed their story multiple times. Clavette initially described his trip to Montana to Agent Earl Kisler as follows. Clavette said that as he was skinning a moose he had killed, he sensed something was wrong. He looked up and saw a seven-or eight-foot bear standing on its hind legs about 25 yards away from him, across a creek that ran past the campsite. He made noises to try to drive the bear away and fired a warning shot with his .7 mm rifle. Then, Clavette said, the bear began to circle the campsite, and Clavette was sure it was going to come forward. He told Kisler that he was terrified. Clavette's wife had retreated into the pickup truck. Clavette stated that when the bear was 40 to 75 yards away from the campsite, he shot it. The first shot hit the bear on the left side and appeared to paralyze its hindquarters, but it kept struggling, trying to get up, and so he emptied his rifle into it, reloaded, and fired more rounds into the bear.

Clavette later stated that there were two bears, although his wife still said that there had been one bear. Then, at trial, Clavette's wife testified that not only were there two bears, but that the second one charged her husband at a dead run. When asked why she had not mentioned two bears before, she explained that only one bear was shot. Clavette himself testified at trial that he had in fact told Agent Kisler about the second bear during their first discussion; Clavette surmised that it must have slipped their minds. He also testified that the second bear charged straight at him and that he crippled it with his first shot at 33 yards. Clavette said he saw the bear spin 180 degrees and dig with its front paws, trying to move away from him. The bear looked as if it was paralyzed in its hindquarters but actually ran another hundred yards away from him, without bleeding, so as to die in the spot where the bear was found by the agents.

Although he could not identify the order in which the shots occurred, Keith Aune, a wildlife laboratory supervisor for the Montana Department of Fish, Wildlife and Parks, testified that the shots Clavette described were inconsistent with his own observations and measurements gathered during the necropsy. No entry wounds appeared on the head, chest or front legs, as would be expected if the bear had been approaching at high speed; all the entry wounds were in the rear portion. The stories were also inconsistent with the physical evidence found by Agent Eicher at the site.

Given the physical evidence and the inconsistencies in the Clavettes' stories, a reasonable person could have found beyond a reasonable doubt that Clavette had not killed the bear in self-defense.

Notes and Questions

1. ESA section 11 provides that no civil or criminal penalty "shall be imposed if it can be shown by a preponderance of the evidence that the defendant committed an act based on a good faith belief that he was acting to protect himself or herself, a member of his or her family, or any other individual from bodily harm, from any endangered or threatened species." ESA §§ 11(a)(3), (b)(3), 16 U.S.C. §§ 1540(a)(3), (b)(3). Suppose that Clavette had shot and wounded a grizzly as it charged him, then stood over the bear and fired six more shots into the bear as it lay motionless on the ground. Or suppose that Clavette killed a grizzly that was attacking his dog. Would he succeed in arguing self-defense?

Idaho Representative Helen Chenoweth objected to efforts to expand the range of the grizzly bear for the following reasons:

> The grizzly bear ... is a huge and dangerous animal, and that is a huge and dangerous problem for us. The grizzly bear is, by its nature, a large predatory mammal that, provoked or unprovoked, can move very quickly to viciously attack a human or an animal ..... An adult grizzly can weigh as much as 450 pounds. It can run up to 40 miles an hour over irregular terrain. It has a keen sense of hearing and an even keener sense of smell. The teeth are large and very, very sturdy, especially the canines, and although they are not particularly sharp, the power of the jaw muscles allow them to readily penetrate deep into soft tissues and to fracture facial bones and bones of the hand and forearm with ease.... The resulting trauma is characteristically a result of punctures with sheering, tearing, and crushing force. Claws on the front pads can be as long as human fingers and can produce significant soft tissue damage in a scraping maneuver that results in deep parallel gashes. The bear paw is capable of delivering powerful forces, resulting in significant blunt trauma, particularly to the head and the neck region, the rib cage and the abdomen.
>
> ... [T]he Fish and Wildlife Service is planning for about one human injury that could result in death due to the grizzly every single year.... [N]ot one human death or injury resulting from a grizzly bear attack is acceptable to this Congressman. In fact, it should not be accepted by anyone who values human life.

143 CONG. REC. H5909 (daily ed. July 28, 1997).

2. While self-defense or defense of others is permissible under the ESA, defense of property is not. In Christy v. Hodel, 857 F.2d 1324 (9th Cir.1988), *cert. denied*, 490 U.S. 1114 (1989), a rancher was fined $2,500 for killing a grizzly bear that had attacked his sheep on a nightly basis. The court upheld the fine and rejected the rancher's argument that the government has effectively taken his property without providing just compensation. The Supreme Court declined to review the case over the dissent of Justice White, who stated that "perhaps a government edict barring one from resisting the loss of one's property is the constitutional equivalent of an edict taking such property in the first place." He then suggested that the situation was analogous to "the government decid[ing] (in lieu of the food stamp program) to enact a law barring grocery store owners from 'harassing, harming, or pursuing' people who wish to take food off grocery shelves without paying for it." 490 U.S. at 1115–16. Would it matter if the government had introduced the grizzly into the area in the first place? Or suppose that the government prohibited a landowner from building a fence that would keep the grizzly bear outside his or her property? *Cf.* New York v. Sour Mtn. Realty, 703 N.Y.S.2d 854 (N.Y.Sup.Ct.1999) (applying state law to enjoin a mining company from constructing a fence that would prevent timber rattlesnakes from traveling across the company's property).

3. There are two instances in which a species listed under the ESA may not receive the protection afforded by the take prohibition of section 9. First, the take prohibition does not apply to endangered or threatened plants. Instead, it is illegal to "remove, cut, dig up, or damage or destroy" any listed plant "in knowing violation of any law or regulation of any state or in the course of any violation of a state criminal trespass law." ESA § 9(a)(2)(B). There are no reported cases involving this provision, which makes plants dependent upon state law for their protection from collectors, vandals and others on private and state lands. On federal lands, it is illegal to take possession of any listed plant or to "maliciously damage or destroy any such species." *Id.* It is also illegal to engage in any commerce involving an endangered or threatened plant.

Professor Jeffrey Rachlinski reports that most states do not protect endangered plants, and most of those that do offer only minimal protection. He concludes that "plants that depend on private property for their habitat do not fare well, and they fare much worse in those states that do not restrict private landowners." Jeffrey J. Rachlinski, *Protecting Endangered Plants Without Regulating Private Landowners: The Case of Endangered Plants*, 8 CORNELL J. L. PUB. POL'Y 1, 3 (1998). Why would Congress fail to protect plants in the same way that it protects fish and wildlife species? And why do states offer limited protections, too?

The second instance in which a species is not automatically protected by the take prohibition of section 9 occurs if the species is listed as threatened instead of endangered. Section 9 excludes threatened species from its scope, but section 4 authorizes the Secretary of the Interior to extend those provisions to threatened species or to otherwise "issue such regulations as he deems necessary and advisable to provide for the conservation" of threatened species. ESA § 4(d). In 1978, the Secretary issued regulations that extend the provisions of section 9 to all threatened species. 50 C.F.R. § 17.31(a). Increasingly, though, the FWS and the NMFS are promulgating rules establishing special provisions regarding the application of the take prohibition to threatened species. These rules are designed to minimize the impact on landowners and other parties caused by the take prohibition while still offering substantial protection to the threatened species. For example, the NMFS has lifted the take prohibition of section 9 for certain threatened salmon and steelhead for a party who is otherwise participating in a program that protects the species. Such programs include fishery management activities, hatchery and genetic management programs, habitat restoration activities, properly screened water diversion devices, ongoing scientific research, and certain forest management activities. Final Rule Governing Take of 14 Threatened Salmon and Steelhead Evolutionary Significant Units (ESUs), 65 Fed. Reg. 42422 (2000).

United States v. Billie

United States District Court for the Southern District of Florida, 1987.
667 F.Supp. 1485.

■ PAINE, DISTRICT JUDGE.

On April 14, 1987, James Billie was charged in a two count information with the taking and subsequent possession, carrying, and transportation of a Florida panther, in violation of the Endangered Species Act, 16 U.S.C. § 1531 *et seq.* (1982) (the Act); *see id.* §§ 1538(a)(1)(B), 1538(a)(1)(D), 1540 (b)(1). The *felis concolor coryi* or Florida panther is a particular subspecies of panther listed as "endangered" pursuant to the Act. The defendant is a member and chairman of the Seminole Indian Tribe, which has approximately 1,700 enrolled members. All of the acts complained of in the information were committed in December 1983 on the Big Cypress Seminole Indian Reservation in the Southern District of Florida.

APPLICABILITY OF ENDANGERED SPECIES ACT TO SEMINOLE INDIAN RESERVATIONS

Billie first moves to dismiss the information on the ground the Act does not apply to non-commercial hunting on the Seminole Indian Reservations. He argues that the Act evinces no Congressional intent to abrogate or modify his traditional right to hunt and fish on the reservation. The Government disagrees, maintaining that the Act is a reasonable, necessary, and nondiscriminatory conservation statute which has limited Indian rights to take or possess species to the extent those rights are inconsistent with the Act. In United States v. Dion, 476 U.S. 734 (1986), the Supreme Court expressly left unresolved the question whether the Act abrogates Indian hunting rights. . . .

A.   *The Endangered Species Act*

The Supreme Court has described the Endangered Species Act as "the most comprehensive legislation for the preservation of endangered species ever enacted by any nation." Tennessee Valley Authority v. Hill, 437 U.S. 153, 180 (1978). . . . The Florida panther, whose historic range is listed as in the United States from Louisiana and Arizona east to South Carolina and Florida, has been listed as endangered since 1967. 50 C.F.R. § 17.11 (1986).

The Act's prohibitions are set forth in 16 U.S.C. § 1538. Included within these prohibitions are the taking of any endangered species within the United States, the possession of any illegally taken endangered species, and the sale or offer for sale of any endangered species in interstate or foreign commerce. Civil and criminal penalties may be imposed for violations of the Act. *Id.* § 1540.

Congress has drawn several extraordinarily narrow exceptions to the Act's prohibitions. Indians, Aleuts, or Eskimos who are Alaskan Natives residing in Alaska and, in some circumstances other non-native permanent residents of Alaskan native villages, may take endangered or threatened species, but only if the taking is primarily for subsistence purposes and only subject to such regulations as the Secretary may issue upon his determination that such takings materially and negatively affect the species. *Id.* § 1539(e). In addition, the Secretary may permit otherwise prohibited acts for scientific purposes, to enhance the propagation or survival of the affected species, or when the taking is incidental to carrying out an otherwise lawful activity. Such permits may be issued only on the basis of stringent statutory procedures designed to assure that any adverse impact on the particular species will be minimized. *Id.* § 1539(a).

B.   *Hunting Rights on the Seminole Indian Reservations*

The Seminole Indian Reservations were established pursuant to an Executive Order by which certain lands were "set aside as a reservation for the Seminole Indians in southern Florida."[2] Although the Executive Order does not expressly mention hunting and fishing rights, those rights were included by implication in the setting aside of the lands as an Indian reservation. The Seminoles' rights to hunt and fish are part of their larger rights of possession.

Although the Congress is empowered to abrogate Indian rights, its intent to do so must be clear and plain.... Billie maintains that the Act and its legislative history lack the evidence of congressional intent necessary to abrogate his hunting rights.

C.   *The Scope of the Right*

Before the court can determine whether the Seminoles' rights have been abrogated, however, it must assess the scope of those rights. As a general rule, treaties with the Indians should be interpreted as the Indians themselves would have understood them. The Supreme Court has stated that Indian treaties "cannot be interpreted in isolation but must be read in light of the common notions of the day and the assumptions of those who drafted them." Oliphant v. Suquamish Indian Tribe, 435 U.S. 191, 206 (1978).

When the Seminole reservations were set aside in 1911, the Florida panther was not endangered. The court has received no evidence showing that in 1911 either the Seminoles or the United States imagined that the Florida panther would be nearly extinct in 1987. Given this historical setting, and as the Government aptly points out in its memorandum, it is inconceivable that the Seminoles would have demanded, and the United States would have conceded, a right to hunt on the lands in question free from regulation by the federal sovereign.

The Supreme Court has confirmed that Indian treaty rights do not extend to the point of extinction.... Indian rights to hunt and fish are not absolute. Where conservation measures are necessary to protect endangered wildlife, the Government can intervene on behalf of other federal interests. The migratory nature of the Florida panther gives Indians, the states, and the federal Government a common interest in the preservation of the species. Where the actions of one group can frustrate the others' efforts at conservation, reasonable, nondiscriminatory measures may be required to ensure the species' continued existence.

---

2. During the nineteenth century, the United States entered into numerous treaties with the Seminoles in attempts to end the Seminole Indian Wars, to convince the Seminoles to settle west of the Mississippi, and to settle land disputes between the Seminole and other tribes that did move west. Not all Seminoles moved west, however. In 1911, President Taft signed an Executive Order No. 1379, creating a reservation for those Seminoles remaining in Florida.

The Endangered Species Act is such a measure. Its general comprehensiveness, its nonexclusion of Indians, and the limited exceptions for certain Alaskan natives, demonstrate that Congress considered Indian interests, balanced them against conservation needs, and defined the extent to which Indians would be permitted to take protected wildlife.... The narrow Alaskan exception, the inclusion of Indians within the Act's definition of "person," the Act's general comprehensiveness, and the evidence that the House committee desired to prohibit Indians from hunting and fishing protected species all provide "clear evidence that Congress actually considered the conflict between its intended action on the one hand and Indian treaty rights on the other, and chose to resolve that conflict by abrogating" the Indian rights. *Dion*, 106 S. Ct. at 2220.

In summary, this court's conclusion that the Endangered Species Act applies to hunting by Indians on the Seminole reservations is based on both the character of their hunting rights and on the Act's abrogation of those rights. On-reservation hunting rights are not absolute when a species such as the Florida panther is in danger of extinction. To the extent that evidence of congressional abrogation of these rights is required, that standard has been met. When Congress passed "the most comprehensive legislation for the preservation of endangered species ever enacted by any nation," *Tennessee Valley Authority*, 437 U.S. at 180, and empowered the Secretary of the Interior to classify a species as "in danger of extinction," 16 U.S.C. § 1532(6), it could not have intended that the Indians would have the unfettered right to kill the last handful of Florida panthers....

RELIGIOUS FREEDOM

Billie last contends that the possession charge violates his right to freedom of religion under the First Amendment. He argues that the Act is overbroad either on its face or as applied to him because, unlike the Bald Eagle Protection Act, 16 U.S.C. § 668 *et seq.* (1982), it does not authorize the Secretary to permit the possession of the species for Indian religious purposes. Thus, according to the defendant, the Act is invalid because it sweeps within its ambit his constitutionally-protected practices.

A.   Facial Overbreadth

A statute is overbroad on its face if it is unconstitutional in every application or if it seeks to prohibit such a broad range of protected conduct that it is unconstitutionally overbroad. The mere fact that some impermissible applications of a statute are conceivable, however, does not render it overbroad.... The Endangered Species Act represents the legislature's attempt to achieve the laudable goal of protecting this country's vulnerable wildlife. The possible constitutional applications of the Act are too numerous to conceive. The possibility that it might be unconstitutionally applied to certain religious practices does not render it void on its face where the remainder of the statute covers a whole range of easily identifiable and constitutionally proscribable conduct. Billie, therefore, must demonstrate that the Act is unconstitutional as applied to him.

B. *Overbreadth as Applied*

Not all burdens on religion are unconstitutional. The Free Exercise Clause "embraces two concepts,—freedom to believe and freedom to act. The first is absolute but, in the nature of things, the second cannot be. Conduct remains subject to regulation for the protection of society." Cantwell v. Connecticut, 310 U.S. 296, 303–04 (1940). . . . Before the court balances competing governmental and religious interests, the governmental action must pass two threshold tests. First, the law must regulate conduct rather than belief. Second, the law must have both a secular purpose and a secular effect. The Endangered Species Act passes both of these tests. The Act regulates conduct, not belief, and is facially neutral in its application. In purpose and effect, the Act protects endangered and threatened wildlife.

The court next "faces the difficult task of balancing governmental interests against the impugned religious interest." The stated purposes of the Act are "to provide a means whereby the ecosystems upon which endangered species and threatened species depend may be conserved" and "to provide a program for the conservation of such endangered species and threatened species." 16 U.S.C. 1531(b). Congress recognized that the "two major causes of extinction" are hunting and destruction of natural habitat. S. Rep. No. 93–307, 93d Cong., 1st Sess. 300, *reprinted in* 1973 U.S. Code Cong. & Admin. News 2989, 2990. In its judgment, the need for strong endangered species efforts was more than aesthetic. It found that many species facing extinction "perform vital biological services to maintain a balance of nature within their environments" and that biological diversity among species was essential for scientific purposes.

The Florida panther was one of the first species protected under the 1966 Endangered Species Preservation Act, the predecessor to the present Endangered Species Act, when it was listed as threatened with extinction. At the evidentiary hearing, the Government presented testimony that population levels are extremely critical. David Wesley, field supervisor of the Jacksonville Office of Endangered Species of the United States Fish and Wildlife Service, estimated that approximately twenty to fifty Florida panthers remain in the wild. A 1985 Interior Department memorandum discussed the species which would probably be most affected were Indians allowed to take endangered and threatened species for religious purposes. The Florida panther was listed as one of those species. The memorandum stated that, although the *coryi* was listed in 1967 as endangered in Louisiana and Arkansas east to South Carolina and Florida, the species now appears to be confined to southern Florida. . . .

The evidence presented by the Government, considered along with the Endangered Species Act and its legislative history, establishes that the governmental interest in protecting the Florida panther on the Seminole Indian Reservations is compelling. Considering the small number of remaining *coryi* and their regular presence on Indian land in South Florida, the cost to the Government of altering its conservation efforts would also be substantial. The governmental interest presented in this case would be substantially harmed by a decision allowing Indians to hunt and possess *coryi* free from regulation on Indian reservations....

According to Seminole tradition, the panther was the first choice of the creator to enter the earth. In Miccosukee, the panther is called *cowachobee*. Buffalo Jim, a Seminole medicine man, testified that panther claws are good for cramps and different ailments. Jim Shore, general counsel to the Seminole tribal council, testified that panther parts are an important part of a medicine man's bundle. It is commonly known that panther claws and tails are used for different ailments, and a medicine man should have them on hand in case they are needed to minister to a particular illness. Sonny Billie, a Seminole and Miccosukee medicine man, testified that the panther is an "important" animal for a medicine man, because they are difficult to find, and a person has to be lucky to kill one. Sonny Billie also stated that panther claws provide "number one" relief from muscle cramps, which cannot be treated as well without the panther part.

The defendant testified that he was initiated into the practice of Seminole medicine in 1983. Although he can practice medicine, he considers himself a "beginner" and will not have the title of medicine man until he has done many deeds or perhaps achieved his first successful operation or healing. The defendant testified that he had no thoughts regarding what he would do with the panther carcass after he shot it until the morning on which it was seized. At that time, it occurred to Billie that he could give it as a gift to a medicine man in order to humble himself. In bestowing such a high honor, Billie hoped that he would be able to learn more medicine.

... Billie has not adequately shown that the possession of panther parts is regular and material to an important religious ceremony or ritual. Although there was testimony that panther parts are important to healing, after having viewed the witnesses, the court is not convinced that panther parts are critical or essential. Sonny Billie stated that panther parts are preferable. They do not appear to be indispensable, however. Furthermore, no evidence was submitted that the *coryi* is the only kind of panther found in Florida today. This lack of evidence gives rise to the inference that other subspecies of panther may be available for Indian religious use and demonstrates Billie's failure to show the gravity of the cost to his religious interest imposed by the Government activity.

Considering the foregoing, the court finds that the evidence has not adequately shown that Billie's religious interest in possessing panther parts[7] should outweigh the compelling governmental interest in protecting the Florida panther.

7. For the first time in his reply memorandum, Billie raises the argument that outlawing possession of *coryi* parts without an exception for Indian religious use if the possession has not been involved in the taking does not serve the compelling governmental purpose of prohibiting the killing of the species. The court is not persuaded. In drafting the Endangered Species Act, Congress made possession, carrying, or transportation of protected species a crime in itself. This demonstrates that Congress believed that preventing protected species from being killed is not the only means to rescue it from endangerment. The court also notes Sonny Billie's testimony that the parts of a panther found

[The court also rejected Billie's claim that he did not "knowingly" take an endangered species, holding that the government did not have to prove that Billie knew that the panther that he shot belonged to the protected subspecies, nor did the government have to prove that Billie knew that it was illegal to kill an endangered species on an Indian reservation.]

Notes and Questions

1. James Billie gained national attention for killing—and then barbequing—a Florida panther in 1983. He reported that the panther tasted "like a cross between a manatee and bald eagle." Efforts to prosecute him failed despite the federal district court's denial of his motion to dismiss the claims in the decision reprinted above. A federal prosecution ended in a mistrial, a state prosecution resulted in an acquittal because the jury questioned whether than animal was a pure Florida panther, and then federal authorities dropped their charges. Billie asked the government for the return of the panther hide and skull which had been used as evidence, but that request was denied because the ESA prohibits the possession of such items. Six years later, Billie was fined $2,000, placed on two years probation, and banned from hunting outside of Seminole tribal reservation after he illegally shot elk in an Idaho national forest.

Billie has been described as "arguably the leading pioneer of Indian gaming in America." Sean Rowe, *Big Chief Moneybags: Part CEO, Part Shaman, Seminole Leader James Billie Has His Tribe Charging Toward Economic Independence*, MIAMI NEW TIMES, Mar. 26, 1998. Billie also operates the tribe's Billie Swamp Safari, where three Florida panthers are on display, and where he lost a finger while wrestling with an alligator. *See* <http://www.seminoletribe.com/safari/>. He served as the Seminole tribal chairman until he was forced out in 2001 amidst charges of financial mismanagement and sexual harassment. *See* Tanya Weingberg, *After 22 Years, Seminole Leader's World Crumbling*, SUN-SENTINEL (Ft. Lauderdale, Fla.), June 10, 2001, at 1B.

2. As discussed in chapter 2, the relationship between Native Americans and the preservation of biodiversity presents a number of difficult problems. Many Native Americans rely upon certain animals, birds and plants for their very subsistence, including food, clothing and shelter. Also, the religious traditions of many tribes rely upon particular animals. Some federal statutes—including the Bald Eagle Act—provide some exceptions from their provisions in recognition of the special needs of Native Americans, but the ESA and other statutes do not. Thus a number of conflicts have occurred between the dictates of the law and the historic practices of tribes and their members. *See generally* MICHAEL J. BEAN & MELANIE J.

dead would not be good enough for use by a medicine man, because there would have to be something wrong with that panther. This testimony tends to show a link between hunting *coryi* and possessing their parts. Congress foresaw such a connection when it drafted the Endangered Species Act.

ROWLAND, THE EVOLUTION OF NATIONAL WILDLIFE LAW 449–67 (3d ed. 1997) (chapter on wildlife and Native Americans).

The ESA does contain an exception for Eskimos and other Alaskan natives who take an endangered or threatened species "primarily for subsistence purposes." ESA § 10(e)(1), 16 U.S.C. § 1539(e)(1). The Secretary of the Interior can regulate the time during which such takings take place if the species are "materially and negatively affect[ed]" by the actions of the Alaskan natives. ESA § 10(e)(4), 16 U.S.C. § 1539(e)(4). Native Hawaiians do not enjoy such treatment, and an equal protection objection to the ESA for failing to extend the exception to Native Hawaiians was unsuccessful. *See* United States v. Nuesca, 945 F.2d 254 (9th Cir.1991).

Should cultural differences result in special treatment under the ESA? Which differences matter? In United States v. Tomono, 143 F.3d 1401 (11th Cir.1998), a Japanese operator of a commercial reptile import and export business was convicted for bringing sixty turtles into the United States in violation of the Lacey Act. The district court reduced his sentence because he did not realize the seriousness of the offense under U.S. law and because of the special role that the turtles play in Japan, where they are quite common. The court of appeals reversed 2–1, concluding that the circumstances of the case were not sufficiently unusual to justify a reduction on the sentence.

3. The Florida panther plays a prominent role in Seminole medicinal practices. Other endangered species are used in traditional Asian medicine. For example, Chinese medicine uses tiger bones (for arthritis and rheumatism), rhino horns (for fevers), and bear gall bladders. Chinese pharmaceutical factories use 1,400 pounds of rhino horns annually, the product of about 650 rhinos. The ESA does not provide an exception for the use of listed (or the parts of listed species) in medicinal practices, but such practices persist, even in the United States. The Fish & Wildlife Service placed advertisements in the Asian language media in Los Angeles that encouraged the use of alternatives to endangered species in Chinese medicine after noting that six of nine Asian product stores visited in the city carried rhino and tiger-based products. *See* Peter Y. Hong, *Remedy to Extinction; Education Effort Targets Use of Tiger, Rhino Parts in Asian Medicines*, L.A. TIMES, Oct. 20, 1995, at B3. Whether such efforts will succeed remains unclear. A 1998 report concluded that "both Hong Kong Chinese and Chinese–Americans have little knowledge of the ingredients in the [traditional Chinese medicines (TCM)] they use, and little interest in obtaining that knowledge before using such products." SAMUEL LEE ET AL., A WORLD APART? ATTITUDES TOWARD TRADITIONAL CHINESE MEDICINE AND ENDANGERED SPECIES IN HONG KONG AND THE UNITED STATES 8 (1998), *available at* <http://www.traffic.org/tcm/ChineseMedicine.pdf>.

4. Like most other courts, *Billie* rejected the claim that application of the ESA violated the rights of Native Americans to practice their religion. Tribal reliance upon the first amendment's free exercise clause is unlikely to work given the Supreme Court's decision in Employment Division v. Smith, 494 U.S. 872 (1990), that the application of a neutral and generally applicable statute does not violate the first amendment. Congress enacted the Religious Freedom Restoration Act (RFRA) to expand the scope of rights to exercise religion, but most Native American claims that the ESA infringes on the rights stated by RFRA have failed, too. *See* United States v. Sandia, 188 F.3d 1215, 1218 (10th Cir.1999) (explaining that "a defendant may not claim First Amendment or RFRA protection for the taking and possession of a protected bird when he subsequently sells it for pure commercial gain"); United States v. Lundquist, 932 F.Supp. 1237 (D.Or. 1996) (holding that RFRA did not provide a defense to a Native American charged with possessing bald eagle feathers); United States v. Jim, 888 F.Supp. 1058 (D.Or.1995) (denying a RFRA challenge to the conviction of a Native American who killed twelve bald eagles); *but see* United States v. Gonzales, 957 F.Supp. 1225 (D.N.M.1997) (upholding a RFRA challenge to the criminal prosecution of a Native American who shot a bald eagle for use in a religious ceremony). RFRA no longer limits the actions of states, *see* City of Boerne v. Flores, 521 U.S. 507 (1997), but the applicability of the statute to *federal* government actions may well remain intact. *See* In re Young, 141 F.3d 854 (8th Cir.1998).

United States v. Winnie

United States Court of Appeals for the Seventh Circuit, 1996.
97 F.3d 975.

■ Terrence T. Evans, Circuit Judge.

Gail Winnie of Harshaw, Wisconsin, was a member of a party that went on a monthlong hunting safari in Africa in 1981. During the safari a cheetah was shot and killed. The cheetah was imported into the United States and found its way into Winnie's home where its skin and skull were mounted and displayed on a basement wall. Eleven years later, in 1992, federal and state wildlife agents descended on Winnie and seized the mounted cheetah, which they claimed he possessed in violation of law.

Three years after the seizure Winnie was charged with a federal misdemeanor—unlawfully possessing a cheetah (it's a good thing its common name is so simple, for a cheetah's formal name is Acinonyx jubatus) traded in contravention of the Convention in International Trade in Endangered Species of Wild Fauna and Flora as prohibited by the Endangered Species Act, 16 U.S.C. § 1538(c)(1).

Winnie possessed the cheetah (at least what was left of it after it was shot) continuously from 1981. But because all elements of the offense were present in 1981, Winnie says the government waited too long to file its charge. It had to act, he said, by 1986. Because he was not charged within five years, Winnie claimed the prosecution was too stale to proceed. Nonsense, said the government. The possession of an illegally imported endangered species is a continuing offense which only stops when the possession stops. Winnie's possession ended in 1992, says the government, so the charge here was filed with plenty of time to spare. The district court bought the government's view, and when Winnie's motion to dismiss came up short he entered a guilty plea to the charge (his penalty; six months probation and a $500 fine), preserving his right to argue his statute of limitations defense on this appeal.

In criminal law, the purpose of a statute of limitations is to limit prosecutions to a fixed period of time after the commission of an offense. The limitation protects individuals from having to defend against stale charges. It also encourages law enforcement officials to move promptly when investigating suspected criminal activity. The "continuing offense" doctrine, which enlarges the permissible time period for bringing criminal charges, is, therefore, applied in only limited circumstances....

We think Winnie's defense must fail without even considering the continuing offense doctrine. The only relevant facts necessary to this appeal are not in dispute. Winnie admits continued possession of the cheetah from 1981 to 1992. His contention that the offense was "committed" in 1981 when he first took possession of the animal is contrary to the plain language of the statute, part of which makes it a crime "to possess" protected wildlife. Winnie's analysis would require a conclusion that the crime defined by Congress was "to take possession of" illegally traded wildlife, which Winnie did in 1981, rather than "to possess" wildlife, which he did through 1992. Congress did not define the crime that way. The statute of limitations thus did not begin to run until Winnie ceased possessing the cheetah. It was only then that he stopped violating the law.

The cheetah was contraband, just like heroin, and the passage of time never made its possession legal. Otherwise, someone like Winnie could hide a cheetah hide for five years and then display it (or even wear it) with impunity. That scenario was not what Congress had in mind when it prohibited the possession of endangered species. So we need not venture into the thicket of the continuing offense doctrine. Winnie was violating the law on the day the cheetah was seized, and the judgment of the district court is affirmed.

■ RIPPLE, CIRCUIT JUDGE.

I concur in the result.

Notes and Questions

1. The statute of limitations issue arises, of course, when an item made from protected wildlife is discovered many years after it was obtained. How do you suppose that federal and state authorities learned of the cheetah eleven years after the animal was shot? Why did Congress criminalize possession of the remains of a dead animal? How does the wrongfulness of possessing a cheetah skin compare to possessing drugs or weapons?

2. ESA section 9(a)(e) makes it illegal to "possess, receive, carry, transport, or ship in interstate or foreign commerce, by any means whatsoever and in the course of a commercial activity" any species that is taken illegally. The law, however, allows the Secretary of the Interior to permit the possession of a species in certain circumstances. They include the

progeny of listed raptors that were held prior to the 1978, articles that are more than 100 years old that were made from a listed species, the possession of a species that was obtained before it was listed, and the possession of a species within one year after it was listed where the owner would suffer undue economic hardship from the application of the ESA's prohibitions. *See* ESA § 10(b), (e), (f), & (h); LITTELL, *supra*, at 65–70.

More commonly, a permit may be obtained to take or possess a species "for scientific purposes or to enhance the propagation or survival of the affected species." ESA § 10(a)(1)(A). The FWS considers a number of factors when evaluating a scientific permit request, including the reason for taking a species from the wild, the impact on the existing wild populations, and the likelihood that the action would reduce the threat to the species. *See* 50 C.F.R. § 17.22(a)(2). This provision enables zoos to obtain and keep a listed species with the government's permission, provided that the zoo is working for the survival of the species. Zoos have become crucial to efforts to save endangered species because they facilitate research into a species and because they work to propagate the species in captivity so that it can be returned to the wild. For example, the Toledo Zoo leads the effort to save the Virgin Islands Tree Boa—discussed above at page 189—through a captive breeding program that has generated over one hundred new snakes. For a description of the work of zoos and aquariums in supporting the captive breeding of endangered species ranging from the Kanab ambersnail to the razorback sucker, *see* Mike Demlong, *Beyond Captive Propagation*, 24 ENDANGERED SPECIES BULLETIN 22 (1999), *available at* <http://endangered.fws.gov/ESB/99/05–06/22–24.pdf>.

3.   A knowing violation of the ESA can also result in a $50,000 fine and one year's imprisonment. *See* ESA § 11(b)(1). The law further provides for the forfeiture of any items illegally possessed or imported—as illustrated by the following case—and the revocation of various hunting, fishing and other permits. But the actual sanctions in most endangered species cases are far more modest. Winnie was fined $500 and placed on six months probation for illegally possessing the cheetah skin. Is that an appropriate penalty? It is the same fine that was imposed on Adriano Teobadlelli, an Italian tourist who was caught with over 250 rare butterflies that he had collected in several western national parks. Another butterfly poacher was fined $3,000 and given three years probation for his role in illegally capturing 2,200 butterflies—some of which were endangered species—after a judge rejected a plea agreement that would have required the defendant to serve time in prison.

United States v. One Handbag of Crocodilus Species

United States District Court for the Eastern District of New York, 1994.
856 F.Supp. 128.

■ HURLEY, DISTRICT JUDGE

[The United States brought a forfeiture action to obtain 55 handbags and two belts made from the yacare—an endangered subspecies of crocodile—and other listed crocodiles and reptiles. The government alleged that the goods were imported into the United States by J.S. Suarez, Inc. in violation of section 11(e)(4)(A) of the ESA. That provision states that ''all fish or wildlife or plants taken, possessed, sold, purchased, offered or sale or purchase, transported, delivered, received, carried, shipped, exported, or imported contrary to the provisions of [the ESA], any regulation made pursuant thereto, or any permit or certificate issued hereunder shall be subject to forfeiture to the United States.'' 16 U.S.C. § 1540(e)(4)(A).]

The claimant maintains that the forfeiture of his merchandise for allegedly containing yacare is a denial of due process, because that subspecies of caiman is not identifiable with reasonable certainty. The Court has already rejected, at least in part, the premises for the present argument by concluding that an expert is capable of identifying products made of yacare with reasonable certainty. However, there are apparently a limited number of such experts, and it may well be that businesspersons in the trade rarely possess the requisite expertise.

Presumably this problem is not limited to commerce in certain caiman skins. Difficulty in identification of many endangered species is to be expected due to the rarity of the product involved, coupled with the fact that trade in the product is prohibited. Such being the case, it appears that few persons would have had an opportunity to gain familiarity with the animal. The situation is further complicated when the endangered wildlife is, as in the present case, but one of a number of subspecies which share similar physical characteristics, so that only experts can draw critical distinctions.

Yacare caimens along the Paraguay River in southern Brazil

Does such difficulty in identification render enforcement of the Endangered Species Act, and concomitant forfeiture actions, violative of the due process clause of the 5th amendment? In this Court's view, the answer to that question is "no." Initially, it should be noted that if the claimant's argument was accepted, it would be virtually impossible to protect endangered species; paradoxically, the degree of difficulty typically would escalate as the number of a protected species in existence declined.

Moreover, and by way of analogy, in the criminal law there are certain crimes that are *malum prohibitum*. In such instances, society has decided that the legislative goal sought to be realized outweighs the unfairness that might befall an individual who unintentionally violates the law. Similarly, those who traffic in products involving endangered species run the risk of having the product forfeited, even if compliance with the law is sometimes difficult because of identification problems. Whether the rationale for this result is based upon the fact that the target of the action is the *res*, and not the person having an interest in the *res*, or whether it is based upon a balancing of the competing interests involved, the result is the same.

In sum, this Court rejects the argument that the identification difficulties associated with yacare render the statutory scheme to protect such animals unconstitutional as violative of due process. . . .

Claimant argues that he is an "innocent owner," thereby insulating his merchandise from forfeiture. Having reviewed the respective positions of the parties, however, as well as the relevant case law, the Court concludes that a good faith defense is not available in a forfeiture proceeding based on violations of the Endangered Species Act. As one court has persuasively explained,

> [T]he application of strict liability in wildlife forfeiture actions is necessary to effect Congressional intent. To permit an importer to recover the property because he or she lacks culpability would lend support to the continued commercial traffic of the forbidden wildlife. Additionally, a foreseeable consequence would be to discourage diligent inquiry by the importer, allowing him or her to plead ignorance in the face of an import violation. Furthermore, it is not unreasonable to expect the importer to protect his or her interest by placing the risk of non-compliance on the supplier in negotiating the sales agreement.

United States v. 1,000 Raw Skins of Caiman Crocodilus Yacare, No. CV–88–3476, 1991 WL 41774, at *4 (E.D.N.Y. Mar.14, 1991); *see also* United States v. 2,507 Live Canary Winged Parakeets, 689 F. Supp. 1106, 1117 (S.D.Fla. 1988) ("The Court is of the opinion that the defense of 'innocent owner' is not available in actions under the Lacey Act. . . . The Act provides for forfeiture of the fish, wildlife and plants on a *strict liability basis*, because the merchandise is, in effect, contraband.") (emphasis in original); United States v. Proceeds From The Sale of Approximately 15,538 Panulirus Argus Lobster Tails, 834 F. Supp. 385 (S.D.Fla.1993). *But see* United States v. 3,210 Crusted Sides of Caiman Crocodilus Yacare, 636 F. Supp. 1281, 1286–87 (S.D.Fla.1986) (court assumed, without analysis, that an innocent owner defense was available in wildlife forfeiture proceeding, but held that defense was unavailable to claimant in that case); Carpenter v. Andrus, 485 F. Supp. 320 (D. Del. 1980) (where owner did not intend to ship leopard hide to United States, but shipping agent accidentally brought the hide into the country, court held that owner was wholly innocent of wrongdoing and denied government's forfeiture application).

Moreover, the Court notes that, as a forfeiture proceeding under the Endangered Species Act, the present case is not one in which some type of nexus has to be established between the property subject to forfeiture and the conduct sought to be controlled, such as those instances in which the government seeks to have currency forfeited as traceable to illegal trafficking in narcotics. Here, the two items are the same: the endangered species is the contraband. Under such circumstances, and given the language of the relevant statutes and the rationale underlying their enactment, a "good faith" or "innocent purchaser" defense is not available in the present case.

Even if, *arguendo*, a claimant's non-culpability is a defense, it should be noted that the Court has significant reservations about Mr. Suarez's innocence. On several prior occasions, he and the same supplier imported products into the United States, including yacare, in violation of the Endangered Species Act. Given the previous illegality, claimant may not close his eyes to the likelihood of a repeat occurrence. Simply relying on that supplier for compliance with the law, as Mr. Suarez testified he did, is not sufficient to trigger a good faith defense, assuming that such a defense exists in this type of forfeiture proceeding.

Notes and Questions

1. How can someone tell if the crocodile handbag that they are buying was made from a protected subspecies like the yacare or from an unprotected subspecies? It is even more difficult to know whether other products are produced from endangered species. That, however, is no defense under the law. *See* Delbay Pharmaceuticals v. Department of Commerce, 409 F.Supp. 637 (D.D.C.1976) (prohibiting the sale of a prescription drug made from a substance derived from the endangered sperm whale).

2. Is forfeiture an appropriate remedy in endangered species cases? Suppose that the government sought forfeiture from a woman who bought a yacare handbag from J.S. Suarez? Note that "[u]pon returning to the United States, unwary tourists are ... forced to surrender their foreign purchases of articles like fur coats and hats, leather pocketbooks, rugs, ivory carvings, or religious statutes." LITTELL, *supra*, at 42 (citing administrative cases).

3. In 2000, the FWS reclassified the Yacare caiman as threatened. At the same time, the FWS issued a special rule permitting strictly regulated trade in the caiman in order "to promote the conservation of the yacare caiman by ensuring proper management of the commercially harvested caiman species in the range countries and, through implementation of trade controls ... to reduce commingling of caiman specimens." Reclassification of Yacare Caiman in South America From Endangered to Threatened, and the Listing of Two Other Caiman Species as Threatened by Reason of Similarity of Appearance, 65 Fed. Reg. 25867 (2000). Why did the FWS allow any trade in the species?

4. Smuggling remains a great threat to many endangered species. Assistant Attorney General Lois Schiffer described the government's successful prosecution of a smuggling ring in Madagascar that used local residents to collect Madagascan tree boas and rare radiated and spider tortoises (each of which is listed under the ESA) and then transported the animals to the United States where reptile dealers could sell them at a 10,000% mark-up on the price paid to collectors. Lois J. Schiffer & Timothy J. Dowling, *Government Works! Case Studies in Environmental Protection*, 32 CREIGHTON L. REV. 781, 784 (1999). The World Wildlife Fund's "Traffic North America" newsletter provides an update on smuggling activities, including a recent report on several cases involving the seizure of large amounts of sea turtle eggs that were smuggled into the United States. *Illegal Egg Trade Threatens Sea Turtles*, 2 TRAFFIC NORTH AMERICA 1 (1999), *available at* <http://www.worldwildlife.org/news/pubs/traffic_sep99.pdf>. Such imports are illegal under the federal Lacey Act and the international Convention on International Trade in Endangered Species of Wild Fauna and Flora (CITES), as well as the ESA. Additional materials on the threat that smuggling and other illegal activities pose to endangered species around the world is contained in chapter 14.

B. HABITAT DESTRUCTION AND INDIRECT TAKING OF ENDANGERED SPECIES

Half of the species listed under the ESA have at least 80% of their habitat on private lands. As detailed in chapter 2, habitat destruction presents the greatest threat to the survival of species in the United States and throughout the world. The drafters of the ESA recognized as much when they stated that the purpose of the law is "to provide a means whereby the ecosystems upon which endangered species and threatened species depend may be conserved." ESA § 2(b). Yet the ESA does not contain a provision that protects all of the habitats of all listed species from destruction or other harms. Instead, the ESA addresses habitat destruction through several different kinds of provisions. Section 4 authorizes the FWS to designate the critical habitat of a species, as described above at page 159. Section 5 directs appropriate federal agencies to acquire land used as habitat by listed species. Most controversially, the take prohibition of section 9 can be read to encompass the modification of the habitat of an endangered species, as discussed in the *Sweet Home* case below.

Besides habitat destruction, a variety of other indirect actions can harm a species. The contested cases have featured disoriented baby sea turtles attracted to beachfront lights, grizzly bears feasting on grain spilled on railroad tracks as another train looms, and the fear that bald eagles would be poisoned by lead shot used to kill deer. Federal, state and local governments often find themselves accused of injuring listed species by failing to take actions necessary to protect them from a variety of threats.

But private parties have been the most critical of the restrictions imposed by the ESA. The extension of the ESA to certain kinds of habitat destruction and other activities that indirectly harm a species has prompted a lively theoretical debate and numerous calls for legislative reform. The same actions that environmentalists view as critical for the survival of a species are often denounced as overzealous government interference with private landowners. The provisions of the ESA itself are at issue in this debate, which shows no signs of ending any time soon.

1.   DEFINING PROHIBITED TAKE OF SPECIES

Babbitt v. Sweet Home Chapter of Communities for a Great Oregon

Supreme Court of the United States, 1995.
515 U.S. 687.

■ JUSTICE STEVENS delivered the opinion of the Court.

The Endangered Species Act of 1973 (ESA or Act), contains a variety of protections designed to save from extinction species that the Secretary of the Interior designates as endangered or threatened. Section 9 of the Act makes it unlawful for any person to "take" any endangered or threatened species. The Secretary has promulgated a regulation that defines the statute's prohibition on takings to include "significant habitat modification or degradation where it actually kills or injures wildlife." This case presents the question whether the Secretary exceeded his authority under the Act by promulgating that regulation.

I

Section 9(a)(1) of the Endangered Species Act provides the following protection for endangered species:

> Except as provided in sections 1535(g)(2) and 1539 of this title, with respect to any endangered species of fish or wildlife listed pursuant to section 1533 of this title it is unlawful for any person subject to the jurisdiction of the United States to ... (B) take any such species within the United States or the territorial sea of the United States[.]

Section 3(19) of the Act defines the statutory term "take": "The term 'take' means to harass, harm, pursue, hunt, shoot, wound, kill, trap, capture, or collect, or to attempt to engage in any such conduct." The Act does not further define the terms it uses to define "take." The Interior Department regulations that implement the statute, however, define the statutory term "harm": "Harm in the definition of 'take' in the Act means an act which actually kills or injures wildlife. Such act may include significant habitat modification or degradation where it actually kills or injures wildlife by significantly impairing essential behavioral patterns, including breeding, feeding, or sheltering." This regulation has been in place since 1975.[8]

A limitation on the § 9 "take" prohibition appears in § 10(a)(1)(B) of the Act, which Congress added by amendment in 1982. That section authorizes the Secretary to grant a permit for any taking otherwise prohibited by § 9(a)(1)(B) "if such taking is incidental to, and not the purpose of, the carrying out of an otherwise lawful activity."

. . . Respondents in this action are small landowners, logging companies, and families dependent on the forest products industries in the Pacific Northwest and in the Southeast, and organizations that represent their interests. They brought this declaratory judgment action against petitioners, the Secretary of the Interior and the Director of the Fish and Wildlife Service, in the United States District Court for the District of Columbia to challenge the statutory validity of the Secretary's regulation defining "harm," particularly the inclusion of habitat modification and degradation in the definition. Respondents challenged the regulation on its face. Their complaint alleged that application of the "harm" regulation to the red-cockaded woodpecker, an endangered species, and the northern spotted owl, a threatened species, had injured them economically.

Respondents advanced three arguments to support their submission that Congress did not intend the word "take" in § 9 to include habitat modification, as the Secretary's "harm" regulation provides. First, they correctly noted that language in the Senate's original version of the ESA would have defined "take" to include "destruction, modification, or curtailment of [the] habitat or range" of fish or wildlife, but the Senate deleted that language from the bill before enacting it. Second, respondents argued that Congress intended the Act's express authorization for the Federal Government to buy private land in order to prevent habitat degradation in § 5 to be the exclusive check against habitat modification on private property. Third, because the Senate added the term "harm" to the definition of "take" in a floor amendment without debate, respondents argued that the court should not interpret the term so expansively as to include habitat modification.

[The District Court upheld the regulation. On appeal, the D.C. Circuit first upheld the regulation 2–1, but on rehearing the court struck down the regulation 2–1 after Judge Williams changed his mind.]

II

Because this case was decided on motions for summary judgment, we may appropriately make certain factual assumptions in order to frame the legal issue. First, we assume respondents have no desire to harm either the red-cockaded woodpecker or the spotted owl; they merely wish to continue logging activities that would be entirely proper if not prohibited by the

---

8. The Secretary, through the Director of the Fish and Wildlife Service, originally promulgated the regulation in 1975 and amended it in 1981 to emphasize that actual death or injury of a protected animal is necessary for a violation.

ESA. On the other hand, we must assume arguendo that those activities will have the effect, even though unintended, of detrimentally changing the natural habitat of both listed species and that, as a consequence, members of those species will be killed or injured. Under respondents' view of the law, the Secretary's only means of forestalling that grave result—even when the actor knows it is certain to occur—is to use his § 5 authority to purchase the lands on which the survival of the species depends. The Secretary, on the other hand, submits that the § 9 prohibition on takings, which Congress defined to include "harm," places on respondents a duty to avoid harm that habitat alteration will cause the birds unless respondents first obtain a permit pursuant to § 10.

The text of the Act provides three reasons for concluding that the Secretary's interpretation is reasonable. First, an ordinary understanding of the word "harm" supports it. The dictionary definition of the verb form of "harm" is "to cause hurt or damage to: injure." In the context of the ESA, that definition naturally encompasses habitat modification that results in actual injury or death to members of an endangered or threatened species.

Respondents argue that the Secretary should have limited the purview of "harm" to direct applications of force against protected species, but the dictionary definition does not include the word "directly" or suggest in any way that only direct or willful action that leads to injury constitutes "harm."[9] Moreover, unless the statutory term "harm" encompasses indirect as well as direct injuries, the word has no meaning that does not duplicate the meaning of other words that § 3 uses to define "take." A reluctance to treat statutory terms as surplusage supports the reasonableness of the Secretary's interpretation.[10]

---

9. Respondents and the dissent emphasize what they portray as the "established meaning" of "take" in the sense of a "wildlife take," a meaning respondents argue extends only to "the effort to exercise dominion over some creature, and the concrete effect of [sic] that creature." This limitation ill serves the statutory text, which forbids not taking "some creature" but "tak[ing] any [endangered] species"—a formidable task for even the most rapacious feudal lord. More importantly, Congress explicitly defined the operative term "take" in the ESA, no matter how much the dissent wishes otherwise, thereby obviating the need for us to probe its meaning as we must probe the meaning of the undefined subsidiary term "harm." Finally, Congress' definition of "take" includes several words—most obviously "harass," "pursue," and "wound," in addition to "harm" itself—that fit respondents' and the dissent's definition of "take" no better than does "significant habitat modification or degradation."

10. In contrast, if the statutory term "harm" encompasses such indirect means of killing and injuring wildlife as habitat modification, the other terms listed in § 3—"harass," "pursue," "hunt," "shoot," "wound," "kill," "trap," "capture," and "collect"—generally retain independent meanings. Most of those terms refer to deliberate actions more frequently than does "harm," and they therefore do not duplicate the sense of indirect causation that "harm" adds to the statute. In addition, most of the other words in the definition describe either actions from which habitat modification does not usually result (e.g., "pursue," "harass") or effects to which activities that modify habitat do not usually lead (e.g., "trap," "collect"). To the extent the Secretary's definition of "harm" may have applications that overlap with other words in the definition, that overlap reflects the broad purpose of the Act.

Second, the broad purpose of the ESA supports the Secretary's decision to extend protection against activities that cause the precise harms Congress enacted the statute to avoid. In *TVA v. Hill*, 437 U.S. 153 (1978), we described the Act as "the most comprehensive legislation for the preservation of endangered species ever enacted by any nation." Whereas predecessor statutes enacted in 1966 and 1969 had not contained any sweeping prohibition against the taking of endangered species except on federal lands, the 1973 Act applied to all land in the United States and to the Nation's territorial seas. As stated in § 2 of the Act, among its central purposes is "to provide a means whereby the ecosystems upon which endangered species and threatened species depend may be conserved...."

Respondents advance strong arguments that activities that cause minimal or unforeseeable harm will not violate the Act as construed in the "harm" regulation. Respondents, however, present a facial challenge to the regulation. Thus, they ask us to invalidate the Secretary's understanding of "harm" in every circumstance, even when an actor knows that an activity, such as draining a pond, would actually result in the extinction of a listed species by destroying its habitat. Given Congress' clear expression of the ESA's broad purpose to protect endangered and threatened wildlife, the Secretary's definition of "harm" is reasonable.

Third, the fact that Congress in 1982 authorized the Secretary to issue permits for takings that § 9(a)(1)(B) would otherwise prohibit, "if such taking is incidental to, and not the purpose of, the carrying out of an otherwise lawful activity," strongly suggests that Congress understood § 9(a)(1)(B) to prohibit indirect as well as deliberate takings. The permit process requires the applicant to prepare a "conservation plan" that specifies how he intends to "minimize and mitigate" the "impact" of his activity on endangered and threatened species, making clear that Congress had in mind foreseeable rather than merely accidental effects on listed species. No one could seriously request an "incidental" take permit to avert § 9 liability for direct, deliberate action against a member of an endangered or threatened species, but respondents would read "harm" so narrowly that the permit procedure would have little more than that absurd purpose. "When Congress acts to amend a statute, we presume it intends its amendment to have real and substantial effect." Congress' addition of the § 10 permit provision supports the Secretary's conclusion that activities not intended to harm an endangered species, such as habitat modification, may constitute unlawful takings under the ESA unless the Secretary permits them.

The Court of Appeals made three errors in asserting that "harm" must refer to a direct application of force because the words around it do.[15] First,

---

15. The dissent makes no effort to defend the Court of Appeals' reading of the statutory definition as requiring a direct application of force. Instead, it tries to impose on § 9 a limitation of liability to "affirmative conduct intentionally directed against a particular animal or animals." Under the dissent's interpretation of the Act, a developer could drain a pond, knowing that the act would extinguish an endangered species of turtles, without even proposing a conservation plan or applying for a permit under the court's premise was flawed. Several of the words that accompany "harm" in the § 3 definition of "take," especially "harass," "pursue," "wound," and "kill," refer to actions or effects that do not require direct applications of force. Second, to the extent the court read a requirement of intent or purpose into the words used to define "take," it ignored § 9's express provision that a "knowing" action is enough to violate the Act. Third, the court employed *noscitur a sociis* to give "harm" essentially the same function as other words in the definition, thereby denying it independent meaning. The canon, to the contrary, counsels that a word "gathers meaning from the words around it." The statutory context of "harm" suggests that Congress meant that term to serve a particular function in the ESA, consistent with but distinct from the functions of the other verbs used to define "take." The Secretary's interpretation of "harm" to include indirectly injuring endangered animals through habitat modification permissibly interprets "harm" to have "a character of its own not to be submerged by its association."

Nor does the Act's inclusion of the § 5 land acquisition authority and the § 7 directive to federal agencies to avoid destruction or adverse modification of critical habitat alter our conclusion. Respondents' argument that the Government lacks any incentive to purchase land under § 5 when it can simply prohibit takings under § 9 ignores the practical considerations that attend enforcement of the ESA. Purchasing habitat lands may well cost the Government less in many circumstances than pursuing civil or criminal penalties. In addition, the § 5 procedure allows for protection of habitat before the seller's activity has harmed any endangered animal, whereas the Government cannot enforce the § 9 prohibition until an animal has actually been killed or injured. The Secretary may also find the § 5 authority useful for preventing modification of land that is not yet but may in the future become habitat for an endangered or threatened species. The § 7 directive applies only to the Federal Government, whereas the § 9 prohibition applies to "any person." Section 7 imposes a broad, affirmative duty to avoid adverse habitat modifications that § 9 does not replicate, and § 7 does not limit its admonition to habitat modification that "actually kills or injures wildlife." Conversely, § 7 contains limitations that § 9 does not, applying only to actions "likely to jeopardize the continued existence of any endangered species or threatened species," and to modifications of habitat that has been designated "critical" pursuant to § 4. Any overlap that § 5 or § 7 may have with § 9 in particular cases is unexceptional, and simply reflects the broad purpose of the Act set out in § 2 and acknowledged in *TVA v. Hill*.

We need not decide whether the statutory definition of "take" compels the Secretary's interpretation of "harm," because our conclusions that Congress did not unambiguously manifest its intent to adopt respondents'

§ 9(a)(1)(B); unless the developer was motivated by a desire "to get at a turtle," no statutory taking could occur. Because such conduct would not constitute a taking at common law, the dissent would shield it from § 9 liability, even though the words "kill" and "harm" in the statutory definition could apply to such deliberate conduct....

view and that the Secretary's interpretation is reasonable suffice to decide this case. *See generally Chevron U.S.A. Inc. v. Natural Resources Defense Council, Inc.*, 467 U.S. 837 (1984). The latitude the ESA gives the Secretary in enforcing the statute, together with the degree of regulatory expertise necessary to its enforcement, establishes that we owe some degree of deference to the Secretary's reasonable interpretation. *See* Breyer, *Judicial Review of Questions of Law and Policy*, 38 ADMIN. L. REV. 363, 373 (1986).

III

Our conclusion that the Secretary's definition of "harm" rests on a permissible construction of the ESA gains further support from the legislative history of the statute. . . . The Senate Report stressed that " '[t]ake' is defined . . . in the broadest possible manner to include every conceivable way in which a person can 'take' or attempt to 'take' any fish or wildlife." The House Report stated that "the broadest possible terms" were used to define restrictions on takings . . . [By contrast, the fact that a proposed endangered species bill included "the destruction, modification, or curtailment of [the] habitat or range of fish and wildlife" does not indicate the take prohibition that was ultimately adopted excludes habitat protection. Additionally, "the history of the 1982 amendment that gave the Secretary authority to grant permits for 'incidental' takings provides further support for his reading of the Act. The House Report expressly states that '[b]y use of the word "incidental" the Committee intends to cover situations in which it is known that a taking will occur if the other activity is engaged in but such taking is incidental to, and not the purpose of, the activity.' "]

IV

When it enacted the ESA, Congress delegated broad administrative and interpretive power to the Secretary. The task of defining and listing endangered and threatened species requires an expertise and attention to detail that exceeds the normal province of Congress. Fashioning appropriate standards for issuing permits under § 10 for takings that would otherwise violate § 9 necessarily requires the exercise of broad discretion. The proper interpretation of a term such as "harm" involves a complex policy choice. When Congress has entrusted the Secretary with broad discretion, we are especially reluctant to substitute our views of wise policy for his. *See Chevron*, 467 U.S. at 865–866. In this case, that reluctance accords with our conclusion, based on the text, structure, and legislative history of the ESA, that the Secretary reasonably construed the intent of Congress when he defined "harm" to include "significant habitat modification or degradation that actually kills or injures wildlife."

In the elaboration and enforcement of the ESA, the Secretary and all persons who must comply with the law will confront difficult questions of proximity and degree; for, as all recognize, the Act encompasses a vast range of economic and social enterprises and endeavors. These questions must be addressed in the usual course of the law, through case-by-case resolution and adjudication.

■ JUSTICE O'CONNOR, concurring.

My agreement with the Court is founded on two understandings. First, the challenged regulation is limited to significant habitat modification that causes actual, as opposed to hypothetical or speculative, death or injury to identifiable protected animals. Second, even setting aside difficult questions of scienter, the regulation's application is limited by ordinary principles of proximate causation, which introduce notions of foreseeability. These limitations, in my view, call into question *Palila v. Hawaii Dept. of Land and Natural Resources*, 852 F.2d 1106 (C.A.9 1988) (*Palila II*), and with it, many of the applications derided by the dissent. Because there is no need to strike a regulation on a facial challenge out of concern that it is susceptible of erroneous application, however, and because there are many habitat-related circumstances in which the regulation might validly apply, I join the opinion of the Court....

■ JUSTICE SCALIA, with whom THE CHIEF JUSTICE and JUSTICE THOMAS join, dissenting.

I think it unmistakably clear that the legislation at issue here (1) forbade the hunting and killing of endangered animals, and (2) provided federal lands and federal funds for the acquisition of private lands, to preserve the habitat of endangered animals. The Court's holding that the hunting and killing prohibition incidentally preserves habitat on private lands imposes unfairness to the point of financial ruin—not just upon the rich, but upon the simplest farmer who finds his land conscripted to national zoological use. I respectfully dissent.

I

... The regulation has three features which ... do not comport with the statute. First, it interprets the statute to prohibit habitat modification that is no more than the cause-in-fact of death or injury to wildlife. Any "significant habitat modification" that in fact produces that result by "impairing essential behavioral patterns" is made unlawful, regardless of whether that result is intended or even foreseeable, and no matter how long the chain of causality between modification and injury. *See, e.g.,* Palila v. Hawaii Dept. of Land and Natural Resources (Palila II), 852 F.2d 1106, 1108–1109 (CA9 1988) (sheep grazing constituted "taking" of palila birds, since although sheep do not destroy full-grown mamane trees, they do destroy mamane seedlings, which will not grow to full-grown trees, on which the palila feeds and nests).

Second, the regulation does not require an "act": the Secretary's officially stated position is that an omission will do.... The third and most important unlawful feature of the regulation is that it encompasses injury inflicted, not only upon individual animals, but upon populations of the protected species. "Injury" in the regulation includes "significantly impairing essential behavioral patterns, including breeding." Impairment of breeding does not "injure" living creatures; it prevents them from propagating, thus "injuring" a population of animals which would otherwise have maintained or increased its numbers. What the face of the regulation shows, the Secretary's official pronouncements confirm. The Final Redefinition of "Harm" accompanying publication of the regulation said that "harm" is not limited to "direct physical injury to an individual member of the wildlife species," and refers to "injury to a population." . . .

II

The Court [argues that] "the broad purpose of the [Act] supports the Secretary's decision to extend protection against activities that cause the precise harms Congress enacted the statute to avoid." I thought we had renounced the vice of "simplistically . . . assum[ing] that whatever furthers the statute's primary objective must be the law." . . . Second, the Court maintains that the legislative history of the 1973 Act supports the Secretary's definition. Even if legislative history were a legitimate and reliable tool of interpretation (which I shall assume in order to rebut the Court's claim); and even if it could appropriately be resorted to when the enacted text is as clear as this, here it shows quite the opposite of what the Court says. I shall not pause to discuss the Court's reliance on such statements in the Committee Reports as " '[t]ake' is defined . . . in the broadest possible manner to include every conceivable way in which a person can 'take' or attempt to 'take' any fish or wildlife." This sort of empty flourish—to the effect that "this statute means what it means all the way"—counts for little even when enacted into the law itself. . . . Both the Senate and House floor managers of the bill explained it in terms which leave no doubt that the problem of habitat destruction on private lands was to be solved principally by the land acquisition program of § 1534, while § 1538 solved a different problem altogether—the problem of takings . . . Habitat modification and takings, in other words, were viewed as different problems, addressed by different provisions of the Act. . . .

III

In response to the points made in this dissent, the Court's opinion stresses two points, neither of which is supported by the regulation, and so cannot validly be used to uphold it. First, the Court and the concurrence suggest that the regulation should be read to contain a requirement of proximate causation or foreseeability, principally because the statute does—and "[n]othing in the regulation purports to weaken those requirements [of the statute]." I quite agree that the statute contains such a limitation, because the verbs of purpose in § 1538(a)(1)(B) denote action directed at animals. But the Court has rejected that reading. The critical premise on which it has upheld the regulation is that, despite the weight of the other words in § 1538(a)(1)(B), "the statutory term 'harm' encompasses indirect as well as direct injuries." Consequently, unless there is some strange category of causation that is indirect and yet also proximate, the Court has already rejected its own basis for finding a proximate-cause limitation in the regulation. In fact "proximate" causation simply means "direct" causation.

. . . The regulation says (it is worth repeating) that "harm" means (1) an act which (2) actually kills or injures wildlife. If that does not dispense with a proximate-cause requirement, I do not know what language would. And changing the regulation by judicial invention, even to achieve compliance with the statute, is not permissible.

The second point the Court stresses in its response seems to me a belated mending of its hold. It apparently concedes that the statute requires injury to particular animals rather than merely to populations of animals. The Court then rejects my contention that the regulation ignores this requirement, since, it says, "every term in the regulation's definition of 'harm' is subservient to the phrase 'an act which actually kills or injures wildlife.'" As I have pointed out, this reading is incompatible with the regulation's specification of impairment of "breeding" as one of the modes of "kill[ing] or injur[ing] wildlife."[5]

... The Endangered Species Act is a carefully considered piece of legislation that forbids all persons to hunt or harm endangered animals, but places upon the public at large, rather than upon fortuitously accountable individual landowners, the cost of preserving the habitat of endangered species. There is neither textual support for, nor even evidence of congressional consideration of, the radically different disposition contained in the regulation that the Court sustains. For these reasons, I respectfully dissent.

Notes and Questions

1. So who is right? Does one "take" an endangered species when one adversely affects its habitat? Did the Congress that enacted the ESA in

---

5. JUSTICE O'CONNOR supposes that an "impairment of breeding" intrinsically injures an animal because "[t]o make it impossible for an animal to reproduce is to impair its most essential physical functions and to render that animal, and its genetic material, biologically obsolete." This imaginative construction does achieve the result of extending "impairment of breeding" to individual animals; but only at the expense of also expanding "injury" to include elements beyond physical harm to individual animals. For surely the only harm to the individual animal from impairment of that "essential function" is not the failure of issue (which harms only the issue), but the psychic harm of perceiving that it will leave this world with no issue (assuming, of course, that the animal in question, perhaps an endangered species of slug, is capable of such painful sentiments). If it includes that psychic harm, then why not the psychic harm of not being able to frolic about—so that the draining of a pond used for an endangered animal's recreation, but in no way essential to its survival, would be prohibited by the Act? That the concurrence is driven to such a dubious redoubt is an argument for, not against, the proposition that "injury" in the regulation includes injury to populations of animals. Even more so with the concurrence's alternative explanation: that "impairment of breeding" refers to nothing more than concrete injuries inflicted by the habitat modification on the animal who does the breeding, such as "physical complications [suffered] during gestation." Quite obviously, if "impairment of breeding" meant such physical harm to an individual animal, it would not have had to be mentioned. The concurrence entangles itself in a dilemma while attempting to explain the Secretary's commentary to the harm regulation, which stated that "harm" is not limited to "direct physical injury to an individual member of the wildlife species." The concurrence denies that this means that the regulation does not require injury to particular animals, because "one could just as easily emphasize the word 'direct' in this sentence as the word 'individual.'" One could; but if the concurrence does, it thereby refutes its separate attempt to exclude indirect causation from the regulation's coverage. The regulation, after emerging from the concurrence's analysis, has acquired both a proximate-cause limitation and a particular-animals limitation—precisely the one meaning that the Secretary's quoted declaration will not allow, whichever part of it is emphasized.

1973 intend for section 9 to apply to habitat modification? Or did that Congress plan to protect habitat through other provisions in the act? Does the addition of section 10's incidental take provision in 1982 suggest that "take" must be read to include habitat modification?

2.   Justice O'Connor's concurrence indicates that she questions the application of the take prohibition in *Palila II*. The palila is a six-inch bird that lives in mamane and naio forests on the slopes of Mauna Kea on the island of Hawaii. That area is also home to the Mauna Kea Game Management Area, where the state introduced wild goats and sheep to facilitate game hunting, an otherwise rare commodity in Hawaii. The goats and the sheep were eating the seedlings, leaves, stems, and sprouts of the mamane and naio trees on which the palila depended for its survival. In *Palila I*, the courts held that the state's act of permitting the goats and the sheep to live in the area constituted a take of the palila. Palila v. Hawaii Dep't of Land & Natural Resources, 471 F.Supp. 985 (D.Haw.1979), *aff'd*, 639 F.2d 495 (9th Cir.1981). *Palila II* reached a similar conclusion regarding the state's introduction of mouflon sheep into the same area, sheep that were prized by hunters but which fed on the mamane trees. Palila v. Hawaii Dep't of Land & Natural Resources, 649 F.Supp. 1070 (D.Haw.1986), *aff'd*, 852 F.2d 1106 (9th Cir.1988). The fact that the goats and sheep were destroying the habitat on which the palila depended led the courts to find a take in both cases, even though there was no evidence of an actual injury to any individual palila and even though the numbers of palilas had not dropped. Subsequently, the state's resulting "[u]ngulate eradication efforts have become so successful that hunters frequently cannot find any sheep to shoot" in the game area. Palila v. Hawaii Dep't of Land & Natural Resources, 73 F.Supp.2d 1181, 1184 (D.Haw.1999). A group of frustrated hunters then moved to dissolve the court's orders, but the court held that the balance of the equities favored the species because "mouflon sheep can always be reintroduced on Mauna Kea," whereas "[p]alila once extinct are gone forever." *Id.* at 1187.

3.   As the Court understands the regulation, which of the following activities constitute a prohibited "take" of an endangered species?

a.   A housing developer knows that a tree contains an endangered spotted owl's nest, but he cuts the tree down anyway—while the bird is perched in the nest.

b.   A housing developer accidentally puts his bulldozer in reverse and knocks over a tree containing a spotted owl's nest.

c.   A housing developer cuts down three acres of trees not knowing that endangered spotted owls live in the forest.

d.   In January, a housing developer cuts down three acres of trees, knowing that endangered black capped vireos (a songbird) live in the forest during the summer, but also knowing that they migrate to Central America during the winter.

e.   A housing developer cuts down three acres of trees that have not been designated as the critical habitat of the black capped vireo, and where no black capped vireos have been seen, but which contain the exact kind of habitat preferred by black capped vireos and which are part of a larger forest in which those birds live.

Alternately, how would you decide the following actual cases?

f. A train spilled grain alongside the railroad tracks, thus attracting hungry grizzly bears to the area. Environmentalists claim that both the failure to remove the grain and the railroad's ongoing operations constitute a take. National Wildlife Fed'n v. Burlington N. R.R., 23 F.3d 1508 (9th Cir.1994).

g. A Tucson school district wants to build a new high school on part of a 73 acre site. A pygmy-owl has been seen on the north part of the property, outside the area where the school will be built but near the construction activities and where the student parking lot will be located. Defenders of Wildlife v. Bernal, 204 F.3d 920 (9th Cir.2000).

h. An irrigation district pumps water from the Sacramento River. Salmon are killed when they crash into a fish screen that was installed by the state wildlife agency, but the district claims that it has not committed a prohibited take because it did not install the fish screen that is killing the fish. United States v. Glenn–Colusa Irrigation Dist., 788 F.Supp. 1126 (E.D.Cal.1992).

i. Dewitt DeWeese wants to build a lounge on ten acres of his land on Perdido Key north of Florida's Highway 182. The critical habitat of the Perdido Key Beach mouse occupies 88 acres south of the highway. A biologist contends that the lounge would result in three different prohibited takes of the mouse because (1) construction activities might actually kill or injure some mice, even though no mouse has been seen on DeWeese's property; (2) the project will degrade the habitat of the mouse; or (3) the lounge will attract cats that will prey upon the mouse. Morrill v. Lujan, 802 F.Supp. 424 (S.D.Ala.1992).

j. A timber company wants to clear cut 93 acres of its old growth forest land. A pair of spotted owls forages on the land, but also on similar forest on neighboring property as well. United States v. West Coast Forest Resources Ltd. Partnership, 2000 WL 298707 (D.Or. 2000).

4. The Court's decision in *Sweet Home* prompted a number of responses. Three of the six bills introduced in Congress in 1995 to amend the ESA would have changed the statutory definition of "harm." For example, H.R. 2275 defined "harm" as "to take a direct action against any member of an endangered species of fish or wildlife that actually injures or kills a member of the species." Would the current FWS regulation be consistent with that definition of harm? Which of the cases listed in note 3 would be decided differently under that definition?

The FWS has tried to respond to the uncertainty that continues to surround the precise scope of the ESA's take prohibition in the aftermath of *Sweet Home*. Whenever it lists a new species, the agency specifies which activities would constitute a take and which would not. For example, when it listed two snails in Alabama as endangered, the FWS advised that it would not view the following activities as a prohibited take: existing permitted discharges into the snail's habitat; typical agricultural and silvicultural practices carried out in compliance with existing state and federal regulations and best management practices; development and construction activities designed and implemented according to state and local water quality regulations; existing recreational activities such as swimming, wading, canoeing, and fishing; and the use of pesticides and herbicides in accordance with the label restrictions within the species' watersheds. But the agency warned that other activities could result in a take, including unauthorized collection or capture of the snails; dredging, channelization, the withdrawal of water, and other unauthorized destruction or alteration of the habitat of the snails; the violation of any discharge or water withdrawal permit; and illegal discharge or dumping of toxic chemicals or other pollutants into waters supporting the snails. Endangered Status for the Armored Snail and Slender Campeloma, 65 Fed. Reg. 10033, 10038 (2000). Is that guidance consistent with *Sweet Home*? Is it likely to be helpful to local residents and landowners? Note that the agency's advice is nonbinding, but the agency promises to answer questions about the application of the take prohibition to particular activities.

2. The Scope of Liability

United States v. Town of Plymouth, Massachusetts

United States District Court for the District of Massachusetts, 1998.
6 F.Supp.2d 81.

■ Patti B. Saris, United States District Judge.

[The piping plover is a small shorebird that nests on sandy coastal beaches from North Carolina to Newfoundland. The 1,200 breeding pairs that live along the Atlantic Coast in the United States are listed as threatened under the ESA. Over one-third of those breeding pairs nest along beaches in Massachusetts. The piping plovers nest on the beaches in Massachusetts from mid-March through May, and the chicks remain there until they are able to fly by late July or early August. Piping plover chicks are particularly vulnerable to off-road vehicles (ORVs) because the chicks leave their nests within hours of hatching and wander some distance along the beach and adjacent areas. The chicks can move quickly, but sometimes when vehicles approach they stand motionless in an attempt to blend into the sand. Thus Dr. Scott Melvin, a rare species zoologist with the Massachusetts Division of Fisheries and Wildlife, documented 25 piping plover chicks and two adults that were found dead in ORV tire ruts on the upper beach in Massachusetts and New York between 1989 and 1997. Dr. Melvin also described how ORVs destroy beach habitat that would be suitable for piping plovers and how ORVs disrupt feeding by scaring chicks and burying insects and other food that would otherwise be available to the piping plovers.

Plymouth Long Beach is 2.8 miles long and is used by both piping plovers and ORVs. Most of the beach is owned by the town of Plymouth, while the remainder is home to eighteen private residences. The town permits ORVs and other vehicles to travel on the beach from 4:00 a.m. until 9:00 p.m daily. The town sold between 1,100 and 2,700 ORV permits in the years from 1991 through 1997, with up to 325 vehicles allowed on the beach at one time. When it became aware of the threat to the piping plovers, the town's park division issued guidelines for protecting the birds, and John Crane—a park division employee who served as the town's natural resources officer (NRO)—worked to install fencing and take other actions to prevent ORVs from harming the birds. Those efforts failed, however, because of opposition from ORV users and pressure from town officials. On June 13, 1996, Crane sought to close the beach while four piping plover chicks were hatched from their eggs, but the town's select-men refused to do so and one of the chicks was found dead in a vehicle track later that same day. On August 5, 1997, irate local residents packed a town meeting concerning the proposed closure of the beach. People wanted to "take our beach back" from the "bird watchers," the crowd booed supporters of the plan to protect the piping plovers, and Crane was hung in effigy. Crane was fired one month later. Later in 1997, Fish and Wildlife Service officials recommended that the town take certain actions to protect the birds, but the town rejected those recommendations in January 1998. The federal government then brought suit against the town and sought a preliminary injunction to prohibit use of the beach by ORVs unless the town were to take numerous specified actions to guard the piping plovers.]

"In ruling on a motion for preliminary injunction, a district court is charged with considering: (1) the likelihood of success on the merits; (2) the potential for irreparable harm if the injunction is denied; (3) the balance of relevant impositions, i.e., the hardship to the nonmovant if enjoined as contrasted with the hardship to the movant if no injunction issues; and (4) the effect (if any) of the court's ruling on the public interest." Strahan v. Coxe, 127 F.3d 155, 160 (1st Cir.1997). "Under the ESA, however, the balancing and public interest prongs have been answered by Congress' determination that the 'balance of the hardships and the public interest tips heavily in favor of protected species'." Id. (quoting National Wildlife Fed'n v. Burlington Northern R.R., 23 F.3d 1508, 1510 (9th Cir.1994)). Examination of the language, history and structure of Section 7 of the ESA indicates "beyond doubt that Congress intended endangered species to be afforded the highest of priorities." Tennessee Valley Authority v. Hill, 437 U.S. 153, 174 (1978) (holding that Congress had foreclosed the exercise of the usual discretion possessed by a court of equity); *see also* Sierra Club v. Marsh, 816 F.2d 1376, 1383 (9th Cir.1987) (holding that the courts may not use equity's scales to strike a difference balance).

1.   The ESA

. . . A " 'take' is construed in the broadest possible manner to include every conceivable way in which a person can 'take' or attempt to 'take' any

The piping plover

fish or wildlife." *Strahan*, 127 F.3d at 162 (quoting S. Rep. No. 93–307, at 7 (1973)); *see also* Babbitt v. Sweet Home Chapter of Communities for a Great Oregon, 515 U.S. 687, 703–04 (1995) (same). Moreover, the ESA's prohibitions contemplate both the actions of individuals who directly take a species and those of a third party authorized by the government to engage in activity resulting in a taking. *See* 16 U.S.C. § 1538(g) ("It is unlawful for any person ... to attempt to commit, solicit another to commit, or cause to be committed, any offense" prohibited by the ESA); *see also Strahan*, 127 F.3d at 163 (holding that "the statute not only prohibits the acts of those parties that directly exact the taking, but also bans those acts of a third party that bring about the acts exacting a taking").

The ESA specifically authorizes the Attorney General, acting on behalf of the United States, to "seek to enjoin any person who is alleged to be in violation of any provision of this chapter or regulation issued under authority thereof." 16 U.S.C. § 1540(e)(6). In order for the United States to prove a likelihood of success on the merits, it must demonstrate that the Town of Plymouth has caused, through action or inaction, the illegal taking of the piping plover on Plymouth Long Beach or that future takes will occur if management of the beach continues on its present course. Proof of a taking requires a showing "that the alleged activity has actually harmed the species or if continued will actually, as opposed to potentially, cause harm to the species." American Bald Eagle v. Bhatti, 9 F.3d 163, 166 (1st Cir.1993) (citing cases); *see also Strahan*, 939 127 F.3d at 162. A movant can make a showing of actual harm by proving that significant modification or damage to the habitat of an endangered or threatened species is likely to occur so as to injure that species. In *TVA v. Hill*, the Supreme Court stated that "some" may find it "curious" that an endangered three-inch fish, the snail darter, could stop a $100 million dam under the ESA, but that "the explicit provisions of the [ESA] require precisely that result." *Id.* at 172–73. Similarly, here a 2.5 inch piping plover can stop a behemoth ORV if the government proves a taking under the ESA.

2.   *Proof of Harm*

The Service has proved a likelihood of success on its claim that current management practices with respect to ORV access to Plymouth Long Beach have actually harmed piping plovers and will continue to cause harm to the species if they remain unchecked. Plymouth's persistent refusal over the last five years to undertake adequate and timely precautionary measures on the initiative of those town officials responsible for management of the beach creates a likelihood that piping plover chicks will be killed and disturbed and that the nesting and feeding habitat will be adversely modified during the upcoming breeding season.

The Town points out that Plymouth Long Beach has experienced an increase in the number of nesting pairs of piping plovers from one in 1991 to nine in 1997, and that the piping plover population is "flourishing" under the Town's "upgraded" management of the beach under the 1992 Plan. Douglass Gray, the Superintendent of Parks and Forestry, claims he is empowered by the by-laws to limit the number of vehicles and by the 1998 Protocol to create "vehicle-free zones for nonessential vehicles where unfledged (flightless) piping plover chicks are present." I do not doubt the good faith or diligence of those employees entrusted with managing Long Beach and with monitoring the piping plover. However, as a matter of unwritten policy and practice, these employees do not have the authority on their own to take necessary measures to create adequate vehicle-free buffer zones.

Rather, there is compelling evidence that certain selectmen and at least one town manager have failed in the past to take prompt measures to close the beach to ORVs to protect nests and unfledged piping plover chicks and have instead declined to authorize a beach closing until state or federal officials intervened and recommended the closures in writing. Moreover, the lack of authority vested in the NRO to close the beach on his own volition to protect piping plovers and the absence of mandatory requirements in the 1998 Protocol with respect to minimum buffer zones for unfledged chicks have restricted the NRO's ability to prevent takes.

This failure to act quickly and decisively resulted, in all probability, in the June 13, 1996 taking of the plover chick. While the Town maintains that no conclusions can be drawn about the cause of the death of the recently hatched plover—observed healthy in the morning and found dead in vehicle tracks just hours later—the reasonable inference to be drawn in this whodunit is that the chick was killed by an ORV. Although the Selectmen perceive Long Beach as a multipurpose beach where ORV users and piping plovers can happily co-exist, the expert evidence proves that the large vehicle-free zones are essential to protect this threatened species from takes, and will often be inconsistent with ORV use during the summer months.

Based on the transcript and videotape of the August 5, 1997 hearing before the Board of Selectmen; the hostility of many residents to "bird watchers;" the roasting of Town Counsel and the Town Manager, who were attempting to work with state and federal authorities; the firing of John Crane, who was doing his best to comply with the federal and state guidelines; and the decision to rescind the Memorandum of Agreement, I conclude that the Selectmen will not authorize the Town Manager or NRO to take appropriate measures to protect the piping plover from "takes" as required by law. The long-standing intransigence of town officials persuades me that without an injunction, the town officials will not act to protect plovers when the pressure from ORV owners intensifies.

Accordingly, I adopt in large part the proposed preliminary relief recommended by the Service based on the Service guidelines and the affidavits of Dr. Melvin and Ms. Von Oettingen, each of whom point out that the Town of Plymouth stands apart from all private and municipally owned beaches in Massachusetts and other New England states in its failure to manage ORV use in accordance with the Service guidelines.

Because the 1992 Plan ceases to be effective on September 25, 1998, I am hopeful that the Town will work with appropriate state and federal officials to reach an acceptable beach management plan which will make the attached court order no longer necessary. Although I have adopted the Service's proposed order verbatim in most respects, I have modified the government's proposed order in the following ways. First, I do not require the Town to report to a Service employee, but instead give it the authority to implement the order itself. If there are violations of the order, the Service may move for contempt sanctions. I expect that there will be adequate consultation so that this will not be necessary. Second, I did not adopt the government's requested definition of "essential" vehicles because it was unclear to me whether residents of Long Beach have other access to their homes. Third, I leave it up to the Town to regulate pets, kites, and fireworks, as there is no evidence that in the past these activities have caused any "takes" or were improperly managed. Fourth, I did leave in the requirement of monitoring by a biologist, but do not expect the Town to hire someone new for this position if a current employee already has adequate training and expertise to do the job; the Town may make another proposal if a biologist is unnecessary. Finally, although there are mandatory buffer zones, the Service suggested that there should be flexibility to permit ORV access where not inconsistent with the breeding, nesting, and feeding needs of the piping plovers. In consultation with state and/or federal officials, the Town may take reasonable measures consistent with the needs of the piping plovers so long as any deviations from the order are reported to the Service and logged. The order requires the Town and the Service to provide full written reports at the end of the summer to ensure compliance.

The order expires in one year unless the Service demonstrates good cause to continue it based on the actions of the Town.

Notes and Questions

1. Judge Saris issued her decision in the battle between the piping plover and ORV users for Plymouth's beach on May 15, 1998. Four days later, the town's selectmen voted 4–1 to comply with the judge's order rather than appealing. The lone dissenter—Selectman Linda Teagan—remarked that she "[c]ontinued to be baffled by a federal judge basically concluding that the oldest town in the United States is unable to govern itself." John O'Keefe, *Plover Dispute Comes to Head; Vehicle Restriction Effective Today*, THE PATRIOT LEDGER (Quincy, Mass.), May 19, 1998, at 5S. The first piping plover chicks hatched on May 23, and the ban on vehicle traffic went into effect. By the end of the summer, the FWS was complimenting the town for doing "a terrific job managing the beach." John O'Keefe, *Off-Road Vehicle Limits May Be Lifted*, THE PATRIOT LEDGER (Quincy, Mass.), July 24, 1998, at 13S. Twenty of the thirty chicks that hatched on Plymouth's beach in 1998 survived. The casualties resulted not from ORVs, but from storms, malnourishment, and predators. Indeed, foxes were the primary reason why only three of the chicks hatched in 1999 lived long enough to fly, even though there were 12 nesting pairs on the beach that summer. The threats that foxes, skunks, crows and other predators posed to piping plovers on nearby beaches caused the Fish and Wildlife Service to summon a sharpshooter to kill a coyote suspected of damaging plover nests, and it prompted speculation that increased enforcement of local dog leash laws would be required by the ESA.

2. The piping plover is one of the few species protected by the ESA that has provoked controversy among residents of the northeastern United States. Besides the dispute in Plymouth, efforts to protect piping plovers have been responsible for modifying Fourth of July fireworks celebrations in Connecticut and slowing beachfront development on Long Island. *See* John Rather, *On Resort Shores, It's People vs. Plovers*, N.Y. TIMES, June 21, 1998, sec. 14LI, at 3; *see also Spielberg Seeks Access to Bluff Point*, N.Y. TIMES, Mar. 15, 1997, at 28 (describing concerns about the possible impact on piping plovers of trucks using a Long Island beach to film the movie *Amistad*). By contrast, most of the controversies about the ESA have involved the Pacific Northwest, southern California, other western states, Texas and Florida. Not surprisingly, support for the ESA is greatest in the northeast while endangered species protections are much more divisive in the west. Imagine, though, what actions could have been contested—and which species could have been saved—if Congress had enacted the ESA in 1800.

3. Colorado Representative Dan Schaeffer has objected to what he views as the litigious radical wing of the environmental movement because of its willingness to employ the ESA's take prohibition to block certain activities. He offered two examples: "In Massachusetts, environmentalists sued the state for merely licensing fishermen who used certain kinds of lobster traps because the traps actually worked. In Florida, one radical environmental group sued in the name of the Loggerhead Turtles because they believed aggressive local actions to curb beach-front lighting were not aggressive enough." 144 Cong. Rec. E2362 (daily ed. Dec. 21, 1998).

Like *Town of Plymouth*, both of the cases mentioned by Representative Schaeffer involve the application of the ESA's take prohibition to the government's failure to regulate private conduct that harms an endangered species. In Strahan v. Coxe, 127 F.3d 155 (1st Cir.1997), the First Circuit held that the Massachusetts state agency responsible for regulating fishing violated the ESA's take prohibition when it issued permits to fishermen whose nets entangled endangered Northern Right whales. Similarly, in Loggerhead Turtle v. County Council of Volusia County, Florida, 148 F.3d 1231 (11th Cir.1998), *cert. denied*, 526 U.S. 1081 (1999), the court held that the defenders of the turtle made a sufficient causal showing to establish standing to contend the county violated the take prohibition when it failed to regulate lighting in homes and businesses along the beach because the lights confuse newly-hatched sea turtles and make them more susceptible to predators and less likely to reach the safety of the water. On remand, however, the district court held that the county's new lighting regulations did not constitute a take because they were designed to help the turtle, and because the county could not be held liable for the unwillingness of private individuals to abide by the regulations. Loggerhead Turtle v. County Council of Volusia County, Fla., 92 F.Supp.2d 1296 (M.D.Fla.2000). Another example (not mentioned by Representative Schaeffer) occurred in Defenders of Wildlife v. Administrator, Environmental Protection Agency, 882 F.2d 1294 (8th Cir.1989), where the court held that EPA's licensing of pesticides that were ultimately ingested by black-footed ferrets constituted a prohibited take. Remember, too, that Hawaii violated the take prohibition when it managed state lands for hunting sheep and goats that injured the habitat of the palila. *See supra* at page 255.

Do such cases satisfy the understanding of the take prohibition articulated in *Sweet Home*? Should state or local officials be held liable for failing to regulate conduct that harms a species? *See* Shannon Petersen, *Endangered Species in the Urban Jungle: How the ESA Will Reshape American Cities*, 19 Stan. Envtl. L. J. 423, 439 (2000) (arguing that *Town of Plymouth*, *Strahan*, and *Loggerhead Turtle* were wrongly decided because (1) "state and local regulatory regimes governing private activities cannot be the proximate cause of an illegal take," (2) "the Tenth Amendment prohibits the federal government from coercing local governments to implement the ESA," and (3) "the doctrine of sovereign immunity should protect state and local governments from ESA citizen suits in some limited situations"). How should local officials respond to residents who ask why they can no longer use ORVs on the beach or turn on their lights at night? *See also* J.B. Ruhl, *State and Local Government Vicarious Liability Under the ESA*, 16 Natural Resources & Env't 70 (2001).

4. *Town of Plymouth* also illustrates the issuance of injunctive relief for future violations of the take prohibition. In Forest Conservation Council v. Rosboro Lumber Company, 50 F.3d 781, 784 (9th Cir.1995), the Ninth Circuit held that "a showing of imminent threat to injury to wildlife suffices" to support an injunction against a future take. In so holding, the court rejected the claim that the regulatory requirement of an "actual" injury was limited to a part or present injury. The court reaffirmed that conclusion in Marbled Murrelet v. Babbitt, 83 F.3d 1060, 1064–65 (9th Cir.1996), rejecting the assertion that the Supreme Court's decision in *Sweet Home* compelled a different result. In both cases the court concluded that injunctions were available against logging that threatened to take endangered birds. A district court has since read the Ninth Circuit test to allow for an injunction in a future take case if the plaintiff proves that (1) the endangered species is located on the property at issue, and (2) the defendant's activities are reasonably certain to constitute a threat of imminent harm to the species. Applying that test, a district court refused to enjoin a private campground project because the two endangered species—the San Francisco garter snake and the red-legged frog—were not present on all parts of the land in question and because the campground and its ancillary activities were not reasonably likely to harm the species. *See* Coastside Habitat Coalition v. Prime Properties, Inc., 1998 WL 231024 (N.D.Cal.1998).

Blackburn's sphinx moth

5. Blackburn's sphinx moth is Hawaii's largest native insect, with a wingspan of five inches. Its natural host plants are a shrub named the popolo and a tree named the aiea, which the moths rely upon for food and shelter. Many other host plants for the moth are not native to Hawaii, including eggplant, tomatoes, Jimson weed, tree tobacco, and *Nicotiana tabacum*—commonly known as commercial tobacco.

The moth is named after the Reverend Thomas Blackburn, a missionary who collected the first specimens in the nineteenth century. Hawaii's landscape has undergone extreme alteration since that time as a result of ranching, agricultural development, the deliberate introduction of alien animals and plants, and tourism. Much of the state's native wildlife and plants have suffered from these events, as illustrated by the fact that Hawaii has more federally listed endangered species than any other state.

The moth was thought to be extinct after an unsuccessful attempt to relocate the species in the late 1970s. In 1984, several moths were found on private and state land on Maui, part of which is within a nature reserve but part of which is used by the national guard for military training. Then two more moths were found on another part of Maui in 1992. This second population was feeding on commercial tobacco on private land that contained none of the native plants favored by the moth. Four more small populations of the moth have been found since then: two on Maui, one on Kahoolawe, and one on the island of Hawaii. These newest populations, numbering less than ten moths each, live in areas where there are few or no aiea trees and few or no commercial tobacco plants, so they rely upon the other host plants listed above instead. The small numbers of these populations and the ongoing threats to all of them persuaded the United States Fish & Wildlife Service to list Blackburn's sphinx moth as endangered on February 1, 2000.

The cultivation of tobacco faces increasing government regulation. The United States Department of Agriculture (USDA) pays millions of dollars to farmers who are willing to switch from growing commercial tobacco to growing other crops instead. The Hawaii legislature recently imposed a $2.00 tax on each package of cigarettes sold in the state. And Hyatt just announced plans to purchase thousands of acres on Maui that are used to grow tobacco—including the land where the two moths were found in 1992 and where a small population of the moths still lives today—and to build a new beachfront resort on that land instead.

Philip Morris, Incorporated, R.J. Reynolds Tobacco Company, and the Gallins Vending Company (a leading seller of cigarette vending machines) worry that the decrease in commercial tobacco cultivation on Maui will harm their businesses. They have brought suit in federal district court alleging that the USDA subsidies, the Hawaiian state cigarette tax, and the Hyatt resort proposal all violate the ESA because they will adversely affect the commercial tobacco plants used by some Blackburn's sphinx moths for food and for shelter. Do any of the three challenged actions violate any provisions of the ESA?

C.   LIMITS ON THE TAKE PROHIBITION

1.   THE SCOPE OF FEDERAL JURISDICTION

National Association of Home Builders v. Babbitt

United States Court of Appeals for the District of Columbia Circuit, 1997.
130 F.3d 1041, *cert. denied*, 524 U.S. 937 (1998).

■ WALD, CIRCUIT JUDGE:

[The saga of the Delhi Sands Flower–Loving Fly is recounted in chapter 1 at page 1. This case arose when the FWS determined that San Bernadino County's plans to redesign an intersection near the new Arrowhead Regional Medical Center would probably cause a take of the fly in violation of section 9 of the ESA. In response, the county, the National Association of Home Builders of the United States and two other builders groups, and the cities of Colton and Fontana brought suit alleging that the application of the take prohibition to protect the fly was beyond the scope of congressional power under the commerce clause. The district court upheld the application of the statute, and the plaintiffs appealed.]

Appellants' Commerce Clause challenge to the application of section 9(a)(1) of the ESA to the Fly rests on the Supreme Court's decision in United States v. Lopez, 514 U.S. 549 (1995). In *Lopez*, the Court held that the Gun–Free School Zones Act of 1990, which made possession of a gun within a school zone a federal offense, exceeded Congress' Commerce Clause authority. Drawing on its earlier Commerce Clause jurisprudence, the *Lopez* Court explained that Congress could regulate three broad categories of activity: (1) "the use of the channels of interstate commerce," (2) "the instrumentalities of interstate commerce, or persons or things in interstate commerce, even though the threat may come only from intrastate activities," and (3) "those activities having a substantial relation to interstate commerce ... i.e., those activities that substantially affect interstate commerce." Possession of a gun within 1000 feet of a school, the Court explained, clearly did not fit the first two categories. In addition, it could not be regulated under the third category as an activity that "substantially affects" interstate commerce because it was not commercial in nature and was not an essential part of a larger regulation of economic activity. Moreover, the Court explained, Congress had made no findings about the effect of gun possession in school zones on interstate commerce. Thus, concluding that Congress had no rational basis for finding that gun possession within school zones had a substantial effect on interstate commerce, the Court declared the statute unconstitutional....

Application of section 9(a)(1) of the ESA to the Fly can be viewed as a proper exercise of Congress' Commerce Clause power over the first category of activity that the *Lopez* Court identified: the use of the "channels of interstate commerce." Although this category is commonly used to uphold regulations of interstate transport of persons or goods, it need not be so limited. Indeed, the power of Congress to regulate the channels of interstate commerce provides a justification for section 9(a)(1) of the ESA for two reasons. First, the prohibition against takings of an endangered species is necessary to enable the government to control the transport of the endangered species in interstate commerce. Second, the prohibition on takings of endangered animals falls under Congress' authority " 'to keep the channels of interstate commerce free from immoral and injurious uses.' "

The ESA's prohibition on takings of endangered species can be justified as a necessary aid to the prohibitions in the ESA on transporting and selling endangered species in interstate commerce. . . . The prohibition on takings of endangered animals also falls under Congress' authority to prevent the channels of interstate commerce from being used for immoral or injurious purposes. This authority was perhaps best described by the Supreme Court in Heart of Atlanta [Motel Inc. v. United States], 379 U.S. 241 [(1964)], which the *Lopez* Court cited and quoted in its reference to Congress' power to regulate the use of the "channels of interstate commerce." In *Heart of Atlanta*, the Supreme Court upheld a prohibition on racial discrimination in places of public accommodation serving interstate travelers against a Commerce Clause challenge. The Court explained that " 'the authority of Congress to keep the channels of interstate commerce free from immoral and injurious uses has been frequently sustained, and is no longer open to question.' " It does not matter if the activities that are regulated are of a "purely local character," the Court elaborated, " 'if it is interstate commerce that feels the pinch, it does not matter how local the operation which applies the squeeze.' " Thus, the power of Congress over interstate commerce "also includes the power to regulate the local incidents thereof, including local activities in both the States of origin and destination, which might have a substantial and harmful effect upon that commerce." . . . . [I]n this case, Congress used this authority [to rid the channels of interstate commerce of injurious uses] to prevent the eradication of an endangered species by a hospital that is presumably being constructed using materials and people from outside the state and which will attract employees, patients, and students from both inside and outside the state. . . . Congress is therefore empowered by its authority to regulate the channels of interstate commerce to prevent the taking of endangered species in cases like this where the pressures of interstate commerce place the existence of species in peril. . . .

The takings clause in the ESA can also be viewed as a regulation of the third category of activity that Congress may regulate under its commerce power. According to *Lopez*, the test of whether section 9(a)(1) of the ESA is within this category of activity "requires an analysis of whether the regulated activity 'substantially affects' interstate commerce." A class of activities can substantially affect interstate commerce regardless of whether the activity at issue—in this case the taking of endangered species—is commercial or noncommercial. As the *Lopez* Court, quoting Wickard v. Filburn, 317 U.S. 111 (1942), noted:

"[E]ven if appellee's activity be local and though it may not be regarded as commerce, it may still, whatever its nature, be reached by Congress if it exerts a substantial economic effect on interstate commerce, and this irrespective of whether such effect is what might at some earlier time have been defined as 'direct' or 'indirect.' "[6]

.... Congress could rationally conclude that the intrastate activity regulated by section 9 of the ESA substantially affects interstate commerce for two primary reasons. First, the provision prevents the destruction of biodiversity and thereby protects the current and future interstate commerce that relies upon it. Second, the provision controls adverse effects of interstate competition.[10]

■ KAREN LECRAFT HENDERSON, CIRCUIT JUDGE, concurring:

I agree with Judge Wald's conclusion that the "taking" prohibition in section 9(a)(1) of the Endangered Species Act (ESA) constitutes a valid exercise of the Congress's authority to regulate interstate commerce under the Commerce Clause. I cannot, however, agree entirely with either of her grounds for reaching the result and instead arrive by a different route.

Judge Wald first asserts that section 9(a)(1) is a proper regulation of the "channels of commerce." In support she cites decisions upholding regulation of commercially marketable goods, such as machine guns and lumber, and public accommodations. In each case, the object of regulation was necessarily connected to movement of persons or things interstate and could therefore be characterized as regulation of the channels of commerce. Not so with an endangered species, as the facts here graphically demonstrate. The Delhi Sands Flower-loving Flies the Department of the Interior seeks to protect are (along with many other species no doubt) entirely *intra*state creatures. They do not move among states either on their own or through human agency. As a result, like the Gun–Free School Zones Act in *Lopez*, the statutory protection of the flies "is not a regulation of the use of the channels of interstate commerce."

6. Indeed, the case at hand is in many ways directly analogous to *Wickard*. In both cases, the appellee's activity, growing wheat for personal consumption and taking endangered species, is local and is not "regarded as commerce." However, in both cases, the activity exerts a substantial economic effect on interstate commerce—by affecting the quantity of wheat in one case, and by affecting the quantity of species in the other.

10. Judge Sentelle asserts that these rationales have "no stopping point." In fact, however, they have very clear and obvious limits. In the case of the first rationale, the argument stops at endangered species. Activities that threaten a species' existence threaten to reduce biodiversity and thereby have a substantial negative effect on interstate commerce. Thus, the biodiversity rationale offered here provides support for the Endangered Species Act only insofar as the Act prevents activities that are likely to cause the elimination of species. In the case of the second rationale, the argument stops at activities are the product of destructive interstate competition. Under this rationale, interstate competition that is likely to produce destructive results, such as elimination of endangered species' habitat, environmental degradation, or exploitation of labor, can be regulated by Congress. Thus, the destructive interstate competition rationale provides support for the Endangered Species Act only insofar as the Act prevents a bidding down of regulatory standards that is likely to result in the elimination of endangered species' habitat.

Judge Wald also justifies the protection of endangered species on the ground that the loss of biodiversity "substantially affects" interstate commerce because of the resulting loss of potential medical or economic benefit. Yet her opinion acknowledges that it is "impossible to calculate the exact impact" of the economic loss of an endangered species. As far as I can tell, it is equally impossible to ascertain that there will be any such impact at all. It may well be that no species endangered now or in the future will have any of the economic value proposed. Given that possibility, I do not see how we can say that the protection of an endangered species has any effect on interstate commerce (much less a substantial one) by virtue of an uncertain potential medical or economic value. Nevertheless, I believe that the loss of biodiversity itself has a substantial effect on our ecosystem and likewise on interstate commerce. In addition, I would uphold section 9(a)(1) as applied here because the Department's protection of the flies regulates and substantially affects commercial development activity which is plainly interstate.

First, I agree with Judge Wald that biodiversity is important to our understanding of ESA and its relation to interstate commerce.... The effect of a species' continued existence on the health of other species within the ecosystem seems to be generally recognized among scientists. Some studies show, for example, that the mere presence of diverse species within an ecosystem (biodiversity) by itself contributes to the ecosystem's fecundity. The Congress recognized the interconnection of the various species and the ecosystems when it declared that the "essential purpose" of ESA, which protects endangered species, is in fact "to protect the ecosystem upon which we and other species depend." H.R. REP. No. 93–412, at 10 (1973); *see also* 16 U.S.C. § 1531 (finding that endangered species "are of aesthetic, *ecological*, educational, historical, recreational, and scientific value") (emphasis added). Given the interconnectedness of species and ecosystems, it is reasonable to conclude that the extinction of one species affects others and their ecosystems and that the protection of a purely intrastate species (like the Delhi Sands Flower-loving Fly) will therefore substantially affect land and objects that are involved in interstate commerce. There is, therefore, "a rational basis" for concluding that the "taking" of endangered species "substantially affects" interstate commerce so that section 9(a)(1) is within the Congress's Commerce Clause authority.

The interstate effect of a taking is particularly obvious here given the nature of the taking the County proposes. In enacting ESA, the Congress expressed an intent to protect not only endangered species but also the habitats that they, and we, occupy. At the same time, the Congress expressly found that "economic growth and development untempered by adequate concern and conservation" was the cause for "various species of fish, wildlife, and plants in the United States having been rendered extinct." 16 U.S.C. § 1531(a)(1). It is plain, then, that at the time it passed ESA the Congress contemplated protecting endangered species through regulation of land and its development, which is precisely what the Department has attempted to do here. Such regulation, apart from the characteristics or range of the specific endangered species involved, has a plain and substantial effect on interstate commerce. In this case the regulation relates to both the proposed redesigned traffic intersection and the hospital it is intended to serve, each of which has an obvious connection with interstate commerce. Insofar as application of section 9(a)(1) of ESA here acts to regulate commercial development of the land inhabited by the endangered species, "it may ... be reached by Congress" because "it asserts a substantial economic effect on interstate commerce." Wickard v. Filburn, 317 U.S. 111, 125, (1942), *quoted in* United States v. Lopez, 514 U.S. 549, 556 (1995). . . . [6]

■ SENTELLE, CIRCUIT JUDGE, dissenting:

... Can Congress under the Interstate Commerce Clause regulate the killing of flies, which is not commerce, in southern California, which is not interstate? Because I think the answer is "no," I can not join my colleagues' decision to affirm the district court's conclusion that it can.

... Though Judge Henderson rejects Judge Wald's "biodiversity" rationale, she relies on a related justification of her own, which is to me indistinguishable in any meaningful way from that of Judge Wald. As I understand her rationale, it depends on "the interconnectedness of species and ecosystems," which she deems sufficient for us "to conclude that the extinction of one species affects others and their ecosystems and that the protection of a purely intrastate species [concededly including the Delhi Sands Flower–Loving Fly] will therefore substantially affect land and objects that are involved in interstate commerce." I see this as no less of a stretch than Judge Wald's rationale. First, the Commerce Clause empowers Congress "to regulate commerce" not "ecosystems." The Framers of the Constitution extended that power to Congress, concededly without knowing the word "ecosystems," but certainly knowing as much about the dependence of humans on other species and each of them on the land as any ecologist today. An ecosystem is an ecosystem, and commerce is commerce.

... There is no showing, but only the rankest of speculation, that a reduction or even complete destruction of the viability of the Delhi Sands Flower–Loving Fly will in fact "affect land and objects that are involved in interstate commerce," let alone do so substantially.[4] Nothing in the statute certainly necessitates such a nexus, nor has my colleague supplied a reason why this basis of regulation would apply to the preservation of a species any more than any other act potentially affecting the continued and stable existence of any other item of a purely intrastate nature upon which one

---

6. The dissent suggests this justification has no "stopping point" as required by *Lopez*. In *Lopez* the Court was concerned that the "theories" offered by the government would authorize regulation of "all activities that might lead to violent crime, regardless of how tenuously they relate to interstate commerce" and "any activity that it found was related to the economic productivity of individual citizens." The rationale on which I rely permits regulation only of activities (including land use) that adversely affect species that affect, or are involved in, interstate commerce.

4. Indeed, there is nothing in either Judge Henderson's opinion or the record to support speculation that the extinction of the Delhi Sands Flower–Loving Fly would have any effect on any other species.

might rest a speculation that its loss or change could somehow affect some other object, land, or otherwise, that might be involved in interstate commerce. In each of those cases, the Supreme Court upheld the relevant statutes, noting that the regulated actors would either destroy other commercial activities or be able to unfairly compete with interstate competitors subject to higher regulatory standards protective of other elements of commerce. In the present case neither Congress nor the litigants, nor for that matter Judge Wald, has pointed to any commercial activity being regulated, any commercial competition being unfairly challenged, or any other sort of commerce being destroyed by the taking of the fly. With reference to her other rationale, I saw no stopping point; here, I am not even sure what the beginning point is, let alone the terminus. I do not think a decision upholding the challenged section of the ESA on this rationale can exist in the same jurisprudence as *Lopez*.

. . . An alternate reading of Judge Henderson's second justification with its stress on the effect of the regulation upon the highway and hospital is that she concludes that Congress may regulate purely intrastate activities—*e.g.*, the habitat modification of the fly—where the *regulation* will then affect items which are arguably in interstate commerce. Again, I do not see the stopping point. Congress is not empowered either by the words of the Commerce Clause or by its interpretation in Lopez to regulate any non-commercial activity where the regulation will substantially affect interstate commerce. The most expansive view of *Lopez* is that Congress can regulate "those activities having a substantial relation to interstate commerce." Nowhere is it suggested that Congress can regulate activities not having a substantial effect on commerce because the regulation itself can be crafted in such a fashion as to have such an effect.

In the end, attempts to regulate the killing of a fly under the Commerce Clause fail because there is certainly no interstate commerce in the Delhi Sands Flower–Loving Fly. . . .

Notes and Questions

1. Judge Wald asked whether there was a sufficient relationship between *endangered species* and interstate commerce. Judge Henderson asked whether there was a relationship between the *hospital* and interstate commerce. Judge Sentelle asked whether there was a relationship between the *fly* and interstate commerce. Who asked the right question? *See* John Copeland Nagle, *The Commerce Clause Meets the Delhi Sands Flower–Loving Fly*, 97 Mich. L. Rev. 174 (1998).

2. Does the fly's role in its ecosystem provide a sufficient justification for congressional action under the commerce clause? The fly serves at least two functions within the Colton Dunes ecosystem: it pollinates two native species of plants, and its disappearance sounds an alarm that the ecosystem itself is in danger. The Colton Dunes is important as one of a decreasing number of isolated ecosystems that are crucial to the development of new species. The preservation of the fly will assure the preservation of the

Colton Dunes. Alternatively, does the fact that the fly might one day offer medical, nutritional or other benefits to people suffice to empower Congress to regulate activities that would harm the fly today?

3.   Judge Henderson and Judge Wald both emphasized that their theories allowed federal regulation only where it was interstate commerce that threatened an endangered species. A subsequent district court case emphasized that the ESA's take prohibition could be applied to the construction of a WalMart, shopping center, and residential homes and apartments because those activities had the requisite connection to interstate commerce, whether or not the five endangered cave species had such a connection as well. *See* GDF Realty Investments, Ltd. v. Norton, 169 F.Supp.2d 648 (W.D.Tex.2001). Could Congress regulate the following activities that threaten the Delhi Sands Flower–Loving Fly and its habitat: (1) the recreational use of off-road vehicles (ORVs) that crush the fly's habitat and interfere with its breeding, (2) the presence of non-native plants that alter the ecological balance upon which the fly depends, and (3) the trampling of land—or flies themselves—by people walking in the area?

4.   The Supreme Court decided not to review the D.C. Circuit's decision despite widespread speculation that the Court would take the case to elaborate on the meaning of *Lopez*. But the issue continues to be raised by those concerned with the effects of the ESA. Several listing decisions have been challenged as beyond the federal government's power under the commerce clause. *See, e.g.,* Endangered and Threatened Wildlife and Plants; Final Rule to List the Illinois Cave Amphipod as Endangered, 63 Fed. Reg. 46,900, 46,902 (1998). The one court of appeals to consider the issue since *NAHB* produced another 2–1 decision upholding the application of the ESA to a landowner who shot a red wolf that had strayed onto his land. *See* Gibbs v. Babbitt, 214 F.3d 483 (4th Cir.2000) (Wilkinson, C.J.) (sustaining the application of the ESA "because the regulated activity substantially affects interstate commerce and because the regulation is part of a comprehensive federal program for the protection of endangered species"); *see also id.* at 506–10 (Luttig, J., dissenting) (denying that the private taking of red wolves has a substantial effect on interstate commerce).

If the courts decide that the application of the ESA in situations like those involving the fly exceeds the scope of the commerce clause, is there any other source of federal power to protect endangered species that live in only one state? And why should it matter whether a species lives in more than one state, anyway?

2.   TAKES OF SPECIES AND TAKINGS OF PROPERTY

Good v. United States

United States Court of Appeals for the Federal Circuit, 1999.
189 F.3d 1355, *cert. denied*, 529 U.S. 1053 (2000).

■ EDWARD R. SMITH, SENIOR CIRCUIT JUDGE.

[In 1973, Lloyd A. Good, Jr. purchased Sugarloaf Shores, (which consists of thirty-two acres of wetlands and freshwater marsh) and eight acres of uplands on Lower Sugarloaf Key, Florida. The sales contract for the land stated that "[t]he Buyers recognize that certain of the lands covered by this contract may be below the mean high tide line and that as of today there are certain problems in connection with the obtaining of State and Federal permission for dredging and filling operations." In 1980, Good hired Keycology, Inc., a land planning and development firm, to obtain the federal, state, and county permits necessary to develop Sugarloaf Shores into a residential subdivision. In their contract, Good and Keycology acknowledged that "obtaining said permits is at best difficult and by no means assured." In March 1981, Good applied to the U.S. Army Corps of Engineers ("Corps") for a permit to fill 7.4 acres of salt marsh and excavate another 5.4 acres of salt marsh in order to create a 54–lot subdivision and a 48–slip marina. The Corps permit was required for dredging and filling navigable waters of the United States, including wetlands adjacent to navigable waters, under the Rivers and Harbors Act of 1899 and under § 404 of the Clean Water Act. In May 1983, Good obtained a permit that was good for five years.

Good was less successful in his efforts to get state and county approval for his plan. His struggles to satisfy state and local environmental regulations took so long that Good had to receive another, slightly revised permit from the Corps in 1988. By the end of 1989, Good had satisfied most of the county's requirements, but he was unable to receive the approval of the South Florida Water Management District (SFWMD) because of the impact of the proposed project on wetlands and on the habitat of the mud turtle and Lower Keys marsh rabbit, both of which were protected under Florida law. So in 1990 Good submitted a new plan to the Corps to build only sixteen homes, a canal, and a tennis court.

Meanwhile, the FWS had listed the Lower Keys marsh rabbit as an endangered species under the ESA. In the ensuing ESA section 7 consultation with the Corps, the FWS concluded that the project proposed in Good's 1990 permit application would not jeopardize the continued existence of the marsh rabbit, though it nonetheless recommended denial of the permit based on the development's overall environmental impact. Next, the FWS listed another species—the silver rice rat—as endangered in April 1991. The consultation triggered by that listing resulted in a December 1991 biological opinion issued by the FWS that determined that both the 1988 and 1990 plans jeopardized the continued existence of both the Lower Keys marsh rabbit and the silver rice rat. FWS recommended that the Corps deny the 1990 application and modify the 1988 permit to include the "reasonable and prudent alternatives" of locating all homesites in upland areas and limiting water access to a single communal dock. The impact of project on the two endangered species caused the Corps to deny Good's 1990 permit application on March 17, 1994, and to notify Good that his 1988 permit had expired. Good then filed suit in the United States Court of Federal Claims alleging that the Corps' denial of his permit application

constituted a taking of his property without just compensation in violation
of the fifth amendment. The court rejected Good's claim, and he appealed.]

The Lower Keys marsh rabbit

ANALYSIS

The Fifth Amendment to the United States Constitution provides that
private property shall not "be taken for public use, without just compensation." U.S. CONST. amend. V. The government can "take" private property
by either physical invasion or regulatory imposition. *See, e.g.,* Loretto v.
Teleprompter Manhattan CATV Corp., 458 U.S. 419 (1982); Lucas [v.
South Carolina Coastal Council], 505 U.S. 1003 [(1992)]. Appellant in this
case alleges a regulatory taking.

It has long been recognized that "while property may be regulated to a
certain extent, if regulation goes too far it will be recognized as a taking."
Pennsylvania Coal Co. v. Mahon, 260 U.S. 393, 415 (1922). The Supreme
Court has set out "several factors that have particular significance" in
determining whether a regulation effects a taking. Penn Central [Transp.
Co. v. New York City], 438 U.S. [104], 124 [(1978)]. These factors are (1)
the character of the government action, (2) the extent to which the
regulation interferes with distinct, investment-backed expectations, and (3)
the economic impact of the regulation. *See id. See also* Loveladies Harbor,
Inc. v. United States, 28 F.3d 1171, 1179 (Fed.Cir.1994); Florida Rock
Inds., Inc. v. United States, 18 F.3d 1560, 1567 (Fed.Cir.1994); Creppel v.
United States, 41 F.3d 627, 632 (Fed.Cir.1994). Because we find the
expectations factor dispositive, we will not further discuss the character of
the government action or the economic impact of the regulation.

REASONABLE, INVESTMENT-BACKED EXPECTATIONS

For any regulatory takings claim to succeed, the claimant must show
that the government's regulatory restraint interfered with his investment-

backed expectations in a manner that requires the government to compensate him. The requirement of investment-backed expectations "limits recovery to owners who can demonstrate that they bought their property in reliance on the non-existence of the challenged regulation." *Creppel*, 41 F.3d at 632. These expectations must be reasonable. *See* Ruckelshaus v. Monsanto Co., 467 U.S. 986, 1005–1006 (1984).

Reasonable, investment-backed expectations are an element of every regulatory takings case. *See Loveladies Harbor*, 28 F.3d at 1179. *See also id. at 1177* ("In legal terms, the owner who bought with knowledge of the restraint could be said to have no reliance interest, or to have assumed the risk of any economic loss. In economic terms, it could be said that the market had already discounted for the risk, so that a purchaser could not show a loss in his investment attributable to it."); *Creppel, 41 F.3d at 632* ("One who buys with knowledge of a restraint assumes the risk of economic loss.").

Good argues that the Supreme Court has eliminated the requirement for reasonable, investment-backed expectations, at least in cases where the challenged regulation eliminates virtually all of the economic value of the landowner's property. In support, Appellant cites *Lucas*, 505 U.S. at 1015, and argues that *Loveladies Harbor* should be reversed as contrary to *Lucas*.

However, we agree with the *Loveladies Harbor* court that the Supreme Court in *Lucas* did not mean to eliminate the requirement for reasonable, investment-backed expectations to establish a taking. It is true that the Court in *Lucas* set out what it called a "categorical" taking "where regulation denies all economically beneficial or productive use of land." 505 U.S. at 1015. The *Lucas* Court, however, clarified that by "categorical" it meant those "categories of regulatory action [that are] compensable without case-specific inquiry *into the public interest advanced in support of the restraint.*" *Id.* (emphasis added). A *Lucas*-type taking, therefore, is categorical only in the sense that the courts do not balance the importance of the public interest advanced by the regulation against the regulation's imposition on private property rights. *See Loveladies Harbor*, 28 F.3d at 1179.

The *Lucas* Court did not hold that the denial of all economically beneficial or productive use of land eliminates the requirement that the landowner have reasonable, investment-backed expectations of developing his land. In *Lucas*, there was no question of whether the plaintiff had satisfied that criterion. *See* 505 U.S. at 1006–1007 ("In 1986, petitioner David H. Lucas paid $975,000 for two residential lots on the Isle of Palms in Charleston County, South Carolina, on which he intended to build single-family homes. In 1988, however, the South Carolina Legislature enacted the Beachfront Management Act, S.C. Code Ann. § 48–39–250 *et seq.* (Supp. 1990), which had the direct effect of barring petitioner from erecting any permanent habitable structures on his two parcels.").

In addition, it is common sense that "one who buys with knowledge of a restraint assumes the risk of economic loss. In such a case, the owner presumably paid a discounted price for the property. Compensating him for a 'taking' would confer a windfall." *Creppel,* 41 F.3d at 632 (citations omitted).

Appellant alternatively argues that he had reasonable, investment-backed expectations of building a residential subdivision on his property. Appellant reasons that the permit requirements of the Rivers and Harbors Act and the Clean Water Act are irrelevant to his reasonable expectations at the time he purchased the subject property, because he obtained the federal dredge-and-fill permits required by those acts three times, and was only denied a permit, based on the provisions of the Endangered Species Act ("ESA"), when two endangered species were found on his property. Therefore, since the ESA did not exist when he bought his land, he could not have expected to be denied a permit based on its provisions.

Appellant's position is not entirely unreasonable, but we must ultimately reject it. In view of the regulatory climate that existed when Appellant acquired the subject property, Appellant could not have had a reasonable expectation that he would obtain approval to fill ten acres of wetlands in order to develop the land.

In 1973, when Appellant purchased the subject land, federal law required that a permit be obtained from the Army Corps of Engineers in order to dredge or fill in wetlands adjacent to a navigable waterway. Even in 1973, the Corps had been considering environmental criteria in its permitting decisions for a number of years. *See* Deltona Corp. v. United States, 228 Ct. Cl. 476, 657 F.2d 1184, 1187 (Ct. Cl. 1981) ("On December 18, 1968, in response to a growing national concern for environmental values and related federal legislation, the Corps [announced that it] would consider the following additional factors in reviewing permit applications: fish and wildlife, conservation, pollution, aesthetics, ecology, and the general public interest."). *See also* 657 F.2d at 1190 ("Since the late 1960's the regulatory jurisdiction of the Army Corps of Engineers has substantially expanded pursuant to § 404 of the [Clean Water Act] and—under the spur of steadily evolving legislation—the Corps has greatly added to the substantive criteria governing the issuance of dredge and fill permits."). By 1973, the Corps had denied dredge-and-fill permits solely on environmental grounds. *See, e.g.,* Zabel v. Tabb, 430 F.2d 199 (5th Cir.1970).

In addition to the federal regulations, development of the subject land required approval by both the state of Florida and Monroe County....

At the time he bought the subject parcel, Appellant acknowledged both the necessity and the difficulty of obtaining regulatory approval. The sales contract specifically stated that "the Buyers recognize that ... as of today there are certain problems in connection with the obtaining of State and Federal permission for dredging and filling operations." Appellant thus had both constructive and actual knowledge that either state or federal regulations could ultimately prevent him from building on the property. Despite his knowledge of the difficult regulatory path ahead, Appellant took no steps to obtain the required regulatory approval for seven years.

During this period, public concern about the environment resulted in numerous laws and regulations affecting land development. For example:

- In December 1973, the Endangered Species Act was enacted. 16 U.S.C. § 1531 *et seq.* (1994). The ESA prohibited federal actions that would be "likely to jeopardize the continued existence of any endangered species," 16 U.S.C. § 1536(a)(2), and made it unlawful to "take" (i.e., kill, harass, etc.) any endangered animal. *See* 16 U.S.C. §§ 1532(19), 1538(a)(1)(B).

- In 1975, the Corps of Engineers issued regulations broadening its interpretation of its § 404 authority to regulate dredging and filling in wetlands. *See* United States v. Riverside Bayview Homes, 474 U.S. 121, 123–124 (1985). In 1977, the Corps further broadened its definition of wetlands subject to § 404's permit requirements. *See id.*

- Also in 1977, Florida enacted its own Endangered and Threatened Species Act, FLA. STAT. ANN. § 372.072 (West 1997), further emphasizing the public concern for Florida's environment. In 1979, the Florida Keys Protection Act was enacted, designating the Keys an Area of Critical State Concern. FLA. STAT. ANN. § 380.0552 (West 1997).

Thus, rising environmental awareness translated into ever-tightening land use regulations. Surely Appellant was not oblivious to this trend.

The picture emerges, then, of Appellant in 1973 acknowledging the difficulty of obtaining approval for his project, then waiting seven years, watching as the applicable regulations got more stringent, before taking any steps to obtain the required approval. When in 1980 he finally retained a land development firm to seek the required permits, he acknowledged that "obtaining said permits is at best difficult and by no means assured."

While Appellant's prolonged inaction does not bar his takings claim, it reduces his ability to fairly claim surprise when his permit application was denied. Appellant was aware at the time of purchase of the need for regulatory approval to develop his land. He must also be presumed to have been aware of the greater general concern for environmental matters during the period of 1973 to 1980. As our predecessor court stated on similar facts: "When Deltona acquired the property in 1964, it knew that the development it contemplated could take place only if it obtained the necessary permits from the Corps of Engineers. Although at that time Deltona had every reason to believe that those permits would be forthcoming when it subsequently sought them, it also must have been aware that the standards and conditions governing the issuance of permits could change. Deltona had no assurance that the permits would issue, but only an expectation." *Deltona*, 657 F.2d at 1193.

Here, as in *Deltona*, Appellant "must have been aware that the standards and conditions governing the issuance of permits could change." *Id.* In light of the growing consciousness of and sensitivity toward environmental issues, Appellant must also have been aware that standards could change to his detriment, and that regulatory approval could become harder to get.

We therefore conclude that Appellant lacked a reasonable, investment-backed expectation that he would obtain the regulatory approval needed to develop the property at issue here. We have previously held that the government is entitled to summary judgment on a regulatory takings claim where the plaintiffs lacked reasonable, investment-backed expectations, even where the challenged government action "substantially reduced the value of plaintiffs' property." Avenal v. United States, 100 F.3d 933, 937 (Fed.Cir.1996). Here, too, Appellant's lack of reasonable, investment-backed expectations defeats his takings claim as a matter of law....

Notes and Questions

1. When did it become unreasonable for Good to expect that he could develop his property in the manner that he had hoped? In 1973, when he bought it? Later that year, when Congress enacted the ESA? In 1979, when the state designated the entire Florida Keys an area of critical state concern? In 1980, when a contractor informed him of the environmental obstacles facing development? In 1990, when the FWS listed the Lower Keys marsh rabbit as endangered under the ESA? Or at some other time?

2. The takings clause has been the subject of intense judicial and scholarly debate in recent years. Governmental regulation of private property has resulted in countless claims that the government must compensate landowners for their inability to use their land as they would like. Only one court, however, has held that the ESA produced a taking for which the government must pay the property owner. *See* Tulare Lake Basin Water Storage Dist. v. United States, 49 Fed.Cl. 313 (2001) (holding that water use restrictions imposed to protect the endangered delta smelt and winter-run chinook salmon worked a taking of the contractual water rights held by two county water districts). Why have the courts found so few takings?

3. The Minnesota Valley National Wildlife Refuge is home to bald eagles and a wide range of other animals and birds. It also lies in the path of a proposed new runway for the Minneapolis–St. Paul airport. The Fish and Wildlife Service concluded that the noise and other activities at the airport would not harm the bald eagle and the other wildlife, but the Federal Aviation Administration agreed to the FWS's request to pay $20 million for a constructive use of the refuge. Property rights advocates contend that the FWS is applying a double standard. Are they right?

4. Property rights advocates have publicized a number of "horror stories" about individuals who suffered economic losses because of the ESA. One of the most sympathetic stories concerns Margaret Rector, who purchased land outside of Austin, Texas in 1973 as an investment for retirement. When she sought to realize her investment two decades later, she discovered that it was "virtually impossible to find a buyer for a tract of land that has been labelled habitat of the golden-cheeked warbler." Endangered Species Act Implementation: Hearing Before the House Committee on

Resources, 104th Cong., 2d Sess. 14–15 (1996). She complained that "I'm 74 years old. I want the money so I can go to a nice rest home.... I think it is unfair for me to bear the burden when the bird belongs to everyone." Ralph Haurwitz, *Who Pays? Warbler Stokes Debate: Landowner Sees Retirement Money Fly Away on Endangered Songbird's Wings*, AUSTIN AMERICAN STATESMAN, Sept. 26, 1994, at A25. Many of the other so-called "horror stories," though, are questioned by supporters of the ESA. *See* Michael Allan Wolf, *Overtaking The Fifth Amendment: The Legislative Backlash Against Environmentalism*, 6 FORDHAM ENVTL. LAW J. 637 (1995).

Such stories have led to claims that fairness dictates that landowners be paid if the presence of an endangered species makes their land less valuable. Before the court decided against his claim, Lloyd Good testified before a congressional committee examining the impact of the ESA on private property rights. He complained that "the designation of this property as the habitat of these two species has in effect rendered it valueless. Real estate brokers won't touch it. Other developers won't touch it. Bankers won't even look at it. Mortgage brokers aren't interested. It's dead." Good concluded that "[t]he Endangered Species Act must, if it is going to continue, contain some provision for the protection of private property rights." *Endangered Species Act Implementation Hearing, supra,* at 16–17. Similarly, when Nebraska Senator Chuck Hagel introduced his proposed Private Property Fairness Act of 1999, he explained that "if the Government condemns part of a farm to build a highway, it has to pay the farmer for the value of his land. But if the Government requires that same farmer [to] stop growing crops on that same land in order to protect endangered species or conserve wetlands, the farmer gets no compensation." 145 CONG. REC. S734 (daily ed. Jan. 20, 1999).

Likewise, Professor Barton Thompson argues that "both incentive and fairness considerations militate in favor of at least partial compensation to landowners injured by the imposition of section 9's constraint on habitat modification." He explains:

> Absent any compensation under section 9, some landowners will destroy the habitat value of their land in order to escape ESA regulation—unnecessarily wasting valuable societal resources to annihilate the very public good that the ESA seeks to preserve. Others will refrain from creating or improving habitat. So long as the listing of a species remains a prerequisite of section 9 regulation, a no compensation rule will also encourage property owners to oppose new listings of endangered species, undermining all recovery efforts for those species. A no compensation rule biases those species-recovery efforts that do occur toward property-focused efforts and, because property owners vary among themselves in the political power they enjoy, distorts which property is used for habitat preservation and which landowners bear the burden of preservation. Finally, a no compensation rule is inequitable. The ESA's impact on particular parcels of land depends not on the behavior of the landowner or even on his ability to finance public

services, but on the happenstance of where the remaining habitat of an endangered species is.

Barton H. Thompson, Jr., *The Endangered Species Act: A Case Study in Takings & Incentives*, 49 STAN. L. REV. 305, 375 (1997).

Are Good, Professor Thompson, and Senator Hagel right? Should landowners be entitled to compensation if the presence of an endangered species limits their ability to use their land? Is Good deserving of compensation? Is Rector? If so, who should pay them?

5. The inability to persuade the courts to extend the fifth amendment's takings clause to force the government to compensate private landowners constrained by wildlife regulations led property rights advocates to propose legislative solutions. A bill considered by Congress in 1995 would have required federal agencies to compensate any private property owner whose land suffers at least a twenty percent reduction of value as a result of restrictions on the use of the property to protect endangered species. The agencies, moreover, would have to pay for such compensation out of their own budgets. Compensation would *not* be required if a landowner's activities are proscribed by state law, constitute a nuisance, or are prohibited by a local zoning ordinance, or if the federal agency is acting to prevent public health or safety hazards, to prevent damage to other property, or pursuant to navigational servitudes. *See* Private Property Protection Act of 1995, H.R. 925, 104th Cong., 1st Sess. (1995); *see also* H.R. Rep. No. 46, 104th Cong., 1st Sess. (1995) (outlining the arguments of supporters and opponents of the bill).

Dean William Treanor has advocated narrowly focused takings legislation that seeks to compensate those most harshly affected by government regulation. He proposes to compensate landowners when "unanticipated regulations destroy a significant portion of the total assets of a property owner." William Michael Treanor, *The Armstrong Principle, the Narratives of Takings, and Compensation Statutes*, 38 WM. & MARY. L. REV. 1151, 1155 (1997). The three qualifications insure that someone who was aware of applicable government regulation or who is a repeat player who loses on one investment but gains on others need not be compensated. Compensation is provided, though, to the small landowners who could lose their life's savings, the individuals most deserving of compensation and those whose predicament has informed much of the current debate. How is that test different from the constitutional formula announced in *Lucas*? From H.R. 925? Would Good receive compensation under the bill or under Dean Treanor's proposal? Would Rector?

6. Are there other ways to encourage private landowners to manage their land in a way to protects biodiversity besides the ESA's prohibition on adversely affecting the habitat of a species? Section 9 operates as a stick to force landowners not to harm an endangered species. Conversely, there is a broad consensus that the law should offer landowners and others affected by the ESA a carrot as well. Congress has considered several proposed bills that would provide a tax incentive to private landowners who act to protect endangered species. The Species Conservation Tax Act of 1999, introduced by Montana Senator Max Baucus, would allow payments to conservation programs to be deducted from federal income taxes, expand the income tax deduction offered for the donation of conservation easements involving endangered species, and create an estate tax exclusion for property subject to a conservation agreement. *See* S. 1392, 106th Cong., 1st Sess. (1999). The National Wildlife Federation has proposed that several tax and financial incentives be extended to landowners and that technical and financial assistance be offered to small private landowners, states, tribes, and local communities. *See* National Wildlife Federation, *Support Landowner Incentives to Promote Endangered Species Conservation*, <http://www.nwf.org/endangered/hcp/lndinc.html>. And as Texas Governor (and presidential candidate), George W. Bush advocated a landowner incentive program that would provide matching grants to states for use in helping private landowners protect rare species while engaging in traditional land management practices. He would also establish a private stewardship grant program to fund for private conservation initiatives. What other incentives should be offered to private landowners and others so that they will act to protect endangered species? *See generally* John F. Turner & Jason C. Rylander, *The Private Lands Challenge: Integrating Biodiversity Conservation and Private Property*, in Private Property and the Endangered Species Act: Saving Habitats, Protecting Homes 116–23 (Jason F. Shogren ed. 1998) (describing possible incentive programs).

D. Authorizing Incidental Take of Protected Species

Friends of Endangered Species, Inc. v. Jantzen

United States Court of Appeals for the Ninth Circuit, 1985.
760 F.2d 976.

■ Pregerson, Circuit Judge:

[In 1975, Visitacion Associates and the Crocker Land Company purchased about 3,000 acres of land on San Bruno Mountain. The owners proposed to build a large residential and commercial development, but they scaled back their plans after environmentalists objected because the land was so rich in wildlife. The U.S. Fish and Wildlife Service then discovered that the endangered Mission Blue butterfly lived on the mountain, and that the butterfly's existence would be seriously threatened by the development. In response, the San Bruno Mountain Steering Committee was formed by the landowners and other prospective developers, environmentalists, and representatives of state and local government agencies. The Committee developed a Habitat Conservation Plan that would preserve 81% of the open space on the mountain as wildlife habitat and require landowners to pay $60,000 annually to finance a permanent habitat conservation program in exchange for permission to proceed with the modified development. To satisfy the ESA, the parties applied for a permit pursuant to ESA section 10(a) that would allow the incidental taking of the endangered Mission Blue butterflies, San Bruno Elfin butterflies, and San Francisco garter snakes in the course of the development. FWS issued the permit, but Friends of Endangered Species, Inc. filed suit alleging that the development violated the ESA and the National Environmental Policy Act (NEPA).]

... Section 10(a) of the ESA allows the Service to permit an applicant to engage in an otherwise prohibited "taking" of an endangered species under certain circumstances. The applicant first must submit a comprehensive conservation plan. The Service then must scrutinize the plan and find, after affording opportunity for public comment, that: (1) the proposed taking of an endangered species will be "incidental" to an otherwise lawful activity; (2) the permit applicant will minimize and mitigate the impacts of the taking "to the maximum extent practicable"; (3) the applicant has insured adequate funding for its conservation plan; and (4) the taking will not appreciably reduce the likelihood of the survival of the species. Appellant challenges the sufficiency of the Permit findings relating to the second and fourth of these section 10(a) requirements.

1. *The field study adequately supported the Service's findings that "the taking will not appreciably reduce the likelihood of the survival of the species."*

The Service went beyond the statutory requirement and concluded that the Permit, coupled with the Plan, was likely to enhance the survival of the Mission Blue butterfly.

Appellant contends that the Service's determination that issuing the permit would not reduce the likelihood of the Mission Blue butterfly's survival was arbitrary and capricious because of alleged scientific shortcomings in the Biological Study upon which the decision was based. Specifically, appellant argues that low recapture rates and allegedly mistaken recaptures by the field crew in the mark-release-recapture phase of the field study invalidated the Study's conclusions, and that the Service abused its discretion in relying upon such data to approve the Permit. With respect to this contention, appellant fails to raise a genuine issue of material fact.

As the district court determined, the legislative history to section 10(a)'s 1982 amendments suggests that Congress viewed appellees' conduct in the present case as the paradigm approach to compliance with section 10(a):

> In some cases, the overall effect of a project can be beneficial to a species, even though some incidental taking may occur. An example is the development of some 3000 dwelling units on San Bruno Mountain near San Francisco. This site is also habitat for three endangered butterflies ... Absent the development of this project these butterfly recovery actions may well have never been developed. The proposed amendment should lead to resolution of potential conflicts between endangered species and the actions of private developers, while at the same time encouraging these developers to become more actively involved in the conservation of these species.

S. REP. No. 97–418, 97th Cong., 2d Sess. 10 (1982) (emphasis added). The House Conference Report indicates, in stronger terms than the Senate Report, that the Service acted properly in relying upon the Biological Study to comply with section 10(a):

> Because the San Bruno Mountain plan is the model for this long term permit and because the adequacy of similar conservation plans should be measured against the San Bruno plan, the Committee believes that the elements of this plan should be clearly understood. . . .
>
> Prior to developing the conservation plan, the County of San Mateo conducted an independent exhaustive biological study which determined the location of the butterflies. . . . The biological study was conducted over a two year period and at one point involved 50 field personnel.
>
> The San Bruno Mountain Conservation Plan is based on this extensive biological study.

H.R. REP. 97–835, 97th Cong., 2d Sess. 31–32 (1982).

In addition, appellant failed to bring many of these purported field data "errors" and "inconsistencies" to the attention of the Service until after the district court denied appellant's motion for summary judgment. Clearly, the Service did not act arbitrarily in failing to consider criticisms not presented to it before issuing the Permit. Review of the reasonableness of an agency's consideration of environmental factors is "limited . . . by the time at which the decision was made."

The Service did, however, extensively solicit and consider expert and public comments on the Biological Study before issuing the Permit. And, the Biological Study itself acknowledged methodological limitations.

Thus, there is no evidence that the Service issued the Permit either in ignorance or deliberate disregard of the Biological Study's limitations. Moreover, the Service responded in good faith in its Permit Findings to the criticisms which it sought out and received concerning the Biological Study, and acted reasonably in relying upon the Biological Study to conclude that the Plan would not reduce the likelihood of survival of the Mission Blue butterfly. Again, the Service cannot be said to have acted arbitrarily by not responding to criticisms not received when it approved the Permit.

We also consider it relevant that the Permit was expressly made subject to revocation and reconsideration based upon data that might be revealed from the continuing monitoring called for under the Plan. Thus, the Service complied with section 10(a)'s mandate by determining that "the incidental taking" of the Mission Blue would enhance the survival of the species. In light of the clear declaration of legislative intent and the Service's efforts to consider all criticisms of the Biological Study before relying upon it, we hold that the district court was correct in concluding that there were no genuine issues of material fact that would preclude it from determining that the Service did not act arbitrarily or capriciously in relying on the Biological Study.

2. *The Service did not act arbitrarily or capriciously in concluding that the Plan complied with section 10(a)'s requirement to minimize and mitigate the impact of the taking upon endangered species.*

The Plan at issue contains various measures to "minimize and mitigate" the impact of the project upon the Mission Blue butterfly. The Plan and the Agreement provide for the permanent protection of 86% of the Mission Blue's habitat. Moreover, funding for the Plan would yield $60,000 annually, which would be used to halt the apparent incursion of brush and gorse into the habitat and permit the re-establishment of grasslands for the butterfly.

In addition to provisions to halt advancing brush and gorse, the Plan contains continuing and comprehensive restrictions on land development and significant financial incentives. Regardless of whether brush and gorse continue to spread, these additional mitigating measures should play a significant role in enhancing the protection of endangered species on the Mountain.

Appellant also contends that the Saddle Area of the Mountain, now publicly owned for parkland and consisting of 75% brush, should be substituted for one of the grassland parcels currently proposed for development. Appellant suggests that this Saddle Area alternative would more effectively mitigate the Plan's effects. The [environmental impact report/environmental assessment (EIR/EA)] authors considered and rejected this Saddle Area development alternative. They concluded, among other things, that development of the Saddle Area would have secondary impacts including a biological impact greater than that produced by the Saddle's proposed use as a county park. The Saddle Area allegedly contains unique wetlands and endangered plants, and its development could meet with stiff environmental opposition.

Thus, the district court correctly concluded that there is simply no genuine factual dispute as to whether the Service acted arbitrarily or unreasonably in determining that the Plan complied with section 10(a)'s mitigation requirement.

3. *Appellant does not raise a genuine issue of material fact in alleging that the Service failed to comply with section 7(a)(2) of the ESA.*

Section 7(a)(2) of the ESA states:

> Each Federal agency shall ... insure that any action authorized, funded, or carried out by such agency ... is not likely to jeopardize the continued existence of any endangered species or threatened species or result in the destruction or adverse modification of habitat of such species which is determined by the Secretary ... to be critical.... In fulfilling [these] requirements ... each agency shall use the best scientific and commercial data available.

An action would "jeopardize" a species if it "reasonably would be expected to reduce the reproduction, numbers, or distribution of a listed species to such an extent as to appreciably reduce the likelihood of the survival and recovery of that species in the wild."

Pursuant to section 7(a)(2), the Service determined that the Permit would not likely jeopardize the continued existence of the Mission Blue butterfly. In so determining, the Service relied upon a variety of information including the Biological Study, the Plan, the Agreement, the EIR/EA, public comments received on the Permit Application, peer reviews of the Biological Study, and file materials on the Mission Blue butterfly.

Appellant erroneously contends that these sources did not represent the best scientific data available because of "the uncontroverted evidence from [the two experts upon whom it relies] revealing major mistakes in the field study."

Again, the low recapture rate realized in the mark-release-recapture phase of the Biological Study was a limitation that the Study itself and the Service acknowledged. And, several peer reviews took note of the limitations inherent in low mark-release-recapture rates. Thus, the Service was aware of all relevant limitations on the Biological Study and the field data, and the Service addressed those limitations in its Permit Findings. During the administrative process, appellant and its two experts did not direct the Service to any better available data. Moreover, the Service considered whatever data and other materials appellants provided. "The issue for review is whether the [agency's] decision was based on a consideration of the relevant factors and whether there has been a clear error of judgment." There is no genuine issue of material fact to dispute the district court's determination that the Service did not act unreasonably or capriciously, or in violation of section 7(a)(2), by considering all the data it received in the present case.

[The court also held that the FWS complied with the requirements of NEPA].

Notes and Questions

1.   The upshot of the habitat conservation plan (HCP) is that some of the butterfly's habitat is protected while the developers are permitted to build amidst other parts of butterfly's habitat. Why would the government agree to that deal? Why would the developer? Note that since the HCP went into effect, 800 housing units have been built on the San Bruno Mountain, while developers and homeowners have paid $70,000 annually to remove non-native plants, replant native species, and monitor the activities of the butterflies. A 1999 study concluded that "[a]t best, the HCP has been a wash. The development has not drastically decreased the population of endangered species on the mountain, but the restoration projects have not dramatically raised the numbers either." Marcus E. Walton, *San Bruno Ecology Faces Threat of Urban Sprawl*, SAN JOSE MERCURY NEWS, June 16, 1999. Most recently, the FWS and the nearby City of Brisbane agreed to spend a total of over one million dollars to purchase a part of the mountain that provides ten percent of the habitat of the callippe silverspot butterfly.

2. Section 10(a)(1)(B), which was part of the 1982 amendments to the ESA, authorizes the FWS to permit an "incidental take" of a listed species. An incidental take is an action that constitutes a "take" within ESA section 9, but the take is "incidental to, and not the purpose of, the carrying out of an otherwise lawful activity." Michael Bean and Melanie Rowland assert that "this provision likely increased the Secretary's leverage over activities that incidentally take endangered species because it substituted a flexible regulatory authority for a threat of prosecution that few found credible." BEAN & ROWLAND, *supra*, at 234. That leverage, however, does face some limits. In Arizona Cattle Growers' Association v. FWS, 273 F.3d 1229 (9th Cir.2001), reprinted above at page 211, the court held that the FWS could not issue an incidental take permit that imposed conditions on federal grazing permits where there was no evidence that the grazing would result in a take of an endangered species in the first place.

3. The developer in *Jantzen* produced the first HCP, and it received the first incidental take permit. Despite the enactment of section 10(a)(1)(B), only three HCPs were adopted between 1982 and 1989. Then the number of HCPs boomed. By September 1995, over 100 HCPs had been approved and about 200 were in various stages of development. As of April 2002, 379HCPs have been approved and many more were being developed. (For a summary of each of the approved plans, visit the FWS's HCP database at <http://ecos.fws.gov/hcp_report/hcp_summary.html?region=9&module=421&view=link>.) So why have HCPs become so popular, especially compared to the lack of interest in them throughout the 1980s?

Wisconsin Statewide Karner Blue Butterfly Habitat Conservation Plan and Environmental Impact Statement

Wisconsin Department of Natural Resources, 1999.
http://www.dnr.state.wi.us/org/land/er/publications/karner/karner.htm

EXECUTIVE SUMMARY

Background

This summary provides a synopsis of the *Wisconsin Statewide Karner Blue Butterfly Habitat Conservation Plan* (HCP) and the associated environmental impact statement (EIS). These two documents have been combined into a single document, with the HCP comprising Chapters I and II of the EIS and several appendices. This approach was taken so that those parties wishing to focus only on the proposed habitat conservation activities can easily extract that material from the overall document.

The HCP is an integral part of an application to the U.S. Fish and Wildlife Service (USFWS) for a statewide incidental take permit covering the federally endangered Karner blue butterfly (*Lycaeides melissa samuelis*). The Wisconsin Department of Natural Resources (DNR) is applying for the permit in collaboration with 26 other private and public partners. The partners are proposing the HCP as a resource management strategy to assure the long-term sustainability of Karner blue butterfly habitat and the persistence of Karner blue butterflies on the Wisconsin landscape.

In addition to the partners, development of the HCP relied heavily on people representing various associations and organizations. These organizations have contributed extensive and continuous time and effort to the process and include groups such as the Sierra Club, the Wisconsin Audubon Council and the Wisconsin Woodland Owners Association. It is anticipated that these organizations, and others, will play an important role in HCP implementation and Karner blue butterfly conservation in Wisconsin.

The HCP describes broad-scale land conservation and outreach/education strategies. Efforts are focused on conservation both in the Karner blue butterfly's high potential habitat area and across the Karner blue butterfly's Wisconsin range, with implementation relying heavily on adaptive management principles. The efficacy of the plan is tied to partner commitments (outlined in legally-binding conservation agreements with the DNR) and a hierarchy of monitoring systems developed to assure the anticipated, positive results of proposed actions.

While the HCP specifically addresses only one species, the Karner blue butterfly, its focus on habitat management is designed to benefit numerous other species that rely on the rare habitats in which the butterfly occurs. Through a state consultation process, implementation of the HCP will factor in considerations for other species listed under the Wisconsin endangered species laws.

The HCP was prepared by the DNR in collaboration with the partners and other participants. Because preparation, approval and implementation of the HCP are actions requiring environmental review, the DNR and the USFWS agreed to prepare a single environmental document (i.e. the EIS) that would comply with the requirements of both the Wisconsin and National Environmental Policy Acts (WEPA and NEPA), as well as other state and federal regulations. Both WEPA and NEPA are intended to help public officials make decisions that are based on an understanding of environmental consequences and to take actions that protect, restore and enhance the environment. Preparation of a joint document is encouraged under both WEPA and NEPA, thereby reducing paperwork and best using limited public resources, while ensuring broad public involvement.

The innovative approach to endangered resources conservation proposed in the HCP is designed to move regulated communities beyond compliance and into efforts to proactively apply conservation measures on the land while engaging in normal land management activities. The U.S. Congress, in establishing the incidental take permit provisions of the Endangered Species Act, expressed the hope that the provision would encourage creative partnerships between the public and private sectors and among governmental agencies in the interest of species and habitat conservation, as well as provide a framework to permit cooperation between the public and private sectors. Those goals are accomplished by the HCP, an effort that arose out of and has been developed through a solid and diverse grassroots effort.

The Karner blue butterfly

The greatest numbers of Karner blue butterflies and some of the best Karner blue butterfly habitat currently occur in Wisconsin. Karner blue butterflies have been extirpated from Ontario, Maine, Pennsylvania, Massachusetts, New Jersey, Ohio and Iowa; appear to be extirpated in Illinois; and persist in only remnant populations in Minnesota, Indiana, New York and New Hampshire. Because of these disappearances and the relative abundance of this species' populations in Wisconsin and Michigan, Wisconsin plays an important role in protecting Karner blue butterflies.

The Karner Blue Butterfly in Wisconsin

Wisconsin supports the largest and most widespread Karner blue butterfly populations worldwide. More than 270 Karner blue butterfly occurrences are known from 23 counties. Most of the occurrences can be grouped in about fifteen large population areas, and most of the larger populations are found on sizable contiguous acreages in central and northwest Wisconsin. . . .

Karner blue butterflies are found in close association with wild lupine (*Lupinus perennis*), the only known host plant for their larvae. Natural habitats that Karner blue butterflies occupy include sandy pine and oak barrens, pine prairies, oak savannas and some lake shore dunes. Current Karner blue butterfly habitat in Wisconsin includes abandoned agricultural fields, mowed utility and road rights-of-way, managed forest lands, military training areas and bombing ranges and managed barrens. Potential habitat at the specific site level can only occur where conditions exist to support wild lupine. Given the knowledge of certain ecological criteria relative to Karner blue butterflies such as the distribution of wild lupine, general soils information and climatic parameters, potential habitat is predictable. . . .

The Need for a Statewide Conservation Effort

While many extant Karner blue butterfly populations occur on public lands, it has become increasingly clear to conservation interests that species conservation cannot occur on public lands alone. There are simply not enough acres in public ownership to provide long-term stewardship. In addition, species like the Karner blue butterfly depend on active land management which results in the perpetuation of particular habitat types.

The amount of conservation that can be accomplished on private lands for which there are economic goals depends on landowner flexibility in time, space and financial strategy. In principle, the most intensive conservation measures for Karner blue butterflies are concentrated on public lands. However, the role of private lands cannot be underestimated. Large scale, multi-level conservation is best accomplished with statewide planning in which the participants accept various levels of responsibility for plan implementation. The *Wisconsin Statewide Karner Blue Butterfly HCP* was formulated with this concept in mind.

The HCP partners envision a *statewide* conservation effort that gains and incorporates the support of landowners and land users throughout Wisconsin. As the applicant for the incidental take permit, the DNR will administer the permit with the cooperation of all the partners, as well as other participants and cooperators. Regardless of their individual roles, each HCP partner has shown the capability and resolve to make significant contributions to the conservation effort through management practices, through public outreach, education and assistance programs, or through both management practices and public outreach and education. The 27 partners own or manage nearly two million acres of land in the state and have agreed to manage their lands in the Karner blue butterfly's high potential range with conservation considerations. Individual partner roles and commitments for both management and outreach are described in each individual partner's conservation agreement ("Species and Habitat Conservation Agreement") with the DNR, in the HCP, or in the DNR's Implementing Agreement with the USFWS.

The Karner blue butterfly is adapted to barrens and other early successional habitats. Because the persistence of these habitats is disturbance-dependent, an important aspect of this HCP is to provide for land management regimes that assure a balance between habitat gain from disturbance and habitat loss from vegetational succession. Stopping land management activities which provide desirable disturbance would be detrimental to maintaining this balance. A conventional "do not touch" regulatory approach, therefore, is inappropriate for the particular considerations presented in the conservation of the Karner blue butterfly. Such an approach would discourage, in many cases, the maintenance of habitat and conservation of the species. Consequently, this statewide conservation program may be distinct in its approach and application.

As applied to statewide landowner involvement in conservation, a participation strategy has been developed to provide incentives for conservation through cooperative partnerships. It includes a review mechanism to ensure that implementation does not adversely affect the species in the long-term. Although outreach is intended to reach any and all parties with the potential to become involved in Karner blue butterfly conservation, geographical areas and activities associated with the greatest potential will be given greater outreach emphasis. For example, broad general efforts will be made statewide and greater efforts will be made in the Karner blue butterfly's high potential range. The participation strategy also includes a notification system designed to inform landowners and users, where possible and feasible, of the opportunities presented under this HCP. Implementation of this strategy is intended to be at the county level, as much as possible, using county and town communication mechanisms and providing information and assistance locally.

Conservation Strategies

Insect conservation efforts are based on different premises than traditional vertebrate conservation efforts. The Karner blue butterfly, like most insect species, has adapted to survive by producing relatively large numbers of eggs and large populations, with short individual lifespans and frequent generation turnovers. Most of the Karner blue butterfly's life is spent in the egg and larval stages. Natural mortality rates during these immature life stages are much greater than mortality rates observed for vertebrate animals. The survival strategy of the Karner blue butterfly centers on the success of overall populations rather than individual organisms. To accommodate this strategy, a focus on habitat conservation and the maintenance of populations—rather than individuals—is key to butterfly preservation.

The long-term viability of Karner blue butterfly populations depends on habitat disturbance. Without periodic disturbance, natural woody succession shades out wild lupine and nectar plants and can passively eliminate Karner blue butterfly populations. Creation of new habitat to replace habitat lost to succession is therefore necessary. This reality underscores the need for managing landscapes for a dynamic, shifting mosaic of habitat and populations. Fortunately, many land management activities, such as those used in forest management and utility right-of-way maintenance, provide such disturbances.

There are two approaches being proposed in the HCP. . . .

Management with consideration for Karner blue butterflies. This management category represents lands owned or managed by partners on which consideration for the Karner blue butterfly and its habitat will be incorporated into land management activities. Partners have committed to management with consideration for Karner blue butterflies on 227,492 acres in the Karner blue butterfly's high potential range. *The long term biological goal on these lands is that Karner blue butterfly habitat gains be equal to or exceed losses occurring through natural succession or otherwise.*

Management to feature, protect or enhance Karner blue butterflies. This management category represents lands that are owned or managed by partners on which one of the primary management goals is to feature the Karner blue butterfly habitat or the broader barrens community that includes it. Partners have committed 37,725 acres in the Karner blue butterfly's high potential range to this level of focus. As with the management with consideration category, *the long term biological goal on these lands is that butterfly habitat gains be equal to or exceed losses occurring through natural succession or otherwise. Additional measures are taken however, to promote viable Karner blue butterfly populations despite potential economic costs.*

Overall, the partners own and maintain 2.03 million acres in Wisconsin. Of these 2.03 million acres, 265,217 acres in the high potential range will be subject to the management categories described above. Incorporating conservation into management activities on lands in the high potential range focuses efforts where they can have the most conservation benefit (the high potential range includes the Karner blue butterfly's documented range, as well as a significant buffer around known Karner blue butterfly occurrences).

Land Management Activities Affecting Karner Blue Butterfly

The partners engage in a variety of activities on the lands they own and manage. These activities have been grouped into five categories for the purposes of developing conservation strategies:

Forest Management
- Timber harvesting
- Stand improvement
- Prescribed fire
- Forest roads
- Forest regeneration

Barrens, Prairie and Savanna Management
- Prescribed fire
- Mechanical treatment
- Herbicide treatment
- Native plant propagation
- Grazing

Recreational Management
- Intensive construction
- Less intensive construction
- Maintenance
- Public use

Transportation Management
- Road development
- Road maintenance
- Vegetation control

Utility Right-of-Way Management
- Construction of electric transmission lines
- Operations and maintenance of electric transmission lines
- Vegetation control
- Construction of new pipelines and underground transmission lines
- Operations and maintenance of pipelines

Participation Strategy and Additional Conservation Measures

. . . The strategy seeks to incorporate conservation into the working landscape. The HCP is built upon the extensive land ownership and conservation commitments of the 27 partners identified in this HCP, but seeks to go beyond those partners to include the assistance and participation of other landowners, nonprofit groups, environmental and industrial organizations and a variety of governmental units.

The HCP, with its biological approach, focuses efforts on geographic areas and activities which provide the highest potential to safeguard or enhance Karner blue butterfly habitat. The strategy goes beyond the initial 27 partners and seeks to reach all landowners and users, but will vary in approach and process. The HCP's inclusion strategy therefore includes:

1. A category of non-voluntary participants that must formally apply for partnership to the DNR and receive a Certificate of Inclusion from the USFWS, because of the value their land and activities provide to conservation of the species;

2. An opportunity for voluntary participants that receive incidental take permit coverage, without further process, so as to encourage land management activities that may benefit the species;

3. An extensive public outreach and education plan to reach all landowners and land users, and others, to describe the effort and encourage their cooperation and participation in this conservation effort; and

4. A review of the participation strategy after three years of implementation to determine its effectiveness, with the option of modifying it should it prove to need changes to provide the anticipated conservation.

Through this HCP, the partnership intends to achieve the endangered species conservation goals while protecting the economic interests of non-federal landowners through this increasing partnership statewide. Strategies that support Karner blue butterfly conservation included in the HCP are as follows:

> *Outreach and Education. Efforts will be made to create awareness by potential mandatory participants, such as county and town highway departments, railroads, electric utilities and others. HCP partners will also encourage Karner blue butterfly conservation by those in the voluntary segment, such as the small private landowners and those in the agricultural community.*

> *Federal Recovery. In 1994, the USFWS appointed a federal recovery team for the Karner blue butterfly, which at the time of this writing has produced a working draft recovery plan. Representatives of three Wisconsin Statewide Karner Blue Butterfly HCP partners participated in the development of the draft recovery plan: the DNR, Consolidated Papers, Inc. and Georgia–Pacific Corporation.*

Two Wisconsin HCP Partners, the DNR and The Nature Conservancy, intend to participate in federal recovery efforts, as funding allows. In addition, Jackson and Eau Claire County Forests (also HCP partners), will consider participating in recovery, pending approval by their respective county boards.

Adaptive Management

The long-term viability of Karner blue butterfly populations depends on habitat disturbance. Without periodic disturbance, natural woody succession shades out wild lupine and nectar plants, eliminating Karner blue butterfly habitat. Given this reality, the absence of management activities that create beneficial disturbances is ultimately detrimental to Karner blue butterflies. Halting on-going management activities to wait until all unanswered questions about Karner blue butterflies are answered is impractical. Therefore, this HCP will be implemented using adaptive management.

Adaptive management is a formal, structured approach to dealing with uncertainty in natural resources management, using the experience of management and the results of research as an on-going feedback loop for continuous improvement. Adaptive approaches to management recognize that the answers to all management questions are not known and that the information necessary to formulate answers is often unavailable. Adaptive management also includes, by definition, a commitment to change management practices when determined appropriate. The HCP partners have committed to using adaptive management.

The adaptive management approach, in part, relies on the management activities and practices that are already in place and for which there is no existing data or obvious reasons that suggest a need for change. This approach allows for up-front conservation measures to be used during routine operations and as management practices are implemented on the landscape. However, under this approach, carefully designed monitoring and research procedures are initiated to determine if there are any effects of the management practices.

Research. Several Karner blue butterfly research efforts are already underway or have recently been completed. A summary of each activity is included in the HCP:

- Effects of Herbicides on the Development of Karner Blue Butterfly Eggs and Larvae Development
- Effects of Herbicide on Seed Germination (wild lupine) and Development
- Effects of Herbicide Application on Lupine and Select Nectar Plants
- Dispersal Research (Karner blue butterflies)

Notes and Questions

1. The Karner blue butterfly was the first insect listed under the ESA. Besides the Wisconsin population, the largest number of butterflies now live near Albany, New York (including a sizeable population amidst the runways of the Saratoga County Airport) and in the Indiana Dunes National Lakeshore along Lake Michigan near the steel mills of Gary,

Indiana. For additional information about the Karner blue butterfly, *see* CHARLES C. MANN & MARK L. PLUMMER, NOAH'S CHOICE: THE FUTURE OF ENDANGERED SPECIES 82–114 (1995); U.S. Fish & Wildlife Service Region 3, *The Karner Blue Butterfly*, <http://www.fws.gov/r3pao/eco_serv/endangrd/news/karnerbl.html>; David Lentz, *Big Plans for a Little Butterfly*, WISCONSIN NATURAL RESOURCES (June 1999), *available at* <http://www.wnrmag.com/stories/1999/jun99/butter.htm>.

2. Consider how much more ambitious the habitat conservation plan (HCP) for the Karner blue butterfly is than the HCP for the Mission blue butterfly. The Karner blue butterfly HCP encompasses 260,000 acres of land throughout central Wisconsin, whereas the Mission blue butterfly plan covers less than 3,000 acres. The Wisconsin HCP reflects the dramatic increase in the scope of HCPs in recent years. The most prominent efforts to develop HCPs that cover entire regions exist in southern California and around Austin, Texas. *See, e.g.*, MANN & PLUMMER, *supra*, at 178–211 (describing the efforts to develop a HCP in Austin); Marc J. Ebbin, *Southern California Habitat Conservation: Is the Southern California Approach to Conservation Succeeding?*, 24 ECOLOGY L.Q. 695 (1997). Over 20 million acres are now covered by HCPs.

The increased scope of HCPs also extends to the number of species. Unlike the Karner blue butterfly plan, many HCPs apply to lots of different species. For example, two of the California plans protect 63 and 47 species, respectively. Collectively, the 379 existing HCPs protect more than 200 endangered or threatened species. Additionally, many plans apply to species that have not yet been listed under the ESA.

3. Interior Secretary Bruce Babbitt trumpeted the development of HCPs as evidence of the success of the ESA. When he signed the Karner blue butterfly HCP, he described it as "the first comprehensive statewide Habitat Conservation Plan and the most inclusive agreement of its kind in the country.... It is an excellent example of how the flexibility of the Endangered Species Act can promote regional habitat conservation planning by states and local governments and is a model for what other states and their partners might consider." More generally, Babbitt described HCPs as "one of the vanguards in a Quiet Revolution in American conservation," adding that "I know these plans work. Businesses know they work. Individual citizens know they work." U.S. Fish & Wildlife Service, *Habitat Conservation Plans: The Quiet Revolution* 5 (1998), *available at* <http://endangered.fws.gov/hcp/Quiet/03–09.pdf>.

Environmentalists are more skeptical. They complain that HCPs are being developed without adequate scientific guidance. They further charge that the FWS is abandoning the safeguards that are supposed to characterize HCPs in its rush to approve more plans. Many of the recent HCPs, they say, are inadequately funded and monitored. They conclude that many HCPs are "habitat giveaways" that cater to the interests of developers rather than furthering the purposes of the ESA. *See, e.g.*, Patrick Parenteau, *Rearranging The Deck Chairs: Endangered Species Act Reforms in an*

Era Of Mass Extinction. 22 WM. & MARY ENVTL. L. & POL'Y REV. 227; John Kostyack, *"Surprise!"*, 15 ENVTL. FORUM 19 (1998).

Some environmental organizations have cautiously approved of HCPs, but only provided that steps are taken to make the plans more protective of the species they are designed to preserve. The National Wildlife Federation (NWF) advocates the inclusion of five additional safeguards in the HCP process: (1) clarify that HCPs must be consistent with an overall recovery strategy before approving them, (2) allow greater participation by concerned citizens and independent scientists, (3) encourage conservation strategies that prevent the need to list species, (4) provide regulatory assurances only when a credible adaptive management strategy is in place, and (5) ensure that HCPs are adequately funded to ensure that they are an effective strategy for conserving biodiversity. National Wildlife Federation, *Habitat Conservation Plans: Safeguards Are Needed to Ensure that the Endangered Species Act's Recovery Goal is Not Undermined,* <http://www.nwf.org/endangered/hcp/hcpsaf.html>. What objections might landowners raise to any of those recommendations? What other actions would improve the HCP process?

For more on the debate about HCPs, see REED F. NOSS, MICHAEL A. O'CONNELL & DENNIS D. MURPHY, THE SCIENCE OF CONSERVATION PLANNING: HABITAT CONSERVATION UNDER THE ENDANGERED SPECIES ACT (1997); *Symposium on Habitat Conservation Plans: Reshaping Habitat Conservation Plans For Species Recovery*, 27 ENVTL. L. 755 (1997); National Wildlife Federation, *Habitat Conservation Plans: The New Movement in Endangered Species Protection,* <http://www.nwf.org/endangered/hcp/index.html>; National Audubon Society, *Report of the National Audubon Society Task Force on Habitat Conservation Plans,* <http://www.audubon.org/campaign/esa/hcp-report.html>.

4. The attractiveness of HCPs to private landowners further increased as a result of the "No Surprises" rule adopted by the FWS, the NMFS, and NOAA. *See* Habitat Conservation Plan Assurances ("No Surprises") Rule, 63 Fed. Reg. 8859 (1998). According to that rule, "[o]nce an HCP permit has been issued and its terms and conditions are being fully complied with, the permittee may remain secure regarding the agreed upon cost of conservation and mitigation. If the status of a species addressed under an HCP unexpectedly worsens because of unforeseen circumstances, the primary obligation for implementing additional conservation measures would be the responsibility of the Federal government, other government agencies, or other non-Federal landowners who have not yet developed an HCP." *Id.* at 8867.

Many environmentalists object to the "No Surprises" rule. They contend that the dynamic nature of ecosystems makes it impossible to predict the needs of a species or the pressures on a species that will occur in future years. The FWS counters that the government is committed to doing whatever it takes to protect a species if the commitments embodied in a HCP prove to be inadequate. It also defends the scientific basis for its plans and its record of monitoring compliance with HCPs. And it emphasizes that the Congress that adopted the HCP provision in 1982 believed that "property owners should be provided economic and regulatory certainty regarding the overall cost of species conservation and mitigation, provided that the affected species were adequately covered by a properly functioning HCP, and the permittee was properly implementing the HCP and complying with the terms and conditions of the HCP permit in good faith." *No Surprises Rule*, 63 Fed. Reg. at 8860 (citing H.R. Rep. No. 835, 97th Cong., 2d Sess. 29 (1982)).

5. The "No Surprises" Rule is only one example of recent efforts to reduce the regulatory impact of the ESA on private parties. The FWS has adopted two regulatory reforms designed to encourage private landowners to protect biodiversity. A safe harbor agreement lifts the restrictions of the ESA from a private landowner whose voluntary habitat creation, restoration or improvement attracts a new species to the land. A candidate conservation agreement offers a similar promise of reduced ESA regulation to a landowner who takes agreed upon steps to conserve the habitat of a species that has not yet been listed under the ESA. For a discussion of each mechanism, *see* Darcy H. Kishida, Note, *Safe Harbor Agreements Under the Endangered Species Act: Are They Right for Hawai'i?*, 23 HAWAI'I L. REV. 507 (2001); Francesca Ortiz, *Candidate Conservation Agreements as a Devolutionary Response to Extinction*, 33 GA. L. REV. 413 (1999); J.B. Ruhl, *The Endangered Species Act and Private Property: A Matter of Timing and Location*, 8 CORNELL J.L. & PUB. POL'Y 37 (1998).

PART THREE

MANAGING ECOSYSTEM DIVERSITY

The Endangered Species Act (ESA) proclaims that one of its purposes is "to provide a means whereby the ecosystems upon which endangered species and threatened species depend may be conserved." 16 U.S.C. 1531(b). In other words, if species biodiversity is to be conserved, the natural settings within which species biodiversity is maintained–ecosystems–must also be conserved. Part II of this book explores how the ESA works toward this ecosystem conservation end. Notably, no other provision of the ESA so much as mentions the word ecosystem. Instead, the statute adopts a species focus as a surrogate for ecosystem-level conservation. The ESA has been criticized for this failure to place ecosystem conservation at the forefront of the planning, regulatory, and enforcement provisions of the statute. On the other hand, as the materials in Part II demonstrate, the species focus has carried the ESA far, albeit indirectly, toward accomplishing biodiversity conservation in many settings.

Of course, there is no law of national scope that could fairly be called the "Ecosystem Conservation Act." (We would have mentioned it by now!) Rather, the ESA and a collection of other laws contain elements and programs that can explicitly or impliedly be advanced toward developing ecosystem-level policies designed to conserve biodiversity. In Part III of this book we examine this bundle of laws under the heading of "ecosystem management." The opening chapter provides an introduction to the concept of ecosystem management as it has progressed in science, policy, and law. In the remaining chapters we survey the law of ecosystem management as it applies in different ecosystem settings: forests, grasslands, freshwater, coastal and marine, and selected "extreme" ecosystems.

CHAPTER 7

INTRODUCTION TO ECOSYSTEM MANAGEMENT LAW

Chapter Outline:

The term "ecosystem management law" includes three components, each of which poses its own set of challenges. First, we must develop some sense of the underlying subject matter—ecosystems. What are they? Where are they? What happens in them? What threatens to disrupt them? In their purest form these are questions of *science*, of the discipline of ecology. But in this case we seek the answers to these questions in order to develop *policy*, that of ecosystem management. Given what we know about ecosystems, what are we trying to accomplish when we think of managing them? How will we know whether we have succeeded? Who gets to decide these questions? As difficult as it is to hash out these policy questions, the third challenge requires that we reduce the decisions to *law*, that is, to legal script. What is the legal definition of an ecosystem? What legal process must be followed for implementing and challenging policy decisions? How will courts review agency decisions implementing ecosystem management policy? Consensus on these questions, as well as the pertinent questions of science and policy, is the extreme exception. This chapter thus delves deeper into the questions surrounding these three components of ecosystem management law in order to provide an overview of the core issues the law faces in the particular ecosystem settings taken up in the remaining chapters of this part of the text.

Before going there, however, some thought must be given to the question, why have *any* law of ecosystem management? The Endangered Species Act may funnel ecosystem protection concerns through a species-specific focus, but why isn't that sufficient to address whatever concerns we have about ecosystems? The answer has much to do with endangered species and much not to do with them. The connection to endangered species is obvious: with habitat loss as the leading cause of endangered status, conservation of the ecosystems within which endangered species exist will be vital to their recovery. Clearly, however, any hope of limiting future additions to the list of endangered species will depend on a habitat conservation policy that is at least in part decoupled from the ESA—that is, which does not require the presence of an endangered species to trigger a response. Moreover, humans may value, and thus wish to manage, ecosystems for reasons completely apart from any concern about species. Even the desire to use an ecosystem primarily for human benefit, such as for timber production or water purification, implies some need for management.

Still, that there may be some potential purposes for ecosystem management law does not answer the question of whether we *need* ecosystem management law now. Other than ecosystems within which endangered species exist, which we could reasonably assume can be managed through the authority of the ESA, is there an "ecosystem problem" in the United States? Most ecologists agree there is, based on any plausible definition of ecosystem. Groundbreaking work on the state of the nation's ecosystems appeared in 1995 through the work of Reed Noss and fellow researchers in a report for the National Biological Service of the U.S. Department of the Interior, *see* REED F. NOSS ET AL., BIOLOGICAL REPORT 28, ENDANGERED ECOSYSTEMS OF THE UNITED STATES: A PRELIMINARY ASSESSMENT OF LOSS AND DEGRADATION (1995), available at <http://biology.usgs.gov/pubs/ecosys.htm>, and in a report prepared for the Defenders of Wildlife, *see* REED F. NOSS AND ROBERT L. PETERS, ENDANGERED ECOSYSTEMS: A REPORT ON AMERICA'S VANISHING HABITAT AND WILDLIFE (1995), available at http://www.defenders.org/pubs/eco00.html. Noss and his colleagues explained that "ecosystems can be lost or impoverished in basically two ways. The most obvious kind of loss is quantitative—the conversion of a native prairie to a corn field or to a parking lot.... The second kind of loss is qualitative and involves a change or degradation in the structure, function, or composition of an ecosystem." Their work assessed ecosystem status in the United States across these two scales. On a national level, the magnitude of quantitative and qualitative ecosystem decline since prior to European settlement is significant. Some indicia of the level of loss and degradation include:

- 177 million acres of wetlands lost—a loss of more than 50 percent
- 25 million acres of ancient forests lost in the Northwest—a 90 percent loss
- 22 million acres of native grassland lost in California

- 360,000 miles of roads in national forests
- 270 million acres of public lands affected by grazing

The study concluded that the overall risk to ecosystems is extreme in seven southeastern states, Texas, California, and Hawaii, high in 25 states, and moderate in only 15 states. Twenty-one ecosystems rank as the nation's most endangered, listed here in descending order based on extent of decline, present area (rarity), imminence of threat, and number of federally listed threatened and endangered species associated with each type:

- South Florida landscape
- Southern Appalachian spruce-fir forest
- Longleaf pine forest and savanna
- Eastern grassland, savanna and barrens
- California native grassland
- Coastal communities in conterminous 48 states
- Southwestern riparian forest
- Hawaiian dry forest
- Large streams and rivers in conterminous 48 states
- Cave and karst systems
- California riparian forest and wetland
- Florida scrub
- Old-growth eastern deciduous forest
- Old-growth forests of Pacific Northwest
- Old-growth red and white pine forest of Great Lakes states
- Old-growth ponderosa pine forest
- Midwestern wetlands
- Southern forested wetlands

Similar findings were made more recently by a multi-nation commission study reported in NORTH AMERICAN COMMISSION FOR ENVIRONMENTAL COOPERATION , THE NORTH AMERICAN MOSAIC: A STATE OF THE ENVIRONMENT REPORT (2002), which found the situation a "widespread crisis." Of course, the presence of many imperiled ecosystems throughout much of the United States does not lead inexorably to a particular design for the law of ecosystem management. People offer many different explanations of the causes of ecosystem decline, and thus what should be done about it in law. Indeed, the design of comprehensive ecosystem management law is a question that has been pondered seriously only since the early 1990s.

The absence of a unified, national ecosystem management statute thus presents both opportunities and obstacles to the development of ecosystem management law. The many federal and state laws dealing explicitly or potentially with facets of ecosystem management allow for experimentation and flexibility in dealing with the challenge of forging policy in varied natural settings. Like ecosystems, ecosystem management law must be diverse, dynamic, and evolving. On the other hand, important overarching issues of ecosystem management may require common understanding and uniform treatment throughout the nation, which is complicated when the law is fractured into many statutes administered by many political units. To assist in the examination of how these laws are or are not evolving toward the ecosystem management theme, this chapter provides a foundational understanding of the three core issues: What is an ecosystem? What is ecosystem management? And what is ecosystem management law?

A. WHAT IS AN ECOSYSTEM?

Like the term "biodiversity," the term "ecosystem" is simply a human invention used to represent what we perceive to be happening in nature. *Eco* represents ecology; *systems* are assemblages of parts forming a complex or unitary whole. What we have in mind with *ecosystems*, therefore, is the sense that the physical and biological parts of nature—water, air, species, etc.—assemble into complex, interacting wholes. From there, however, agreement on what constitutes an ecosystem, particularly for purposes of formulating ecosystem management law, is elusive. In its *People's Glossary of Ecosystem Management Terms*, the United States Forest Service defines ecosystem as:

> An arrangement of living and non-living things and the forces that move among them. Living things include plants and animals. Non-living parts of ecosystems may be rocks and minerals. Weather and wildfire are two of the forces that act within ecosystems.

That is a fairly broad sweep, but some definitions—and many have been offered—are even broader. Testifying before Congress as Secretary of the Department of the Interior, Bruce Babbitt once observed that "an ecosystem is in the eye of the beholder." That may be true for purposes of conceptualizing ecosystems, but that will not suffice for making ecosystem management *law*.

The challenge, therefore, is in defining the concept of ecosystem in such a way as to provide a metric for formulating management objectives and promoting the development of policy and law. There is no obvious spatial or functional unit to select. As the U.S. Fish and Wildlife Service has observed, an ecosystem could reasonably be defined as anything "from a drop of water to the North American continent to the entire biosphere." U.S. FISH AND WILDLIFE SERVICE, AN ECOSYSTEM MANAGEMENT APPROACH TO FISH AND WILDLIFE CONSERVATION: AN APPROACH TO MORE EFFECTIVELY CONSERVE THE NATION'S BIODIVERSITY 6 (1994). Some critics have used this trait to question the wisdom of attempting to forge policy and law around any concept that is so amorphous. Sound natural resources policy, they say, requires the ability to designate unambiguous ecosystem boundaries on maps, so that the jurisdiction and application of government authority can be clearly delineated. *See* ALLAN K. FITZSIMMONS, DEFENDING ILLUSIONS: FEDERAL PROTECTION OF ECOSYSTEMS 34–39 (1999). On the other hand, few people would mistake the Everglades for the Rocky Mountains, suggesting that there are some unambiguous distinguishing characteristics with which to work after all. Nevertheless, the demand for a metric of ecosystems that is not only scientifically accurate, but also politically and legally viable, cannot be ignored.

The law of ecosystem management thus will require some sort of framework for delineating and describing the ecosystems being managed. In these materials we explore five different approaches for that purpose. First, we must have some way of identifying where the ecosystem is located. One approach is to use *geographic units*, such as an island, a forest, or a lake. Several federal agencies have sliced up the ecosystem landscape using this approach, as in the U.S Fish and Wildlife Service's map of watershed-based ecosystem geography (see Figure 1).

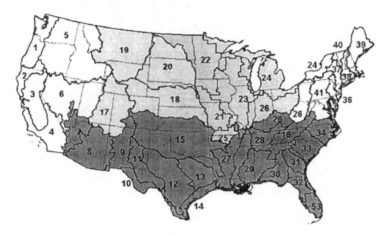

The United States Fish and Wildlife Service has identified and defined boundaries for 53 ecosystem units in the contiguous states according to watershed geography units.
U.S. FWS

But there are other ways to slice up the landscape, as the Bureau of Land Management has done by using its conception of "ecoregions:"

The Bureau of Land Management's ecoregion-based division of the lower 48 states
U.S. BLM

And as the U.S. Environmental Protection Agency has suggested with its "Large Ecosystem" map:

USEPA Watershed Information Network Water Atlas

As these maps illustrate, moreover, the geography of ecosystems does not always correspond to the way the political and social landscape is divided. We may want to describe ecosystems, therefore, according to *management units*, such as a park, a county, or a farm. Once we have identified where the ecosystem is located, we are likely to be interested in what is happening within and to its physical and biological systems. In this sense, ecosystems consist of *functional units*, such as nutrient cycles, energy transfers, and chemical processes. These functional processes support the biodiversity found in the ecosystem, and for some people the primary goal of ecosystem management is to ensure the continued integrity of the processes that maximize biodiversity. The processes in some ecosystems, however, are managed for other purposes. A private timber forest, for example, is managed to maximize the *commodity units* of timber value. A public water supply company, by contrast, might manage the forest in the watershed of its water supply reservoir to take advantage of the *ecosystem service units* of water purification.

These five ways of thinking about ecosystems are not mutually exclusive, but neither are they entirely compatible. Management units such as parks may contain all of a geographic ecosystem unit, or only part of several units. Functional processes in ecosystems often transcend both geographic and management boundaries. Managing a forest for commodity value of the timber may result, incidentally, in maximizing some ecosystem service values beneficial to human populations, but may sacrifice some functional processes beneficial to biodiversity. In different settings, therefore, there may be scientific and policy-based advantages and disadvantages to emphasizing one or a particular combination of the five approaches in the law of ecosystem management. The following materials explore the challenges that await the law of ecosystem management as it searches for the appropriate mix of these approaches to defining ecosystems in different contexts.

1. SCIENTIFIC PERSPECTIVES

John M. Blair, Scott L. Collins and Alan K. Knapp, Ecosystems As Functional Units In Nature

14 Natural Resources & Environment 150 (2000).

In the last two decades or so, the word "ecosystem" has become increasingly used by the media and in nonscientific public and private sectors to denote some portion of the natural world, ranging in size and scope from a rotting log to the entire planet on which life as we know it exists. It also has become associated with environmental issues, and with certain land use practices and management approaches (e.g., ecosystem management). This variable and sometimes vague use of the term ecosystem has led to some confusion regarding the meaning of this important ecological concept. Our objectives in this essay are to define the ecosystem concept and briefly describe ecosystem ecology as a scientific discipline. We then provide some examples of ecosystem research and how human activities can impact ecosystems. Finally, we link ecosystem research to the recently conceived concept of ecosystem management and discuss the potential for using knowledge of how ecosystems function to achieve desired management goals.

The ecosystem concept has been used in many contexts. In the scientific realm, the term ecosystem is used most often to describe a relatively discrete unit of nature, such as a lake, a grassland, or a forested mountain valley. Such a view is consistent with much current ecosystem-based research in which the units of study are often well-delineated watersheds, or catchment basins, which are areas defined by topographic features and a common hydrologic drainage. Ecosystem research also includes studies of processes within these entities, such as the flow of energy from plants to consumers (e.g., animals, fungi, bacteria), or the processes determining the amount of soil nitrogen available for use by plants. Research concerned with these processes is a cornerstone of ecosystem science, but such research is not necessarily constrained by a need for ecosystems with distinct physical boundaries. A third use of the ecosystem concept is based on the application of principles derived from ecosystem ecology to achieve certain management goals. An example of this usage would be the Greater Yellowstone Ecosystem that includes Yellowstone and Grand Teton National Parks, surrounding public and private lands, human settlements, and multiple types of other land use. Thus the ecosystem concept can apply to the description and study of a distinct entity, such as a grassland or forested watershed; a process or collection of processes, such as nitrogen cycling; or a management unit like the Greater Yellowstone Ecosystem. Use of the ecosystem concept in all of these contexts is appropriate, when defined explicitly.

Ecosystem ecology as a scientific discipline has relatively recent origins. Although the term ecosystem was coined in 1935 by the British ecologist Arthur Tansley, ecosystem ecology did not begin to flourish as a scientific discipline until the 1960s. Several factors, including an influx of new funding for ecological research, an increase in academic positions at universities, the development of computer technology, and an increased awareness of the effects of human activities on the environment, combined to enhance the development of ecosystem ecology as an important new area of ecological study. * * * Together, these factors enhanced the growth of ecosystem science in the U.S. and the relevance of this discipline for addressing important environmental issues. Indeed, ecosystem science was one of the first of ecological disciplines to explicitly include the impact of human populations and activities as part of its comprehensive research agenda.

Just What Is an Ecosystem?

An ecosystem can be described in simple terms as a biological community (all of the organisms in a given area) plus its abiotic (nonliving) environment. In fact, the word ecosystem was first used by Tansley to describe natural systems in a way that encompassed all of the living organisms occurring in a given area and the physical environment with which they interact. In this sense, the ecosystem is the first level in the traditional hierarchical arrangement of biological systems. It explicitly includes both living organisms and the abiotic environment as integral parts of a single system. This is one reason that ecosystem studies often focus on quantifying transfer of energy and materials between living organisms and the physical environment.

Tansley originally described the ecosystem as part of a continuum of physical systems in nature, and in fact the concept of "system" and much of the language used to describe ecosystem structure and function are borrowed from physics. In very general terms, a system is any collection of components, interacting in an organized manner to form a unit, through which may flow materials, energy and information. An ecosystem is a particular kind of system, defined by Eugene P. Odum, a leading proponent of ecosystem ecology, as any unit in nature that includes all the organisms that function together in a given area and their abiotic environment, interacting so that energy flows lead to biotic structure and material cycles. ODUM, BASIC ECOLOGY (1983).

The definition above implies that ecosystems occupy a given area, and that they therefore have boundaries (at least conceptually). Indeed, some ecosystems have fairly obvious, distinct boundaries (a lake, an urban forest, a distinct watershed); but more often the boundaries of an ecosystem are much less distinct and may be user-defined. The concept of the ecosystem as a real physical entity sometimes presents problems, especially in terrestrial habitats, where it is often difficult to say where one ecosystem ends and another begins. Further, boundaries which might be appropriate for quantifying hydrologic fluxes or nutrient budgets may not apply to more mobile or migratory organisms. However, the difficulty in defining precise boundaries does not negate the value of the ecosystem concept, and many ecosystem ecologists simply define their ecosystems based on the area under study. Indeed the same difficulties in defining precise boundaries apply to natural populations and communities; yet these organizational levels are widely accepted and used by ecologists and provide the basis for many management-related decisions.

* * *

A Process–Based View of Ecosystems

An alternate way of viewing an ecosystem is based on the processes that comprise ecosystem functioning. These processes include productivity, energy flow among trophic levels, decomposition, and nutrient cycling. In this way, an ecosystem can be defined based on the collection of processes required to continue normal functioning, as opposed to the physical boundaries within which those processes operate. Defining what constitutes "normal" functioning is a critical issue, with important implications for preserving and managing ecosystems. Some early proponents of the ecosystem concept viewed ecosystems as highly integrated collections of organisms which would achieve a stable equilibrium (the "balance of nature") in the absence of outside disturbances. However, this view has largely been replaced by a non-equilibrium perspective in which disturbances are seen as natural phenomena. Thus, defining a natural disturbance regime and the boundaries of ecosystem functioning are challenges for both ecosystem scientists and ecosystem managers.

Ecosystems function by processing energy and materials, operating under certain internal constraints or rules and producing certain outputs. Systems, by definition, have boundaries that delimit them from their surroundings, or their environment. An important aspect of ecosystems, however, is that they are "open" systems with respect to both materials and energy. They are subject to an input environment (beyond the system boundaries), which includes inputs of energy (e.g., sunlight) and materials (e.g., water and nutrients). These inputs can cause changes in the internal components of an ecosystem (compartments called state variables) over time. The interactions of inputs and state variables affect ecosystem processes and result in ecosystem outputs, or an output environment. For example, inputs of water and carbon dioxide in combination with state variables such as soil nutrients and plants leads to primary productivity— an ecosystem process—that may be consumed by grazers and transported to another ecosystem. Inputs that play a major role in altering ecosystem components and outputs are also referred to as forcing functions. For example, sunlight, water and nutrients are important forcing functions that affect the structure and functioning of most ecosystems. Natural disturbances, such as periodic fires or hurricanes, also play a role in affecting ecosystem structure and function. It is also important to note that the open nature of ecosystems means that the output environment of one ecosystem becomes part of the input environment of another. Thus, flows of material and energy link ecosystems in the biosphere.

Although ecosystems are open with respect to flows of both energy and materials, the behaviors of energy and materials, once they are in ecosystems, fundamentally differ. Energy transformations are essentially one-way flows, and thus energy cannot be recycled or reused within an ecosystem. Nutrients, however, can circulate or cycle within an ecosystem, and the flow of nutrients among different compartments is referred to as nutrient cycling. Nutrients include substances such as phosphorus, nitrogen, potassium, and calcium that are required by living organisms. Because the flows of chemical nutrients typically involve both biological (organic) and geological (inorganic) pools, nutrient cycles also are referred to as biogeochemical cycles. Ecosystem ecologists are often interested in describing and quantifying the storage and movement of nutrients among these pools, as well as understanding the factors that regulate these patterns of movement. This is important from both theoretical and practical perspectives. Unwanted, and unintended, changes in nutrient cycles can have profound consequences for environmental quality and human populations (e.g., groundwater contamination, acid rain, global climate change, eutrophication of lakes).

* * *

Ecosystem management

Ecosystem management has been defined as "management driven by explicit goals, executed by policies, protocols, and practices, and made adaptable by monitoring and research based on our best understanding of the ecological interactions and processes necessary to sustain ecosystem composition, structure, and function." Norman L. Christensen et al., *The Report of the Ecological Society of America Committee on the Scientific Basis for Ecosystem Management*, 6 ECOLOGICAL APPLICATIONS 665 (1996). This definition includes both common uses of the term ecosystem discussed earlier—the ecosystem as a functional and identifiable unit in nature—and ecosystem processes (such as nitrogen cycling). In this case the goal is to manage a defined ecosystem, often to provide a product (e.g., water, wood products, beef) while sustaining natural processes that include ecosystem services, such as purification of air and water, regeneration of soil nutrients, and maintenance of biodiversity. As noted by Costanza, d'Arge and de Groot in *The Value of the World's Ecosystem Services*, 387 NATURE 253–260 (1997), the monetary value of these services is tremendous. As our understanding of the processes occurring within and among ecosystem components increases, ecosystem management may provide a viable strategy for sustainable use of natural resources. For this to be realized, knowledge of the inputs, processes and cycles, and outputs of a managed ecosystem is required. This may be most feasible in clearly defined ecosystems, ... where the natural boundaries are relatively well defined and encompass most of the major components necessary for ecosystem functioning. Ecosystem management of less well delineated or more fragmented ecosystems (e.g., a forest fragment surrounded by agricultural fields) will be far more challenging because of the increased interactions with surrounding ecosystems and the potential for greater movement of materials and organisms from the surrounding landscape into and out of the ecosystem of interest.

Such challenges should not be viewed as impediments to ecosystem management, however. The benefits of ecosystem versus species-level management are numerous. Attempts to manage species as entities isolated from their biotic and abiotic environment (the ecosystem) are doomed to failure and the costs associated with the loss of ecosystem services due to mismanagement can be astounding. Clearly, the open nature of ecosystems and their lack of discrete boundaries should not be used as excuses for failing to design policies to manage them. Our legal system is replete with regulations that cross boundaries (e.g., interstate laws) and regulate processes and phenomena that are not discrete entities (e.g., air quality).

* * *

The ecosystem concept is often applied to well-defined and relatively small geographic entities in nature, where the input and outflow of energy and materials is reasonably well delineated. The main goal of ecosystem ecology as a research discipline is to understand, in detail, the processes of energy capture and the conversion of matter (e.g., nitrogen, carbon) and energy (e.g., sunlight) from one state to another within ecosystems, and the rules that govern these processes. The ecosystem concept, however, can also be applied to systems in nature that are less clearly bounded, including larger geographic areas that often contain several smaller ecosystems in whole or in part. These ecosystems may be more difficult to manage, unless management is scaled to definable subunits. This explains, perhaps, why ecosystem management is better developed in some ecosystems (forested watersheds) than others (coastal ocean areas, large rivers). In fact, it may be more difficult to apply ecosystem management approaches to large ecosystems that interact in many complex ways with the surrounding landscape, as is the case with coastal estuaries. Perhaps that is why many of the most difficult and unpredictable environmental problems (e.g., outbreaks of the toxic marine dinoflagellate *Pfeisteria*) occur in such complex, open systems. The challenges associated with ecosystem management highlight the importance of basic research in ecosystem ecology for understanding how ecosystems work at many different spatial scales. Ecosystem components represent the basic building blocks of complex food webs, of which humans are a part, and ecosystem processes provide many services upon which humans rely. As the impacts of human activities on ecosystem structure and function increase through time, a better understanding of the workings of both natural and managed ecosystems can be combined with predictive models to help manage ecosystems in a sustainable way.

Notes and Questions

1. One problem in using geographic units to define ecosystem boundaries is that there is more than ample disagreement over which unit to use. The principal units discussed in the scientific literature are watersheds, biomes, and ecoregions. Their working definitions are:

> watershed: The entire topographic region within which apparent surface water runoff drains to a specific point on a stream or to a body of water such as a lake or bay.

> biome: A major regional community of plants and animals with similar life forms and environmental conditions. It is the largest geographical biotic unit, named after the dominant type of life form, such as rain forest, grassland, or coral reef.

> ecoregion: Regions of relative homogeneity of biotic, abiotic, terrestrial, and aquatic ecological components (e.g., physiography, soils, vegetation, geology, climate, and land use) within which the mosaic of characteristics is holistically different than adjacent regions.

None of these is fully satisfying for purposes of defining ecosystem boundaries. Watersheds are hard to define in flat and arid areas. Ecoregions can require sophisticated data and mapping analysis to delineate. Biomes can be large areas containing many watershed and ecoregions. See generally G. E. Griffith et al., *Ecoregions, Watersheds, Basins, and HUCs: How State and Federal Agencies Frame Water Quality*, 54 JOURNAL OF SOIL AND WATER CONSERVATION 666 (1999); James M. Omernik and Robert G. Bailey, *Distinguishing Between Watersheds and Ecoregions*, 33 JOURNAL OF THE AMERICAN WATER RESOURCES ASSOCIATION 935 (1997); M. LYNNE CORN, CONGRESSIONAL RESEARCH SERVICE REPORT FOR CONGRESS, ECOSYSTEMS, BIOMES, AND WATERSHEDS: DEFINITIONS AND USE, 93–655 ENR (1993). Some commentators argue that all three units must be used in combination to effectively manage ecosystems, while others argue that doing so will impede the formation of coherent ecosystem management policy, thus one unit should be adopted as the dominant unit. *Compare* Griffith et al., *supra*, and Omernik and Bailey, *supra* (use combinations of units), *with* J. B. Ruhl, *The (Political) Science of Watershed Management in the Ecosystem Age*, 35 JOURNAL OF THE AMERICAN WATER RESOURCES ASSOCIATION 519 (1999) (adopt watersheds as the dominant unit). What do you think?

2. As Blair et al. suggest, most scientific definitions of ecosystems, once they go beyond the obvious statement that ecosystems are interacting collections of biological and physical parts, focus on the *process* characteristics of an ecosystem as a·means of defining its boundaries. It is what happens in an ecosystem that makes it more than just the sum of its parts or a set of lines on a map. Consider, however, whether this manner of describing ecosystems provides a politically and legally useful metric. How would you draft the *legal* definition of "ecosystem functions" or "ecosystem processes" in such a way as to describe clearly for agencies, courts, and citizens where the ecosystems are?

3. The Department of Interior's Bureau of Land Management recently outlined the ecosystem functions it believes should be managed on federal public rangeland, which is located primarily in states west of the Mississippi and leased extensively to private ranchers for grazing cattle and sheep. *See* 60 Fed. Reg. 9894 (Feb. 22, 1995). Some of the function-management objectives the agency specified include:

> Maintaining or promoting adequate amounts of vegetative ground cover, including standing plant material and litter, to support infiltration, maintain soil moisture storage, and stabilize soils.

> Maintaining or promoting subsurface soil conditions that support permeability rates appropriate to climate and soils.

> Maintaining, improving or restoring riparian-wetland functions including energy dissipation, sediment capture, groundwater recharge, and stream bank stability.

> Maintaining or promoting stream channel morphology (e.g., gradient, width/depth ratio, channel roughness and sinuosity) and functions appropriate to climate and landform.

> Maintaining or promoting the appropriate kinds and amounts of soil organisms, plants and animals to support the hydrologic cycle, nutrient cycle, and energy flow.

Do these adequately translate Blair's description of ecosystem functions into administrative standards capable of being easily understood and implemented without difficult interpretation? How would an agency, or an interest group watching over the agency's management decisions, conclude whether a decision the agency made maintains, degrades, or promotes any of these specified functions?

4. Blair contends that "the benefits of ecosystem versus species-level management are numerous." Do you agree? What implications does this view have for the future of the Endangered Species Act in the formation of ecosystem management law? Should an ecosystem approach replace or supplement the species approach of the ESA? Most conservation biologists, not surprisingly, argue that both approaches are needed. As Noss et al. argue:

> Ecosystem conservation offers several advantages over a species-by-species approach for the protection of biodiversity: it directly addresses the primary cause of many species declines (habitat destruction), it offers a meaningful surrogate to surveying every species, and it provides a cost-effective means for simultaneous conservation and recovery of groups of species. The species-by-species approach—although extremely important to our efforts of saving biodiversity—is inefficient.... As the public becomes more familiar with the evidence that entire ecosystems or groups of species have declined and that saving

individual species under the Endangered Species Act of 1973 does not solve all conservation problems and does not necessarily prevent the need for future listings, the rationale for ecosystem management becomes more apparent.

REED F. NOSS ET AL., BIOLOGICAL REPORT 28, ENDANGERED ECOSYSTEMS OF THE UNITED STATES: A PRELIMINARY ASSESSMENT OF LOSS AND DEGRADATION 6 (1995).

5. Sir Arthur George Tansley (1871–1955), mentioned in Blair's article as the founder of the concept of ecosystems, was educated at Highgate School, University College, London, and Trinity College, Cambridge. He returned to University College as assistant to Professor F. W. Oliver until 1906, when he moved to Cambridge as a lecturer. In 1901 he founded the *New Phytologist*, an influential botanical journal which he continued to edit for thirty years. Tansley took a prominent part in the development of plant ecology in Britain. He was instrumental in founding the British Ecological Society in 1913, and edited its *Journal of Ecology* for many years. In 1927 he was appointed Sherardian Professor of Botany at Oxford from which he retired with the title of Professor Emeritus in 1937. His classic book *The British Islands and their Vegetation* appeared in 1939. The last fifteen years of his life were largely devoted to working for nature conservation in Britain, a subject which had long been of concern to him. He was knighted in 1950 while serving as the first chairman of the Nature Conservancy.

6. Tansley's contemporary in the United States, Barrington Moore, described the dawn of ecology sciences in the 1920 inaugural issue of *Ecology*, the American counterpart to Britain's *Journal of Ecology*:

> There have been three stages in the development of the biological sciences: first, a period of general work, when Darwin, Agassiz and others amassed and gave their knowledge of such natural phenomena as could be studied with the limited methods at hand; next, men specialized in different branches, and gradually built up the biological sciences which we know today; and now has begun the third or synthetic stage. Since the biological field has been reconnoitered and divided into its logical parts, it becomes possible to see the interrelations and to bring these related parts more closely together. Many sciences have been developed to the point where, although the field has not yet been fully covered, contact and cooperation with related sciences are essential to full development. Ecology represents the third phase.

Barrington Moore, *The Scope of Ecology*, 1 ECOLOGY 3 (1920). Although the "methods at hand" for scientific discourse have progressed tremendously since Moore wrote those words, principally through the computer, his theme of interdisciplinary synthesis remains at the forefront of the biological sciences. *See* EDWARD O. WILSON, CONSILIENCE: THE UNITY OF KNOWLEDGE (1998).

2.   POLICY PERSPECTIVES

The system of metrics chosen for defining ecosystems should have scientific validity, but it also has much to do with the underlying question of policy objective. For example, one concern some people have with the view of ecosystems as functional process units in nature is that it can lead to a primarily biocentric approach to ecosystem management—i.e., we should manage ecosystems above all else to preserve natural ecosystem functions where and as they operate. Of course, for many people that is precisely the approach that is needed to rebut the anthropocentric view of nature as primarily a source of commodities humans can sever from nature and use to serve human needs, such as lumber, minerals, and water. Given their vastly different objectives, a commodity-based view of ecosystems is likely to adopt a much different metric for ecosystem management than will a preservation-based view. This form of economy versus environment debate has plagued environmental law for decades. As Senator Hatfield suggested when he introduced the Ecosystem Management Act of 1994 (reproduced *infra*), developing ecosystem management policy will require that we "look beyond the polarized positions of 'economic growth' and 'environmental protection' which have crippled our system of land management planning and implementation in recent years." 141 Cong. Rec. S 320 (Jan. 4, 1995).

The functional view of ecosystems, however, does not necessarily align with only a biocentric set of policy objectives for resources management. Likewise, even a commodity-oriented view of ecosystems must acknowledge that ecosystem functions are essential to the goal of continued delivery of an ecosystem's market-valued goods. But the functions themselves also provide economic value to humans in the form of natural *services*. These include:

- mitigation of droughts and floods
- purification of air and water
- generation and preservation of soils and renewal of their fertility
- detoxification and decomposition of wastes
- pollination of crops and natural vegetation
- dispersal of seeds
- cycling and movement of nutrients
- control of the vast majority of potential agricultural pests
- maintenance of biodiversity
- protection of coastal shores from erosion by waves
- protection from the sun's harmful ultraviolet rays
- partial stabilization of climate
- moderation of weather extremes and their impacts

Estimates of the value of these services to humans are staggering at the global scale. In 1997, the ecologist Robert Costanza and a team of researchers gained headline news when they published an article in *Nature* magazine estimating the global value of nature's services at over $33 trillion. In the book on the topic published later that year, NATURE'S

SERVICES: SOCIETAL DEPENDENCE ON NATURAL ECOSYSTEMS (Gretchen Daily ed., 1997), the biologist Gretchen Daily and other authors from the fields of ecology and economics fused decades worth of path-breaking research and scholarship on the economic value *to humans* of ecosystem functions.

There being no more unambiguous yardstick of value than money, legal scholars quickly adapted the concept of nature's services as an answer to the quest for a scientifically *and* politically viable metric for defining ecosystems and the objectives of ecosystem management. For example, in his review of *Nature's Services*, the first part of which is excerpted below (the remaining parts are reproduced *infra*), law professor James Salzman provides a framework for beginning to think of ecosystem functions in anthropo*metric*, and thus potentially more policy-viable, terms:

James Salzman, Valuing Ecosystem Services

24 Ecology Law Quarterly 887 (1997).

INTRODUCTION

Beneath the Arizona desert sun on September 26, 1991, amid reporters and flashing cameras, eight men and women entered a huge glass-enclosed structure and sealed shut the outer door. Their 3.15 acre miniature world, called Biosphere II, was designed to re-create the conditions of the earth (modestly named Biosphere I). Built at a cost of over $200 million, Biosphere II boasted a self-sustaining environment complete with rain forest, ocean, marsh, savanna, and desert habitats. The eight "Bionauts" intended to remain inside for two years. Within sixteen months, however, oxygen levels had plummeted thirty-three percent, nitrous oxide levels had increased 160–fold, ants and vines had overrun the vegetation, and nineteen of the twenty-five vertebrate species and all the pollinators had gone extinct. Eden did not last long.

What went wrong? With a multi-million dollar budget, the designers of Biosphere II had sought to re-create the level of basic services that support life itself—services such as purification of air and water, pest control, renewal of soil fertility, climate regulation, pollination of crops and vegetation, and waste detoxification and decomposition. Together, these are known as "ecosystem services," taken for granted yet absolutely essential to our existence, as the inhabitants of Biosphere II ruefully learned. Created by the interactions of living organisms with their environment, ecosystem services provide both the conditions and processes that sustain human life. Despite their obvious importance to our well-being, recognition of ecosystem services and the roles they play rarely enters policy debates or public discussion.

The general ignorance of ecosystem services is partly the result of modern society's dissociation between computers, cars and clothing on the one hand and biodiversity, nutrient cycling, and pollination on the other. It is perhaps not surprising that many children, when asked where milk comes from will reply without hesitation, "from the grocery store." The primary reason that ecosystem services are taken for granted, however, is that they are free. We explicitly value and place dollar figures on "ecosystem goods" such as timber and fish. Yet the services underpinning these goods generally have no market value—not because they are worthless, but rather because there is no market to capture and express their value directly.

* * *

Perhaps the most fundamental policy challenge facing ecosystem protection is that of valuation—how to translate an ecosystem's value into common units for assessment of development alternatives. The tough decisions revolve not around whether protecting ecosystems is a good thing but, rather, how much we should protect and at what cost. For example, how would the flood control and water purification services of a particular forest be diminished by the clearcutting or selective logging of 10%, 20% or 30% of its area? At what point does the ecosystem's net value to humans diminish, and by how much? Can the degradation of these services (in addition to ecosystem goods) be accurately measured? And, if so, how can partial loss of these services be balanced against benefits provided by development or pollution?

One might argue that ecosystem services cannot be evaluated, but this is clearly incorrect. We implicitly assess the value of these services every time we choose to protect or degrade the environment. The fundamental question is whether our implicit valuation of ecosystem services is accurate, and if not, what should be done about it. Indeed, studies such as *Nature's Services* indicate that our valuations are grossly and systematically understated. This essay explores the importance—and the challenges—of integrating ecosystem services research with the law. The potential is exciting, for a focus on ecosystem services would significantly change the way we understand and apply environmental law.

I. ECOSYSTEM SERVICES

Nature's Services addresses two basic questions: what services do natural ecosystems provide society, and what is a first approximation of their monetary value? Separate chapters describe the range of services and physical benefits provided by climate, biodiversity, soil, pollinators, pest control, the major biomes (oceans, freshwater, forests, and grasslands), and offer case studies of ecosystem services whose values are particularly well-known. The authors do not attempt to measure non-use values such as aesthetic or existence values, arguing that such work has already been done elsewhere. Instead, the authors determine lower-bound estimates of monetary value, using replacement costs where possible. Such information, it is hoped, will provide a basis for better incorporation of ecosystem services in decisionmaking.

The chapter on soil provides a useful example of the book's specific findings. More than a clump of dirt, soil is a complex matrix of organic and inorganic constituents transformed by numerous tiny organisms. The level of biological activity within soil is staggering. Under a square meter of pasture soil in Denmark, scientists identified over 50,000 worms, 48,000 small insects, and ten million nematodes. This living soil provides six ecosystem services: buffering and moderation of the hydrological cycle (so precipitation may be soaked up and metered out rather than rushing off the land in flash floods); physical support for plants; retention and delivery of nutrients to plants; disposal of wastes and dead organic matter; renewal of soil fertility; and regulation of the major element cycles. What are these services worth in the aggregate?

Take, for example, soil's service of providing nitrogen to plants. Nitrogen is supplied to plants through both nitrogen-fixing organisms and recycling of nutrients in the soil. As mentioned above, the authors rely primarily on replacement costs to estimate the value of ecosystem services. If nitrogen were provided by commercial fertilizer rather than natural processes, the lowest-cost estimate for its use on crops in the U.S. would be $45 billion, the figure for all land plants $320 billion. Most of the services identified in the book, however, such as breaking down dead organic material, are not valued in dollars because no technical substitutes are available.

Overall, *Nature's Services* reaches four conclusions. First, the services that ecosystems provide are both wide-ranging and critical. The question, "where would we be without ecosystem services?" is nonsensical, for we simply would not exist without them. Second, as Biosphere II's failure showed, the substitute technologies for most ecosystem services are either prohibitively expensive or non-existent. Massive hydroponic gardening in the absence of soil is at least conceivable, if unfeasible. Substitutes for climate regulation are neither conceivable nor feasible. Third, our overall understanding of ecosystem services—the contributions of individual species, threshold effects, synergies, etc.—is poor. Finally, even taking into account the inevitable imprecision of such valuation exercises, ecosystem services have extraordinarily high values. A recent study in the journal *Nature* estimated their aggregate value at between $16–54 trillion per year. The global GNP is $18 trillion.

Whether such a total estimate is precisely accurate is beside the point. The sheer magnitude of their dollar figures dictates that ecosystem services cannot be treated as merely add-on considerations. Nor can they be shunted aside as soft numbers (as often occurs with scenic beauty or existence value) when assessing the impacts of development or pollution. Tastes may differ over beauty, but they are in universal accord over fertile soil. If the goal of ecologists is to wake people up with big numbers, *Nature's Services* delivers. But are these numbers a convertible currency?

The greatest shortcoming of *Nature's Services*, one openly admitted by its authors, is the macro-scale of the analysis. The fact that pollinators annually provide Americans up to $1.6 billion of service or that soil fertility is worth $45 billion is important to know for general policy direction, but that fact does not help to inform specific land-use or pollution permitting decisions. One cannot divide the $45 billion value of soil fertility by the nation's total agricultural acreage to determine the value of the services of five acres of land threatened by development. Thus, the greatest need for ecosystem service valuation is at the margins. Few policy decisions, thankfully, will involve obliterating an ecosystem service. Rather, policy decisions tend to be incremental. What is the extent of degradation to these services at various points along a continuum of impacts? Given the complexity of ecosystem services, the responses are almost certainly nonlinear.

This problem, the assessment and valuation of services at the margin, is at once the most useful and most difficult challenge for economists and ecologists. *Nature's Services* establishes the range of ecosystem services and their great significance. The next step is to pick up where the book leaves off and identify how ecosystem services should be explicitly considered in real-life decisions, for ecosystem services are rarely, if ever, considered in current agency cost-benefit analyses.

The ideal method to assess development alternatives would be to give local ecosystem services an accurate monetary value. As a complement to the more subjective and controversial non-use measures such as existence and option values, dollar figures for ecosystem services would reflect practical benefits delivered to society. More important, this method would also permit direct comparisons between investments in physical capital and investments in natural capital as well as projections of future costs and benefits. Beyond ensuring wiser development, this method would respond to the regulatory mandates of wetlands mitigation banking, environmental impact statements, and natural resource damages that specifically request such figures.

Notes and Questions

1. The concept of ecosystem services suggests that even an individual tree provides some service value to humans—that it provides economic utility other than its commodity value as cut timber. Salzman suggests that the challenge facing efforts to make the ecosystem services concept a policy-viable tool is that it is much easier to determine a single tree's commodity value than its service value. In short, the theory and methodology of valuing ecosystem services at micro levels are far behind the abilities of the market to generate commodity values. On the other hand, we know with certainty that ecosystem values are extremely high at macro levels. Is knowing that enough to justify moving ahead with ecosystem service values as a policy-shaping tool before we refine the microeconomic understanding?

2. Dailey's and Costanza's work on ecosystem service valuation has opened up a sizable rift among ecologists, many of whom argue that it is immoral to attempt to assign anthropometric values to biometric processes. On the one hand, many ecologists contend that it is morally wrong to value the ecological attributes of a resource, as valuation implies that these attributes are only of relative and not absolute importance. Even under an anthropocentric view, they would contend that ecological considerations, such as the maintenance of ecological integrity, are of such fundamental importance to future generations of people that they should never be compromised no matter how large any assessment of the economic benefits of changing an ecosystem may be. On the other hand, respond Costanza, Daily, and others advocating ecosystem service valuation, both as a society and as individuals we make choices and trade-offs about ecosystems every day. These choices necessarily imply valuations. To say that we shouldn't try to perform more precise, quantifiable valuation of ecosystems is to simply deny these trade-offs are already occurring, and are occurring in the absence of reliable data.

3. The economic theory of ecosystem service values was actually first laid out comprehensively, as far as your authors can tell, in Edward Farnworth et al., *The Value of Natural Ecosystems: An Economic and Ecological Framework*, 8 ENVIRONMENTAL CONSERVATION 275 (1981). Farnworth et al. described the purpose of their work as "to describe the values of goods and services of natural ecosystems, and to structure these values according to a functional framework," which they believed "provides recognition of total value of natural ecosystems, and provides also a mechanism to translate values into action decisions." Their article lays out an ecological and economic framework for understanding ecosystem service values that corresponds closely with Costanza's and Daily's work. Ironically, they were perhaps too far ahead of their time—their article is not cited in Costanza's work or any of the chapters of *Nature's Services*, and seems to have been largely forgotten, if ever noticed.

B. WHAT IS ECOSYSTEM MANAGEMENT?

As we refine our understanding of ecosystems and design a politically and legally viable set of units for representing what they are and what they mean to us, the next threshold question on the way to formulating a law of ecosystem management is deciding the specific management goals and methods. After all, ideally the law will simply put the goals and methods into motion. Thus, at the most fundamental level:

> Ecosystem management is management driven by explicit goals, executed by policies, protocols, and practices, and made adaptable by monitoring and research based on our best understanding of the ecological interactions and processes necessary to sustain ecosystem composition, structure, and function.

Norman L. Christensen, *The Report of the Ecological Society of America Committee on the Scientific Basis for Ecosystem Management*, 6 ECOLOGICAL APPLICATIONS 665 (1996). Of course, this only begs the question: What are the goals, policies, protocols, and practices of ecosystem management?

Indeed, there is yet a threshold question to consider before debating goals and methods. Assuming we have agreed on where an ecosystem is and whether to use natural function preservation, commodity extraction value, service delivery value, or some combination thereof as the yardstick of ecosystem management, the initial management challenge is in devising the "currency" with which to define ecosystem management goals and measure the performance of the policies and practices used to achieve them. The specification of currency, goals, and methods, in that order, thus are the three central challenges to devising a coherent ecosystem management policy.

1. Currency Selection

As challenging as it is to manage a business, the goal is usually quite straightforward (make a profit) and the method, at least in theory, can be textbook simple (sell more product). There is also no shortage of readily available measurements for expressing business goals and measuring performance. Business and economic performance indicators abound, and management can use them to make strategic decisions about employment, capital expenditure, product line expansion, and so on. Market forces ruthlessly weed out bad decisions and reward the good ones. Every newspaper has pages of business and economic data available to investors. CEO employment packages now routinely use one or a combination of precisely measurable performance indicators to set compensation and bonuses. In short, although figuring out how to boost profits can be quite a challenge, knowing where profits stand is not.

Ecosystem management does not have it so easy. To be sure, an ecosystem managed purely for its commodity value units (e.g., a farm, a fishery, a timber forest) can take advantage of business management performance indicators, but management for other purposes becomes far more complicated to evaluate. For example, if we were to prescribe "no net loss of habitat" as the goal for management of a large park or forest, what would that mean? We could use acres as the yardstick for determining net position over time. But we know that not all habitat is created equal—that some habitat is "better" than other habitat. Merely keeping the number of acres constant could lead to a significant degradation of habitat quality over time. So, we could use total habitat quality as the index and forget about counting acres. But quality based on what? Quality for rodents? For birds? High quality bird habitat may make low quality rodent habitat, so which species wins? And how do we quantify habitat quality in any event? We could avoid those questions altogether by focusing on ecosystem processes as the measure of habitat, as in no net loss of ecosystem processes. This would leave it to whatever happens under those processes to decide what the habitat looks like and which species occupy it. While this seems to be the direction most ecosystem management literature takes, it simply raises a different set of questions about measurement, such as which processes to count and how to count them.

Alas, ecosystem management law cannot get very far off the ground until these very basic questions are answered. It is relatively easy to describe ecosystem processes and declare abstract goals for them, such as "protect," "maintain," or "restore," but those are not legally viable concepts until agencies, courts, and interest groups have a common language for determining whether management decisions comply with the law.

Returning to Professor Salzman's review of *Nature's Services*, we see that achieving agreement on this fundamental currency selection issue for ecosystem management, in this case taking the ecosystem services perspective of ecosystems, is no mean feat:

James Salzman, Valuing Ecosystem Services

24 Ecology Law Quarterly 887 (1997).

* * *

II. VALUING ECOSYSTEMS

So how does one value an ecosystem? Assume our object of study is a wetland along the banks of the Potomac. The first step lies in defining the ecosystem's contribution to human well-being. An ecosystem may be characterized by its physical features (site-specific characteristics such as landscape context, vegetation type, salinity), its goods (vegetation, fish), its services (nutrient cycling, water retention) or its amenities (recreation, bird-watching). These four aspects may not be complementary. For example, one could manage a wetland for cranberry production at the expense of primary productivity and services. Furthermore, the location of the system will be a critical factor of its net utility because location determines the distribution of goods and services. An ecosystem's carbon sequestration and biodiversity will be valuable even if distant from human populations, but its role in pollination and flood control likely will not. Thus two identical ecosystems may have very different values depending on their landscape context.

Economists classify these characteristics using four categories. The most obvious category includes consumable ecosystem goods such as cranberries and crabs that are exchanged in markets and easily priced (direct market uses). Activities such as hiking and fishing (direct non-market uses) as well as more intangible existence and option values (non-market, non-use) are not exchanged in markets. As a result, their values must be determined indirectly by shadow pricing techniques such as hedonic pricing, travel-cost methodologies, or contingent valuation. Ecosystem services are categorized as indirect non-market uses, for while they provide clear benefits to humans they are neither directly "consumed" nor exchanged in markets. These are also classic public goods because their use cannot be exclusively controlled.

* * *

Currently, there are three challenges to incorporating benefits of ecosystem services more directly into decisionmaking: identifying services on a local scale, measuring the value of these services, and projecting their future value. First, ecologists must understand the services provided by a specific ecosystem. For example, wetlands provide an important service in nutrient trapping, which retards and prevents eutrophication. The capacity to trap nutrients depends on the biophysical capacity of the site (e.g., its vegetation, benthic community, size, slope) and on the in-flow from adjacent water sources. Data on these factors can be provided in great detail. One can make empirically sound predictions that actions on a gross scale, such as clearcutting, will affect nutrient flows and services, or that a loss of populations reduces ecosystem resiliency. In aggregate, such knowledge can provide policy guidance in warning against extreme actions. But in most cases, our scientific knowledge is inadequate to predict with any certainty how specific local actions affecting these factors will impact the local ecosystem services themselves.

This lack of knowledge is due both to the lack of relevant data and to the multivariate complexity of the task. Analysis of how ecosystems provide services has proceeded slowly because ecosystem level experiments are difficult, costly, and lengthy. More important, research to date has focused much more on understanding ecosystem processes than determining ecosystem services, and how an ecosystem works is not the same as the services it provides. This focus has been reinforced, and partly driven, by regulatory requirements. Federal and state wetland regulations assess the adequacy of wetland mitigation on the basis of the site's functional capacity, not on the basis of the services actually provided and their extant benefits to humans. For this reason alone, publications such as *Nature's Services* are valuable: they increase awareness of the need to shift the focus of ecological research toward provision of services.

* * *

As noted above, an ecosystem's benefit to humans is not a straightforward biophysical measure, for identical ecosystems in different locations will have very different values. The value of a wetland's nutrient trapping service, for instance, depends on the location of its out-flow. Does it flow to shellfish beds (high value) or a fast-flowing ocean current (low value)? In valuing each ecosystem service, and indirectly the "cost" of its diminution, substitutes become important. Will the threatened service be replaced by other natural processes? Is it redundant or scarce? To what extent can technology overcome or mitigate these harms? If the loss of a service is important and non-linear, when will it become asymptotically more valuable approaching the point of collapse? None of these questions can be answered without intricate, localized knowledge of the ecosystem service itself.

Despite these difficulties, let us assume we understand fully the ecosystem service and have determined its current value. Even then, we face a third challenge when we try to determine in dollars the future stream of services flowing from the current biophysical features and landscape of the ecosystem. This figure is important because the net present value of most proposed actions that will degrade an ecosystem, such as shopping mall developments, take into account future streams of income. To ensure a full accounting of costs and benefits, the future "income" flow of the ecosystem service should be factored into its current value as well, since that value may change over time due to land-use patterns, weather, pollution, etc. How then does one link a site's current ecological characteristics with future ecosystem services?

* * *

[T]here is no clear method yet for valuing or measuring ecosystem services, much less future services. Differing expectations also contribute to the disparity. In the environmental context, uncertain values are often set at zero. In contrast, when assessing corporate acquisitions, financial analysts routinely provide credible values greater than zero. Uncertainty is an accepted part of the profession. Finally, the lack of information plays a role. Making accurate projections of ecosystem services (or future stock prices) requires a great deal of robust data. Yet basic research and regulatory compliance have focused far more on the biophysical capacities of ecosystems than on their services. Moreover, information on ecosystem services is expensive to collect because the benefits of ecosystem services are a public good. Since there is no financial gain from "investing" in services (unlike with IBM's stock) there is no secondary market generating relevant data.

The combination of methodological difficulty, inherent complexity, and lack of data makes placing absolute dollar figures on local ecosystem services unfeasible in many cases. At the same time, the current research and regulatory focus on ecosystems' biophysical measures is too removed from valuation of services. Is there a middle ground to inform decisionmakers?

Wall Street and IBM's stock price may provide some guidance. As noted above, many of the sources on which analysts rely to value stocks are not, in fact, monetary. They are composite indicators such as market strength, consumer confidence, and housing starts. Similarly, some of the most advanced work in wetlands valuation is now focusing on non-monetary indicators. This research area combines traditional biophysical measures (i.e., the capacity to provide ecosystem services and goods) with landscape context to determine the opportunity and impact of providing these services to people. Such indicators do not provide dollar figures for ecosystem services, but they do provide more accurate bases for assessing relative qualities of different ecosystems.

* * *

Notes and Questions

1. The installment of Salzman's article used in this section pinpoints why it is that we can more easily determine a tree's commodity value than its service value: unlike commodities such as lumber, ecosystem services are indirect non-market uses, for while they provide clear benefits to humans they are neither directly "consumed" nor exchanged in markets, and their use cannot be exclusively controlled. But is that necessarily the case? Could ecosystem services be made a form of private property that owners could withhold or remove from distribution? If so, a direct use market in ecosystem services would arise as the beneficiaries of the service values pay to have them restored. Salzman's point, of course, is that it is difficult to withhold the service of, say, air purification. Some ecosystem services, however, can be withheld or removed by property owners, such as water purification from riparian vegetation. But who "owns" those services? And why is there presently no market in them?

2. Another problem with ecosystem services that Salzman's work reveals is that it may be too expensive, given current knowledge, to "mint" the currency. Measuring ecosystem service value delivery at local levels takes time and extensive, expensive research. Less elegant, less precise, but far less expensive currencies thus threaten to crowd out ecosystem service values as the measure of policy performance. In the context of wetland conservation, for example, a cheap, fast, and easy way to measure policy performance is to count acres of wetlands under a "no net loss" policy goal. Measuring wetland process integrity is far more difficult, and measuring wetland service value delivery is yet more difficult. Not surprisingly, therefore, actual wetlands conservation policy has stuck closely to acres as the currency. *See* J. B. Ruhl and R. Juge Gregg, *Integrating Ecosystem Services Into Environmental Law: A Case Study of Wetlands Mitigation Banking*, 20 STANFORD ENVIRONMENTAL LAW JOURNAL 365 (2001) (excerpted *infra* in Chapter 10).

3. Hence the dilemma: we know that ecosystem functions provide services that are valuable to humans, but we do not know how either (1) efficiently to measure those values or (2) efficiently to exchange the values in markets even if we could measure them. So when a landowner decides to convert riparian vegetation areas to other uses such as crop production or suburban housing, what currency *are* we supposed to use in order to weigh the ecosystem effects of that decision?

4. Does this currency problem explain why the species-by-species approach of the Endangered Species Act remains essentially the default approach for much of what we call ecosystem management, particularly on non-public lands? In effect, haven't we used endangered species as the currency for much of ecosystem management policy?

5. For examples of how ecosystem services in different settings might be valued using monetary and non-monetary methods, see Ecosystem Valuation, available at <http://www.ecosystemvaluation.org/uses.htm>. For a comprehensive discussion of the importance of developing measures or indicators of ecological performance and integrity, see NATIONAL ACADEMY OF SCIENCES/NATIONAL RESEARCH COUNCIL, ECOLOGICAL INDICATORS FOR THE NATION (2000).

2. GOAL SELECTION

In its *People's Glossary of Ecosystem Management Terms*, the United States Forest Service defines ecosystem management as "an ecological approach to natural resource management to assure productive, healthy ecosystems by blending social, economic, physical, and biological needs and values." As does this rather breezy effort, almost every definition offered of ecosystem management includes some normative goals-statement dimension. Indeed, the very notion of ecosystem management generally invokes debate cast in normative terms.

At one extreme, some people decry ecosystem management *on any basis* as an unwarranted human interference with nature. This "nature knows best" and "leave only footprints" camp follows the preservation principle of "nondestruction, noninterference, and generally, nonmeddling." Tom Regan, *The Nature and Possibility of an Environmental Ethic*, 3 ENVIRONMENTAL ETHICS 31 (1981). Wildness is their currency, and "many who value wildness, although unable to say exactly what it is, are nonetheless positive that less management is better management." Peter Alpert, *Incarnating Ecosystem Management*, 9 CONSERVATION BIOLOGY 952 (1995). For them, therefore, ecosystem management is an unwarranted intrusion on the wildness of ecosystems.

Others, however, condemn ecosystem management as an unwarranted intrusion on private property rights. Allan K. Fitzsimmons, for example, contends that ecosystem management, "if fully implemented, would greatly intrude on the property rights of all Americans." ALLAN K. FITZSIMMONS, DEFENDING ILLUSIONS 16 (1999). It is a form of "national land use planning wherein nature protection takes precedence over improvements in human well-being." *Id*. For him and others of this view, therefore, ecosystem management is a form of "nature worship," overly biocentric in perspective and relying too heavily on management of private land to achieve its goals.

Yet another view, this one in support of an active framework of ecosystem management, is that it is a form of beneficial interference, designed usually to undo or impede all or many of the results of past human destructive interference, and based on some sense of what the ecosystem's "natural" conditions should be. As the environmental ethicist Mark Michael has pointed out, "to claim that a human action interferes with an ecosystem or species is to say nothing more than that its presence or occurrence makes a difference in terms of what happens to the ecosystem or species." Mark A. Michael, *How to Interfere With Nature*, 23 ENVIRONMENTAL ETHICS 135 (2001). Interference, in other words, is simply a causal state; whether the effects are desirable is a different matter. And, as Michael posits, the effects of properly designed interference, even in the eyes of the most ardent preservationist, may be healthy for the ecosystem.

The latter view, of course, is what justifies publishing the remainder of this part of the text! There will be no law of ecosystem management under extreme versions of the wildness principle or the property rights principle, except by default. Moreover, given the rather unavoidable realities that humans are on the planet and that part of the planet consists of private property, some exploration of how to manage the whole package of ecosystems, humans, and property seems worth undertaking.

Yet with that normative debate out of the way for our immediate pedagogical purposes (to be taken up again *infra*), only more numerous and complicated normative debates are opened. In short, what are the goals of a "beneficial interference" framework of ecosystem management? While consensus on that question remains elusive, the following landmark article by Edward Grumbine has become a reference point for virtually every person who has made a stab at an answer.

R. Edward Grumbine, What Is Ecosystem Management?

8 Conservation Biology 27 (1994).

Introduction

Deep in a mixed conifer forest on the east side of the Washington Cascades, a U.S. Forest Service silviculturalist, responding to a college student's query, suggests that ecosystem management means snag retention and management of coarse woody debris on clear cut units.

In northern Florida on a US. Department of Defense reservation, a team of biologists and managers struggles with the design of a fire management plan in longleaf pine (Pinus palustris) forests that mimics natural disturbance regimes while minimizing the risk of burning adjacent private lands.

To avert what he calls "national train wrecks," Interior Secretary Bruce Babbitt announces that the Clinton Administration plans to shift federal policy away from a single species approach to one that looks "at entire ecosystems."

Commenting on a draft federal framework for the Greater Yellowstone Ecosystem that proposes increased interagency cooperation, a lawyer claims that "Congress does not intend that national forests should be managed like national parks" and that there exists no need to create ecosystem management.

Other observers in the Greater Yellowstone region contend that an ecosystem approach could provide just the holistic management necessary for sustaining resources in a complex ecological/political landscape.

What is ecosystem management?

The above vignettes portray but a few of the various interpretations of ecosystem management that can be found in the conservation biology, resource management, and popular literature. Since the ecosystem approach is relatively new, and still unformed, this is not surprising. As any concept evolves, debates over definition, fundamental principles, and policy implications proceed apace. Yet the discussion surrounding ecosystem management is not merely academic. Nor is it limited to those scientists, resources professionals, and policymakers who work directly with federal management issues. The debate is raising profound questions for most people who are concerned with the continuing loss of biodiversity at all scales and across many administrative boundaries and ownerships. Along with defining the ecosystem management approach as a new policy framework there appears to be a parallel process of redefining the fundamental role of humans in nature.

* * *

Dominant Themes of Ecosystem Management

Ecosystem management has not been uniformly defined or consistently applied by federal or state management agencies. Yet consensus is developing, at least within the academic literature. Using standard keyword search techniques focused on "ecosystem management," "ecosystem health," "biodiversity" "management," "adaptive management," etc., I surveyed papers published on ecosystem management in peer reviewed journals (Conservation Biology, Environmental Management, Ecological Applications, Society and Natural Resources etc.) up through June 1993 to determine where agreement exists on the subject. Articles came from a broad spectrum of disciplines including conservation biology, resource management, and public policy. I also reviewed books with substantive accounts of ecosystem management, lay environmental publications, and several federal and state-level documents that discuss ecosystem level policymaking.

Ten dominant themes of ecosystem management emerged from my review. Dominant themes were those attributes that authors identified explicitly as critical to the definition, implementation, or overall comprehension of ecosystem management. The ten dominant themes emerged repeatedly throughout the literature. I believe the following themes faithfully represent areas of agreement. * * *

1. **Hierarchical Context.** A focus on any one level of the biodiversity hierarchy (genes, species, populations, ecosystems, landscapes) is not sufficient. When working on a problem at any one level or scale, managers must seek the connections between all levels. This is often described as a "systems" perspective.

2. **Ecological Boundaries.** Management requires working across administrative/political boundaries (i.e., national forests, national parks) and defining ecological boundaries at appropriate scales.* * *

3. **Ecological Integrity**. Norton defines managing for ecological integrity as protecting total native diversity (species, populations, ecosystems) and the ecological patterns and processes that maintain that diversity. Most authors discuss this as conservation of viable populations of native species, maintaining natural disturbance regimes, reintroduction of native, extirpated species, representation of ecosystems across natural ranges of variation, etc.

4. **Data Collection.** Ecosystem management requires more research and data collection (i.e., habitat inventory/classification, disturbance regime dynamics, baseline species and population assessment) as well as better management and use of existing data.

5. **Monitoring.** Managers must track the results of their actions so that success or failure may be evaluated quantitatively. Monitoring creates an ongoing feedback loop of useful information.

6. **Adaptive Management.** Adaptive management assumes that scientific knowledge is provisional and focuses on management as a learning process or continuous experiment where incorporating the results of previous actions allows managers to remain flexible and adapt to uncertainty.

7. **Interagency Cooperation.** Using ecological boundaries requires cooperation between federal, state, and local management agencies as well as private parties. Managers must team to work together and integrate conflicting legal mandates and management goals.

8. **Organizational Change.** Implementing ecosystem management requires changes in the structure of land management agencies and the way they operate. These may range from the simple (forming an interagency committee) to the complex (changing professional norms, altering power relationships).

9. **Humans Embedded In Nature.** People cannot be separated from nature. Humans are fundamental influences on ecological patterns and processes and are in turn affected by them.

10. **Values.** Regardless of the role of scientific knowledge, human values play a dominant role in ecosystem management goals.

These ten dominant themes form the basis of a working definition: *Ecosystem management integrates scientific knowledge of ecological relationships within a complex sociopolitical and values framework toward the general goal of protecting native ecosystem integrity over the long term.*

* * *

Ecosystem Management Goals

Most of the authors cited in this review agree that setting clear goals is crucial to the success of ecosystem management. Within the overall goal of sustaining ecological integrity, five specific goals were frequently endorsed:

1. Maintain viable populations of all native species in situ.

2. Represent, within protected areas, all native ecosystem types across their natural range of variation.

3. Maintain evolutionary and ecological processes (i.e., disturbance regimes, hydrological processes, nutrient cycles, etc.).

4. Manage over periods of time long enough to maintain the evolutionary potential of species and ecosystems.

5. Accommodate human use and occupancy within these constraints.

The first four of these goals are value statements derived from current scientific knowledge that aim to reduce (and eventually eliminate) the biodiversity crisis. The fifth goal acknowledges the vital (if problematic) role that people have to play in all aspects of the ecosystem management debate.

These fundamental goals provide a striking contrast to the goals of traditional resource management. Though different agencies operate under a variety of federal and state mandates, current resource management in the U.S. is based on maximizing production of goods and services, whether these involve number of board feet (commodities) or wilderness recreational visitor days (amenities). Managers and lawmakers have always been careful to speak of "balance" and "sustained yield" but this language is obfuscatory—balance has never been defined in any U.S. environmental law and sustained yield has often been confused with sustainability.

If ecosystem management is to take hold and flourish, the relationship between the new goal of protecting ecological integrity and the old standard of providing goods and services for humans must be reconciled. Much of the oft-complained "fuzziness" or lack of precision surrounding ecosystem management derives from alternative views on this point. Kessler et al., for example, suggest that ecosystem management represents a further evolution of multiple use, sustained yield policy where managers "must not diminish the importance of products and services, but instead treat them within a broader ecological and social context." These authors envision ecosystem-level management as an expansion of focus from particular resource outputs to the ecosystem as "life support system [for humans]." Kessler et al. fail to see that expanding the scale of concern by itself does not address the fact that there are certain ecological limits in any system which constrain human use. Ecological integrity as expressed by the five specific goals explicitly considers all resource use as a managerial artifact that may flow sustainably from natural systems only if basic ecosystem patterns and processes are maintained.

Echoing Kessler et al., the most detailed Forest Service working definition of ecosystem management exemplifies lack of clarity over the key policy problem of defining ecosystem management goals. The report defines the philosophy of ecosystem management as sustaining "the patterns and processes of ecosystems for the benefit of future generations, while providing goods and services for each generation." The study characterizes the main limiting factors to ecosystem management as defining societal expectations, integrating these expectations with the sustainable capabilities of ecosystems, and filling information gaps in baseline data describing historical ecosystem variability and disturbance regimes. The Forest Service prescribes adaptive management as a process to blend ecosystem sustainability and human concerns. Specific solutions offered, however, are problematic. If societal goals conflict with ecosystem sustainability, cost/benefit analyses are offered as the standard for solutions. Adaptive management is described as an ongoing experiment yet "landscapes can be restored," managers are said to already be capable of mimicking natural disturbance regimes successfully, and there is speculation that future experiments may reveal new sustainable ecosystem states that may differ from evolutionary and historical states. In short, the Forest Service defines the goals of ecosystem management narrowly within the old resource management paradigm ("for the benefit of future generations") and seeks to operationalize this goal within a positivistic scientific framework. These characterizations of ecosystem management are also found in the other government policy documents in this review.

As several analysts point out, however, it takes more than scientific knowledge to reframe successfully complex policy problems. Knowledge of organizational structure and behavior as well as the policy process itself are equally important. Yet none of the five government treatments of the ecosystem management concept reviewed here mention substantive organizational change, nor do they discuss the policy process as it is defined by policy scientists. This emphasis on science is an artifact of the training and professional norms of the major group writing about ecosystem management—scientists. But defining ecosystem management goals is also a political process; those authors advocating a new vision of ecological integrity are more often employed independently or in academia. Authors affiliated with government agencies tend to support the Forest Service version of ecosystem management. As policy analyst Tim Clark (personal communication) has pointed out, " 'the ecosystem management debate' is really a complex, competitive, conflictual social process about whose values will dominate, it is not about science."

Management goals are statements of values—certain outcomes are selected over others. Choosing the management goal of maintaining ecological integrity along with the five specific goals may be debated, but in the academic and popular literature there is general agreement that maintaining ecosystem integrity should take precedence over any other management goal. This may be due partially to the fact that, given the rate and scale of environmental deterioration along with our profound scientific ignorance of ecological patterns aid processes, we are in no position to make judgments about what ecosystem elements to favor in our management efforts. An increasing number of people also believe that humans do not have any privileged ethical standing from which to arbitrate these types of questions.

Conclusion

History tells us that change does not always come easily, peacefully, or in a planned manner. Implementing the short-term scientific aspects of ecosystem management is daunting enough. For the moment, however, ecosystem management provides our best opportunity to describe, understand, and fit in with nature. We know that the risk of extinction increases under certain conditions, that wildfires cannot long be suppressed without significant successional consequences, that political power must somehow become less centralized, that whales and spiders must also be allowed to vote. We are also coming to realize that resourcism has for so long prevented us from putting our ecological knowledge to work that we are facing the limits of life on Earth for many species. Where once we thought endangered species were the problem, we now face the loss of entire ecosystems.

* * *

Ecosystem management, at root, is an invitation, a call to restorative action that promises a healthy future for the entire biotic enterprise. The choice is ours—a world where the gap between people and nature grows to an incomprehensible chasm, or a world of damaged but recoverable ecological integrity where the operative word is hope.

Notes and Questions

1. The unmistakable unifying theme of Grumbine's treatment of ecosystem management is "the general goal of protecting native ecosystem integrity over the long term." Is it self-evident what a "native ecosystem" is, or what its "integrity" involves? Is Grumbine suggesting that any ecosystem management approach necessarily must adopt this as its goal to be considered true ecosystem management, or simply that the weight of academic literature at the time—around the mid–1990s—had identified this as the goal? Could ecosystem management chart protection of something other than "native ecosystem integrity" as its goal and still be ecosystem management? If so, what would that goal be?

2. Adhering to his "native ecosystem integrity" model, Grumbine criticizes Kessler et al. for positing as the goal of ecosystem management protecting "the patterns and processes of ecosystems for the benefit of future generations, while providing goods and services for each generation." Grumbine suggests that this is an overly anthropocentric perspective and the wrong "philosophy" of ecosystem management. How different are Grumbine's and Kessler's philosophies of ecosystem management? With which do you agree more?

3. The disagreement between Grumbine's and Kessler's approaches is covered thoroughly in Thomas R. Stanley Jr., *Ecosystem Management and the Arrogance of Humanism*, 9 CONSERVATION BIOLOGY 255 (1995). The debate permeates ecosystem management literature. Compare these two definitions of ecosystem management:

> A collaborative process that strives to reconcile the promotion of economic opportunities and livable communities with the conservation of ecological integrity and biodiversity. THE KEYSTONE NATIONAL POLICY DIALOGUE ON ECOSYSTEM MANAGEMENT 6 (1996).

> Management of natural resources using systemwide concepts to ensure that all plants and animals in ecosystems are maintained at viable levels in native habitats and basic ecosystem processes are perpetuated indefinitely. U.S. FISH AND WILDLIFE SERVICE, AN ECOSYSTEM APPROACH TO FISH AND WILDLIFE CONSERVATION 4 (1994).

Are these close in theme? Which is closer to Grumbine? To Kessler? Does lack of agreement over the basic objectives of ecosystem management suggest that fundamental normative decisions have yet to be made about ecosystem management policy?

4. Consider the Grumbine–Kessler debate in terms of the five ecosystem unit frameworks discussed in the previous section of the text—geographic, management, functions, commodities, and services. What mix of these comes closest to capturing Grumbine's philosophy of ecosystem management? Kessler's? Are they much different? Would any particular mix of units preclude using Grumbine's vision of ecosystem management policy, or Kessler's?

5. Grumbine also suggests the need for taking a "systems approach" to ecosystem management, which he defined as requiring that "when working on a problem at any one level or scale, managers must seek the connections between all levels." Notice that the U.S. Fish and Wildlife Service's definition of ecosystem management quoted in Note 3 above also relies on "management of natural resources using systemwide concepts." What are "systemwide" concepts? In large part these references are to the system component of eco*system*, and to the development in the discipline of ecology of a greater understanding and appreciation of the complex, adaptive, dynamic nature of ecological settings. While once thought of as reaching constant equilibrium states, as in so-called climax forests, today's understanding of ecosystems is that they are governed by forces of short-term disturbance (e.g., fire, flood, and drought) and long-term change (e.g., decline of one species, gradual change in climate), and that what was taken as "equilibrium" and "the balance of nature" is actually a state of constant and resilient adaptation at a *system* level to these perturbations. As one of the leaders of this "new ecology" has put it:

> Throughout the 20th century, most theoretical and empirical research attempted to understand the structure and dynamics of populations, communities, and ecosystems by identifying their components and studying their relations in isolation from the complicating influences of larger systems. This research strategy was successful in elucidating fundamental ecological processes: response to stresses of extreme abiotic conditions; limiting resources of food, water, and inorganic nutrients; and the biotic interactions of competition, mutualism, predation, parasitism, and disease. It was less successful in revealing the complex patterns of temporal and spatial variation in abundance, distribution, and diversity of species or the complicated roles of species in ecosystems. By the 1980s, it was becoming apparent that more holistic, synthetic approaches were needed. To understand realistically complex ecological systems, it is necessary to study how the components affect and are affected by the larger, more complicated systems within which they are located.

James H. Brown et al., *Complex Species Interactions and the Dynamics of Ecological Systems: Long–Term Experiments*, 293 Science 643 (2001). Always a step or two behind the advances in science, legal scholars have only recently begun to ponder the effect of systems-based ecology on our overall approach to environmental law and policy. *See, e.g.,* Symposium, *Beyond the Balance of Nature: Environmental Law Faces the New Ecology*, 7 Duke Environmental Law Journal 1 (1996); Symposium, *Ecology and the Law*, 69 Chicago-Kent Law Review 847 (1994); Jonathan Baert Wiener, *Law and the New Ecology: Evolution, Categories, and Consequences*, 22 Ecology Law Quarterly 325 (1995).

6.   Just as the "old ecology" had its model of equilibrium and nature in balance, so too must the "new ecology" develop models of complex ecosystem dynamics. *See* Norman L. Christiansen et al., *The Report of the Ecological Society of America Committee on the Scientific Basis for Ecosystem Management*, 6 Ecological Applications 665 (1996) ("Ecosystem management should be rooted in the best current models of ecosystem functioning."). But modeling equilibrium is easier than modeling disequilibrium. The effort requires access to vast repositories of data and sophisticated computational capacity. The scientific community is in agreement that substantial progress is needed on both fronts. *See* James S. Clark et al., *Ecological Forecasts: An Emerging Imperative*, 293 SCIENCE 657 (2001). Moreover, the upshot of such modeling efforts is that the models show exactly what the new ecology posits—uncertainty, nonlinearity, and un-predictability. For example, in their efforts to model the effects of different controlled fire management strategies in the longleaf pine ecosystem at Eglin Air Force Base in Florida, researchers found that their model "hypothesized that some kinds of disturbance may send ecosystems on a trajectory of nearly irreversible change" and that such "degradation might not appear for two or more decades, by which time reversing it would be difficult and expensive." *See* Jeff Hardesty et al., *Simulating Management with Models*, 1 CONSERVATION BIOLOGY IN PRACTICE 26 (Spring 2001). As it turned out, however, these findings suggested that the traditional approach the base had been advised to use for fire management "would lead to long-term landscape-scale ecological degradation." *Id.* By adopting a more adaptive fire management strategy tested through the model, the base appears to have averted that outcome. The upshot is that "expert opinion alone would have steered Eglin in some ecologically damaging and costly wrong directions." *Id.* For more on modeling in ecosystem management contexts, particularly the need to fuse ecological with economic models, see Robert Costanza, *Ecological Economics: Reintegrating the Study of Humans and Nature*, 6 ECOLOGICAL ECONOMICS 978 (1996).

7. Grumbine later took stock of the development of ecosystem management policy after his landmark article, in R.E. Grumbine, *Reflections on "What is Ecosystem Management,"* 11 CONSERVATION BIOLOGY 41 (1997), and concluded based on interviews and a literature review that "many [re-source] managers and academics believe that the 10 themes described in 'What is Ecosystem Management' remain useful as a framework for specific applications of EM. No major additions or subtractions have been proposed, though much new information has come into light."

8. If Grumbine is correct that his version of ecosystem management continues to make sense to most resource managers and academics in terms of policy goals selection, is it also important to know whether it has a sound scientific basis? A committee of the Ecological Society of America recently considered that issue, concluding that there is a good match between the current scientific understanding of ecosystems dynamics and the ecosystem management approach Grumbine outlined, particularly in light of the threats to ecosystem integrity researchers like Reed Noss have begun to identify. *See* Norman L. Christiansen et al., *The Report of the Ecological Society of America Committee on the Scientific Basis for Ecosystem Management*, 6 ECOLOGICAL APPLICATIONS 665 (1996). The committee's 27–page report, possibly the most compact and accessible comprehensive treatment of ecosystem management available still five years after its publication, can be obtained for a nominal fee from the Ecological Society of America, 2010 Massachusetts Ave., NW, Suite 400, Washington, DC 20036.

9. One implicit scientific assumption of ecosystem management policy, regardless of which side of the Grumbine–Kessler debate one takes, is that ecosystem integrity is essential to promoting high levels of species biodiversity. But are high levels of species biodiversity essential to promoting ecosystem integrity? Ecologists disagree on this question, quite intensely. *See* David Tilman, *Diversity and Production in European Grasslands*, 286 SCIENCE 1099 (1999); Paul L. Angermeier and James R. Karr, *Biological Integrity versus Biological Diversity as Policy Directives*, 44 BIOSCIENCE 690 (1994). Advocates of ecosystem management quite naturally contend that "biological diversity is central to the productivity and sustainability of the earth's ecosystems." Norman L. Christiansen et al., *The Report of the Ecological Society of America Committee on the Scientific Basis for Ecosystem Management*, 6 ECOLOGICAL APPLICATIONS 665 (1996). To a large extent, however, this assertion rests on an extrapolation from ample evidence of the contribution of *individual* species to ecosystem functioning. But other researchers argue that there is insufficient evidence that, beyond some minimum level needed for basic ecosystem functioning, higher and higher species diversity continues inexorably to enhance overall ecosystem health through synergistic properties. To resolve the debate, experiments are underway to test for "overyielding" effects—evidence that the total productivity of a species grown in mixture is greater than the total that is produced when each species in the mixture is grown in isolation—which might point in the direction of confirming biodiversity's contribution to ecosystem integrity. The jury is still out. *See* M. Loreau et al., *Biodiversity and Ecosystem Functioning: Current Knowledge and Future Challenges*, 294 SCIENCE 804 (2001); Jocelyn Kaiser, *Rift Over Biodiversity Divides Ecologists*, 289 SCIENCE 1282 (2000). For an early but still quite valuable study of the possible linkages between species biodiversity and ecosystem integrity, see COUNCIL ON ENVIRONMENTAL QUALITY, 21ST ANNUAL REPORT ON ENVIRONMENTAL QUALITY 135 (1990).

10. When all is said and done, is Grumbine's vision of ecosystem management—indeed, *any* vision of ecosystem management—just a fantasy? Is it really possible to accomplish most, or even a few, of what Grumbine suggests are the core goals of ecosystem management? At least one commentator believes not, suggesting that all forms of ecosystem management rely unrealistically on the ability of humans, through policy and technology, to manipulate ecological conditions. Thus

> the assumptions underlying ecosystem management are presumptuous and false. Ecosystem management cannot deliver what it has promised, and to deny this is to set a destructive course destined to fail.... As currently espoused, ecosystem management is a magical theory ... that promises the impossible—that we can have our cake and eat it too. Worse, however, it addresses only the symptoms of the problem and not the problem itself. The problem is not how to maintain current levels of resource output while also maintaining ecosystem integrity; the problem is how to control population growth and constrain resource consumption.

Thomas R. Stanley, Jr., *Ecosystem Management and the Arrogance of Humanism*, 9 CONSERVATION BIOLOGY 255 (1995). Under this view, the Grumbine–Kessler debate is rather moot. Naturally, however, your authors hope you will forge ahead in these materials even if Stanley has you convinced that ecosystem management is folly!

11. Define nature. What is it? Where is it? Can nature exist where ecosystem management has tread? Can you define nature without reference to humans? Try it! For more, see Steven Vogel, *Environmental Philosophy after the End of Nature*, 24 ENVTL. ETHICS 23 (2002), and BILL MCKIBBEN, THE END OF NATURE (1989).

3. METHOD SELECTION

As described by Grumbine, ecosystem management refers to a set of policy *goals* directed at ensuring the sustainability of natural resource qualities within ecologically functional units. Grumbine mentions another concept, "adaptive management," as a set of policy *tools* intended to move decision making from a process of incremental trial and error to one of experimentation using continuous monitoring, assessment, and recalibration. Ecosystem management and adaptive management are not interchangeable, but are nearly inseparable. Successful ecosystem management usually requires a heavy dose of adaptive management, and using adaptive management in natural resources conservation contexts generally leads to expressing goals in term of ecosystem management. As the General Accounting Office summarized in its recent assessment of the status of ecosystem management in federal agencies:

> Just as ecosystems are continually changing over time, so, too, will the understanding of their ecology and, by implication, the management choices based on this understanding. Scientists and policy analysts generally recognize that their understanding of how different ecosystems function and change and how they are affected by human activities is incomplete. For this reason, they see a need for continually researching, monitoring, and evaluating the ecological conditions of ecosystems and, where necessary, modifying management on the basis of new information to better accommodate socioeconomic considerations while ensuring the minimum or desired ecological conditions are being achieved.

> This process, sometimes known as "adaptive management," has been identified as a requirement for ecosystem management by ... the federal interagency team tasked to examine ecosystem management in the old-growth forests of the Pacific Northwest. It is also reflected in the administration's principle to "use monitoring and assessment and the best science available." Thus, applying this principle will require (1) continually researching, monitoring, and assessing ecological conditions as well as the effects of activities on ecosystems and (2) modifying prior management choices on the basis of this new information. This ... underscores the continuing, iterative nature of ecosystem management.

General Accounting Office, Ecosystem Management: Additional Actions Needed to Test a Promising Approach (August 1994).

Adaptive management theory traces its origins to C. S. "Buzz" Holling's seminal book, *Adaptive Environmental Assessment and Management*, published in 1978. During the two decades following the book's publication, Holling's basic framework for adaptive management found followers in environmental policy and law circles, eventually becoming one of the coordinating principles for ecosystem management implementation in many contexts today. When the term "adaptive management" rolls off people's tongues with reference to ecosystem management policy, they mean essentially what Holling outlined over 20 years ago, but the concept is often described in terms too brief to be meaningful. For example, in its "People's Glossary of Ecosystem Management Terms," <http://www.fs.fed.us/land/emterms.html>, the United States Forest Service defines adaptive management as:

> A type of natural resource management that implies making decisions as part of an on-going process. Monitoring the results of actions will provide a flow of information that may indicate the need to change a course of action. Scientific findings and the needs of society may also indicate the need to adapt resource management to new information.

To see if there is any more to it than that, it is worth examining what Holling had in mind in more detail.

Holling edited *Adaptive Environmental Assessment and Management* while at the Institute of Animal Resource Ecology at the University of British Columbia. He and a team of hand-picked scientists met over a period of two years to hammer out an alternative to conventional environmental assessment methods, such as the Environmental Impact Statement (EIS) process of the National Environmental Policy Act (NEPA). The group perceived the EIS and similar assessment methods as involving "fixed review of an independently designed policy" which over time would "inhibit laudable economic enterprises as well as violate critical environmental constraints." They called their alternative "adaptive management and policy design, which integrates environmental with economic and social understanding at the very beginning of the design process, in a sequence of steps during the design phase and after implementation."

The team introduced adaptive management first by debunking twelve principles—what they called myths—which they contended formed the premises of conventional environmental assessment theory. These statements, in other words, define what adaptive management theory explicitly *rejects*. Four of these myths centered around the premises of environmental management policy:

> *Myth 1* The central goal for design is to produce policies and developments that result in stable social, economic, and environmental behavior.
>
> *Myth 2* Development programs are fixed sets of actions that will not involve extensive modification, revision, or additional investment after the development occurs.

> *Myth 3*   Policies should be designed on the basis of economic and social goals with environmental concerns added subsequently as constraints during a review process.
>
> *Myth 4*   Environmental concerns can be dealt with appropriately only by changing institutional constraints.

To these the group added eight myths relating to the premises underlying conventional approaches to environmental assessment:

> *Myth 5*   Environmental assessment should consider all possible impacts of the proposed development.
>
> *Myth 6*   Each new assessment is unique. There are few relevant background principles, information, or even comparable past cases.
>
> *Myth 7*   Comprehensive "state of the system" surveys (species lists, soil conditions, and the like) are a necessary step in environmental assessment.
>
> *Myth 8*   Detailed descriptive studies of the present condition of system parts can be integrated by systems analysis to provide overall understanding and predictions of systems impact.
>
> *Myth 9*   Any good scientific study contributes to better decision making.
>
> *Myth 10*   Physical boundaries based on watershed area or political jurisdictions can provide sensible limits for impact investigations.
>
> *Myth 11*   Systems analysis will allow effective elections of the best alternative from several proposed plans and programs.
>
> *Myth 12*   Ecological evaluation and impact assessment aim to eliminate uncertainty regarding the consequences of proposed development.

One does not have to agree that all these statements are flatly wrong and misguided, or even that they undergird EIS and similar assessment methods, in order to appreciate the alternative view adaptive management theory introduced. Holling's group found the 12 conventional "myths" at odds with four basic properties of ecological systems. First, although the parts of ecological systems are connected, not all parts are strongly or intimately connected with all other parts. It cannot possibly be the case, for example, that every species in an ecosystem depends for its survival on the survival of every other species. The connections within ecosystems are themselves selective and variable, meaning what should be measured will depend on our understanding of the way the system as a whole works. Second, events are not uniform over space, meaning that impacts of development do not gradually dilute with distance from the development. In particular, induced effects of developments such as pipelines and water reservoirs may be of greatest magnitude at distant points. Third, ecological systems exhibit multi-equilibrium states between which the system may move for unpredictable reasons, in unpredictable manners, and at unpredictable times. Small variations in conditions such as temperature, nutrient content, or species composition can "flip" ecosystems into vastly different behavioral states, sometimes well after the event that started the reaction. The upshot is that the unexpected can happen, and it will be difficult to predict when, where, and to what degree. Finally, Holling's group observed that because ecosystems are not static but in continual change, environmental quality is not achieved by eliminating change. Flood, fire, heat, cold, drought, and storm continually test ecosystems, enhancing resilience through system "self-correction." Efforts to suppress change are thus not only futile, but counter-productive.

Ecologists today generally agree that these four properties define important characteristics of ecosystems. Holling's point was that environmental assessment methods thus ought to take these four properties into account, and that the 12 "myths" of conventional environmental assessment theory are unsuitable premises for doing so. In their place, Holling's adaptive management theory builds on eight key premises:

1. Since everything is not intimately connected to everything else, there is no need to measure everything. There is a need, however, to determine the significant connections.

2. Structural features (size distribution, age, who connects to whom) are more important to measure than numbers.

3. Changes in one variable (e.g., a population) can have unexpected impacts on variables at the same place but several connections away.

4. Events at one place can re-emerge as impacts at distant places.

5. Monitoring of the wrong variable can seem to indicate no change even when drastic change is imminent.

6. Impacts are not necessarily immediate and gradual; they can appear abruptly some time after the event.

7. Variability of ecological systems, including occasional major disruptions, provides a kind of self-monitoring system that maintains resilience. Policies that reduce variability in space or time, even in an effort to improve environmental "quality," should always be questioned.

8. Many existing impact assessment methods (e.g., cost-benefit analysis, input-output, cross-impact matrices, linear models, discounting) assume none of the above occurs or, at least, that none is important.

So the central question Holling's group confronted was: If none of the existing assessment methods takes these critically important premises into account, what kind of assessment protocol would? Their answer was the creation of the adaptive management method, which they formed through integration of several key design features and process steps:

1. Environmental dimensions should be introduced at the very beginning of the development or policy design process and should be integrated as equal partners with the economic and social dimensions.

2. Thereafter, in the design phase, there should be periods of intense, focused innovation involving significant outside constituencies, followed by periods of stable consolidation.

3. Part of the design should include benefits attached to increasing information on unknown or partially known social, economic, and

environmental effect. Information can be given a value just as jobs, income, and profit can.

4. Some of the experiments designed to produce information can be part of an integrated research plan, but others should be designed into the actual management activities. Managers as well as scientists learn from change.

5. An equally integral part of the design are the monitoring and remedial mechanisms. They should not simply be post hoc additions after implementation.

6. In the design of those mechanisms there should be careful analysis of the economic trade-offs between structures and policies that presume that the unexpected can be designed into insignificance and less capital intensive mechanisms that monitor and ameliorate the unexpected.

Although it is seldom described with the preceding level of detail, when most people today use the term adaptive management they mean what was laid out in Holling's seminal work. No one since has been able to improve upon it, and because of it, adaptive management has been joined at the hips with ecosystem management as the tool with which to implement the policy. For example, the U.S. Fish and Wildlife Service and National Marine Fisheries Service have placed adaptive management front and center as the approach to carrying out the agencies' new focus on ecosystem management as a unifying approach for implementing the Endangered Species Act. For example, in the habitat conservation plan (HCP) program (discussed *supra* in Chapter 6), under which the agencies can authorize actions that incidentally take protected species provided the action is designed to minimize and mitigate adverse effect to the species, the agencies have portrayed adaptive management as an important practical tool that "can assist the Services and the applicant in developing an adequate operating conservation program and improving its effectiveness." Notice of Availability of a Final Addendum to the Handbook for Habitat Conservation Planning and Incidental Take Permitting Process, 64 Fed. Reg. 35242, 35252 (June 1, 2000).

Notes and Questions

1. There appears to be a general consensus among resource managers and academics that adaptive management is the only practical way to implement ecosystem management. *See* Ronald D. Brunner and Tim W. Clark, *A Practice–Based Approach to Ecosystem Management*, 11 CONSERVATION BIOLOGY 48 (1997); Paul L. Ringold et al., *Adaptive Management Design for Ecosystem Management*, 6 ECOLOGICAL APPLICATIONS 745 (1996); Anne E. Heissenbuttel, *Ecosystem Management–Principles for Practical Application*, 6 ECOLOGICAL APPLICATIONS 730 (1996). Indeed, the Ecological Society of America's comprehensive study of ecosystem management treats the use of adaptive management methods as a given. *See* Norman L. Christensen et al., *The Report of the Ecological Society of America Committee on the*

Scientific Basis for Ecosystem Management, 6 Ecological Applications 665 (1996).

2. Perhaps adaptive management is the only *practical* way to implement ecosystem management, but is adaptive management as Holling describes it *politically* feasible? Holling himself has recognized that adaptive management requires flexible institutions, as in "ones where signals of change are detected and reacted to as a self-correcting process." C.S. Holling, *Surprise for Science, Resilience for Ecosystems, and Incentives for People*, 6 Ecological Applications 733 (1996). But how, politically and legally, would this "self-correcting process" be constructed within resource management agencies? Will the public have a right to participate in the "self-correction process," and how will judicial review be conducted of the "self-correction process" decisions? Is our political context willing to give agencies the discretion they would need to engage in constant "self-correction" free of legislative, judicial, and citizen oversight and challenge? As Grumbine has observed, agencies "have not often been rewarded for flexibility, openness, and their willingness to experiment, monitor, and adapt." *Reflections on "What is Ecosystem Management*,*"* 11 Conservation Biology 41 (1997). Why is that so?

3. Even if adaptive management is politically viable, can we afford it—is it *financially* viable? Who will pay for the constant flow of monitoring, measuring, modeling, and assessing that adaptive management contemplates? Are you concerned that if we adopt adaptive management as the method for implementing ecosystem management, but then under-fund the process itself, we risk making irreparably damaging decisions? *See* D. James Baker, *What Do Ecosystem Management and the Current Budget Mean for Federally Supported Environmental Research?*, 6 Ecological Applications 712 (1996).

4. Draft a legal definition of adaptive management—one that could be inserted in a new law instructing a federal resource management agency how to implement ecosystem management. Does your definition address any concerns discussed above about political and financial viability?

C.   What is Ecosystem Management Law?

As law professor Robert Keiter has observed, "until Congress speaks, ecosystem management can only claim a tenuous legitimacy, which also leaves the concept undefined for legal purposes." Robert Keiter, *Toward Legitimizing Ecosystem Management on the Public Domain*, 6 Ecological Applications 727 (1996). He elaborates:

> An antagonistic and recalcitrant Congress can impede and even reverse agency policies with which it disagrees. Congressional funding for key ecosystem initiatives can be stopped through appropriations riders without full debate over the merits of the policy. Administrative regulations that are not statutorily mandated can always be revised by a subsequent, unsympathetic administration, just as policies lacking

congressional support can be abandoned or reformulated. And courts inclined to defer to legislative or administrative discretion are unlikely to intervene in the absence of an express ecosystem mandate.

Yet the prospect of enacting a comprehensive, mandate-stating national ecosystem management law in the foreseeable future is essentially nil. The last bill to attempt anything close to that goal was introduced in Congress in 1994 and quickly died. That bill was not very ambitious. It did not even define ecosystem management! Rather, it would have established a federal commission to study how to coordinate an ecosystem management approach for federal land management policy. Nevertheless, the bill's sponsor, Senator Hatfield, introduced the bill with an impassioned plea for Congress to do *something* about ecosystem management, and the bill did contain the seeds of an ecosystem management approach:

ECOSYSTEM MANAGEMENT ACT OF 1995 (S. 2189)

104th Congress, 1st Session, 141 Cong. Rec. S 320.
Wednesday, January 4, 1995.

MR. HATFIELD. Mr. President, the last proposal I will introduce today relates to ecosystem management and watershed protection. These are the "buzz words" for a new generation of land management philosophies and techniques. A number of federal land management agencies are now working to implement ecosystem management on a landscape levels, including the Bureau of Land Management, the Forest Service and the Bureau of Reclamation.

* * *

Unfortunately, we as legislators and appropriators understand little about this new and innovative land management technique. Each federal government agency, state agency, interest group and Congress-person has his or her own idea of what ecosystem management means for the people and ecology of their particular state or region. As appropriators, we are required to fund these actions with little more than faith that the agencies' recommendations are based on sound science and a firm understanding of the needs of ecosystems and the people who live there.

Numerous additional questions surround not only the integrity but the functionality of the ecosystem management boat we have already launched. For example, what is ecosystem management, how should it be implemented and who should be implementing it? How does the ecosystem oriented work of the federal agencies, states, municipalities, counties, and interest groups mesh? And is the existing structure of our government agencies adequate to meet the requirements of managing land across which state and county lines have been drawn? Finally, with a decreasing resource production receipt base, how shall we pay for ecosystem management? Direct federal appropriations? Consolidation of federal, state, local and private funds? And if we determine how to pay for ecosystem management, who coordinates collection of these funds and how are they distributed?

I do not disagree with the theory that holistic, coordinated management of our natural resources is necessary. On the contrary, I and many of my Senate colleagues are prepared to move in that direction. It makes eminent sense to manage resources by the natural evolution of river basins and watersheds rather than according to the artificial boundaries established by counties, states and nations. Nevertheless, as our nation's funding resources become more scarce and our government agencies, states, localities and private interests seek to coordinate their ecosystem restoration efforts, Congress and the Executive Branch need to avail themselves of the best information in order to make educated, informed decisions about how ecosystem management will affect our nation's people, environment and federal budget.

To help answer these questions, I am introducing legislation today to create an Ecosystem Management Study Commission.... The Commission will submit a report to Congress 1 year after enactment which: defines ecosystem management; identifies constraints and opportunities for coordinated ecosystem planning; examines existing laws and Federal agency budgets to determine whether any changes are necessary to facilitate ecosystem management; identifies incentives, such as trust funds, to encourage parties to engage in the development of ecosystem management strategies; and identifies, through case studies representing different regions of the United States, opportunities for and constraints on ecosystem management.

It is time to look beyond the polarized positions of "economic growth" and "environmental protection" which have crippled our system of land management planning and implementation in recent years. Instead we must work toward the creation of cooperative, regionally-based, incentive-driven planning for the management of our water, air, land and fish and wildlife resources in perpetuity.

* * *

S. 93 Be it enacted by the Senate and House of Representatives of the United States of America in Congress assembled,

SECTION 1. SHORT TITLE.

This Act may be cited as the "Ecosystem Management Act of 1995."

SEC. 2. ECOSYSTEM MANAGEMENT.

(a) Definitions. Section 103 of the Federal Land Policy and Management Act of 1976 (43 U.S.C. 1702) is amended by adding at the end the following new subsections:

* * *

"(r) The term 'systems approach', with respect to an ecosystem, means an interdisciplinary scientific method of analyzing the ecosystem as a whole that takes into account the interconnections of the ecosystem."

(b) Ecosystem Management. Title II of the Federal Land Policy and Management Act of 1976 (43 U.S.C. 1711 et seq.) is amended by adding at the end the following new sections:

"ECOSYSTEM MANAGEMENT

Sec. 216. It is the policy of the Federal Government to carry out ecosystem management with respect to public lands in accordance with the following principles:

(1) Human populations form an integral part of ecosystems.

(2) It is important to address human needs in the context of other environmental attributes-

(A) in recognition of the dependency of human economies on viable ecosystems; and

(B) in order to ensure diverse, healthy, productive, and sustainable ecosystems.

(3) A systems approach to ecosystem management furthers the goal of conserving biodiversity.

(4) Ecosystem management provides for the following:

(A) The promotion of the stewardship of natural resources.

(B) The formation of partnerships of public and private interests to achieve shared goals for the stewardship of natural resources.

(C) The promotion of public participation in decisions and activities related to the stewardship of natural resources.

(D) The use of the best available scientific knowledge and technology to achieve the stewardship of natural resources.

(E) The establishment of cooperative planning and management activities to protect and manage ecosystems that cross jurisdictional boundaries.

(F) The implementation of cooperative, coordinated planning activities among Federal, tribal, State, local, and private landowners."

"ECOSYSTEM MANAGEMENT COMMISSION

Sec. 217. (a) Establishment. There is established an Ecosystem Management Commission (referred to in this section as the 'Commission').

(b) Purposes of the Commission. The purposes of the Commission are as follows:

(1) To advise the Secretary and Congress concerning policies relating to ecosystem management on public lands.

(2) To examine opportunities for and constraints on achieving cooperative and coordinated ecosystem management strategies that provide for cooperation between the Federal Government and Indian tribes, States and

political subdivisions of States, and private landowners to incorporate a multijurisdictional approach to ecosystem management.

* * *

(e) Duties of the Commission. The duties of the Commission are as follows:

(1) To conduct studies to accomplish the following:

(A) To develop, in a manner consistent with section 216, a definition of the term 'ecosystem management'.

(B) To identify appropriate geographic scales for coordinated ecosystem-based planning.

(C) To identify, with respect to the Federal Government, the governments of Indian tribes, States and political subdivisions of States, and private landowners, constraints on, and opportunities for, ecosystem management in order to facilitate the coordination of planning activities for ecosystem management among the governments and private landowners.

(D) To identify strategies for implementing ecosystem management that recognize the following:

(i) The role of human populations in the operation of ecosystems.

(ii) The dependency of human populations on sustainable ecosystems for the production of goods and the provision of services.

(E) To examine this Act, and each other Federal law or policy that directly or indirectly affects the management of public lands, including Federal lands that have been withdrawn from the public domain, to determine whether any legislation or changes to administrative policies, practices, or procedures are necessary to facilitate ecosystem management by the Federal Government in accordance with section 216.''

* * *

Notes and Questions

1. Senator Hatfield's bill reduced ecosystem management to six core topics of policy development:

- stewardship of natural resources
- partnerships of public and private interests
- public participation in decisions and activities
- use of the best available scientific knowledge and technology
- cooperative planning and management activities across jurisdictional boundaries
- cooperative, coordinated planning activities among Federal, tribal, State, local, and private landowners

Do these correspond well with Grumbine's suggested approach to ecosystem management outlined *supra*? Are there other key topics that must be considered?

2. Rather than defining ecosystem management, Senator Hatfield's bill defined the "systems approach" and emphasized its importance to developing an effective ecosystem management policy. Senator Hatfield defined systems approach as "an interdisciplinary scientific method of analyzing the ecosystem as a whole that takes into account the interconnections of the ecosystem." Recall that Grumbine's work on ecosystem management also acknowledges the importance of taking this kind of approach, which he defined as requiring that "when working on a problem at any one level or scale, managers must seek the connections between all levels." Also, Holling's theory of adaptive management, detailed *supra*, is based on the conception of ecosystems as dynamic systems. Was Senator Hatfield's reference to the "systems approach" thus an endorsement of the Grumbine–Hollings approach to ecosystem management?

3. Senator Hatfield's proposed law would have required the Ecosystem Management Commission to define ecosystem management consistent with the systems approach and the six specified policy components. Assume you are a member of the Commission. What definition would you propose?

4. Note the reference in the bill to "the dependency of human populations on sustainable ecosystems for the production of goods and the provision of services." Are the "services" to which the bill refers the same as the "ecosystem services" Salzman, Costanza, and Daily propose as a unit of ecosystem measurement? What other "services" could the bill have had in mind?

1. CHANGING THE FOCUS OF EXISTING LAWS

While neither Senator Hatfield's Ecosystem Management Act nor anything like it stands much chance of being enacted in the foreseeable future, the existing array of environmental laws presents an opportunity to forge ecosystem management through reinvention rather than invention. Indeed, the guru of government reinvention during the 1990s, Vice President Al Gore, suggested just this approach in his September 1993 National Performance Review, in which he directed federal environmental agencies to develop "a proactive approach to ensuring a sustainable economy and a sustainable environment through ecosystem management."

Recall, however, Professor Keiter's description of the ways in which any such effort can be thwarted: new administrations may change course; legislatures may enact new law to preempt the direction of reinvention; courts may find existing law does not authorize where reinvention has led. One way to conceptualize these possibilities is to think of existing law as defining a box of "policy space." Statutory text rarely defines the policy options in such a way that the box is extremely small. Agencies charged with implementing the statute thus often have plenty of space within which to develop alternative policies. Indeed, often the box is sufficiently large that we don't know how large it is until a court finds that an agency has adopted a policy that falls outside the box. Of course, at any time the legislature could alter the dimensions of the box through statutory amendments, putting the point in space representing the current administrative policy position outside or inside of the new box.

The Clinton Administration's call for reinvention of environmental policy through ecosystem management thus was an effort to move within the policy space box from one point—the point where previous administrations had left policy—to another. The materials in this section chart the administration's initial efforts in that regard, taking place primarily in the time period from 1995 through 2000. Keiter's point, of course, is clearly put front and center by the events of politics and law since that time: a new administration; a new Congress; and time for courts to test whether ecosystem management violates the boxes of existing law. For example, soon after taking office, the Bush Administration announced that it would revisit many of the rulemakings the Clinton Administration made in furtherance of ecosystem management. For contrasting views on the wisdom of such seesaw politics, compare OFFICE OF MANAGEMENT AND BUDGET, OFFICE OF INFORMATION AND REGULATORY AFFAIRS, MAKING SENSE OF REGULATION: 2001 REPORT TO CONGRESS ON THE COSTS AND BENEFITS OF REGULATIONS AND UNFUNDED MANDATES ON STATE, LOCAL, AND TRIBAL ENTITIES (2001), with NATURAL RESOURCES DEFENSE COUNCIL, REWRITING THE RULES: THE BUSH ADMINISTRATION'S UNSEEN ASSAULT ON THE ENVIRONMENT (2002). The specific consequences of these dynamics, repeated at state and local levels, are examined in the chapters that follow this one.

In general, though, fields of law and policy may tend to ossify over time—it becomes harder to change the box of policy space. Court interpretations of statutes become not news but rather long-held precedent. Legislatures become less willing, or able, to reopen statutes hashed out in the past. Administrations come and go with less effect on the course of long-term policy. Occasionally, of course, there is chaos in the form of new Supreme Court jurisprudence or a wave of "reform" fever in the legislature. By and large, however, ecosystem management law is simply too young to have ossified. For purposes of law and policy, it began yesterday, and is still finding its feet. It is useful, therefore, to examine its origins, for they continue to have a profound influence on its future.

a. EARLY ASSESSMENTS

Soon after Senator Hatfield introduced his bill, House Natural Resources Committee Chair Rep. George Miller instructed the Congressional Research Service (CRS) and the Government Accounting Office (GAO) to examine what federal agencies were doing to move toward ecosystem management. After surveying and examining the work of a group of federal agencies that had formed an Interagency Ecosystem Management Coordinating Group (IEMCG), CRS and GOA concluded as follows:

Congressional Research Service, Ecosystem Management: Federal Agency Activities

(April 19, 1994).

Many Federal agencies appear to be very aggressive in redirecting current efforts or initiating ecosystem management activities. These efforts and activities are diverse in scale, focus, institutional relationships, goals, and accomplishments. When they are viewed together, two themes seem most common. One is improved communication and coordination. Improvement is based on building new partnerships, sharing information, and reaching agreements or definitions for key terms and data collection. The second is trying to improve the condition of resources, which some are calling protection of biodiversity. These efforts might grow out of working to protect individual species and their habitats, restoring ecological processes and ecosystem services, or restoring degraded resources in an area. Agencies have listed many examples of these approaches in these summaries.

The current focus seems centered on the communication and coordination aspects of ecosystem management, perhaps because results here are visible far more quickly than for significant changes in resources. Improved resource conditions are decades away in many cases, even if ecosystems are managed in a scientifically sound and programmatically consistent manner. And these improved conditions are hard to measure, because ecosystems involve so many interrelated components, but the crucial variable is the health of the overall system rather than any single component. The [Interagency Ecosystem Management Coordinating Group] plays a key role in these efforts by supporting activities to improve communication and coordination and to change resource conditions. In a recent brochure, its listed efforts included:

- forming committees and working groups to address the complex of issues arising as agencies adopt ecosystem management;
- facilitating the standardization of data sharing techniques and ecosystem mapping techniques;
- establishing general implementation guidelines on ecosystem management for assimilation by participating agencies;
- identifying training needs and instituting collaborative training programs;
- adopting a Memorandum of Understanding among participating agencies to formally establish the Coordination Group and its role;
- serving as a resource to ensure each agency is current with ongoing research and information including legislative concerns;
- facilitating the adoption of standard terminology for participating agencies to use in ecosystem management and relevant aspects of research and operation;
- maintaining an education and outreach component; and
- encouraging existing and new partnerships for management on an ecosystem basis on a multi-agency basis.

These papers also illustrate the wide variety of missions by reporting agencies. The agencies responsible for Federal lands and the resources on those lands have a very different mix of efforts than the agencies that deal with resources that are primarily on private lands. Some agencies deal more with the tools of resource analysis and with the development of resource information while others are more interested in the actual management of resources. These differences in missions are cause for a remarkable diversity in the approaches taken, in the reasons for interest in ecosystem management, and in the ways that agencies have integrated these efforts with ongoing activities and approaches to problem solving or resource management.

The intensity of the effort is also striking. All these agencies are devoting significant resources, both staff and financial, to this effort at a time of budget constraints and Federal downsizing. But the benefits of this approach, as they are defining it, seem to outweigh any costs, assuming successful implementation. Equally striking is the faith that the many government employees who participated in this demonstration appear to place in an ecosystem approach as a more rational way of serving the public good. One of the more difficult challenges for these agencies will be to maintain this intensity in continuing to work together successfully toward a consistent approach over time and in the face of changes in priorities, authorizations, and appropriations from year to year.

General Accounting Office, Ecosystem Management: Additional Actions Needed to Test a Promising Approach

(August 1994).

Purpose

Even though many laws have been enacted to protect individual natural resources—air, water, soils, plants, and animals, including forests, rangelands, threatened and endangered species, wetlands, and wilderness areas—ecological conditions on many federal lands have declined. As a result of these declines and the recognition that some historic levels of natural resource-commodity production and other natural resource uses cannot be sustained indefinitely; federal land managers have had to substantially decrease production of some renewable commodities, such as timber, and other uses, such as recreational activities, on some land units. These reductions have, in some instances, disrupted local economies and communities, contributing to intractable conflicts between ecological and economic values and concerns.

Since the late 1980s, many federal agency officials, scientists, and natural resource policy analysts have advocated a new, broader approach to managing the nation's lands and natural resources called "ecosystem management.'' This approach recognizes that plant and animal communities are interdependent and interact with their physical environment (soil, water, and air) to form distinct ecological units called ecosystems that span federal and nonfederal lands. In response to congressional requests, GAO identified (1) the status of federal initiatives to implement ecosystem management, (2) additional actions required to implement this approach, and (3) barriers to governmentwide implementation.

* * *

Results in Brief

Over the past 2 years, all four of the primary federal land management agencies have independently announced that they are implementing or will implement an ecosystem approach to managing their lands and natural resources, and each has been working to develop its own strategy primarily within its existing framework of laws and land units. In addition, the administration is proposing in its fiscal year 1995 budget, among other things, to fund the initial stage of a governmentwide approach to ecosystem management, including four ecosystem management pilot projects. It is also considering various principles for its governmentwide approach, including managing along ecological rather than political or administrative boundaries.

Implementing the initial stage of a governmentwide approach to ecosystem management will require clarifying the policy goal for ecosystem management and taking certain practical steps to apply the principles being considered by the administration. These steps include (1) delineating ecosystems, (2) understanding their ecologies, (3) making management choices, and (4) adapting management on the basis of new information. In taking these steps, the federal government will have to make difficult policy decisions about how it can best fulfill its stewardship responsibilities.

The administration's initiatives to implement ecosystem management governmentwide face several significant barriers. For example, although ecosystem management will require greater reliance on ecological and socioeconomic data, the available data collected independently by various agencies for different purposes, are often noncomparable, and insufficient, and scientific understanding of ecosystems is far from complete. While ecosystem management will require unparalleled coordination among federal agencies, disparate missions and planning requirements set forth in federal land management statutes and regulations hamper such efforts. And although ecosystem management will require collaboration and consensus-building among federal and nonfederal parties within most ecosystems, incentives, authorities, interests, and limitations embedded in the larger national land and natural resource use framework—many beyond the ability of the federal land management agencies individually or collectively to control or affect—constrain these parties' efforts to work together effectively.

Notes and Questions

1. Do the CRS and GAO reports differ significantly in overall perspective on the progress of ecosystem management policy in the federal agencies at the time of their assessments? Doesn't CRS seem more positive in its comments? For example, CRS marvels at the "remarkable diversity in the approaches taken," whereas GAO expresses the concern that "while ecosystem management will require unparalleled coordination among federal agencies, disparate missions and planning requirements set forth in federal land management statutes and regulations hamper such efforts." Looking back over the materials in the prior sections of this Chapter, which of the two reports strikes you as more attuned to the issues that Blair raises about ecosystem functions, that Grumbine discusses in his exposition on ecosystem management policy, and that Holling's team revealed in their work on adaptive management?

2. The GAO refers to the "four ... primary federal and management agencies." These agencies administer vast expanses of federally-owned land and natural resources. Three are within the Department of the Interior: the Bureau of Land Management (264 million acres); the National Park Service (84 million acres); and the United States Fish and Wildlife Service (93 million acres). The fourth agency is the United States Forest Service, a branch of the Department of Agriculture, which oversees 191 million acres of national forests and grasslands. As the GAO points out, how these agencies decide to manage their combined total of over 630 million acres of federal land has played and will continue to play a prominent role in the development of ecosystem management law.

b. INITIAL EFFORTS

As the CRS and GAO reports clearly demonstrate, notwithstanding the congressional endorsement Professor Keiter contends will be necessary for moving ecosystem management from de facto to de jure status, there is room for steering the ship of existing environmental law toward ecosystem management. It matters, of course, what basic objectives have been set, currencies selected, and management goals and methods installed. But as those policy decisions are made, even existing law can be used to put them into action.

In that regard, the laws that could be part of the reshaping movement toward ecosystem management can be thought of as falling into five categories. One set of laws has general conservation of environmental quality as the principle focus already, but the laws lack explicit reference to ecosystem management as a conservation objective. The Endangered Species Act is an example. A second set of laws, including most prominently the National Environmental Policy Act, uses impact assessment procedures as an opportunity to improve government decision making. If impact assessment were to be defined to include ecosystem-level impacts, these law could focus attention on the need for ecosystem management. Another set of laws, typified by the Clean Water Act and Clean Air Act, has pollution control as the central subject matter and protection of public health as the primary concern. Improving environmental quality is an additional objective of many of these laws, but it is not accomplished in the statutory schemes through anything like ecosystem management. A fourth type of law consists of those concerned primarily with management of defined public land or resource units, such as forests, parks, or fisheries. This variety of laws provides the most promising foundation from which to build an ecosystem management approach, as the laws already have ecosystem units or functions as their core organizing principle. Many of these laws, however, were enacted before ecosystem management became a dominant policy theme, and thus the degree to which it can be incorporated for the future is uncertain. Moreover, these laws are limited in operative effect to public (primarily federal) lands. Finally, the body of law with perhaps the most influence over how land is used has long been local land use controls such as zoning and growth management. While the emphasis of zoning has been form and function, and growth management has emerged largely in response to the nuisances of urban sprawl, there is no reason why local land use laws could not move toward ecosystem management as an additional objective. Yet, being local in nature and subject ultimately to state authority, the question is whether state and local political will are primed to do so. As the following materials suggest, therefore, each of these five types of laws presents opportunities and challenges for development of a law of ecosystem management.

(1) Environmental Conservation Laws

How far has ecosystem management come? How much does existing law box it in? If any existing statute provides the test case for asking those questions, it is the Endangered Species Act. As noted in the opening to this Part of the text, the ESA proclaims that one of its purposes is "to provide a means whereby the ecosystems upon which endangered species and threatened species depend may be conserved." 16 U.S.C. § 1531(b). More than any other federal statute, the agencies charged with administering the ESA, the U.S. Fish and Wildlife Service (FWS) and the National Marine Fisheries Service (NMFS), have embraced ecosystem management and attempted to make it the dominant theme of law and policy. Did they succeed?

To examine that question, one of your authors has proposed a set of questions designed to test how far "big" policy ideas have advanced toward formulation as hard law to apply. The seven "degrees of relevance" through which concepts such as ecosystem management progress from mere idea to law are:

    level one: The idea has become widely expressed through a generally accepted norm statement.

    level two: Advocating the opposite of the norm is no longer a tenable policy position.

    level three: The charge of acting contrary to the norm no longer can be left unaddressed.

level four: Failure affirmatively to portray an action as being consistent with the norm is seen as a significant deficiency.

level five: Important governmental authorities have established the norm as an explicit policy goal.

level six: Proposed actions are being denied or delayed on the basis of perceived failure to facilitate the norm.

level seven: The norm is fully transformed into law to apply.

The materials presented thus far in this chapter go a long way toward establishing that ecosystem management has in general passed at least as far as the fourth level—ecosystem management has been associated with specified normative goals and, while pockets of resistance exist, it has become a dominant theme of academic and policy dialogue. But is it *law*? Is it relevant *to lawyers*? The following excerpt picks up the analysis of that question at the fifth level:

J. B. Ruhl, Ecosystem Management, The Endangered Species Act, and the Seven Degrees of Relevance

14 Natural Resources & Environment 156 (2000).

Level 5: Have important governmental authorities established the norm as an explicit policy goal? Has anyone employed by FWS or NMFS *not* heard of ecosystem management? The story of these two agencies' aggressive, and at times bold, effort to infuse the ESA with ecosystem management policy begins in March 1994 with FWS's publication of *An Ecosystem Approach to Fish and Wildlife Conservation: An Approach to More Effectively Conserve the Nation's Biodiversity* (March 1994). The agency portrayed this policy document as its road map for applying "the concept of managing and protecting ecosystems to everything the Service does." *Id.* at 5. The agency thereby explicitly endorsed the emerging body of conservation biology literature and research advocating that biological diversity—the variety and number and distribution of species across the earth—is the primary index of the health of the environment, and that the way to sustain that diversity is through protection and management of whole ecosystems. Each species is part of a dynamic, co-adapted assemblage of species dependent on and interacting with their surrounding habitat. It is that total package that must be managed, not just some of the bits and pieces. To do so, FWS has divided the country into watershed-based planning units representing major ecosystems of the nation, around which it will orient implementation of its various regulatory authorities. FWS announced through this publication that, where it can, it will attempt to use its powers to manage on the ecosystem level, for protection of the ecosystem dynamics, and thereby promote conservation of all the assembled species and environmental qualities. The agency promised that specific ecosystem-based reform measures for the ESA would follow.

FWS soon lived up to its promise through other initiatives, and often hand-in-hand with NMFS. Shortly after FWS published the general "Ecosystem Approach" agenda, FWS and NMFS adopted two significant policies designed to take the new focus on ecosystem dynamics straight to the ESA. The magic behind the agencies' approach was the realization that whereas the agencies do not have the discretion to transform the ESA into an ecosystem protection statute—the Act is fundamentally species-by-species in its orientation—nothing in the statute prevents the agencies from considering ecosystem factors in making species-specific decisions. Thus, in July 1994 the agencies announced that they would "promote healthy ecosystems through activities undertaken by the Service under authority of the Endangered Species Act." Notice of Interagency Cooperative Policy for the Ecosystem Approach to the Endangered Species Act, 59 Fed. Reg. 34273 (July 1, 1994). The agencies declared that henceforth they would incorporate ecosystem-level considerations into species listings and recovery planning under section 4 of the Act and for interagency cooperation under section 7. The agencies thus will "develop and implement recovery plans for communities or ecosystems where multiple listed and candidate species occur." *Id.* That policy recognizes that, in many ways, all the species are in the same boat, and our efforts to bring about their recovery must be approached with that reality in mind. Even more specifically, FWS and NMFS announced in a separate policy that the "method to be used for recovery plan preparation shall be based on several factors, including the range or ecosystem of the species," *id.*, and that recovery planning teams would be assembled with the species' ecosystem in mind. An expert in the biology of a species, for example, may not be an expert in the ecosystem functions that have been degraded and pose the threat to the species. If recovering the species requires repairing the ecosystem, the recovery planning effort must use expertise from both disciplines.

FWS and NMFS have acted to cement their ecosystem management policies for the ESA in a number of subsequent publications and announcements. For example, in 1997 the two agencies jointly published a policy statement emphasizing how the emerging ecosystem management approach would guide their ESA implementation in a variety of specific programs. *See Making the ESA Work Better: Implementing the 10 Point Plan ... and Beyond* (June 1997). More recently, the [then] Director of FWS proclaimed that she has "challenged the agency to pioneer the practical implementation of an ecosystem approach," citing habitat conservation plan permits as a focal point for the ESA in that respect, and she invited the scientific community to join that endeavor. *See* Jamie Rappaport Clark, *The Ecosystem Approach from a Practical Point of View*, 13 CONSERVATION BIOLOGY 679 (1999). Through the efforts of FWS and NMFS, therefore, ecosystem management has become a new defining model for how the ESA will be implemented as a matter of official *policy*.

Level 6: Are proposed actions being denied or necessary authorization delayed on the basis of perceived failure to facilitate the norm? Notwithstanding all the momentum behind ecosystem management in the scientific literature, in the rhetoric of environmental and preservation groups, and in the official policies of FWS, NMFS, and other federal, state, local, and tribal agencies . . ., the practically-minded lawyer will still want to know what his or her burden of proof is—i.e., will anything about the ecosystem management dimension of a proposed project make or break the project's approval. An attorney hired to oppose a project will want to know this just as much as the attorney hired to advocate on behalf of the project. In the long run, all the rah-rah support for ecosystem management means very little in the right here, right now context in which real-world lawyers operate if it does not have the capacity directly and overtly to change their clients' outcomes.

In the ESA setting, this question focuses attention on the fact that nothing in the ESA advances ecosystem management beyond an implied consequence of the stated purpose of conserving ecosystems upon which protected species rely. Rather, virtually everything that counts under the ESA is species specific: *species* are listed and their critical habitat designated under section 4; federal agencies must conserve listed *species* under section 7(a)(1); federal agencies must avoid jeopardizing *species* and destroying *species'* critical habitat under section 7(a)(2); all persons must avoid taking animal *species* under section 9; permits may be granted for incidental take of listed *species* under section 10. Hence, FWS and NMFS are left to channel all their ecosystem management policy into a law that is species-specific in implementation, a tactic they know can go only so far before it is open to attack as *ultra vires*. For example, the agencies declined recently to develop a policy of designating and listing distinct breeding populations based on the importance of the population to its ecosystem. The adopted policy observed that "despite its orientation toward conservation of ecosystems, the Services do not believe that the Act provides authority to recognize a potential [population] as significant on the basis of the importance of its role in the ecosystem in which it occurs." *Policy on Recognition of Distinct Vertebrate Population Segments Under the Endangered Species Act*, 61 Fed. Reg. 4722, 4723 (1996).

Does this mean ecosystem management can make no difference to outcomes under the ESA? Not at all. But it does mean that it is difficult to know when it will make a difference, as it is difficult for the agencies to pin up or down decisions directly on ecosystem management goals and criteria. Some actions are easier in this respect, such as grouping species listings when the species are found in the same ecosystem, or developing recovery plans designed around ecosystem restoration. In those cases the nexus between ecosystem and the ESA action is often direct—i.e., a significant component in the explanation for the species' decline and a key factor in designing the species' recovery.

By contrast, actions that are species-specific and *project*-specific, such as small-scale habitat conservation plan permits under section 10 and federal agency consultations under section 7(a)(2), make it more difficult for the agencies to base decision making on ecosystem management policy. The ESA's evaluation criteria for such actions do not expressly adopt ecosystem management standards, but do leave enough room for the agencies to steer outcomes based on ecosystem management policy. FWS and NMFS have done so principally through the "take" prohibition found in section 9 of the Act, or more specifically, the so-called "harm" rule. Section 9(a)(1)(B) of the ESA makes it unlawful for any person to "take" endangered species of fish or wildlife. 16 U.S.C. 1538(a)(1)(B). Section 3(19) defines take to mean "to harass, harm, pursue, hunt, shoot, wound, kill, trap, capture, or collect, or to attempt to engage in any such conduct." *Id.* 1532(19). Few single words have been the subject of as much litigation, administration, and consternation as has "harm" in that definition. As Congress left it undefined, FWS defined it by rule to include "significant habitat modification or degradation where it actually kills or injures wildlife by significantly impairing essential behavioral patters, including breeding, feeding or sheltering." *See* 50 C.F.R. 17.3; *see also* 63 Fed. Reg. 24148 (1998) (NMFS proposed rule adopting similar definition). Hello ecosystem management!

After a lengthy series of court battles, the Supreme Court endorsed the rule as consistent with congressional intent, but only when read strictly to limit harm to cases in which the challenged action is the foreseeable, proximate cause of actual death or injury to identifiable individuals of the species. *See Babbitt v. Sweet Home Chapter of Communities for a Great Oregon*, 515 U.S. 687 (1995). The lower federal courts are trying to divine *Sweet Home's* exact meaning.

* * *

While these issues filter through the courts ... FWS in particular has effectively used the harm rule to police habitat destruction and modification and thereby serve as a surrogate for background ecosystem protection goals. For example, in Austin, Tucson, and southern California, FWS has attempted to extend the harm rule's umbrella of protection to so-called "suitable habitat," or "potential habitat," or "contiguous habitat." These loose terms, which lack any official definition, serve to describe areas where none of the relevant protected species are found but which seem like awfully nice habitat nevertheless. Sounds like ecosystem management to me.

Level 7: Is the norm fully transformed into law to apply? Even under the harm rule, FWS and NMFS must have a listed endangered animal species relying on the ecosystem to provide the necessary leverage to shape the ESA's species-specific and project-specific outcome with ecosystem management policy. The most die-hard ecosystem management advocate thus must concede that ecosystem management has not attained the status of law to apply under the ESA. Although its policy star is rising, ecosystem management remains in the background of the ESA. No ecosystem management rules are to be found in the agencies' ESA regulations. No citizen suits are brought for violation of ecosystem management. No court has enjoined a development project because it impedes ecosystem management. The ESA's species-specific focus will limit ecosystem management in this respect at every turn, preventing it from becoming more than a driving policy force in species-specific legal actions. But perhaps that is not much of a distinction to the practically-minded real-world lawyer interested more in the outcome than in how one gets there. As FWS increasingly uses the harm rule as a surrogate for ecosystem management, it matters less and less that ecosystem management itself is not in the black letter law.

Indeed, the greatest challenge to a lawyer is when an idea has made it to Level 6—where it affects legal outcomes—and is stuck there, for in that setting the lawyer knows there is lawyering to be done to deal with the issue but the lawyer cannot point to clear legal standards to guide clients, to argue to agencies, and to appeal to courts. Despite all the efforts of FWS and NMFS to push ecosystem management to the forefront, and despite all the scientific journal and law review articles advocating ecosystem management, this is, I am afraid, where ecosystem management is stuck under the ESA, and it will remain so in the absence of legislative reform.

Notes and Questions

1. Given this view of the ESA, shaped largely by its species-by-species approach, is it possible that the ESA may eventually hinder rather than support the formulation of ecosystem management law? The point of the article seems to be that ecosystem management law fundamentally *cannot* be based on a species-by-species focus, so eventually the authority of the ESA to bring about ecosystem management will run out, as will, therefore, its effectiveness in achieving the goals of ecosystem management. Do you agree? How would you change the ESA to eliminate these concerns? For more on this theme, see Jacqueline Lesley Brown, *Preserving Species: The Endangered Species Act Versus Ecosystem Management Regime, Ecological and Political Considerations, and Recommendations for Reform*, 12 Journal of Envtl. Law and Litigation 151 (1997).

2. How close did FWS and NMFS under the Clinton Administration come to crossing their line of authority? Assume you have been hired by an industry trade group to challenge FWS and NMFS ecosystem management policy on the ground that it violates the ESA. What arguments would you make?

(2) Environmental Assessment Laws

The Grumbine–Holling approach to ecosystem management outlined previously in this chapter relies heavily on the adaptive management process of monitoring, modeling, assessing, and adjusting. Holling forged that approach in response to what he saw as the shortcomings of conventional environmental assessment procedures in use in the 1970s. The genesis of those procedures was the National Environmental Policy Act (NEPA) and its infamous "Environmental Impact Statement" procedure.

Although Holling roundly criticized the assumptions and methods of the EIS process and his substitute of adaptive management has been widely endorsed as the only way to go in ecosystem management, NEPA continues to have its supporters as a tool for promoting ecosystem management. *See* Dinah Bear, *The Promise of NEPA*, in Biodiversity and the Law 178 (William Snape III ed., 1996). As the following excerpt reveals, however, the promise of NEPA in this regard may be quite dim.

David R. Hodas, NEPA, Ecosystem Management and Environmental Accounting

14 Natural Resources & Environment 185 (2000)

If law were to reorient our analytical framework so that each decision were to include, to the greatest extent possible, adverse environmental consequences, we could institutionalize a process of making sound ecosystem management decisions. One law that was supposedly designed to break decision-making out of its narrow, economically focused box was the National Environmental Policy Act (NEPA), 42 U.S.C. §§ 4321–4370d. Predicated on the idea that governmental decisions should not be made without full consideration of adverse environmental implications of the decisions, NEPA suggests that the more environmentally realistic our expectations, the greater the opportunity to reduce poverty, increase wealth, and diminish environmental degradation. Unfortunately, NEPA does not advance the cause of sound ecosystem management or the related concept of sustainable development, but, as will become apparent, allows decisions affecting ecosystem development to be whitewashed with a thin coat of "apparent" soundness or sustainability. In other words, NEPA, as it has evolved, lets us feel comforted by the illusion that our decisions are environmentally sensitive; as a society we willingly pretend that environmental impact statements are important, thorough, reliable analyses, when in most cases they are mere formalities based on data and predictions made by people who have no accountability for error.

Enacted January 1, 1970, NEPA was the first environmental law of the modern environmental age, and is now the model for a law that has been adopted worldwide. Although it does not use the phrase ecosystem management or sustainable development in its text, the purpose of NEPA was to achieve that which is now referred to as sustainable development, namely, "... [t]o declare a national policy which will encourage productive and enjoyable harmony between man and his environment; to promote efforts which will prevent or eliminate damage to the environment and biosphere and stimulate the health and welfare of man." *Id.* at § 4321. To implement this policy, Congress directed that agencies "insure that presently unquantified environmental amenities and values ... be given appropriate consideration in decision-making along with economic and technical considerations;" and that they "include in every recommendation or report on proposals for legislation and other major federal actions significantly affecting the quality of the human environment, a detailed statement . . . on . . . the environmental impact of the proposed action." *Id.* at § 4332(2)(B), (C).

As with most new requirements, compliance with the new beast known as the environmental impact statement (EIS) requirement was slow. Although NEPA § 102 requires that each federal agency must prepare an EIS when making a decision that could significantly affect the human environment, many agencies, particularly those with a mandate to promote development projects, such as the Atomic Energy Commission, vigorously resisted. In response, the early litigation under NEPA, such as *Calvert Cliffs' Coordinating Comm., Inc. v. U.S. Atomic Energy Comm'n*, 449 F.2d 1109 (D.C.Cir.1971), addressed the fundamental failure of the agency to include environmental impacts in its evaluation and approval of licenses to construct nuclear power plants. In that now famous decision, the court declared that "Congress did not intend [NEPA] to be . . . a paper tiger," and ruled that all agencies of the federal government "must—to the fullest extent possible under its statutory obligations—consider alternatives to its actions which would reduce environmental damage." After it became clear that NEPA applied broadly to all agencies of the government, the litigation shifted to more lawyerly gamesmanship in which agencies tried to avoid significant environmental evaluation by narrowly defining the statutory requirements so that hard issues might not be subject to the EIS mandate.

The early litigation that arose under NEPA fell into two large categories: threshold questions and adequacy questions. In cases raising threshold questions, agencies asserted that no EIS was needed because one of the elements of the statute's requirement ("proposals . . . for . . . major federal action significantly affecting the quality of the human environment") for an EIS was not present. The second category addressed whether the EIS adequately evaluated the adverse environmental consequences of a project, as well as alternatives that could avoid or mitigate the harm. In 1978, the President's Council on Environmental Quality (CEQ) promulgated regulations, 40 C.F.R. parts 1500–1508, that substantially standardized the way that federal agencies approach the NEPA process.

The courts, however, significantly narrowed the practical impact of the mandate that agencies think deeply about the environmental consequences of their actions, that they seriously explore alternatives, and that they consider the larger, long-term picture of accommodating development with ecological soundness, even though these requirements remain in the words of NEPA and the CEQ regulations. In a series of decisions, the U.S. Supreme Court has bleached out the meaning of NEPA and the CEQ regulations by its "crabbed interpretation of NEPA" and its dismissal of the goals of NEPA in § 101(b) as "largely rhetorical." Lynton Caldwell, *NEPA Revisited: A Call for a Constitutional Amendment*, The Environmental Forum, Nov.–Dec. 1989, at 18.

The Supreme Court's Evisceration of NEPA

* * *

Since the late 1970s, the Supreme Court has been unwilling to read the substantive goals of NEPA into its interpretations, especially when major government policy issues were at stake. It did this, in part, by narrowing the remedies it would permit under NEPA. First, in *Kleppe v. Sierra Club*, 427 U.S. 390 (1976), the Court, defined "proposal" in the most narrow, legalistic sense possible, on the theory that an agency could avoid preparing an EIS so long as it was only contemplating action.... According to the Court, an EIS need not be prepared until the eleventh hour: "the moment at which an agency must have a final statement ready is the time at which it makes a recommendation or report on a proposal for federal action." *Id.* at 405–06.... By looking at NEPA solely as a procedural requirement devoid of any substantive value, the Court signaled its hostility toward NEPA's advancement of any of its sustainable development goals, even in a requirement as minor as allowing courts to order agencies to begin early preparation of EISs.

Shortly after its *Kleppe* decision, ... the Court narrowed the vision called for by NEPA. In *Vermont Yankee Nuclear Power v. NRDC*, 435 U.S. 519 (1978), public interest groups challenged the issuance of a nuclear power plant construction license on the grounds, *inter alia*, that the agency had failed to consider alternative sources of electricity, including energy conservation. The Court in *Vermont Yankee* stated explicitly that although NEPA established "significant substantive goals for the Nation," the duties it imposed on agencies was "essentially procedural."

* * *

Just two years after *Vermont Yankee*, the Court announced summarily that an agency was "free under NEPA to reject an alternative acknowledged to be environmentally preferable solely on the ground that any change in [plans] would cause delay." *Strycker's Bay Neighborhood Council v. Karlen*, 444 U.S. 223, 230 (1980) (Marshall, J., dissenting). As a result, NEPA does not require an agency either to develop or implement a plan to mitigate environmental damage, so long as the agency considers mitigation in general terms as an option; nor does NEPA require an agency to perform a "worst-case analysis" to assess the effects of catastrophe. *Robertson v. Methow Valley Citizens Council*, 490 U.S. 332 (1989).

* * *

Thus, NEPA now merely requires a relatively narrow document that accompanies files reflecting foregone conclusions. At best, NEPA may marginally improve narrow decisions affecting the environment, but NEPA does not provide even marginal ecological or sustainable security. Unfortunately, NEPA, the most widely copied environmental law in the world, now provides the means to thoroughly wallpaper serious structural flaws in our decisions, so that decisions appear to be sustainable when in reality they are no more than mirages of environmental concern.

* * *

Placing the Risk of Mistake on the Predictor

NEPA's tragic devolution can be traced to the lack of any criteria to measure conduct and to hold actors accountable for their decisions. The lack of post-EIS review and monitoring not only makes the promises of mitigation hollow, but decision-makers and project advocates have also learned the short-term lessons of NEPA litigation in meeting NEPA's technical requirements without hindering a project by asking important questions. This short-term approach results in no post-project monitoring and deprives us of the feedback needed to improve future decisions. Under the current law, project proponents know there is no consequence from underestimating adverse environmental effects. Because there is no liability for inaccuracy, there is no need for post-project review that would check the accuracy of the predictions.

* * *

To improve decision-making under NEPA, the decision-makers and project proponents must be held accountable for their predictions, mitigation promises, and the residual environmental consequences of the projects. The EIS must be required to identify objective, measurable criteria that can be used to judge the ultimate accuracy of the prediction. These criteria should reflect the quantity and quality of the external environmental impacts created by the project, and be translatable into specific, dollar-based valuations that can be incorporated into project valuation, and secured against, as with any other financial risk.

Fortunately, the emerging discipline of environmental externality valuation will allow us to use law to define ecosystem management principles and sustainable development in economic terms. At a macroeconomic level, externality valuation has been pursued by the emergence of natural resource accounting, under which national income accounts (e.g., Gross Domestic Product) are adjusted directly or indirectly to reflect environmental degradation associated with a nation's economic activity. For example, if the wealth of a country selling off its forests were reduced by the value of the topsoil lost from the clear-cutting, the country's annual income would be revealed to be not the result of sustainable production, but of liquidation of capital. Natural resource accounting more accurately reflects a country's true annual income and net worth than the current system of financial accounting.

* * *

In the United States, evaluation expertise has emerged in two areas: natural resource damage assessment and state public utility commission integrated resource planning. Damages to natural resources from oil spills or from hazardous wastes must now be paid for under oil pollution and hazardous waste laws. Natural Resource Damages Assessment, 15 C.F.R. pt. 990. In the area of energy regulation, expertise is emerging in establishing preliminary values for the external damages caused by residual emissions from utilities, after all environmental regulations are met. Social Costs and Sustainability: Valuation and Implementation in the Energy and

TRANSPORT SECTOR (Olav Hohmeyer *et al.*, eds., 1997). Thus, both after-the-fact liability for external damages and before-the-fact evaluation of environmental externalities can be monetized and directly incorporated into a decision-maker's calculus.

Securing Against the Risk: Externality Insurance

* * *

The core feature of this proposal is the mandate that the potential liability for external environmental harm be secured. Although the method of securitization will vary depending upon the nature of the project, a central tenet of sustainable development, that environmental externalities be internalized into routine decision-making, requires that the risk of uncertain adverse consequences be placed upon project proponents and direct beneficiaries. One method of achieving this goal of creating accountability for externalities is to combine the concept of a performance bond with insurance.

In its simplest form, the environmental externality bond would be a condition of project approval, with the bond being in the amount of the environmental externalities the project is estimated to create. Alternatively, the bond could be set in the amount of damages greater than estimated as part of the public approval process. In either case, the operation of the environmental externality bond system would be similar. At the conclusion of the project, the bond would be returned, less the amount of the environmental harm actually caused by the project, or less the amount of excess externalities experienced. This approach would motivate both the government and the project proponent to investigate more thoroughly the environmental impacts of the project, mitigation measures available to reduce project impacts, innovative alternatives to the project, and methods to measure impact.

Another method of holding project proponents accountable could be to require that the proponents insure against the cost of future abandonment or modification of a project to mitigate adverse environmental consequences that become apparent or regulated.... To do this, it would be necessary that potential external costs be secured so that the risk shifting cannot be avoided by insolvency or bankruptcy.... This risk shifting could be "collateralized" by combining modified offsets with adequate, relatively liquid, identifiable loss reserves. To insure the public and itself against these risks, the project proponent could be required to create a segregated environmental damage reserve account that will insure against reasonably foreseeable new environmental requirements such as greenhouse gas taxes or emissions limitations. The project proponent would either deposit annual insurance premiums into this reserve fund or purchase environmental externality insurance from private insurers. Self-insurance premiums would be calculated by multiplying the environmental externality value set by the government ($/pound or $/ton) times annual emissions. Premiums for private externality insurance would be set by the private insurer.

Ultimately, the best insurance against future environmental harm and the best method to reduce the risk of future regulation is to reduce emissions and adverse environmental consequences now. Therefore, in addition to the market-leveling approach of environmental externality insurance, the system also should have a direct incentive to reduce emissions and adverse environmental effects. The project proponent should be entitled to reduce its risk and pay all or part of its environmental externality insurance premium by implementing an environmentally beneficial emission reduction project deemed to be a satisfactory equivalent reduction of the externality caused by the future emissions. In other words, "buying" environmentally beneficial projects that reduce emissions or adverse ecosystem impacts would be deemed equivalent to buying actual insurance or funding a self-insurance account. Any expenditures on these emission reduction or ecosystem protection activities deemed environmentally equivalent to externalities would reduce the utility's annual insurance obligations.

* * *

When NEPA was first enacted in 1970 it commanded each agency to "identify and develop methods and procedures . . . which will insure that presently unquantified environmental amenities and values may be given appropriate consideration in decision-making along with economic and technical considerations." NEPA § 102(2)(B). NEPA § 102(1) requires that, to the "fullest extent possible," the federal government shall administer regulations and public laws "in accordance with" the policies set forth in the statute. As a first step in finally implementing this requirement, all that is needed is for the CEQ to amend its regulations to require externality valuation and securitization for all decisions for which NEPA documents are prepared. . . . Each federal agency would then be required to make corresponding changes to its NEPA regulations or implementing procedures.

* * *

This proposal would combine the substantive goals of NEPA with the traditional risk shifting skills of lawyers and the emerging discipline of environmental externality valuation. From post-project monitoring and feedback, ecological lessons can be learned, and externality measurement criteria can be refined, which in turn will result in better information for project evaluators considering project designs and alternatives. In so doing, law will be the engine that both defines and implements sound ecosystem management for our nation and the world.

Notes and Questions

1. Professor Hodas's rather gloomy view of NEPA's usefulness in ecosystem management illustrates one of the difficulties of trying to stretch existing environmental laws into ecosystem management laws—the pre-existing "box" of judicial interpretations that may limit how far the stretching can go. Yet, is even Professor Hodas underestimating how small the NEPA box really is? He posits that to implement his proposals "all that is needed is for the CEQ to amend its regulations to require externality valuation and securitization for all decisions for which NEPA documents are prepared." Nothing in the Supreme Court's NEPA jurisprudence seems to prevent CEQ from demanding externality valuation analysis in NEPA documents, as that component of the proposal easily fits within the scope of effects analysis the NEPA procedure already requires. By contrast, doesn't the securitization component of the proposal impose a substantive financial burden on federal agencies (or state and private entities seeking federal funding or approvals) that is unrelated to the completion of the effects analysis *procedure*. Does Professor Hodas' description of the Supreme Court cases under NEPA justify the conclusion that CEQ could demand that part of his proposal as well? Is it enough to say, as Professor Hodas seems to suggest, that the securitization requirement will improve the quality of how agencies fulfill the effects analysis procedure, and thus is acceptable under NEPA notwithstanding the non-procedural burdens it imposes?

2. Complying with NEPA's assessment and documentation procedure is the responsibility of the federal agency proposing the action, which raises concerns as to the objectivity of the process. The U.S. Environmental Protection Agency provides some level of independent review under its authority, found in the Clean Air Act, to review and comment on the environmental impacts of "any matter relating to duties and responsibilities granted" to EPA under federal law, and to refer to the White House Council on Environmental Quality any proposed action EPA finds "is unsatisfactory from the standpoint of public health or welfare or environmental quality." 42 U.S.C. 7609(a). EPA uses this authority to review and comment on NEPA EISs it and other agencies prepare. Recently, EPA released a guidance designed to focus its NEPA review efforts on the degree to which the EIS reflects consideration of the impact the proposed action will have on biodiversity conservation and ecosystem functions. *See* USEPA, CONSIDERING ECOLOGICAL PROCESSES IN ENVIRONMENTAL IMPACTS ASSESSMENTS (1999). The guidance, which builds on previous guidance documents that focused the use of terrestrial environment and habitat evaluations in NEPA reviews, explicitly adopts conservation biology as the review framework and specifies ten ecosystem processes that its NEPA reviewers should consider when assessing the impact of proposed actions. While EPA cannot mandate that other federal agencies adopt conservation biology or an ecosystem processes approach to their NEPA duties, the fact that EPA will use those frameworks in its NEPA review authority is likely to help promote the development of ecological economics as a discipline and to develop substantive measurements of ecological processes. *See generally* Robert L. Fischman, *EPA's NEPA Duties and Ecosystem Services*, 20 STANFORD ENVTL. LAW JOURNAL 497 (2001).

(3) Pollution Control Laws

The U.S. Environmental Protection Agency (EPA) is the nation's leading pollution control authority, administering dozens of statutes designed to control industrial, municipal, agricultural, and other sources of environmental media pollutants. EPA's primary media-specific pollution control authorities include the Clean Water Act (water pollution), Clean Air Act (air pollution), Resources Conservation Recovery Act (solid and hazardous waste), Comprehensive Environmental Response, Compensation, and Liability Act (contaminated media remediation), Federal Insecticide, Rodenticide, and Fungicides Act (pesticides), and Toxic Substances Control Act (chemicals). Many conventional courses and casebooks on environmental law use a "tour" of these EPA authorities as their core subject matter. But EPA's administration of its pollution control authorities has not been particularly focused on ecosystem-level concerns. Indeed, the laws themselves for the most part do not build explicitly on such an approach. Rather, technology-based and health-based standards are the chief concern of EPA as it regulates discrete sources of air, water, and soil pollution. The authority to regulate based on environmental quality performance that does exist in some of EPA's statutes had been largely untapped through the 1980s. With the emergence of ecosystem management as a broad federal policy objective in the 1990s, however, the agency began to focus on that latent potential:

U.S. Environmental Protection Agency, Ecosystem Protection Workgroup, Toward a Place–Driven Approach: The Edgewater Consensus on an EPA Strategy for Ecosystem Protection

(March 15, 1994 Draft).

Background

To date, EPA has accomplished a great deal, addressing many major sources of pollution to the nation's air, water and land. Yet, even as we resolve the more obvious problems, scientists discover other environmental stresses that threaten our ecological resources and general well-being. Evidence of these problems can be seen in the decline of the salmon populations in the Pacific Northwest and the oyster stock in the Chesapeake Bay, the decline in migratory bird populations, and degraded coral reef systems.

The causes of these problems are as varied as human activity itself: the way we farm, work, travel, and spend our leisure hours. Although many federal, state, and local regulations address these problems, past efforts have been as fragmented as our authorizing statutes. Because EPA has concentrated on issuing permits, establishing pollutant limits, and setting national standards, the Agency has not paid enough attention to the overall environmental health of specific ecosystems. In short, EPA has been "program-driven" rather than "place-driven."

Recently, we have realized that, even if we had perfect compliance with all our authorities, we could not assure the reversal of disturbing environmental trends. We must collaborate with other federal, state, and local agencies, as well as private partners to reverse those trends and achieve our ultimate goal of healthy, sustainable ecosystems that provide us with food, shelter, clean air, clean water and a multitude of other goods and services. We therefore should move toward a goal of ecosystem protection.

Goal of Ecosystem Protection

The goal of EPA's ecosystem protection approach is to help improve the Agency's ability to protect, maintain, and restore the ecological integrity of the nation's lands and waters (which includes the health of humans, as well as plant and animal species) by moving toward a place-driven focus. This approach will integrate environmental management with human needs, consider long-term ecosystem health, and highlight the positive correlations between economic prosperity and environmental well-being. An ecosystem protection approach will create a framework within which to discuss numerous other issues facing the Agency and our country, such as environmental justice, unfunded mandates, flexibility, state capacity, sustainable development, use of science and data, and measuring environmental results.

The Edgewater Consensus

On March 5, 1994, EPA's Ecosystem Protection Workgroup met in Edgewater, Maryland to develop a strategy for realizing that goal. The Workgroup described a vision for reorienting the Agency toward a "place-driven" orientation; that is, the work of the Agency would be driven by the environmental needs of communities and ecosystems. For any given "place," EPA would establish a process for determining long-term ecological, economic, and social needs and would reorient its work to meet those needs. Although this approach is being demonstrated in a number of places, the Workgroup envisioned that, over time, the entire country would benefit from this approach. To realize that vision, the Workgroup determined that systems must be established to move toward a place-driven approach that would:

1. establish a process (with appropriate partners) for picking places, and for

   - developing steps for implementing ecosystem protection that:
   - identifies environmental goals and indicators;
   - identifies ways to support sustainable development and communities
   - develops and implements a joint action plan based on sound science; and,
   - measures progress and adapts management to new information over time;

2. coordinates different programs within our Agency, and collaborates with our external partners and defines roles and responsibilities at each identified place; and,

3. identifies tools and support that could be provided at a national level.

Critical Success Factors

Three critical success factors exist for EPA to make ecosystem protection a reality. First, government activities must be driven by the issues faced by particular ecosystems and the economies founded upon them. The protection of ecosystems cannot be viewed by the Agency as a separate task existing on the margins of environmental protection, or as a special initiative imposed from above. The issues threatening the sustainability of ecosystems must drive the Agency's agenda, and policy, planning, budgeting, and information systems must be developed accordingly. This will involve "changing the unit of work" from piecemeal program mandates to the imperatives of a specific place. It will require EPA to devise programs that respond to the needs of specific geographic areas, not simply statutory mandates. Success will be achieved with greater integration and teamwork among environmental and natural resource agencies, and commerce, trade, and economic development programs.

Second, the ecosystem approach requires coordinated, integrated action by federal, state, tribal, and local agencies; between government and private enterprises (e.g., NGOs and industry); and, most importantly, between government and the people for whom services are being offered and provided. Environmental problems are almost always beyond the purview of any one program or organization. EPA is committed to working more effectively in an interagency, intergovernmental process. EPA will enlist the support of a spectrum of participants in the priority-setting and decision-making processes. EPA will increase it efforts to support the ecosystem protection efforts of states and local agencies that have recently proliferated throughout the country.

Third, information is a key to empowerment that moves communities to action. The availability of quality information on the resources to be protected is essential and, in many cases, is primarily available at the local level. Whereas the traditional approach spawned rules for an agency to follow, ecosystem protection, restoration and management at EPA will be dedicated to strategically responding to the best information on the needs of the resource and adapting management over time based on careful monitoring and new information. This means we will improve and integrate the information we gather and make it more accessible through the Agency's information systems, and forge a much stronger link between the Agency's scientific community and the information technology community, so that work is aligned with the needs of protecting entire ecosystems. EPA will also provide leadership in conducting public education and outreach programs to explain the importance of ecosystem protection to the public.

Making it Happen

Recognizing that we must move to a place-driven approach, the Edgewater Conference identified existing barriers to progress. These barriers include a lack of information on specific ecosystems; inadequate ecological endpoints for specific places; staff lacking a focus toward which to orient their work; historical single-media focus on programs and not places; the

Agency's lack of a central system for planning, budgeting, and accountability; and staff not trained or hired for the right skills.

To address the critical success factors and barriers discussed above, the Agency will respond on several paths. We will align our policy, regulatory, institutional, and administrative infrastructure to support ecosystem protection; we will develop information and tools to facilitate the approach; and we will reorient the Agency's culture to facilitate a place-driven approach. We also will promote our ecosystem protection activities within and outside the Agency and measure our success as we proceed.

Notes and Questions

1. Recall Grumbine's goal for ecosystem management of protecting "native ecological integrity." EPA says its goal for ecosystem management is "to protect, maintain, and restore the ecological integrity of the nation's lands and waters (which includes the health of humans, as well as plant and animal species)." Are these consistent? Assume that we could, for any particular ecosystem, identify its "native ecological integrity" and measure how close its current condition is to that state. Would using EPA's goal justify attempting to restore the ecosystem to that native state? Would using EPA's goal justify a decision *not* to restore the ecosystem to that native state?

2. To follow through with the general theme of the "Edgewater Consensus," EPA later identified the provisions in its media-specific pollution control authorities that could be shaped toward ecosystem management, mainly by linking EPA's primary regulatory authority with its coordinate species protection responsibilities under the Endangered Species Act. *See* ENVIRONMENTAL LAW INSTITUTE (FOR EPA), USING POLLUTION CONTROL AUTHORITIES TO PROTECT THREATENED AND ENDANGERED SPECIES AND REDUCE ECOLOGICAL RISK. EPA also produced an inventory of ongoing projects that presented some ecosystem focus component, though admitting that very few involved comprehensive ecosystem assessment or management. *See* USEPA, A PHASE I INVENTORY OF CURRENT EPA EFFORTS TO PROTECT ECOSYSTEMS (1995), available at http://www.epa.gov/docs/ecoplaces. The projects are taking place at many levels. Examples include:

large scale: Great Lakes Program; Colorado River Program; Gulf of Mexico Program; Pacific Northwest Forest Plan; Prairie Pothole Region Ecosystem Assessment

local scale: Merrimack River, N.H.; Lake Champlain, N.Y.; Tampa Bay, Fla.; Corpus Christi Bay, Tex.; Malibu Creek, Haw.

multi-site: Wetlands Restoration Research Project; Pacific Salmon Habitat Recovery Project; Clean Lakes Program

As of 1995, these projects appear to have blanketed most of the United States:

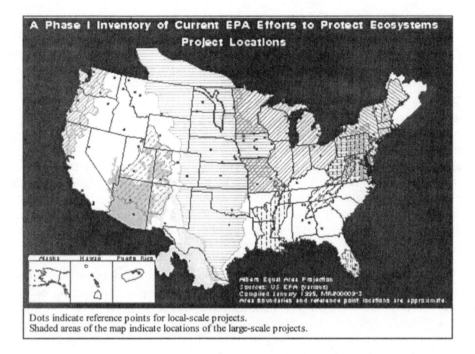

A Phase I Inventory of Current EPA Efforts to Protect Ecosystems
Project Locations

Dots indicate reference points for local-scale projects.
Shaded areas of the map indicate locations of the large-scale projects.

EPA has not, however, developed a comprehensive proposal for redrafting its pollution control regulations into an ecosystem management regime.

3. The states play a large role in implementation of pollution control authorities, both through administration of state pollution control laws and through a process by which EPA delegates to states the authority to implement federal law. Over a dozen states have moved their pollution control authorities increasingly toward ecosystem management goals, principally through watershed-based initiatives. *See* R. Steven Brown and Karen Marshall, *Ecosystem Management in State Government*, 6 ECOLOGICAL APPLICATIONS 712 (1995); Jessica Bennett, *State Biodiversity Planning*, THE ENVTL. FORUM, July/August 1998, at 19. For example, even before the EPA undertook its ecosystem management initiative, the Florida Department of Environmental Protection (FDEP) undertook a similar effort to reorient its pollution control (and other) authorities with ecosystem management as a coordinating purpose. In legislation consolidating several state environmental and resources agencies into the unified FDEP, the Florida legislature and Governor Lawton Chiles required FDEP to develop and implement measures to "protect the functions of entire ecological systems through the enhanced coordination of public land acquisition, regulatory, and planning programs." FDEP convened an Ecosystem Management Work Group to develop means of using ecosystem management to fulfill that legislative directive. In the series of reports that emanated from the Work Group, FDEP outlined an approach for integrating ecosystem management into its existing authorities, including principally its pollution control authorities, that relied heavily on the principles R.E. Grumbine laid out in his landmark article *What Is Ecosystem Management?*, which is reproduced in the previous section of this book. *See* FLORIDA DEP, BEGINNING ECOSYSTEM MANAGEMENT (1994); FLORIDA DEP, ECOSYSTEM MANAGEMENT AT WORK IN FLORIDA (1998). Still, like EPA's Edgewater Consensus, FDEP's vision lacked concrete legislative and regulatory proposals, and the effort has become largely moribund with the change of administration in the Governor's office.

4. EPA's movement toward "place-based" ecosystem management approaches actually began several years before the *Edgewater Consensus* meeting. In 1990, EPA's Science Advisory Board recommended as one of the top ten priorities for the agency that "EPA should attach as much importance to reducing ecological risk as it does to reducing human health risk." USEPA, SCIENCE ADVISORY BOARD, REDUCING RISK: SETTING PRIORITIES AND STRATEGIES FOR ENVIRONMENTAL PROTECTION (1990). Why did it take four more years for EPA's policy side to formulate the *Edgewater Consensus* position?

(4) Public Land and Resource Management Laws

The Congressional Research Service and General Accounting Office assessments of ecosystem management policy that open this section of the text focus on four federal public land management agencies: The U.S. Fish and Wildlife Service; the National Park Service; the Bureau of Land Management; and the U.S. Forest Service. The Property Clause of the U.S. Constitution provides that "Congress shall have power to dispose of and make all needful Rules and Regulations respecting the Territory or other Property belonging to the United States." U.S. const. Art IV, § 3, cl. 2. The Supreme Court has held that this authority provides Congress essentially unfettered power to regulate actions and to protect the resources on federal public land. Kleppe v. New Mexico, 426 U.S. 529 (1976). So, if Congress wished to make ecosystem management the law for all federal public land, it could. But we know it hasn't. The following excerpt explores how far the federal land management agencies have used existing public land and resource management authorities to do so in the absence of congressional initiative.

Rebecca W. Thomson, "Ecosystem Management"— Great Idea, But What Is It, Will It Work, and Who Will Pay?

9 Natural Resources & Environment 42 (Winter 1995).

Existing Law and Litigation

Ecosystem management's *potential* significance is enormous—if incorporated into agency practice and law, it could act as an overlay over *all environmental laws*. Generally, federal land management and environmental laws are not organized around either biodiversity or ecosystem concepts. The public lands (30 percent of the land mass of the United States) are essentially divided up along jurisdictional boundaries between four public land agencies: Bureau of Land Management (BLM), Forest Service, Fish and Wildlife Service (FWS), and Park Service. The legal mandates of these agencies are largely directed to different purposes—the Forest Service and BLM lands are to be managed for multiple uses, including commodity production (trees, grass, and minerals), while the Park Service and FWS lands mainly serve conservation goals. Traditionally, the scientists who worked in each agency were trained to accomplish the agency's primary mandate with little interdisciplinary exchange.

* * *

In 1993, the White House Office on Environmental Policy (White House Task Force) established an Interagency Ecosystem Management Task Force to focus the administration's ecosystem management efforts. The administration's fiscal year 1995 budget includes $700 million for ecosystem management initiatives, over 70 percent of which will fund four "pilot ecosystem management projects" in the Pacific Northwest, Prince William Sound, the Everglades, and the Anacostia River in the District of Columbia.

The White House Task Force has also developed "draft" principles that would guide federal implementation of ecosystem management:

- Manage along ecological, rather than political or administrative boundaries.

- Ensure coordination among federal agencies and increased collaboration with state, local, and tribal governments; the public; and Congress.

- Use monitoring, assessment, and the best science available.

- Consider all natural and human components and their interactions.

Independently, the four public land agencies have also begun to move toward ecosystem management. Most notable, for the amount of controversy it engendered, was the Department of the Interior (DOI) proposal for [a National Biological Survey (NBS)] to inventory the nation's natural resources. DOI Secretary Bruce Babbitt described the NBS as a "very basic instrument for getting at these basic issues of inventory classification . . . to get, at least, a first cut of how we relate to these ecosystems." After property rights proponents stalled the NBS legislation in Congress, DOI implemented the NBS administratively in November 1993 by using existing funding, authority, and personnel.

On December 14, 1993, BLM issued a report on ecosystem management that contains its three principles: (1) sustaining the productivity and diversity of *viable ecological processes and functions*; (2) adopting an interdisciplinary approach to land management in which *program advocacy will yield to ecosystem advocacy*; and (3) basing management on long-term horizons and goals. FWS issued a report in March 1994 describing how to apply ecosystem management to fish and wildlife conservation. FWS has developed a map of fifty-two ecosystems, based on watersheds, in 50 states and, most significantly, reinforced its February 1992 policy decision to take a multispecies approach (rather than a species-by-species approach) to protecting plants and animals. To that end, FWS will list *groups* of species in a particular area to move to protection of ecosystems upon which several species depend. The National Park Service established an ecosystem working group and plans to promote comprehensive regional ecosystem restoration and management throughout the park system.

In 1989, the Forest Service announced "New Perspectives for Managing the National Forest System." The Forest Service stated that it would move from a multiple-use management approach—where forest land is viewed as a "place to produce commodities and amenities"—to an ecological vision in which living systems on the land "have importance beyond traditional commodity and amenity uses." Kessler, *New Perspectives for Sustainable Natural Resource Management*, 2 ECOLOGICAL APPLICATIONS 221, 222 (1992).

Accordingly, in 1992, the chief of the Forest Service announced a commitment to the use of ecosystem management to "blend[ ] the needs of people and environmental values in such a way that the national forests and grasslands represent diverse, healthy, productive, and sustainable ecosystems." Significantly, at the same time, the Forest Service announced its intent to reduce clear-cutting as a standard commercial timber harvest practice by over 70 percent from its 1988 level.

A Forest Service report describes four fundamental principles that guide its approach to ecosystem management: the use of an ecological approach to multiple-use management; application of the best scientific knowledge and technologies to decisionmaking; encouragement of partnerships with state agencies and private landholders; and the promotion of grass-roots participation in the planning process. FOREST SERVICE, A NATIONAL FRAMEWORK: ECOSYSTEM MANAGEMENT (1994). And, on June 13, 1994, Chief Thomas told regional foresters in a letter that "new forest plan standards and guidelines and program levels will be fully based on and carried out in accordance with the evolving concepts of ecosystem management."

The Forest Service announced its intent to revise its forest planning regulations, 36 C.F.R. § 219, to incorporate ecosystem management in accord with its four principles. These proposed rules have been slated for publication since August 1994, but apparently are on hold until the White House Task Force sorts out its administrationwide ecosystem management efforts.

The Forest Service also holds the distinction of implementing one of the most high-profile efforts at ecosystem management—the Federal Ecosystem Management Assessment Team (FEMAT) created to "solve" the spotted owl/timber crisis through ecosystem management. The President's 1993 Timber Summit directed this team of scientists to recommend an alternative based on analysis of biological and socioeconomic science that would balance the need for a healthy ecosystem with the needs of timber dependent communities in the Pacific Northwest. However, "Option 9," FEMAT's preferred alternative that calls for a 70 percent reduction in timber harvests, immediately resulted in suits from both the environmentalists and timber interests.

The Barriers and the Future

* * *

[T]he fundamental barrier to implementing ecosystem management identified by GAO is the failure of the White House Task Force to address and answer the tough public policy question of "who wins" in a conflict between socioeconomic needs and ecosystem health. The Report argues that a "healthy" ecosystem should take priority and buttresses its argument with references to NEPA. The Report also notes that the principles developed by BLM and FWS "leave no doubt" that greater priority will have to be given to maintaining or restoring minimum levels of ecosystem integrity "over unsustainable commodity production and other uses." How the White House Task Force will address this politically sensitive policy issue is not known at this time. If FEMAT is any example, it looks like the ecosystem "wins."

The hurdles described by the GAO report will not be easily cleared. Nonetheless, there is a great deal of momentum from many disparate groups to adopt ecosystem management. Agencies praise ecosystem management as having the potential to resolve the multiple, piecemeal battles over individual species or commodity uses. Timber industry groups have expressed guarded optimism about ecosystem management, for its inclusion of human needs, reliance on sound science, adaptive management, and emphasis on a cooperative, open process. Environmentalists, though they fear that the inclusion of human concerns and consensus building will invite compromises detrimental to the environment, are supportive of management seemingly directed to "healthy" ecosystems rather than only commodities production. As the CRS report aptly noted, "there is not enough agreement on the meaning of the concept to hinder its popularity."

Senator Hatfield asked how ecosystem management should be defined, implemented, and funded. Without really knowing the answers to any of these questions, the federal government is forging ahead. What we do know, is that for the near future, the policy of ecosystem management will be yet another overlay on top of existing environmental and public land law and processes.

Notes and Questions

1. Thompson refers to different "mandates" Congress has given the different federal public land agencies. It is important to recognize that there are many different units of federal public lands—parks, refuges, forests, recreation areas, etc.—serving many different purposes. There is, in other words, no single unifying federal land policy. Rather, different units of federal land are managed according to one of three principal types of mandate. *Multiple use* lands are managed to optimize a bundle of prescribed uses. *Dominant use* lands have a primary use specified, but may accommodate other uses so long as the primary use is not impeded. *Single use* lands have, as the name suggests, just one prescribed use. The vast majority of federal land is subject to a multiple use mandate of one form or another, and usually some of the uses involve an intensive human use, such as timber production, grazing, recreation, or mineral extraction, while others are more consistent with the goals of ecosystem management, such as wildlife conservation. Congress usually leaves the agencies a fairly large policy space "box," meaning a wide variety of options is available to the agency. Nevertheless, given constant oversight by Congress, industry, conservation groups, and the public in general, the perpetual challenge federal land agencies face under multiple use mandates is to balance the multiple uses, particularly balancing the intense human uses against the resource conservation uses.

2.   Thompson's perspective clearly is that by adopting ecosystem management principles, federal agencies are changing the balance of multiple uses, or even functionally transforming multiple use lands into dominant use lands under the ecosystem management "overlay." This raises the question, explored in detail in the remaining chapters of Part III, as to whether the agencies have reached the outer edges of the "box"—that is, have taken ecosystem management beyond their legal authority.

3.   Thompson mentions another federal agency, the National Biological Service, now known as the Biological Resources Division (BRD) of the Department of the Interior's U.S. Geological Survey (USGS). The BRD is a bio-geographical research agency, not a public land management agency. The BRD's roots go back to 1885, when Congress allocated funds to the Division of Entomology of the U.S. Department of Agriculture to conduct biological studies of birds in their environment. The program was expanded to include mammals the next year, and in 1890 the funding language referred specifically to investigations of species' geographic distributions. The Division was elevated to Bureau status in 1905, and for the next three decades the Bureau of Biological Survey conducted numerous important studies. In 1939 the Bureau was transferred to the Department of Interior as part of the Fish and Wildlife Service, where it remained until 1993. That year, the Department of Interior consolidated the biological research components of all of its various branches into the newly organized National Biological Survey, which was designed to operate as a nonadvocacy biological science research program. Funding for creation of a new branch of the Department was controversial in Congress, however, and in 1995 Secretary Bruce Babbitt administratively moved the program to operate within the Department's existing USGS and renamed it the National Biological Service. Congress later renamed and funded the agency as the Biological Resources Division of the USGS. One of its primary missions, besides providing biological research capacity to all of the Department's branches, is to coordinate the nation's ongoing National Biological Information Infrastructure (NBII) that emerged from the recommendations of Vice President Gore's 1997 National Performance Review of federal agencies. For current information on the BRD and the NBII see the agency's homepage <http://biology.usgs.gov/>. For background on its precursor agencies, see Milton Friend, *Conservation Landmarks: Bureau of Biological Survey and National Biological Service*, in NATIONAL BIOLOGICAL SERVICE, DEPARTMENT OF the Interior, Our Living Resources 7 (1995); O. J. Reichman and H. Ronald Pulliam, *The Scientific Basis for Ecosystem Management*, 6 Ecological Applications 694 (1996). Why do you suppose funding for a science-oriented, non-regulatory biological research agency was controversial in Congress?

4. Federal lands are held in a wide variety of forms of conservation. A partial listing of the different federal land units and their basic authorities shows this tremendous variation:

UNIT	AUTHORITY	AGENCY
Bureau of Land Management Land	Taylor Grazing Act, 43 U.S.C. 315—315o; Federal Land Policy and Management Act, 43 U.S.C. 1701—1782	BLM
Game and Bird Preserves	Game and Bird Preserves Act, 16 U.S.C. 671—698t	various
National Conservation Recreation Area	Refuge Recreation Act, 16 U.S.C. 460k	Dept. of Interior
National Forest	Forest Service Administration Act, 16 U.S.C. 473—478, 479—482; National Forest Management Act, 16 U.S.C. 1600—1614	Forest Service
National Grassland	Bankhead–Jones Farm Tenant Act, 7 U.S.C. 1000—1013; National Forest Management Act, 16 U.S.C. 1600—1614	Forest Service
National Lakeshore	National Park Service Act, 16 U.S.C. 1—18f–1	Park Service
National Marine Sanctuary	Marine Protection, Research, and Sanctuaries Act, 16 U.S.C. 1431—1445a	Dept. of Commerce
National Monument	Antiquities Act, _16 U.S.C . 431;_ National Park Service Act, 16 U.S.C. 1—18f–1	Park Service
National Park	National Park Service Act, 16 U.S.C. 1—18f–1	Park Service
National Preserve	National Park Service Act, 16 U.S.C. 1—18f–1	Park Service
National Recreation Area	various	various
National Scenic Trail	National Trail Systems Act, 16 U.S.C. 1241—1251	various
National Wildlife Refuge	National Wildlife Refuge Administration Act, 16 U.S.C. 668dd—668ee	FWS
Primitive Area	Wilderness Act, 16 U.S.C. 1131—1138	Forest Service
Wild and Scenic River	Wild and Scenic Rivers Act, 16 U.S.C. 1271—1287	various
Wilderness Area	Wilderness Act, 16 U.S.C. 1131—1136	various

(5) Local Land Use Laws

Habitat loss on non-federal lands is the leading cause of species decline in the United States. *See* William Stolzenberg, *Habitat Is Where It's At*, Nature Conservancy, Nov–Dec 1997, at 6. Although under five percent of the nation's land mass is "built up" in the urban sense, conversion of undeveloped and agricultural land to urban uses nonetheless is associated with species and ecosystem decline in many areas, leading to "hot spots" of concern. *See* A.P. Dobson et al., *Geographic Distribution of Endangered Species in the United States*, 275 SCIENCE 550 (1997); T. Adler, *Mapping Out Endangered Species' Hot Spots*, 150 SCIENCE NEWS 101 (1996). Much of that land development activity is subject to regulation through local land use controls such as zoning and growth management. Traditionally, however, cities and counties have not used their land use authorities to effectuate any meaningful ecosystem-level conservation policy. Local governments face several constraints to doing so. First, their boundaries often are too limited to provide effective management of ecosystem-level conditions. Second, fiscal concerns may deter one local government from investing in protecting ecosystems that transcend political boundaries, lest neighboring localities "free ride" by forgoing similar investments or, worse, scoop up the businesses and industry that may no longer be welcome in the locality that exercises ecosystem management. These boundary and fiscal concerns could be solved through regional cooperation, but that requires divesting local control over issues or even succumbing to preemptive regional and state government authority.

On the other hand, local governments have often exhibited a distaste for federal control of matters that are profoundly local in impact. And ecosystem management is, in the end, profoundly local in impact given its focus on landscape-level implementation. There is, moreover, little question that local governments could exercise considerable power in the field of ecosystem management. *See* A. Dan Tarlock, *Local Government Protection of Biodiversity: What Is Its Niche?*, 60 U.CHI.L.REV. 555 (1993). There is also the concern that the federal government at some point will exhaust its authority to implement ecosystem management when it affects decisions that are fundamentally within the scope of local land use authority. The federalism of ecosystem management at the landscape level is thus quite complex. *See* A. Dan Tarlock, *Federalism Without Preemption: A Case Study in Bioregionalism*, 27 PAC. L. J. 1629 (1996). As one observer sums up:

> Supposing local governments have the incentives, either through federal and state encouragement or from the bottom up, to tackle critical issues of biodiversity protection, what tools are likely to be at their disposal? There is a wide web of real and potential legal authorities in contemporary American local government law with which to address some of the problems of resource use and ecosystem management. Some of these authorities require institutional and doctrinal innovations; that is, they require a change in local government law to better serve the aims of ecosystem management. At the same time, other sources of local government authority are currently serviceable. While other demands and exigencies may make these innovations undesirable, there are good reasons to consider the potential of local institutions and local government law to grapple with the needs of modern ecosystems.

Daniel B. Rodriguez, *The Role of Legal Innovation In Ecosystem Management: Perspectives from American Local Government Law*, 24 ECOLOGY L. Q. 745, 755 (1997). Local governments thus find themselves in one sense in much the same position as the federal and state governments: some authority already exists to shape an ecosystem management policy, but legal reform would be required to do so comprehensively.

2. DESIGNING NEW HARD LAW TO APPLY

As you may have noticed by now, thus far the law of ecosystem management has failed to take hold widely in concrete legal forms—in ''hard law'' that can be applied to discrete factual settings. In the absence of a comprehensive substantive ecosystem management legislation, federal and state agencies have tried to recraft specific conservation, impact assessment, pollution control, and land management laws into ecosystem management tools by administratively-led policy initiatives. But, as the case of the ESA demonstrates, that approach can only go so far before the connection between the text of the statutes involved and explicit policy adoption of ecosystem management goals becomes too tenuous to withstand challenge. At some point, in other words, Congress and the state legislatures, if they become committed to ecosystem management in a particular setting, will need to design substantive statutory authority for explicit adoption of ecosystem management goals under the relevant laws. The final segment of Professor Salzman's review of *Nature's Services* provides a vision of how substantially amended or entirely new laws could be designed to fulfill a greater ecosystem management purpose, in this case with the objective of ecosystem service delivery values in mind as the common policy theme:

James Salzman, Valuing Ecosystem Services

24 Ecology Law Quarterly 887 (1997).

III. ECOSYSTEM SERVICES AND THE LAW

In addition to its ecological and economic analyses, *Nature's Services* is fascinating because it recognizes a key role for ecosystem services in environmental law and policy. In fact, as potential symbiotic partners, both environmental law and research on ecosystem services have much to offer: together, they provide a new way to view environmental law, beyond command-and-control mandates and single-species protections.

How can environmental law promote our understanding of ecosystem services? It can do so through the creation of information markets that drive scientific research. Our understanding of groundwater chemistry and hydrology has increased tremendously in recent years, due primarily to markets created for this information as a result of CERCLA actions. Potentially responsible parties require a sophisticated understanding of local groundwater conditions to design the most efficient remediation strategies, and now-wealthy consulting businesses have arisen to meet these needs. Indeed, the role of regulation in creating secondary information markets is an important pillar of economics of information theory.

Ecosystem services have real value, yet they are not understood well enough to be valued monetarily. Could current regulations spur the creation of secondary information markets without the liability hammer of CERCLA? To a large extent, current wetlands regulations have already created information markets for wetlands vegetation and hydrology data. A great deal more is known today than just ten years ago, largely because the assessment models used to comply with wetlands regulations have focused on biophysical characteristics. But such emphasis is misplaced if ecosystem services are as valuable as current research indicates.

If government officials explicitly required significant data on ecosystem services for natural resource damage assessments and environmental impact statements, then a secondary information market likely would develop. Some regulations have begun to make these demands in the areas of groundwater hydrology and wetlands vegetation.* * * If ecosystem services are significantly undervalued, and such undervaluation therefore leads to misallocation of resources, then the use of regulations to create a profitable secondary market in ecosystem service data and indicators could prove an efficient intervention for improved management of resources.

What does ecosystem service research offer in return to environmental law? [One possibility] is specificity of indicators. For some services, benefits are too diffuse and monetary valuation is no more than a guess. Here, the law can use indicators of ecosystem services as a surrogate for economic value. While ecosystem management has become a catchword in government, a recent study by Professor Oliver Houck indicates serious shortcomings. He makes a strong case that, despite the trumpeting of an ecosystem approach to conservation, "[e]cosystem management, as currently promoted, is politics with a strong flavor of law-avoidance."[30] He argues that the only effective legal standards to ensure protection of an ecosystem rely on assessments of keystone or indicator species:

> Why is it that indicator species work? Granted, they are by no means perfect surrogates for ecosystems and, granted again, the proof of their requirements can be complex and demanding for scientists operating at the far edge of data and predictability and trained to conclude nothing until all possible alternative hypotheses, however remote, have been disproved. Nonetheless, indicators work because, in the end, they produce specifics.[31]

Robust, quantified indicators of ecosystem services could serve a similar role, providing an additional legal standard on which to base ecosystem management strategies. Much as the [National Forest Management Act] currently requires conservation of indicator species as a surrogate measure for ecosystem health, one could imagine a legal standard requiring mainte-

30. Oliver Houck, On the Law of Biodiversity and Ecosystem Management, 81 Minn. L. Rev. 869, 975 (1997).

31. *Id.* at 976–977.

nance of a specified, measurable level of local ecosystem services. Thus indicators assessing water flow into and out of a wetland might, for example, include dynamic measures of water retention, nutrient trapping, or water quality. These indicators, at least on the local level, could mandate management of ecosystems based on functional standards, i.e., maintaining the provision of baseline levels of services. Moreover, the direct benefit to humans of such conservation actions would be more obvious than the current focus on indicator species.

* * *

Perhaps the greatest value that increased understanding of ecosystem services offers to environmental policy, however, is its persuasive argument that biodiversity and habitat protection provide important benefits in ways not normally considered. Wheeling out the rosy periwinkle and charismatic megafauna every time the Endangered Species Act or wetlands protections come under threat goes only so far. *Nature's Services* takes a different, potentially more effective tack, calling for explicit recognition of ecosystem services because of the direct, tangible benefits they provide. Such recognition could provide a more integrated and compelling basis for action than those suggested by a focus on single-species or biodiversity protection for the simple reason that the impacts of these services on humans are more immediate and undeniably important. Indeed, a focus on ecosystem services has the potential to unify disparate parts of environmental law, linking the conservation goals in laws such as the Endangered Species Act and National Forest Management Act more closely with the human health goals in seemingly unconnected laws such as the Clean Air Act and Safe Drinking Water Act.

These developments in environmental law are at once speculative and foreseeable consequences of future research on the production and delivery of ecosystem services. The study of ecosystem services is a new and very promising area of interdisciplinary research with the potential to create a significant shift in how we address environmental protection. Just as *Nature's Services* provides a valuable bridge linking ecologists and economists to policymakers, so, too, is it important for environmental lawyers to engage themselves in this research effort, both to explore the role ecosystem services should play in the law's development and to influence the direction of research so that the services provided by nature may be accorded their proper value.

Notes and Questions

1. How would you implement Salzman's vision of using ecosystem services as the organizing principle of ecosystem management law? Along with a team of biologists and economists, Salzman has since outlined a proposal for the creation of Ecosystem Service Districts (ESDs) and empowering them with land use and taxing authority to gather information about ecosystem services, assess their flow through the environment and economy, and manage their use and distribution. *See* Geoffrey Heal et al.,

Protecting Natural Capital Through Ecosystem Service Districts, 20 STAN-
FORD ENVIRONMENTAL LAW JOURNAL 333 (2001). Naturally this would involve
some shifting of power from political units arranged more along political
boundaries, such as counties, to units arranged more in line with environmentally relevant boundaries, such as watersheds. Do you think ESDs
could be both a politically and environmentally viable arrangement?

2. The article on ESDs discussed in the note above appears in an issue of
the *Stanford Environmental Law Journal* devoted entirely to the concept of
ecosystem services. Symposium, 20 STANFORD ENVIRONMENTAL LAW JOURNAL
309 (2001). The *Journal* hosted a conference in 2000 at which economists,
ecologists, lawyers, and representatives from industry, government, and
non-governmental organizations discussed the viability of ecosystem services as a foundation for the formation of ecosystem management and
environmental policy generally. The articles in the symposium issue cover
that topic in a wide variety of settings. Consider some of the questions
debated at the conference: How viable do you find the concept of ecosystem
services as a core principle for purposes of developing ecosystem management law? Should any new law of ecosystem management use it as a core
principle? If not, around what core principles would you fashion new hard
law to apply in ecosystem management?

3. In his preamble to the Ecosystem Management Act of 1994 (excerpted
above at the beginning of this part of the Chapter), Senator Hatfield asks a
series of questions about ecosystem management law and policy:

- What is ecosystem management, how should it be implemented and
  who should be implementing it?

- How does the ecosystem oriented work of the federal agencies,
  states, municipalities, counties, and interest groups mesh?

- And is the existing structure of our government agencies adequate to
  meet the requirements of managing land across which state and
  county lines have been drawn?

- Finally, with a decreasing resource production receipt base, how
  shall we pay for ecosystem management? Direct federal appropriations? Consolidation of federal, state, local and private funds?

- And if we determine how to pay for ecosystem management, who
  coordinates collection of these funds and how are they distributed?

After studying and discussing the materials presented thus far in this
chapter, do you feel any more comfortable than Senator Hatfield apparently did that we have good answers to these questions? Do we need good
answers to these questions in order to formulate good ecosystem management law and policy?

4. In his strident attack on the ecosystem management movement, *Defending Illusions: Federal Protection of Ecosystems*, Allan K. Fitzsimmons
seizes on Senator Hatfield's questions as the reason why any effort to steer
existing law toward ecosystem management or to forge altogether new law

is misguided. His overall take on ecosystem management as it was implemented in the Clinton Administration:

> Empowering the federal government to manage and protect ecosystems across the country is bad public policy. Advocates of ecosystem arguments use sectarian arguments built on false claims of pending environmental doom. They cloak themselves in a garment woven from ill-defined scientific buzzwords that are valueless for the purpose of guiding government decision making. They employ religious arguments that confuse the Creator with creation. New paradigmists would lessen human liberty and improvements in human well-being through assaults on constitutionally protected property rights. They would establish a vast new centralized bureaucracy to manage land use throughout the nation, using a maze of rules and regulations that would greatly surpass in complexity and cost any set of environmental regulations previously seen in the United States.

DEFENDING ILLUSIONS, *supra*, at 251. Do you find these qualities in the materials included in this chapter that advocate use of ecosystem management? Does Grumbine, for example, base his description of ecosystem management on "false claims of pending environmental doom"? Is Holling's adaptive management method "woven from ill-defined scientific buzzwords" and "valueless for the purpose of guiding government decision making"? Is Salzman proposing "assaults on constitutionally protected property rights"? Have the Environmental Protection Agency and Fish and Wildlife Service ever mentioned the Creator or creation in any of their materials on ecosystem management that you have reviewed? To whom *is* Fitzsimmons referring?

D. CASE STUDY: NATIONAL WILDLIFE REFUGE SYSTEM POLICY

The National Wildlife Refuge System is "a national network of lands and waters for the conservation, management, [and] appropriate restoration of the fish, wildlife, and plant resources and their habitats within the United States for the benefit of present and future generations of Americans." 16 U.S.C. § 668dd(a)(2). The system today comprises over 520 refuges and related wildlife conservation units. It covers 93 million acres of federal public land in total—an area larger than the National Park System—and including a wide variety of ecosystem types. Every state has at least one national refuge. Current administration of these refuge lands through the Department of the Interior's U.S. Fish and Wildlife Service (FWS) is controlled principally under the terms of the National Wildlife Refuge System Administration Act of 1966, as amended by the National Wildlife Refuge System Improvement Act of 1997. But the history of national wildlife refuges begins far prior to those laws.

The first instance of setting aside federal public land specifically for wildlife protection came in 1869 when Congress protected the Pribilof

Islands in Alaska as a reserve for the northern fur seal. Later initiatives led to national parks and national forests, but not until 1903, when President Theodore Roosevelt used a presidential proclamation to set aside the three-acre Pelican Island in south Florida as a place for wildlife protection, were any lands designated as a wildlife refuge. Quickly thereafter, many other islands were given the same status, and eventually Congress and the Executive began routinely to establish wildlife refuges throughout the nation. The enactment of the Migratory Bird Treaty Act in 1918 fueled the process given the Federal government's new role in migratory bird protection, which was carried out often through Executive Order designating bird refuges.

To bring some order to the hodge-podge of refuges established through the collection of legislative and executive actions, FWS published a "Refuge Manual" in 1942 to guide individual refuge managers. Nevertheless, by the 1960s, although tens of millions of acres had been placed in refuge status, the ad hoc process of establishing refuge units and purposes resulted in numerous, uncoordinated establishing acts and executive orders and no coherent management standards, leading one commentator to describe it as the National "Bunch" of Wildlife Refuges. *See* Cam Tredennick, *The National Wildlife System Improvement Act of 1997: Defining the National Wildlife Refuge System for the Twenty–First Century*, 12 Fordham Environmental Law Journal 41, 46 (2000).

Congress made a stab at bringing order to the program in the Refuge Recreation Act of 1962, which allowed refuge managers to prohibit forms of recreation on refuge land that are not directly related to the primary purposes of the refuge *unless* such uses would not interfere with those primary purposes. This "non-interference" standard provided a system-wide test for a set of allowable uses, in this case recreational uses, but in the end only compounded the management confusion by failing to provide meaningful clarifying standards for determining when interference was present.

The 1966 legislation was a more concerted effort to manage the various refuge units as a single coordinated network, and thus can be regarded as an organic act for the refuge program, or at least the beginnings thereof. The 1966 legislation consolidated the various refuge units into a unified "system." The legislation failed to provide specific guidelines and directives for administration and management of areas in the system, but did authorize the Secretary to "permit the use of any area within the System for any purpose, including but not limited to hunting, fishing, public recreation and accommodations, and access whenever he determines that such uses are compatible with the major purposes for which such areas were established." 16 U.S.C. § 668dd(d)(1). This so-called "compatibility" standard placed refuge lands in the "dominant use" model of public lands management—i.e., although multiple uses are allowed, protection of wildlife dominates over the others. Beyond that, however, the statute did not define compatibility or attach specific planning or substantive duties to it.

The FWS defined compatibility in its refuge manual to mean that a use "does not materially interfere with or detract from the purpose(s) for which the refuge was established." This not especially demanding or clarifying standard helped little to avoid inconsistent applications of the compatibility requirement across the System. But a series of events moved the Congress ever closer to additional legislative intervention on the subject. First, the District of Columbia federal district court held that refuge managers bear the burden of proving that a secondary use is "incidental to, compatible with and does not interfere with the primary purpose of a refuge." Defenders of Wildlife v. Andrus, 11 Env't Rep. Cas. (BNA) 2098, 2101 (D.D.C.1978); *see also* Defenders of Wildlife v. Andrus, 455 F.Supp. 446 (D.D.C.1978). This placed pressure on the agency to articulate and defend tests for compatibility. Then, declining migratory bird populations in the 1980s prompted Congress to direct the General Accounting Office (GAO) to study refuge management practices. GAO's report was less than glowing to say the least, criticizing FWS for allowing numerous harmful secondary uses. See U.S. GENERAL ACCOUNTING OFFICE, NATIONAL WILDLIFE REFUGES: CONTINUING PROBLEMS WITH INCOMPATIBLE USES CALL FOR BOLD ACTION, GAO/T–RCED–89–196 (1989). Several years later FWS settled a lawsuit brought in 1992 that tracked the GAO report in challenging numerous secondary use decisions at specific refuges around the nation. FWS agreed to review each secondary use and continue only those deemed compatible with the refuge. National Audubon Society v. Babbitt, No. C92–1641 (W.D. Wash) (settlement agreement filed Oct. 22, 1993).

By this time support had grown in Congress, led by Senator Bob Graham of Florida, for a legislative overhaul of the program. But Senator Graham's bills failed sufficiently to protect hunting and fishing as refuge uses in the eyes of those interest groups, and thus failed to pass. *See, e.g.,* S. 823, 103d Cong., 1st Sess. (1993). The Clinton Administration then thwarted efforts to enact legislation that would have strengthened the prominence of hunting and fishing uses in refuges, and in 1996 President Clinton adopted an Executive Order specifically making wildlife conservation the primary purpose of the entire National Wildlife Refuge System, relegating hunting, fishing, and other "wildlife-dependent public uses" to subordinate use priority. *See* Executive Order No. 12,996, 61 Fed. Reg. 13647 (March 28, 1996). This opened the door to the 1997 legislation.

The 1997 legislation endorsed many facets of the Executive Order, focusing in particular on articulating the compatibility standard by incorporating detailed planning and substantive duties for the agency. The legislation defined "compatible use" to mean "a wildlife-dependent recreational use or any other use of a refuge that, in the sound professional judgment of the Director, will not materially interfere with or detract from fulfillment of the mission of the System or the purposes of the refuge." 16 U.S.C. § 668ee(1). The mission of the System, and thus the defining dominant use, is conservation of wildlife, plants, and their habitats. *Id.* § 668dd(a)(2). To the extent they are compatible with that dominant use, wildlife-dependent human uses, which Congress defined to include hunting and fishing as well as less intrusive activities such as photography and observation, are the highest priority for human uses. *Id.* § 668dd(a)(4)(D). Other "general public uses," such as grazing and oil exploration, are allowed only where they are deemed appropriate and where they are compatible not only with the wildlife protection purpose, but also all higher-priority human uses. The tiered use structure can be summarized as follows:

priority of use	type of use
	conservation of wildlife, plants, and their habitats (System mission) plus specific purposes for which the particular Refuge was created
first subordinate	a wildlife-dependent recreational use, such as hunting, fishing, photography, and observation, or any other use of the Refuge that, in the sound professional judgment of the Director, will not materially interfere with or detract from fulfillment of the mission of the System or the purposes of the Refuge
second subordinate	general public uses, such as grazing and oil exploration, but only where they are deemed appropriate and where they are compatible not only with the System and Refuge purposes, but also all higher-priority human uses

FWS has issued policies outlining how it will permit wildlife-dependent and other general public uses in refuge lands. *See* 66 Fed. Reg. 3668 (Jan. 16, 2001). With respect to the *dominant* mission of wildlife conservation, the 1997 legislation added the directive that the agency "ensure that the biological integrity, diversity, and environmental health of the System are maintained." *Id.* § 668dd(a)(4)(B). In October 2000, USFWS issued a draft policy for implementing this core directive, which, not surprisingly, incorporated many principles of ecosystem management and adaptive management. The agency summarized its proposal as follows:

We (U.S. Fish and Wildlife Service) propose to establish an internal policy to guide personnel of the National Wildlife Refuge System (Refuge System) in implementing the clause of the National Wildlife Refuge System Improvement Act of 1997 (Refuge Improvement Act) that calls for maintaining the "biological integrity, diversity, and environmental health" of the Refuge System. The holistic integration of these three qualities constitutes ecological integrity. The concept of ecological integrity requires a frame of reference for natural conditions. Our frame of reference extends from 800 AD to 1800 AD. The former date marked the beginning of an ecological transformation associated with higher temperatures; the latter approximates the advent of the industrial era, including drastic and widespread habitat loss. In areas where pre-industrial European settlement was particularly intensive, however, our frame of reference may be shorter. Natural conditions also include those that would have persisted or evolved to the present time if European settlement and industrialization had not occurred. At each refuge, we ascertain natural conditions, assess current conditions, and strive to decrease the difference. However, we are especially concerned with ecological integrity of the Refuge System as a whole, which can conflict with the maintenance of ecological integrity at individual refuges. In some cases, we may compromise the ecological integrity of a refuge for the sake of maintaining a higher level of ecological integrity at the Refuge System scale.

Department of the Interior, Draft Policy on Maintaining the Ecological Integrity of the National Wildlife Refuge System Fish and Wildlife Service, 65 Fed. Reg. 61356 (October 17, 2000).

To summarize, the agency outlined a three-step process to be carried out at each Refuge in order to fulfill the System-wide mandate of protecting biological integrity, diversity, and environmental health:

Step 1   determine the natural conditions of the Refuge that existed in the time frame of 800 to 1800 AD

Step 2   determine the current conditions of the Refuge and any extent to which they deviate from natural conditions

Step 3   design management actions that will eliminate the deviation of current conditions from natural conditions

This proposed approach received sharp criticism in the public comment process, motivating the agency to recraft certain terms and conditions in the final policy. Many of these criticisms resonate in the general themes described in the preceding section as running through the law of ecosystem management. Consider how FWS responds to these issues and criticisms in arriving at a final policy.

Department of the Interior, Policy on Maintaining the Biological Integrity, Diversity, and Environmental Health of the National Wildlife Refuge System

66 Fed. Reg. 3810 (Jan. 16, 2001).

* * *

The proposed policy was derived from Section 5(a)(4)(B) of the Refuge Improvement Act that the Secretary of the Interior "ensure that the biological integrity, diversity, and environmental health of the System are maintained * * * " The policy presented in this notice is a final policy that has been modified after consideration of public comment. The finalized policy will constitute part 601 Chapter 3 of the Fish and Wildlife Service Manual.

Purpose of This Policy

The purpose of the policy is to provide guidance for maintaining, and restoring where appropriate, the biological integrity, diversity, and environmental health of the National Wildlife Refuge System.

Response to Comments Received

The combined comment periods totaled 60 days. We received 106 comments from the following sources: Non-governmental organizations (36); State agencies or commissions (31); Federal agencies or facilities (9); local or county governmental agencies (3); and individuals (24). The key points raised by these comments fell into 10 general categories:

— Creation of the term "ecological integrity" and its definition:

— Definition of the term "natural conditions" and application of the concept in management;

— Impact of the policy on the ongoing refuge management activities;

— Impact of the policy on recreational use of refuges, primarily hunting and fishing;

— Concern that the policy would not meet specific refuge purpose(s) in favor of the System mission or some other management direction;

— Concern that the policy might adversely affect private property rights of refuge neighbors, and does not adequately recognize the State interests in how we manage refuges;

— Confusion regarding management for biological integrity, diversity, and environmental health at various landscape scales;

— Concern that the policy contains too many exceptions;

— General support either for the entire policy or significant elements of it; and

— A collection of other issues.

Issue 1: The Term "Ecological Integrity"

Comment: Most of the commenters (9 of 14) who cited this term stated that it went beyond the Refuge Improvement Act by creating a term that was not contained in the law or legislative history. Another stated it provided managers too much latitude to threaten private landowners. Still others stated it was too academic and basically unnecessary to meet the requirements of the Refuge Improvement Act. One commenter supported the term but stated the definition needed further refinement pursuant to scientific literature and that we should provide more guidance as to how to measure it.

Response: We never intended for the term "ecological integrity" to be more than a convenient means of referencing the terms biological integrity, diversity and environmental health. We agree, however, that as we used the term throughout the policy it appeared to take on meaning beyond the reference to the three terms. We abandoned the term in the final policy and substitute its appearance with the three specific terms as they appear in the law.

Issue 2: The Definition of the Term "Natural Conditions" and Its Application in Management

Fifty-nine of 106 commenters made specific references to the definition of natural conditions. Of these, 14 generally favored the concept and the remainder expressed concern about the concept and/or its application in management. An additional 9 commenters indicated general support for the policy overall, thus indicating support for the concept as well. However, even the 14 commenters who specifically endorsed the concept did so with various qualifications or suggestions. Overall, the commenters raised the following concerns:

Comment: A reference period is unnecessary, since the Refuge Improvement Act merely requires us to maintain the biological integrity, diversity, and environmental health necessary to meet refuge purposes.

Response: We believe the use of a reference point is pivotal to compliance with the mandate of the Refuge Improvement Act to ensure the maintenance of biological diversity, integrity, and environmental health. To implement the Refuge Improvement Act mandate, we needed definitions for the three terms. We believe a reference period is a critical element in these definitions and thus critical to the assessment of current habitat and wildlife conditions. * * * In using the term "natural conditions" relative to a specific period (i.e., 800 to 1800 AD), we chose an approach with scientific underpinnings. . . . Our intent in using the period was not to suggest a return to some particular community or habitat but, in fact, to reference something within the historic range of variability as found within that time frame. . . . Notwithstanding, the way the draft policy presents this concept clearly created a catalyst for controversy among reviewers, and while nine commenters supported the concept with some variation, the great majority expressed strong concern. Thus, we agree that the term "natural conditions" and the implications for management in the framework we have described should be removed from the policy. Instead, we adopted the more general and open-ended term, "historic conditions," which we refer to as the condition of the landscape in a particular area before the onset of significant, human-caused change. See final policy Section 3.12. On that basis, we refined the definitions of biological integrity and environmental health to mean composition, structure and functioning of ecosystems "comparable to historic conditions." The intent is to emphasize not a particular point in time, but the range of ecosystem processes and functions that we believe would have occurred historically.

Issue 3: Implications for Refuge Purposes and System Mission

* * *

Comment: One commenter felt that the Ecological Integrity Policy and Refuge Improvement Act should take precedence over, or replace refuge purpose(s).

Response: The fulfillment of refuge purpose(s) is a nondiscretionary statutory duty of the Service. However, the law also requires that we ensure that the biological integrity, diversity, and environmental health of the System is maintained, and therefore, this is an additional duty which we must fulfill as we endeavor to achieve refuge purpose(s) and System mission.

Issue 4: Impacts on Public Use, Especially Hunting and Fishing

We received 34 letters that addressed the relationship between the draft policy and its relationship to public uses on refuges and public use as mandated under the Refuge Improvement Act.

Comment: More than half of these letters (17) were concerned that the policy, as drafted, would interfere with or eliminate hunting and fishing on refuges while another 13 letters were concerned that this policy would affect or find all public uses incompatible with ecological integrity.

Response: We did not write the draft policy with the intent or direction to eliminate hunting, fishing, or other priority public uses recognized by the Refuge Improvement Act. This draft policy rarely mentions public use, but where it does, the purpose is for refuge managers to consider impacts on wildlife and habitat (i.e., biological integrity, diversity, and environmental health) when implementing public uses. The authority for this draft policy is the Refuge Improvement Act, which also clearly identifies hunting and fishing as priority public uses. Section 2.(6) of the Refuge Improvement Act states, "When managed in accordance with principles of sound fish and wildlife management and administration, fishing, hunting * * * in national wildlife refuges have been and are expected to continue to be generally compatible uses."

* * *

Comment: A few letters thought that hunting, fishing and trapping should not be permitted on refuges because they interfere with ecological integrity, while one letter wanted "trapping" added to Section 3.14 where hunting and fishing are encouraged in cooperation with State fish and wildlife management agencies.

Response: The six priority wildlife-dependent uses are given special status by the Refuge Improvement Act, which specifically recognizes hunting, fishing, wildlife observation, photography, interpretation, and environmental education. Refuges must facilitate these uses when compatible. The Refuge Improvement Act does not similarly recognize trapping.

Issue 5: Implications for States and Other Partnerships

Comment: Various States commented that the policy should place emphasis on cooperation and coordination with States in the management of wildlife populations on refuges.

Response: Strong partnerships with the respective States are an essential part of all refuge planning and management, including the maintenance of biological integrity, diversity, and environmental health of refuges. We encourage and expect managers to forge effective partnerships with States through cooperation and coordination in the management of wildlife habitats and populations found on refuges. We have changed the language in the final policy, Section 3.14, to more clearly state this expectation.

Issue 6: Implications for Private Property Rights

Comment: Several commenters were concerned that the policy was not mindful of the property rights of others and encouraged managers to seek resolutions to problems injuring resources on refuges through litigation.

Response: We changed Section 3.20 of the final policy to emphasize that the preferred course of action for managers in cases of injury to refuge resources from outside sources is first to seek cooperative resolution to such conflicts through neighborly discussion, negotiation, and consultation. This includes working with State or local agencies and other third party interests to seek solutions of mutual satisfaction. The revised policy offers several steps for a manager to take in this regard. Ultimately, however, and with full respect of private property rights, we recognize our responsibility to protect the property and resources of the American public, and state the responsibility to do so.

Issue 7: Implications for Wildlife and Habitat Management on Refuges

Comment: We received many comments which expressed concern about the role of active management on refuges under the proposed policy. These comments noted that active management is often necessary to achieve refuge purpose(s). Some felt management for natural conditions basically implied an absence of management and would, therefore, conflict with achieving refuge purpose(s). Comments also noted that numerous refuges are located in highly altered landscapes where active management is needed to maintain wildlife values of the refuge. A few comments identified that active management actions are required to maintain desirable wildlife populations where habitats surrounding the refuge have been degraded.

Response: We acknowledge that active management is often critically important to achieve refuge purpose(s). We also acknowledge that at some refuges very intensive management actions are required to maintain high densities of some wildlife species. We will continue active management where needed. However, we will evaluate management practices on all refuges to ensure that we take appropriate management action to achieve refuge purpose(s), while at the same time addressing the guidelines identified in the final policy.

* * *

Issue 8: Implications of Policy at Different Landscape Scales

Comment: There were 12 letters that raised issues of scale and the definitions and references to landscapes.

Response: Use of the term "local landscape" in the draft policy caused some confusion among these commenters. We intended the term to describe the refuge and its immediate surroundings. In the final policy, we dropped the "landscape" part of the term and use "local scale" or "refuge scale" to refer to a refuge and the area around it.

* * *

The text of the final policy follows:

* * *

3.3 What Is the Biological Integrity, Diversity, and Environmental Health Policy?

The policy is an additional directive for refuge managers to follow while achieving refuge purpose(s) and System mission. It provides for the consideration and protection of the broad spectrum of fish, wildlife, and habitat resources found on refuges and associated ecosystems. Further, it provides refuge managers with an evaluation process to analyze their refuge and recommend the best management direction to prevent further degradation of environmental conditions; and where appropriate and in concert with refuge purposes and System mission, restore lost or severely degraded components.

3.4 What Are the Objectives of This Policy?

A. Describe the relationships among refuge purposes, System mission, and maintaining biological integrity, diversity, and environmental health.

B. Provide guidelines for determining what conditions constitute biological integrity, diversity, and environmental health.

C. Provide guidelines for maintaining existing levels of biological integrity, diversity, and environmental health.

D. Provide guidelines for determining how and when it is appropriate to restore lost elements of biological integrity, diversity, and environmental health.

E. Provide guidelines to follow in dealing with external threats to biological integrity, diversity, and environmental health.

* * *

3.6 What Do These Terms Mean?

A. Biological diversity. The variety of life and its processes, including the variety of living organisms, the genetic differences among them, and communities and ecosystems in which they occur.

B. Biological integrity. Biotic composition, structure, and functioning at genetic, organism, and community levels comparable with historic conditions, including the natural biological processes that shape genomes, organisms, and communities.

C. Environmental health. Composition, structure, and functioning of soil, water, air, and other abiotic features comparable with historic conditions, including the natural abiotic processes that shape the environment.

D. Historic conditions. Composition, structure, and functioning of ecosystems resulting from natural processes that we believe, based on sound professional judgment, were present prior to substantial human related changes to the landscape.

E. Native. With respect to a particular ecosystem, a species that, other than as a result of an introduction, historically occurred or currently occurs in that ecosystem.

3.7 What Are the Principles Underlying This Policy?

A. Wildlife First

The Refuge Administration Act, as amended, clearly establishes that wildlife conservation is the singular National Wildlife Refuge System mission. House Report 105–106 accompanying the National Wildlife Refuge System Improvement Act of 1997 states " * * * the fundamental mission of our System is wildlife conservation: wildlife and wildlife conservation must come first." Biological integrity, diversity, and environmental health are critical components of wildlife conservation.

B. Accomplishing Refuge Purposes and Maintaining Biological Integrity, Diversity, Environmental Health of the System

The Refuge Administration Act states that each refuge will be managed to fulfill refuge purpose(s) as well as to help fulfill the System mission, and we will accomplish these purpose(s) and our mission by ensuring that the biological integrity, diversity, and environmental health of each refuge is maintained, and where appropriate, restored. We base our decisions on sound professional judgment.

C. Biological Integrity, Diversity, and Environmental Health in a Landscape Context

Biological integrity, diversity, and environmental health can be described at various landscape scales from refuge to ecosystem, national, and international. Each landscape scale has a measure of biological integrity, diversity, and environmental health dependent on how the existing habitats, ecosystem processes, and wildlife populations have been altered in comparison to historic conditions.* * *

D. Maintenance and Restoration of Biological Integrity, Diversity, Environmental Health

We will, first and foremost, maintain existing levels of biological integrity, diversity, and environmental health at the refuge scale. Secondarily, we will restore lost or severely degraded elements of integrity, diversity, environmental health at the refuge scale and other appropriate landscape scales where it is feasible and supports achievement of refuge purpose(s) and System mission.

E.   Wildlife and Habitat Management

Management, ranging from preservation to active manipulation of habitats and populations, is necessary to maintain biological integrity, diversity, and environmental health. We favor management that restores or mimics natural ecosystem processes or function to achieve refuge purpose(s). Some refuges may differ from the frequency and timing of natural processes in order to meet refuge purpose(s) or address biological integrity, diversity, and environmental health at larger landscape scales.

F.   Sound Professional Judgment

Refuge managers will use sound professional judgment when implementing this policy primarily during the comprehensive conservation planning process to determine: The relationship between refuge purpose(s) and biological integrity, diversity, and environmental health; what conditions constitute biological integrity, diversity, and environmental health; how to maintain existing levels of all three; and, how and when to appropriately restore lost elements of all three. These determinations are inherently complex. Sound professional judgment incorporates field experience, knowledge of refuge resources, refuge role within an ecosystem, applicable laws, and best available science including consultation with others both inside and outside the Service.

G.   Public Use

The priority wildlife-dependent public uses, established by the National Wildlife Refuge System Improvement Act of 1997, are not in conflict with this policy when determined to be compatible. The directives of this policy do not generally entail exclusion of visitors or elimination of public use structures, e.g., boardwalks and observation towers. However, maintenance and/or restoration of biological integrity, diversity, and environmental health may require spatial or temporal zoning of public use programs and associated infrastructures. General success in maintaining or restoring biological integrity, diversity, and environmental health will produce higher quality opportunities for wildlife-dependent public use.

* * *

3.9   How Do We Implement This Policy?

The Director, Regional Directors, Regional Chiefs, and Refuge Managers will carry out their responsibilities specified in Section 3.8 of this chapter. In addition, refuge managers will carry out the following tasks.

A.   Identify the refuge purpose(s), legislative responsibilities, refuge role within the ecosystem and System mission.

B.   Assess the current status of biological integrity, diversity, and environmental health through baseline vegetation, population surveys and studies, and any other necessary environmental studies.

C.   Assess historic conditions and compare them to current conditions. This will provide a benchmark of comparison for the relative intactness of ecosystems' functions and processes. This assessment should include the opportunities and limitations to maintaining and restoring biological integrity, diversity, and environmental health.

D. Consider the refuge's importance to refuge, ecosystem, national, and international landscape scales of biological integrity, diversity, and environmental health. Also, identify the refuge's roles and responsibilities within the Regional and System administrative levels.

E. Consider the relationships among refuge purpose(s) and biological integrity, diversity and environmental health, and resolve conflicts among them.

G. Through the comprehensive conservation planning process, interim management planning, or compatibility reviews, determine the appropriate management direction to maintain and, where appropriate, restore, biological integrity, diversity, and environmental health, while achieving refuge purpose(s).

H. Evaluate the effectiveness of our management by comparing results to desired outcomes. If the results of our management strategies are unsatisfactory, assess the causes of failure and adapt our strategies accordingly.

* * *

Notes and Questions

1. Overall, how different is the final policy from the proposed policy? Does the final policy remove agency discretion that may have existed under the proposed policy? Can you think of a decision with respect to management of a particular refuge that could have been made under the proposed policy but cannot be made under the final policy?

2. More to the point, is there any meaningful difference between the "natural conditions" concept as outlined in the proposed policy and the "historic conditions" concept adopted in the final policy? The basic three-step process the agency outlined in the proposed rule is the same in the final rule, except the concept of historical conditions replaces that of natural conditions. Can you envision any set of circumstances in which a manager of a particular refuge would be legally constrained by the historic conditions policy but would not have been under the natural conditions policy? Do you suppose those who criticized the natural conditions policy are satisfied with the agency's final historic conditions policy?

3. Can you articulate the difference between "ecological integrity" and the combination of "biological integrity, diversity, and environmental health?" What particular refuge management action could be accomplished under the former but not the latter? Do you believe the agency's response that it intended the ecological integrity term to serve merely as shorthand for the longer statutory list of three objectives, or do you think the agency had some ulterior motive?

4. What is "sound professional judgment," and how will we know when the agency has failed to apply it? The 1997 refuge system legislation defined it to mean "a finding, determination or decision that is consistent with principles of sound fish and wildlife management and administration, available science and resources, and adherence to the requirements of this act and other applicable laws." 16 U.S.C. § 668ee(3). Does that help? Is this just another way of saying the agency should be left free to exercise adaptive management as best as it can? But if so, does this not also illustrate the difficulty of legislative, judicial, and citizen oversight of adaptive management? It is difficult for the legislature to define ahead of time what is and is not sound judgment in the exercise of adaptive management, because the challenges of adaptive management arise unpredictably. Citizens who contest agency decisions thus bear the burden of proving them to be "unsound," which courts are loathe to find in the absence of strong evidence that the agency acted no less than arbitrarily.

5. Is the process FWS outlines for refuge management consistent with adaptive management theory?

6. Why are hunting and fishing allowed on *wildlife refuges*? Ironically, hunting has been perhaps the best friend of wildlife refuges. Under the Migratory Bird Conservation Stamp Act of 1929, 16 U.S.C. §§ 715–715r, and the Migratory Bird Hunting Stamp Act of 1934, 16 U.S.C. §§ 718–718h, all waterfowl hunters must purchase a federal waterfowl "stamp" and attach it to their state hunting licenses. FWS deposits revenue from the stamp sales into the Migratory Bird Conservation Fund, which funds acquisition of refuge lands. This source of funding has contributed to all or part of the lands acquisition of over 345 refuges. *See* MIGRATORY BIRD CONSERVATION COMMISSION, U.S. FISH AND WILDLIFE SERVICE, REPORT FOR THE FISCAL YEAR 1999 (2000). This steady source of funding, however, has "almost assured a refuge system keyed principally to the production of migratory waterfowl." MICHAEL J. BEAN AND MELANIE J. ROWLAND, THE EVOLUTION OF NATIONAL WILDLIFE LAW 284 (3d ed., 1997).

7. What private property rights are threatened by the proposed or final policy for management of public wildlife refuge lands? How might the exercise of private property rights threaten public wildlife refuge interests?

8. For what is by far the most comprehensive, up to date, and insightful overview of the wildlife refuge system, see Robert L. Fischman, *The National Wildlife Refuge System and the Hallmarks of Modern Organic Legislation*, 29 ECOLOGY L.Q.—(forthcoming 2002). We used the wildlife refuge system as the case study for the introductory chapter on ecosystem management law because, unlike many other federal land conservation units, it cuts across many ecosystem types, thus forcing a more generic management approach. The remaining chapters of this part of the text divide the materials by ecosystem type, and other federal land conservation units correspond more closely to that division (e.g., National Forests with the forest ecosystems materials; BLM Land and National Grasslands with the grasslands ecosystems materials). The wildlife refuge policy presents a template of the issues and controversies explored in these chapters in more particularized and often more intensified settings. As has the refuge policy debate, those controversies often focus on the issue of use versus restoration, and on what, precisely, restoration means. For a broad take on the concept of restoration in environmental law, see Alyson C. Flournoy, *Restoration Rx: An Evaluation and Prescription*, 42 ARIZONA L. REV. 187 (2000).

CHAPTER 8

FOREST ECOSYSTEMS

Chapter Outline:
A. Biodiversity and Forest Ecosystems
    Note on Fire in Forest Ecosystems
B. Primer on the Law of Forests
    1.   Historical Background
    2.   Current Scene
C. Biodiversity Management in Forest Ecosystems
    1.   Public Lands: National Forests and the Biodiversity Mandate
        a.   Conservation Biology Science
        b.   Indicator Species
        c.   Species Viability Analysis
        d.   Ecological Sustainability
    Note on the Endangered Species Act in National Forests
    Note on Roads and Roadless Areas in National Forests
    Note on Ecosystem Management in State–Owned Forest Lands
    2.   Private Lands: Integrated Approaches
        a.   The Policy Grab Bag
        b.   Dealing with the Takings Issue
D. Case Study: The Shawnee National Forest in Southern Illinois

---

A. BIODIVERSITY AND FOREST ECOSYSTEMS

Nearly half the forests that once covered the earth are gone. Notwithstanding the loss of biodiversity values suffered as a result of that depletion of forest resources, forests today provide our planet's largest terrestrial storehouse of biodiversity, covering 27 percent of the ice-free land surface of the earth. How the earth's forest resources are managed in the future—in particular, whether the rate of loss can be slowed or even reversed—thus will play an important role in biodiversity conservation at the global level.

In addition to storing biodiversity resources, forests provide critical ecosystem services such as shelter for countless species, protection of soils from rainfall erosion, flood control, carbon sequestration, and water filtration. Of course, humans value forests also for purely economic reasons, namely for lumber, paper, and fuel production. About 55 percent of the wood cut today is used for fuel (primarily in India, China, Brazil, Indonesia, and Nigeria), with the remainder going to production for lumber, paper, and other industrial products (primarily in the United States, Canada, and Russia). Forest lands have also been valuable to humans for conversion to other uses, principally agriculture, urban development, and recreational uses. One of the challenges for management of forests thus has been how to take into account the biodiversity storage and ecosystem service values of forests when making decisions about how to take advantage of their more direct economic values.

If the biodiversity and ecosystem service values of forest parcels could be precisely monetized, so that their losses could be factored into forest management decisions in direct economic terms, timber and paper production or conversion of forested areas to urban development may make much less economic sense in many settings. Even though we lack the technical and financial resources to perform this sort of perfect monetization, the knowledge that forest ecosystems deliver significant value to humans suggests that forest management decisions could factor those values into the decision making process in some manner. Yet only a few countries have begun to consider the biodiversity storage and ecosystem service values of forests as core decision making criteria. In the last two decades of heightened global awareness of biodiversity losses, total global forest losses experienced in that period would cover an area larger than Mexico. Still, forests cover more than one quarter of the earth's land area (excluding Antarctica and Greenland), with the largest holdings in Russia, Brazil, Canada, the United States, China, Indonesia, and the Democratic Republic of the Congo, in that order.

Even in countries that have in recent decades halted or reversed forests losses, such as the United States, biodiversity values in forested areas are generally on the decline. In many countries the primary forest stands have been largely depleted and production is mostly from secondary stands or tree plantations. In the United States, for example, about 46 of the land was forested at the time of European settlement, and about 32 percent is today, but virtually none of the original trees—the so-called old growth forests—remain today. The biodiversity values of the secondary stands and managed plantations that replaced them, which do not count in most studies as deforested acres, are generally less than for primary stands. Moreover, the biodiversity capacity of many standing forests has been imperiled by industrial and agricultural pollution of air and water, invasive species introduced through inadvertent or purposeful human import, raging wildfires caused, ironically, by fuel buildup resulting from prior fire suppression, and other results of human management decisions both within and external to forests. And as overall forest health declines, natural threats such as insects and disease become more acute. For example, over 90 million acres of standing forests in the United States are threatened by seven major pests and diseases.

Thus, although headlines focus on loss of the tropical rainforests to agriculture and timber extraction, the United States has been no stranger to forest resource depletion and continues to suffer serious declines in health of its remaining forest stands. Our nation's appetite for wood is voracious: we consume one fourth of the world's industrial timber; ten percent of the world's industrial timber is used in the U.S. construction industry, mostly for home building; we produce 30 percent of the world's paper. These demands have taken their toll on American forests. Today, about 33 percent of the U.S. land area—747 million acres—is forest lands, but that is two-thirds of our forested area in 1600. More than 75 percent of the 307 million acres lost to other uses were converted in the 19th century, mostly to agricultural uses and mostly in the Midwest and lower Mississippi Valley. The area of forested land has remained relatively stable since 1920, as the introduction of mechanized farm machines reduced the need to devote cropland to growing feed for farm draft animals. In short, we have little need for converting more forest land into farmland. Yet only a small fraction of what remains of our forested lands is primary old growth forest.

About two-thirds of the nation's remaining forested lands are classified as timberland suitable for production of lumber, paper, and other industrial goods. The federal government owns and manages about one-third of U.S. forested land. The rest is owned by nonfederal public agencies, forest industry firms, farmers, and some 6 million other private individuals. Forest lands, types, and ownership patterns are distributed unevenly. North Dakota, for example, is only one percent forested land, while Maine has the most at 89 percent. Most of the privately owned forest land is east of the Mississippi River, while most forest land to the west is publicly owned. Most hardwood timber stock is in the east; western timber is mostly softwood. Private forest lands in the eastern states, particularly plantation forests in the southeast, will be our nation's primary source of timber in this century. This diversity of distribution, ownership, and type of forest lands makes formation of a national management policy for our forest ecosystems a difficult political and scientific undertaking.

Notes and Questions

1. The marketplace assigns values to timber products and non-forest uses of forested land, but often overlooks the value of ecosystem services that standing forests provide. *See* Robert Bonnie et al., *Counting the Cost of Deforestation*, 288 SCIENCE 1763 (2000). Norman Myers lists among the ecosystem services forests supply with real value to humans: landscape stabilization; soil nutrient cycling; water flow regulation; energy balance; siltation removal; sunlight reflectance (albedo), atmospheric moisture regulation; and a host of others. *See* Norman Myers, *The World's Forests and Their Ecosystem Services*, in NATURE'S SERVICES: SOCIETAL DEPENDENCE ON NATURAL ECOSYSTEMS 215–26 (Gretchen Daily ed. 1997). Assigning monetized values to these services is difficult, but rough estimates can be derived in numerous contexts from real-world examples, and they are staggering in magnitude. *See id.* at 226–28. For example, if a forest buffer could reduce pollution of a fishery from upland runoff, one could consider the added fishery yield as a value gained from the forest, and the cost of duplicating that effect through technological means as an avoided cost. While leaving the forest buffer in that use may impose an opportunity cost in terms of foregone development of that land to "higher" uses, ought not the cost-benefit calculus include the values gained and costs avoided as a result of the forest's delivery of the identified services to other economically valuable land and resource uses? On the other hand, the value of cut timber is subject to fairly precise quantification, whereas the ecosystem service values of standing timber are not. Should the cost-benefit decision making be required to use both valuation components given that one side suffers from far inferior measurement accuracy?

2. A related problem in deciding how to value timber as cut versus as uncut is the lack of knowledge we have about the long term effects of cutting timber on overall ecosystem integrity and productivity. The value of cut timber may be precisely quantifiable, but that market value does not account for potential loss of ecological and economic productivity the removal of timber imposes. For example, cutting roads into tropical forests has been shown to begin a process of forest transformation that fundamentally reduces the forest's value even in terms of cut timber, and the long term effects on forest ecosystem service dynamics are even more severe. *See* Claude Gascon, *Receding Forest Edges and Vanishing Reserves*, 288 Science 1356 (2000). Timber harvesting also has been shown to have devastating long term effects on timber resources through increased fire risk, and on other commercial activities, such as seed harvesting, through the depletion of vegetative diversity. *See* Gary Hartshorn and Nora Bynum, *Tropical Rainforest Synergies*, 286 Science 2093 (2000). Even putting aside the question whether ecosystem service values can be sufficiently quantified to include in forest management decision making, does research such as this suggest we should be less confident that we know the true value of cut timber?

3. Some argue that, regardless of whether it focuses narrowly on market value or broadly to include ecosystem service value, ecosystem *productivity* should not be used as the criterion in deciding how to manage a forest ecosystem. Rather, in this alternative view, ecosystem *restoration* is the correct policy goal. But restoration to what condition? This question has triggered a burning debate among ecologists. *See* Keith Kloor, *Returning America's Forests to Their "Natural" Roots*, 287 Science 573 (2000). While acting as Secretary of the Interior, Bruce Babbitt suggested that ecosystem restoration means returning lands "to a presettlement equilibrium." In conventional use this means prior to European settlement, but some ecologists contend that Native Americans were significantly shaping and altering the land, particularly through fire manipulation and clearing on forested lands, well before European settlers arrived. Indeed, picking any particular date for restoration also fails to account for a variety of naturally-induced factors, such as drought, fire, temperature swings, that caused long term variations in healthy forest ecosystems. Trying to "lock" a forest ecosystem into a particular "equilibrium" state in which it existed in the past thus may not truly "restore" it to "presettlement" conditions. Instead, some ecologists advocate working to rehabilitate an "ailing" or "out of control" forest back into its natural "envelope of variability." Others eschew any notion of "stasis" and advocate use of past conditions simply as "reference points" designed to guide in the restoration of a "natural flux" in "dynamic processes" such as fire, flood, predation, and regeneration. Clearly, there is a lack of agreement among scientists as to where and how to do forest restoration, how much to use the past as the guide, and which past to use. And if the scientists can't agree, what should the policy-makers do?

4. Speaking of dynamic processes, what may be the impact of anthropo-genically-induced climate change on forest ecosystems? There is evidence, for example, that regional climate change brought on by alterations in land use patterns can have a profound effect on local forest ecosystem dynamics. *See* R. O. Lawton et al., *Climatic Impact of Tropical Lowland Deforestation on Nearby Montane Cloud Forests*, 294 SCIENCE 584 (2001). What could happen as *global* climate changes? According to the U.S. EPA, the projected 2°C (3.6°F) warming could shift the ideal range for many North American forest species by about 300 km (200 mi.) to the north. If the climate changes slowly enough, warmer temperatures may enable the trees to colonize north into areas that are currently too cold, at about the same rate as southern areas became too hot and dry for the species to survive. If the earth warms this much in 100 years, however, the tree species would have to migrate about 2 miles every year. Trees whose seeds are spread by birds may be able to spread at that rate. But neither trees whose seeds are carried by the wind, nor such nut-bearing trees such as oaks, are likely to spread by more than a few hundred feet per year. Poor soils may also limit the rate at which tree species can spread north. Thus, the range over which a particular species is found may tend to be squeezed as southern areas become inhospitably hot. The net result is that some forests may tend to have a less diverse mix of tree species.

Several other impacts associated with changing climate further compli-cate the picture. On the positive side, CO_2 has a beneficial fertilization effect on plants, and also enables plants to use water more efficiently. These effects might enable some species to resist the adverse effects of warmer temperatures or drier soils. On the negative side, forest fires are likely to become more frequent and severe if soils become drier. Changes in pest populations could further increase the stress on forests. Managed forests may tend to be less vulnerable than unmanaged forests, because the managers will be able to shift to tree species appropriate for the warmer climate.

Perhaps the most important complicating factor is uncertainty whether particular regions will become wetter or drier. If climate generally becomes wetter, then forests are likely to expand toward rangelands and other areas that are dry today; if climate generally becomes drier, then forests will retreat away from those areas. Because of these fundamental uncertainties, existing studies of the impact of climate change have ambiguous results.

The potential impacts of climate change on forest wildlife are also poorly understood. If habitats simply shift to cooler areas (i.e., higher latitudes or higher altitudes), many forms of wildlife could potentially adapt to global warming, just as they have adapted to the changes in climate that have occurred over the last several million years. Unlike previous climatic shifts, however, roads, development, and other modifications to the natural environment may block the migration routes. For more analysis of these issues, see www.epa.gov/globalwarming/impacts.

5. Which is more important to manage in the forest ecosystem, the vegetation or the animals? Of course, the theme of ecosystem management is that the relationship between such components of ecosystems is complex, and management cannot favor one over the other. Forest ecosystems regulate energy flow through "bottom up" processes such as chemical uptake by vegetation, and through "top down" processes such as predator control of other species. Indeed, even these processes, while starting at different "ends" of the ecosystem dynamics, are not discrete. For example, research conducted on islands created when a large reservoir in Venezuela was impounded showed that many islands lost most or all of the apex vertebrate predators. On those islands, herbivore species densities quickly skyrocketed and, not surprisingly, seed and seedling densities depleted. Over the long run, in other words, the loss of apex predators could spell doom for the vegetative regime on these islands. *See* John Terborgh et al., *Ecological Meltdown in Predator–Free Forest Fragments*, 294 SCIENCE 1923 (2001). What does this suggest for forest ecosystem management on larger scales?

6. Many of the statistics discussed in the text are described in more detail in the WorldWatch Institute's annual *State of the World* series, which provides additional information on the biodiversity values of forests, the historical losses of and current threats to them, and the effects of human economy on forests. *See* Janet N. Abramovitz and Ashley T. Mattoon, *Recovering the Paper Landscape*, in LESTER R. BROWN ET AL., STATE OF THE WORLD 2000 101 (2000); Janet N. Abramovitz and Ashley T. Mattoon, LESTER R. BROWN ET AL., *Reorienting the Forest Products Economy*, in STATE OF THE WORLD 1999 60 (1999); Janet N. Abramovitz, *Sustaining the World's Forests*, in LESTER R. BROWN ET AL., STATE OF THE WORLD 1998 21 (1998).

Note on Fire in Forest Ecosystems

One ecosystem management issue that cuts across all forest ownership regimes, but is utterly confounded by the patchwork of modern forest land ownership patterns, is fire. The importance of fire to forest ecosystem dynamics is indisputable. A variety of natural causes of fire, particularly lightning, lead to a disturbance regime of periodic fires to which forest ecosystems adapt and, eventually, upon which they come to depend. For example, the heat from such fires releases seeds in many pine species without burning them, and it clears out underbrush vegetation to control species distribution. Forest ecosystems respond to these periodic fires through a balance between understory and forest crown vegetation that keep fires generally closer to the surface and thus not a threat to the majority of biomass in the forest.

Native Americans and early European settlers understood fire's role in forest and grassland ecosystems, using it to manage vegetative regimes to serve their purposes, usually for agricultural clearing and game management. Where human populations were relatively sparse and agriculture was not yet intensively developed, these fire *manipulation* techniques posed little threat either to forest ecosystems or to human settlements. *See* U.S FOREST SERVICE, FACES OF FIRE (1998).

With increased urbanization and conversion of land to agricultural uses, forest fires increasingly became a threat to human lives and property. In 1910, for example, fires in Idaho and Montana swept through millions of acres, killing dozens of firefighters and destroying extensive private holdings. Thus was born the national policy of fire *suppression*. Fires in forests were to be stopped at all costs, as quickly as possible. Icons of popular culture such as Bambi and Smokey the Bear instilled a "fire as bad" mentality in the public, and the job of forest managers was to keep fire at bay.

The problem with fire suppression, however, is that it does not work and, indeed, can make matters worse for humans and ecosystems in many settings. As fire suppression policies entered forest ecosystems adapted to natural fire regimes, it disrupted the balance of forest floor litter, undergrowth, and mature crown growth in the forest regime. The fuel available for fires when they started was far more abundant as a result, making suppression that much more difficult. In short, over time the fire suppression policy fueled catastrophic wildfire conditions. But allowing natural fire regimes to return to forested lands was not a viable option near urbanized areas or, for that matter, in forests that were used for timber harvesting, recreation, and other human uses. Indeed, even in remote forests designated for wilderness, the understory is often so developed from decades of fire suppression that any fire is likely to rage catastrophically and potentially alter basic ecosystem dynamics permanently.

The answer to this dilemma was the policy of fire *prescription*, or prescribed burns. Prescribed burning involves intentional setting of fire under planned and controlled conditions to meet specified management objectives, such as heat release and removal of understory fuel. The National Park Service adopted prescribed burning as its fire management policy in 1967, followed in 1978 by the Forest Service, and then in 1995 by a unified policy for all federal land management agencies. *See* NATIONAL PARK SERVICE ET AL., WILDLAND AND PRESCRIBED FIRE MANAGEMENT POLICY IMPLEMENTATION PROCEDURES REFERENCE GUIDE (1998). Many state governments adopted prescribed burn policies as well, including measures to provide private landowners limited liability protection for prescribed burns if proper procedures are followed. Florida, for example, describes prescribed burning as "a land management tool that benefits the safety of the public, the environment, and the economy of the state," and protects certified burners that follow specified prescribed burn procedures from common law negligence and nuisance liabilities. *See, e.g.*, Florida Prescribed Burning Act of 1990, Fla. Stat. § 590. On the other hand, there is no liability, common law or otherwise, for failing to conduct prescribed burning. *See* Department of Transportation v. Whiteco Metrocom, 1999 WL 1486516 (Fla. Div. Admin. Hrgs.).

Prescribed burning has proven to be no panacea, however, as escaped fire is always a concern and even well-managed prescribed fires pose inconveniences to nearby human populations. Moreover, prescribed fire policies could not be implemented in all forests quickly enough to counter the buildup of understory fuel, creating the risk that prescribed burns simply could not be adequately controlled and might trigger more intensive wildfires. Indeed, the years after prescribed burning was adopted as official national and state policy have included some of the worst fire years on record. In 1998, 2000 wildfires raged across Florida, burning 500,000 acres to a crisp. And in 2000, the nation suffered 123,000 fires, burning a total of more than 8 million acres, and the federal government alone spent over $2 billion fighting them. The most visible of wildfires in many years occurred in May 2000, when a prescribed burn set in the Bandelier National Monument in New Mexico escaped into the Santa Fe National Forest and triggered a wildfire that burned 17,000 acres and destroyed 380 structures in the Los Alamos area. *See Cerro Grande Fire, 2000: Hearings Before the Senate Committee on Energy and Natural Resources* 106th Cong., 2d. Sess. (2000).

With over 11,000 communities located precariously close to federal lands considered at risk of wildland fire, it seemed an appropriate time to develop a more concerted national fire management policy for forest and range lands. President Clinton and Congress thus directed the Secretaries of Interior and Agriculture to work with the states to develop a comprehensive approach to the management of wildland fires, hazardous fire conditions, and ecosystem restoration for federal and adjacent state, tribal, and private forest lands. *See* Pub. L. 106–291 (2000). The agencies soon developed a National Fire Plan with four key principles: (1) improve prevention and suppression; (2) reduce hazardous fuels; (3) restore fire adapted ecosystems; and (4) promote community assistance. *See* U.S. FOREST SERVICE ET AL., A COLLABORATIVE APPROACH FOR REDUCING WILDLAND FIRE RISKS TO COMMUNITIES AND THE ENVIRONMENT: 10–YEAR COMPREHENSIVE STRATEGY (Aug. 2001).

If you were in charge of designing a National Fire Plan, what would you recommend for management of fire in a public forest that is near an urban community and experiences high levels of understory fuels as a result of decades of fire suppression? Continue fire suppression? Take a chance on prescribed burns? Manually remove or thin the understory? What about the policy for a forest far from urbanized and rural land uses, but popular with recreational users? Would you let it "go natural" entirely?

For more information on the National Fire Plan, see http://www.fire-plan.gov. For background on the history of fire management policy and its current applications, see Stephen J. Pyne, *The Perils of Prescribed Fire: A Reconsideration*, 41 NATURAL RESOURCES J. 1 (2001); Laura Sweedo, *Where There Is Fire, There Is Smoke: Prescribed Burning in Idaho's Forests*, 8 DICKINSON J. ENVTL. LAW & POLICY 121 (1999); David Steinau, *Legal Issues Affecting Forestry Practices and Wildfires in Florida*, FLORIDA BAR SECTION OF ENVIRONMENTAL AND LAND USE LAW REPORTER (Jan. 2002).

B.   PRIMER ON THE LAW OF FORESTS

1.   HISTORICAL BACKGROUND

Although the early North American colonists found a virtually boundless supply of forests, it was not long before the common practice of setting fire to forests as a means of opening the land to other uses became regarded as economically wasteful and environmentally unsound. For example, a 1626 ordinance of the Plymouth Colony in Massachusetts banned setting of open forest fires on the ground that an inconvenience might result from a depletion of the timber supply, and much later a North Carolina act of 1777 proscribed unlawful firing of the woods as being extremely prejudicial to the soil. By the early eighteenth century many colonial governments also forbade unnecessary cutting of timber on common lands, and shortly after the Revolutionary War concluded Massachusetts enacted a law requiring licenses for removal of large white pines from state lands.

The newly formed national government entered the forest policy scene quickly after the war as a result of repeated pirate attacks on American merchant vessels. Congress, realizing the need for a national navy, also realized the need for a national timber supply to build the navy's ships. By an act of February 25, 1799, 1 Stat. 622, Congress appropriated money to purchase a timber supply, which later was used to acquire Grover's and Blackbeard's Islands off the coast of Georgia to secure about 2000 acres of prime timber property. Later, after the navy proved instrumental to victory against the British in the War of 1812, an act of March 1, 1817, 3 Stat. 347, authorized the Secretary of the Navy to select vacant and unappropriated tracts on federal lands with prime timber resources and recommend that the President reserve them from sale.

Timber policy through 1850 thus was defined largely by ad hoc state laws regulating open fires and federal reservation of timber supplies on federal lands to meet the navy's shipbuilding needs. Reports of widespread theft of timber from federal, state, and private lands suggest that these laws were difficult to enforce. The development of iron shipbuilding technology in the mid–1800s, moreover, led Congress to return most of its naval timber reserves to the public domain, meaning they were once again open for sale. The federal and state land settlement policies of the 1800s then surged the transfer of title to federal and state lands from public to private ownership by sale, homestead laws, railroad laws, and other means, shifting most eastern forest lands to private hands by the late 1800s.

The first glimmer of forest conservation policy was lit when the American Forestry Congress of 1882, meeting in Cincinnati, created the American Forestry Association to cooperate with the federal and state governments toward formulating a definite policy for managing public forest lands. In the following two decades, several states created state-level forestry agencies to regulate fires, encourage timber culture, and, in some cases, promote forest conservation, preservation, and extension. While in most cases the motivating force behind these laws was security of timber supply, laws such as one New York enacted in 1885, Session Laws, N.Y., 1885 ch. 283, p. 482, were among the first truly comprehensive forest management policies in America. The New York law established a system for designating, maintaining, and protecting state forests, complete with a state forest commission, wardens, forest inspectors, foresters, and other staff.

Federal policy witnessed a similar trend after the 1873 meeting of the American Society for the Advancement of Science appointed a committee to present to Congress a plan for the extension and preservation of forests, providing the impetus for a flurry of additional studies and proposals and even several federal laws promoting timber culture and the collection of forest statistics. The seeds of today's Forest Service also were planted in the Department of Agriculture through creation of the Division of Forestry, which began with one employee and an annual budget of $2000. By the late 1800s, though, federal forest policy remained a complete muddle—while promoting the extension of forest lands by subsidized plantings and other culture programs, albeit often in areas not suited to trees of any kind, the federal government was at the same time disposing of vast tracts of prime forest lands into private possession.

To resolve this inconsistency, a "rider" provision to the 1891 General Revision Act, known as the Forest Reserve Act of 1891, authorized the President to establish forest reservations on federal lands. *See* Act of March 3, 1891, ch. 561, § 26 Stat. 1095, 1103 (repealed 1976). Fortunately, not all the federal lands had yet been given away or sold, and Presidents Harrison and Cleveland withdrew extensive areas in the Western states from sale or entry and declared them national forest reserves. But much remained uncertain: Western interests were quite bitter over the turn of events, there was debate over the actual authority of the president under the 1891 law, and there were no monies appropriated to manage what by 1896 amounted to millions of acres of national forests. Congress resolved the situation with the passage of the Forest Service Organic Act of 1897, 30 Stat. 11, 34–35 (Organic Act), which ratified the presidential reservations and authorized administration of the national forests, then called forest reserves, through a federal agency.

The Organic Act marks the beginning of the development of comprehensive federal forest policy and administration. Although it made no mention of biodiversity, ecosystems, or even wildlife, it provided that the national forests should be established "to improve and protect the forest within the boundaries, or for the purposes of securing favorable water flows, and to furnish a continuous supply of timber for the use and necessities of citizens of the United States." 15 U.S.C. § 473. For the latter purpose, the Organic Act authorized the Department of the Interior, then after 1905 the Forest Service in the Department of Agriculture, to

> cause to be designated and appraised so much of the dead, matured or large growth trees found upon such national forests as may be compatible with the utilization of the forests thereon, and may sell the same.... Such timber, before being sold, shall be marked and designated, and shall be cut and removed under the supervision of some person appointed for that purpose by the Secretary....

Gifford Pinchot, who became head of the Forestry Service in 1898, envisioned the national forests as primarily a timber supply resource, and for nearly seventy years the Forest Service interpreted the Organic Act as allowing widespread extraction of timber. Indeed, after World War II, housing construction demands placed tremendous new pressure on the nation's timber supply, and on the Forest Service. Clearcutting became a common practice on private forest lands, thus depleting private timber supplies and causing the timber industry to pressure the Forest Service to increase the yield from national forests. The agency met this demand, but by doing so fueled a conflict between timber harvesting and another demand that boomed after the war—recreation. As clearcutting became common in the national forests, so too did the previously uncommon instance of public criticism of Forest Service decisions.

Congress nevertheless gave the Forest Service basically a free hand in all such matters of national forest policy, intervening only once to enact the Multiple Use and Sustained Yield Act of 1960 (MUSY). MUSY expanded the purposes of national forest management from water flows and timber supply to include "outdoor recreation, range, timber, watershed, and wild-life and fish purposes." 16 U.S.C. § 528. Recognizing that "some land will be used for less than all the resources," MUSY required that the five multiple uses, which Congress deliberately named in alphabetical order, be treated as co-equal and managed "with consideration being given to the relative values of the various resources." *Id*. The statute described the core mandate of multiple use as meaning

> The management of all the various renewable surface resources of the national forests so that they are utilized in the combination that will best meet the needs of the American people; making the most judicious use of the land for some or all these resources or related services over areas large enough to provide sufficient latitude for periodic adjustments in use to conform to changing needs and conditions. . . .

Id. § 531.

Conservation and recreation interests opposed the legislation while the agency and timber industry actively supported it. Critics of the Forest Service charged that the agency had elevated timber extraction above the other uses and exercised widespread clearcutting without due regard to the Organic Act, and would continue to do both under MUSY. Indeed, for all practical purposes MUSY codified precisely the policy discretion the agency sought (and argued it had even without MUSY). After MUSY, the law of national forests explicitly recognized the breadth of the agency's discretion. While courts demanded that the agency give "due consideration" to each of the multiple use components, *See* Parker v. United States, 307 F.Supp. 685 (D.Colo.1969), in the final analysis most courts agreed that "the decision as to the proper mix of uses within any particular area is left to the sound discretion and expertise of the Forest Service." Sierra Club v. Hardin, 325 F.Supp. 99, 123 (D.Alaska 1971). MUSY's multiple use mandate was essentially rendered directionless, leaving it to the agency to decide where to go and providing no meaningful legislative or judicial check on the path chosen.

As Congress and the courts continued to afford the Forest Service wide latitude in setting policy after MUSY, the agency began experimenting with planning as a way to resolve multiple use conflicts, requiring each national forest to develop a land use plan. Yet the agency used the end products as a vehicle to portray its policy of clearcutting as not merely a capitulation to the powerful timber industry, but the result a rational, scientifically sound policy decision making process. Neither congressional appropriations nor agency will would have supported any other outcome, thus leaving the growing recreational and conservation interests on the outside of the forest.

The first chink in this armor came in 1964 with passage of the Wilderness Act. 16 U.S.C. §§ 1131–36. Since 1929 the Forest Service had regulations in place in one form or another for designating "primitive," "wilderness," and "wild" areas to be removed from timber harvesting and other resource extraction uses. Although this administrative policy had produced over 2 million mainly roadless acres of preserve acres and was well within the agency's discretion before and after MUSY became part of national forest law, the agency could not satisfy conservationists that the agency would retain such designations in the face of increased demand for timber. After all, it would have been well within the agency's discretion and financial capability to withdraw the protected status, build roads, and open the lands to harvesting. After a decade of lobbying, these interests finally succeeded in convincing Congress to pass the Wilderness Act, in which the Forest Service was forced to place the protected lands into "The National Wilderness System" of wilderness and pristine lands, and thus out of the path of the timber industry. And although the Wilderness Act left MUSY untouched for the remaining lands under Forest Service control, the new law required the agency to examine the eligibility of certain roadless national forest lands to be added to the National Wilderness System, a process that has embroiled the agency ever since and is discussed in more detail later in this chapter.

An incremental step toward wrestling the agency under control came with the Forest and Rangelands Renewable Resources Planning Act of 1974, Pub. L. No. 93–378, 88 Stat. 476. This law required the Forest Service to prepare a system-wide five-year plan for the national forests, known as the Renewable Resources Program, but also directed the agency to develop "land and resource management plans for units of the National Forest System." Yet the statute provided no direction as to procedure or content of such plans, leaving the agency in a position to continue the forest plan process it had already devised as a means of supporting the clearcutting program.

The next significant blow came in 1975, as the clearcutting age came to a screeching halt when the court in West Virginia Division of Izaak Walton League v. Butz, 522 F.2d 945 (4th Cir.1975), used plain English meanings to interpret the Organic Act's "designated," "marked," "dead," "mature," and "large" terms to prohibit widespread clearcutting in most circumstances. The court simply noted that the Organic Act referred only to "dead, matured, or large growth" trees as eligible for harvesting, and that the statute required the Forest Service to designate and mark the trees before removal. MUSY did not alter that basic starting point, so, the court concluded, clearcutting is illegal. In modern terms, this simply did not "compute" for the agency or the industry. As environmental groups seized the moment and filed suits around the nation to extend the Fourth Circuit's reasoning, the Forest Service and timber industry immediately sought congressional action to clear up what the agency could and could not do with respect to timber extraction policy.

Of course, what the agency sought and fully expected it would receive was legislative nullification of the court's opinion, but by this time criticism of the Forest Service had crept into Congress, focused initially through the so-called Bolle Report, commissioned by Senator Lee Metcalf of Montana, and the Church Commission hearings held in 1971 before the Senate Subcommittee on Public Lands of the Committee on Interior and Insular Affairs. At the request of Senator Metcalf, Arnold Bolle led a team of academics from the University of Montana in 1970 to study Forest Service practices. Their report, entitled *A University View of the Forest Service*, was sharply critical of the Forest Service's land management practices, concluding that the agency overemphasized timber production and thus undermined the multiple use mandate. At the Church Commission hearings the next year, numerous distinguished witnesses testified in those hearings as to the environmental harm Forest Service policies had caused. Amidst the emerging broad attention to environmental affairs that took hold in Congress during the early 1970s, this testimony proved critical in convincing Congress that the agency required more explicit direction. The result was the National Forest Management Act of 1976 (NFMA), Pub. L. No. 94–588, 90 Stat 2949, codified at 16 U.S.C. §§ 1601–14.

Adding to rather than replacing the Organic Act and MUSY, the NFMA prescribed a set of substantive standards and planning requirements for the Forest Service, the details of which are explored in the principal cases included in this chapter. Generally it restricts timber harvests to only those national forest lands where "soil, slope, or other watershed conditions will not be irreversibly damaged" and which could "be adequately restocked within five years after harvest." 16 U.S.C. § 1604(g)(3)(E). In particular, clearcutting and other even-aged management techniques are specifically addressed and restricted by standards which, while loose, were more than had appeared in law previously. *Id.* § 1604(g)(3)(F). Also, making NFMA relevant to the scope of this text, the statute requires the Forest Service to "provide for diversity of plant and animal communities." *Id.* § 1604(g)(3)(B). These and other standards are to be coordinated for each national forest though individual "land and resources management plans" that require public input and are subject to judicial review. *Id.* §§ 1601–1604. Hence, although it is not without its detractors, the NFMA unquestionably charted a new direction for national forest policy, one in which, for the first time in Forest Service history, biodiversity values had to be taken into account.

2. CURRENT SCENE

Although the NFMA provided vastly more detail to guide Forest Service policy than had the Organic Act and MUSY, it left many more questions than it answered. A rich history of litigation helps fill in the details of such issues as where timber can be harvested under the "irreversible damage" standard, when clearcutting is allowed, and whether forest plans have been properly compiled. In this chapter we focus on the "diversity of plant and animal communities" standard in particular, showing how it has been the subject of varying judicial and administrative interpretations and had emerged by the close of the century as a potential force of change in future Forest Service policy.

There are other laws, of course, that play an important role in forest management, both within national forests and on state, local, and private forest lands. In the years preceding enactment of NFMA, Congress passed a flurry of other environmental laws, many with general application to federal agencies and some with general application to the public at large. Although the impact of these laws on forest management was not immediately apparent or felt substantially in the 1970s and 1980s, by the early 1990s it had become clear, primarily through citizen initiated litigation, that a web of federal environmental laws profoundly affects forest management decisions taken by federal, state, local, and even private forest managers. Indeed, as the materials in this chapter illustrate, in many circumstances statutes such as the Endangered Species Act, the National Environmental Policy Act, and the Clean Water Act have had a far more prescriptive effect on Forest Service policy than have the agency's trio of primary statutes—the Organic Act, MUSY, and NFMA. Their effect on state, local, and private forest lands has been no less substantial.

At the same time as the web of environmental laws affecting forest management is being more tightly woven, another trend has emerged to complicate forest policy—the so-called property rights movement. Property rights advocates focus on the Supreme Court's famous statement in Pennsylvania Coal Co. v. Mahon, 260 U.S. 393 (1922), that "while property may be regulated to a certain extent, if regulation goes too far it will be recognized as a taking" and thus violate the Fifth Amendment's protection against taking of private property without just compensation. Sparked by the Supreme Court's renaissance in takings jurisprudence begun in the late 1980s, in which the Court has given some meaning to when environmental and other land use regulations "go too far," takings claims have abounded in federal and state courts and spawned enactment or proposal of private property protection legislation, or "takings laws," in many states as well as in Congress. Indeed, the Endangered Species Act has been a primary target of that effort, though to date with little success. In this chapter we revisit the takings question to illustrate how difficult it may be to carry forest policy as far toward preservation of biodiversity and ecosystem values on private lands as it may be headed on many public lands.

Notes and Questions

1. Why have national forests? If the Forest Service identifies an area of federally-owned forest as most suitable for timber harvesting, why not auction the land—not just the timber—for sale to timber companies? If another area is deemed best suited to recreation, why not sell it to the Disney Corporation? Having paid for the land, won't these private interests have more incentive to manage it appropriately than does the federal government? Indeed, rather than having the federal government decide what use is suitable for particular parcels of forest, wouldn't a better way to determine "highest and best use" be to let the market decide through open auctions? Won't the successful bidder have the most incentive to manage the land in such a way as to maximize its realization of highest and best use, thus maximizing total social wealth? These are not idle questions. For decades federal land policy was one of disposition, not permanent retention. Only recently has the principal issue of federal land management become one of use rather than to whom to dispose and when. *See* GEORGE C. COGGINS ET AL., FEDERAL PUBLIC LAND AND RESOURCES LAW 39–148 (4th ed. 2001). And serious questions still are raised about the wisdom of the permanent retention approach. Some economists, for example, advocate privatizing the national forests to end many of the management issues that plague the Forest Service by replacing the incentives of negative-sum political dynamics with those of positive-sum private entrepreneurship. *See* RICHARD L. STROUP & JOHN A. BADEN, NATURAL RESOURCES: BUREAUCRATIC MYTHS AND ENVIRONMENTAL MANAGEMENT 118–27 (1983). As you delve into the materials on public forest lands, keep in mind that alternative ownership frameworks exist and consider whether they offer superior avenues to ecological management of forested lands.

2. Why have a Forest Service? Consider that the Forest Service has a workforce of over 30,000 employees. The agency is divvied up into 9 regions, within which are distributed 155 national forests, which together are comprised of a total of more than 600 ranger districts ranging in size from 50,000 to 1 million acres. Each district has a staff of 10 to 100 people. Overseeing the whole structure is the Chief, who reports directly to the USDA's Under Secretary for Natural Resources and Environment. Even if we maintain federal public lands as "national forests," are there alternatives we might consider for implementing national objectives for those lands other than through a large, centralized, hierarchical federal agency? Indeed, in his recent book, Professor Robert Nelson has characterized the Forest Service as bloated, anachronistic, and lacking a coherent vision, choosing instead to hop from one fashionable environmental solution to the next—most recently ecosystem management. He advocates abolishing the agency and replacing it with a decentralized system to manage protection of our national forests. *See* ROBERT NELSON, A BURNING ISSUE: A CASE FOR ABOLISHING THE U.S. FOREST SERVICE (2000). Other observers propose models that retain a role for the Forest Service in setting national forest management policy, but vesting greater power in local and private decision making bodies. *See* ROGER A. SEDJO, THE NATIONAL FORESTS: FOR WHOM AND FOR WHAT?,

PERC POLICY SERIES NO. PS–23 (2001). Indeed, in 1987 New Zealand, which at one time managed its national forests under a system explicitly modeled after the Forest Service, replaced its centralized agency with a system of dominant-use management by a conservation agency and a state-owned forestry corporation. *See* Robert L. Fischman and Richard L. Nagle, *Corporatisation: Implementing a Forest Management Reform in New Zealand*, 16 ECOLOGY L.Q. 719 (1989). For a spectrum of views about the role of national forests and the Forest Service, see A VISION FOR THE U.S. FOREST SERVICE (Roger A. Sedjo ed., 2000).

3. Given that, for now and the foreseeable future, we do have national forests and a Forest Service, a few words are merited about the mechanics of one of the agency's principal functions—selling timber. Timber sale contracts are governed by a blend of traditional contract law and statutory and regulatory provisions that alter normal rules of contract interpretation in some circumstances. The Forest Service must publicly advertise all sales, providing information about the quality and age of the timber and methods of bidding and payment. Bidding must be conducted through open and fair competition. The timber sale contract must incorporate the successful bidder's plan of operation designed to ensure orderly harvesting of the timber consistent with the NFMA. At the end of the contract term—usually ten years—the purchaser must replant and take measures to reduce erosion. Rules of contract modification and termination, particularly in the context of changing environmental constraints, have often been tested in the courts. *See Everett Plywood Corp. v. United States*, 651 F.2d 723 (Ct.Cl.1981); *Scott Timber Co. v. United States*, 40 Fed.Cl. 492 (1998). For more on the timber sale contracting process, *See* GEORGE C. COGGINS ET AL., FEDERAL PUBLIC LAND AND RESOURCES LAW 628–55 (4th ed. 2001); Thomas R. Lundquist, *Providing the Timber Supply from National Forest Lands*, 5 NATURAL RESOURCES & ENV'T 6 (Winter 1991). The Forest Service's multiple-use mandate also requires it to administer rules for mineral resource extraction on national forest lands. *See* Lyle K. Rising, *Public Land and National Forest Access for Mining*, 5 NATURAL RESOURCES & ENV'T 16 (Winter 1991); Jan G. Laitos, *Oil and Gas Leasing on Forest Service Lands*, 5 NATURAL RESOURCES & ENV'T 23 (1991).

4. The early history of federal and state forest policy is thoroughly explored in J. CAMERON, THE DEVELOPMENT OF GOVERNMENTAL FOREST CONTROL IN THE UNITED STATES (1972); J. P. KINNEY, THE DEVELOPMENT OF FOREST LAW IN AMERICA (1917); and James L. Huffman, *A History of Forest Policy in the United States*, 8 ENVTL. L. 239 (1978). Excellent summaries of Forest Service policies prior to, leading to, and after the enactment of NFMA are found in MICHAEL J. BEAN AND MELANIE J. ROWLAND, THE EVOLUTION OF NATIONAL WILDLIFE LAW 340–56 (3rd ed.1997); Lawrence Ruth, *Conservation on the Cusp: The Reformation of National Forest Policy in the Sierra Nevada*, 18 J. ENVTL. L. 1 (2000); Frederico Cheever, *Four Failed Forest Standards: What We Can Learn From the History of the National Forest Management Act's Substantive Timber Management Provisions*, 77 OR. L. REV. 601 (1998); and Oliver A. Houck, *On the Law of Biodiversity and Ecosystem Management*, 81 MINN. L. REV. 869, 883–929 (1997).

5. International law has also entered into the domestic law development scene in recent years. In June 1992, at the United Nations Conference on Environment and Development (UNCED) in Rio de Janeiro, participating nations adopted the so-called Rio Declaration to formally endorse the concept of sustainable development and to commit to formulating national plans of action for achieving it. *See* Rio Declaration on Environment and Development, UNCED, U.N. Doc. A/CONF.151/Rev. 1, 31 I.L.M. 874 (1992). UNCED's blueprint for what national sustainable development plans should contain is known as the Agenda 21 document, which covers an array of topics and provides guiding principles of sustainable development. *See* UNCED Agenda 21, U.N. Doc. A/CONF.151.26 (1992). Although no single chapter of Agenda 21 addresses sustainable forestry, the topic is covered in an amalgam of Chapters 10 (management of land resources), 11 (combating deforestation), and 15 (conservation of biological diversity). Also, the Principle 8 of the Rio Declaration urges nations to integrate forest management with large scale land planning and to protect "ecologically viable representative or unique examples of forests." Many of the domestic U.S. forest management initiatives discussed in this chapter have been adopted since Agenda 21 was promulgated. As you review them and the laws under which they were promulgated (most of which were enacted prior to Agenda 21), consider how our domestic law of forest management lives up to the expressed goals of sustainable forestry. For a comprehensive assessment in that regard, notable not only for its breadth of programmatic coverage but also its depth of analysis of each program, see Robert L. Fischman, *Stumbling to Johannesburg: The United States' Haphazard Progress Toward Sustainable Forestry Law*, 32 ENVTL. L. REP. (Envtl. L. Inst.) 10291 (2002).

C. BIODIVERSITY MANAGEMENT IN FOREST ECOSYSTEMS

As the foregoing history of forest law suggests, two difficult sets of issues have plagued U.S. forest policy at all levels of government in recent decades: how much to take into account the biodiversity storage and ecosystem service values of forests in forest management decisions, and how to formulate a national forest policy given the diverse geography of ownership of forested lands. The two sets of issues are, of course, interrelated, as public agencies will have more discretion in establishing a biodiversity-ecosystem values policy for lands they own than for lands that are privately owned.

1. PUBLIC LANDS: NATIONAL FORESTS AND THE BIODIVERSITY MANDATE

As the nation's largest single owner of forest lands, the federal government's decisions about forest management have lasting effects on the biodiversity and ecosystem values of our nation's forests. Of the federal land management agencies, the U.S. Forest Service controls the largest holding of federal forests—192 million acres of land in 42 states, the Virgin Islands, and Puerto Rico—through its jurisdiction over the National Forest System. The system is composed of 155 national forests, 20 national grasslands, and various other lands under the jurisdiction of the Secretary of Agriculture (the Secretary). The vast majority of the national forests acres are located west of Texas and the Great Plains states, though some other states have significant holdings.

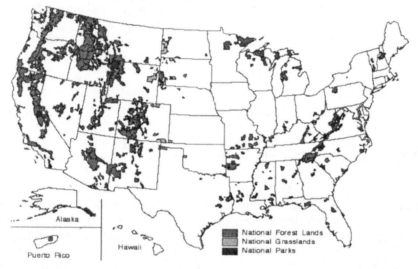

National Forest lands are located predominantly in the western states and Appalachian Mountains
U.S. Forest Service

According to the combined mandates of the Multiple–Use Sustained–Yield Act (MUSY) and the National Forest Management Act (NFMA), the National Forest System lands are to be managed for a variety of uses on a sustained-yield basis to ensure a continued supply of products and services in perpetuity. (Further detail on the MUSY and NFMA mandates is provided in the first principal case presented below). Exactly what that management mandate means has been the subject of intense debate in Congress, the Forest Service, and the courts. The NFMA injects a biodiversity factor into the mix of overlapping and conflicting goals the Forest Service serves, requiring that the agency

> provide for diversity of plant and animal communities based on the suitability and capability of the specific land area in order to meet overall multiple-use objectives, and within the multiple-use objectives of a land management plan adopted pursuant to this section, provide, where appropriate, to the degree practicable, for steps to be taken to preserve the diversity of tree species similar to that existing in the region controlled by the plan.

16 U.S.C. § 1604(g)(3)(B). The Forest Service has implemented that statutory provision, complete with all its escape valves, qualifiers, and sources of discretion, through its regulation at 36 C.F.R. pt. 219.

Pursuant to that regulation, the diversity mandate, as well as all the other concerns the Forest Service must consider under its multiple use mandate, is factored into each national forest's land and resource management plan, or LRMP. Preparation of an LRMP is the first step in resource allocation within a national forest. In the case of timber harvesting, the LRMP outlines where, when, and under what conditions harvesting generally can occur. The Forest Service then authorizes harvesting in particular locations by selecting a timber sale area and preparing an environmental assessment subject to public review and comment. The agency must consider the environmental consequences of each sale and must determine that a decision to sell in a particular area complies with the LRMP. Only then can the agency award a timber harvest contract.

During the 1990s in particular, environmental groups pressed hard on the Forest Service to emphasize this biodiversity side of the agency's forest management mandate and de-emphasize the use of national forests for timber extraction. The groups initiated litigation challenging numerous LRMPs resulting in the trilogy of influential cases presented in this section, through which they tried to force the agency to (1) use a new brand of science known as conservation biology as the guiding light for all forest management decisions; (2) manage forests to ensure the long-term viability of sensitive species above all else; and (3) perform rigorous population viability analyses for forest species before making forest management decisions. Although the suits largely failed in achieving the intended overhaul of Forest Service policy by judicial decree, the effort scored some modest successes in some courts and the relentless pressure the groups placed on the agency, coupled with recommendations from an independent body of experts, had by the end of the decade produced a proposal for change by the agency under the heading of "ecological sustainability." The expert committee's recommendations and the agency's response in the form of a new forest planning rule adopted in late 2000 are excerpted at the end of this section. As a hint of where the agency landed, consider this broad policy description from the final rule:

> The concept of sustainability has become an internationally recognized objective for land and resource stewardship. In 1987, the Brundtland Commission Report (The World Commission on Environment and Development) articulated in "Our Common Future" the need for intergenerational equity in natural resource management. The Commission defined sustainability as meeting the needs of the present without compromising the ability of future generations to meet their own needs. During the last twenty years, the world has increasingly come to recognize that the functioning of ecological systems is a necessary prerequisite for strong productive economies, enduring human communities, and the values people seek from wildlands.

> Similarly, the Forest Service and scientific community have developed the concepts of ecosystem management and adaptive management. Scientific advances and improved ecological understanding support an approach under which forests and grasslands are managed as

ecosystems rather than focusing solely on single species or commodity output. Indeed, ecosystem management places greater emphasis on assessing and managing broad landscapes and sustaining ecological processes. Ecosystem management focuses on the cumulative effects of activities over time and over larger parts of the landscape. Planning and management under ecosystem management also acknowledge the dynamic nature of ecological systems, the significance of natural processes, and the uncertainty and inherent variability of natural systems. Ecosystem management calls for more effective monitoring of management actions and their effects to facilitate adaptive management, which encourages changes in management emphasis and direction as new, scientific information is developed. In accord with ecosystem management, regional ecosystem assessments have become the foundation for more comprehensive planning, sometimes involving multiple forests and other public land management units....

* * *

Taken together, ecosystem management, scientific reviews, and collaboration enable the Forest Service to identify key scientific and public issues and to target its limited resources on trying to resolve those issues at the most appropriate time and geographic scale. Based on these changes in the state of scientific and technical knowledge, the Forest Service's extensive experience, and a series of systematic reviews, the Forest Service has concluded that 36 CFR part 219 must be revised in order to better reflect current knowledge and practices and to better meet the conservation challenges of the future. Indeed, while the 1982 planning rule was appropriate for developing the first round of plans from scratch, it is no longer well suited for implementing the NFMA or responding to the ecological, social, and economic issues currently facing the national forests and grasslands.

65 Fed. Reg. 67514 (Nov. 7, 2000).

Only time will tell if the agency's shift of policy is a sea change or semantics. For a sense of how far the agency has purported to change course, we suggest that you read the three principal cases together (notes and questions are reserved until the conclusion of the trilogy) and then pause to ask what the Forest Service's duty is under the case law with respect to conserving biodiversity. Then read the materials on "ecological sustainability" that follow the cases and ask yourself whether the agency's duty has become more or less clear and enforceable under the new rule.

a. CONSERVATION BIOLOGY SCIENCE

In Chapter 7 of the text we explored the three critical decisions for formulation of ecosystem management policy: (1) choosing the framework for describing ecosystems; (2) selecting a currency for describing policy goals and performance; and (3) adopting a method for implementing the policy goals and measuring policy success. The first case in this trilogy focuses on step one, illustrating how what is at one level purely a matter of scientific discourse—the ecosystem biology principles used in forest management—has profound implications for the Forest Service's multiple use policy decisions.

Sierra Club v. Marita

United States Court of Appeals, Seventh Circuit, 1995.
46 F.3d 606.

FLAUM, CIRCUIT JUDGE:

Plaintiffs Sierra Club, Wisconsin Forest Conservation Task Force, and Wisconsin Audubon Council, Inc. (collectively, "Sierra Club") brought suit against defendant United States Forest Service ("Service") seeking to enjoin timber harvesting, road construction or reconstruction, and the creation of wildlife openings at two national forests in northern Wisconsin. The Sierra Club claimed that the Service violated a number of environmental statutes and regulations in developing forest management plans for the two national forests by failing to consider properly certain ecological principles of biological diversity. The district court determined that the plaintiffs' claims were justiciable but then granted the Service summary judgment on the merits of those claims. We affirm.

I.

The National Forest Management Act ("NFMA") requires the Secretary of Agriculture, who is responsible for the Forest Service, to develop "land and resource management plans" to guide the maintenance and use of resources within national forests. 16 U.S.C. §§ 1601–1604. In developing these plans the Secretary must determine the environmental impact these plans will have and discuss alternative plans, pursuant to the National Environmental Policy Act ("NEPA"), 42 U.S.C. § 4321 et seq. The Secretary must also consider the "multiple use and sustained yield of the several products and services obtained" from the forests, pursuant to the Multiple–Use Sustained Yield Act ("MUSYA"), 16 U.S.C. §§ 528–531. The process for developing plans is quite elaborate. The Service must develop its management plans in conjunction with coordinated planning by a specially-designated interdisciplinary team, extensive public participation and comment, and related efforts of other federal agencies, state and local governments, and Indian tribes. 36 C.F.R. §§ 219.4–219.7. Directors at all levels of the Service participate in the planning process for a given national forest. The Forest Supervisor, who is responsible for one particular forest, initially appoints and then supervises the interdisciplinary team in order to help develop a plan and coordinate public participation. The Supervisor and team then develop a draft plan and draft environmental impact statement ("EIS"), which is presented to the public for comment. 36 C.F.R. §§ 219.10(a), 219.10(b). After a period of comment and revision, a final plan and final EIS are sent to the Regional Forester, who directs one of four national forest regions, for review. If the Regional Forester approves them, she issues both along with a Record of Decision ("ROD") explaining her reasoning. 36 C.F.R. § 219.10(c). An approved plan and final EIS may be appealed to the Forest Service Chief ("Chief") as a final administrative decision. 36 C.F.R. §§ 219.10(d), 211.18.

The final plan is a large document, complete with glossary and appendices, dividing a forest into "management areas" and stipulating how resources in each of these areas will be administered. The plans are ordinarily to be revised on a ten-year cycle, or at least once every fifteen years. 36 C.F.R. § 219.10(g). The present case concerns management plans developed for two forests: Nicolet National Forest ("Nicolet") and Chequamegon (She–WA-me-gon) National Forest ("Chequamegon"). Nicolet spreads over 973,000 acres, of which 655,000 acres are National Forest Land, in northeastern Wisconsin, while Chequamegon encompasses 845,000 publicly-owned acres in northwestern and north-central Wisconsin. Collectively, the Nicolet and the Chequamegon contain hundreds of lakes and streams, thousands of miles of roads and trails, and serve a wide variety of uses, including hiking, skiing, snowmobiling, logging, fishing, hunting, sightseeing, and scientific research. The forests are important for both the tourism and the forest product industries in northern Wisconsin.

A scenic view from the Chequamegon-Nicolet National Forest

Until the mid–1800s, both the Nicolet and Chequamegon were old-growth forests consisting primarily of northern hardwoods. Pine logging around 1900, hardwood logging in the 1920s, and forest fires (caused by clear cutting) significantly affected the landscape. Government replanting and forest-fire control efforts beginning in the 1930s have reclaimed much of the land as forest. The forests now contain a mixture of trees that markedly differs from the forests' pre–1800 "natural" conditions but is also more diverse in terms of tree type and age.

[In the late 1970s and early 1980s, the Nicolet and Chequamegon Forest Supervisors and interdisciplinary teams each began drafting a forest management plan for their respective forests. These plans were expected to guide forest management for ten to fifteen years beginning in 1986. The Regional Forester issued final drafts of both plans on August 11, 1986, as well as final environmental impact statements (FEIS) and RODs explaining the final planning decisions. The Sierra Club brought an action against the Service in the district court on April 2, 1990, over the Nicolet plan and on October 10, 1990, over the Chequamegon plan. Suing under the Administrative Procedure Act (APA), 5 U.S.C. § 701–06, the Sierra Club argued in both cases that the Service had acted arbitrarily or capriciously in developing these forest management plans and FEISs.]

The Sierra Club's primary contention concerned the Service's failure to employ the science of conservation biology, which failure led it to violate a number of statutes and regulations regarding diversity in national forests. Conservation biology, the Sierra Club asserted, predicts that biological diversity can only be maintained if a given habitat is sufficiently large so that populations within that habitat will remain viable in the event of disturbances. Accordingly, dividing up large tracts of forest into a patchwork of different habitats, as the Nicolet and Chequamegon plans did, would not sustain the diversity within these patches unless each patch were sufficiently large so as to extend across an entire landscape or regional ecosystem. See, generally, Reed F. Noss, Some Principles of Conservation Biology, As They Apply to Environmental Law, 69 Chi.-Kent L. Rev. 893 (1994). Hence, the Sierra Club reasoned, the Service did not fulfil its mandates under the NFMA, NEPA and MUSYA to consider and promote biological diversity within the Nicolet and the Chequamegon.

On February 9, 1994, the district court denied the Sierra Club's motion for summary judgment and granted the Service's with regard to the Nicolet. The court held that the Sierra Club had standing to challenge the forest management plan without attacking any specific action under the plan and that the plan was ripe for judicial review. The court then found for the Service on the merits, holding that because of the uncertain nature of application of many theories of conservation biology, the Service had not erred in failing to apply it and so had not violated the NFMA, NEPA, or MUSYA. Sierra Club v. Marita, 843 F.Supp. 1526 (E.D.Wis.1994) ("Nicolet"). The court issued a similar opinion with regard to the Chequamegon plan on March 7, 1994. Sierra Club v. Marita, 845 F.Supp. 1317 (E.D.Wis. 1994) ("Chequamegon"). This consolidated appeal of the two cases followed.

II.

[The court determined that the Sierra Club had standing to challenge the plans and that the claims were ripe for judicial review]

III.

The Sierra Club claims that the Service violated the NFMA and NEPA by using scientifically unsupported techniques to address diversity concerns in its management plans and by arbitrarily disregarding certain principles of conservation biology in developing those plans. The Sierra Club asserts that the Service abdicated its duty to take a "hard look" at the environmental impact of its decisions on biological diversity in the forests on the erroneous contentions that the Sierra Club's proposed theories and predictions were "uncertain" in application and that the Service's own methodology was more than adequate to meet all statutory requirements. According to the Sierra Club, the Service, rather than address the important ecological issues the plaintiffs raised, stuck its head in the sand. The result, the Sierra Club argues, was a plan with "predictions about diversity directly at odds with the prevailing scientific literature."

A.

Several statutes and regulations mandate consideration of diversity in preparing forest management plans. Section 6(g) of the NFMA, the primary statute at issue, directs the Secretary of Agriculture in preparing a forest management plan to, among other things,

> provide for diversity of plant and animal communities based on the suitability and capability of the specific land area in order to meet overall multiple-use objectives, and within the multiple-use objectives of a land management plan adopted pursuant to this section, provide, where appropriate, to the degree practicable, for steps to be taken to preserve the diversity of tree species similar to that existing in the region controlled by the plan[.]

16 U.S.C. § 1604(g)(3)(B).

A number of regulations guide the application of this statute. The most general one stipulates that:

> Forest planning shall provide for diversity of plant and animal communities and tree species consistent with the overall multiple-use objectives of the planning area. Such diversity shall be considered throughout the planning process. Inventories shall include quantitative data making possible the evaluation of diversity in terms of its prior and present condition. For each planning alternative, the interdisciplinary team shall consider how diversity will be affected by various mixes of resource outputs and uses, including proposed management practices.

36 C.F.R. § 219.26. Another regulation addresses the substantive goals of the plan:

> Management prescriptions, where appropriate and to the extent practicable, shall preserve and enhance the diversity of plant and animal communities, including endemic and desirable naturalized plant and animal species, so that it is at least as great as that which would be expected in a natural forest and the diversity of tree species similar to that existing in the planning area. Reductions in diversity of plant and animal communities and tree species from that which would be expected in a natural forest, or from that similar to the existing diversity in the planning area, may be prescribed only where needed to meet overall multiple-use objectives....

36 C.F.R. § 219.27(g); see also 36 C.F.R. § 219.27(a)(5) (requiring that all management prescriptions "provide for and maintain diversity of plant and animal communities to meet overall multiple-use objectives"). Diversity is defined for the purposes of these regulations as "[t]he distribution and abundance of different plant and animal communities and species within the area covered by a land and resource management plan." 36 C.F.R. § 219.3.

Regulations implementing the NFMA with regard to the management of fish and wildlife resources are more specific still. First,

> [f]ish and wildlife habitat shall be managed to maintain viable populations of existing native and desired non-native vertebrate species in the planning area.... In order to ensure that viable populations will be maintained, habitat must be provided to support, at least, a minimum number of reproductive individuals and that habitat must be well distributed so that those individuals can interact with others in the planning area.

36 C.F.R. § 219.19. In order to perceive the effects of management on these species, the Service must monitor the populations of specially selected "management indicator species" ("MIS"). 36 C.F.R. § 219.19(a)(1). The selection of MIS must include, where appropriate, "endangered and threatened plant and animal species" identified on state and federal lists for the area; species with "special habitat needs that may be influenced significantly by planned management programs; species commonly hunted, fished or trapped, non-game species of special interest; and additional ... species selected because their population changes are believed to indicate the effects of management activities on other species ... or on water quality." Id.

The NFMA diversity statute does not provide much guidance as to its execution; "it is difficult to discern any concrete legal standards on the face of the provision." ... However, "when the section is read in light of the historical context and overall purposes of the NFMA, as well as the legislative history of the section, it is evident that section 6(g)(3)(B) requires Forest Service planners to treat the wildlife resource as a controlling, co-equal factor in forest management and, in particular, as a substantive limitation on timber production."

* * *

B.

The Service addressed diversity concerns in the Nicolet and Chequamegon in largely similar ways, both of which are extensively detailed in the district court opinions issued below. See Nicolet, 843 F.Supp. at 1533–40; Chequamegon, 845 F.Supp. at 1322–28. The Service defined diversity as "[t]he distribution and abundance of different plant and animal communities and species within the area covered by the Land and Resource Management Plan." The Service assumed that "an increase in the diversity of habitats increases the potential livelihood of diverse kinds of organisms."

The Service focused its attention first on vegetative diversity. Diversity of vegetation was measured within tree stands as well as throughout the forest, noting that such diversity is "desirable for diverse wildlife habitat, visual variety, and as an aid to protecting the area from wildfire, insects, and disease." The Service assessed vegetative diversity based on vegetative types, age class structure of timber types, within-stand diversity of tree species, and the spacial distribution pattern of all these elements across the particular forest. The Service also factored in other considerations, including the desirability of "large areas of low human disturbance" and amount of "old-growth" forest, into its evaluations. Using these guidelines, the Service gathered and analyzed data on the current and historical composition of the forests to project an optimal vegetative diversity.

The Service assessed animal diversity primarily on the basis of vegetative diversity. Pursuant to the regulations, the Service identified all rare and uncommon vertebrate wildlife species as well as those species identified with a particular habitat and subject to significant change through planning alternatives. The Service grouped these species with a particular habitat type, identifying 14 categories in the Nicolet and 25 (reduced to 10 similar types) in the Chequamegon. For each of these habitat types, the Service selected MIS (33 in the Nicolet and 18 in the Chequamegon) to determine the impact of management practices on these species in particular and, by proxy, on other species in general. For each MIS, the Service calculated the minimum viable population necessary in order to ensure the continued reproductive vitality of the species. Factors involved in this calculation included a determination of population size, the spatial distribution across the forest needed to ensure fitness and resilience, and the kinds, amounts and pattern of habitats needed to support the population.

Taking its diversity analysis into consideration, along with the its numerous other mandates, the Service developed a number of plan alternatives for each of the forests (eight in the Nicolet and nine in the Chequamegon). Each alternative emphasized a different aspect of forest management, including cost efficiency, wildlife habitat, recreation, and hunting, although all were considered to be "environmentally, technically, and legally feasible." In the Nicolet, the Service selected the alternative emphasizing resource outputs associated with large diameter hardwood and softwood vegetation; in the Chequamegon an alternative emphasizing recreational opportunities, quality saw-timber, and aspen management was chosen.

C.

The Sierra Club argues that the diversity statute and regulations ... required the Service to consider and apply certain principles of conservation biology in developing the forest plan. These principles, the Sierra Club asserts, dictate that diversity is not comprehensible solely through analysis of the numbers of plants and animals and the variety of species in a given area. Rather, diversity also requires an understanding of the relationships between differing landscape patterns and among various habitats. That understanding, the Sierra Club says, has led to the prediction that the size of a habitat—the "patch size"—tends to affect directly the survival of the habitat and the diversity of plant and animal species within that habitat.

A basic generalization of conservation biology is that smaller patches of habitat will not support life as well as one larger patch of that habitat, even if the total area of the smaller patches equals the total area of the large patch. This generalization derives from a number of observations and predictions. First, whereas a large-scale disturbance will wipe out many populations in a smaller patch, those in a larger patch have a better chance of survival. Second, smaller patches are subject to destruction through "edge effects." Edge effects occur when one habitat's environment suffers because it is surrounded by different type of habitat. Given basic geometry, among other factors, the smaller the patch size of the surrounded habitat, the greater the chance that a surrounding habitat will invade and devastate the surrounded habitat. Third, the more isolated similar habitats are from one another, the less chance organisms can migrate from one habitat to another in the event of a local disturbance. Consequently, fewer organisms will survive such a disturbance and diversity will decline. This third factor is known as the theory of "island biogeography." Thus, the mere fact that a given area contains diverse habitats does not ensure diversity at all; a "fragmented forest" is a recipe for ecological trouble. On the basis of these submissions, the Sierra Club desires us to rule that to perform a legally adequate hard look at the environmental consequences of landscape manipulation across the hundreds of thousands of hectares of a National Forest, a federal agency must apply in some reasonable fashion the ecological principles identified by well accepted conservation biology. Species-by-species techniques are simply no longer enough. Ecology must be applied in the analysis, and it will be used as a criterion for the substantive results.

As a way of putting conservation biology into practice, the Sierra Club suggested that large blocks of land (at least 30,000 to 50,000 acres per block), so-called "Diversity Maintenance Areas" ("DMAs"), be set aside in each of the forests. The Sierra Club proposed and mapped three DMAs for the Nicolet and two for the Chequamegon. In these areas, which would have included about 25% of each forest, habitats were to be undisturbed by new roads, timber sales, or wildlife openings. Neither forest plan, however, ultimately contained a DMA; the Chequamegon Forest Supervisor initially did include two DMAs, but the Regional Forester removed them from the final Chequamegon plan.

The Sierra Club contends that the Service ignored its submissions, noting that the FEISs and RODs for both the Nicolet and the Chequamegon are devoid of reference to population dynamics, species turnover, patch size, recolonization problems, fragmentation problems, edge effects, and island biogeography. According to the Sierra Club, the Service simply disregarded extensive documentary and expert testimony, including over 100 articles and 13 affidavits, supporting the Sierra Club's assertions and thereby shirked its legal duties.

The Service replies that it correctly considered the implications of conservation biology for both the Nicolet and Chequamegon and appropriately declined to apply the science. The Service asserts that it duly noted the "concern [of the Sierra Club and others] that fragmentation of the . . . forest canopy through timber harvesting and road building is detrimental to certain plant and animal species." The Service decided that the theory had "not been applied to forest management in the Lake States" and that the subject was worthy of further study. However, the Service found in both cases that while the theories of conservation biology in general and of island biogeography in particular were "of interest, . . . there is not sufficient justification at this time to make research of the theory a Forest Service priority." Given its otherwise extensive analysis of diversity, as well as the deference owed its interpretation of applicable statutory and regulatory requirements, the Service contends that it clearly met all the "diversity" obligations imposed on it.

IV.

The case now turns to whether the Service was required to apply conservation biology in its analysis and whether the Service otherwise complied with its statutory mandates and regulatory prescriptions regarding diversity in national forests. We hold that the Service met all legal requirements in addressing the concerns the Sierra Club raises.

* * *

The Sierra Club's arguments regarding the inadequacy of the Service's plans and FEISs can be distilled into five basic allegations, each of which we address in turn. First, the Sierra Club asserts that the law "treats ecosystems and ecological relationships as a separately cognizable issue from the species by species concepts driving game and timber issues." The Sierra Club relies on the NFMA's diversity language to argue that the NFMA treats diversity in two distinct respects: diversity of plant and animal communities and diversity of tree species. See 16 U.S.C. § 1604(g)(3)(B). The Sierra Club also points to NEPA's stipulations that environmental policy should focus on the "interrelations of all components of the natural environment," 42 U.S.C. § 4331, and regulations which require an EIS to include an analysis of "ecological" effects. See 40 C.F.R. § 1508.8. The Sierra Club concludes from these statutes and regulations that the Service was obligated to apply an ecological approach to forest management and failed to do so. In the Sierra Club's view, MISs and population viability analyses present only half the picture, a picture that the addition of conservation biology would make complete.

The Sierra Club errs in these assertions because it sees requirements in the NFMA and NEPA that simply do not exist. The drafters of the NFMA diversity regulations themselves recognized that diversity was a complex term and declined to adopt any particular means or methodology of providing for diversity. Report of the Committee of Scientists to the Secretary of Agriculture Regarding Regulations Proposed by the United States Forest Service to Implement Section 6 of the National Forest Management Act of 1976, 44 Fed. Reg. 26,599, 26,609 (1979). We agree with the district court that "[i]n view of the committee's decision not to prescribe a particular methodology and its failure to mention the principles that plaintiffs claim were by then well established, the court cannot fairly read those principles into the NFMA. . . ." Nicolet, 843 F.Supp. at 1542; Chequamegon, 845 F.Supp. at 1330. Thus, conservation biolog[is not a necessary element of diversity analysis insofar as the regulations do not dictate that the service analyze diversity in any specific way.

Furthermore, the Sierra Club has overstated its case by claiming that MIS and population viability analyses do not gauge the diversity of ecological communities as required by the regulations. Except for those species to be monitored because they themselves are in danger, species are chosen to be on an MIS list precisely because they will indicate the effects management practices are having on a broader ecological community. Indeed, even if all that the Sierra Club has asserted about forest fragmentation and patch size and edge effects is true, an MIS should to some degree indicate their impact on diversity. See Report of the Committee of Scientists, 44 Fed. Reg. at 26,627 (noting that MIS are chosen "because they indicate the consequences of management on other species whose populations fluctuate in some measurable manner with the indicator species"); Judy L. Meyer, The Dance of Nature: New Concepts in Ecology, 69 Chi.-Kent L. Rev. 875, 885 (1994) (noting that the most sensitive indicator of environmental stress is the population level). While the NFMA would not permit the Service to limit its choices to either enhancing diversity or protecting a particular species, see Seattle Audubon Society v. Evans, 952 F.2d 297, 301–02 (9th Cir.1991), such is not the case here. The Sierra Club may have wished the Service to analyze diversity in a different way, but we cannot conclude on the basis of the records before us that the Service's methodology arbitrarily or capriciously neglected the diversity of ecological communities in the two forests.

In a second and related argument, the Sierra Club submits that the substantive law of diversity necessitated the set-aside of large, unfragmented habitats to protect at least some old-growth forest communities. The Sierra Club points out that 36 C.F.R. § 219.27(g) requires that "where appropriate and to the extent practicable" the Service "shall preserve and enhance the diversity of plant and animal communities . . . so that it is at least as great as that which would be expected in a natural forest. . . ." Furthermore, "[r]eductions in diversity of plant and animal communities and tree species from that which would be expected in a natural forest or from that similar to the existing diversity in the planning area[ ] may be prescribed only where needed to meet overall multiple-use objectives." Id. Diversity, the Sierra Club asserts, requires the Service to maintain a range of different, ecologically viable communities. Because it is simply not possible to ensure the survival of any old-growth forest communities without these large, undisturbed patches of land, the Service has therefore reduced diversity. The Service was thus bound to protect and enhance the natural forest or explain why other forest uses prevented the Service from doing so. The Sierra Club believes the Service did neither.

The Sierra Club asserts that the diversity regulations require a certain procedure and that because the substantive result of the Service's choices will produce, in the Sierra Club's view, results adverse to "natural forest" diversity, the Service has violated its mandate. However, as the Service points out, the regulations do not actually require the promotion of "natural forest" diversity but rather the promotion of diversity at least as great as that found in a natural forest. The Service maintains that it did provide for such diversity in the ways discussed above. Additionally, the Service did consider the maintenance of some old-growth forest, even though the Sierra Club disputes that the Service's efforts will have any positive effects. And to the extent the Service's final choice did not promote "natural diversity" above all else, the Service acted well within its regulatory discretion. See Sierra Club v. Espy, 38 F.3d 792, 800 (5th Cir.1994) ("That [NFMA diversity] protection means something less than the preservation of the status quo but something more than eradication of species suggests that this is just the type of policy-oriented decision Congress wisely left to the discretion of the experts—here, the Forest Service."); cf. Methow Valley, 490 U.S. at 350 ("If the adverse environmental effects of the proposed action are adequately identified and evaluated, the agency is not constrained by NEPA from deciding that other values outweigh the environmental costs.").

* * *

V.

The creation of a forest plan requires the Forest Service to make trade-offs among competing interests. See Sierra Club v. Espy, 38 F.3d at 802. The NFMA's diversity provisions do substantively limit the Forest Service's ability to sacrifice diversity in those trades, and NEPA does require that decisions regarding diversity comply with certain procedural requirements. However, the Service neither ignored nor abused those limits in the present case. Thus, while the Sierra Club did have standing to challenge the choices made by the Service, the Service made those choices within the boundaries of the applicable statutes and regulations.

For the foregoing reasons, we affirm the decisions of the district court.

AFFIRMED.

b.   INDICATOR SPECIES

Regardless of whether the Forest Service chooses conservation biology or some other brand of biology as the guiding scientific perspective for ecosystem management, it must identify some "currency" for formulating policy goals and measuring the success of its multiple use decisions. The agency has done so by designating "indicator species" and basing conservation decisions around them, on the premise that by conserving these species the ecosystems within which they exist (and thus the other species in those ecosystems) will be adequately conserved. The selection of the indicator species thus becomes a critical decision in national forest planning, making it subject to tremendous pressure and debate.

Oregon Natural Resources Council v. Lowe

United States Court of Appeals, Ninth Circuit, 1997.
109 F.3d 521.

PER CURIAM

OVERVIEW

Appellants Oregon Natural Resources Council ("ONRC") and other environmental groups filed this action against the United States Forest Service, alleging that the Forest Service failed to comply with the National Forest Management Act ("NFMA") and the National Environmental Policy Act ("NEPA") in developing and amending the Winema National Forest Land and Resource Management Plan ("LRMP" or "Forest Plan").

* * *

FACTUAL AND PROCEDURAL BACKGROUND

In this suit, the ONRC challenges two Forest Service planning decisions relating to the management of old growth forests on the Winema National Forest, located in south-central Oregon. It challenges the Winema LRMP and Amendment 3 to that plan. Both the LRMP and Amendment 3 were developed pursuant to section 6(a) of the NFMA, which directs the Secretary of the Forest Service to develop, maintain, and revise resource plans for units of the National Forest Service. 16 U.S.C. § 1604(a) . . .

The Winema proposed Forest Plan and DEIS were published in December 1987 for a 100–day public review and comment period. On the basis of comments received regarding the DEIS and proposed plan, the Forest Service issued a final Forest Plan and an FEIS in 1990. The Winema Forest Plan breaks the forest into a number of Management Areas ("MAs"), which are characterized by different types of forest and objectives for use. One of these, MA7, is devoted to the provision and maintenance of old growth forest and old growth associated species. The Forest Service designated five [Management Indicator Species (MIS)] associated with old growth forest, which were to be managed so as to ensure general species viability in MA7: the pileated woodpecker, northern goshawk, three-toed woodpecker, pine marten, and northern spotted owl.

* * *

The ONRC and other plaintiffs administratively appealed both the LRMP and Amendment 3. The Chief of the Forest Service denied both appeals. Having exhausted its administrative remedies, the ONRC and six other organizations filed this suit in the district court on September 9, 1992. They alleged that the LRMP and Amendment 3 failed to insure viable populations of old growth associated wildlife, failed to identify adequate old growth MIS, and failed to comply with NEPA. The Thomas Lumber Company and others intervened as defendants. Both the ONRC and the Forest Service moved for summary judgment.

On September 28, 1993, the district court entered a final judgment granting the Forest Service's motion and denying the ONRC's motion, and dismissing the ONRC's complaint. Oregon Natural Resources Council v. Lowe, 836 F.Supp. 727 (D.Or.1993).

* * *

ANALYSIS

I. ONRC's Claims Under the NFMA

* * *

B. Failure to Designate the White–Headed Woodpecker as an MIS

The ONRC argues that in failing to designate the white-headed woodpecker as an MIS, the Forest Service acted arbitrarily and capriciously and left "the most critical and imperiled forest type," old growth ponderosa pine, entirely unprotected. In support of this argument, the ONRC points to an October 1989 letter to the Forest Service Supervisor from a group of Forest Service biologists, including one sent from the Winema National Forest, which concluded that the only animals closely associated with old growth ponderosa pine were the flammulated owl and the white-headed woodpecker. The ONRC also notes that both the Klamath Tribe and the Oregon Department of Fish and Wildlife ("ODFW") criticized the Forest Service's failure to designate the white-headed woodpecker as an MIS in comments on the draft EIS.

The Forest Service argues that the white-headed woodpecker is adequately protected by the [minimum management requirements (MMRs)] for pileated woodpeckers and goshawks and by the forest-wide standards for cavity nesters such as snag retention requirements. Further, it asserts that although neither the pileated woodpecker nor the goshawk are closely associated with old-growth ponderosa pine, there is enough overlap between the habitats of the whiteheaded woodpecker and the goshawk and pileated woodpecker that the Winema LRMP provides adequate protection for the whiteheaded woodpecker. Because the ONRC has not shown that these justifications are arbitrary and capricious, we hold that the Forest Service did not violate the NFMA in failing to designate the whiteheaded woodpecker as an MIS in the LRMP ...

c. SPECIES VIABILITY ANALYSIS

Once the Forest Service has designated the MISs for an area of national forest, it must develop a method of measuring the status of the MIS "currencies" in order to develop and evaluate its multiple use planning. Here again we find ample room for debate over the appropriate method.

Inland Empire Public Lands Council v. United States Forest Service

United States Court of Appeals, Ninth Circuit, 1996.
88 F.3d 754.

CYNTHIA HOLCOMB HALL, CIRCUIT JUDGE:

The United States Forest Service proposed eight timber sales in the Upper Sunday Creek Watershed region of the Kootenai National Forest in northwest Montana. The environmental impact statement it prepared in anticipation of the sales evaluated the project's impact on a number of "sensitive species" living in that region. Plaintiffs, a number of environmental groups, challenged the sale first in administrative hearings and ultimately in district court, claiming that the Service's analysis of the sale's impact on seven species—the lynx, boreal owl, flammulated owl, black-backed woodpecker, fisher, bull charr, and wet-sloped cutthroat trout—was inadequate under both the National Forest Management Act, 16 U.S.C. §§ 1600, et seq., and the National Environmental Policy Act of 1969, 42 U.S.C. §§ 4321 et seq. The district court concluded that the Service's analysis was sufficient and thereafter granted summary judgment for the Service and refused to enjoin the sales. In this expedited appeal, Plaintiffs now argue: (1) that the Service failed to comply with 36 C.F.R. § 219.19, which requires a minimum level of population viability analysis; and (2) that the Service violated the National Environmental Policy Act because the viability analysis it did perform only examined the effect of the timber sales on wildlife populations living within the project boundaries.

* * *

The Kootenai National Forest is a 2.2 million acre tract of land nestled against the Salish Range of the Northern Rockies, in northwestern Montana. The Forest Service ("the Service") completed its stage-one, forest-wide plan for the Forest in 1987 (hereinafter "Kootenai Forest Plan"). Five years later, the Service entertained notions of selling timber from a 28,485 acre area of the Forest known as the Sunday Creek Watershed. By late 1992, the Service refined its plans and proposed eight timber sales from a 12,374 acre tract in the upper portion of the Watershed (hereinafter "Upper Sunday").

* * *

Inland Empire Public Lands Council and other environmental groups (hereinafter "Plaintiffs") filed suit in district court on August 25, 1994. Plaintiffs alleged that the Service's Upper Sunday EIS was deficient and violated both NFMA and NEPA. Plaintiffs first contended that the EIS did not conduct a proper population viability analysis for the seven "sensitive" species living in the area: the lynx, boreal owl, black-backed woodpecker, flammulated owl, fisher, bull charr, and the wet-sloped cutthroat trout. Plaintiffs claimed that the Service fell short of what the NFMA required because it never examined the species' population size, their population trends, or their ability to interact with other groups of the species living in neighboring patches of forest. The district court rejected this argument on summary judgment, reasoning that Plaintiffs were quibbling over the choice of scientific methodologies, a decision to which a reviewing court should defer.

* * *

When the district court denied Plaintiffs' motion for a preliminary injunction to enjoin the timber sales, Plaintiffs filed this expedited appeal.

II. Population Viability Analysis

Plaintiffs first claim that the district court erred in granting summary judgment on their claim that the Forest Service's Upper Sunday EIS violated the National Forest Management Act. We review de novo the district court's grant of summary judgment. Nevada Land Action Ass'n v. United States Forest Serv., 8 F.3d 713, 716 (9th Cir.1993).

As noted above, the NFMA imposes substantive duties on the Forest Service, one of which is the duty to "provide for diversity of plant and animal communities." 16 U.S.C. § 1604(g)(3)(B). Regulation 219.19, one of the many regulations promulgated to ensure such diversity, states in relevant part that:

> Fish and wildlife habitat shall be managed to maintain viable populations of existing native and desired non-native vertebrate species in the planning area. For planning purposes, a viable population shall be regarded as one which has the estimated numbers and distribution of reproductive individuals to insure its continued existence is well distributed in the planning area. In order to insure that viable populations will be maintained, habitat must be provided to support, at least, a minimum number of reproductive individuals and that habitat must be well distributed so that those individuals can interact with others in the planning area.

36 C.F.R. § 219.19. This duty to ensure viable, or self-sustaining, populations, applies with special force to "sensitive" species. Oregon Natural Resources Council v. Lowe, 836 F.Supp. 727, 733 (D.Or.1993) ("The Forest Service has interpreted the viable population provision [Regulation 219.19] as requiring additional attention to certain 'sensitive species.' "); Forest Conservation Council v. Espy, 835 F.Supp. 1202, 1206 (D.Idaho 1993) ("In keeping with [Regulation 219.19], the Forest Service at times designates 'sensitive species' ... within a particular planning area."), aff'd, 42 F.3d 1399 (9th Cir.1994). Because neither party disputes the Service's ultimate obligation to ensure viable populations, the key to this appeal is deciding what type of population viability analysis the Service must perform in order to comply with Regulation 219.19.

Each party suggests its own answer. The Forest Service proposes that its "habitat viability analyses" were sufficient. For four of the species (the black-backed woodpecker, lynx, fisher, and boreal owl), the Service did the following: It consulted field studies that disclosed how many acres of territory an individual of each species needed to survive and the percentage of that acreage that was used for nesting, feeding, denning, etc. (e.g. a lynx needs a 200 acre territory, 20 acres—or 10%—of which must be suitable for denning). The Service then assumed that these percentages would hold true regardless of the size of the individual's territory (e.g. that a lynx would need 10% of whatever acreage of territory it inhabited to be denning habitat). The Service examined each proposed alternative to see how many acres of each type of relevant habitat would remain after the timber was harvested (e.g. Alternative 1 would leave 2,000 acres of denning habitat). It next determined what percentage of the decision area that the remaining types of habitat constituted (e.g. decision area was 10,000 acres so that remaining denning habitat is 20% of the decision area). The Service concluded a species would remain viable as long as the threshold percentage of each type of habitat remaining in the chosen alternative was greater than the percentage required for that species to survive (e.g. the lynx population would remain viable because Alternative 1 left 20% denning habitat and a lynx needs only 10% of its territory to be suitable for denning).

The Service's analysis of the remaining species was not as detailed. As to the flammulated owl, the Service noted that the Upper Sunday area contained 366 acres of habitat suitable for nesting and feeding, enough for three potential owl territories; it noted that the timber sales would reduce the size of one of those territories. As to the bull charr trout, the Service stated that the trout only marginally used the streams within the decision area, but that the timber sales would not appreciably raise the sediment or carbonate levels in those streams.

Plaintiffs contend that the Service's manifold "habitat viability analyses" are insufficient. They argue that Regulation 219.19 also requires the Service to examine: (1) the population of each species; (2) the population dynamics (trends, etc.) of each species; and (3) whether the species could travel between different patches of forest ("linkages"). Plaintiffs claim that their form of analysis is the minimum required by law.

In deference to an agency's expertise, we review its interpretation of its own regulations solely to see whether that interpretation is arbitrary and capricious. 5 U.S.C. § 706(2)(A); Oregon Natural Resources Council v. Marsh, 52 F.3d 1485, 1488 (9th Cir.1995). This is especially true when questions of scientific methodology are involved. Inland Empire Public Lands Council v. Schultz, 992 F.2d 977, 981 (9th Cir.1993) ("We defer to agency expertise on questions of methodology unless the agency has completely failed to address some factor, 'consideration of which was essential to a truly informed decision whether or not to prepare an EIS.'") (citation omitted); see Sierra Club v. Marita ("Marita II"), 46 F.3d 606, 619–20 (7th Cir.1995) (holding that Forest Service's failure to employ "conservation biology" methodology when conducting population viability analysis was not arbitrary or capricious). Thus, we will uphold the Forest Service's interpretation "unless it is plainly erroneous or inconsistent with the regulation." Nevada Land Action Ass'n, 8 F.3d at 717 (citations and internal quotations omitted).

We start, as we must, with the plain language of the Regulation. Idaho First Nat'l Bank v. Commissioner of Internal Revenue, 997 F.2d 1285, 1289 (9th Cir.1993) ("If the [statutory] language . . . is unambiguous, and its literal application does not conflict with the intentions of its drafters, the plain meaning should prevail."). The Regulation specifically provides that the Forest Service may discharge its duties though habitat management as long as "habitat [is] provided to support, at least, a minimum number of reproductive individuals and that habitat [is] well distributed so that those individuals can interact with others in the planning area." 36 C.F.R. § 219.19.

We do not believe that the habitat management analysis conducted in this case for the black-backed woodpecker, lynx, fisher, and boreal owl was in any way "plainly erroneous" or "inconsistent" with this regulatory duty. Regulation 219.19 ultimately requires the Forest Service to maintain viable populations. In this case, the Service's methodology reasonably ensures such populations by requiring that the decision area contain enough of the types of habitat essential for survival. In applying this methodology, the Service recognizes that decision areas are artificial boundaries that change depending on the project at issue, and that the species inhabiting these areas pay no attention to such boundaries.

We recognize that the Service's methodology necessarily assumes that maintaining the acreage of habitat necessary for survival would in fact assure a species' survival. The Service is entitled to rely on reasonable assumptions in its environmental analyses. See, e.g., Sierra Club v. Marita ("Marita I"), 845 F.Supp. 1317, 1331 (E.D.Wis.1994) (finding it permissible to assume that population trends affecting one species in a particular habitat will similarly affect other species in the same habitat), aff'd, 46 F.3d 606 (7th Cir.1995); Greenpeace Action v. Franklin, 14 F.3d 1324, 1335–36 (9th Cir.1992) (finding it permissible for Service to assume that declines in the Stellar sea lion population would be the same for the harbor seal population, given their similarities). We find the above-stated assumption eminently reasonable and therefore do not find that the Forest Service's habitat analyses for the black-backed woodpecker, lynx, fisher, and boreal owl were arbitrary or capricious.

Nor do we believe that the less rigorous analysis performed for the remaining three species—the flammulated owl, the bull charr trout, and the wet-sloped cutthroat trout—was arbitrary and capricious. The Service's failure to engage in a more intensive analysis for the bull charr trout and the wet-sloped cutthroat trout is understandable, as neither species would be affected by the timber sales. See Final EIS at IV:64–67 (noting that the bull charr trout's habitat within the decision area would not be affected by the timber sales); Final EIS at III:37 (listing the 12 sensitive species that "possibly occur[ ]" in the decision area and omitting the wet-sloped cutthroat trout from that list).

The Service's treatment of the flammulated owl is also reasonable. In its EIS, the Service determined that the Upper Sunday decision area contained habitat to support three potential flammulated owl territories and concluded that Alternative E–Modified would shrink the size of the smallest of these territories from 40 to 35 acres. The Service did not engage in a more extended analysis of the owl's nesting and feeding habitat requirements because such data were unavailable. See Richard T. Reynolds & Brian D. Linkhart, "The Nesting Biology of Flammulated Owls in Colorado," Biology & Conservation of Northern Forest Owls 259 (1987) ("It spite of its wide distribution, little is known of the flammulated owl's nesting biology and population status."). We believe that an analysis that uses all the scientific data currently available is a sound one. See Seattle Audubon Soc'y v. Moseley, 80 F.3d 1401, 1404 (9th Cir.1996) (upholding a viability analysis that was "based on the current state of scientific knowledge"). We therefore find no fault with the Service's analysis of the flammulated owl.

* * *

We therefore affirm the district court's conclusion that the Service's population viability analysis was not "arbitrary and capricious."

Notes and Questions

1. Based on the three principal cases excerpted above, what *is* the Forest Service's substantive duty under the MUSY and NFMA with respect to biodiversity and ecosystem management, and how much latitude does the agency have in choosing the method of implementing that duty? If the agency had settled the cases by agreeing to do everything the plaintiffs argued was necessary to comply with the statutes and regulations, could timber industry interests have challenged the agency's action as arbitrary and capricious? Indeed, the outcomes in the cases are of little surprise, as one would be hard-pressed to read the MUSY and NFMA as making clear choices about which science to use, which species to designate as indicator species, and how to measure species viability. Who should make these decisions, Congress, the Forest Service, or the courts?

2. Why all the fuss? How much difference does it make? In other words, how altered would the Forest Service's planning process and decision making outcomes be for the national forests if it (1) had to apply conservation biology; (2) had to adopt more indicator species; and (3) had to conduct population surveys to evaluate indicator species viability? Clearly, the plaintiffs in the cases believe the difference would be substantial, and apparently so did the Forest Service given how steadfastly it opposed the plaintiffs' arguments. But what would be the bottom line impact on Forest Service decisions for national forests?

3. One commentator, critical in general of Forest Service policy with respect to biodiversity protection, has pointed to this line of cases as evidence that the Forest Service "has avoided judicial decisions that would restrict its discretion—by crafting an issue as one of scientific methodology over substantive choice, thereby taking advantage of the arbitrary and capricious standard of review." Greg D. Corbin, *The United States Forests*

Service's Response to Biodiversity Science, 29 Envtl. L. 377, 400 (1999). But why shouldn't the agency seek to avoid judicial decisions that would restrict its discretion? Would this commentator complain about the agency's strategic behavior had the agency adopted the positions the *plaintiffs* advocated and prevailed in litigation brought by timber industry interests? As in many administrative law settings, the combination of Congress delegating broad discretion to the Forest Service with the practice of judicial deference to administrative implementation of that discretion makes for a double-edged sword.

4. The same commentator argues that the *Marita* decision "allows the Forest Service to ... insulate itself from scientific advances in the name of uncertainty." *Id.* at 406. Assuming an agency has discretion to choose the methodology to implement a substantive statutory mandate, when is a "scientific advance" so untested that for the agency to adopt it would be arbitrary and capricious, versus being sufficiently tested to allow the agency to adopt it or not, versus becoming so dominant in the field that the agency *must* adopt it to the exclusion of all other scientific views? The criticism also begs the question of whether conservation biology is so "advanced" that the Forest Service has no choice but to adopt it. Has conservation biology become the *only* science of ecosystem management? Is it even clear from *Marita* what conservation biology is? The leading journal of the discipline, *Conservation Biology*, is in only its 14th volume, so the field cannot be described as seasoned. But it has quickly become a leading force in ecosystem management policy. With its heavy emphasis on preservation of large contiguous tracts of habitat, however, is conservation biology well-suited to the multiple use mandate under which the Forest Service operates?

5. The extensive discretion the Forest Service enjoyed in the *Marita/Lowe/Inland Empire* line of cases was neither an accident nor a recent phenomenon. As a leading expert in national forest law and policy explains, Gifford Pinchot, the progenitor of the national forests and first leader of the Forest Service, worked hard to steer Congress toward a statutory text for the Organic Act that appeared on its surface to impose mandates, but which had no depth of content or sense of direction. In other words, "Pinchot received *carte blanche*." *See* Frederico Cheever, *The United States Forest Service and National Park Service: Paradoxical Mandates, Powerful Founders, and the Rise and Fall of Agency Discretion*, 74 Denv. U. L. Rev. 625, 638 (1997). Congress since then has resisted efforts to enact prescriptive reform legislation, instead expanding the multiple use mandate through the MUSY and adding planning layers through the NFMA.

6. The cases and the preceding notes evidence an inherent tension in national forest policy between use and protection of biodiversity. As Professor Cheever explains, this also was by design, as Pinchot extended the offer of a protection mandate to soften the use mandate that was his central goal in shaping the Organic Act. *See id.* at 631–35. A wealthy man used to acting as a strong leader, Pinchot used the Forest Service's wall of discretion to build agency prestige and pride. But over time the combination of multiple use mandate and extensive discretion became the agency's downfall. The environmental protection movement of the 1970s forged a powerful set of interest groups who clamored for more emphasis of the protection mandate. The agency thus was pulled between its use and protection mandates more forcefully than it had been in the past, though how to strike the balance was no more clear even after the NFMA was enacted. Gradually, each of the multiple uses became increasingly associated with strong interest groups demanding that their use be the dominant use. The agency's discretion then became more a burden than a benefit, for the open-ended statutes "allow those of us interested in public land management to project our vision and values onto the language Congress used to instruct [the agency]. This almost insures that some significant part of the interested public will believe that the [agency's] conduct is not only wrong but illegal." *Id.* at 629. It is no surprise that after 20 years of such battering from both sides, Forest Service prestige and morale have eroded.

7. Several courts have departed from the conclusion in *Inland Empire* that *habitat* viability analysis satisfies the agency's *species* viability evaluation regulations, describing the NFMA as requiring "on-the-ground inventorying and monitoring." *See* Sierra Club v. Martin, 168 F.3d 1, 5–6 (11th Cir.1999); Sierra Club v. Peterson, 185 F.3d 349, 372 (5th Cir.1999), *vacated and remanded on other grounds*, 228 F.3d 559 (2000) (en banc); Forest Guardians v. U.S. Forest Service, 180 F.Supp.2d 1273 (D.N.M.2001). The Tenth Circuit has taken a middle ground, agreeing with *Inland Empire* when there is no substantial ex ante evidence of species presence in the area in question, but suggesting that such evidence could at some point trigger a duty to engage in species surveys and compile "hard population data." *See* Colorado Environmental Coalition v. Dombeck, 185 F.3d 1162, 1169–70 (10th Cir.1999).

8. For further background on the three principal cases and other important NFMA jurisprudence, see Frederico Cheever, *Four Failed Forest Standards: What We Can Learn From the History of the National Forest Management Act's Substantive Timber Management Provisions*, 77 OR. L. REV. 601 (1998); Greg D. Corbin, *The United States Forests Service's Response to Biodiversity Science*, 29 ENVTL. L. 377 (1999); Michael A. Padilla, *The Mouse that Roared: How National Forest Management Act Diversity of Species Requirement is Changing Public Timber Harvesting*, 15 UCLA J. Envtl. L. & Pub. Pol'y 113 (1996–97); Jack Tuholske and Beth Brennan, *The National Forest Management Act: Judicial Interpretation of a Substantive Environmental Statute*, 15 PUB. LAND. L. REV. 53 (1994); Julie A. Weis, *Eliminating the National Forest management Act's Diversity Requirement as a Substantive Standard*, 27 ENVTL. L. 641 (1997); Charles F. Wilkinson, *The National Forest Management Act: The Twenty Years Behind and Twenty Years Ahead*, 68 U. COLO. L. REV. 659 (1997). For an overview of conservation biology accessible to the non-scientist, see RICHARD B. PRIMACK, A PRIMER OF CONSERVATION BIOLOGY (2000).

9. The materials in this text focus primarily on the biodiversity and ecosystem management aspects of the different legal regimes considered. For a broad overview of the full scope of the NFMA, see *The National Forest Management Act: Law of the Forest in the Year 2000*, 21 J. Land, Res., & Envtl. L. 151 (2001). For a thorough description of all that the Forest Service must cover in its land and resource management plans for national forests, see Michael J. Gippert and Vincent L. DeWitte, *The Nature of Land and Resource Management Planning Under the National Forest Management Act*, 3 Envtl. Law. 149 (1996).

d. ECOLOGICAL SUSTAINABILITY

As noted above, the *Marita/Inland Empire/Lowe* trilogy largely stymied the environmental groups' efforts to overhaul Forest Service policy toward biodiversity and ecosystem values through judicial interpretation of the agency's then-existing regulations. Congress during this period was by no means friendly to the groups' positions either, which left only direct pressure on the agency itself as a means to effect change. Ironically, it was there, after decades of fighting in the courts, that the groups had had the most visible success by the close of the 1990s. With the help of the auspicious sounding Committee of Scientists, many of the policies the environmental groups had been pursuing in court surfaced in the form of a proposed regulation and a proposed strategic plan the Forest Service published late in 1999, then a final rule the agency adopted late in 2000. The following excerpt from the Forest Service's final regulation explains the agency's controversial and erratic evolution through the previous decade, which began with promulgation of the first set of planning rules in 1982 and culminated in the recent adoption of a new set of planning rules focused on ecosystem management as a unifying theme:

> The Forest Service has undertaken two systematic reviews of the planning process mandated by the 1982 rules. The first began in 1989, when it conducted a comprehensive review of its land management planning process in cooperation with the Conservation Foundation and Purdue University's Department of Forestry and Natural Resources.... Based in part on this review, the Forest Service published an Advance Notice of Proposed Rulemaking (56 FR 6508, Feb. 15, 1991) regarding possible revisions to the 1982 planning rule. The agency conducted four public meetings to explain and discuss ideas for revising the planning rule; and received comments from over 600 individuals and groups. These comments were used in the development of a proposed rule published on April 13, 1995 (60 FR 18886). However, due to comments received on the 1995 proposed rule and lessons learned from experiences in developing the Northwest Forest Plan, regional assessments, and other regional ecosystem management strategies, the Secretary elected not to proceed with that proposal.

> The second systematic review was undertaken in December 1997, when the Secretary of Agriculture convened a 13–member Committee of Scientists to review the Forest Service planning process and offer

recommendations for improvements within the statutory mission of the Forest Service and the established framework of environmental laws. The members of this Committee of Scientists represented a diversity of views, backgrounds, and academic expertise. The Committee's charter was to "provide scientific and technical advice to the Secretary of Agriculture and the Chief of the Forest Service on improvements that can be made in the National Forest System Land and Resource Management Planning Process and to address such topics as how to consider the following in land and resource management plans: biological diversity, use of ecosystem assessments in land and resource management planning, spatial and temporal scales for planning, public participation processes, sustainable forestry, interdisciplinary analysis, and any other issues that the Committee identifies that should be addressed in revised planning regulations."

* * *

[T]he Committee of Scientists issued a final report on March 15, 1999, entitled Sustaining the People's Lands. The Committee found that, through careful management, National Forest System lands can continue to provide many diverse benefits to the American people in perpetuity. These benefits include clean air and water, productive soils, biological diversity, a wide variety of products and services, employment, community development opportunities, and recreation. The Committee recognized that many Forest Service managers have developed innovative ways to commingle science and collaborative public processes to improve land management decisions, and that these innovative strategies provided a good starting point for developing a more integrated, long-lasting, and flexible planning framework. The Committee concluded that the Forest Service can improve its planning and decisionmaking by relying on the concepts and principles of sustainable natural resource stewardship; by applying the best available scientific knowledge to management choices; and by effectively collaborating with a broad array of citizens, other public servants, and governmental and private entities.

* * *

Based on scientific advances in forestry, forest management and range science, the 1990 Critique of land and resource management planning and the Committee of Scientists' findings and recommendations as contained in its 1999 report, the various laws and regulations that guide National Forest System planning and management, and over 17 years of experience in developing and implementing the existing 127 land and resource management plans, a team of Forest Service employees from national, regional, and local offices, aided by an interagency steering committee, prepared the October 5, 1999, proposed rule to comprehensively revise the land and resource management planning regulation at 36 CFR part 219 (60 FR 18886 Oct. 5, 1999).

65 Fed. Reg. 67,514 (Nov. 9, 2000)

What follows are excerpts from the Committee of Scientists' recommended rule for the ecological sustainability component of the new planning approach it outlined, then additional excerpts from the Forest Service's recent final rule in response.

Committee of Scientists, Sustaining the People's Lands: Recommendations for Stewardship of the National Forests and Grasslands Into the Next Century

March 15, 1999.

CHAPTER SIX

Implementing the Laws and Policies Governing the National Forests and Grasslands in the Context of Sustainability

The previous chapters have developed a framework for management of the national forests and grasslands to achieve ecological, economic, and social sustainability. In this chapter, we apply the concepts from those chapters in suggesting planning principles for implementing the environmental laws and policies under which the Forest Service operates: the National Forest Management Act, Multiple–Use Sustained–Yield Act, Organic Act, Endangered Species Act, Clean Water Act, Clean Air Act, and related legislation. We use the suite of legislation that influences the management of the national forests and grasslands, rather than focus solely on the National Forest Management Act, in keeping with our overall goal of assisting in the development of an integrated planning process.

* * *

The Committee recognizes that its role is not to dictate specific management approaches for the Forest Service but to provide advice that the Secretary and Chief may act on as they deem appropriate. Nonetheless, the Committee recognizes that such concepts as focal species, ecological integrity, and the use of scientific information may involve technical issues and that the Committee thus has an obligation to the Secretary and the Chief to provide some insight on how this framework for ecological sustainability might be converted from concept to application. Therefore, while our approach has not been field-tested, the Committee has drafted the following regulatory language, that, we believe, provides a useful approach to this issue.

Committee's Proposed Regulation on Ecological Sustainability

36 CFR Sec. 219. Ecological Sustainability.

A. Goals. Nature provides many goods, services, and values to humans. These ecological benefits occur as two, major, interdependent forms: the variety of native plants and animals and the products of ecological systems, such as clean water, air, and fertile soil. The most fundamental goal of the National Forest System is to maintain and restore ecological sustainability, the long-term maintenance of the diversity of native plant and animal communities and the productive capacity of ecological systems. Ecological sustainability is the foundation of national forest stewardship and makes it possible for the national forests and grasslands to provide a wide variety of benefits to present and future generations.

B. Diversity. Ecosystems are inherently dynamic; changes regularly result from natural events, such as floods, fires, or insect outbreaks. Human intervention, such as through forest cutting and water diversions, is often substantial. Thus, because species must have the capability and opportunity to respond adaptively to changes in their environment, species diversity and ecological processes can only be sustained if the essential elements of the natural dynamics of ecosystems are recognized and accommodated when human intervention occurs. Planners and managers must apply the best available scientific information and analysis so that the diversity and adaptive capability of ecosystems will be maintained and restored.

1. Levels of diversity. Ecological diversity must be considered at three hierarchical levels: ecosystems, species, and genes, all of which are necessary parts of a strategy to sustain species values and ecological goods and services. Ecosystem diversity, including landscape diversity, is the coarsest level of resolution in this hierarchy. Ecosystem are physical environments and the associated communities of interacting plants and animals. Ecosystem diversity can be described by the variety of components, structures, and processes within an ecosystem and the variety among ecosystem types and functions across broad areas, such as watersheds, landscapes, and regions. Ecosystem diversity provides essential elements for sustaining individual species and the productive capacity of ecosystem. Species diversity refers to variation in the number and relative abundance of species (including subspecies and distinct populations) within a given area. To maintain species diversity, individual species must have the capability and opportunity to respond adaptively to their environment. Genetic diversity, the finest level of resolution in this hierarchy, refers to the degree of variation in heritable characteristics (including life histories) within and among individual organisms and populations.

2. Use of surrogate approaches. Ecological diversity is expressed at a variety of spatial and temporal scales. Explicitly describing and managing all elements of diversity and their interconnections within a single assessment or planning effort is beyond the capacity of the agency. Thus, planners must identify surrogate approaches that rely on a subset of ecological measurements that are sensitive to management and indicative of overall diversity. Although all three levels of diversity are essential to providing ecological sustainability, the most developed scientific knowledge and assessment strategies relevant to broad-scale resource management occur at the ecosystem (especially landscape scales) and species levels. Accordingly, this section primarily addresses ecosystem and species diversity.

C. Ecosystem Diversity. The first step in providing for ecological sustainability is to sustain the variety and functions of ecosystems across multiple spacial scales, from microsites to large landscapes, to maintain the diversity of native plant and animal communities and the productive capacity of ecological systems.

1. Management standards: ecological integrity. The decisions of resource managers must be based upon the best available scientific information and analysis to provide for conditions that support ecological integrity sufficient to meet the goals of this section. The ecological integrity of an ecosystem can be defined as the completeness of the composition, structure, and processes that are characteristic of the native states of that system.... Ecological integrity should be analyzed at appropriate spatial and temporal scales and consider the cumulative effects of human and natural disturbances.

2. Assessment and planning. Measures of ecosystems integrity shall be developed in regional assessments based on scientific principles and knowledge of local conditions. As natural forests and grasslands may comprise only a portion of the landscape under consideration, coordination with other landowners and institutions concerning probable future conditions is critical. Planning documents must explicitly set forth the constraints and opportunities for sustaining ecological systems presented by jurisdictional patterns and varying land-management objectives. In general, in assessing and planning for ecosystem integrity, the planning process must address the larger physical landscape (its historical legacy, its current condition, its biological potential, and its expected changes over successional time) both within and the national forests and grasslands.

3. Validation. The assumption that coarse-filter elements can serve as a basis of sustaining native species diversity shall be validated through monitoring and research. The best available scientific information and analysis shall be used to assess this assumption in a timely manner. If this assumption is invalid, then additional coarse-filter elements will be required, or modification of the coarse-filter approach will be needed, and appropriate management action shall be taken to meet the goals of this section.

D. Species Diversity. A second step in providing for ecological sustainability is to sustain the diversity of native plant and animal communities through maintaining and restoring the viability of the species that comprise them. The goal of this section is to provide the ecological conditions needed to protect and, as necessary, restore the viability of native species.

1. Focal species. The primary obligation in the selection of focal species is to provide for the diversity of native species. However, since it is not feasible to assess the viability of all species, this section will employ focal species to provide for plant and animal diversity. The status of a single species, or group of species, such as a functional guild of species, can convey information about the status of the larger ecological system in which it resides or about the integrity of specific habitat or ecosystem processes. Regional assessments shall select an appropriate number of focal species that represent the range of environments within the planning area, serve an umbrella function in terms of encompassing habitats needed for

many other species, play key roles in maintaining community structure or processes, and are sensitive to the changes likely to occur.

2.   Management standards: species viability. The decisions of resource managers must be based upon the best available scientific information and analysis to provide ecological conditions needed to protect and, as necessary, restore the viability of focal species and of threatened, endangered, and sensitive species. A viable species is defined as consisting of self-sustaining populations that are well distributed throughout the species's range. Self-sustaining populations are those that are sufficiently abundant and have sufficient diversity to display the array of life-history strategies and forms that will provide for their persistence and adaptability in the planning area over time.

3.   Validation. The assumption that focal species are providing reliable information about the status and trend of species not being directly monitored shall be validated through monitoring and research. The best available scientific information and analysis shall be used to assess this assumption in a timely fashion. If this assumption is invalidated for a given focal species, then such focal species shall be augmented or replaced by species that better meet the criteria, and appropriate management action shall be taken to meet the goals of this section.

E.   Implementation. The determinations required regarding ecosystem integrity and species viability shall be made at the appropriate planning level. Decisions at each level must be consistent with such determinations for wide-ranging species are best made at the regional scale. Planners and managers must then demonstrate consistency with this determination in all subsequent decisions made at finer scales of planning, including the project level.

F.   Monitoring. Effective monitoring is a critical aspect of achieving ecological sustainability. Monitoring, which must be an ongoing process, provides a better understanding of how to sustain ecosystems and serves as an "early warning system" to detect declines in ecosystem integrity and species viability before irreversible loss has occurred. The monitoring program must select indicators of ecosystem integrity and species viability, develop methods for measuring such indicators, designate critical indicator values that would trigger changes in management practices, obtain data to determine whether such critical values are being approached, and interpret those data in relation to past and potential management decisions. If analysis and assessment concludes that some critical values are being approached, then the appropriate plan must be reevaluated to determine whether amendments are necessary to comply with the provisions of this section.

G.   Development of Viability Assessment Methods and Conservation Strategies. Regional assessments shall develop methods for assessing ecosystem diversity and species diversity, including methods for assessing ecological integrity and the viability of focal, threatened, endangered, and sensitive species, and shall apply them to estimate the likely condition of ecosystems and species. These assessments shall also propose strategies for use in testing the effectiveness of plans in conserving ecosystem diversity and species diversity.

H.   Evaluation of Plans. The following evaluations shall occur during planning: (1) an evaluation of the plan's capability to provide for the ecological conditions necessary to support ecosystem diversity and species diversity and (2) an independent review, before publication of the plans, by Forest Service and other scientists of the effectiveness of the plan in meeting the goals of this section. The results from this work shall be made available to the public.

———

After public notice and comment on a proposed rule incorporating many of the Committee of Scientists proposals, *See* 64 Fed. Reg. 54073 (Oct. 5, 1999), the Forest Service promulgated the following final rule.

U.S. Forest Service, Department of Agriculture, National Forest System Land and Resource Management Planning; Final Rule

65 Fed. Reg. 67,514 (Nov. 9, 2000).

SUMMARY: This final rule describes the framework for National Forest System land and natural resource planning; reaffirms sustainability as the overall goal for National Forest System planning and management; establishes requirements for the implementation, monitoring, evaluation, amendment, and revision of land and resource management plans; and guides the selection and implementation of site-specific actions. The intended effects are to simplify, clarify, and otherwise improve the planning process; to reduce burdensome and costly procedural requirements; to strengthen and clarify the role of science in planning, and to strengthen collaborative relationships with the public and other government entities.

* * *

Today's Final Rule

Today's final rule will help the Forest Service improve forest planning and on-the-ground management and enable the agency to improve the long-term health of the national forests and grasslands while better meeting the needs of the American people. The final rule affirms ecological, social, and economic sustainability as the overall goal for managing the National Forest System lands and makes the maintenance and restoration of ecological sustainability a first priority for management of the national forests and grasslands so these lands can contribute to economic and social sustainability by providing a sustainable flow of uses, values, products, and services.

* * *

Another key element in the final rule is greater emphasis on the use of science in planning. The final rule requires the use of the best available science to give the Forest Service and the people, communities, and organizations involved in the planning process sound information on which to make recommendations about the resource conditions and outcomes they desire. The final rule incorporates science in the planning and decisionmaking process in a number of ways.

First, the rule recognizes the lessons learned in recent years about developing and analyzing information at the regional ecosystem level. Regional ecosystem assessments have proven to be an extremely valuable and efficient means of understanding the scientific, ecological, social, and economic issues and trends affecting national forests and grasslands and generating baseline data for use in planning and decisionmaking.

Second, consistent with the 1990 Critique and the Committee of Scientists report, the final rule emphasizes monitoring and evaluation of resource conditions and trends over time so that management can be adapted as conditions change. Specifically, the required monitoring and evaluation will assist in determining if desired outcomes are being achieved and how to adapt if they are not. This emphasis is in keeping with NFMA's direction to ensure research on evaluation of the effects of each management system, based on continuous monitoring and assessment in the field, to the end that it will not produce substantial and permanent impairment of the productivity of the land (16 U.S.C. § 1604(g)(3)(C)). As noted by the Committee of Scientists, "Monitoring is a key component of planning * * * Monitoring procedures need to be incorporated into planning procedures and should be designed to be part of the information used to inform decisions. Adaptive management and learning are not possible without effective monitoring of actual consequences from management activities."

Third, the final rule provides for the establishment of science advisory boards to improve decision makers' and planners' access to current scientific information and analysis. It also provides for an independent scientific review of the effectiveness of land management plans in meeting the goal of sustainability during the revision process, and, when appropriate, science consistency evaluations to determine whether the planning process is consistent with the best available science. As the Committee of Scientists observed, "To ensure public trust and support innovation, scientific and technical review processes need to become essential elements of management and stewardship * * * The more that conservation strategies and management actions are based on scientific findings and analysis, the greater the need for an ongoing process to ensure that the most current and complete scientific and technical knowledge is used."

Fourth, the proposed rule affirms the Forest Service's commitment to the viability of all species in accordance with the NFMA requirement to provide for the diversity of plant and animal communities and recognizes the unique contributions national forest and grassland stewardship can make in maintaining species viability. At the same time, the rule recognizes the limits of our scientific understanding and financial and technical capability to conduct viability assessments. To assess the viability of appropriate species of flora and fauna, the rule calls for the use of focal species as indicators of ecological conditions and the best available science and information, including professional opinion and the principles of conservation biology.

Finally, the final rule provides a planning framework that facilitates the identification and responsive resolution to emerging problems. The final rule simplifies required planning steps to enable responsible officials to more readily address emerging issues than is possible under the 1982 rule. For example, the final rule would clarify that, where appropriate, multiple planning activities of one or more national forests or grasslands can be combined along administrative boundaries. Additionally, current requirements for detailed analyses, such as those required for benchmark analyses, would be streamlined or eliminated. Moreover, planning would be done at the most appropriate scale in order to address key issues, and forest and grassland plans and projects would use the same planning framework. The final rule also allows the steps in the planning framework to be coordinated with the scoping requirements under the Forest Service NEPA procedures when appropriate. This will reduce duplication in the preparation of environmental documents associated with management of the National Forest System.

In summary, the final rule will enable the Forest Service to make better decisions about the National Forest System and guide Forest Service planning and management clearly and effectively well into the 21st Century. Grounded in law and experience, the final rule affirms sustainability as the overall goal for national forest and grassland management, requires greater cooperation and collaboration with the public and other private and public entities, and more effectively integrates science into Forest Service planning and management. At the same time, the rule also includes the essential features of National Forest System planning that Chief Gifford Pinchot established almost a century ago and that the Forest Service has used throughout the history of the agency. These features include detailed inventories, monitoring of forest conditions, determination of sustainable levels of uses, and exclusion of uses, where necessary, to protect watershed and other resources (1906 Use Book).

* * *

PART 219—PLANNING

Subpart A—National Forest System Land and Resource Management Planning

* * *

Ecological, Social, and Economic Sustainability

§ 219.19   Ecological, social, and economic sustainability.

Sustainability, composed of interdependent ecological, social, and economic elements, embodies the Multiple–Use Sustained–Yield Act of 1960 (16 U.S.C. 528 et seq.) without impairment to the productivity of the land and is the overall goal of management of the National Forest System. The first priority for stewardship of the national forests and grasslands is to maintain or restore ecological sustainability to provide a sustainable flow of uses, values, products, and services from these lands.

§ 219.20 Ecological sustainability.

To achieve ecological sustainability, the responsible official must ensure that plans provide for maintenance or restoration of ecosystems at appropriate spatial and temporal scales determined by the responsible official.

(a) Ecological information and analyses. Ecosystem diversity and species diversity are components of ecological sustainability. The planning process must include the development and analysis of information regarding these components at a variety of spatial and temporal scales. These scales include geographic areas such as bioregions and watersheds, scales of biological organization such as communities and species, and scales of time ranging from months to centuries.... For plan revisions, and to the extent the responsible official considers appropriate for plan amendments or site-specific decisions, the responsible official must develop or supplement the following information and analyses related to ecosystem and species diversity:

(1) Characteristics of ecosystem and species diversity. Characteristics of ecosystem and species diversity must be identified for assessing and monitoring ecological sustainability....

(i) Ecosystem diversity. Characteristics of ecosystem diversity include, but are not limited to:

(A) Major vegetation types. The composition, distribution, and abundance of the major vegetation types and successional stages of forest and grassland systems; the prevalence of invasive or noxious plant or animal species.

(B) Water resources. The diversity, abundance, and distribution of aquatic and riparian systems including streams, stream banks, coastal waters, estuaries, groundwater, lakes, wetlands, shorelines, riparian areas, and floodplains; stream channel morphology and condition, and flow regimes.

(C) Soil resources. Soil productivity; physical, chemical and biological properties; soil loss; and compaction.

(D) Air resources. Air quality, visibility, and other air resource values.

(E) Focal species. Focal species that provide insights to the larger ecological systems with which they are associated.

(ii) Species diversity. Characteristics of species diversity include, but are not limited to, the number, distribution, and geographic ranges of plant and animal species, including focal species and species-at-risk that serve as surrogate measures of species diversity. Species-at-risk and focal species must be identified for the plan area.

(2) Evaluation of ecological sustainability. Evaluations of ecological sustainability must be conducted at the scope and scale determined by the responsible official to be appropriate to the planning decision. These evaluations must describe the current status of ecosystem diversity and species diversity, risks to ecological sustainability, cumulative effects of human and natural disturbances, and the contribution of National Forest System lands to the ecological sustainability of all lands within the area of analysis.

(i) Evaluation of ecosystem diversity. Evaluations of ecosystem diversity must include, as appropriate, the following:

(A) Information about focal species that provide insights to the integrity of the larger ecological system to which they belong.

(B) A description of the biological and physical properties of the ecosystem using the characteristics identified in paragraph (a)(1)(i) of this section.

(C) A description of the principal ecological processes occurring at the spatial and temporal scales that influence the characteristic structure and composition of ecosystems in the assessment or analysis area. . . .

(D) A description of the effects of human activities on ecosystem diversity. These descriptions must distinguish activities that had an integral role in the landscape's ecosystem diversity for a long period of time from activities that are of a type, size, or rate that were not typical of disturbances under which native plant and animal species and ecosystems developed.

(E) An estimation of the range of variability of the characteristics of ecosystem diversity, identified in paragraph (a)(*l*)(i) of this section, that would be expected under the natural disturbance regimes of the current climatic period. . . .

(F) An evaluation of the effects of air quality on ecological systems including water.

(G) An estimation of current and foreseeable future Forest Service consumptive and non-consumptive water uses and the quantity and quality of water needed to support those uses and contribute to ecological sustainability.

(H) An identification of reference landscapes to provide for evaluation of the effects of actions.

(ii) Evaluations of species diversity. Evaluations of species diversity must include, as appropriate, assessments of the risks to species viability and the identification of ecological conditions needed to maintain species viability over time based on the following:

(A) The viability of each species listed under the Endangered Species Act as threatened, endangered, candidate, and proposed species must be assessed. Individual species assessments must be used for these species.

(B) For all other species, including other species-at-risk and those species for which there is little information, a variety of approaches may be used, including individual species assessments and assessments of focal species or other indicators used as surrogates in the evaluation of ecological conditions needed to maintain species viability.

* * *

(b) Plan decisions. When making plan decisions that will affect ecological sustainability, the responsible official must use the information developed under paragraph (a) of this section. The following requirements must apply at the spatial and temporal scales that the responsible official determines to be appropriate to the plan decision:

(1) Ecosystem diversity. Plan decisions affecting ecosystem diversity must provide for maintenance or restoration of the characteristics of ecosystem composition and structure within the range of variability that would be expected to occur under natural disturbance regimes of the current climatic period in accordance with paragraphs (b)(1)(i) through (v) of this section.

(i) Except as provided in paragraph (b)(1)(iv) of this section, in situations where ecosystem composition and structure are currently within the expected range of variability, plan decisions must maintain the composition and structure within the range.

(ii) Except as provided in paragraph (b)(1)(v) of this section, where current ecosystem composition and structure are outside the expected range of variability, plan decisions must provide for measurable progress toward ecological conditions within the expected range of variability.

(iii) Where the range of variability cannot be practicably defined, plan decisions must provide for measurable progress toward maintaining or restoring ecosystem diversity. The responsible official must use independently peer-reviewed scientific methods other than the expected range of variability to maintain or restore ecosystem diversity. The scientific basis for such alternative methods must be documented in accordance with (§§ 219.22–219.25).

(iv) Where the responsible official determines that ecological conditions are within the expected range of variability and that maintaining ecosystem composition and structure within that range is ecologically, socially or economically unacceptable, plan decisions may provide for ecosystem composition and structure outside the expected range of variability. In such circumstances, the responsible official must use independently peer-reviewed scientific methods other than the expected range of variability to provide for the maintenance or restoration of ecosystem diversity....

(v) Where the responsible official determines that ecological conditions are outside the expected range of variability and that it is not practicable to make measurable progress toward conditions within the expected range of variability, or that restoration would result in conditions that are ecologically, socially or economically unacceptable, plan decisions may provide for ecosystem composition and structure outside the expected range of variability. In such circumstances, the responsible official must use independently peer-reviewed scientific methods other than the expected range of variability to provide for the maintenance or restoration of ecosystem diversity. . . .

* * *

§ 219.21   Social and economic sustainability.

To contribute to economic and social sustainability, the responsible official involves interested and affected people in planning for National Forest System lands (§§ 219.12–219.18), provides for the development and consideration of relevant social and economic information and analyses, and a range of uses, values, products, and services.

(a) Social and economic information and analyses. To understand the contribution national forests and grasslands make to the economic and social sustainability of local communities, regions, and the nation, the planning process must include the analysis of economic and social information at variable scales, including national, regional, and local scales. Social analyses address human life-styles, cultures, attitudes, beliefs, values, demographics, and land-use patterns, and the capacity of human communities to adapt to changing conditions. Economic analyses address economic trends, the effect of national forest and grassland management on the well-being of communities and regions, and the net benefit of uses, values, products, or services provided by national forests and grasslands. Social and economic analyses should recognize that the uses, values, products, and services from national forests and grasslands change with time and the capacity of communities to accommodate shifts in land uses change. Social and economic analyses may rely on quantitative, qualitative, and participatory methods for gathering and analyzing data.

* * *

The Contribution of Science

§ 219.22   The overall role of science in planning.

(a) The responsible official must ensure that the best available science is considered in planning. The responsible official, when appropriate, should acknowledge incomplete or unavailable information, scientific uncertainty, and the variability inherent in complex systems.

(b) When appropriate and practicable and consistent with applicable law, the responsible official should provide for independent, scientific peer reviews of the use of science in planning. Independent, scientific peer reviews are conducted using generally accepted scientific practices that do not allow individuals to participate in the peer reviews of documents they authored or co-authored.

* * *

Notes and Questions

1. As the rule's preamble describes, the Forest Service's November 2000 final rule was the culmination of a decades-long effort by the agency to define its biodiversity protection mandate more clearly than did the 1982 planning rule, but without substantially tying its hands. The genesis of the effort was an informal policy directive known as "New Perspectives in Forestry," which launched a public relations effort designed to convey the agency's approach to biodiversity protection planning. *See* Harold Salawasser, *New Perspectives for Sustaining Diversity in the U.S. National Forest Ecosystems*, 5 CONSERVATION BIOLOGY 567, 567–69 (1991). But "New Perspectives" was quickly, and rightly, criticized as primarily a presentation of vague policy goals with little practical effect. *See* Oliver Houck, *On the Law of Biodiversity and Ecosystem Management*, 81 MINN. L. REV. 869, 923 (1997). By the early 1990s the agency was faced with the reality that the time was due for an overhaul of its 1982 rule.

2. A key ingredient leading to the Forest Service's new forest planning rules was the work of the Committee of Scientists. Congress created the Committee when it enacted the NFMA in 1976 to guide the Forest Service in drafting regulations to implement the Act. *See* 16 U.S.C. § 1604(h)(1). The original seven-member Committee met in eighteen public meetings around the country by the end of 1978 and issued a final report in late 1979. *See Final Report of the Committee of Scientists*, 44 Fed. Reg. 26599 (1979). The Forest Service adopted substantially the Committee's recommendations in September 1979, and reconvened the Committee to assist in the 1982 amendments to the rules. At every stage in this process, the Committee emphasized the importance of science, the central role the diversity protection mandate must play, and the need for specificity. *See* Oliver Houck, *On the Law of Biodiversity and Ecosystem Management*, 81 MINN. L. REV. 869, 887–89 (1997). In an effort to replicate these good experiences, and no doubt seeing the writing on the wall, Agriculture Secretary Dan Glickman appointed a second Committee in 1997 to make recommendations for improving the forest planning process. The thirteen-member Committee of experts was racked by disputes over how much of a role to give ecological goals and how much of a role science should play in shaping policy. *See* Charles C. Mann and Mark L. Plummer, *Call for "Sustainability" In Forests Sparks a Fire*, 283 SCIENCE 1996 (1999). After several drafts, in 1999 the second Committee issued the report excerpted above. Law professor Charles Wilkinson, a member of the second Committee, provides a synopsis of its deliberations and a summary of its perspectives in *A Case Study in the Intersection of Law and Science: The 1999 Report of the Committee of Scientists*, 42 ARIZ. L. REV. 307 (2000). For further background on the development of the Forest Service's ecosystem management policy and the events leading to its new rule, *See* Susan Bucknum, *The U.S. Commitment to Agenda 21: Chapter 11 Combating Deforestation–The Ecosystem Management Approach*, 8 Duke Envtl. L. & Pol'y F. 305, 332–42; Greg D. Corbin, *The United States Forests Service's Response to Biodiversity Science*, 29 Envtl. L. 377 (1999); Heidi J. McIntosh, *National Forest Management: A New Approach Based on Biodiversity*, 16 J. of Energy, Natural Resources, and Envtl. Law 257 (1996).

3. Are the second Committee of Scientists' and the Forests Service's visions of ecosystem management compatible? Both emphasize the importance of ecosystem and species diversity, the use of focal species as a surrogate measurement of ecosystem health, and the management process of monitoring and assessment. What does the Forest Service's rule include that the Committee of Scientists does not discuss in its proposal?

4. The questions in Note 3 assume that the Forest Service's vision of ecosystem management is comprehensible. Summarize in 25 words or less what you believe that vision is. One of the more vocal critics of ecosystem management law and policy does so as follows:

> To summarize, the FS seeks to oversee the national forest system in order to sustain undefined conditions on undefined landscape units that exist in limitless numbers in undefined locations and that are dynamic and constantly changing over time and space in unclear ways. Moreover, there are no standards by which land managers are to measure the undefined landscape conditions that the rule is intended to guarantee.

Allan K. Fitzsimmons, Defending Illusions: Federal Protection of Ecosystems 185 (1999). Does that accurately sum it up? Does your summary suggest there is more precision to the Forest Service's rule? Is it necessary to have more precision in order to accomplish effective ecosystem management? Is it *possible* to have more precision?

5. If the Forest Service's new rule had been in effect in the early 1990s, would the results in *Marita*, *Lowe*, and *Inland Empire* have been different? In other words, would the rule have *required* the agency to use conservation biology science? Would the Forest Service have been *required* to designate the white-headed woodpecker an indicator species (or, in the new parlance, a focal species)? And would the Forest Service have been *required* to conduct population surveys to satisfy the viability analysis requirement? Examine the standards of review the courts applied to those issues. Does the new rule change that? If the answer is that the new rule would not have required the agency to act differently then, what does the new rule change for the future?

6. If the new rule does significantly alter the Forest Service's substantive duties, one question is whether it goes too far. Professor Wilkinson, the legal expert member of the second Committee, has noted that the Committee's recommendations were criticized on the ground that the Forest Service would lack authority to adopt such regulations because they would create a new, single-purpose mission for the agency. Is that a valid concern?

If so, did the Forest Service ameliorate any such problems in its final rules? Based on this response to comments on the 1999 proposed rule, the agency seems to think so:

> Comment: Ecological Sustainability and Compliance with the Multiple–Use Sustained–Yield Act of 1960 and the Organic Administration Act. Many respondents felt that the agency erred in placing ecological sustainability as the first priority. They felt that the agency was ignoring its legislative mandates for multiple-use and had slighted the importance of humans and their needs in the management of National Forest System lands. According to some respondents, changing the emphasis of planning to ecological sustainability would virtually make it impossible to comply with the MUSYA. They were concerned that the MUSYA requirement, to ensure a continued supply of products and services in perpetuity, would be jeopardized. Additional public comments expressed concern that provisions of the Organic Administration Act of 1897 could not be achieved with ecological sustainability as the primary objective.
>
> Response: The proposed rule's focus on sustaining ecosystems is fully compatible with the Forest Service's underlying statutes. In order to ensure that the multiple-uses can be sustained in perpetuity, decisions must be made with sustainability as the overall guiding principle. Ecological sustainability lays a necessary foundation for national forests and grasslands to contribute to the economic and social needs of citizens. . . .
>
> It is the Department's view that the rule is consistent with the Forest Service's conservation and legislative mandates. Contrary to some comments received, the proposed rule did not change the overarching purpose for planning. Rather, it affirmed the direction in the MUSYA. As used in the final rule, sustainability embodies the congressional mandates of multiple-use and sustained-yield without impairing the productivity of the land. In the final rule, sustainability is described as comprising three intricately linked elements that integrate the ecological, social, and economic aspects of our world. It is virtually impossible to separate one element from the other. . . .
>
> Under the Organic Administration Act of 1897, the forest reserves were set aside and protected from exploitation, with the intention to embrace a system of sustainable forest reserves that would protect water resources and ensure a continuous supply of timber for the benefit of the American public.

65 Fed. Reg. 67,514, 67,520–21 (2000). If what the agency says is true about how ecosystem management fits into the MUSY and NFMA mandates, has it left itself any room for turning back from ecosystem management? On the other hand, in its environmental assessment of the rule's impacts, the agency concluded that

> [p]romulgation of the final rule would not result in any immediate changes in the management of any particular national forest or grass-

land or in activities permitted or conducted on those lands. Thus, the adoption of the final rule would not have a direct impact on the quality of the human environment. However, future implementation of the final rule on individual national forests or grasslands could affect decisions that are made for those lands.

USDA, Forest Service, Finding of No Significant Impact For the National Forest System Land and Resource Management Planning Final Rule (July 21, 2000). Does this suggest that the rule dictates nothing substantive, is purely procedural?

7. If the answers to the questions posed in Notes 5 and 6 do not jump out at you from the text of the new rule, you are not alone. When asked to opine on the new rule, a panel of experts on forest policy representing a diverse array of perspectives produced a startling spectrum of opinions on what the rule means, its legal authority, and its practical effect. Views ranged from that of one environmental activist concerned that the Bush administration will use the Clinton administration's rule to *increase* timber output, to the view that the new rule illegally subjugates the multiple-use mandate to a dominant use regime of ecological integrity. *See The Forum*, THE ENVIRONMENTAL FORUM, May/June 2001, at 60. Indeed, a coalition of environmental organizations filed litigation challenging the rule as falling short of what they perceived is mandated in the NFMA on behalf of the ecological sustainability objective, *see* Citizens for Better Forestry v. U.S. Forest Service (N.D. Cal.), while a coalition of timber and grazing interests filed a separate action arguing essentially the opposite, *see* American Forest and Paper Association v. Veneman (D.D.C.).

8. The Forest Service contends in the portion of the preamble to the final rule excerpted above that "sustainability is described as comprising three intricately linked elements that integrate the ecological, social, and economic aspects of our world." Consistent with that theme, the agency published its *Draft USDA Forest Service Strategic Plan* (2000 Revision) in November 1999, in which it identifies "ecosystem health" and "multiple benefits for people" as two of the agency's four strategic goals and describes its mission statement as "to sustain the health, diversity, and productivity of the Nation's forests and grasslands to meet the needs of present and future generations." The *Draft Strategic Plan* details four objectives to meet its goal of promoting ecosystem health and conservation:

> Improve and protect watershed conditions to provide the water quality and quantity and the soil productivity necessary to support ecological functions and intended beneficial water uses.

> Increase the amount of habitat capable of sustaining viable populations of all native species and support desirable levels of selected species.

> Increase the amount of forests and rangelands restored or maintained in a healthy condition with reduced risk and damage from fires, insects and diseases, and invasive species.

Increase collaboration with, and the participation of, a greater diversity of people and members of underserved and low-income populations in planning and implementing programs and activities.

Who would argue with these as policy goals? Do they provide any meaningful standards by which to measure agency success in meeting the objectives? Why are "multiple benefits" and "productivity" never too far away from "ecosystem health" in the Forest Service's descriptions of its forest management policy? Recall the MUSY and NFMA statutory mandates.

9. *Epilogue.* Surely you noticed that the agency's new final rule was published in the last days of the Clinton administration. In addition to the litigation filed almost immediately thereafter (see note 7 above), Congress and the incoming Bush administration took many shots at the new rule. In May 2001, the Forest Service, under new direction, extended for one year the date specified in the rule by which all land and resource management plans must comply with its terms. The agency concluded that it is "not sufficiently prepared to fully implement the rule agencywide," and that the additional time would allow the agency to resolve "serious concerns [that] have arisen regarding some of the provisions of the new planning rule." 66 Fed. Reg. 27552 (May 17, 2001). No revised rule had been issued at the time of this writing.

Note on the Endangered Species Act in National Forests

As the three principal cases illustrate, at the heart of Forest Service ecosystem management practice are *planning* and *assessment*. The NFMA, through land resource management plans, and the National Environmental Policy Act (NEPA), through environmental assessments and impact statements, impose significant process duties on the agency and result in countless written forest plans and environmental impact assessments. The Forest Service's recent forest planning rule promulgation illustrates that the agency's movement toward ecosystem management builds on that core, expanding planning and assessment deep into the agency's environmental decision making process. In the final analysis, however, planning and assessment duties do not dictate substantive outcomes, and a multiple use mandate remains the background for all of the agency's planning focus.

Dissatisfied with the agency's environmental policy direction, and often turned away at the courthouse when making their case as a challenge to agency planning and assessment under the NFMA and NEPA, advocates of an "environment-first" policy for the national forest have seized on the Endangered Species Act (ESA) as a way to cut through the multiple use muddle. The listing as endangered of two remarkably unadaptable species of birds—the Northern Spotted Owl and the Marbled Murrelet—played center stage in court battles that raged across western forests in the 1990s, and yet a third bird—the Red Cockaded Woodpecker—served that position in southeastern forests. With the additional listing of several endangered runs of salmon in rivers of the Pacific Northwest running through the heart of several national forests, it is no exaggeration to suggest that by the end of the 1990s the ESA had brought the Forest Service to its knees.

The reason the ESA could prove so effective for the environmental advocates, in precisely the settings where the NFMA and NEPA had not, has to do with the ESA's fundamentally different structure. The ESA is anything but a multiple use planning statute. As the materials in Chapter 5 explain, federal agencies face numerous substantive constraints under the ESA which they cannot plan around or balance with countervailing policy objectives. Section 9(a)(1) of the ESA prohibits any person, including federal agencies, from taking a protected species, which can include destruction of the species' habitat. *See* Babbitt v. Sweet Home Chapter of Communities for a Great Oregon, 515 U.S. 687 (1995). Section 7(a)(2) of the ESA prohibits federal agencies from jeopardizing the continued existence of protected species through actions the agency carries out, funds, or authorizes, which includes the "granting of licenses, contracts, easements, rights-of-way, permits, or grants-in-aid." 50 C.F.R. 402.02. This "no jeopardy" constraint is one of the most unyielding provisions of environmental law. *See* Tennessee Valley Authority v. Hill, 437 U.S. 153 (1978).

Timber management plans and timber sale contracts fall squarely into these two regulatory proscriptions. Notwithstanding that a timber sale contract and all the Forest Service's planning leading up to it may comply in all procedural and substantive respects with the NFMA and NEPA, the ESA adds an additional layer of requirements that fall outside the agency's multiple use mandate and planning discretion. In short, if the harvesting authorized in a timber sale contract may result in take of a listed species, the Forest Service must defend its decision to let the contract under not only the NFMA and NEPA, but also the ESA. *See* Murray Feldman, *National Forest Management under the Endangered Species Act*, NAT. RESOURCES & ENV'T, Winter 1995, at 32.

Beginning in the 1980s, environmental advocates started using this added factor of the ESA to challenge Forest Service decisions about specific projects in national forests. *See, e.g.*, Thomas v. Peterson, 753 F.2d 754 (9th Cir.1985) (challenge to construction of road in Nez Perce National Forest in Idaho). Following some success in that setting, ESA cases began taking on broader forest management planning decisions in particular forests, such as the decision to use even-aged timber management. *See* Sierra Club v. Yeutter, 926 F.2d 429 (5th Cir.1991). By the early1990s, the focus of the ESA litigation effort had expanded to challenging large-scale programmatic decisions regarding entire national forest plans and even national forest management policies affecting large regions and many national forests. *See, e.g.*, Resources Ltd., Inc. v. Robertson, 35 F.3d 1300 (9th Cir.1994) (challenge to forest plan for the Flathead National Forest). The grand slam came in a series of cases enjoining timber sales in the range of the spotted owl for failure to satisfy the procedures for fulfilling the "no jeopardy" restriction of the ESA. *See, e.g.*, Seattle Audubon Soc'y v. Espy, 998 F.2d 699 (9th Cir.1993).

With the stakes this high, the Clinton Administration responded in 1993 with the Northwest Forest Plan (NFP), an ecosystem management planning process designed to produce land management plans for 24.5 million acres in nineteen national forests covering the range of owl habitat. Against challenges from environmental and industry groups, the federal courts upheld FEMAT's compromise. *See* Seattle Audubon Society v. Lyons, 871 F.Supp. 1291 (W.D.Wash.1994), *aff'd*, 80 F.3d 1401 (9th Cir.1996). A similar effort followed in owl habitat further to the south, in the national forests throughout California's Sierra Nevada region, through an effort known as the Sierra Nevada Ecosystem Project (SNEP). *See* Lawrence Ruth, *Conservation on the Cusp: The Reformation of National Forest Policy in the Sierra Nevada*, 18 J. ENVTL L. 1, 58–81 (1999–2000). Simultaneously, President Clinton directed the Forest Service to develop a plan for national forests in the Columbia River Basin to avoid the same conflicts in salmon habitat that prompted the need for the NFP in owl country. Suits involving the salmon had already had an impact on timber sales in some forests, so the writing was on the wall that attacks on a more programmatic level were to come. *See* Pacific Rivers Council v. Robertson, 854 F.Supp. 713 (D.Or.1993), *aff'd in part, rev'd in part*, 30 F.3d 1050 (9th Cir.1994) (enjoining timber sales while ESA compliance procedures for salmon were underway). The agency's resulting Interior Columbia Basin Ecosystem Management Project (ICBEMP) eventually reached over 70 million acres of federal lands including thirty-two national forests. *See also* Susan Bucknum, *The U.S. Commitment to Agenda 21: Chapter 11 Combating Deforestation–The Ecosystem Management Approach*, 8 DUKE ENVTL. L. & POL'Y F. 305, 332–42 (1998) (summarizing other regional applications).

Nevertheless, few observers of national forest policy are surprised that the NFP and ICBEMP spawned more litigation than they avoided. Indeed, the ESA continues to play a central role in national forest policy in the affected areas, with courts maintaining the role of policing timber sales, forest plans, and national forest policy in general. *See, e.g.*, Pacific Coast Federation of Fishermen's Associations v. National Marine Fisheries Service, No. 1757R (W.D. Wash. Dec. 7, 2000) (enjoining 150 Forest Service timber sales for failure to satisfy ESA consultation procedures). While it has by no means entirely displaced the multiple-use mandate, the ESA's single-use focus has squeezed its way into national forest policy, in many cases completely overshadowing and controlling the NFMA multiple use planning and NEPA environmental impact assessment processes.

Consider, for example, how the NFP controls Forest Service planning decisions. One of the key components of the NFP is the Aquatic Conservation Strategy (ACS), a comprehensive plan designed to maintain and restore the ecological health of the waterways in the federal forests. There are four components to the ACS: (1) key watersheds (the best aquatic habitat, or hydrologically important areas), (2) riparian reserves (buffer zones along streams, lakes, wetlands and mudslide risks), (3) watershed analysis (to document existing and desired watershed conditions), and (4) watershed restoration (a long-term program to restore aquatic ecosystems and watershed health). The ACS also has binding standards and guidelines that restrict certain activities within areas designated as riparian reserves or key watersheds. Additionally, ACS has nine objectives designed to maintain or restore properly functioning aquatic habitats.

When a timber sale or other project is proposed for the NFP region, it is initially subject to the Forest Service's internal planning process under the NFMA. The Forest Service then creates a team of biologists and other resource management specialists to incorporate the NFP requirements, including ACS standards and guidelines. A biologist on the team uses a Matrix of Pathways and Indicators (the "MPI") and a checklist developed by The National Marine Fisheries Service (NMFS) to assess the project's effect on listed species. The MPI and checklist help the biologist to analyze 18 different habitat indicators and determine whether they are properly functioning, at risk, or not properly functioning. The biologists also determine whether the proposed action is likely to restore, maintain, or degrade the indicator. Projects that receive either zero or only one degrade checkmark are considered "not likely to adversely affect" listed species.

Those projects determined "likely [to] adversely affect" listed species, i.e., those that received one or more degrade checkmarks, are referred to a Level 1 Team. This team is made up of biologists from various agencies. It reviews the proposed project for ACS consistency. The team can suggest changes in the plan to bring it into ACS compliance.

If the Level 1 Team agrees that the project complies with ACS, it then forwards the project to the NMFS for formal consultation. Otherwise, the team elevates the review to a Level 2 Team, and the project undergoes the same review process. Failure to reach a consensus elevates the project to a Level 3 Team. Once one of these three teams approves the project, it goes to NMFS for ESA consultation. The NMFS must review the project pursuant to Section 7 of ESA, which, as described in Chapter 5 of your text, requires federal agencies to "insure that any action authorized, funded, or carried out by such agency ... is not likely to jeopardize the continued existence of" any species listed as threatened or endangered under the ESA. 16 U.S.C. § 1536(a)(2). Finally, NMFS issues a Biological Opinion for Forest Service use in going forward. *See* Pacific Coast Federation of Fishermen's Associations v. NMFS, 253 F.3d 1137 (9th Cir.2001).

Drawing from this history and current experience, Professor Oliver Houck suggests that the ESA has thus served to "convene the meeting and draw a bottom line. It has acted as the therapist for conduct we all knew was harmful and had limits, but could not bring ourselves to admit was a problem, much less begin to solve." Oliver Houck, *On the Law of Biodiversity and Ecosystem Management*, 81 MINN. L. REV. 869, 959 (1997). But along with this "muscle" comes fewer policy options and more litigation to enforce the "bottom line," leading another commentator to suggest that the ESA has "reduced agencies' flexibility to deal with forest health in the integrated fashion that ecosystem management envisioned." Rebecca W. Watson, *Ecosystem Management in the Northwest: "Is Everybody Happy?"*, 14 NATURAL RESOURCES & ENV'T 173, 178 (2000).

Has the ESA become the tail wagging the NFMA dog? Is the tradeoff between regulatory muscle and policy flexibility inevitable? If not, how are the two harmonized? If the tradeoff is inevitable, which approach serves ecosystem management objectives more effectively over the long run?

Note on Roads and Roadless Areas in National Forests

Most of the preceding discussion of ecosystem management in national forests has focused on how the Forest Service has adapted its timber harvest policies to changes in law and administrative goals. Yet the multiple-use mandate means, of course, that activities other than timber harvesting will take place in national forests with frequency. Hardrock mining, energy resource extraction, grazing, water uses, recreation, and conservation are among the other uses competing for access to national forest lands, and must necessarily be a component of any national forest plan based on an ecosystem management philosophy. Construction and maintenance of roads in national forests is a critical infrastructure issue that dictates in large part which of these uses can claim a stake in the multiple-use competition for a given area of national forest.

The Forest Service maintains over 370,000 miles of roads in national forests, which carry 9000 Forest Service vehicles, 15,000 timber harvest vehicles, and 1.7 million vehicles involved in recreation each day. Yet vast areas of national forests have long been legally off limits to roads—lands designated as wilderness under the Wilderness Act of 1964 and subsequent congressional designations amount today to 35 million acres within which roads, motor vehicles, motorized equipment, and mechanical transport are prohibited. 43 U.S.C. § 1133. Rather, the roads debate has focused on the much larger area of national forest lands—over 58 million acres, or 31 percent of the national forest lands—that are not legally off limits to roads as designated wilderness areas, but have yet to be opened to road access. Over 33 million of those acres consist of blocks of over 5000 acres of contiguous, undisturbed forest land, eight million acres of which are classified as suitable for timber production. In short, there is a lot of land potentially up for grabs in the national forests, the fate of which depends on whether new roads are constructed into their presently undisturbed interiors.

Debate over these so-called "de facto wilderness" lands has been brewing for decades. In 1967, the Forest Service engaged in a Roadless Area Review and Evaluation (RARE I) to evaluate the suitability of roadless areas for wilderness designation. RARE I inventoried the 5000–acre blocks of roadless non-wilderness lands, but accomplished little more. Litigation over the fate of the inventoried roadless areas resulted in an injunction against timber harvesting until the Forest Service could complete environmental impact review under NEPA. *See* Wyoming Outdoor Coordinating Council v. Butz, 484 F.2d 1244 (10th Cir.1973). The Forest Service published an EIS on the inventoried areas in 1973, selecting about 12 million acres for further study as potential wilderness areas. In 1977, the new Administration geared up another study, known as RARE II, which defined the divide between non-wilderness and potential wilderness areas for over 60 million identified acres of national forests. Subsequent litigation again successfully enjoined the Forest Service from changing the character of designated non-wilderness areas without NEPA review. *See* California v. Block, 690 F.2d 753 (9th Cir.1982). After that blow, the Forest Service began a new review, known unoriginally as RARE III. In 1984, however, Congress intervened in the seemingly endless series of RAREs with 19 bills designating a total of nine million acres as wilderness protected under the Wilderness Act. This nevertheless left unclear the status of tens of millions of roadless areas RARE II had designated as nonwilderness, as well as those areas RARE II had designated as potential wilderness but which Congress had not so declared. *See, e.g.,* Smith v. United States Forest Service, 33 F.3d 1072 (9th Cir.1994) (holding that Forest Service must consider character of area as 5000–acre roadless area in EIS authorizing timber sale).

In 1998, the Forest Service sought once again to bring closure to the roadless area issue by proposing to temporarily suspend road construction and reconstruction in the roadless areas while devising a new policy for their future. Following a series of agency proposals and presidential directives, the Forest Service promulgated a final rule on January 11, 2001 designed "to protect and conserve inventoried roadless areas on the National Forest System lands." 66 Fed. Reg. 3244 (Jan. 12, 2001). The agency identified eight key values and amenities the roadless areas convey more strongly and efficiently than other lands in the national forests:

- High quality or undisturbed soil, water, and air.
- Sources of public drinking water.
- Diversity of plant and animal communities.
- Habitat for threatened, endangered, proposed, candidate, and sensitive species and for those species dependent on large, undisturbed areas of land.
- A place for primitive, semi-primitive non-motorized, and semi-primitive motorized classes of dispersed recreation (e.g., mountain bikes).
- Reference landscapes for research.
- Natural appearing landscape with high scenic value.
- Traditional cultural properties and sacred sites.

To protect these values, the rule prohibits new road construction and reconstruction in inventoried roadless areas on national forest lands, except in limited circumstances. The rule also substantially restricts cutting, sale, and removal of timber in inventoried roadless areas.

The rule and its timing—less than ten days before the inauguration of President George W. Bush—engendered tremendous controversy, with many Republican lawmakers vowing to fight it in every possible forum. Not surprisingly, the timber industry declared the rule the beginning of the industry's demise and filed litigation to block the rule. In May 2001, a federal district court in Idaho preliminarily enjoined USDA from implementing the roadless area rule because the agency failed to follow necessary procedures for environmental impact assessment and public participation. *See* Kootenai Tribe of Idaho v. Veneman (D. Id.), and Idaho v. U.S. Forest Service (D. Id.). Although the Bush administration had previously stated its commitment to the overall objective of conservation in the roadless areas, the Forest Service had already begun a process of revisiting the rule. *See* 66 Fed. Reg. 8899 (Feb. 5, 2001) (delaying effective date of the rule by 60 days, until May 12, 2001). The court's injunction accelerated that process: in July 2001, the agency announced that it "is studying whether to amend the Roadless Area Conservation Rule ... or to provide further administrative protections" and invited public comment on that decision. *See* 66 Fed. Reg. 35918 (July 10, 2001). Thereafter, in August and December 2001 the agency promulgated for public comment a series of "interim directives" governing roadless areas, both of which signaled that a retreat from the Clinton Administration rule is likely. *See* 66 Fed. Reg. 44590 (Aug. 1, 2001); 66 Fed. Reg. 65796 (Dec. 20, 2001). No revised final rule had been issued at the time of this writing.

Regardless of how the litigation and administrative examination of the roadless areas rule turn out, the basic gist of the roadless areas rule fits seamlessly with the Forest Service's new (but suspended) national forest planning regulation and the *Draft Strategic Plan* as another manifestation of the agency's movement during the 1990s toward ecosystem management as a unifying policy. If the Bush Administration adheres to this body of policy and to the specific directives of the roadless area rule, the relevant question for the litigation will be whether the rule is within the agency's discretion. Does anything in the NFMA and the cases interpreting it suggest that the agency cannot legally put these areas off limits to roads provided it has jumped through the proper planning and procedure hoops?

For more on the background and status of the roadless areas rule, check out the Forest Service's special website on the topic at www.roadless.fs.fed.us. For a more thorough discussion of the history and mechanics of the RARE processes, see GEORGE C. COGGINS ET AL., FEDERAL PUBLIC LAND AND RESOURCES LAW 1178–98 (4th ed. 2001); Mary Katherine Ishee, *Review and Management of Roadless Lands in Wilderness Planning*, 5 NATURAL RESOURCES & ENV'T 3 (Winter 1991). For information on the ecological effects of roads on forest ecosystems in general, see *Special Section: Ecological Effects of Roads*, 14 CONSERVATION BIOLOGY 16 (2000).

Note on Ecosystem Management in State–Owned Forest Lands

The materials in this section of the chapter focus on federal administration of forest policy in national forests. As the previous historical overview explained, however, many states have also had a long history of administering forest policy on state-owned lands. Lest we leave the impression that state forest policies are unimportant to the question of how to describe the law of forest ecosystem management, a few words on state-owned forest lands are merited here.

All states have established agencies charged with duties similar to those of the Forest Service, and many states own substantial areas of land designated as state forests or devoted to similar purposes. In total, state and local governments own almost 70 million acres of forest land, or 8.75 percent of total national forested land. While this pales by comparison to federal and private holdings, much of the state and local forest lands provide valuable recreational opportunities or were acquired specifically to preserve sensitive resources.

Most states, however, have adopted forest management policies that put their forest management agencies in the same predicament as the Forest Service—having to satisfy a multiple-use mandate on public lands while responding to the increased focus on ecosystem management for protection of biodiversity. Florida, for example, maintains a state forest system covering over 830,000 acres in 36 state forests, as well as significant additional public land holdings devoted to forestry management. The Division of Forestry, an arm of the Florida Department of Agriculture and Consumer Services, oversees these lands under the following mandate:

> The Division of Forestry shall provide direction for the multiple-use management of forest lands owned by the state; serve as the lead management agency for state-owned land primarily suited for forest resource management; and provide to other state agencies having land management responsibilities technical guidance and management plan development for managing the forest resources on state-owned lands managed for other objectives. Multiple-use purposes shall include, but is not limited to, water-resource protection, forest-ecosystems protection, natural-resource-based low-impact recreation, and sustainable timber management for forest products.

Fla. Stat. 589.04(3). Other state agencies with oversight of Florida's state public lands must prepare a multiple-use analysis for tracts over 1000 acres, which must assess

> the feasibility of managing timber resources on the parcel for resource conservation and revenue generation purposes through a stewardship ethic that embraces sustainable forest management practices if the lead resource agency determines that the timber resource management is not in conflict with the primary objectives of the parcel. For purposes of this section, practicing sustainable forest management means meeting the needs of the present without compromising the ability of future generations to meet their own needs by practicing a land stewardship ethic which integrates the reforestation, managing, growing, nurturing, and harvesting of trees for useful products with the conservation of soil, air and water quality, wildlife and fish habitat, and aesthetics.

Fla. Stat. 253.036. Do these provisions give a clearer mandate to the Division of Forestry than the MUSY and NFMA give the Forest Service?

Does the Florida Division of Forestry enjoy less discretion in determining how to implement the "stewardship ethic" than does the Forest Service in its implementation of the NFMA's biodiversity principle?

2.   PRIVATE LANDS: INTEGRATED APPROACHES

Private lands make up over 57 percent of forest cover and 73 percent of the nation's timberland forests, divided among over six million owners. In eastern states the ratio between public and private forests is even more unbalanced. Florida, for example, has 830,000 acres in state forests and another two million acres of public lands devoted to forest management, but over 13 million acres, or *90 percent* of the state's forested land and 42 percent of the state's *total* land area, is privately-owned timberland. Although the lion's share of that land is held by nonindustrial private forest (NIPF) landowners, the forest industry controls about one-third. Forestry is among the state's major agricultural industries, contributing over $8 billion to the state economy annually. The leading product is nothing glamorous—plywood.

Clearly, private forest lands in states such as Florida present tremendous potential as economic and environmental resources. Yet there is no national forest management law other than the rules governing forests on federal public lands. The question, therefore, is how to integrate private forest lands into a coordinated forest ecosystem management policy without running directly into the same multi-objective problems faced on public lands and, more to the point, without running afoul of constitutional limitations on regulation of private land. This section addresses those two concerns.

a.   THE POLICY GRAB BAG

In June 2000, Maryland, Pennsylvania, Virginia, the District of Columbia, the Chesapeake Bay Commission, and several federal agencies entered into a renewed Chesapeake Bay Agreement, a multi-government compact designed to manage the resources of the Chesapeake Bay and its watershed of 64,000 square miles. The first Bay Agreement, signed in 1983, was "a huge step forward in the restoration of the Bay. It brought together the principal leaders in the watershed, committed them to action, and created an organized structure dedicated to the systematic and scientific analysis of the Bay's problems and the refinement of a science-based, consensus driven plan for the Chesapeake's cleanup." Harry R. Hughes and Thomas W. Burke, Jr., *The Cleanup of the Chesapeake Bay: A Test of Political Will*, NAT. RESOURCES & ENV'T, Fall 1996, at 30, 31. A second agreement, signed in 1986, was more specific and established a set of goals in six areas: living resources, water quality, population, public information, public access, and governance. The second agreement also adopted measurable performance standards and placed political responsibility, and authority, in the hands of the elected leaders of the state signatories. *See id.* at 31–33. The new Chesapeake Bay Agreement adds to that structure, among other things, a commitment to take actions that will "promote the expansion and connection of the contiguous forests" in the Bay watershed.

Forests remain the primary land cover in the Bay watershed, covering 24.1 million acres of the watershed's 41.2 million acres in the three signatory states. Over 80 percent of that forested area is in private ownership. The parcel size of the forested land is decreasing over time, NIPF ownership is increasing, and the patches of forest are increasingly surrounded by urban land uses. Clearly, therefore, any effort to expand and connect contiguous forests must enlist the cooperation of a multitude of private forest land owners or be doomed to failure. The question is, how?

To answer that question, the Environmental Law Institute (ELI) recently laid out a multi-faceted plan of integrated policy approaches for forest management in the Chesapeake Bay watershed. *See* ENVTL. L. INST., FORESTS FOR THE BAY (2000). Like many proposals evaluating options in similar circumstances around the nation, ELI gravitated to five basic policy tools: (1) tax reform; (2) voluntary conservation programs; (3) land and development rights acquisition; (4) targeted subsidies and incentives; and (5) conventional regulation. Each of these tools itself offers many options, and finding the right configuration for each and blend of all for private forest land conservation will prove to be at least as challenging as has been the experience for designing public forest policy. *See* Lee P. Breckenridge, *Reweaving the Landscape: The Institutional Challenges of Ecosystem Management for Lands in Private Ownership*, 19 VT. L. REV., 363 (1995) (covering the northern forests of New England and New York). Consider how ELI and others have proposed using these tools in the private forest land context.

Tax Reform. Forests and forest lands are taxed at federal, state, and local levels in a variety of ways that do not always favor conservation and stewardship goals. For example, a small private commercial timber operation faces several tax policies that run counter to conservation goals:

- Federal capital gains taxes penalize investment in growing and holding older growth timber resources, because the taxed inventory value increases over time with no revenue yield until the timber is cut.

- Annual local property taxes, while lower in most states than for developed land, impose a carrying cost on land value that places pressure on the landowner to generate revenues through timber harvests.

See CONSTANCE BEST AND LAURIE A. WASHBURN, AMERICA'S PRIVATE FORESTS 109–112 (2001). Reversing these and other tax policy perversions is a chief goal of private forest conservation advocates. Indeed, the three Chesapeake Bay states have programs in place to reduce the property tax burden on private forest land owners who meet certain conditions. For example, Maryland's Forest Conservation Management Agreement program allows land owners to agree not to develop forest land for nonforest uses for 15 years and thereby freeze property valuation assessments at $100/acre for the duration of the agreement term. Md. Code Ann., Tax–Property 8–209. Many other states have similar property tax reduction programs. *See* FORESTS FOR THE BAY, *supra*, at 16–20.

While removing these tax barriers would go a long way toward relieving anti-conservation pressures, ELI and others have proposed using incentive-based tax policy to *promote* private forest land conservation, such as tax deductions or credits for investment in replantings and forest management practices. Virginia, for example, allows a state income tax credit for owners of forest land who forego timber harvesting along rivers and streams. Va. Code 58.1–512. The federal tax code also allows a blend of tax credits and deductions for qualifying reforestation expenses. Internal Revenue Code 631(b). ELI recommends that such programs link the tax benefits to preparation of and adherence to forest management plans, the costs of which would also benefit from a tax credit. *See* FORESTS FOR THE BAY, *supra*, at 20–26.

Voluntary Conservation Programs. Tax policy reform may remove disincentives and create incentives for conservation of privately-owned forest land, but in neither case does it direct the land owner's behavior. Other government and private programs also attempt to lead private landowners in the direction of selected forest conservation goals without commanding them to do so. Many of these programs are designed to link landowners and timber operators with viable markets, or to increase efficiency through cooperative efforts. For example, the Maine Low Impact Forestry Project consists of loggers, foresters, and land owners practicing and promoting low impact forestry who have cooperatively worked to develop and access markets for sustainably harvested forest products. *See* <http://www.acadia.net/hcpc/>. Similar sustainable forestry cooperatives operate in a number of states. *See* FORESTS FOR THE BAY, *supra*, at 75–80. Many forest products companies provide free forest management advice to assist NIPFs located within the relevant "woodshed" supplying mills and other manufacturing centers in forest management planning. Large organizations of NIPFs and other commercial forest operations, such as the American Forest Foundation and the National Woodland Owners Association, also provide valuable repositories of information about forest management practices and offer regular educational programs. *See* AMERICA'S PRIVATE FORESTS, *supra*, at 146–48.

Timber and Wood Product Certification Programs. An emerging nongovernmental voluntary program that taps more overtly into market forces involves efforts to "certify" timber and finished wood products as ecologically "friendly." This is part of a larger trend in which businesses commit to meet environmental standards set by an independent certifying body, banking on the consuming public to pay premiums for the positive effects associated with purchase of certified products. *See* Errol E. Meidenger, *Environmental Certification Programs and Environmental Law: Closer than You May Think*, 31 ENVTL. L. REP. (Envtl. L. Inst.) 10162 (2001). For forest products, the Forest Stewardship Council (FSC) is an international, non-profit, non-governmental organization founded in 1993 to act as the certifying body. The FSC has issued its Principles and Criteria for Forest

Management, a set of certification standards developed by highly respected individuals representing FSC's expertise in environmental, economic, and social factors involving forests and forest management. *See* http:// www.fscus.org. Criteria 6.3 states that "Ecological functions and values shall remain intact, enhanced, or restored." Using this and other criteria, FSC approves regional certification standards tailored to the type of forests in the region and accredits certifiers to work in that region. Two FSC accredited certification organizations, SmartWood and Scientific Certification Systems, are located in the United States and have issued management certificates to forest operations with combined forest holdings of more than 4.8 million acres. Forest products coming from FSC certified timber operations, which is determined through "chain-of-custody" certification, can carry the FSC product label. A similar program is the American Forest and Paper Association's Sustainable Forest Initiative, the ultimate goal of which is to have 100 percent of the loggers supplying material to the Association's member companies verify their adherence to a set of sustainable forestry standards and principles. *See* http://www.afandpa.org/iinfo/iinfo.html. Other programs like these are described in Forests for the Bay, *supra*, at 67–69. Some conservation biologists remain skeptical, however, of the efficacy of these programs toward promotion of their version on ecosystem management. *See* Conservation Forum, *Timber Certification*, 15 Conservation Biology 308 (2001) (collection of articles critiquing timber certification programs).

Land and Development Rights Acquisition. One of the more obvious policy options for ecosystem management in private forest lands is to convert them to public forest lands through direct acquisition by the government. For example, the Forest Service administers the Forest Legacy Program, through which state foresters can receive generous federal funding matches for acquisition of "environmentally important forest areas that are threatened by conversion to nonforest uses." 16 U.S.C. § 2103(c). *See* http:// www.fs.fed.us/loa/coop. The Chesapeake Bay Agreement states all have state funded programs for acquisition of threatened forest land as well. *See* Forests for the Bay, *supra*, at 32–36. Maryland's Rural Legacy program, for example, has acquired woodlands for the protection of watersheds, streams, and wetlands. *See* http://www.dnr.state.md.us/rurallegacy.html.

A variant of direct acquisition is the purchase of a conservation easement that limits the landowner's development options for the land. It may not be economically or socially desirable, or environmentally necessary, to move all or most of a region's private timber land into public preserve status. Where private forestry practices are consistent with ecosystem management objectives, the main conservation threat may be the potential for conversion to urban nonforest uses. Buying those development rights through purchase of conservation easements allows the landowner to retain ownership and continue deriving economically productive use of the land, but precludes the potential for deleterious future uses. Such conservation easement payments can also be conditioned on the landowner's use of prescribed sustainable forest practices, thus enhancing the value of continued forest use of the land. *See* Forests for the Bay, *supra*, at 38–39.

Either of the direct acquisition or conservation easement options can be accomplished by private entities as well, or through cooperation of public governments and private organizations. Indeed, the largest conservation easement in U.S. history was announced in 1999, when the Pingree family agreed to sell development rights to over 760,000 acres of forest land in northern Maine to the New England Forestry Foundation. The Foundation raised $28 million in private funding over the next two years to finance the acquisition, and closed the deal in March 2001. The Pingrees, which had amassed a reputation for environmentally sensitive timber management, retained the right to continue limited timber harvests pursuant to guidelines spelled out in the easement, and the public retained recreational access to much of the land (a longstanding Maine custom), but conversion to nonforest uses was taken off the table through the purchase of the development rights. *See* http://www.neforestry.org.

Targeted Subsidies and Incentives. An extrapolation of the tax incentive policies discussed above is the use of more aggressive subsidy and incentive funding options designed to promote forest conservation. The Forest Service's Forest Stewardship Program, for example, provides technical support through state foresters to landowners who develop Forest Stewardship Plans designed to manage private forests for timber, wildlife, watersheds, and other benefits. 16 U.S.C. § 2103a. Cost-sharing support for development and implementation of the plans is available through the Forest Service's Stewardship Incentive Program. 16 U.S.C. § 1603b. *See* http://www.fs.fed.us/loa/coop. The Chesapeake Bay Agreement states and many other states offer more generous incentives for reforestation and riparian forest preservation. *See* FORESTS FOR THE BAY, *supra*, at 46–49. The California Forest Improvement Program, for example, provides cost-share assistance to private forest landowners (20—5000 acres) to develop management plans, replant, improve timber stands, improve fish and wildlife habitat, and engage in forest conservation management practices. *See* http://www.fire.ca.gov/ResourceManagement/CFIP.asp

Conventional Regulation. All of the policy options discussed above rely on some relatively passive mechanism—the market, peer pressure, financial incentives—to lead private forest land owners toward ecosystem management goals the government or some private entity has selected. The most obvious option available for ecosystem management in private forest contexts, however, is plain vanilla government regulation. Federal laws such as the Endangered Species Act and Clean Water Act clearly have significant regulatory effects on private forest lands. *See* Jan S. Pauw and James R. Johnston, *Habitat Planning Under the ESA on Commercial Forestlands*, 16 NATURAL RESOURCES & ENVIRONMENT 102 (2001). Other than through the incidental effects of such environmental conservation or pollution control laws, however, there is no body of federal forestry law for private lands. By contrast, Maryland's Forest Conservation Act, Md. Code Ann., Nat. Res., 5-1601 et seq., requires each unit of local government having planning and zoning authority to adopt a forest conservation program applicable to subdivision developments. Plans must provide for forest retention and reforestation generally, and for forest retention in sensitive areas. Most states, however, do not have nearly as aggressive regulation of development activities as does Maryland, although the combination of land use, zoning, and environmental quality regulatory authorities in many states provides a foundation for such regulation. *See* FORESTS FOR THE BAY, *supra*, at 83–99.

Some states also regulate beyond the question of conversion of forest land to nonforest uses by regulating forestry practices on private lands. The variety of approaches, however, is remarkable. Within the Chesapeake Bay Agreement states, for example, Virginia and Pennsylvania have no law directly covering private forest management and harvesting, while Maryland has enacted a relatively comprehensive effort in its Forest Conservancy District Law. Md. Code Ann., Nat. Res. 5–601 et seq. The Maryland law allows the state to "administer forest conservation practices on privately owned forest land" through rules implemented by appointed forest conservancy district boards. *See* FORESTS FOR THE BAY, *supra*, at 61–64. Moreover, like the federal system, most states have environmental protection laws that incidentally regulate forestry practices. *See id.* at 64–67. The multiple-use dilemma that plagues public forest land policy is absent from this private forestry regulation context of course, but the challenge of prescribing and proscribing private forestry practices is nonetheless challenging. The principal case in the following section provides an example of state regulation in this vein.

Notes and Questions

1. How confident are you that ecosystem management goals like those Grumbine and other conservation biologists have prescribed can be achieved through the non-regulatory tools discussed above? Virginia is an example of a state that has pressing private forest issues, little regulatory authority on the books, and a moderate commitment to voluntary and incentive programs. To more effectively address the goals of the Chesapeake Bay Agreement, should Virginia (a) increase regulation of private forestry and conversion of forest land to nonforest uses, or (b) increase subsidies and incentives, or (c) increase both but to a lesser degree?

2. Assuming you were charged with drafting a state law to comprehensively regulate private forestry practices and conversion of forest land to nonforest uses, how useful would you find the Forest Service's recent series of policy initiatives as a guide for what to do with respect to private lands? Would you ban new roads in forested areas? Would you require private landowners to restore private forests to some reference point of ecological composition or process integrity?

3. In your law, how useful would the ecosystem services concept be to designing regulations or subsidies and incentives? Whom would you vest with ownership of the services? The public? The forest owner? The recipient of the service values? What would you do with the ecosystem service "property" rights once you assigned ownership? Could they be traded, sold, and banked?

4. For more information on the Chesapeake Bay Program, primarily its aquatic resources, see Chapter 11 *infra*. For additional discussion of the various regulatory and other policy tools available to advance private forest ecosystem management, see Robert L. Fischman, *Stumbling to Johannesburg: The United States' Haphazard Progress Toward Sustainable Forestry Law*, 32 ENVTL. L. REP. (Envtl. L. Inst.) 10291, 10299–10305 (2002).

b. DEALING WITH THE TAKINGS ISSUE

The Forest Service's proposed ecological sustainability policy may trigger debate over the extent to which it satisfies the agency's statutory mandate, but one thing is clear: Congress has tremendous latitude is setting forest policy on federal lands, and can be expected to continue to delegate considerable discretion to the Forest Service to implement that policy. Indeed, 6 percent of all U.S. public forest lands—47 million acres—is reserved from commercial timber harvest in wilderness, parks, and other land classifications. And there is no constitutional mandate to prevent further reservation on a larger scale. If Congress wanted to make all federally-owned forests a huge biodiversity reserve, it could. State and local governments have similar latitude when formulating policy for forest lands they own. As the following case illustrates, however, there are limits to how far federal and state governments could make the same policy go with respect to private lands.

Boise Cascade Corporation v. State of Oregon

Oregon Court of Appeals, 1999.
991 P.2d 563.

DE MUNIZ, PRESIDING JUDGE:

The state appeals from a jury verdict in favor of plaintiff Boise Cascade (Boise) on its claim for a temporary taking of a stand of timber in which a pair of northern spotted owls were nesting. For the following reasons, we reverse and remand.

In 1988, Boise acquired 1,770 acres of commercial timberlands in Clatsop County and conducted some logging activities on its property. Also in 1988, the Oregon Department of Fish and Wildlife designated the northern spotted owl as a threatened species. In 1990, the State Forester adopted an administrative policy precluding timber harvesting within a 70–acre area around known spotted owl nesting sites, ultimately codified as *former* OAR 629–24–809.

In 1991, Boise sold all of those commercial timberlands except for a 64–acre parcel (the Walker Creek site), which the buyer refused to accept due to the presence of a northern spotted owls' nest on the site. The present dispute centers around the state's refusal to permit logging on the Walker Creek site during the period that the spotted owls were nesting there.

A spotted owl had been seen on the Walker Creek site in 1986, and a pair of spotted owls nested on the site in 1990, hatching two offspring.

Another spotted owl was seen on the site in 1991. A breeding pair was present on the site in 1992. In early 1992, Boise sought approval from the State Forester of its plan to harvest the timber on the site. The State Forester did not approve Boise's harvesting plan because the plan did not identify for protection 70 acres of suitable spotted owl habitat encompassing the nesting site at Walker Creek. The Board of Forestry upheld the denial of Boise's plan on the ground that the proposed plan failed to comply with *former* OAR 629-24-809. A subsequent plan permitted Boise to log several acres of the Walker Creek site but only during time frames when no owls were nesting on the site.

Boise initiated this action for inverse condemnation, arguing that the refusal to permit it to log the Walker Creek site constituted a taking under Article I, section 18, of the Oregon Constitution, as well as under the Fifth Amendment, as applied to the states through the Fourteenth Amendment, of the United States Constitution. Boise further alleged that the restriction on logging the other acres during the owl nesting season was a temporary taking under both constitutions. The trial court dismissed the complaint. On appeal, we reversed, *Boise Cascade Corp. v. Board of Forestry*, 131 Or.App. 538, 886 P.2d 1033 (1994), and the Supreme Court allowed review. On review, the Supreme Court affirmed in part and reversed in part, concluding that, although Boise failed to state a claim for a temporary taking of the small amount of timber that the Board of Forestry permitted to be logged, Boise did state a claim for a taking of the remainder of the Walker Creek site. *Boise Cascade Corp. v. Board of Forestry*, 325 Or. 185, 935 P.2d 411 (1997).

On remand, Boise dropped its claim under the Oregon Constitution and proceeded only on its federal constitutional claim. Boise moved for partial summary judgment, and the trial court ruled as a matter of law that a regulatory taking had occurred. The question of damages was tried to a jury, as was a question as to whether a taking by "physical invasion" had occurred. Meanwhile, one of the spotted owls on the Walker Creek site had died and the other had left the site, and all restrictions on logging the site were lifted. The jury returned a verdict for Boise, and the trial court entered judgment for Boise in the amount of $2,279,223 in damages for the temporary restriction on its logging of the Walker Creek site. This appeal ensued.

The state ... argues that the trial court erred in denying its motion to dismiss for failure to state a claim under either of two theories: the *Lucas* theory (deprivation of all beneficial use of property) and the *Loretto* theory (physical occupation of property). *See generally Lucas v. South Carolina Coastal Council, 505 U.S. 1003 (1992); Loretto v. Teleprompter Manhattan CATV Corp., 458 U.S. 419 (1982)*. To state a claim under the Fifth Amendment for a taking under a *Lucas* theory, the property owner must allege that a governmental action has deprived the owner of all economically beneficial use of the property. Both this court and the Oregon Supreme Court concluded that plaintiff had stated a claim under this type of theory in the course of the previous appeal. *See Boise Cascade*, 131 Or App at 551 ("plaintiff alleges, in essence, that the government has regulated its property in such a way that productive uses are unavailable and all viable economic and beneficial use has been eliminated. Those allegations suffice to state regulatory taking claims under * * * *Lucas*"); *Boise Cascade*, 325 Or at 198 (Plaintiff's allegations were "sufficient to meet the 'deprivation of all economically viable use of the property' standard. The Court of Appeals was correct in so holding."). Although plaintiff amended its pleadings after remand, the amended pleadings, insofar as the *Lucas* theory of recovery is concerned, are much the same as its pleadings discussed in the previous appellate decisions in this case. Plaintiff alleged that the state has regulated its property in such a way that all viable economic and beneficial use of the property was eliminated. The trial court properly denied the state's motion to dismiss on the ground that plaintiff failed to state a claim under the *Lucas* theory.

* * *

The state argues that the trial court erred in striking its defense that Boise's proposed logging would constitute a nuisance and that the state cannot be liable for refusing to permit Boise to perform acts that constitute a nuisance and violate the law. In *Lucas*, the Court noted that there would be no taking if "the proscribed use interests were not part of [the property owner's] title to begin with." 505 US at 1027. The Court further noted that such limitations "must inhere in the title itself, in the restrictions that background principles of the State's law of property and nuisance already place upon land ownership." *Id.* at 1029. The state's defense at issue here appears to rest on this part of the *Lucas* case.

The state offers no authority for the proposition that knocking down a bird's nest on one's property has ever been considered a public nuisance. The case on which it relies, *Columbia Fishermen's Union v. St. Helens*, 160 Or. 654, 87 P.2d 195 (1939), concerned a suit by fishermen to enjoin the city of St. Helens from dumping raw sewage into the Columbia River—a practice which, needless to say, was detrimental to the fish population on which the fishermen depended for their livelihoods. That case concluded that the fisherman had standing to maintain an action against the city. *Id.* at 666.

However, any analogy to the present case is less than clear. The court in *Columbia* indicated that the state could protect its navigable waters from pollution because it had an interest in *ferae naturae* (the fish) in the waters. It does not follow, as the state seems to posit, that any act taken by the state to protect *ferae naturae* on private property is the equivalent to an abatement of a public nuisance or, alternatively, any act by a private party to destroy *ferae naturae* on private property constitutes a public nuisance. *Cf. State Dept. of Env. Qual. v. Chem. Waste*, 19 Or.App. 712, 719, 528 P.2d 1076 (1974) ("defendant has been 'operating' the site in violation of the Environmentally Hazardous Wastes Statutes from the time they became effective in early 1972, [but] that continuing violation does not require a finding that the site constitutes a public nuisance"). The trial court correctly struck the state's defense that plaintiff's proposed logging constituted a public nuisance.

In its final assignment of error concerning the pleadings, the state argues that the trial court erred in striking its defense that it labeled as "failure to exhaust its administrative remedies." The state argues that Boise's regulatory takings claim is unripe because it did not make an effort to obtain an "incidental take" permit pursuant to 16 USC section 1539(a). *Former* OAR 660–24–809(5) provided that "[e]xceptions to the requirements for protecting northern spotted owl nesting sites may be approved by the State Forester if the operator has obtained an incidental take permit from federal authorities under the federal Endangered Species Act." The state argues that, because plaintiff did not attempt to avail itself of this exception, it may not yet assert a takings claim. [The court agreed with the state on this ground, reversed the award of compensation, and remanded the case for further proceedings.]

Notes and Questions

1. What do you think of the court's assertion that "the state offers no authority for the proposition that knocking down a bird's nest on one's property has ever been considered a public nuisance"? If injury to a specific individual of an endangered species is not a nuisance, what about widespread depletion of a forest's biodiversity value generally? Has "biodiversity destruction" ever been a basis for a nuisance claim? Should it be? How would we measure biodiversity destruction and when would we know that it had reached the level of public nuisance? For a discussion of the topic, see Hope M. Babcock, *Should Lucas v. South Carolina Coastal Council Protect Where the Wild Things Are? Of Beavers, Bob-o-Links, and Other Things that Go Bump in the Night*, 85 IOWA L. REV. 849 (2000).

2. What if we could reliably quantify the ecosystem service values a forest delivers to humans, such as the value of flood control, denitrification of runoff waters, and carbon sequestration? Could destruction of those values reasonably be described as a public nuisance, or something akin to it, making regulation of actions leading to loss of such values more immune to takings challenges?

3. The *Boise Cascade* court gave little credence to the *Loretto*-based taking argument that the protection of the owl on the timber company's private property imposes a *per se* taking through physical occupation of the property without just compensation. The court denied the claim on the ground that "[t]he state did not cause or induce the spotted owls to breed on plaintiff's property. The state simply regulated plaintiff's use of the property based on the presence of the spotted owls there." Yet the argument that government protection of wildlife produces a physical invasion of private property has recurred throughout the development of conservation law. Recall Barrett v. State of New York, reproduced in chapter 3. For an elaboration on the related topic of whether species protection laws violate the third amendment's protection against quartering of soldiers, see Andrew P. Morriss and Richard L. Stroup, *Quartering Species: The "Living Constitution," the Third Amendment, and the Endangered Species Act*, 30 ENVTL. LAW 769 (2000).

4.   Perhaps the most incendiary private forest use issue in decades, the so-called Headwaters Forest controversy in California, also revolved around the threat of a takings claim. The Pacific Lumber Company (PALCO) owned the largest remaining stand of privately-owned, old-growth redwoods in the world, including some trees well over 2000 years old. Charles Hurwitz, the CEO of PALCO's parent company MAXAAM, made it clear that to PALCO the 3,500–acre stand of redwoods was essentially a valuable crop ready for harvest. To ardent environmentalist groups, however, the trees held nearly religious values. They argued that PALCO had no right to log the forest given the spiritual value of the redwoods. As habitat for the endangered marbled murrelet, a small bird, the Endangered Species Act (ESA) held the chain saws at bay. But PALCO, which had holdings of over 210,000 acres of redwood forest land on which it hoped to harvest, used the takings issue as its leverage in the affair through an inverse condemnation lawsuit it filed against the United States. A long, arduous series of negotiations between PALCO, the Department of Interior, and the State of California brokered a deal involving the $380 million purchase of 7,500 acres of PALCO's land, including the old growth stands, and the issuance of an ESA permit to conduct harvesting on PALCO's remaining lands under specified conservation conditions. The environmental advocacy groups were incensed, working tirelessly to block the deal, and PALCO and Interior ran into many last-minute negotiation impasses, but on March 1, 1999, the deal was finalized.

Putting aside their status as endangered species habitat, did PALCO have the right to remove these 2000–year old trees from the face of the planet? If so, would any of the policy tools discussed in the previous section other than outright acquisition have been useful in dissuading PALCO from doing so? Assuming, as PALCO asserted, that the commercial timber value of the disputed redwoods was well over the $380 million purchase price, did the government use the ESA to take the land for less than the just compensation it would have had to pay under an exercise of eminent domain? For two sharply contrasting insider views of these and other questions swirling in Headwaters Forest controversy, see Gideon Kanner, *Redwoods, Junk Bonds, and Tools of Cosa Nostra: A Visit to the Dark Side of the Headwaters Controversy*, 30 ENVTL. LAW REPORTER (ELI) 10756 (2000) (attorney for PALCO in the takings litigation); David J. Hayes, *Saving the Headwaters Forest: A Jewel That Nearly Slipped Away*, 30 ENVTL. LAW REPORTER (ELI) 10131 (2000) (attorney for Interior in the acquisition negotiation).

D.   CASE STUDY: THE SHAWNEE NATIONAL FOREST IN SOUTHERN ILLINOIS

The Shawnee National Forest in southern Illinois is a 268,000–acre patch-work national forest which, like many in the eastern United States, the national government purchased when agricultural uses failed and soil erosion was a serious problem. In pre-European settlement times the Shawnee area was covered predominantly with a thriving hardwood forest and open glades or barrens communities. Early settlers cleared the forest to make way for croplands, but the combination of erosion of poor soils and agricultural market failures during the Depression wiped out the hard-scrabbled farming communities. The federal government purchased many abandoned and marginal tracts of land in the 1930s and planted short-leaf pines to control soil erosion and to establish a timber source. President Roosevelt designated the Shawnee National Forest in 1939.

Today the Shawnee protects at least 7 federally-listed endangered species and provides numerous outdoor recreation opportunities. But it is not what it once was. The planted pine forest, though thriving and covering 16 percent of the forest, provides only a fraction of the habitat value and species diversity the pre-settlement hardwood forest provided. The Forest Service thus devised an "ecological restoration program" to return approximately 10,000 acres of the Shawnee to pre-settlement hardwood and mixed hardwood/pine forest conditions. By allowing timber companies to selectively cut pines, the Forest Service hoped to open the forest canopy sufficient to promote hardwood growth and, eventually, to support a hardwood forest of superior ecological value than the existing pine forest.

A scenic view from cliffs overlooking the Shawneee National Forest
U.S. Forest Service

One area the Forest Service chose for the project, however, had grown dear to many environmental activists, not to mention many pine forest species. Known as the Bell Smith Springs area, it is filled with deep untouched pine stands, crystal clear streams, and miles of peaceful walking trails. Species don't know which trees are supposed to be where, and so over the years since the planting of the pines, species that favor pine ecosystems had made Bell Smith Springs their home. This group of species, anchored by a small bird known as the pine warbler, was known as the "pine warbler guild." The Forest Service did not dispute that removal of the pines would severely affect the pine warbler guild, most likely extirpating the species from the Bell Smith Springs area. As one Friends of Bell Smith Springs activist put it, "if they take those trees out now, it will set the area back 60 years," which was, ironically, precisely the agency's objective. Notwithstanding the agency's strong case that the hoped-for hardwood forest would supply higher biodiversity, activists fought hard to protect what they facetiously called the "offending pines."

The central portion of the Shawnee National Forest showing the Bell Smith Springs area and the division between federal land (dark) and non-federal land (light).

Notes and Questions

1. Using the *Marita/Inland Empire/Lowe* trio of cases and the Forest Service statutes and regulations discussed therein, what would you tell the Friends of Bell Smith Springs when they seek your advice on how to proceed in a lawsuit seeking to enjoin the agency's restoration plan timber cuts? Do they have a chance? What are their best arguments?

2. Now put yourself in the role of the Forest Service's attorney. How can the agency portray its plan so as to minimize vulnerability to the activists' case? In the end, although the Forest Service has a substantive duty to protect biodiversity, is its discretion so broad that it has the power to choose *which version* of biodiversity it protects?

3. Are your answers affected by the additional fact that Forest Service has designated the pine warbler as a management indicator species for pine forest areas in the Shawnee National Forest? If so, is it also important to know whether the pine warbler is a native or non-native species in the forest? How long does a species have to be "in" a forest for it to be "native" to that forest?

4. Does the description of conservation biology in *Marita* suggest an approach a conservation biologist would take to the Forest Service's "ecological restoration program" for the Shawnee? Can you tell whether the Forest Service's scientists were practicing conservation biology when they proposed the forest "restoration" program for the Shawnee? How important is it to your answer to any of these questions that the pine forest itself was not native to the Bell Smith Springs area, but rather was introduced following agricultural clearing of the pre-settlement hardwood forest? How long do the pines have to be in the forest before they are no longer "offending" but rather the ecosystem to protect?

5. Consider how the courts that decided these questions ruled. The litigation, Glisson v. U.S. Forest Service, 138 F.3d 1181 (7th Cir.1998), aff'g 876 F.Supp. 1016 (S.D.Ill.1993), culminated in this opinion by Chief Judge Posner of the Seventh Circuit:

> Before us is an appeal from a grant of summary judgment to the Forest Service in a suit to enjoin an "ecological restoration" project in a 10,500 acre tract ("Opportunity Area 6") of the Shawnee National Forest in southern Illinois. The litigation has a long history unnecessary to recount, as the issues presented by the appeal are narrow ones. The first concerns the Service's compliance with a federal regulation that requires it "to maintain viable populations of existing native" species. 36 C.F.R. § 219.19. The project, although designed to promote a variety of fauna and flora native to the area, will have adverse effects on shortleaf pines and pine warblers. The Forest Service, however, interprets the word "native" in the regulation to mean "native to the project area," that is, to OA6, rather than native to the state or even to the national forest; and it interprets "native" to mean existing in a "natural" state in the area rather than introduced by man in recent times. On the basis of these interpretations, the Service held that the regulation does not bar the restoration project. The pines, which in turn provide a habitat for the warblers, were first planted in OA6 during the 1930s and 1940s, and that is too recent to satisfy the Service's conception of what it means for a plant or animal to be "native" to an area.

> An agency is entitled to broad latitude in interpreting its regulations. E.g., Thomas Jefferson University v. Shalala, 512 U.S. 504, 512 (1994); Arkansas v. Oklahoma, 503 U.S. 91, 112 (1992); Lyng v. Payne, 476 U.S. 926, 939 (1986). Since the regulation in question contains no definition of "native," we are required to uphold the Forest Service's interpretation unless it is unreasonable. E.g., Thomas Jefferson University v. Shalala, supra, 512 U.S. at 512; Bradvica v. INS, 128 F.3d 1009, 1014 (7th Cir.1997). It is not.

Does this outcome surprise you? Recall that this is the same court that decided *Marita*.

6. Would either the Committee of Scientists' proposal, the Forest Service's 2001 rule, or the ecosystem health objectives in the Forest Service's *Draft Strategic Plan* have required a different ruling in the *Glisson* case?

7. The map of the central portion of the Shawnee National Forest reproduced above shows how federal and non-federal lands co-exist in close proximity, with non-federal lands often "intruding" on the integrity of the national forest (and vice-versa!). Assume that private lands jut into portions of the Shawnee pine forest area in which the Forest Service would like to see its "ecological restoration project" take hold. The private lands contain thriving pine forest habitat which, because of its adjacency to the project area, could impede the effects of the project within the Shawnee. The Forest Service cannot include those private lands in the timber leases designed to cull pine trees. What can the agency do to avoid having private land management activities thwart the effects of the project within the national forest?

8. Private and state lands virtually slice the Shawnee National Forest in half. Does this geographic distribution of land ownership constrain Forest Service decision options for the Shawnee? What techniques could federal or state law use to coerce or provide incentives for owners of private forested lands near the Shawnee to "cooperate" with Forest Service policy goals for the national forest?

9. For a history of the formation of the Shawnee National Forest from the depleted agricultural lands of southern Illinois, see U.S. FOREST SERVICE, THE CREATION OF THE SHAWNEE NATIONAL FOREST 1930–1938, available at <http://www.fs.fed.us/r9/shawnee/snfchis1.html>.

CHAPTER 9

GRASSLAND ECOSYSTEMS

Chapter Outline:
A. Biodiversity and Grassland Ecosystems
B. Primer on the Law of Grasslands
   1. Historical Background
   2. Current Scene
C. Biodiversity Management in Grassland Ecosystems
   1. BLM's Rangeland Health and Rangeland Planning
     a. The Fundamentals of Rangeland Health
     b. BLM's Other Rangeland Management Planning Requirements
   Note on Forest Service Management of the National Grasslands System
   2. Restoring Grassland Ecosystems on Private Lands
D. Case Study: The Little Darby Prairie National Wildlife Refuge

---

A. BIODIVERSITY AND GRASSLAND ECOSYSTEMS

The Great Plains of North America comprise more than 500 million acres of grasslands or former grasslands west of the 100th meridian, extending from central Alberta, Canada, to the Texas panhandle. Grasslands are composed of a rich complex of grasses and forbs–few trees or shrubs are found in intact grasslands. Most grasslands are also associated with low moisture and seasonal extremes in temperature. Two general varieties exist in the world: temperate grasslands are found primarily in North America, Europe, and in the plains of Asia; savanna grasslands dominate in the hotter climates of South America, Africa, and the Australian Outback. In the United States, grasslands east of the continental divide reaching to western Indiana and Kansas were dominated by tall, mixed, and shortgrass prairie (French for meadow), with the more arid lands being in the "rain shadow" of the Rocky Mountains. West of the continental divide were found the sagebrush grasslands of such Intermountain West states as Wyoming, Montana, Colorado, and New Mexico.

Much of the world's temperate grassland areas are well-suited to agriculture and have been converted to row crops or pasture. Some of this continent's prairie grasslands are among the world's best agricultural lands. In 1820, for example, at least 60 percent of Illinois' land area was high quality grassland of one type or another. Today, 99.99 percent of that habitat is gone, most of it converted to crop production, making the state known as the Prairie State a bit of an imposter. Worse, the remaining 2352 acres of original Illinois grasslands are fragmented into 253 remnants, four out of five of which are smaller than ten acres and one third of which are smaller than one acre–a size too small to function as a self-sustaining ecosystem. Over two hundred of these remnant sites are not protected as

dedicated preserves. *See* ILLINOIS DEPARTMENT OF ENERGY AND NATURAL RE-
SOURCES, THE CHANGING ILLINOIS ENVIRONMENT: CRITICAL TRENDS (1994).

To be sure, manipulation of grasslands began well before European
settlers introduced large-scale agriculture, as Native Americans set many
small fires throughout the central grasslands, possibly to manage grassland
species composition and to prevent establishment of forests, but also
possibly simply to improve travel corridors or as an act of warfare. Later, as
settlers suppressed fire, species composition moved in the other direction–
toward woody plants and other species introduced from adjacent forests
that were unchecked by periodic fires, natural or otherwise. Today, less
than one percent of the original grasslands of the American Great Plains
are undisturbed by human activities. Much of the area that is used for
rangeland is owned and administered by the federal government through
the Department of Interior's Bureau of Land Management (BLM). Most of
the private land is irrigated for agricultural uses. In short, no person who
walked the land of the Great Plains in 1800 would recognize what is there
today as a grassland.

The combination of agricultural encroachment, fire suppression, and
elimination of keystone species such as bison and prairie dogs (see Chapter
4) has led to a general decline in native grassland biodiversity in the United
States. Native grassland birds are among the most imperiled avian species
in North America. Three bird species once common on Illinois prairies are
completely extirpated from the state. Native prairie fishes have also experienced losses as impoundments on rivers and streams have altered habitat
and fragmented populations. Even insects are finding it hard to adapt–the
Karner blue butterfly, native to northern Illinois, had not been seen in the
state for over a century until a single sighting event in 1992.

There is promise in efforts to rehabilitate grassland ecosystems, however. The Great Plains comprise 20 percent of the land area of the lower 48
states, but house only 2 percent of the nation's population. The rural
nature of the area provides opportunities for land acquisition and restoration efforts. On the other hand, the rural population, though sparse, is
committed to its lifestyle, and that lifestyle relies largely on access to large
stretches of land. Efforts to acquire or regulate areas for grassland ecosystem management may threaten the economic and cultural viability of these
rural communities. As the materials in this chapter attest, the history and
future of land management on the Great Plains thus involves much
controversy.

Notes and Questions

1. The degree to which Native Americans altered grassland ecosystems
prior to European settlement is controversial. There is a raging debate
among ecological historians as to the impact early human inhabitants of
the continent had on the surrounding flora and fauna. *See Did Human
Hunting Cause Mass Extinction?*, 294 SCIENCE 1459 (2001) (series of letters
debating the question). In particular, they disagree as to causes of the rapid

extinction of the rich array of mega-fauna that were found on the continent around the time the first human migrations to North America are believed to have occurred (about 13,000 years ago). Was it the deadly Clovis point, used expertly by early native hunters? Or was it climate change, or species competition? And while there is ample evidence of Native American manipulation of vegetative regimes through fire, how widespread was it, and how lasting were its effects? In any event, there is little debate that the largest set of impacts on the Great Plains has been the post-European settlement introductions of widespread irrigated agriculture, intensive domestic cattle and sheep grazing, and concerted fire suppression. Background on the ongoing degrading condition of North American grasslands attributable to these factors is available in U.S. DEPT. OF INTERIOR, OUR LIVING RESOURCES 295–307 (1995), and WORLD RESOURCES INST., PEOPLE AND ECOSYSTEMS 119–135 (2000). For epic accounts of the ecological history of our continent, including the interaction of early human inhabitants and their surroundings, see TIM FLANNERY, THE ETERNAL FRONTIER: AN ECOLOGICAL HISTORY OF NORTH AMERICA AND ITS PEOPLES (2001); DAVID S. WILCOVE, THE CONDOR'S SHADOW: THE LOSS AND RECOVERY OF WILDLIFE IN AMERICA (1999). An excellent history of the social, physical, and ecological conditions associated with grazing on the American rangelands is found in DEBRA L. DONAHUE, THE WESTERN RANGE REVISITED 1–160 (1999).

2. One reason to focus on grazing policy as the medium for studying the law of grassland ecosystems is that the impact of grazing on the environment is a matter of degree. Grazing is simply one form of "herbivory," which is a natural process in any ecosystem where flora and fauna co-exist. A leaf-chewing grasshopper is an herbivore. The now extinct mega-fauna of the pre-human continent were herbivores. Herbivory contributes to nutrient cycling, species control, and other natural ecosystem processes. Grazing by mega-fauna ungulate (hooved) species plays an important role in maintaining soil composition and mediating vegetative species abundance. So, is grazing of cattle and sheep a "natural" ecosystem process? That depends on how it is conducted. By most ecological accounts, it has been conducted too intensively on many of our nation's public lands to sustain the conditions believed to be necessary to support a thriving grassland ecosystem. Overstocking the lands (placing too many cattle or sheep in an area at once) or overgrazing the lands (allowing the stock to graze an area too much) can lead to extensive trampling and compaction of soil, deposition of urine and feces, devegetation, and introduction of nonnative grasses and weeds. Yet, many ranching interests, and even a fair number of range ecologists, maintain that there is a level of grazing that is sustainable ecologically. The challenge for grassland ecosystem management policy is finding that level. For detailed, albeit contrasting, accounts of the impacts grazing has on grassland ecosystems, compare Thomas L. Fleischner, *Ecological Costs of Livestock Grazing in Western North America*, 8 CONSERVATION BIOLOGY 629 (1994) (grazing is detrimental), with James H. Brown and William McDonald, *Livestock Grazing and Conservation on Southwest Rangelands*, 9 CONSERVATION BIOLOGY 1644 (1995) (grazing is not detrimental). Whatever the effects of early Native American and modern grazing practices may be, it is clear that grasslands offer significant ecosystem service values, including sequestering carbon, methane, and nitrogen and, of most direct value, reducing soil erosion, and that improperly managed cultivation and grazing can significantly reduce the ability of grasslands to deliver those values. *See* Osvaldo E. Sala and Jose M. Paruelo, *Ecosystem Services in Grasslands*, in NATURE'S SERVICES 237 (Gretchen C. Daily ed., 1997).

3. The potential effects of global climate on North American grassland areas is uncertain. According to the U.S. Environmental Protection Agency, the impacts of climate change on grasslands have not been studied in the same detail as the implication for forests. Nevertheless, the existing research suggests a number of likely outcomes. Perhaps most importantly, because of its effects on water availability, climate change could harm grazing activities on both federal and private lands. The decline in western water availability suggested by several studies would seriously decrease the economic viability of grazing on these lands. Because grazing on the open range accounts for a small and declining fraction of U.S. cattle, however, national beef production would not be seriously impaired. Changing climate is also likely to alter both the geographical extent and the plant composition of rangelands. If a drier climate causes some areas of the Southeast or Midwest to lose their ability to sustain a forest, the terrain in those areas may come to resemble the landscape of the open range. A wetter climate, by contrast, might enable forests to grow in areas that are now grasslands, while also enabling range grasses to grow in areas that are deserts today. Within existing rangelands, elevated levels of carbon dioxide may induce a shift from grasses toward shrubs and other woody plants. For more information on these climate change effects, see www.epa.gov/globalwarming/impacts.

B. PRIMER ON THE LAW OF GRASSLANDS

1. HISTORICAL BACKGROUND

The importance of history, both social and legal, to the development of ecosystem management law could not be more plainly illustrated than through the lens of the following recent United States Supreme Court opinion. Before addressing the subtle merits of a challenge to new "Rangeland Reform" regulations the Bureau of Land Management (BLM) adopted for administration of public rangelands (discussed *infra*), the Court described the unfolding of the federal law of rangelands and its gradual movement toward an ecosystem management model.

This history has witnessed the evolution of two themes headed on a collision course. The first involves the definition of grazing privileges on the 176 million acres of public rangelands BLM administers in the western states (excluding Alaska), 90 percent of which are approved for livestock grazing. The second involves the ecological condition of the rangelands and the degree to which BLM approval of grazing levels takes those conditions into account.

While grazing on public lands has always been a privilege and nothing more in the strict legal sense, a culture of property rights emerged around those privileges based on how BLM administered them and on how ranchers used them. To understand how this transpired, read carefully the Court's description of the "grazing preference" method of rangeland grazing privilege allocation and the related concepts of "base property" and "suspended Animal Unit Months." As the Court describes how this history unfolded, track also the Court's description of the ecological condition of the rangelands through time.

Public Lands Council v. Babbitt

Supreme Court of the United States, 2000.
529 U.S. 728.

■ Justice Breyer delivered the opinion of the Court.

This case requires us to interpret several provisions of the 1934 Taylor Grazing Act, 48 Stat. 1269, 43 U.S.C. § 315 et seq. The Petitioners claim that each of three grazing regulations, 43 CFR §§ 4100.0–5, 4110.1(a), and 4120.3–2 (1998), exceeds the authority that this statute grants the Secretary of the Interior. We disagree and hold that the three regulations do not violate the Act.

I

We begin with a brief description of the Act's background, provisions, and related administrative practice.

A

The Taylor Grazing Act's enactment in 1934 marked a turning point in the history of the western rangelands, the vast, dry grasslands and desert that stretch from western Nebraska, Kansas, and Texas to the Sierra Nevada. Ranchers once freely grazed livestock on the publicly owned range as their herds moved from place to place, searching for grass and water. But the population growth that followed the Civil War eventually doomed that unregulated economic freedom.

A new era began in 1867 with the first successful long drive of cattle north from Texas. Cowboys began regularly driving large herds of grazing cattle each year through thousands of miles of federal lands to railheads like Abilene, Kansas. From there or other towns along the rail line, trains carried live cattle to newly opened eastern markets. The long drives initially brought high profits, which attracted more ranchers and more cattle to the land once home only to Indian tribes and buffalo. Indeed, an early–1880's boom in the cattle market saw the number of cattle grazing the Great Plains grow well beyond 7 million. See R. White, "It's Your Misfortune and None of My Own": A History of the American West 223

(1991); see generally, E. Osgood, The Day of the Cattleman 83–113 (1929); W. Webb, The Great Plains 205–268 (1931).

But more cattle meant more competition for ever-scarcer water and grass. And that competition was intensified by the arrival of sheep in the 1870's. Many believed that sheep were destroying the range, killing fragile grass plants by cropping them too closely. The increased competition for forage, along with droughts, blizzards, and growth in homesteading, all aggravated natural forage scarcity. This led, in turn, to overgrazing, diminished profits, and hostility among forage competitors–to the point where violence and "wars" broke out, between cattle and sheep ranchers, between ranchers and homesteaders, and between those who fenced and those who cut fences to protect an open range. See W. Gard, Frontier Justice 81–149 (1949). These circumstances led to calls for a law to regulate the land that once was free.

The calls began as early as 1878 when the legendary southwestern explorer, Major John Wesley Powell, fearing water monopoly, wrote that ordinary homesteading laws would not work and pressed Congress to enact "a general law . . . to provide for the organization of pasturage districts." Report on the Lands of the Arid Region of the United States, H. Exec. Doc. No. 73, 45th Cong., 2d Sess., 28 (1878). From the end of the 19th century on, Members of Congress regularly introduced legislation of this kind, often with Presidential support. In 1907, President Theodore Roosevelt reiterated Powell's request and urged Congress to pass laws that would "provide for Government control of the public pasture lands of the West." S. Doc. No. 310, 59th Cong., 2d Sess., 5 (1907). But political opposition to federal regulation was strong. President Roosevelt attributed that opposition to "those who do not make their homes on the land, but who own wandering bands of sheep that are driven hither and thither to eat out the land and render it worthless for the real home maker"; along with "the men who have already obtained control of great areas of the public land . . . who object . . . because it will break the control that these few big men now have over the lands which they do not actually own." Ibid. Whatever the opposition's source, bills reflecting Powell's approach did not become law until 1934.

By the 1930's, opposition to federal regulation of the federal range had significantly diminished. Population growth, forage competition, and inadequate range control all began to have consequences both serious and apparent. With a horrifying drought came dawns without day as dust storms swept the range. The devastating storms of the Dust Bowl were in the words of one Senator "the most tragic, the most impressive lobbyist, that ha[s] ever come to this Capitol." 79 Cong. Rec. 6013 (1935). Congress acted; and on June 28, 1934, President Franklin Roosevelt signed the Taylor Grazing Act into law.

B

The Taylor act seeks to "promote the highest use of the public lands." 43 U.S.C. § 315. Its specific goals are to "stop injury" to the lands from "overgrazing and soil deterioration," to "provide for their use, improvement and development," and "to stabilize the livestock industry dependent on the public range." 48 Stat. 1269. The Act grants the Secretary of the Interior authority to divide the public range lands into grazing districts, to specify the amount of grazing permitted in each district, to issue leases or permits "to graze livestock," and to charge "reasonable fees" for use of the land. 43 U.S.C. §§ 315, 315a, 315b. It specifies that preference in respect to grazing permits "shall be given ... to those within or near" a grazing district "who are landowners engaged in the livestock business, bona fide occupants or settlers, or owners of water or water rights." § 315b. And, as particularly relevant here, it adds:

> "So far as consistent with the purposes and provisions of this subchapter, grazing privileges recognized and acknowledged shall be adequately safeguarded, but the creation of a grazing district or the issuance of a permit ... shall not create any right, title, interest, or estate in or to the lands." Ibid.

C

The Taylor Act delegated to the Interior Department an enormous administrative task. To administer the Act, the Department needed to determine the bounds of the public range, create grazing districts, determine their grazing capacity, and divide that capacity among applicants. It soon set bounds encompassing more than 140 million acres, and by 1936 the Department had created 37 grazing districts, see Department of Interior Ann. Rep. 15 (1935); W. Calef, Private Grazing and Public Lands 58–59 (1960). The Secretary then created district advisory boards made up of local ranchers and called on them for further help. See 2 App. 809–811 (Rules for Administration of Grazing Districts (Mar. 2, 1936)). Limited department resources and the enormity of the administrative task made the boards "the effective governing and administrative body of each grazing district." Calef, supra, at 60; accord, P. Foss, Politics and Grass 199–200 (1960).

By 1937 the Department had set the basic rules for allocation of grazing privileges. Those rules recognized that many ranchers had long maintained herds on their own private lands during part of the year, while allowing their herds to graze farther afield on public land at other times. The rules consequently gave a first preference to owners of stock who also owned "base property," i.e., private land (or water rights) sufficient to support their herds, and who had grazed the public range during the five years just prior to the Taylor act's enactment. See 2 App. 818–819 (Rules for Administration of Grazing Districts (June 14, 1937)). They gave a second preference to other owners of nearby "base" property lacking prior use. Ibid. And they gave a third preference to stock owners without base property, like the nomadic sheep herder. Ibid. Since lower preference categories divided capacity left over after satisfaction of all higher preference claims, this system, in effect, awarded grazing privileges to owners of land or water. See Foss, supra, at 63 (quoting Grazing Division Director F.R. Carpenter's remarks that grazing privileges are given to ranchers "not as individuals, nor as owners of livestock," but to "build up [the] lands and give them stability and value").

As grazing allocations were determined, the Department would issue a permit measuring grazing privileges in terms of "animal unit months" (AUMs), i.e., the right to obtain the forage needed to sustain one cow (or five sheep) for one month. Permits were valid for up to 10 years and usually renewed, as suggested by the Act. See 43 U.S.C. § 315b; Public Land Law Review Commission, One Third of the Nation's Land 109 (1970). But the conditions placed on permits reflected the leasehold nature of grazing privileges; consistent with the fact that Congress had made the Secretary the landlord of the public range and basically made the grant of grazing privileges discretionary. The grazing regulations in effect from 1938 to the present day made clear that the Department retained the power to modify, fail to renew, or cancel a permit or lease for various reasons.

First, the Secretary could cancel permits if, for example, the permit holder persistently overgrazed the public lands, lost control of the base property, failed to use the permit, or failed to comply with the Range Code. See, e.g., 43 CFR 160.26(a)-(f) (1938); Department of Interior, Federal Range Code §§ 6(c)(6), (7), (10) (1942) (hereinafter 1942 Range Code); 43 CFR §§ 161.6(c)(6)-(7), (10)-(12) (1955); 43 CFR §§ 4115.2–1(d), (e)(7)-(11) (1964); 43 CFR §§ 4115.2–1(d), (e)(7)-(11) (1977); 43 CFR 4170.1–2 (1994); 43 CFR 4170.1–2 (1998). Second, the Secretary, consistent first with 43 U.S.C. § 315f, and later the land use planning mandated by 43 U.S.C. § 1712, was authorized to reclassify and withdraw land from grazing altogether and devote it to a more valuable or suitable use. See, e.g., 43 CFR § 160.22 (1938); 1942 Range Code § 6(c)(4); 43 CFR § 161.6(c)(5) (1955); 43 CFR §§ 4111.4–2(f), 4115.2–1(e)(6) (1964); 43 CFR §§ 4111.4–3(f), 4115.2–1(e)(6) (1977); 43 CFR § 4110.4–2(a) (1994); 43 CFR § 4110.4–2(a) (1998). Third, in the event of range depletion, the Secretary maintained a separate authority, not to take areas of land out of grazing use altogether as above, but to reduce the amount of grazing allowed on that land, by suspending AUMs of grazing privileges "in whole or in part," and "for such time as necessary." 43 CFR § 4115.2–1(e)(5) (1964); see also 43 CFR § 160.30 (1938) (reservation (b)); 1942 Range Code § 6(c)(8); 43 CFR § 161.4(8) (1955); 43 CFR §§ 4111.4–3, 4115.2–1(e)(5) (1977); 43 CFR § 4110.3–2 (1994); 43 CFR § 4110.3–2 (1998).

Indeed, the Department so often reduced individual permit AUM allocations under this last authority that by 1964 the regulations had introduced the notion of "active AUMs," i.e., the AUMs that a permit initially granted minus the AUMs that the department had "suspended" due to diminished range capacity. Thus, three ranchers who had initially received, say, 3,000, 2,000, and 1,000 AUMs respectively, might find that they could use only two-thirds of that number because a 33% reduction in the district's grazing capacity had led the Department to "suspend" one-third of each allocation. The "active/suspended" system assured each rancher, however, that any capacity-related reduction would take place proportionately among permit holders, see 43 CFR § 4111.4–2(a)(3) (1964), and that the Department would try to restore grazing privileges proportionately should the district's capacity later increase, see § 4111.4–1.

In practice, active grazing on the public range declined dramatically and steadily (from about 18 million to about 10 million AUMs between 1953 and 1998).... Despite the reductions in grazing, and some improvements following the passage of the Taylor act, see App. 374–379 (Department of Interior, 50 Years of Public Land Management 1934–1984), the range remained in what many considered an unsatisfactory condition. In 1962, a congressionally mandated survey found only 16.6% of the range in excellent or good condition, 53.1% in fair condition, and 30.3% in poor condition. Department of Interior Ann. Rep. 62 (1962). And in 1978 Congress itself determined that "vast segments of the public rangelands are ... in an unsatisfactory condition." 92 Stat. 1803 (codified as 43 U.S.C. § 1901(a)(1)).

D

In the 1960's, as the range failed to recover, the Secretary of the Interior increased grazing fees by more than 50% (from 19 cents to 30 cents per AUM/year), thereby helping to capture a little more of the economic costs that grazing imposed upon the land. Department of Interior Ann. Rep. 66 (1963). And in 1976, Congress enacted a new law, the Federal Land Policy and Management Act of 1976 (FLPMA), 90 Stat. 2744, 43 U.S.C. § 1701 et seq., which instructed the Interior Department to develop districtwide land use plans based upon concepts of "multiple use" (use for various purposes, such as recreation, range, timber, minerals, watershed, wildlife and fish, and natural and scenic, scientific, and historical usage), § 1702(c), and "sustained yield" (regular renewable resource output maintained in perpetuity), § 1702 (h). The FLPMA strengthened the Department's existing authority to remove or add land from grazing use, allowing such modification pursuant to a land use plan, §§ 1712, 1714, while specifying that existing grazing permit holders would retain a "first priority" for renewal so long as the land use plan continued to make land "available for domestic livestock grazing." § 1752(c).

In 1978, the Department's grazing regulations were, in turn, substantially amended to comply with the new law. See 43 Fed. Reg. 29067. As relevant here, the 1978 regulations tied permit renewal and validity to the land use planning process, giving the Secretary the power to cancel, suspend, or modify grazing permits due to increases or decreases in grazing forage or acreage made available pursuant to land planning. See 43 CFR §§ 4110.3–2(b), 4110.4–2 (1978); see also 43 CFR § 4110.4–2 (1994); 43 CFR § 4110.4–2 (1998).

That same year Congress again increased grazing fees for the period 1979 to 1986. See Public Rangelands Improvement Act of 1978, 43 U.S.C. § 1905. However neither of the two Acts from the 1970's significantly modified the particular provisions of the Taylor Act at issue in this case.

Notes and Questions

1. What led to the emergence of the property rights culture in BLM grazing administration? The short answer is that it grew from the early decision to link grazing preferences to tracts of land through the "base property" concept. Preferences for grazing privileges, expressed as Animal Unit Months (AUMs), fluctuated according to BLM's decisions about the impact of total grazing on the relevant rangeland unit, but did so proportionately among all affected ranchers and without ever having the so-called "suspended AUMs" expire. Ranchers, over time, began treating their preferences, including suspended AUMs, as one of the sticks in their bundle of property rights associated with their base property. As one of the leading experts on grazing law explains:

> Preference for grazing privileges, then, was given to lands and not people. Because the preference right virtually guaranteed perpetual renewal of the 10–year grazing permits, the permits came to be treated by the market as appurtenances to base properties, with the market price of the base property, and its assessed value for mortgage purposes, reflecting the value of the permit.

Joseph M. Feller, *Back to the Present: The Supreme Court Refuses to Move Public Range Law Backward, but Will the BLM Move Public Range Law Forward?*, 31 Envtl. L. Rep. (ELI) 10021 (2001). As we shall see in the remaining materials, however, the property rights status of the grazing preferences was far less stable than many ranching interests believed.

2. If ranchers using public rangelands don't have a property interest in their grazing preferences, what *do* they own, other than their livestock and private ranch holdings? That question was central in an arduously long litigation brought by ranching interests in Nye County, Nevada, whose grazing permits BLM had suspended and cancelled. They argued that the BLM's actions took several property interests without just compensation as required by the Fifth Amendment to the U.S. Constitution. In its final opinion in the case, the Court of Federal Claims explained that ranchers: (1) have no property interest under state or federal law in the surface estate of the federal lands on which their grazing allotments applied, and thus suffered no taking in that respect as a result of the reduced grazing allotments, but (2) do have vested rights in certain water rights and rights of way that were perfected according to state and federal law, and thus could suffer takings of those interests as a result of BLM's actions. *See* Hage v. United States, 51 Fed.Cl. 570 (2002). For the history of the litigation, which took several twists and turns, see Hage v. United States, 35 Fed.Cl. 147 (1996) (denying United States motions to dismiss); Hage v. United States, 35 Fed.Cl. 737 (1996) (granting intervention to state and environmental interests); Hage v. United States, 42 Fed.Cl. 249 (1998) (ruling that takings claims required factual evidence).

3. Much like the history of national forest law discussed in Chapter 8, federal rangeland law evolved through three major statutes representing three phases of development. First, BLM's organic statute, the Taylor

Grazing Act (TGA), provided the basic authority for BLM to administer grazing privileges, with some attention to environmental impact. Next, the Federal Lands Planning and Management Act (FLPMA) emphasized BLM's multiple-use mandate and overlaid a more sophisticated planning regime on BLM's administration of grazing. Finally, the Public Rangelands Improvement Act (PRIA) tweaked the system toward a greater focus on economic and ecological factors. For more extensive histories of these three primary BLM laws and related legal developments, see George Cameron Coggins & Margaret Lindeberg–Johnson, *The Law of Public Rangeland management II: The Commons and the Taylor Act*, 13 ENVTL. L. 1 (1982); George Cameron Coggins, *The Law of Public Rangeland Management IV: FLPMA, PRIA, and the Multiple Use Mandate*, 14 ENVTL. L. 1 (1983).

2.   CURRENT SCENE

The controversy in *Public Lands Council v. Babbitt* was over the effect Rangeland Reform regulations the Bureau of Land Management (BLM) adopted in 1995 would have on BLM's public grazing preferences allocation system by altering the historically entrenched "property" status of the privileges. Elsewhere in the new Rangeland Reform rules, BLM also changed its criteria for adjusting the quantity of allotments based on newly-formulated ecological factors. Known as the Fundamentals of Rangeland Health regulations, these criteria lay out a new ecosystem management focus and make subtle but nonetheless potentially far-reaching changes to the way BLM allocates grazing rights. In its preamble to the final adoption of the rules, BLM provided its take on the ecological condition of the rangelands and its justification of both the need and authority for legal reform directed at improving those conditions.

Department of the Interior, Bureau of Land Management, Department Hearings and Appeals Procedures; Cooperative Relations; Grazing Administration—Exclusive of Alaska

60 Fed. Reg. 9894 (Feb. 22, 1995).

* * *

SUMMARY: This final rule amends the regulations that govern how the Secretary of the Interior, through the Bureau of Land Management (BLM), administers livestock grazing. This rule applies to all lands on which BLM administers livestock grazing. This rule also amends the Department of the Interior's appeals regulations pertaining to livestock grazing to provide consistency with administrative remedies provided for in the grazing regulations, increases public participation in the management of the public grazing lands, and amends the regulations on cooperative relations to reflect changes in the organization of certain advisory committees. The changes will improve the management of the Nation's public rangeland resources.

* * *

Numerous comments addressed the overall rulemaking. These comments asserted several central themes which crosscut different sections of the rulemaking. Accordingly, BLM has decided to address these central issues in this portion of the preamble. Within the context of such discussion, particular sections of the proposed and final rules will be referred to as necessary. Nevertheless, in these responses, BLM focuses upon central issues that were of concern to commenters throughout the proposal.

* * *

Some commenters took the position that general rangeland improvement is unnecessary. Their view was that current legislation, regulations, and procedures provide enough latitude and capability for the government to administer the public rangelands properly, therefore there is no justification for designing and implementing the rangeland improvement program. They stated that the initiative should be dropped or abandoned immediately. They asserted that the government has not shown that the proposal will benefit the western range and many of the elements of the rule are more appropriately dealt with in manuals, instruction memos, and policy guidance.

In addition, the comment was often made that the National Research Council study commissioned by the National Academy of Sciences reports that the conditions of rangeland health in the West are largely unknown. If the conditions are unknown, stated the commenters, it is impossible to demonstrate a need for the proposed rule. Some commenters stated that the entire proposal and EIS were politically driven and did not relate to the resource protection issues of public land administration.

The Department believes that there is a need for changes in public rangeland grazing administration. The Department has been collecting data on the condition of the rangelands for over 60 years. The Department does have considerable information on all BLM lands, based on these years of data collection, although the same level of detailed knowledge may not be available on every allotment. The information available is sufficient to identify trends in rangeland health across the western rangelands.

The status and trends of the western rangelands upon passage of the Public Rangelands Improvement Act (PRIA) in 1978 indicated that western rangelands were producing below their potential and that rangelands would remain in unsatisfactory condition or decline further unless the unsatisfactory conditions could be addressed and corrected by intensive public rangelands maintenance, management and improvement. Congress articulated its view in PRIA that such unsatisfactory conditions on public rangeland present a risk for soil loss, siltation, desertification, water loss, loss of wildlife and fish habitat, loss of forage for livestock and other grazing animals, degradation of water quality, flood danger, and threats to local economies. In addition, BLM National Public Lands Advisory Council recommended in 1992 that " * * * foremost consideration needs to be given to protecting the basic components of soil, water and vegetation. Without assurances for the future well-being of these basic natural resources, there is little to squabble about."

BLM's research has concluded that in the long term under current management practices 22 million acres of BLM uplands would be functioning but susceptible to degradation, and about 20 million acres would be nonfunctioning. The vegetation in some areas would change because of overgrazing, fire, or drought. Conditions would be worse in riparian and wetland areas. The overall trends would be a slow, steady, long-term decline in conditions. Approximately 466,000 acres of riparian areas (43 percent of the total) on BLM land would be functioning but susceptible to degradation, and 219,000 acres (21 percent) would be nonfunctioning.... These studies show that without some changes in the current program conditions in critical riparian areas would continue to decline.

* * *

Some commenters asserted that rangeland improvement is unnecessary because it will not improve the condition of the public rangelands. The Department disagrees. Commenters argued that few permittees or lessees are poor stewards of the public rangelands. They stated that the program will alienate many conscientious ranchers. The commenters asserted that the agencies and public may lose the service and support of these users in maintaining and improving the conditions of the public rangelands, and that rangeland conditions are likely to degrade. Therefore, they claimed, the initiative should be abandoned. However, the Department believes that improving administration of public rangelands will improve their condition, which will benefit all uses, including livestock grazing....

The standards and guidelines in the final rule are aimed at improving the ecological health of the rangelands.... The Department recognizes that the majority of public land grazing permittees and lessees are conscientious stewards. However, it also notes that line managers need clear authority and guidance to help correct problems in grazing use and to improve the degraded condition of some areas expeditiously. This program is intended to facilitate cooperation between BLM employees and public land users in making those improvements. Also, by making BLM and Forest Service management more similar, it will be easier for permittees and lessees to comply with land use requirements. Good stewards will not be adversely affected by this initiative and will have an opportunity to work with the Department to sustain the economic vigor of their industry while maintaining or improving the ecological health of the public lands. The Department recognizes that it is in the best interests of the users, the public, and BLM to cooperate in meeting these objectives.

* * *

A number of comments questioned whether the proposed amendments to the grazing rule conflict directly with TGA, FLPMA, PRIA and other related Federal laws. The BLM's main statutory authorities for regulating grazing on the public lands are TGA, FLPMA and PRIA. In TGA Congress directed the Secretary to bring order to the management of the public rangelands and improve range conditions.

Specifically, Section 2 of TGA provides:

The Secretary of the Interior shall make provision for the protection, administration, regulation, and improvement of such grazing districts * * * and he shall make such rules and regulations * * * and do any and all things necessary to accomplish the purposes of this Act * * * namely to regulate their occupancy and use, to preserve the land and its resources from destruction or unnecessary injury, to provide for the orderly use, improvement, and development of the range * * *.

The TGA authorizes the Secretary to, among other things, establish fees, issue permits and leases and prescribe terms and conditions for them, issue range improvement permits, and provide for local hearings on appeals. The emphasis on disposal of Federal lands changed with the Classification and Multiple Use Act in 1964 and FLPMA in 1976. In FLPMA Congress articulated the national policy that "the public lands be retained in Federal ownership." 43 U.S.C. § 1701. FLPMA also directs that land management be on the basis of multiple use and sustained yield, thus clarifying that other uses of public lands are equally appropriate. FLPMA did not repeal TGA, but did provide additional management direction. For example, section 402 of FLPMA provides that grazing permits and leases shall be:

[S]ubject to such terms and conditions the Secretary concerned deems appropriate and consistent with the governing law, including, but not limited to the authority of the Secretary concerned to cancel, suspend, or modify a grazing permit or lease for any violation of a grazing regulation or of any term or condition of such grazing permit or lease.

In 1978 Congress again focused on the public rangelands when it passed PRIA. In Section 2 of that Act Congress found that "vast segments" of the public rangelands were "producing less than their potential for livestock, wildlife habitat, recreation, forage and water and soil conservation benefits," and so were considered to be in an unsatisfactory condition. Congress went on in Section 2 to reaffirm a national commitment to "manage, maintain and improve the condition of the public rangelands so that they become as productive as feasible for all rangeland values." The Department has concluded that the amendments to the grazing rule are within the statutory authority granted by Congress to the Secretary to administer the public lands under TGA, FLPMA, PRIA, and related acts.

Notes and Questions

1. BLM unequivocally states that the Fundamentals of Rangeland Health and their associated standards and guidelines set forth in the final rule are aimed at improving the *ecological health* of the rangelands. Before delving further into what BLM thinks ecological health entails, are you convinced from BLM's description of the current state of the rangelands that they are ecologically unhealthy? Based on its account of history, does the Supreme Court in *Public Lands Council v. Babbitt* seem convinced?

2. BLM describes the majority of ranching interests using BLM lands as "conscientious stewards" and notes that "good stewards will not be adversely affected by this initiative." But if they majority of the ranchers are good stewards, why the need for rules to improve the ecological health of the rangelands?

3. The BLM is responsible for managing 264 million acres of land–about one-eighth of the land in the United States–and about 300 million additional acres of subsurface mineral resources. The agency is also responsible for wildfire management and suppression on 388 million acres. Most of the lands the BLM manages are located in the western United States, including Alaska, and are dominated by extensive grasslands, forests, high mountains, arctic tundra, and deserts. The BLM manages a wide variety of resources and uses, including energy and minerals; timber; forage; wild horse and burro populations; fish and wildlife habitat; wilderness areas; archaeological, paleontological, and historical sites; and other natural heritage values.

Public Lands Managed by the Bureau of Land Management (BLM)

The BLM's roots go back to the Land Ordinance of 1785 and the Northwest Ordinance of 1787. These laws provided for the survey and settlement of the lands that the original 13 colonies ceded to the Federal government after the War of Independence. As additional lands were acquired by the United States from Spain, France, and other countries, Congress directed that they be explored, surveyed, and made available for settlement. In 1812, Congress established the General Land Office in the Department of the Treasury to oversee the disposition of these Federal lands. As the 19th century progressed and the Nation's land base expanded further west, Congress encouraged the settlement of the land by enacting a wide variety of laws, including the Homesteading Laws and the Mining Law of 1872. These statutes served one of the major policy goals of the young country–settlement of the Western territories. With the exception of the Mining Law of 1872 and the Desert Land Act of 1877 (which was amended), all have since been repealed or superseded by other statutes.

The late 19th century marked a shift in Federal land management priorities with the creation of the first national parks, forests, and wildlife refuges. By withdrawing these lands from settlement, Congress signaled a shift in the policy goals served by the public lands. Instead of using them to promote settlement, Congress decided that they should be retained in public ownership because of their other resource values.

In the early 20th century, Congress directed the Executive Branch to manage activities on the remaining public lands–the lands now known as BLM Lands–not already devoted to a specific purpose such as park or refuge. The Mineral Leasing Act of 1920 allowed leasing, exploration, and production of selected commodities such as coal, oil, gas, and sodium to take place on public lands. The Taylor Grazing Act of 1934 established the U.S. Grazing Service to manage the public rangelands. In 1946, the Grazing Service was merged with the General Land Office to form the Bureau of Land Management within the Department of the Interior. When the BLM was initially created, there were over 2,000 unrelated and often conflicting laws for managing the public lands. The BLM had no unified legislative mandate until Congress enacted the Federal Land Policy and Management Act of 1976 (FLPMA). As noted in *Public Lands Council v. Babbitt*, FLPMA declared that the lands known today as the BLM lands would remain in public ownership and directed the agency to implement "multiple use" management, defined as "management of the public lands and their various resource values so that they are utilized in the combination that will best meet the present and future needs of the American people." 43 U.S.C. § 1702(c).

4. Maybe the problem is not the ranchers, but BLM, or, more specifically, public lands operated under a multiple-use mandate. Wouldn't the "health" of the public rangelands be most improved if they were sold to the highest bidder? It is unlikely that a private rancher would degrade lands he or she owns by overgrazing. And if the highest bidder were a land conservation trust, it would hardly need BLM to help it manage the land for grassland values. The lands least suited to grazing or grassland conservation may go to recreational, mining, or urban uses. So what? How, as a matter of national policy that "best meet[s] the present and future needs of the American people," is it preferable to have BLM manage these lands, given that it has done so in a manner that, by its own description, is *neither* economically efficient *nor* ecologically sound? For contrasting views on this topic, *compare* Charles H. Callison, *The Fallacies of Privatization*, 25 ENVIRONMENT 18 (Oct. 1983), *with* John Baden and Laura Rosen, *The*

Environmental Justification, 25 ENVIRONMENT 7 (Oct. 1983), and Delores T. Martin, *Divestiture and the Creation of Property Rights in Public Lands: A Comment*, 2 CATO JOURNAL 687 (1992).

C.   BIODIVERSITY MANAGEMENT IN GRASSLAND ECOSYSTEMS

As the preceding materials suggest, by far the most influential force in national grassland management policy is the Department of Interior's Bureau of Land Management (BLM). To a far lesser extent the U.S. Forest Service, through the 10 million-acre National Grasslands program, and the U.S. Fish and Wildlife Service, through its management of the National Wildlife Refuge System, also have a hand in grassland management policy. And the substantial amount of private lands in the Great Plains used largely for irrigated agriculture presents another opportunity for development of national, state, and local grassland management policy. The following materials provide background on each of those fronts of grassland ecosystem management.

1.   BLM's RANGELAND HEALTH AND RANGELAND PLANNING POLICY

a.   THE FUNDAMENTALS OF RANGELAND HEALTH

We now come to the substantive provisions of the Rangeland Reform rules BLM adopted in 1995 that triggered *Public Lands Council v. Babbitt* and which represent BLM's turn toward ecosystem management. Recall that we have traced the development of two themes in BLM grazing policy: (1) describing the status of grazing preferences as property rights or personal privileges; and (2) the regulation of grazing allotments pursuant to ecological factors. The materials in this section bring closure to the history of those two themes (at least for the time being), first through the BLM's description of its ecosystem management criteria, and then through the Supreme Court's endorsement of BLM's new way of describing grazing preferences.

In its Fundamentals of Rangeland Health and Standards and Guidelines for Grazing Administration, BLM took explicit steps toward use of ecosystem management principles in the determination of range grazing levels. The rules establish nationally uniform "fundamentals" and then require the adoption by field offices of the agency of state and local "standards and guidelines" consistent with the national criteria. BLM's field level decisions, moreover, are to be made in consultation with Resource Advisory Councils (RACs) composed of people representing a variety of interests, including grazing interests, environmental interests, and state and local political interests.

Department of the Interior, Bureau of Land Management, Department Hearings and Appeals Procedures; Cooperative Relations; Grazing Administration—Exclusive of Alaska

60 Fed. Reg. 9894 (Feb. 22, 1995).

* * *

Subpart 4180—Fundamentals of Rangeland Health and Standards and Guidelines for Grazing Administration

Section 4180.1, The fundamentals of rangeland health (titled National Requirements for Grazing Administration in the proposed rule) for grazing administration, are added to establish fundamental requirements for achieving functional, healthy public rangelands. These fundamentals address the necessary physical components of functional watersheds, ecological processes required for healthy biotic communities, water quality standards, and habitat for threatened or endangered species or other species of special interest.

Where it is determined that existing grazing management needs to be modified to ensure that the conditions of healthy rangelands set forth in § 4180.1, Fundamentals of rangeland health, are met or significant progress is being made to meet the fundamentals, the authorized officer must take appropriate action as soon as practical, but not later than the start of the next grazing season. This may include actions such as reducing livestock stocking rates, adjusting the season or duration of livestock use, or modifying or relocating range improvements.

Section 4180.2, Standards and guidelines for grazing administration, is added to direct that standards and guidelines will be developed for an entire State or for an area encompassing portions of more than one State, except where the geophysical or vegetal character of an area is unique and the health of the rangelands will not be ensured by using standards and guidelines developed for a larger geographical area. The geographical area covered will be determined by BLM State Directors in consultation with affected RACs. Once standards and guidelines are in effect, the authorized officer shall take appropriate action as soon as practical, but not later than the start of the next grazing year upon determining that existing grazing management practices are significant factors in failing to ensure significant progress toward the fulfillment of the standards and toward conformance with the guidelines. The preparation of standards and guidelines will involve public participation and consultation with RACs, Indian tribes, and Federal agencies responsible for the management of lands within the affected area.

Section 4180.2(d) lists factors that, at a minimum, must be addressed in the development of State or regional standards. The guiding principles for the development of standards pertain to the factors needed to help achieve rangeland health. More specifically, the factors relate to watershed function, threatened or endangered species and candidate species, habitat for native plant and animal populations, water quality and the distribution of nutrients and energy flow. Section 4180.2(e) lists guiding principles to be addressed in the development of guidelines.

* * *

Subpart 4180—Fundamentals of Rangeland Health and Standards and Guidelines for Grazing Administration

§ 4180.1   Fundamentals of rangeland health.

The authorized officer shall take appropriate action under subparts 4110, 4120, 4130, and 4160 of this part as soon as practicable but not later than the start of the next grazing year upon determining that existing grazing management needs to be modified to ensure that the following conditions exist.

(a) Watersheds are in, or are making significant progress toward, properly functioning physical condition, including their upland, riparian-wetland, and aquatic components; soil and plant conditions support infiltration, soil moisture storage, and the release of water that are in balance with climate and landform and maintain or improve water quality, water quantity, and timing and duration of flow.

(b) Ecological processes, including the hydrologic cycle, nutrient cycle, and energy flow, are maintained, or there is significant progress toward their attainment, in order to support healthy biotic populations and communities.

(c) Water quality complies with State water quality standards and achieves, or is making significant progress toward achieving, established BLM management objectives such as meeting wildlife needs.

(d) Habitats are, or are making significant progress toward being, restored or maintained for Federal threatened and endangered species, Federal Proposed, Category 1 and 2 Federal candidate and other special status species.

§ 4180.2   Standards and guidelines for grazing administration.

(a) The Bureau of Land Management State Director, in consultation with the affected resource advisory councils where they exist, will identify the geographical area for which standards and guidelines are developed. Standards and guidelines will be developed for an entire state, or an area encompassing portions of more than 1 state, unless the Bureau of Land Management State Director, in consultation with the resource advisory councils, determines that the characteristics of an area are unique, and the rangelands within the area could not be adequately protected using standards and guidelines developed on a broader geographical scale.

(b) The Bureau of Land Management State Director, in consultation with affected Bureau of Land Management resource advisory councils, shall develop and amend State or regional standards and guidelines. The Bureau of Land Management State Director will also coordinate with Indian tribes, other State and Federal land management agencies responsible for the management of lands and resources within the region or area under consideration, and the public in the development of State or regional standards and guidelines. Standards and guidelines developed by the Bureau of Land Management State Director must provide for conformance with the fundamentals of § 4180.1. . . .

(c) The authorized officer shall take appropriate action as soon as practicable but not later than the start of the next grazing year upon determining that existing grazing management practices or levels of grazing use on public lands are significant factors in failing to achieve the standards and conform with the guidelines that are made effective under this section. . . .

(d) At a minimum, State or regional standards developed under paragraphs (a) and (b) of this section must address the following:

(1) Watershed function;

(2) Nutrient cycling and energy flow;

(3) Water quality;

(4) Habitat for endangered, threatened, proposed, Candidate 1 or 2, or special status species; and

(5) Habitat quality for native plant and animal populations and communities.

(e) At a minimum, State or regional guidelines developed under paragraphs (a) and (b) of this section must address the following:

(1) Maintaining or promoting adequate amounts of vegetative ground cover, including standing plant material and litter, to support infiltration, maintain soil moisture storage, and stabilize soils;

(2) Maintaining or promoting subsurface soil conditions that support permeability rates appropriate to climate and soils;

(3) Maintaining, improving or restoring riparian-wetland functions including energy dissipation, sediment capture, groundwater recharge, and stream bank stability;

(4) Maintaining or promoting stream channel morphology (e.g., gradient, width/depth ratio, channel roughness and sinuosity) and functions appropriate to climate and landform;

(5) Maintaining or promoting the appropriate kinds and amounts of soil organisms, plants and animals to support the hydrologic cycle, nutrient cycle, and energy flow;

(6) Promoting the opportunity for seedling establishment of appropriate plant species when climatic conditions and space allow;

(7) Maintaining, restoring or enhancing water quality to meet management objectives, such as meeting wildlife needs;

(8) Restoring, maintaining or enhancing habitats to assist in the recovery of Federal threatened and endangered species;

(9) Restoring, maintaining or enhancing habitats of Federal Proposed, Category 1 and 2 Federal candidate, and other special status species to promote their conservation;

(10) Maintaining or promoting the physical and biological conditions to sustain native populations and communities;

(11) Emphasizing native species in the support of ecological function; and

(12) Incorporating the use of non-native plant species only in those situations in which native species are not available in sufficient quantities or are incapable of maintaining or achieving properly functioning conditions and biological health.

Notes and Questions

1. Consider the 12 criteria the state and regional guidelines must address under section 4180.2(e) of BLM's rule. Compare them to Grumbine's list of ingredients in his 1994 description of ecosystem management, *What Is Ecosystem Management?*, reproduced in Chapter 7. Does BLM's list seem complete? Does it seem reasonable to expect the BLM to be able to afford to satisfy the criteria throughout its public land domain?

2. The obvious implication of BLM's rule is that ecological criteria will be used to determine grazing levels. As a leading grazing law expert explains, ranchers thus feared

> that the BLM, acting in response to concerns raised by environmentalists or other public land users, will use its land planning process to substantially reduce authorized grazing levels, and possibly eliminate grazing altogether in some areas, in order to favor competing interests such as preservation and improvement of wildlife habitat, enhancement of recreational opportunities, or protection of the natural character of wilderness areas.

Joseph M. Feller, *Back to the Present: The Supreme Court Refuses to Move Public Range Law Backward, but Will the BLM Move Public Range Law Forward?*, 31 Envtl. L. Rep. (ELI) 10021 (2001). Do you think the BLM's rules could support such an effort? Will they *necessarily* lead to such a result? How much discretion does BLM retain under the new rules?

3. To evaluate the effect of the Fundamentals of Rangeland Health rule in terms of its potential to change grazing allotment decision outcomes, fill in the following chart. The chart uses the three components of ecosystem management policy articulated in Chapter 7 as a way of identifying differences in approach between the pre-and post-rule policies. Use the two excerpts from *Public Lands Council v. Babbitt* to guide your description of BLM's "before 1995 rule" policy approaches.

	Currency	**Goals**	**Methods**
Before 1995 rule			
After 1995 rule			

4. Are you convinced that BLM has the legal authority, under the three relevant statutes, to manage the rangelands based on improving their ecological health?

5. Before adopting its 1995 rule, BLM had issued two policy statements that pointed clearly in the direction of an ecological health focus. First, in 1994 the agency published *Ecosystem Management in the BLM: From Concept to Commitment*, wherein it defined nine operating principles of ecosystem management. Two of the nine principles included "sustain the productivity and diversity of ecological systems" and "minimize or repair impacts to the land." The agency also advocated using the adaptive management methods described in Chapter 7. Later that year, the agency released its *Blueprint for the Future*, in which it stated its policy focus would include "becoming more aware of the status, trend, and overall health of the land." These precursors to the 1995 rule laid the groundwork for the ecological health focus of the rule. *See also* Michael P. Dombeck, Thinking Like a Mountain: BLM's Approach to Ecosystem Management, 6 Ecological Applications 699 (1996) (Director of BLM).

6. Some commenters asserted that various sections of BLM's proposed rule, which was substantially the same as the final rule, raised the possibility of a "taking" of private property rights without "just compensation" as required by the Fifth Amendment. BLM pointed out that, as a threshold matter, the United States Constitution gives Congress the "Power to dispose of and make all needful Rules and Regulations respecting the Territory or other Property belonging to the United States." Article IV, § 3, cl. 2. This power includes authority to control the use and occupancy of Federal lands, to protect them from trespass and injury and to prescribe the conditions upon which others may obtain rights in them. *See* Utah Power & Light Co. v. United States, 243 U.S. 389, 405 (1917). BLM also pointed to the series of laws through which Congress has delegated primary responsibility and authority to manage livestock grazing on public lands to the Secretary, acting through BLM. The basic laws, again, are the Taylor Grazing Act, Federal Land Planning and Management Act, and the Public Rangelands Improvement Act. As BLM explained:

> In authorizing the issuance of grazing permits in TGA, Congress expressly provided that the "issuance of a permit * * * shall not create any right, title, interest, or estate in or to the [public] lands." 43 U.S.C. § 315b. In FLPMA, Congress authorized the Secretary to "cancel, suspend, or modify a grazing permit or lease, in whole or in part, pursuant to the terms and conditions" of the permit or lease. 43 U.S.C.

§ 1752(a). The same section also authorizes the Secretary to "cancel or suspend a grazing permit or lease for any violation of a grazing rule or of any term or condition of such permit or lease." These statutes are implemented by BLM's regulations at 43 CFR Part 4100 et seq., including the amendments adopted here.

The Fifth Amendment to the United States Constitution provides in relevant part that no person shall be denied property without due process of law, and no private property shall be taken for public use, without just compensation. This Amendment protects private property. Because Congress made clear in TGA that grazing permits create no private property interest in public lands, the Fifth Amendment's protection is not implicated. The Courts have long held that no taking of private property occurs in the course of lawful administration and regulation of Federal grazing lands because the grazing permit represents a benefit or privilege bestowed by the Federal government upon a private individual and not a compensable property interest under the Fifth Amendment.

Thus, an authorized officer's decision to change permitted use (§ 4110.3), decrease permitted use (§ 4110.3–2), implement a reduction in permitted use (§ 4110.3–3), decrease land acreage (§ 4110.4–2), approve an AMP (§ 4120.2), or approve a cooperative range improvement agreement (§ 4120.3–2) does not give rise to a takings claim.

Some commenters asserted that permittees and lessees should be compensated for any indirect adverse impact that cancellation, nonrenewal, suspension or modification of grazing permits might have on the permittee's base property. While base property is private property protected by the Fifth Amendment, the United States Supreme Court, in an opinion by Chief Justice Rehnquist, specifically considered and rejected the argument that the increment of value added to a private ranch by a public land grazing permit is a compensable property interest, United States v. Fuller, 409 U.S. 488 (1973).

Even if, in other words, cancellation, nonrenewal, suspension, or changes in the terms and conditions of a grazing permit might have some negative effect on the value of the base property, the Supreme Court has made clear this is not a "taking."

Does this resolve the matter to your satisfaction?

7. Why allow *any* grazing of cattle and sheep in public lands that were formerly grasslands? Many people are surprised to learn that, while grazing is authorized on most of the western lands under BLM's jurisdiction, (1) the livestock produced from those lands accounts for a small fraction of the nation's beef supply; (2) the direct and indirect ranching economies associated with public land grazing, including down to the coffee shops where the ranchers meet, is insignificant to the national economy; and (3) revenues from the grazing fees BLM charges fail to cover even the agency's costs of administration. *See* Thomas M. Power, Lost Landscapes and Failed Economies (1996); U.S. Department of the Interior, Rangeland Reform '94, Draft and Final Environmental Impact Statements (1994 and 1995); General Accounting Office, Rangeland Management: BLM's Hot Desert Grazing Program Merits Reconsideration (Nov. 1991). As currently administered, therefore, BLM's grazing program thus is both economically and ecologically a losing proposition on a national level. Why squander these irreplaceable and diminishing national public resources on such an activity? Why indeed, asked Professor Deb Donahue, in her recent book calling for an end to grazing on the public rangelands, *The Western Range Revisited: Removing Livestock from Public Lands to Conserve Native Biodiversity*. Focusing on the more arid rangelands where the negative impact of grazing is even more pronounced, she argues:

> Management of public lands to produce livestock is vulnerable to challenge on several grounds. It is "out of sync" with changed public values and demands, with policies currently expressed in federal law, and with historical legislative and popular intent regarding use of public rangelands. Moreover, it is beyond cavil that for decades neither ranchers nor land managers have understood range ecology; as a consequence, rangelands that never should have been grazed were, and continue to be, used by domestic livestock.

See also Debra L. Donahue, *Justice for the Earth in the Twenty–First Century*, 1 WYO. L. REV. 373 (2001) (summarizing her thesis). This unquestionably is a controversial position for anyone to take. What made it all the more so for Professor Donahue is that she is on the law faculty at the University of Wyoming, one of the more cattle-friendly states in the nation. The reaction was predictable: letters to the editor condemning Donahue filled Wyoming's newspapers; the Wyoming Senate entertained legislation to close the law school; livestock interests demanded an investigation and an overhaul of faculty selection procedures; and the University's president proclaimed the institution's official position of "support for those industries—including production agriculture—that have brought this state from its status as a territory in 1886 to its promise in the new millennium." *See* Tom Kenworthy, *A Discouraging Word in Tome on the Range*, USA TODAY, Mar. 3, 2000, at 3A. Ironically, the fallout of the 1999 publication of her book brought her thesis national media attention. Regardless of what you think of her idea, could BLM adopt her recommendation and still comply with its regulations and authorizing statutes? Given what you know about the state of the rangelands and the causes thereof, *must* BLM adopt her proposal? How, using the statutory and regulatory text, would you construct the legal argument leading to Professor Donahue's policy recommendation?

8. In the final analysis, the Fundamentals of Rangeland Health regulations only open the door to ecosystem management. They do not preclude or mandate grazing anywhere within BLM's domain. How BLM—now under a new administration—implements them will be the real test of their effect. For more on the rules and their prospects for implementation, *see* Karl N. Arruda and Christopher Watson, *The Rise and Fall of Grazing Reform*, 32 LAND & WATER L. REV. 413 (1997); Joseph H. Feller, *'Til the Cows*

Come Home: The Fatal Flaw in the Clinton Administration's Public Lands Grazing Policy, 25 Envtl. L. 703 (1995); Todd M. Olinger, *Public Rangeland Reform: New Prospects for Collaboration and Local Control Using the Resource Advisory Councils*, 69 U. Colo. L. Rev. 633 (1998); Bruce M. Pendery, *Reforming Livestock Grazing on the Public Domain: Ecosystem Management–Based Standards and Guidelines Blaze a New Path for Range Management*, 27 Envtl. L. 513 (1997).

9. In the introduction to *Public Lands Council v. Babbitt* at the beginning of the section on the history of grasslands law, we asked you to read carefully the Court's description of the grazing preference allotment system. Nothing in BLM's new Fundamentals of Rangeland Health criteria alters the way that system operates. But implementing the ecosystem management focus of the new rule may have been difficult had the precise status of the preferences remained fuzzy. Read on to see how BLM brought the preference system terminology into the age of ecosystem management.

Public Lands Council v. Babbitt

Supreme Court of the United States, 2000.
529 U.S. 728.

■ Justice Breyer delivered the opinion of the Court.

[The first portion of the opinion, providing the historical background of the Taylor Grazing Act and the subsequent law of grazing on BLM's federal public lands, is reproduced at the beginning of this chapter.]

E

This case arises out of a 1995 set of Interior Department amendments to the federal grazing regulations. 60 Fed. Reg. 9894 (1995) (Final Rule). The amendments represent a stated effort to "accelerate restoration" of the rangeland, make the rangeland management program "more compatible with ecosystem management," "streamline certain administrative functions," and "obtain for the public fair and reasonable compensation for the grazing of livestock on public lands." 58 Fed. Reg. 43208 (1993) (Proposed Rule). The amendments in final form emphasize individual "stewardship" of the public land by increasing the accountability of grazing permit holders; broaden membership on the district advisory boards; change certain title rules; and change administrative rules and practice of the Bureau of Land Management to bring them into closer conformity with related Forest Service management practices. See 60 Fed. Reg. 9900–9906 (1995).

Petitioners Public Lands Council and other nonprofit ranching-related organizations with members who hold grazing permits brought this lawsuit against the Secretary and other defendants in Federal District Court, challenging 10 of the new regulations. The court found 4 of 10 unlawful. 929 F.Supp. 1436, 1450–1451 (D.Wyo.1996). The Court of Appeals reversed the District Court in part, upholding three of the four. 167 F.3d 1287, 1289 (C.A.10 1999). Those three (which we shall describe further below) (1) change the definition of "grazing preference"; (2) permit those who are not "engaged in the livestock business" to qualify for grazing permits; and (3) grant the United States title to all future "permanent" range improvements. One judge on the court of appeals dissented in respect to the Secretary's authority to promulgate the first and the third regulations. See 167 F.3d, at 1309–1318. We granted certiorari to consider the ranchers' claim that these three regulatory changes exceed the authority that the Taylor act grants the Secretary.

II

A

The ranchers attack the new "grazing preference" regulations first and foremost. Their attack relies upon the provision in the Taylor act stating that "grazing privileges recognized and acknowledged shall be adequately safeguarded...." 43 U.S.C. § 315b. Before 1995 the regulations defined the term "grazing preference" in terms of the AUM-denominated amount of grazing privileges that a permit granted. The regulations then defined "grazing preference" as "the total number of animal unit months of livestock grazing on public lands apportioned and attached to base property owned or controlled by a permittee or lessee." 43 CFR § 4100.0–5 (1994).

The 1995 regulations changed this definition, however, so that it now no longer refers to grazing privileges "apportioned," nor does it speak in terms of AUMs. The new definition defines "grazing preference" as "a superior or priority position against others for the purpose of receiving a grazing permit or lease. This priority is attached to base property owned or controlled by the permittee or lessee." 43 CFR § 4100.0–5 (1995).

The new definition "omits reference to a specified quantity of forage." 60 Fed. Reg. 9921 (1995). It refers only to a priority, not to a specific number of AUMs attached to a base property. But at the same time the new regulations add a new term, "permitted use," which the Secretary defines as "the forage allocated by, or under the guidance of, an applicable land use plan for livestock grazing in an allotment under a permit or lease and is expressed in AUMs." 43 CFR § 4100.0–5 (1995).

This new "permitted use," like the old "grazing preference," is defined in terms of allocated rights, and it refers to AUMs. But this new term as defined refers, not to a rancher's forage priority, but to forage "allocated by, or under the guidance of an applicable land use plan." Ibid. (emphasis added). And therein lies the ranchers' concern.

The ranchers refer us to the administrative history of Taylor Act regulations, much of which we set forth in Part I. In the ranchers' view, history has created expectations in respect to the security of "grazing privileges"; they have relied upon those expectations; and the statute requires the Secretary to "safeguar[d]" that reliance. Supported by various farm credit associations, they argue that defining their privileges in relation to land use plans will undermine that security. They say that the content of land use plans is difficult to predict and easily changed. Fearing that the resulting uncertainty will discourage lenders from taking mortgages on ranches as security for their loans, they conclude that the new regulations threaten the stability, and possibly the economic viability, of their ranches, and thus fail to "safeguard" the "grazing privileges" that Department regulations previously "recognized and acknowledged." Brief for Petitioners 22–23.

We are not persuaded by the ranchers' argument for three basic reasons. First, the statute qualifies the duty to "safeguard" by referring directly to the Act's various goals and the Secretary's efforts to implement them. The full subsection says:

"So far as consistent with the purposes and provisions of this subchapter, grazing privileges recognized and acknowledged shall be adequately safeguarded, but the creation of a grazing district or the issuance of a permit pursuant to the provisions of this subchapter shall not create any right, title, interest or estate in or to the lands." 43 U.S.C. § 315b (emphasis added).

The words "so far as consistent with the purposes ... of this subchapter" and the warning that "issuance of a permit" creates no "right, title, interest or estate" make clear that the ranchers' interest in permit stability cannot be absolute; and that the Secretary is free reasonably to determine just how, and the extent to which, "grazing privileges" shall be safeguarded, in light of the Act's basic purposes. Of course, those purposes include "stabiliz[ing] the livestock industry," but they also include "stop[ping] injury to the public grazing lands by preventing overgrazing and soil deterioration," and "provid [ing] for th[e] orderly use, improvement, and development" of the public range. 48 Stat. 1269; see supra, at 1819.

Moreover, Congress itself has directed development of land use plans, and their use in the allocation process, in order to preserve, improve, and develop the public rangelands. See 43 U.S.C. §§ 1701(a)(2), 1712. That being so, it is difficult to see how a definitional change that simply refers to the use of such plans could violate the Taylor Act by itself, without more. Given the broad discretionary powers that the Taylor Act grants the Secretary, we must read that Act as here granting the Secretary at least ordinary administrative leeway to assess "safeguard[ing]" in terms of the Act's other purposes and provisions. Cf. §§ 315, 315a (authorizing Secretary to establish grazing districts "in his discretion" (emphasis added), and to "make provision for protection, administration, regulation, and improvement of such grazing districts").

Second, the pre–1995 AUM system that the ranchers seek to "safeguard" did not offer them anything like absolute security—not even in respect to the proportionate shares of grazing land privileges that the "active/suspended" system suggested. As discussed above, the Secretary has long had the power to reduce an individual permit's AUMs or cancel the permit if the permit holder did not use the grazing privileges, did not use the base property, or violated the Range Code. See supra, at 1820 (collecting CFR citations 1938–1998). And the Secretary has always had the statutory authority under the Taylor act and later FLMPA to reclassify and withdraw range land from grazing use, see 43 U.S.C. § 315f (authorizing Secretary, "in his discretion, to examine and classify any lands ... which are more valuable or suitable for the production of agricultural crops ... or any other use than [grazing]"); §§ 1712, 1752(c) (authorizing renewal of permits "so long as the lands ... remain available for domestic livestock grazing in accordance with land use plans "(emphasis added)). The Secretary has consistently reserved the authority to cancel or modify grazing permits accordingly. See supra, at 1820–1821 (collecting CFR citations). Given these well-established pre-1995 Secretarial powers to cancel, modify, or decline to review individual permits, including the power to do so pursuant to the adoption of a land use plan, the ranchers' diminishment-of-security point is at best a matter of degree.

Third, the new definitional regulations by themselves do not automatically bring about a self-executing change that would significantly diminish the security of granted grazing privileges. The Department has said that the new definitions do "not cancel preference," and that any change is "merely a clarification of terminology." 60 Fed. Reg. 9922 (1995). It now assures us through the Solicitor General that the definitional changes "preserve all elements of preference" and "merely clarify the regulations within the statutory framework." See Brief in Opposition 13, 14.

The Secretary did consider making a more sweeping change by eliminating the concept of "suspended use"; a change that might have more reasonably prompted the ranchers' concerns. But after receiving comments, he changed his mind. See 59 Fed. Reg. 14323 (1994). The Department has instead said that "suspended" AUMs will "continue to be recognized and have a priority for additional grazing use within the allotment. Suspended use provides an important accounting of past grazing use for the ranching community and is an insignificant administrative workload to the agency." Bureau of Land Management, Rangeland Reform '94: Final Environmental Impact Statement 144 (1994).

Of course, the new definitions seem to tie grazing privileges to land-use plans more explicitly than did the old. But, as we have pointed out, the Secretary has since 1976 had the authority to use land use plans to determine the amount of permissible grazing, 43 U.S.C. § 1712. The Secretary also points out that since development of land use plans began nearly 20 years ago, "all BLM lands in the lower 48 states are covered by land use plans," and "all grazing permits in those States have now been issued or renewed in accordance with such plans, or must now conform to them." Brief for United States 26. Yet the ranchers have not provided us with a single example in which interaction of plan and permit has jeopardized or might yet jeopardize permit security. An amicus brief filed by a group of Farm Credit Institutions says that the definitional change will "threate[n]" their "lending policies." Brief for Farm Credit Institutions as Amicus Curiae 3. But they do not explain why that is so, nor do they state that the new definitions will, in fact, lead them to stop lending to ranchers.

We recognize that a particular land use plan could change pre-existing grazing allocation in a particular district. And that change might arguably lead to a denial of grazing privileges that the pre–1995 regulations would have provided. But the affected permit holder remains free to challenge such an individual effect on grazing privileges, and the courts remain free to determine its lawfulness in context. We here consider only whether the changes in the definitions by themselves violate the Taylor act's requirement that recognized grazing privileges be "adequately safeguarded." Given the leeway that the statute confers upon the Secretary, the less-than-absolute pre-1995 security that permit holders enjoyed, and the relatively small differences that the new definitions create, we conclude that the new definitions do not violate that law.

* * *

The judgment of the Court of Appeals is

Affirmed.

Notes and Questions

1. Have you pieced together what made this portion of BLM's 1995 rule controversial? Consider this summary by an expert in grazing law:

> The Rangeland Reform amendments sought to clear away some of the mythological haze associated with "suspended" AUMs and with the use of the term "preference" over the previous two decades. In order to clarify that the preferences established under the Taylor Act determine only who is permitted to place livestock on the range and not how many cattle or sheep are allowed, the new regulations define "grazing preference" to mean simply "a superior priority position against others for the purpose of receiving a grazing permit or lease." To emphasize that number of livestock is a resource management issue governed by BLM's land use plans, the amendments added a new regulatory term, "permitted use," defined as "the forage allocated by, or under the guidance of, an applicable land use plan for livestock grazing in an allotment under a permit or lease."
>
> To further clarify that grazing permittees have no permanent entitlement to graze a particular number of livestock, the Rangeland Reform Amendments prospectively discontinued, for the most part, the terminology of "suspending" rather than reducing grazing privileges when a permit is scaled down.

Joseph M. Feller, *Back to the Present: The Supreme Court Refuses to Move Public Range Law Backward, but Will the BLM Move Public Range Law Forward?*, 31 Envtl. L. Rep. (ELI) 10021 (2001). Couple these changes with the new Fundamentals of Rangeland Health rules, which emphasize ecological factors in the determination of grazing levels, and one has the recipe for ecosystem management on BLM lands. For more on this theme, see Erik Schlenker–Goodrich, *Moving Beyond Public Lands Council v. Babbitt: Land Use Planning and the Range Resource*, 16 J. Envtl. Law & Litigation

139 (2001). For the counterview, that *Public Lands Council v. Babbitt* was wrongly decided and expanded BLM authority beyond congressional intent, see Julie Andersen, *Public Lands Council v. Babbitt: Herding Ranchers Off Public Land?*, 2000 BYU L. REV. 1273.

2. Of course, emphasizing ecosystem factors does not bring them to fruition. The Court in *Public Lands Council v. Babbitt* states that a particular land use plan could lead to a denial of grazing privileges that the pre–1995 regulations would have provided. How, substantively and procedurally, does the 1995 rule make that possible?

3. As the Court also noted, the adoption of the "permitted use" definition to describe what grazing privileges really are did not change what they had been all along–privileges, not property rights. Yet BLM must have felt it was important to make that clear through the new definition as a means of facilitating its ecosystem management policy and the Fundamentals of Rangeland Health. What if grazing privileges *were* property rights? Could BLM still apply its Fundamentals of Rangeland Health criteria to determine grazing levels?

b. BLM'S OTHER RANGELAND MANAGEMENT PLANNING REQUIREMENTS

In addition to the substantive provisions of its organic statutes, BLM is subject to planning requirements found principally in FLPMA and the National Environmental Policy Act. The following administrative opinion illustrates the steps BLM must take to satisfy its FLPMA planning obligations. In particular, note how the opinion deftly handles the question whether BLM must engage in a rigorous cost-benefit or public interest analysis when reaching its grazing allotment decisions.

National Wildlife Federation v. Bureau of Land Management

Interior Board of Land Appeals, 1997.
IBLA 94–264.

■ OPINION BY DEPUTY CHIEF ADMINISTRATIVE JUDGE HARRIS.

On December 20, 1993, District Chief Administrative Law Judge John R. Rampton, Jr., issued a Decision involving appeals relating to grazing in the Comb Wash Allotment in southeastern Utah. In his Decision, Judge Rampton precluded BLM from allowing any further grazing of cattle on certain public lands in the allotment, specifically five canyons (Arch, Mule, Fish Creek, Owl Creek, and Road Canyons), pending compliance with the environmental review mandate of section 102(2)(C) of the National Environmental Policy Act of 1969 (NEPA), as amended, 42 U.S.C. § 4332(2)(C) (1994), and the requirements of sections 302(a) and 309(e) of the Federal Land Policy and Management Act of 1976 (FLPMA), 43 U.S.C. §§ 1732(a) and 1739(e) (1994), for consideration of principles of "multiple use" and public participation.

The American Farm Bureau Federation (AFBF) and Utah Farm Bureau Federation (UFBF), the Ute Mountain Ute Indian Tribe (Tribe), and BLM each filed appeals from that Decision. Together with its notice of appeal, the Tribe filed a petition for stay. Neither of the other Appellants sought a stay. National Wildlife Federation, Southern Utah Wilderness Alliance, and Joseph M. Feller (collectively referred to as NWF) filed a document, inter alia, opposing the petition for stay and requesting that the Decision be put into full force and effect.

In a Decision dated March 1, 1994, the Board denied the petition for stay and granted the request to put Judge Rampton's Decision into effect, pursuant to 43 C.F.R. § 4.477(b) (1994), thereby prohibiting grazing in the canyons pending resolution of the appeals on their merits. National Wildlife Federation v. BLM, 128 IBLA 231 (1994). In a subsequent Decision, the Board granted a motion to dismiss the appeal of AFBF, but granted AFBF amicus curiae status. National Wildlife Federation v. BLM, 129 IBLA 124 (1994).

I. Factual and Procedural Background

The Comb Wash Allotment encompasses nearly 72,000 acres, of which approximately 63,000 acres are public lands, the remainder being state and private. The allotment is located southwest of Blanding, Utah, and northwest of Bluff, Utah. Within the allotment boundaries is the geographic feature from which the allotment derives its name, Comb Wash, a narrow valley that runs north-south just west of the Comb Ridge for about 20 miles. Draining into Comb Wash from the west are five canyons, Arch, Mule, Fish Creek, Owl Creek, and Road, sections of which, ranging from approximately 4 to 7 miles in length, are also within the allotment boundaries. The canyon bottoms are narrow, generally less than a half-mile wide and in places no more than 200 yards wide. The canyons encompass about 7,000 acres, or 10 percent of the allotment land. Each canyon contains a perennial or ephemeral stream, with an associated riparian area. The canyons provide recreational opportunities for camping, hiking, photography, sightseeing, and the viewing of archaeological sites, including many remnants of the ancient Anasazi culture.

* * *

The White Mesa Community, whose members are enrolled to the Tribe, owns the White Mesa Cattle Company, which conducts grazing operations on the allotment. The Comb Wash Allotment is one of six allotments on approximately 250,000 acres of public and national forest lands used by the White Mesa Community for its grazing operations. From 1987 to 1991, the White Mesa Community employed about four tribal members per year in its grazing operations. Despite limited employment opportunities for tribal members, the operation provides income for the White Mesa Community, which the community uses to make home repairs for senior citizens, purchase clothing for school children, and maintain a small store.

In 1966, BLM established an active grazing preference of 3,796 animal unit months (AUM's) for BLM-administered public lands in the allotment for the annual grazing season from October 16 to May 31. That preference allowed 506 cattle to graze the allotment. Prior to 1985, the number of cattle using the canyons was unknown because cattle were allowed to graze the entire allotment, including the canyon areas, during the annual 7 ½ month grazing season.

In May 1986, BLM issued a draft resource management plan (RMP) and environmental impact statement (EIS) to establish general management standards and guidelines concerning grazing and other permitted resource uses of the 1.8 million acres of public land in its San Juan Resource Area. In developing that document, BLM grouped the Comb Wash Allotment with other allotments for purposes of considering the environmental consequences of alternative grazing management plans for the resource area. In September 1987, BLM issued a proposed RMP and final EIS (FEIS) for the resource area. Following another comment period, BLM reissued the proposed RMP in April 1989. However, prior to reissuance, on February 20, 1989, the San Juan Resource Area Manager issued a 10–year grazing permit to the Tribe for the Comb Wash Allotment authorizing grazing in the amount of 3,791 AUM's. The Utah State Director, BLM, approved the RMP in a March 18, 1991, Record of Decision and Rangeland Program Summary.

During the development of the RMP, BLM changed its grazing scheme for the Comb Wash Allotment. In the fall of 1986, it began to manage the canyon areas, as well as other land in the allotment, as separate pastures, regulating the movement of cattle among the pastures by the use of existing or newly-constructed fences. The BLM authorized the Tribe to drive cattle into the canyons and hold them there for a month at a time with fences constructed across the canyon mouths. It also began to monitor forage utilization. It established trend study plots and established objectives for increasing the frequency of key plant species in the allotment. However, none of those study plots was in the canyons or in any riparian area on the Comb Wash Allotment. Following a number of years of monitoring, BLM found that key species in the study plots had not changed or had decreased in frequency. In response to that finding, BLM adjusted its objectives downward to reflect the status quo in 1991.

The BLM developed a plan for a 4–year grazing cycle for the allotment in order to improve the vegetation and in response to increased recreational use of the canyons. The features of that plan were that the season of use would be changed so that the canyons would be grazed only during the dormant season, from November 1 to the middle of March; only 50 cattle would be grazed in each canyon for a period of only 1 month; at least one canyon would be rested every fourth or fifth year; pastures outside the canyons would be grazed each year in an alternating pattern of spring grazing for two seasons followed by winter grazing for two seasons; and forage utilization in the canyons would be limited to 40 percent and outside the canyons to 50 percent in spring and 60 percent in the winter.

The BLM never formally adopted the 4–year grazing cycle, but it nevertheless implemented the plan through annual grazing authorizations with the 1990–91 season being the first year of the cycle. However, BLM did not adhere strictly to the features of the system. There was evidence at the hearing of the authorization of more than 50 cattle to graze in certain canyons, scheduling grazing in a canyon that was to be rested, and overutilization of certain key species in certain canyons.

The annual grazing authorizations, two of which were challenged in this case, were issued by the San Juan Resource Area Manager, Edward Scherick, who believed he had discretion under the RMP to determine whether grazing should take place in the canyons. However, in exercising that discretion, he relied exclusively on the recommendations provided to him by Paul Curtis, the San Juan Resource Area range conservationist. Scherick did not conduct any independent analysis of the effects of grazing on other resources. He testified that he always accepted Curtis' recommendations concerning grazing.

Curtis stated that it had been decided in the RMP that the canyons were available for livestock use. He came to that conclusion because the RMP did not preclude grazing in the Comb Wash Allotment. However, he acknowledged that he did not know if there was specific information in the RMP about the impacts of grazing in the canyons.

Curtis testified that he "monitor[s] the grazing in the San Juan resource area, and presently that covers approximately two million acres. And I deal with the biggest percent of that two million acres, and approximately 66 different allotments and 66 different permittees, give or take a few." Given the scope of his duties, it is not surprising that Curtis stated that he did not have the time or personnel to conduct the necessary monitoring in the canyons. Under those circumstances, he is generally left to rely on the permittee to adhere to the grazing schedule in the Comb Wash Allotment. He stated: "I try to call them, you know, periodically to make sure that things are going close to right." It is clear from the record that things did not always go right. . . .

IV. Discussion

    * * *

B. FLPMA Violation

The next issue for resolution is whether Judge Rampton properly held that BLM violated FLPMA. After citing section 302(a) of FLPMA, 43 U.S.C. § 1732(a) (1994), which requires the Secretary to "manage the public lands under principles of multiple use and sustained yield," Judge Rampton quoted the FLPMA definition of "multiple use." That term is defined as the management of the public lands and their various resource values so that they are utilized in the combination that will best meet the present and future needs of the American people; * * * the use of some land for less than all of the resources; a combination of balanced and diverse resource uses that takes into account the long-term needs of future generations for renewable and non-renewable resources, including, but not limited to, recreation, range, timber, minerals, watershed, wildlife and fish, and natural scenic, scientific and historical values; * * * with consideration given to the relative values of the resources and not necessarily to the combination of uses that will give the greatest economic return or greatest unit production. 43 U.S.C. § 1702(c) (1994). He then cited a statement by the court in Sierra Club v. Butz, 3 Envtl. L. Rep. 20,292, 20,293 (9th Cir.1973), that the multiple-use principle "requires that the values in question be informedly and rationally taken into balance." He concluded that an agency is required to engage in such a balancing test in order to determine whether a proposed activity is in the public interest.

However, in applying those standards to the facts in this case, Judge Rampton held that "BLM violated FLPMA by failing to make a reasoned and informed decision that the benefits of grazing the canyons outweigh the costs." It is with this highlighted language that all Appellants disagree, asserting that FLPMA does not require an economic cost/benefits analysis.

It is not clear that Judge Rampton intended that BLM engage in an economic cost/benefits analysis. A reading of his Decision at pages 23–25 discloses that the sentence quoted above is the only place in his Decision where he uses the word "costs," other than in the heading to the discussion. Later in his Decision when he is addressing the appropriate relief for the various violations, he describes the violation as the failure to make a reasoned and informed decision to graze the canyons in violation of FLPMA. He mentions neither costs nor benefits. He described the appropriate relief, as follows: "Because BLM may choose to prohibit grazing in the canyons in the future, BLM is not compelled to make a reasoned and informed decision that grazing the canyons is in the public interest. However, until a decision is made, BLM is prohibited from allowing grazing in the canyons." Again, no mention is made of benefits and costs.

* * *

To the extent Judge Rampton's Decision may be construed as requiring an economic cost/benefit analysis, it is modified to make it clear that no such analysis is required.

On appeal, BLM makes no argument that it satisfied FLPMA's multiple-use mandate in authorizing grazing in the canyons. Instead, it agrees that the actions it takes, including authorizing grazing on the public lands, are required to be "in the public interest," but it asserts that if Judge Rampton intended to impose a "specific public interest determination," such as is found in section 206(a) of FLPMA, 43 U.S.C. § 1716(a) (1994), dealing with exchanges, "he has clearly overstepped his authority." It contends that "FLPMA simply does not require a specific public interest finding in the grazing context."

We agree with BLM that FLPMA does not require a "specific" public interest determination for grazing. However, FLPMA's multiple-use mandate requires that BLM balance competing resource values to ensure that public lands are managed in the manner "that will best meet the present and future needs of the American people." 43 U.S.C. § 1702(c) (1994). Indeed, all parties agree that BLM must conduct some form of balancing of competing resource values in order to comply with the statute. Counsel for BLM states that "we agree that all BLM decisions should be in the public interest as that interest is defined by Congress in law * * *." The UFBF also recognizes that "[c]learly, management for multiple use does require a balancing and review of the relative resource values." The Tribe also "does not dispute that under FLPMA the BLM must give consideration to the relative values of the resources in the Comb Wash canyons. Moreover, we agree that those values must be rationally considered." And NWF concurs that "BLM must informedly and rationally balance competing values."

What is important in this case, and what we affirm, is Judge Rampton's finding that BLM violated FLPMA, because it failed to engage in any reasoned or informed decisionmaking process concerning grazing in the canyons in the allotment. That process must show that BLM has balanced competing resource values to ensure that the public lands in the canyons are managed in the manner that will best meet the present and future needs of the American people.

Notes and Questions

1. The *NWF v. BLM* case illustrates the important role of administrative litigation in environmental law, and for matters within the jurisdiction of the Department of the Interior (DOI) in particular. Within DOI's Office of the Secretary resides the Office of Hearings and Appeals (OHA). OHA is the Secretary's delegated representative for matters subject to secretarial review, which comprises a wide variety of topics in the case of DOI. OHA is divided into a Hearings Division and three standing appeals boards, one of which is the Interior Board of Land Appeals (IBLA) involved in *NWF v. BLM*. IBLA is the largest of the appeals boards and has jurisdiction over the variety of public lands issues within the jurisdiction of such agencies as BLM and the Fish and Wildlife Service. When DOI agencies such as BLM make decisions, an appeal to the Secretary generally can involve, first, a hearing before an Administrative Law Judge (ALJ) of the OHA Hearings Division, and then an appeal to IBLA, the decision of which is final agency action for purposes of judicial review. The rules of jurisdiction and procedure are, however, quite complex and depend on the subject matter and underlying administrative circumstances. No lawyer wishing to practice ecosystem management law can avoid having to wade through these and other procedural and substantive niceties of administrative law. For a thorough explanation of the IBLA and administrative litigation in DOI generally, see Michael C. Hickey, *Litigation Before the Department of the Interior*, 11 Nat. Resources & Env't 20 (Summer 1996).

2. FLPMA requires BLM to make grazing decisions that "will best meet the present and future needs of the American people." 43 U.S.C. § 1702(c). IBLA's interpretation of that provision in *NWF v. BLM* is DOI's final interpretation of the statute. What, then, did IBLA require BLM to do to satisfy the statutory standard? It is clear, at least, what IBLA believes this provision does *not* require. First, it rejected the argument that FLPMA requires any sort of "economic cost/benefit analysis." IBLA also rejected the argument that FLPMA requires "a 'specific' public interest determination." Why are neither of those two analyses required in order for an agency to make decisions that "best meet the present and future needs of the American people?" Don't a weighing of cost and benefit consequences and an analysis of impacts to competing public interests seem relevant inquiries to any effort to make such a decision? Does IBLA satisfy you that FLPMA does not require these detailed analysis? What would the BLM decision making process, and litigation over BLM grazing decisions, look like if BLM had to compile detailed cost-benefit and public interest analyses?

3. IBLA does interpret FLPMA to require that BLM demonstrate that it engaged in a "reasoned [and] informed decisionmaking process concerning grazing in the ... allotment," a process, IBLA explains, that "must show that BLM has balanced competing resource values to ensure that the public lands ... are managed in the manner that will best meet the present and future needs of the American people." Does that clarify things for you? Does this standard *require* that BLM adopt ecosystem management principles? Does it *allow* BLM to do so?

4. Weren't the San Juan Resource Area Manager, Edward Scherick, and his field conservationist assistant, Paul Curtis, practicing adaptive management as that method of ecosystem management is described in chapter 7? Scherick believed he had discretion to determine whether grazing should take place in the canyons, and he relied on the recommendations Curtis provided to him based on his field investigations and discussions with ranchers. To be sure, Curtis did not have the resources to conduct his assessments as thoroughly as he (or IBLA) would have liked, but isn't the basic understanding Scherick and Curtis had of their discretion, and the manner in which they executed it, what adaptive management theory has in mind?

5. Like the other federal land management agencies, a collection of laws outside of BLM's three primary organic statutes necessarily imposes substantive and procedural requirements on the agency, most notably the Endangered Species Act and the National Environmental Policy Act. For a discussion of how these ancillary laws have affected BLM, see George Cameron Coggins, *The Law of Public Rangeland Management III: A Survey of Creeping Regulation at the Periphery*, 13 ENVTL. L. 295 (1983).

Note on Forest Service Management of the National Grasslands System

The United States acquired most of the Great Plains from France with the Louisiana Purchase of 1803. Until the late 1860s, the Great Plains region remained a true "frontier." But the Homestead Act of 1862 brought almost six million settlers by 1890, many of whom tried to replace grass with crops more beneficial to economic aspirations. The settlers soon discovered, however, that while these vast grasslands were productive in wet years, they were also subject to serious drought and bitter winters. Plowed land exposed topsoil to incessant dry winds. Above parts of Oklahoma, Texas, Wyoming, Nebraska, Kansas, Colorado and the Dakotas, dust clouds rose to over 20,000 feet. Ten-foot drifts of fine soil particles piled up, burying fences and closing roads. During the same time, bison were largely eliminated by westward expansion. Ranchers filled the large open ranges of the plains with cattle and sheep. By the early 1930s, the broad midsection of America was in trouble, not only because of the Dust Bowls, but because the Great Depression was reaching its economic depths.

Emergency measures taken to provide relief to the farmers and settlers included the National Industrial Recovery Act of 1933 and the Emergency Relief Appropriations Act of 1935, which allowed the federal government to purchase and restore damaged lands and to resettle destitute families. Under the direction of the Department of Agriculture's Soil Conservation Service, the purchased lands were rehabilitated and enrolled in Land Utilization Projects for use as summer pasture lands by the farms and ranches that had withstood the economic and agricultural collapse of the prior decade.

From these roots, one hundred years after the Homestead Act, the National Grasslands were born when on June 23, 1960, the Secretary of Agriculture designated 3.8 million acres of the Land Utilization Project lands as the new National Grasslands System. Today, there are 20 National Grasslands, 17 of which are located in the Great Plains region east of the Rockies (the other three are in California, Oregon, and Idaho). The Little Missouri National Grassland in North Dakota is the biggest with 1,028,051 acres. The smallest National Grassland is McClelland Creek in Texas with 1,449 acres. Biological resources on national grasslands can be rich. For

Entry to the Cimarron National Grassland in Kansas

example, the Comanche National Grassland has approximately 275 different species of birds and the longest dinosaur track-way in the world.

The Department of Agriculture's Forest Service manages the National Grasslands System for sustainable multiple uses as part of and under the same legal authorities applicable to the National Forest System. These laws, and the complex legal issues arising under them, are the focus of Chapter 8 of the text. The multiple use mandate leads, indeed, to multiple uses. For example, the Caddo and southwestern LBJ National Grasslands in Texas, which are within a four-hour drive of four million people, provide forage for more than 1,584 head of cattle on 3,050 acres of improved pasture and 19,600 acres of native pasture. The largest coal producing mine in the world (Thunder Basin) is on the Thunder Basin National Grassland in Wyoming.

One of the most controversial issues facing the Forest Service on several large National Grasslands involves the black-tailed prairie dog and the impact its possible listing as an endangered species could have on grazing and other activities in and around the National Grasslands (the potential listing of the prairie dog is covered in Chapter 4 of the text). Even without a listing of the prairie dog, the Forest Service recently initiated fundamental changes in its management policy for several large national grasslands designed to improve, among other things, conservation of prairie dog habitat. *See* U.S. Forest Service, Northern Great Plains Revised Management Plans and Final Environmental Impact Statement (July 2001). The plans are highly controversial among local ranchers who question the scientific bases of the plans and fear reduced grazing opportunities on the public lands. *See* Paul Thacker, *A New Wind Sweeps the Plains*, 292 SCIENCE 2427 (2001).

Only a few cases involve Forest Service management of national grasslands, *see, e.g.*, Sharps v. U.S. Forest Service, 823 F.Supp. 668 (D.S.D. 1993), and no specialized doctrine has been applied. For a more complete understanding of National Grassland System management, therefore, refer to the public lands forest ecosystem law materials in Chapter 8 of the text. *See also* Coby C. Dolan, *The National Grasslands and Disappearing Biodiversity: Can the Prairie Dog Save Us From an Ecological Desert?*, 29 ENVTL. L. 213 (1999).

2.   RESTORING GRASSLAND ECOSYSTEMS ON PRIVATE LANDS

Ted Turner, the founder of CNN, is the nation's largest private landowner. Is he turning his 1.7 million acres of personal land holdings into telecommunications centers and shopping malls? Quite the opposite. Ted Turner is a ecosystem management dream come true for the grasslands of the far western states. On over 15 huge ranches throughout Montana, New Mexico, South Dakota, Nebraska, and Kansas, Turner is tearing down barbed wired fences, replacing cattle and sheep with buffalo, reintroducing fire-based vegetation management, restocking streams and rivers with native trout, raising and releasing endangered wolves, and doing just about everything else many ecologists say BLM ought to be doing on the public rangeland.

Before you get inspired to join in Turner's private grassland restoration effort, however, consider that just one of the ranches he has acquired, the 580,000–acre Vermejo Park Ranch in New Mexico, cost him over $80 million. So, what can be done to expand private grassland ecosystem restoration efforts short of winning the lottery? Congress has grappled with that question recently, searching for a model that treads lightly on property rights but yields some meaningful ecological results. One approach suggested in connection with debate over farm policy has been to initiate a program for purchase of easements on private property that would allow the owner to continue use of the land (usually for agricultural purposes, given where most land suitable for grassland restoration is located) in return for instituting a program of grassland conservation measures. The following is an example of such a program Congress adopted in 2002 in connection with adoption of the 2002 the Farm Bill:

Farm Security and Rural Investment Act of 2002

H.R.2646 (2002).

Subtitle E—Grassland Reserve

SEC. 2401. GRASSLAND RESERVE PROGRAM.

Chapter 2 of the Food Security Act of 1985 (as amended by section 2001) is amended by adding at the end the following:

Subchapter C—Grassland Reserve Program

SEC. 1238N. GRASSLAND RESERVE PROGRAM.

(a) ESTABLISHMENT—The Secretary shall establish a grassland reserve program (referred to in this subchapter as the "program") to assist owners in restoring and conserving eligible land described in subsection (c).

(b) ENROLLMENT CONDITIONS—

(1) MAXIMUM ENROLLMENT—The total number of acres enrolled in the program shall not exceed 2,000,000 acres of restored or improved grassland, rangeland, and pastureland.

(2) METHODS OF ENROLLMENT—

(A) IN GENERAL—Except as provided in subparagraph (B), the Secretary shall enroll in the program from a willing owner not less than 40 contiguous acres of land through the use of—

(i) a 10–year, 15–year, or 20–year rental agreement;

(ii)(I) a 30–year rental agreement or permanent or 30–year easement; or (II) in a State that imposes a maximum duration for easements, an easement for the maximum duration allowed under State law.

* * *

(c) ELIGIBLE LAND—Land shall be eligible to be enrolled in the program if the Secretary determines that the land is private land that is—

(1) grassland, land that contains forbs, or shrubland (including improved rangeland and pastureland); or

(2) land that—

(A) is located in an area that has been historically dominated by grassland, forbs, or shrubland; and

(B) has potential to serve as habitat for animal or plant populations of significant ecological value if the land is—

(i) retained in the current use of the land; or

(ii) restored to a natural condition;* * *

SEC. 1238O. REQUIREMENTS RELATING TO EASEMENTS AND AGREEMENTS.

(a) REQUIREMENTS OF LANDOWNER—

(1) IN GENERAL—To be eligible to enroll land in the program through the grant of an easement, the owner of the land shall enter into an agreement with the Secretary—

(A) to grant an easement that applies to the land to the Secretary;

(B) to create and record an appropriate deed restriction in accordance with applicable State law to reflect the easement;

(C) to provide a written statement of consent to the easement signed by persons holding a security interest or any vested interest in the land;

(D) to provide proof of unencumbered title to the underlying fee interest in the land that is the subject of the easement; and

(E) to comply with the terms of the easement and restoration agreement.

(2) AGREEMENTS—To be eligible to enroll land in the program under an agreement, the owner or operator of the land shall agree—

(A) to comply with the terms of the agreement (including any related restoration agreements); and

(B) to the suspension of any existing cropland base and allotment history for the land under a program administered by the Secretary.

(b) TERMS OF EASEMENT OR RENTAL AGREEMENT—An easement or rental agreement under subsection (a) shall—

(1) permit—

(A) common grazing practices, including maintenance and necessary cultural practices, on the land in a manner that is consistent with maintaining the viability of grassland, forb, and shrub species common to that locality;

(B) subject to appropriate restrictions during the nesting season for birds in the local area that are in significant decline or are conserved in accordance with Federal or State law, as determined by the Natural Resources Conservation Service State conservationist, haying, mowing, or harvesting for seed production; and

(C) fire rehabilitation and construction of fire breaks and fences (including placement of the posts necessary for fences);

(2) prohibit—

(A) the production of crops (other than hay), fruit trees, vineyards, or any other agricultural commodity that requires breaking the soil surface; and

(B) except as permitted under this subsection or subsection (d), the conduct of any other activity that would disturb the surface of the land covered by the easement or rental agreement; and

(3) include such additional provisions as the Secretary determines are appropriate to carry out or facilitate the administration of this subchapter.

(c) EVALUATION AND RANKING OF EASEMENT AND RENTAL AGREEMENT APPLICATIONS—

(1) IN GENERAL—The Secretary shall establish criteria to evaluate and rank applications for easements and rental agreements under this subchapter.

(2) CONSIDERATIONS—In establishing the criteria, the Secretary shall emphasize support for—

(A) grazing operations;

(B) plant and animal biodiversity; and

(C) grassland, land that contains forbs, and shrubland under the greatest threat of conversion.

(d) RESTORATION AGREEMENTS—

(1) IN GENERAL—The Secretary shall prescribe the terms of a restoration agreement by which grassland, land that contains forbs, or shrubland that is subject to an easement or rental agreement entered into under the program shall be restored.

(2) REQUIREMENTS—The restoration agreement shall describe the respective duties of the owner and the Secretary (including the Federal share of restoration payments and technical assistance).

SEC. 1238P. DUTIES OF SECRETARY.

(a) IN GENERAL—In return for the granting of an easement, or the execution of a rental agreement, by an owner under this subchapter, the Secretary shall, in accordance with this section—

(1) make easement or rental agreement payments to the owner in accordance with subsection (b); and

(2) make payments to the owner for the Federal share of the cost of restoration in accordance with subsection (c).

(b) PAYMENTS—

(1) EASEMENT PAYMENTS—

(A) AMOUNT—In return for the granting of an easement by an owner under this subchapter, the Secretary shall make easement payments to the owner in an amount equal to—

(i) in the case of a permanent easement, the fair market value of the land less the grazing value of the land encumbered by the easement; and

(ii) in the case of a 30–year easement or an easement for the maximum duration allowed under applicable State law, 30 percent of the fair market value of the land less the grazing value of the land for the period during which the land is encumbered by the easement.* * *

(c) FEDERAL SHARE OF RESTORATION—The Secretary shall make payments to an owner under this section of not more than—

(1) in the case of grassland, land that contains forbs, or shrubland that has never been cultivated, 90 percent of the costs of carrying out measures and practices necessary to restore functions and values of that land; or

(2) in the case of restored grassland, land that contains forbs, or shrubland, 75 percent of those costs.

SEC. 1238Q. DELEGATION TO PRIVATE ORGANIZATIONS.

(a) IN GENERAL—The Secretary may permit a private conservation or land trust organization (referred to in this section as a "private organization") or a State agency to hold and enforce an easement under this subchapter, in lieu of the Secretary, subject to the right of the Secretary to conduct periodic inspections and enforce the easement, if—

(1) the Secretary determines that granting the permission will promote protection of grassland, land that contains forbs, and shrubland;

(2) the owner authorizes the private organization or State agency to hold and enforce the easement; and

(3) the private organization or State agency agrees to assume the costs incurred in administering and enforcing the easement, including the costs of restoration or rehabilitation of the land as specified by the owner and the private organization or State agency.

Notes and Questions

1. The Natural Resources Conservation Service (NRCS), formerly the Soil Conservation Service, is the division of the Department of Agriculture that focuses primarily on conservation practices on private lands. Soil conservation divisions of the USDA were formed in response to the Dust Bowl catastrophe of the mid–1930's. The agency's first chief, Hugh Hammond Bennett, convinced the Congress that soil erosion was a national threat; that a permanent agency was needed within the USDA to call landowners' attention to their land stewardship opportunities and responsibilities; and that nationwide coordination between federal agencies and local communities was needed to help farmers and ranchers conserve their land. Today, more than six decades later, NRCS is USDA's lead conservation agency, partnering with state and federal agencies, NRCS Earth Team volunteers, agricultural and environmental groups, professional societies, and the nation's 3,000 local soil and water conservation districts that are at the heart of the NRCS conservation delivery system. Most NRCS employees serve in USDA's network of local, county-based offices. The rest are at state, regional, and national offices, providing technology, policy, and administrative support. NRCS employees have technical expertise and field experience in many scientific and technical specialties, including soil science, soil conservation, agronomy, biology, agroecology, range conservation, forestry, engineering, geology, hydrology, cultural resources, and economics. Nearly three-fourths of the technical assistance provided by the agency goes to helping farmers and ranchers develop conservation systems uniquely suited to their land and individual ways of doing business. The agency also provides assistance to rural and urban communities to reduce erosion, conserve and protect water, and solve other resource problems. For more on the agency, see www.nrcs.usda.gov. Based on this description, do you find that NRCS is an appropriate agency to implement the Grassland Reserve Program?

2. Note that the easement agreements executed under the new Grasslands Reserve Program would permit "grazing on the land in a manner that is consistent with maintaining the viability of natural grass, shrub, forb, and wildlife species indigenous to that locality." Based on the readings in the previous sections of this chapter, are you comfortable that BLM or the Forest Service has determined what satisfies that criterion on public rangelands? Do you anticipate that NRCS would apply a different approach in the context of easements on private lands? Can you, in 50 words or less, draft a regulation defining what conditions should satisfy this criterion in that context?

3. NRCS is also required to "prescribe the terms of a restoration agreement by which grassland that is subject to an easement or rental agreement entered into under the program shall be restored." To what extent should NRCS do so by general regulation versus through specific, tailor-made provisions of each easement? Notice that the legislation does not define "restored." If this type of legislative provision were adopted, should NRCS provide a definition in general regulations, or leave the concept undefined, to be worked out on a case-by-case basis? What should be the ecological reference point for knowing when a grassland has been restored?

4. The new program allows NRCS to permit a private conservation or land trust organization to hold and enforce an easement under this subchapter, in lieu of the Secretary. Several land conservation organizations already have placed stunning grasslands into permanent preservation status. The Nature Conservancy, for example, recently acquired the 42–square-mile Zumwalt Prairie Preserve in eastern Oregon, home to a rich diversity of bunchgrasses, and a 25,000–acre tallgrass prairie preserve in Minnesota, part of the organization's larger tallgrass prairie preserve project, that is home to a variety of rare birds, flowers, and native grasses. What do you think about having NRCS turn over management and enforcement of its grassland preserves to organizations like The Nature Conservancy?

5. NRCS already engages in private grassland conservation outreach through its participation with state resource agencies and private agricultural interests in the Grazing Lands Conservation Initiative (GLCI), which is designed to deliver technical, educational, and related assistance to owners of private grazing lands through a voluntary "Prescribed Grazing" program. Based on the premise that grazing can be a component of ecosystem management in grassland settings, GLCI focuses its program on improved grazing management through soil conservation methods, invasive weed controls, drought preparation and other water conservation practices, and wildlife habitat protection. In 2001, over 28,000 individuals participated in the program, covering over 18 million acres of private grazing land. *See* http://www.glci.org.

D.   CASE STUDY: THE LITTLE DARBY PRAIRIE NATIONAL WILDLIFE REFUGE

Another federal public land management agency with a hand in grassland ecosystem management is the U.S. Fish and Wildlife Service (FWS), which administers the 93 million-acre National Wildlife Refuge System (see Chapter 7, Part D). The difference between BLM and the Forest Service on the one hand, and FWS on the other, is that FWS generally has to acquire new land suitable for grassland restoration to add to its refuge land inventory, whereas BLM and the Forest Service are devising ecosystem management policies for lands already under their jurisdiction (albeit under the horribly complex multiple-use mandate). As the following examples show, FWS has had some warm and some cold receptions when it has pulled into town with grassland restoration as the goal of establishing new refuge lands.

U.S. Fish and Wildlife Service, Region 3 For Immediate Release: Midwest Celebrates Its Newest National Prairie Refuge Public Invited To Dedication Friday, Aug. 10, 2001 at 4 p.m.

Today, less than one percent of the northern tallgrass prairie, which once blanketed the upper Midwest, still remains. Once stretching from horizon to horizon, this native prairie now exists only in scattered patches tucked away in pasture corners or atop rock-strewn rises. These parcels—forgotten, neglected or accidental survivors—are the tracts that Ron Cole, a U.S. Fish and Wildlife Service manager, wants to protect before they disappear forever. His ultimate goal: to protect up to 77,000 acres of native tallgrass prairie in Minnesota and Iowa, half through protective easements and half through outright purchase from willing sellers. On Aug. 10, he'll purchase his first tract of land for the new Northern Tallgrass Prairie

National Wildlife Refuge, a 360–acre tract near Luverne, Minn. It will be the first step in an effort which may take decades to reach fruition.

"We're going to celebrate," Cole said. "This is a very important first step for the new refuge. It's taken several years of hard work by refuge staff, our Service realty folks, our local Friends of Prairie group, The Nature Conservancy, and the Brandenburg Prairie Foundation, to make this first acquisition for the refuge a reality."

The public celebration of the new Northern Tallgrass Prairie NWR will kick-off at 4:00 p.m. on Friday, Aug. 10, on the site of the new prairie tract near Luverne, Minn. Transportation to the site will be provided by the Luverne Chamber of Commerce—buses will depart from Luverne High School shortly before 4:00 p.m. Following the dedication, participants are invited to a reception at the Brandenburg Gallery in Luverne, followed by dinner and a slide show of prairie images presented by internationally-renowned photographer Jim Brandenburg. Contact the Chamber for more information at 507/283–4061.

Cole notes the National Wildlife Refuge System will be celebrating its centennial anniversary in 2003 and sees the establishment of the Northern Tallgrass Prairie refuge as a timely, and fitting, gesture. "A hundred years ago, the prairie took care of us; this rich soil has fed generations of Americans." he said. "Now it's time for us to take care of the prairie, at least a small part of what's left."

Things were not nearly as rosy for the Fish and Wildlife Service on the other side of the Great Lakes, in Ohio, where the proposed Little Darby Prairie National Wildlife Refuge met vociferous local landowner opposition and has been ground into a state of political limbo. FWS explained the rationale for the refuge as follows:

U.S. Fish and Wildlife Service Notice of Intent to Prepare an Environmental Impact Statement (EIS) for the Proposed Establishment of the Little Darby National Wildlife Refuge

65 Fed. Reg. 36711 (June 9, 2000).

* * *

Purpose of Action.

The general purpose of the refuge would be "for the development, advancement, management, conservation, and protection of fish and wild-life resources" (Fish and Wildlife Act of 1956). More specifically, the Service's interests include preservation and restoration of Federal threatened and endangered species and migratory birds and their habitats in the Little Darby Creek Watershed, ensuring that the overall Darby Creek watershed biodiversity and Federal wildlife trust resources are protected and enhanced, while providing opportunities for wildlife-dependent public uses consistent with preservation and restoration of the natural resources.

* * *

Need for Action

Big and Little Darby Creeks, located 20 miles west of downtown Columbus, are the major streams in a 580–square mile watershed encompassing portions of 6 counties in central Ohio. The Darby watershed is one of the healthiest aquatic systems of its size in the Midwest and is ranked among the top five warm freshwater habitats in Ohio by the Ohio Environmental Protection Agency. Land use in the drainage basin has historically been agriculture, with appropriately 80 percent of the land area in fields, row-cropped, in a corn-soybean rotation. The project area was the location of the easternmost extension of the mid-continent tallgrass prairie. The following eight points help explain the need to preserve this area:

(1) Existing and threatened conversion of the watershed, from agriculture to urban land uses, presents an increased risk to the health of this aquatic system.

(2) Scientists (Ohio EPA surveys) place the number of fish species in the Darby Creek System at 94 and 60 + in the Little Darby Creek sub watershed. The number of mollusk species, including the federally endangered Northern riffle shell and the Northern club shell, is 35 (Dr. Tom Watters). They are reported to be declining.

(3) There are 3 federally endangered, 1 threatened, 1 candidate, and 10 monitored species confirmed in the original project area or likely to be in the original project area.

(4) Collectively, 44 species are designated as being state threatened or endangered throughout the watershed. Another 36 species are identified as potentially threatened or of special interest in the state. A total of 38 (24 percent) species listed in the Service's regional conservation priorities would be affected potentially by the project as proposed in the draft Environmental Assessment.

(5) While the Refuge project area encompasses only 14–15 percent of the entire Little Darby Creek Watershed, it includes almost 50 percent of all stream miles and important aquatic habitat that is in the watershed.

(6) The Ohio Department of Natural Resources, the National Park Service and the Nature Conservancy have all given special designations to the Big Darby and Little Darby Creeks. The Nature Conservancy identified this watershed as one of the "Last Great Places" in the Western Hemisphere.

(7) A 1996 report (Swanson, D.) found that the population trend in Ohio for 10 species of nongame grassland migratory birds exhibited declines in populations from 30 to 84 percent.

(8) The Service's Regional Wetlands Concept Plan, November 1990, identified the Big Darby Creek Watershed that includes Little Darby Creek as, "One of the last remaining watersheds in Ohio with excellent biological diversity." Under threat from development for water use and urban development, the area was listed as a potential wetland acquisition site.

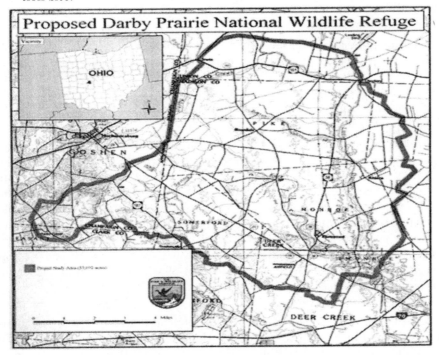

The only party held over the Darby refuge, however, was a September 2000 rally of several hundred angry people in London, Ohio, protesting the federal government's proposed "land grab." Here is an example of the local reaction to the Darby refuge:

Stewards of the Darby (SOD) and Citizens Against Refuge Proposal (CARP), Declaration:"Our Land Is Our Responsibility"

www.nodarbyrefuge.org/stewards_of_the_darby.htm

We, the residents of the area publicized as the "Darby Prairie National Wildlife Refuge Study Area" want our voices heard! We, who live and work in this farming community, believe the impact to area businesses would jeopardize their very existence. The Madison County Auditor's Office projects the affected region generates $300 per acre, which turns over 6–7 times (in buying power) before leaving the community. This translates into a potential deficit of $90 million dollars to our area businesses.

* * *

The State Forest Department manages and protects 7.1 million acres of forest land in Ohio, for the benefit of all Ohio citizens. One hundred eighty one thousand acres of State-owned forest land are available for multiple benefits, including wildlife, recreation, timber products, and soil and water protection. In addition, there are 72 State Parks in Ohio where the public can interact with nature at its leisure. With this great abundance of parks and wildlife areas, all supported by our tax dollars, is there really a need for more public land?

Actual area land auctions show that a 500–acre farm is worth $1.5 million dollars. To this initial cost, add a reasonably priced home at a cost of $85,000, and minimal equipment at approximately $641,000, and the combined start-up cost totals $2.226 MILLION DOLLARS. After committing to an investment of such magnitude, why would our astute, agriculturally-and family-minded farmers want to sell?

In the case of the proposed Darby Prairie National Wildlife Refuge, most of the 53,692 acres is land that has been acquired by our farmers over many generations. This "ownership endurance" enables us to continue our conservation-accredited farming skills, thus growing with our investments. At an average of 4.5 persons per home, this equates to the possible residential displacement of over 7,500 people from the Study Area alone, with a loss of approximately 4,000 taxpayers to the community. We have a proven track record of providing Americans with a diversity of products in the global marketplace, with a combination of wheat, corn, and soybeans; there would be a loss of over 3 million bushels of grain from the Study Area!

With well over 50,000 acres lost to food production, how many non-farmers would be willing to relinquish their combined homes and yards to replace the fertile soil that presently feeds so many, that would be permanently lost by the introduction of a National Wildlife Refuge? At some point, we will no longer have the abundance of high-quality, reasonably priced food that we now take for granted at our supermarkets.

The growing of food to nourish our citizens is certainly as much a consideration as re-establishing a tall grass prairie. Eating is not going to go out of style, and we are not willingly going to yield our bountiful land to either developers or Federal Agencies who say they are 'protecting us' from development.

Those of us who have been entrusted with the privilege of caring for the land, know well the proper care and nurturing required to maintain, protect and preserve our farmlands, and sustain a well-established wildlife habitat through conservation management. With an eye to the future, and the experience of almost two hundred years, we know that Our Land Is Our Responsibility!

Notes and Questions

1. What explains the starkly different reactions to the Northern Tallgrass Prairie refuge and the Little Darby refuge? Both, clearly, are worthwhile projects from an ecological perspective. Neither involves regulation of private land, relying instead on federal acquisition at fair market value. Is the difference that the Darby refuge is proposed in an area heavily dominated by agricultural land uses and in a state that has little previous FWS land management presence, whereas the Northern Tallgrass Prairie refuge is proposed in a state with a long history of national wildlife refuge presence? Perhaps the farming community in Ohio is simply fearful of what a new federal land presence will mean to its future. Yet the Darby refuge proposal definitively points out that:

- The federal government will not use condemnation powers to acquire refuge land, relying on willing sellers at negotiated prices.

- The refuge would make revenue sharing payments to the county to make up for lost tax revenues.

- The refuge cannot regulate agricultural practices on non-refuge lands.

- The refuge, at its largest proposed extent, would retire only 50,000 of the area's 860,000 acres of active farmland.

So, what are the farmers in the Darby refuge area afraid of? Is it really credible for them to pose food scarcity as a reason not to support the refuge? Or is the main issue one of the preservation of their culture, their way of living? Is that element truly at risk as a result of the Darby proposal? If so, how would you account for that in refuge design and management?

2. As noted previously in the chapter, most of the nation's original prairie lands have been converted to agricultural uses, mostly on public lands but much on private lands as well, as the Darby refuge story attests. This suggests that most land suitable for grassland restoration will be in or near agricultural lands. Hence, can grassland restoration, and grasslands ecosystem management generally, get very far without coming to grips with the agriculture question, that is, the question of what to do with the agricultural interests affected by grassland restoration? Efforts to restore grasslands in agriculture-intensive areas, such as Deb Donahue's call for an end to grazing on public lands and the Darby refuge proposal, meet stiff local, agriculturally-based opposition, but often receive broad national support and the ringing endorsement of ecologists. Should local agricultural interests be given a "veto" over national resource goals such as grassland ecosystem restoration? Should national goals such as grassland restoration take priority over local community preferences?

3. Instead of purchasing lands for the Darby refuge, couldn't FWS purchase easements and allow continued limited agricultural uses on the refuge lands? Indeed, FWS has proposed doing so for a ring of land around the central refuge preserve land, stating that it prefers to see the lands bounding the central preserve remain in agricultural uses as opposed to being converted to urban uses. But could the entire refuge be handled in this manner, as is the approach of the Grassland Reserve bill discussed in the previous section of the chapter? What concerns would you have in doing so?

4. Given the degree of local opposition, should FWS give up on the Darby refuge? Or is it time for the farmers to make way for the return of the grasslands?

Epilogue: In the end, the farmers won. On March 12, 2002, Craig Manson, Assistant Secretary of the Interior for Fish and Wildlife, sent letters to U.S. Representatives Deborah Pryce and David Hobson informing them that "conservation of the agricultural and natural resources of the Darby watershed is important," and that "the strong interest expressed in protecting the rural nature of this area indicates it is best that all levels of government work with local citizens to find a preferred approach to conserving those resources." The refuge proposal, he wrote, would be officially withdrawn. Obviously, it was not the preferred approach.

CHAPTER 10

FRESHWATER ECOSYSTEMS

A. BIODIVERSITY AND FRESHWATER ECOSYSTEMS

The water in all of Earth's rivers, lakes, and wetlands makes up just 0.01 percent of the world's water and only one percent of the earth's surface. Fortunately, the United States has bountiful supplies of freshwater resources: 3.6 million miles of rivers and streams, 41 million acres of lakes, 275 million acres of wetlands, and 33 trillion gallons of groundwater. These resources provide tremendous value to biodiversity conservation, as well as to human economies. Globally, for example, over 12 percent of all wildlife species live in freshwater ecosystems, and most other species depend on freshwater resources in some way for their survival. Wetlands, in particular, harbor many species and provide valuable shoreline protection, inland flood control, and sediment and chemical filtration services. Humans use freshwater resources to generate over 20 percent of the world's electricity through hydropower, to irrigate 40 percent of the world's crops, and to supply 12 percent of the world's fish consumption. Half of the population of the United States, and virtually all its rural population, depends on groundwater for its drinking water supply.

How well are we managing these freshwater resources for humans and nature? Wherever that question is asked, on whatever scale it is posed, the answer boils down to two interrelated but distinct issues: quantity and quality. Freshwater ecosystems depend, obviously, on water to function. And where a sufficient quantity of water is present to allow functions to

523

transpire naturally, very often the quality of the water will be a factor in determining whether those functions actually do so.

On the quantity issue, much depends on location. The United States uses over 500,000 gallons of water per person, per year. That sounds like a lot to most people. Nationally, however, our renewable water supply is 4 times the amount withdrawn and 15 times the amount actually consumed. So why all the headlines about water shortages? The answer, of course, is that this pattern of use and replenishment is not uniform geographically or temporally. Many areas of our nation are close to or beyond their physical capacity to supply suitable water to meet human and environmental needs. This condition is particularly acute in the West and Southwest, where low water renewal plus high water consumption have combined to spell economic and ecological trouble in recent decades. Ironically, irrigated agriculture, which consumes four times as much freshwater as all other uses in the nation combined, is concentrated in the 19 western states and is subsidized extensively through federal reclamation projects.

1990 TOTAL WATER WITHDRAWALS
(excluding power)

A topographic representation of water withdrawals, showing disproportionate removal of water from freshwater systems in the western states, primarily as a result of irrigated agriculture.
U.S. Geological Survey

But water scarcity is increasingly becoming a serious economic and environmental issue for states east of the Mississippi River as well. Georgia, Alabama, and Florida, for example, have been locked in battle for several years over the distribution of water in the Apalachicola–Chattahoochee–Flint River basin, which many ecologists consider one of the world's most stunning and imperiled hot spots of biodiversity (more on the "ACF" in Part D of this chapter). In times of drought, which often last for years, these and similar water battles become more heated, but also only more difficult to resolve. As of this writing, for example, drought is persistent throughout many of the eastern states and has triggered intrastate and interstate controversies. *See* Traci Watson and Paul Overberg, *Rivers Down to Barest of Levels*, USA Today, Mar. 28, 2002, at 1A. In short, water quantity is becoming an ecosystem management issue in many parts of the United States as human consumption and ecosystem needs find themselves increasingly in competition.

The functions of freshwater ecosystems, of course, are not limited to the water itself. The *habitat* that freshwater resources support can be reduced in quantity without impairing the quantity of *water* available in the system. Perhaps no topic has presented this dimension of the water resources quantity issue as the history and status of the nation's wetlands–areas such as swamps, bogs, marshes, and other areas saturated by surface or ground water sufficiently to support vegetation adapted for saturated soil conditions (more on these qualities later). Wetlands can be broadly divided into two categories: coastal wetlands and inland wetlands. Coastal wetlands are mostly marshes and swamps that are flooded by the tides. Non-tidal inland wetlands include wetlands along rivers and lakes, as well as isolated wetlands that are not directly connected to a major body of water but which are nonetheless valuable ecological resources. An important example of the latter variety is the "prairie pothole," which provides habitat for about half of the U.S. waterfowl population. Among the most biologically productive of land types, wetlands cover approximately 4 to 6 percent of the Earth's land surface. Their high productivity results from the essential characteristic of a wetland: an area that is flooded part of the time but not all of the time. This flooding ensures that the wetlands have ample supplies of water, minerals, or both. In addition to their high biological productivity, wetlands are important habitat for birds, fish, and other species. Wetlands are also important cleansing mechanisms for preventing pollutants from farms and other activities from running off and degrading water quality in rivers, lakes, and streams.

Two recent federal studies document the massive loss of wetland resources our nation has witnessed over the past 200 years. *See* U.S. Fish and Wildlife Service, Status and Trends of Wetlands in the Coterminous United States 1986 to 1997 (2000); USDA, Natural Resources Conservation Service, National Resources Inventory (2000). At the time of European settlement, the area that is today the coterminous United States had approximately 221 million acres of wetlands. Today about 105 million acres remain in these lower 48 states. Six states have lost over 85 percent of their wetlands, and 22 have lost more than 50 percent. None of the lower 48 states has lost less than 20 percent of its original wetlands. Florida has lost the most acres of wetlands–9.3 million acres–though it also is the state with the most remaining acres of wetlands, other than Alaska. Alaska adds 170 million acres of wetlands to the total, only a small fraction of which have been lost in the last 200 years.

Most of this history of loss is attributable to drainage of wetlands for conversion to agriculture, a practice that began in earnest with the earliest European settlers and was the official federal and state policy throughout most of the 1800s. Wetlands were considered undesirable, swampy, mosqui-to-infested wastelands that truly were wasted if not converted to some better use, usually agriculture but increasingly for urban development. This attitude prevailed well into the 1900s. Consider Gene Zion's 1957 children's book, *Dear Garbage Man*, which tells the supposedly happy story of how garbage should be used:

> That night as the city slept, the tugboats chugged and whistled softly as they pulled the barges down the river. The trash and ashes they carried would be used to fill in swampland. Then parks and play-grounds would be built there.

And many such parks and playgrounds were built. But as the ecological value of wetlands was increasingly understood and appreciated, protection of wetlands became a major public policy objective by the mid–1970s. That trend, plus the gradual decline in conversion of land to agricultural uses significantly dampened the rate of wetland losses. During the 1990s, we lost on average only about 58,500 acres annually to a combination of urban development (30%), agriculture (26%), forestry (23%), and rural development (21%), which is a dramatic reduction from the previous decade. While this does not satisfy the "no net loss" goal that federal policy has set, it is a significant improvement over historical experience.

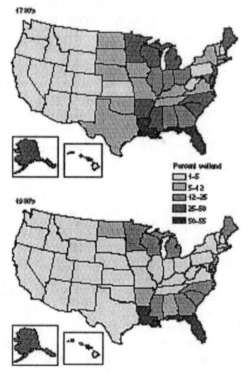

Percentage of the areas of each state occupied by wetlands in the 1780's (top map) and the 1980's.(bottom map).
U.S. Geological Survey, Biological Resources Division, Status and Trends of the Nation's Biological Resources (1999)

Water and habitat quantity in rivers, streams, lakes, and wetlands is one primary metric of freshwater ecosystem management policy. The other is water quality. A variety of factors can impair water quality for human and ecological uses. The chief threats to ecological processes include low dissolved oxygen, excessive nutrients, unsuitable temperature, sediment and siltation, bacteria and other pathogens, toxic organic chemicals and dissolved metals, unsuitable pH levels, and loss or degradation of habitat. All of these factors continue to pose threats to our nation's waters, though in many areas the magnitude of the threat has decreased measurably after decades of concerted water quality protection regulation.

The U.S. Environmental Protection Agency recently prepared two comprehensive assessments of national water quality, evaluating data on a significant portion of our nation's freshwater ecosystems. *See* USEPA, OFFICE OF WATER, ATLAS OF AMERICA'S POLLUTED WATERS (May 2000); USEPA, OFFICE OF WATER, NATIONAL WATER QUALITY INVENTORY: 1994 REPORT TO CONGRESS (Dec. 1995). For rivers and streams, EPA evaluated data for 17 percent of total river miles, finding 36 percent of that portion to be impaired. For lakes, ponds, and reservoirs, EPA surveyed data for 42 percent of total surface acres, finding 37 percent of those waters to be impaired. The leading source of water quality impairment in both categories of water bodies, by far, was agriculture, which is a significant source of nutrients, sediment, and pathogens. EPA also assessed available data on wetland quality, finding sediment, flow alterations, and habitat alterations as leading threats to habitat quality, with agriculture, again, the leading source of those threats.

Water quality and water quantity are intricately related factors in overall functioning of freshwater ecosystems. Recently, for example, the National Academy of Scientists/National Research Council concluded that higher spring flows and lower summer flows–i.e., a flow regime closer to natural conditions–are necessary along the Missouri River in order to reverse the widespread degradation that ecosystem has suffered in recent decades. *See* NATIONAL ACADEMY OF SCIENTISTS/NATIONAL RESEARCH COUNCIL, THE MISSOURI RIVER ECOSYSTEM: EXPLORING THE PROSPECTS FOR RECOVERY (2002). The Missouri River has dozens of dams and levees controlling water flow, and has over 735 miles "channelized" to promote barge navigation. As a result, over 3 million acres of riparian habitat have been substantially degraded. Farmers, however, are fearful that high spring flows will flood croplands, and barge operators contend low summer flows will stall barge traffic. And the U.S. Army Corps of Engineers, which maintains navigability of the river, and the U.S. Fish and Wildlife Service, which oversees habitat quality, have disagreed hotly over how to manage the flow regime. Clearly, the quantity and timing of river flows have much to do with both the ecological and the economic values a river delivers.

Indeed, dams are perhaps the human-induced change that has had the most adverse overall impact on freshwater ecosystems, largely because they affect both quantity and quality. Large dams–those over 15 meters high– now impound over 14 percent of the freshwater runoff in the world. Over

60 percent of the globe's major rivers are fractured, often significantly, by dams. While dams increase surface area of reservoirs, they also enable larger withdrawal of water from the system. For example, most of the world's irrigation is made possible by dams. Also, the interruption of water and silt flow that dams cause, combined with the disruption of thermal conditions caused by their altered outflow water temperatures, has wreaked ecological havoc on almost every dammed river system.

The impact of these and other degradations of freshwater ecosystems on biodiversity is palpable: in the United States, 37 percent of freshwater fish species, 67 percent of mussel species, 51 percent of crayfish species, and 40 percent of amphibian species are considered threatened, endangered, or extinct. Overall, therefore, while much progress has been made on both the quantity and quality fronts for freshwater ecosystems, much room for improvement remains.

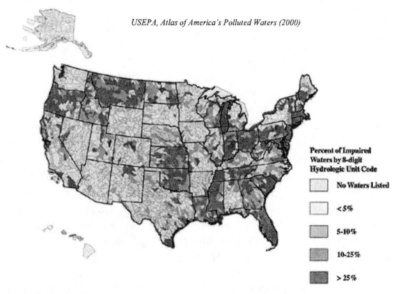

USEPA, Atlas of America's Polluted Waters (2000)

Percent of Impaired Waters by 8-digit Hydrologic Unit Code

- No Waters Listed
- <5%
- 5-10%
- 10-25%
- >25%

Notes and Questions

1. Filling wetlands for playgrounds has obvious impacts to wetland resources, and further losses attributable to such obvious causes can be easily halted. But more often the causes of wetlands degradation are complex and not easily solved. For example, the nation's two large wetlands complexes, the Everglades in Florida and the Mississippi River Delta in Louisiana, face demise from contrasting and seemingly intractable conditions. The Everglades are a naturally low-lying freshwater system that is suffering from nutrient overload caused by agricultural and urban runoff that is "plugged up" by canals and dikes restricting outflow from the system. Efforts to "re-plumb" the system and control nutrient inflow hold some promise, but at the expensive price tag of over $10 billion. *See* Mary Doyle and Donald E. Jodrey, *Everglades Restoration: Forging New Law in Allocating Water for the Environment*, 8 ENVTL. LAWYER 255 (2002); Alfred Light, *Ecosystem*

Management in the Everglades, 14 NATURAL RESOURCES & ENV'T 166 (2000). The Delta wetlands, twice the size of the Everglades, are actually starving for nutrients as the nutrient-rich Mississippi outflow has been diked and channeled so as to send the water out into the Gulf of Mexico well beyond the reach of Delta wetlands. And the Delta is sinking—actually, slumping outward and downward into the Gulf—due to geologic faults that run throughout the area. The Delta now loses over 60 square kilometers of wetlands annually. Plans to reverse these losses, known as Coast 2050, will cost over $14 billion to implement. *See Louisiana's Vanishing Wetlands: Going, Going ...* , 289 SCIENCE 1860 (2000). We never said ecosystem management would be cheap!

2. Not all nations turned the corner on appreciation of wetlands as soon as did the United States. In 1983, for example, Romanian dictator Nicolae Ceausescu decreed that the Romanian portion of the Danube River delta, one of the largest wetland complexes in Europe, would be diked and converted to agricultural and other land uses. His regime managed to transform 15 percent of the delta into marginal cropland–the delta's soil is too salty to support viable agriculture–before his government's fall saved not only Romania's future, but also the delta's. The current government has proclaimed a policy of restoring and protecting the delta that appears thus far to be having meaningful results. *See* Karen F. Schmidt, *A True–Blue Vision for the Danube*, 294 SCIENCE 1444 (2001).

3. The impact of climate change on freshwater resources depends on changes in the amount of rainfall, as well as when it occurs, which scientists are unable to forecast under current models of climate change dynamics. Recent studies suggest that the influence of human population increases will place far greater stress on the adequacy of freshwater supply than will climate change. *See* Union of Concerned Scientists and the Ecological Society of America, Confronting Climate Change in the Gulf Coast Region (2001), *available at* http://www.ucsusa.org. But when climate change is added to the scenario, many parts of the world that have never or rarely been concerned with water availability, including the eastern United States, could experience dramatic increases in water supply stress. *See* Charles J. Vorosmarty et al., *Global Water Resources: Vulnerability from Climate Change and Population Growth*, 289 SCIENCE 284 (2000).

Climate change could also affect the quality and prevalence of wetlands. In those areas where the climate becomes drier, drought will tend to lower water tables more than the lowering that occurs during droughts today. The resulting effect on prairie potholes would be similar to what happens today when farmers drain wetlands. The open water ponds, which are critical habitat for waterfowl, would be replaced by relatively damp land, although some form of wetland vegetation would remain. Lower water tables would also leave areas that currently have some form of wetland vegetation dry for a longer part of the year, which would reduce biological productivity and in some cases leave the land too dry to be considered a wetland. A drier climate could also lead farmers to increase their use of irrigation, which could further lower groundwater tables and indirectly drain the prairie potholes. A wetter climate would have the opposite effect.

The eventual effect of climate change on freshwater systems may also depend on how people respond to the increased flood risk. If people respond by moving out of hazardous areas, there will be more undeveloped land in the floodplain where wetlands can form. If people respond by building dams, river levees, or other structures to prevent floods, both the structures and the decline in flooding will decrease the total area where floodplain wetlands can form. In other words, as is often the cases with ecological dynamics of this scale, it's too early to tell what will happen, where, and when! To help improve our ability to model and predict these outcomes, EPA has established a research grant program to explore the effects of climate change on water quality and aquatic ecosystems. *See* 66 Fed. Reg. 35240 (July 3, 2001) (notice of availability of grants). For more information on the effects of climate change on wetlands and other aquatic ecosystems, at least as we understand them today, see www.epa.gov/global-warming/impacts.

4. For more information on the extent and value of freshwater resources, see Sandra Postel and Stephen Carpenter, *Freshwater Ecosystem Services, in* Nature's Services: Societal Dependence on Natural Ecosystems (Gretchen C. Daily ed., 1997); Nels Johnson et al., *Managing Water for People and Nature*, 292 Science 1071 (2001), as well as the material excerpted in Chapter 2 of this text. For detailed statistical presentation of water use in the United States, see U.S. Geological Survey, Estimated Use of Water in the United States in 1990: Total Water Use, *available at* http://water.usgs.gov/watuse/wuto.html. For additional information on the history of wetland losses in the United States, see Thomas E. Dahl and Gregory J. Allord, Technical Aspects of Wetlands, History of Wetlands in the Coterminous United States, U.S. Geological Survey Water Supply Paper 2425 (1997); Ralph E. Heimlich, Wetlands and Agriculture: Private Interests and Public Benefits, USDA Agricultural Economic Report No. 765 (Sept. 1998); Jonathan Tolman, Gaining More Ground: Analysis of Wetland Trends in the United States (Competitive Enterprise Inst. 1994).

B.   Primer on the Law of Freshwater Resources

The law of freshwater resources–of lakes, rivers, wetlands, and their watersheds–is an incredible patchwork of federal, state, and local initiatives. The major force in that body of law, and our major focus in this chapter, is the Federal Water Pollution Control Act, also known by the name of its 1972 amendatory legislation, the Clean Water Act. A brief description of the history of that law and its current policy direction follows.

1.   Historical Background

In a recent case involving the jurisdictional limits of the Clean Water Act with respect to wetlands, the U.S. Supreme Court took the occasion to explore the legislative history of that statute. Because he disagreed with the majority's interpretation of congressional intent, Justice Stevens, writing for the dissenting Justices, delved even more deeply into the history of federal regulation of water resources. His opinion provides an eloquent capsule history of that topic. (The portions of the opinions in the case addressing the jurisdictional limits of the Clean Water Act are reproduced *infra.*)

Solid Waste Agency of Northern Cook County v. United States Army Corps of Engineers

Supreme Court of the United States, 2001.
531 U.S. 159.

■ JUSTICE STEVENS, with whom JUSTICE SOUTER, JUSTICE GINSBURG, and JUSTICE BREYER join, dissenting.

* * *

Federal regulation of the Nation's waters began in the 19th century with efforts targeted exclusively at "promot[ing] water transportation and commerce." Kalen, Commerce to Conservation: The Call for a National Water Policy and the Evolution of Federal Jurisdiction Over Wetlands, 69 N.D.L.Rev. 873, 877 (1993). This goal was pursued through the various Rivers and Harbors Acts, the most comprehensive of which was the RHA of 1899. Section 13 of the 1899 RHA, commonly known as the Refuse Act, prohibited the discharge of "refuse" into any "navigable water" or its tributaries, as well as the deposit of "refuse" on the bank of a navigable water "whereby navigation shall or may be impeded or obstructed" without first obtaining a permit from the Secretary of the Army. 30 Stat. 1152.

During the middle of the 20th century, the goals of federal water regulation began to shift away from an exclusive focus on protecting navigability and toward a concern for preventing environmental degradation. Kalen, 69 N.D.L.Rev., at 877–879, and n. 30. This awakening of interest in the use of federal power to protect the aquatic environment was helped along by efforts to reinterpret § 13 of the RHA in order to apply its permit requirement to industrial discharges into navigable waters, even when such discharges did nothing to impede navigability. See, e.g., United States v. Republic Steel Corp., 362 U.S. 482, 490–491 (1960) (noting that the term "refuse" in § 13 was broad enough to include industrial waste). Seeds of this nascent concern with pollution control can also be found in the [Federal Water Pollution Control Act (FWPCA)], which was first enacted in 1948 and then incrementally expanded in the following years.

The FWPCA of 1948 applied only to "interstate waters." § 10(e), 62 Stat. 1161. Subsequently, it was harmonized with the Rivers and Harbors Act such that–like the earlier statute–the FWPCA defined its jurisdiction with reference to "navigable waters." Pub.L. 89–753, § 211, 80 Stat. 1252. None of these early versions of the FWPCA could fairly be described as establishing a comprehensive approach to the problem, but they did contain within themselves several of the elements that would later be employed in the [Clean Water Act (CWA)]. Milwaukee v. Illinois, 451 U.S. 304, 318, n. 10 (1981) (REHNQUIST, J.) (Congress intended to do something "quite different" in the 1972 Act); 2 W. Rodgers, Environmental Law: Air and Water § 4.1, pp. 10–11 (1986) (describing the early versions of the FWPCA).

The shift in the focus of federal water regulation from protecting navigability toward environmental protection reached a dramatic climax in 1972, with the passage of the CWA. The Act, which was passed as an amendment to the existing FWPCA, was universally described by its supporters as the first truly comprehensive federal water pollution legislation. The "major purpose" of the CWA was "to establish a comprehensive long-range policy for the elimination of water pollution." S.Rep. No. 92–414, p. 95 (1971), reprinted in 2 Legislative History of the Water Pollution Control Act Amendments of 1972 (Committee Print compiled for the Senate Committee on Public Works by the Library of Congress), Ser. No. 93–1, p. 1511 (1971) (hereinafter Leg. Hist.) (emphasis added). And "[n]o Congressman's remarks on the legislation were complete without reference to [its] 'comprehensive' nature.... " Milwaukee v. Illinois, 451 U.S. 304, 318 (1981) (REHNQUIST, J.). A House sponsor described the bill as "the most comprehensive and far-reaching water pollution bill we have ever drafted," 1 Leg. Hist. 369 (Rep. Mizell), and Senator Randolph, Chairman of the Committee on Public Works, stated: "It is perhaps the most comprehensive legislation that the Congress of the United States has ever developed in this particular field of the environment." 2 id., at 1269. This Court was therefore undoubtedly correct when it described the 1972 amendments as establishing "a comprehensive program for controlling and abating water pollution." Train v. City of New York, 420 U.S. 35, 37 (1975).

Section 404 of the CWA resembles § 13 of the RHA, but, unlike the earlier statute, the primary purpose of which is the maintenance of navigability, § 404 was principally intended as a pollution control measure. A comparison of the contents of the RHA and the 1972 Act vividly illustrates the fundamental difference between the purposes of the two provisions. The earlier statute contains pages of detailed appropriations for improvements in specific navigation facilities, 30 Stat. 1121–1149, for studies concerning the feasibility of a canal across the Isthmus of Panama, id., at 1150, and for surveys of the advisability of harbor improvements at numerous other locations, id., at 1155–1161. Tellingly, § 13, which broadly prohibits the discharge of refuse into navigable waters, contains an exception for refuse "flowing from streets and sewers ... in a liquid state." Id., at 1152.

The 1972 Act, in contrast, appropriated large sums of money for research and related programs for water pollution control, 86 Stat. 816–833, and for the construction of water treatment works, id., at 833–844. Strikingly absent from its declaration of "goals and policy" is any reference to avoiding or removing obstructions to navigation. Instead, the principal objective of the Act, as stated by Congress in § 101, was "to restore and maintain the chemical, physical, and biological integrity of the Nation's waters." 33 U.S.C. § 1251. Congress therefore directed federal agencies in § 102 to "develop comprehensive programs for preventing, reducing, or eliminating the pollution of the navigable waters and ground waters and improving the sanitary condition of surface and underground waters." 33 U.S.C. § 1252. The CWA commands federal agencies to give "due regard," not to the interest of unobstructed navigation, but rather to "improvements which are necessary to conserve such waters for the protection and propagation of fish and aquatic life and wildlife [and] recreational purposes." Ibid.

Because of the statute's ambitious and comprehensive goals, it was, of course, necessary to expand its jurisdictional scope. Thus, although Congress opted to carry over the traditional jurisdictional term "navigable waters" from the RHA and prior versions of the FWPCA, it broadened the definition of that term to encompass all "waters of the United States." § 1362(7). Indeed, the 1972 conferees arrived at the final formulation by specifically deleting the word "navigable" from the definition that had originally appeared in the House version of the Act.

* * *

2. CURRENT SCENE

Justice Stevens' dissenting opinion in *SWANCC* traces the transformation of federal water resources policy from its focus on protection of navigation to its modern focus on water quality. Yet, his opinion does not reveal the aspect of federal water quality policy that has plagued it in terms of effectiveness in ecosystem management–the distinction between "point source" and "nonpoint source" pollution and the disparate regulatory treatment the two categories receive.

Under the CWA, a point source is a discharge of regulated pollutants from "any discernable, confined and discrete conveyance, including but not limited to any pipe, ditch, channel, tunnel, conduit, well, discrete fissure, container, rolling stock, concentrated animal feeding operation, or vessel or other floating craft, from which pollutants are or may be discharged." 33 U.S.C. § 1362(14). As all-encompassing as this definition seems, it misses a huge component of water pollution sources. First, by exclusion, any regulated pollutant not so discharged is a so-called nonpoint source of pollution. This includes, significantly, overland runoff from rain and snowmelt that is not collected in storm sewers. Second, by definition, most agricultural pollution, including irrigation return flows carried, believe it or not, *in ditches and pipes*, is not point source pollution. See 33 U.S.C. § 1362(14). Therein lies the problem: As noted previously, EPA has identified agricultural pollution as the leading cause of impairment to our nation's rivers, lakes, and wetlands, and not far behind in all of those cases is urban runoff. In other words, today's leading causes of impairment to freshwater resources are the two significant sources of water pollution that are *not* regulated as point sources under the CWA.

So how has federal law managed these nonpoint sources? Not much. Efforts to address nonpoint source water pollution in the CWA and other statutes have been feeble, unfocused, and underfunded. Section 208 of the CWA requires states to develop areawide waste treatment management plans that include a process for identifying nonpoint sources and establishing feasible control measures. *See* 33 U.S.C. § 1288(a). With high expectations for this program, Congress used it as the rationale for moving irrigation return flows from the point source side of the CWA to the nonpoint source side of the CWA. *See* S. Rep. No. 370, at 35 (1977), *reprinted in* 1977 U.S.C.C.A.N. 4326, 4360 ("All such sources, regardless of the manner in which the flow was applied to the agricultural lands, and regardless of the discrete nature of the entry point, are more appropriately treated under the requirements of section 208"). Similarly, in the 1987 amendments, Congress added section 319 to the statute to require states to prepare "state assessment reports" that identify waters that cannot reasonably be expected to meet water quality standards because of nonpoint source pollution. 33 U.S.C. § 1329(a). States must prepare "state management programs" prescribing the "best management practices" to control sources of nonpoint pollution. When EPA approves a state's assessment reports and the management plans, the state is eligible for financial assistance to implement its programs.

In the absence of any concrete, enforceable federal blueprint for addressing nonpoint source pollution, the efficacy of the section 208 and section 319 programs depended largely on state initiative to develop new approaches. It is little surprise, then, that neither section 208 nor section 319 led anywhere meaningful. An EPA Federal Advisory Committee summed up the weakness of the section 208 and 319 programs by explaining that "EPA had no 'hammer' provision for states not adopting programs and no ability to establish a program if a State chose not to." EPA TMDL Federal Advisory Committee, Discussion Paper, Nonpoint Source–Only Waters 5 (1997).

Congress thus took a more aggressive step in section 6217 of the Coastal Zone Act Reauthorization Amendments of 1990. Pub. L. No. 101–508, Title VI, § 6217 (1990). This legislation amended the Coastal Zone Management Act (CZMA) (discussed more fully in Chapter 11) to add a requirement that any state with a federally approved coastal zone management program must develop a Coastal Nonpoint Pollution Program subject to federal review and approval. States must identify land uses leading to nonpoint pollution and develop measures to apply "best available nonpoint pollution control practices, technologies, processes, siting criteria, operating methods, or other alternatives." 16 U.S.C. § 1455b(g)(5). When EPA and the National Oceanic and Atmospheric Administration approve a state's Coastal Nonpoint Pollution Program, the federal government agrees not to fund, authorize, or carry out projects inconsistent with the state's plan. *Id.* 1455b(k). For coastal states, this requirement can serve as an impetus for more aggressive regulation of nonpoint source pollution generally, though funding assistance from the federal government is woefully short of the expected cost of Coastal Nonpoint Pollution Program plan preparation and implementation for the states (for example, the appropriation for fiscal year 1998 was $12 million).

By the mid–1990s, therefore, it had become clear that the primary objective of ecosystem management for freshwater resources had to be getting a handle on nonpoint source water pollution, but that significant obstacles would need to be overcome, not the least of which is the gaping hole in the CWA. Congress remained essentially inert on such matters through the 1990s (and into the 2000s). Thus, on October 18, 1997, during the Clean Water Act's 25th Anniversary celebration, then Vice President Gore requested that the Secretary of Agriculture and the Administrator of the United States Environmental Protection Agency (EPA), in consultation with affected federal agencies, devise an action plan to address enhanced protection from public health threats posed by water pollution; more effective control of polluted runoff; and promotion of water quality protection on a watershed basis. The result, the *Clean Water Action Plan: Restoring and Protecting America's Waters* (CWAP), outlined an ambitious agenda for coordinating existing federal water conservation authorities toward, among other things, ecosystem management. A principal focus of this component of the CWAP agenda was bringing nonpoint source pollution more fully within the federal water policy umbrella, but its scope also extended to wetland conservation policy and the broader ambition of developing a watershed-based framework for resource management. The remainder of this chapter traces the development of those three themes.

Notes and Questions

1. In addition to the several programs discussed in the text that were initiated specifically to address nonpoint source pollution, over 30 other federal programs deal with nonpoint source pollution in some way or another. Many of these programs provide funding assistance and incentives to states or private actors to improve nonpoint source pollution management—for example, to improve farm runoff control practices. Some of the programs address actions on federal lands, where the solution is more amenable to direct regulation of private actors or altering federal agency action. Overall, the federal government spends over $3 billion per year implementing this array of programs, with EPA, USDA, and the Department of the Interior as the lead agencies. For an overview of the programs, see U.S. GENERAL ACCOUNTING OFFICE, FEDERAL ROLE IN ADDRESSING—AND CONTRIBUTING TO—NONPOINT SOURCE POLLUTION, GAO/RCED–99–45 (Feb. 1999).

2. Knowing what we now know about their impact on freshwater ecosystems, why hasn't Congress regulated nonpoint sources more aggressively? One factor is that they are fundamentally different from point sources in terms of amenability to regulation. Point sources such as chemical plants and wastewater treatment facilities generally are highly regulated enterprises. They discharge water pollutants in discrete, easily monitored locations and events. The discharge is usually a byproduct or waste of a technological process that can be altered with technological solutions. Regulation of point sources, in other words, is primarily a technology issue, not a land use issue. Most categories of nonpoint sources, by contrast, are diffuse and diverse land uses. Take farms for example. There are one million farms in the United States today, using over 900 million acres for crop and livestock production. Differences in weather, soil, and other local factors lead to tremendous diversity of farming practices around the nation. While modern farming is more technologically intensive than in the past, it is still primarily a local land use practice, and the federal government has traditionally been reluctant to regulate local land uses directly. Still, given how significant the nonpoint source pollution problem is, does it surprise you that Congress has responded how it has?

3.   Perhaps it is understandable that Congress has not tried to undertake comprehensive nation-wide regulation of nonpoint sources such as farms. Have the states closed that gap? Generally, no. Most states follow the federal lead and focus regulatory clout almost entirely on point sources. Some states have authorities in place that could, arguably, be used to regulate nonpoint sources, but have not used them in any concerted effort to do so. And the few states that have ventured into more aggressive regulation of nonpoint sources generally leave the worst offender–farms–relatively untouched. *See* Environmental Law Institute, Enforceable State Mechanisms for the Control of Nonpoint Source Water Pollution (1997); Environmental Law Institute, Almanac of Enforceable State Laws to Control Nonpoint Source Water Pollution (1998); James M. McElfish, *State Enforcement Authorities for Polluted Runoff*, 28 Envtl. L. Rep. (ELI) 10,181 (1998). Recently, however, several states have received federal approval of a Nonpoint Source Pollution Control Program under Section 6217 of the Coastal Zone Management Act. In July 2000, California became the first state to receive such approval. Its 400–page, state-wide plan addresses nonpoint source pollution from agriculture, forestry, urban areas, marinas, and other sources over a 15–year planning horizon. It contemplates an initial period of voluntary improvement by each source category followed by aggressive regulation and enforcement if improvements are not brought about voluntarily. California also plans to use state implementation of a controversial federal program–the total maximum daily load program–as the cornerstone of nonpoint source management, including agricultural nonpoint source pollution (the so-called TMDL program is the subject of the materials in the next section of the text). It may be, therefore, that states will use the opportunity of compiling their coastal nonpoint source plans to reverse the trend of general neglect of nonpoint source pollution.

4.   The historical discussion presented in this section reviews laws governing water quality in lakes, rivers, and wetlands. What about the 33 trillion gallons of groundwater? Even more so than for nonpoint source pollution, the Clean Water Act leaves management and regulation of groundwater pollution to the states. Some courts have held that the regulatory arm of the CWA–the NPDES permit program–covers discharges of pollutants to groundwater that is hydrologically connected to jurisdictional surface waters. *See, e.g.*, Idaho Rural Council v. Bosma, 143 F.Supp.2d 1169 (D.Idaho 2001); Sierra Club v. Colorado Refining Co., 838 F.Supp. 1428 (D.Colo. 1993). Most courts, however, have held that the statute does not reach that far, *see, e.g.*, Village of Oconomowoc Lake v. Dayton Hudson Corp., 24 F.3d 962 (7th Cir.1994); Exxon Corp. v. Train, 554 F.2d 1310 (5th Cir.1977), and no court has applied the CWA to isolated groundwater. *See generally* Jason R. Jones, *The Clean Water Act: Groundwater Regulation and the National Pollutant Discharge Elimination System*, 8 DICKINSON J. ENVTL. LAW & POLICY 93 (1999).

Some states, however, regulate to protect groundwater quality more comprehensively. In the central Texas "Hill Country" around San Antonio and Austin, for example, the Edwards Aquifer is a highly productive karst (limestone) aquifer providing a bountiful and valuable source of high-quality water for residential and industrial purposes. It is the principal water supply for San Antonio and a major source of agricultural irrigation water in counties farther west. Through natural spring openings, the aquifer also supplies water to many surface streams and rivers that are home to a host of endangered aquatic species. Given the importance of this resource to the state, the Texas Natural Resources Conservation Commission and its predecessor agencies have for many years regulated development over the aquifer's recharge zone to prevent intrusion of pollutants and contaminants. *See* 30 Tex. Admin. Code ch. 213, *available at* http://www.tnrcc.state.tx.us/EAPP/index.html. In addition, more recently Texas took the controversial step of regulating withdrawals from the aquifer. Landowners previously were entitled, under the so-called "rule of capture," to use whatever groundwater they could withdraw from the aquifer free of quantitative regulation. The Edwards Aquifer Authority Act of 1993 required landowners to document historical uses and seek a permit from the Edwards Aquifer Authority to continue withdrawals, subject to additional restrictions. *See* 31 Tex. Admin. Code chs. 701–711, *available at* http://www.edwardsaquifer.org/. For a current overview of the Texas system for groundwater regulation in the Edwards Aquifer region, see Gregory Ellis and Jace A. Houston, *Senate Bill 2: "Step Two" Towards Effective Water Resource Management and Development for Texas*, 32 ST. BAR OF TEXAS ENVTL. L.J. 53 (2002).

5. International law has also entered into the domestic law development scene in recent years. In June 1992, at the United Nations Conference on Environment and Development (UNCED) in Rio de Janeiro, participating nations adopted the so-called Rio Declaration to formally endorse the concept of sustainable development and to commit to formulating national plans of action for achieving it. *See* Rio Declaration on Environment and Development, UNCED, U.N. Doc. A/CONF.151/Rev. 1, 31 I.L.M. 874 (1992). UNCED's blueprint for what national sustainable development plans should contain is known as the Agenda 21 document, which covers an array of topics and provides guiding principles of sustainable development. See UNCED Agenda 21, U.N. Doc. A/CONF.151.26 (1992). Chapter 18 of

Agenda 21 addresses protection of the quality and supply of freshwater resources, advocating the use of integrated, watershed-based policies. Many of the domestic U.S. laws and initiatives discussed in this chapter have been adopted or amended since Agenda 21 was promulgated. For a comprehensive assessment of how they live up to the Agenda 21 vision, notable not only for its breadth of programmatic coverage but also its depth of analysis of each program, see Robert W. Adler, *Fresh Water—Toward a Sustainable Future*, 32 Envtl. L. Rep. (Envtl. L. Inst.) 10169 (2002).

C. BIODIVERSITY MANAGEMENT IN FRESHWATER ECOSYSTEMS

We have divided the discussion of the law of freshwater ecosystem management into three topics: (1) lakes and rivers; (2) wetlands; and (3) watersheds. We have two reasons for taking this approach. One was simple convenience–the legal framework follows essentially the same approach. The other reason, however, was to emphasize the complexity of the task. When asked to describe a freshwater ecosystem–where it begins and ends, what is in it, what influences it–the simplest unit is a lake or river. They have geographically-defined and well-known boundaries. Clearly, however, lakes and rivers are influenced by events outside their boundaries, even from areas infrequently or never inundated by the water body surface that defines the lake or river. One significant influence is from the quantity and quality of adjacent wetlands, which purify water entering the lake or river and provide habitat and nutrients for the wildlife associated with the lake or river. But we cannot stop even there. A lake or river feels influences from well beyond the geographic line where wetlands end and dry uplands begin. Any lake or river has its watershed, from which it draws water, nutrients, sediments, and other inputs vital to the processes of the water body itself.

So what should be the focus of freshwater ecosystem management policy? Should we set the scale at the larger, holistic watershed level, or at the smaller, more easily managed scale of individual lakes and rivers? Should the approach be regulatory or incentive-based? Should we focus on national concerns through federal initiatives, or keep things local? Given the patchwork of federal, state, and local laws having to do with conservation of water quality and water quantity, the answer, thus far, has been a little bit of everything. The following materials provide a taste of the major initiatives in that regard. We leave it to you to decide whether anything coherent has emerged from them that we can call the law of freshwater ecosystem management.

1. LAKES AND RIVERS

a. THE TOTAL MAXIMUM DAILY LOAD PROGRAM

Given the relative impotence of CWA programs designed specifically for nonpoint sources, EPA looked elsewhere in the 1990s for a hook upon which to hang the nonpoint source program hat. It found it, not entirely voluntarily, in a previously little-noticed program of the CWA known as total maximum daily loads (TMDL).

The TMDL program is a bridge between the two primary CWA programs for water pollution control. One program is the National Pollutant Discharge Elimination Permit System (NPDES), through which regulated point sources obtain permits to discharge water pollutants pursuant to nationally-uniform, technology-based discharge limits that are devised for each major industry. The other major branch of the CWA is the Water Quality Standards (WQS) program, under which states designate ambient water quality goals and criteria to arrive at expressions of maximum tolerable contaminant concentrations in a waterbody. In essence, the TMDL program answers the question of what happens when a particular WQS for a waterbody is not met even though all the NPDES-regulated discharges into the waterbody are complying with the most stringent technology-based limits applicable to the relevant industries. Because NPDES limits are based on end-of-the-pipe effluent concentrations, and WQS are based on ambient conditions in the entire waterbody, no obvious connection exists by which to adjust the NPDES discharges. The TMDL program is designed to produce that connection, albeit in ways that are far less than obvious. The following materials provide a background of the TMDL program and demonstrate how even it may be ineffective in the long run for dealing with nonpoint source pollution as an ecosystem management issue.

June F. Harrigan–Lum and Arnold L. Lum, Hawaii's Total Maximum Daily Load Program: Legal Requirements Versus Environmental Realities

15 Natural Resources & Environment 12 (2000).

In order to restore the environmental integrity of our nation's waters, Congress in 1972 enacted the Federal Water Pollution Control Act, commonly referred to as the Clean Water Act (CWA). To achieve this objective, Congress declared as a "national goal" that "the discharges of pollutants into the navigable waters [of the United States] be eliminated by 1985." 33 U.S.C. § 1251(a)(1). Where pollution cannot be abated by limiting discharges of point source pollutants through the National Pollution Discharge Elimination System (NPDES) permit process, the CWA requires that states designate water quality limited segments (WQLSs) for waters within their jurisdiction that have water quality below the state's water quality standards; establish a priority ranking of such waters; and adopt more stringent pollution limits, called total maximum daily loads (TMDLs). 33 U.S.C. § 1313(d)(1)(A) and (C). TMDLs set an absolute upper limit on the amount of a pollutant that a WQLS can receive from NPDES-permitted point sources and nonpoint sources. This section briefly discusses the implementation of TMDL programs by the states, viewed against the backdrop of the CWA's congressional mandate that water pollution be eliminated by 1985.

The Clean Water Act establishes a procedure whereby states must submit lists of their water quality limited segments and TMDLs to EPA at periodic intervals. The first submission of such lists was due on June 26, 1979. *Idaho Sportsmen's Coalition v. Browner*, 951 F.Supp. 962, 965 (D.Wash.1996). Many states failed to make initial submissions before the deadline expired. Both Illinois and Indiana failed to meet this deadline, and a citizens' suit, *Scott v. City of Hammond, Ind.*, 741 F.2d 992 (7th Cir.1984), was subsequently brought, calling into question the remedy that should be imposed, if any, for a state's inaction in initiating a TMDL program. In the *Scott* case, the court of appeals held that the CWA requires the EPA Administrator to approve a state-proposed TMDL not later than thirty days following submission. If the Administrator disapproves the state submission, EPA must then establish TMDLs as necessary to implement applicable state water quality standards, pursuant to 33 U.S.C. § 1313(d)(2). Reversing the holding of the district court, which had ruled that EPA had no duty to establish TMDLs in situations where a state failed to submit TMDLs to the agency for approval, the Court of Appeals for the Seventh Circuit held that "prolonged failure [to submit a TMDL] may amount to the 'constructive submission' by that state of no TMDL's [*sic*]." * * * Therefore, after the 1984 decision in *Scott*, it appeared that mere inaction by a state in initiating a TMDL program would trigger a mandatory duty on the part of EPA to promulgate TMDLs for that state's waters.

* * *

While EPA is proposing a rule change that will obligate states to develop comprehensive schedules for establishing TMDLs within fifteen years from the date a water quality limited segment is listed in Part 1 of a state's WQLS list, 64 Fed. Reg. 46,050 (Aug. 23, 1999), it may be prudent for states to take note of the fact that the district court in *Idaho Sportsmen's Coalition v. Browner*, 951 F. Supp. 962, 967 (D.Wash.1996), has cautioned EPA that "Congress prescribed early deadlines for the TMDL process [and] 'long-term' [development of TMDLs] *at most can mean months and a few years, not decades*" (emphasis added). Although a decade and a half per WQLS is less than "decades," given the courts' expressed impatience with the states' and EPA's delay in submitting and approving TMDLs, respectively, it is questionable whether a fifteen-year TMDL schedule can survive judicial scrutiny.

* * *

TMDL costs [for Hawaii] are unclear, but preparation costs alone are likely approach $1 million if all streams draining into listed water bodies require multiple TMDLs. In view of the cost and effort required by law, the state would, in our view, have three main points of contention with the federal government.

The first issue includes TMDLs as a regulation. CWA § 303(d), as applied to allocating wasteloads among a number of independent point source discharges along rivers and streams or into lakes, with an allocation for background or nonpoint sources, is an effective regulation for controlling cumulative impacts of discharges regulated under the NPDES and CWA §§ 404 and 401 permit programs. These regulations do not make sense when applied to isolated high subtropical islands with small land areas, such as the Hawaiian Islands, and will not accomplish local environmental management goals. Because all of Hawaii's proposed TMDLs will be for streams impaired by nonpoint sources, with regulatory coverage limited to that presently required under the NPDES stormwater permit program, the state lacks authority to compel TMDL implementation to the degree needed to show measurable improvements in water quality. The TMDL program was not designed to accomplish water quality improvement in streams in small Hawaiian watersheds impaired mainly by nonpoint source pollutants, habitat destruction, dewatering of streams, and growth of dense stands of introduced vegetation in stream channels.

Although states do not want to be burdened with more regulations, resistance from agricultural interests to implementation of nonpoint source pollutant load limits will hamper water quality improvement in many streams flowing through agricultural lands. State laws are needed to provide regulatory and enforcement backup for voluntary programs and may eventually be required under a separate federal coastal zone management requirement, the Coastal Zone Act Reauthorization Amendments § 6217 process. As Hawaii's large sugar cane plantations have closed, lands have been subdivided and leased to small farmers who lack the environmental management expertise and funds previously available to the larger growers. In addition to new requirements at either the federal or state level aimed at controlling nonpoint source loads originating on agricultural lands, increased funding for federal agricultural support programs may be needed to assist farmers with TMDL implementation in their watersheds.

The second issue relates to TMDLs as public policy. The CWA may be too narrow a legal authority to accomplish community goals for restoration of impaired streams. Limiting loads of material pollutants entering highly impaired stream channels does accomplish the Act's regulatory goal of pollutant reduction but will not result in water quality improvement in cases where these loads will continue to accumulate in stream channels under low flow conditions, albeit at a slower rate than existed before the TMDL was implemented. Combining some degree of habitat restoration . . . in order to open up the channels and improve flows will aid in water quality improvement by allowing a slow discharge of stream water containing sediments and nutrients to the bay at levels that can be assimilated. Other benefits of habitat and flow restoration include provision of a persistent surface hydrological connection between the ocean and the stream for migration of native biota dependent on both the ocean and stream for life cycle completion, and will provide flood control capacity not now present due to dense vegetation in the channels.

Community interest in an aesthetically pleasing stream that supports native species and provides flood control capacity far outweighs interests focused solely on reduction of material pollutants. Unless TMDL preparation is more closely tied to restoration of waterways at the federal level, there will continue to be inadequate public support for implementation of TMDLs, and monies spent on TMDL preparation in Hawaii will accomplish little beyond satisfying federal paperwork requirements. At present, the projected cost of the TMDL program outweighs its practical accomplishments.

The final issue involves TMDLs as they relate to realistic expectations for success in the physical environment. Failure to implement TMDLs within a reasonable period of time after preparation wastes the resources expended on preparation. However, limited funding, the difficulty of obtaining cooperation from agricultural interests due to the strictly voluntary nature of the state's nonpoint source control efforts, and frequent conflicts over land use in Hawaii's small drainage basins inhibit implementation and monitoring activities needed to realize water quality improvements.

The authors must reluctantly conclude that, although the intent of the TMDL program is reduction of loads of material pollutants and temperature so that water quality standards are attained and maintained in states' surface waters, in practice, additional elements are needed (primarily habitat restoration and adequate dry weather flows) before our state's impaired perennial streams will meet water quality goals. As long as runoff from agricultural lands is not regulated under either state or federal law, and as long as a body of federal law is not in place that requires consideration of habitat and water quantity in TMDL preparation and implementation, the TMDL program will not accomplish a noticeable degree of water quality restoration in inland waters of the Hawaiian islands.

Pronsolino v. Nastri

U.S. Court of Appeals for the Ninth Circuit, 2002.
291 F.3d 1123.

■ BERZON, CIRCUIT JUDGE.

The United States Environmental Protection Agency ("EPA") required California to identify the Garcia River as a water body with insufficient pollution controls and, as required for waters so identified, to set so-called "total maximum daily loads" ("TMDLs")—the significance of which we explain later—for pollution entering the river. Appellants challenge the EPA's authority under the Clean Water Act ("CWA" or the "Act") § 303(d), 33 U.S.C. § 1313(d), to apply the pertinent identification and TMDL requirements to the Garcia River. The district court rejected this challenge, and we do as well.

CWA 303(d) requires the states to identify and compile a list of waters for which certain effluent limitations are not stringent enough to implement the applicable water quality standards for such waters. 303(d)(1)(A). Effluent limitations pertain only to point sources of pollution; point sources of pollution are those from a discrete conveyance, such as a pipe or tunnel. Nonpoint sources of pollution are non-discrete sources; sediment run-off from timber harvesting, for example, derives from a nonpoint source. The Garcia River is polluted only by nonpoint sources. Therefore, neither the effluent limitations referenced in 303(d) nor any other effluent limitations apply to the pollutants entering the Garcia River.

The precise statutory question before us is whether the phrase "are not stringent enough" triggers the identification requirement both for waters as to which effluent limitations apply but do not suffice to attain water quality standards and for waters as to which effluent limitations do not apply at all to the pollution sources impairing the water. We answer this question in the affirmative, a conclusion which triggers the application of the statutory TMDL requirement to waters such as the Garcia River.

I. STATUTORY BACKGROUND

Resolution of the statutory interpretation question before us, discrete though it is, requires a familiarity with the history, the structure, and, alas, the jargon of the federal water pollution laws. Natural Res. Def. Council v. EPA, 915 F.2d 1314, 1316 (9th Cir.1990). We therefore begin with a brief overview of the Act.

A. The Major Goals and Concepts of the CWA

Congress enacted the CWA in 1972, amending earlier federal water pollution laws that had proven ineffective. EPA v. California, 426 U.S. 200, 202 (1976). Prior to 1972, federal water pollution laws relied on "water quality standards specifying the acceptable levels of pollution in a States interstate navigable waters as the primary mechanism ... for the control of water pollution." Id. The pre–1972 laws did not, however, provide concrete direction concerning how those standards were to be met in the foreseeable future.

In enacting sweeping revisions to the nation's water pollution laws in 1972, Congress began from the premise that the focus on the tolerable effects rather than the preventable causes of pollution constituted a major shortcoming in the pre 1972 laws. Oregon Natural Desert Assoc. v. Dombeck, 172 F.3d 1092, 1096 (9th Cir.1998) (quoting EPA v. State Water Resources Control Board, 426 U.S. 200, 202–03 (1976)). The 1972 Act therefore sought to target primarily the preventable causes of pollution, by emphasizing the use of technological controls. Id.; Oregon Natural Res. Council v. United States Forest Serv., 834 F.2d 842, 849 (9th Cir.1987).

At the same time, Congress decidedly did not in 1972 give up on the broader goal of attaining acceptable water quality. CWA § 101(a), 33 U.S.C. § 1251(a). Rather, the new statute recognized that even with the application of the mandated technological controls on point source discharges, water bodies still might not meet state-set water quality standards, Natural Res. Def. Council, 915 F.2d at 1316–17, and therefore put in place mechanisms other than direct federal regulation of point sources designed to "restore and maintain the chemical, physical, and biological integrity of the Nation's waters." 101(a).

In so doing, the CWA uses distinctly different methods to control pollution released from point sources and those that are traceable to nonpoint sources. Oregon Natural Res. Council, 834 F.2d at 849. The Act directly mandates technological controls to limit the pollution point sources may discharge into a body of water. *Dombeck*, 172 F.3d at 1096. On the other hand, the Act "provides no direct mechanism to control nonpoint source pollution but rather uses the 'threat and promise' of federal grants to the states to accomplish this task," id. at 1907 (citations omitted), thereby "recogniz[ing], preserv[ing], and protect[ing] the primary responsibilities and rights of States to prevent, reduce, and eliminate pollution, [and] to plan the development and use ... of land and water resources...." § 101(b).

B.   The Structure of CWA 303, 33 U.S.C. 1313

1. *Water Quality Standards*

Section 303 is central to the Act's carrot-and-stick approach to attaining acceptable water quality without direct federal regulation of nonpoint sources of pollution. Entitled Water Quality Standards and Implementation Plans, the provision begins by spelling out the statutory requirements for water quality standards: "Water quality standards" specify a water body's "designated uses" and "water quality criteria," taking into account the water's "use and value for public water supplies, propagation of fish and wildlife, recreational purposes, and agricultural, industrial, and other purposes...." 303(c)(2). The states are required to set water quality standards for all waters within their boundaries regardless of the sources of the pollution entering the waters. If a state does not set water quality standards, or the EPA determines that the state's standards do not meet the requirements of the Act, the EPA promulgates standards for the state. §§ 303(b), (c)(3)–(4).

2. *Section 303(d): "Identification of Areas with Insufficient Controls; Maximum Daily Load"*

Section 303(d)(1)(A) requires each state to identify as "areas with insufficient controls" "those waters within its boundaries for which the effluent limitations required by section [301(b)(1)(A)] and section [301(b)(1)(B)] of this title are not stringent enough to implement any water quality standard applicable to such waters." Id. The CWA defines "effluent limitations" as restrictions on pollutants "discharged from point sources." CWA § 502(11), 33 U.S.C. § 1362(11). Section 301(b)(1)(A) mandates application of the "best practicable control technology" effluent limitations for most point source discharges, while § 301(b)(1)(B) mandates application of effluent limitations adopted specifically for secondary treatment at publicly owned treatment works. § 301(b)(1), 33 U.S.C. § 1311(b)(1).

For waters identified pursuant to § 303(d)(1)(A)(the § 303(d)(1) list), the states must establish the "total maximum daily load" ("TMDL") for pollutants identified by the EPA as suitable for TMDL calculation. § 303(d)(1)(C). "A TMDL defines the specified maximum amount of a pollutant which can be discharged or 'loaded' into the waters at issue from all combined sources." Dioxin/Organochlorine Center v. Clarke, 57 F.3d 1517, 1520 (9th Cir.1995). The TMDL "shall be established at a level necessary to implement the applicable water quality standards...." § 303(d)(1)(C).

Section 303(d)(2), in turn, requires each state to submit its § 303(d)(1) list and TMDLs to the EPA for its approval or disapproval. If the EPA approves the list and TMDLs, the state must incorporate the list and TMDLs into its continuing planning process, the requirements for which are set forth in § 303(e). § 303(d)(2). If the EPA disapproves either the § 303(d)(1) list or any TMDLs, the EPA must itself put together the missing document or documents. Id. The state then incorporates any EPA-set list or TMDL into the states continuing planning process. Id.

Each state must also identify all waters not placed on its § 303(d)(1) list (the "§ 303(d)(3) list") and "estimate" TMDLs for pollutants in those waters. § 303(d)(3). There is no requirement that the EPA approve the § 303(d)(3) lists or the TMDLs estimated for those waters. Id.

The EPA in regulations has made more concrete the statutory requirements. Those regulations, in summary, define "water quality limited segment[s]"—those waters that must be included on the § 303(d)(1) list—as"[a]ny segment where it is known that water quality does not meet applicable water quality standards, and/or is not expected to meet applicable water quality standards, even after the application of the technology-based effluent limitations required by sections 301(b) and § 306 [33 U.S.C. § 1316]." 40 C.F.R. § 130.2(j) (2000). The regulations then divide TMDLs into two types: "load allocations," for nonpoint source pollution, and "wasteload allocations," for point source pollution. § 130.2(g)-(i); see also pp. 7919, infra. Under the regulations, states must identify those waters on the § 303(d)(1) lists as "still requiring TMDLs" if any required effluent limitation or other pollution control requirement (including those for nonpoint source pollution) will not bring the water into compliance with water quality standards. § 130.7(b) (2000).

3. *Continuing Planning Process*

The final pertinent section of § 303, § 303(e), requiring each state to have a continuing planning process, gives some operational force to the prior information-gathering provisions. The EPA may approve a state's continuing planning process only if it "will result in plans for all navigable waters within such State" that include, inter alia, effluent limitations, TMDLs, areawide waste management plans for nonpoint sources of pollution, and plans for "adequate implementation, including schedules of compliance, for revised or new water quality standards." § 303(e)(3).

The upshot of this intricate scheme is that the CWA leaves to the states the responsibility of developing plans to achieve water quality standards if the statutorily-mandated point source controls will not alone suffice, while providing federal funding to aid in the implementation of the state plans. See *Dombeck*, 172 F.3d at 1097; § 303(e); see also § 319(h), 33 U.S.C. § 1329(h) (providing for grants to states to combat nonpoint source pollution). TMDLs are primarily informational tools that allow the states to proceed from the identification of waters requiring additional planning to the required plans. See Alaska Center for the Environment v. Browner, 20 F.3d 981, 984–85 (9th Cir.1994). As such, TMDLs serve as a link in an implementation chain that includes federally-regulated point source controls, state or local plans for point and nonpoint source pollution reduction, and assessment of the impact of such measures on water quality, all to the end of attaining water quality goals for the nation's waters.

II. FACTUAL AND PROCEDURAL BACKGROUND

A. The Garcia River TMDL

In 1992, California submitted to the EPA a list of waters pursuant to § 303(d)(1)(A). Pursuant to § 303(d)(2), the EPA disapproved California's 1992 list because it omitted seventeen water segments that did not meet the water quality standards set by California for those segments. Sixteen of the seventeen water segments, including the Garcia River, were impaired only by nonpoint sources of pollution. After California rejected an opportunity to amend its § 303(d)(1) list to include the seventeen sub-standard segments, the EPA, again acting pursuant to § 303(d)(2), established a new § 303(d)(1) list for California, including those segments on it. California retained the seventeen segments on its 1994, 1996, and 1998 § 303(d)(1) lists.

California did not, however, establish TMDLs for the segments added by the EPA. Environmental and fishermen's groups sued the EPA in 1995 to require the EPA to establish TMDLs for the seventeen segments, and in a March 1997 consent decree the EPA agreed to do so. See Pacific Coast Fishermens Assocs. v. Marcus, No. 95–4474. According to the terms of the consent decree, the EPA set March 18, 1998, as the deadline for the establishment of a TMDL for the Garcia River. When California missed the deadline despite having initiated public comment on a draft TMDL and having prepared a draft implementation plan, the EPA established a TMDL for the Garcia River. The EPA's TMDL differed only slightly from the states' draft TMDL.

The Garcia River TMDL for sediment is 552 tons per square mile per year, a sixty percent reduction from historical loadings. The TMDL allocates portions of the total yearly load among the following categories of nonpoint source pollution: a) "mass wasting" associated with roads; b) "mass wasting" associated with timber-harvesting; c) erosion related to road surfaces; and d) erosion related to road and skid trail crossings.

B. The Appellants

In 1960, appellants Betty and Guido Pronsolino purchased approximately 800 acres of heavily logged timber land in the Garcia River watershed. In 1998, after re-growth of the forest, the Pronsolinos applied for a harvesting permit from the California Department of Forestry ("Forestry").

In order to comply with the Garcia River TMDL, Forestry and/or the state's Regional Water Quality Control Board required, among other things, that the Pronsolinos' harvesting provide for mitigation of 90% of controllable road-related sediment run-off and contain prohibitions on removing certain trees and on harvesting from mid-October until May 1. The Pronsolino's forester estimates that a large tree restriction will cost the Pronsolinos $750,000.

Larry Mailliard, a member of the Mendocino County Farm Bureau, submitted a draft harvesting permit on February 4, 1998, for a portion of his property in the Garcia River watershed. Forestry granted a final version of the permit after incorporation of a 60.3% reduction of sediment loading, a requirement included to comply with the Garcia River TMDL. Mr. Mailliard's forester estimates that the additional restrictions imposed to comply with the Garcia River TMDL will cost Mr. Mailliard $10,602,000.

Bill Barr, another member of the Mendocino County Farm Bureau, also applied for a harvesting permit in 1998 for his property located within the Garcia River watershed. Forestry granted the permit after incorporation of restrictions similar to those included in the Pronsolino's permit. A forester states that these additional restrictions, included to comply with the TMDL, will cost Mr. Barr at least $962,000.

* * *

III. ANALYSIS

* * *

B. Plain Meaning and Structural Issues

1. *The Competing Interpretations*

Section 303(d)(1)(A) requires listing and calculation of TMDLs for "those waters within [the states] boundaries for which the effluent limitations required by section [301(b)(1)(A)] and section [301(b)(1)(B)] of this title are not stringent enough to implement any water quality standard applicable to such waters." § 303(d) (emphasis added). The precise statutory question before us is whether, as the Pronsolinos maintain, the term "not stringent enough to implement . . . water quality standard[s]" as used in § 303(d)(1)(A) must be interpreted to mean both that application of effluent limitations will not achieve water quality standards and that the waters at issue are subject to effluent limitations. As only waters with point source pollution are subject to effluent limitations, such an interpretation would exclude from the § 303(d) listing and TMDL requirements waters impaired only by nonpoint sources of pollution.

The EPA, as noted, interprets "not stringent enough to implement . . . water quality standard[s]" to mean "not adequate" or "not sufficient . . . to implement any water quality standard," and does not read the statute as implicitly containing a limitation to waters initially covered by effluent limitations. According to the EPA, if the use of effluent limitations will not implement applicable water quality standards, the water falls within

§ 303(d)(1)(A) regardless of whether it is point or nonpoint sources, or a combination of the two, that continue to pollute the water.

2. *The Language and Structure of § 303(d)*

Whether or not the appellants' suggested interpretation is entirely implausible, it is at least considerably weaker than the EPA's competing construction. The Pronsolinos' version necessarily relies upon: (1) understanding "stringent enough" to mean "strict enough" rather than "thorough going enough" or "adequate" or "sufficient"; (2) reading the phrase "not stringent enough" in isolation, rather than with reference to the stated goal of implementing "any water quality standard applicable to such waters." Where the answer to the question "not stringent enough for what?" is "to implement any [applicable] water quality standard," the meaning of "stringent" should be determined by looking forward to the broad goal to be attained, not backwards at the inadequate effluent limitations. One might comment, for example, about a teacher that her standards requiring good spelling were not stringent enough to assure good writing, as her students still used bad grammar and poor logic. Based on the language of the contested phrase alone, then, the more sensible conclusion is that the § 303(d)(1) list must contain any waters for which the particular effluent limitations will not be adequate to attain the statute's water quality goals.

* * *

Nothing in § 303(d)(1)(A) distinguishes the treatment of point sources and nonpoint sources as such; the only reference is to the "effluent limitations required by" § 301(b)(1). So if the effluent limitations required by § 301(b)(1) are "as a matter of law" "not stringent enough" to achieve the applicable water quality standards for waters impaired by point sources not subject to those requirements, then they are also "not stringent enough" to achieve applicable water quality standards for other waters not subject to those requirements, in this instance because they are impacted only by nonpoint sources.* * *

3. *The Statutory Scheme as a Whole*

The Pronsolinos' objection to this view of § 303(d) ... is, in essence, that the CWA as a whole distinguishes between the regulatory schemes applicable to point and non-point sources, so we must assume such a distinction in applying §§ 303(d)(1)(A) and (C). We would hesitate in any case to read into a discrete statutory provision something that is not there because it is contained elsewhere in the statute. But here, the premise is wrong: There is no such general division throughout the CWA.

Point sources are treated differently from nonpoint sources for many purposes under the statute, but not all. In particular, there is no such distinction with regard to the basic purpose for which the § 303(d) list and TMDLs are compiled, the eventual attainment of state-defined water quality standards. Water quality standards reflect a state's designated uses for a water body and do not depend in any way upon the source of pollution. See § 303(a)-(c).

* * *

True, there are, as the Pronsolinos point out, two sections of the statute as amended, § 208 and § 319, that set requirements exclusively for nonpoint sources of pollution. But the structural inference we are asked to draw from those specialized sections—that no other provisions of the Act set requirements for waters polluted by nonpoint sources—simply does not follow. Absent some irreconcilable contradiction between the requirements contained in §§ 208 and 319, on the one hand, and the listing and TMDL requirements of § 303(d), on the other, both apply.

There is no such contradiction. Section 208 provides for federal grants to encourage the development of state areawide waste treatment management plans for areas with substantial water quality problems, § 208(a), (f), and requires that those plans include a process for identifying and controlling non-point source pollution "to the extent feasible." § 208(b)(2)(F). Section 319, added to the CWA in 1987, directs states to adopt "nonpoint source management programs;" provides grants for nonpoint source pollution reduction; and requires states to submit a report to the EPA that "identifies those navigable waters within the State which, without additional action to control nonpoint sources of pollution, cannot reasonably be expected to attain or maintain applicable water quality standards or the goals and requirements of this chapter." § 319(a)(1)(A). This report must also describe state programs for reducing nonpoint source pollution and the process "to reduce, to the maximum extent practicable, the level of pollution" resulting from particular categories of nonpoint source pollution. § 319(a)(1)(C), (D).

The CWA is replete with multiple listing and planning requirements applicable to the same waterways (quite confusingly so, indeed), so no inference can be drawn from the overlap alone. See, e.g., § 208(b); § 303(d)(1)(A), (d)(1)(B), (d)(3), (e); CWA § 304(*l*), 33 U.S.C. § 1314(*l*); CWA § 314, 33 U.S.C. § 1324(a); § 319(a). Nor are we willing to draw the more discrete inference that the § 303(d) listing and TMDL requirements cannot apply to nonpoint source pollutants because the planning requirements imposed by § 208 and § 319 are qualified ones—"to the extent feasible" and "to the maximum extent practicable"—while the § 303(d) requirements are unbending. For one thing, the water quality standards set under § 303 are functional and may permit more pollution than it is "feasible" or "practicable" to eliminate, depending upon the intended use of a particular waterway. For another, with or without TMDLs, the § 303(e) plans for attaining water quality standards must, without qualification, account for elimination of nonpoint source pollution to the extent necessary to meet those standards. § 303(e)(3)(F).

The various reporting requirements that apply to nonpoint source pollution are no more impermissibly redundant than are the planning requirements. Congress specifically provided that in preparing the § 319 report, states may rely on information from § 303(e), which incorporates the TMDLs. § 319(a)(2). Moreover, states must produce a § 319 report only once, but must update the § 303(d)(1) list periodically. § 319; § 303(d)(2). Also, the § 319 report requires the identification of a plan to reduce nonpoint source pollution, without regard to the attainment of water quality standards, while the plans generated using the § 303(d)(1) lists and TMDLs are guided by the goal of achieving those standards. § 319; § 303(d), (e).

Essentially, § 319 encourages the states to institute an approach to the elimination of nonpoint source pollution similar to the federally-mandated effluent controls contained in the CWA, while § 303 encompasses a water quality based approach applicable to all sources of water pollution. As various sections of the Act encourage different, and complementary, state schemes for cleaning up nonpoint source pollution in the nation's waterways, there is no basis for reading any of those sections—including § 303(d)—out of the statute.

There is one final aspect of the Act's structure that bears consideration because it supports the EPA's interpretation of § 303(d): The list required by § 303(d)(1)(A) requires that waters be listed if they are impaired by a combination of point sources and nonpoint sources; the language admits of no other reading. Section 303(d)(1)(C), in turn, directs that TMDLs shall be established at a level necessary to implement the applicable water quality standards.... Id. (emphasis added). So, at least in blended waters, TMDLs must be calculated with regard to nonpoint sources of pollution; otherwise, it would be impossible "to implement the applicable water quality standards," which do not differentiate sources of pollution. This court has so recognized. *Browner*, 20 F.3d at 985 ("Congress and the EPA have already determined that establishing TMDLs is an effective tool for achieving water quality standards in waters impacted by non-point source pollution.").

Nothing in the statutory structure—or purpose—suggests that Congress meant to distinguish, as to § 303(d)(1) lists and TMDLs, between waters with one insignificant point source and substantial nonpoint source pollution and waters with only nonpoint source pollution. Such a distinction would, for no apparent reason, require the states or the EPA to monitor waters to determine whether a point source had been added or removed, and to adjust the § 303(d)(1) list and establish TMDLs accordingly. There is no statutory basis for concluding that Congress intended such an irrational regime.

Looking at the statute as a whole, we conclude that the EPA's interpretation of § 303(d) is not only entirely reasonable but considerably more convincing than the one offered by the plaintiffs in this case.

C. Federalism Concerns

The Pronsolinos finally contend that, by establishing TMDLs for waters impaired only by nonpoint source pollution, the EPA has upset the balance of federal-state control established in the CWA by intruding into the state's traditional control over land use. See Solid Waste Agency of

Northern Cook County v. United States Army Corps of Eng'rs, 531 U.S. 159, 172–73 (2001). That is not the case.

The Garcia River TMDL identifies the maximum load of pollutants that can enter the Garcia River from certain broad categories of nonpoint sources if the river is to attain water quality standards. It does not specify the load of pollutants that may be received from particular parcels of land or describe what measures the state should take to implement the TMDL. Instead, the TMDL expressly recognizes that "implementation and monitoring" "are state responsibilities" and notes that, for this reason, the EPA did not include implementation or monitoring plans within the TMDL.

Moreover, § 303(e) requires—separately from the § 303(d)(1) listing and TMDL requirements—that each state include in its continuing planning process "adequate implementation, including schedules of compliance, for revised or new water quality standards" "for all navigable waters within such State." § 303(e)(3). The Garcia River TMDL thus serves as an informational tool for the creation of the state's implementation plan, independently—and explicitly—required by Congress.

California chose both if and how it would implement the Garcia River TMDL. States must implement TMDLs only to the extent that they seek to avoid losing federal grant money; there is no pertinent statutory provision otherwise requiring implementation of § 303 plans or providing for their enforcement. See CWA § 309, 33 U.S.C. § 1319; CWA § 505, 33 U.S.C. 1365.

* * *

IV. CONCLUSION

For all the reasons we have surveyed, the CWA is best read to include in the § 303(d) listing and TMDLs requirements waters impaired only by nonpoint sources of pollution.... We therefore hold that the EPA did not exceed its statutory authority in identifying the Garcia River pursuant to § 303(d)(1)(A) and establishing the Garcia River TMDL, even though the river is polluted only by nonpoint sources of pollution.

Notes and Questions

1. The TMDL process outlined in *Pronsolino*, in essence, is the following:

    - States identify specific waters where problems exist or are expected as a result of point *and/or nonpoint* sources.

    - States allocate pollutant loadings among point *and nonpoint* sources contributing to the impairment, and EPA approves State actions or acts in lieu of the State if necessary.

    - Point *and nonpoint* sources then reduce pollutants to achieve the pollutant loadings established by the TMDL through a wide variety of Federal, State, Tribal, and local authorities, programs, and initiatives.

As it relates to nonpoint source pollution, then, application of the TMDL program breaks down into three discrete questions. First, can impairment resulting entirely or mostly from nonpoint sources be considered when deciding whether a water body must be listed as impaired for purposes of section 303? Second, if a water body impaired in part by nonpoint sources can be listed, can a TMDL waste load allocation be made that includes nonpoint sources contributing to impairment of the water body? Finally, if a TMDL waste load allocation can be made to nonpoint sources, how can it be enforced against them? *Pronsolino* is seen by many as a victory on the first two issues for advocates of using the TMDL program as the hook for controlling nonpoint source pollution. Plaintiffs argued that EPA lacked authority to require TMDLs for waters of the Garcia River in Northern California impaired solely by nonpoint sources of pollution due to sediment from soil erosion caused by timber harvesting along the river's banks. The court disagreed, ruling that EPA has authority to require states to list waters receiving only nonpoint source pollutants on a state's Section 303(d) list and subsequently to prepare TMDLs for the listed waters.

2.   But what about the third question–enforcement of the TMDL against nonpoint sources? Here *Pronsolino* is less helpful to those who would use the TMDL program as the basis of nonpoint source regulation. The court observed that EPA could not regulate land use because that matter was reserved for the states. The court further concluded that, although section 303(e) of the CWA requires states to include TMDLs such as the Garcia River sediment TMDL in their continuing planning processes, California is free to decide whether to implement the TMDL by regulating nonpoint sources such as farms and timber operations through its land use practices, and is free to risk possible loss of federal environmental grant funds if it decides not to implement the TMDL. So where does that leave matters? Do you think most states will accept the invitation to regulate nonpoint sources? Or will they bite the bullet, leave their nonpoint sources alone, and forego federal monies? Will it depend on who the nonpoint sources are? And, if the states do refuse to go after the nonpoint sources, what happens to the TMDL?

3.   In an effort to entice states into including nonpoint sources in their TMDL implementation programs, EPA ingeniously (some have said devi- ously) unveiled a plan to take advantage of the point source side of the TMDL program as leverage for the nonpoint source side. Point sources, of course, have long been subject to direct technology-based regulation under the CWA NPDES permitting program. Nonpoint sources have not. It is reasonable to assume, therefore, that the marginal cost of satisfying TMDL waste load allocations–i.e., the cost of reducing a unit of pollutant load–will be more for point sources than it will for nonpoint sources. Hence, as an alternative to direct regulation of nonpoint sources, EPA suggested that states could allow NPDES dischargers to pay for nonpoint source dischar- gers' reductions in discharge loads and thereby avoid additional load restrictions in their NPDES permits. *See* Revisions to the Water Quality Planning and Management Regulation and Revisions to the National Pollu- tant Discharge Elimination System Program in Support of Revisions to the

Water Quality Planning and Management Regulation, 65 Fed. Reg. 4365 (2000) (amending various provisions of 40 C.F.R. pts. 9, 122, 123, 124, 130). Provided the state can demonstrate with reasonable assurance that the nonpoint source load reduction measures will actually lead to load reductions, such trading can take advantage of the disparity in marginal cost of load unit of reduction–i.e., point sources could actually save money by paying for nonpoint source reductions. Of course, this plan won't work for waters impaired entirely by nonpoint sources, such as those involved in *Pronsolino* and, according to the Lums, all the impaired waters of Hawaii.

4. There is far from universal agreement, however, that EPA has the authority the *Pronsolino* court believes it does. The history of EPA's efforts to establish its position in this regard is a story in bizarre politics, though one many would regard as politics as usual. In 1999, EPA published proposed revisions to its TMDL regulations that would have brought the program squarely in line with *Pronsolino* and the strategy for nonpoint sources the court endorsed. *See* 64 Fed. Reg. 46,057 (1999); 64 Fed. Reg. 46,011 (1999); *see generally* Lisa E. Roberts, *Is the Gun Loaded This Time? EPA's Proposed Revisions to the Total Maximum Daily Load Program*, 6 ENVTL. LAWYER 635 (2000). These proposals immediately attracted controversy. EPA received over 34,000 comments on the proposed rules. Farming groups initiated litigation challenging EPA's authority to implement the TMDL program so as to assign allocations to nonpoint sources. *See* American Farm Bureau Federation v. Browner, Nos. 00–1320 and consolidated cases (D.C. Cir). The National Governors Association immediately sought federal legislation to add funding and flexibility to the TMDL program. Some members of Congress also questioned EPA's authority in this regard and took measures to block implementation of the final rules. Indeed, prior to EPA's promulgation of the final rules, Congress adopted a rider to the 2001 Military Construction/Supplemental Appropriations bill (the bill dealt with funding of U.S. Forces in Kosovo and of Columbian anti-drug efforts) that prevented EPA not only from enforcing new TMDL rules before fiscal year 2002, but also from even finishing its work on the proposed rules and adopting them as final. In a political gambit to thwart Congress, EPA adopted the final rules, which retain most of the relevant structure of the proposed rules, before President Clinton signed the appropriations bill restricting the agency from doing so. *See* Revisions to the Water Quality Planning and Management Regulation and Revisions to the National Pollutant Discharge Elimination System Program in Support of Revisions to the Water Quality Planning and Management Regulation, 65 Fed. Reg. 4365 (2000) (amending various provisions of 40 C.F.R. pts. 9, 122, 123, 124, 130). In other words, the rules went on the books, but were "not effective until 30 days after the date that Congress allows EPA to implement this regulation." 65 Fed. Reg. at 53586. Since then, of course, a new administration has taken command of EPA. In July 2001, one year after the final rules were "adopted," the Bush administration announced that it would further delay implementation of the rules so that it may reconsider them in light of concerns raised by stakeholders, with spring 2002 as a projected target date for proposing any changes to the rules. *See* 66 Fed. Reg. 41,817

(2001). The court in American Farm Bureau Federation v. Browner then stayed the challenge to the "old" rules pending issuance of any revised rules.

5. Assuming the TMDL rules that are eventually put into effect retain the basic features EPA adopted in July 1999, and do apply to nonpoint sources at least as far as *Pronsolino* says they do, to what extent can the TMDL program become a useful tool in the ecosystem management effort? The Lums do not appear too optimistic in this respect. Some of their reasons are politically-based: states will be reluctant to take on regulation of agricultural interests; costs of TMDL calculation, not to mention implementation, are staggering; local land use decisions are highly complex and must be coordinated with state TMDL decisions; and so on. But some of their concerns are physical: TMDLs may be very difficult to accomplish in small watersheds; TMDLs are too narrow in focus, failing to take into account habitat quality and water quantity issues; and nonpoint source pollution often is caused in part by physical environment conditions beyond the control of regulators or dischargers. *See also* Jory Ruggiero, *Toward a Law of the Land: The Clean Water Act as a Mandate for the Implementation of an Ecosystem Approach to Land Management*, 20 PUBLIC LAND AND RE-SOURCES L.J. 31 (1999). For a contrary view based on the experience of TMDL implementation in Texas, see Margaret Hoffman, *Integrating TMDLs into Watershed–Based Water Quality Management*, 31 ST. BAR OF TEXAS ENVTL. L.J. 193 (2001). Were the EPA under the Clinton administration, as well as the environmental advocates that sued the agency into action, asking too much of the TMDL program?

6. Another action Congress took after EPA proposed its TMDL rule in 2000 was to commission the National Research Council (an arm of the National Academy of Sciences) to evaluate the scientific basis of the rule and the TMDL program in general. Consistent with the Lums' assessment, the Council's report concluded, among other things, that the TMDL program narrowly focuses on pollutant loads, whereas habitat degradation and stream channel modifications should play a key role in any water quality restoration program. Also, consistent with much of the literature on ecosystem management, the Council recommended that the TMDL program employ adaptive management techniques so that TMDLs can evolve as better data are collected and assessed. *See* NATIONAL ACADEMY OF SCI-ENCES/NATIONAL RESEARCH COUNCIL, ASSESSING THE TMDL APPROACH TO WATER QUALITY MANAGEMENT (2001).

7. The *Pronsolino* court noted that, because EPA has no authority to regulate nonpoint sources directly, the only leverage EPA really has over states that refuse to regulate nonpoint sources is withdrawing federal environmental grant funds. As the court explained, while this is not federal regulation, the withdrawal effects may be too severe for states that have become dependent on the federal dollars to help run state environmental programs. They may knuckle under, as the court put it, and apply the federal program uncritically and contrary to other legitimate state policy objectives. What do you think of this technique of "federalizing" environmental policy? For an overview of how Congress can use the spending power to influence state environmental policy, see Denis Binder, *The Spending Clause as a Positive Source of Environmental Protection: A Primer*, 4 Chapman L. Rev. 147 (2001).

8. The Lums predict that TMDL compliance will be costly in the State of Hawaii, and the *Pronsolino* opinion drives the cost point home by attaching some whopping compliance cost figures to individual landowner contexts in California–in some cases to the tune of millions of dollars each. Why is it so expensive to reduce pollutant loads from their lands? Is it that their forestry and other land management practices were so bad, or is it simply that controlling nonpoint source pollution is so difficult? If it will cost landowners tens of millions of dollars to comply with TMDLs in just this small watershed, imagine the total national price tag for controlling nonpoint source pollution! Is it worth it? Should the cost be borne entirely by the private landowners?

9. For more on the TMDL program, particularly as it relates to nonpoint sources, see Oliver A. Houck, The Clean Water Act TMDL Program: Law, Policy, and Implementation (Envtl. L. Inst. 1999); Robert W. Adler, *Controlling Nonpoint source Pollution: Is Help on the Way (From the Courts or EPA?)*, 31 Envtl. L. Rep. (ELI) 10270 (2001); Sarah Birkeland, *EPA's TMDL Program*, 28 Ecology L.Q. 297 (2001); James Boyd, *The New Face of the Clean Water Act: A Critical Review of EPA's New TMDL Rules*, 11 Duke Envtl. L. & Pol'y F. 39 (2000).

b. WILD AND SCENIC RIVERS

The Lums' criticism of the TMDL program as an ecosystem management tool boils down to the observation that it is too powerful an authority channeled into too narrow a mandate. An alternative approach would be a broader mandate–protect the overall integrity of the river or the lake–within a more flexible regulatory framework. One model of such a program is the Wild and Scenic Rivers Act of 1968 (WSRA), 16 U.S.C. § 1271–1287. Enacted in advance of the flurry of environmental legislation that began in the 1970s, WSRA protects "free-flowing" rivers that possess "outstandingly remarkable value." 16 U.S.C. § 1271. The designation and boundary delineation of such rivers is made by Congress on its own initiative, by congressional approval of a federal agency nomination, or by a state with the approval of the Secretary of the Interior.

WSRA classifies designated rivers into three categories based on their degree of naturalness (i.e., lack of evidence of human influence): wild, scenic, and recreational. Depending on the classification of a designated river, different management mandates apply, to be implemented by the appropriate federal or state agency. The responsible agency must prepare a management plan designed to implement its duty to "protect and enhance the values which caused [the river] to be included in said system without, insofar as is consistent therewith, limiting other uses that do not substantially interfere with public use and enjoyment of those values." *Id.* 1281(a). WSRA thus is organized around the functions of designating, classifying, managing, and protecting free-flowing rivers for their identified outstanding resource values.

WSRA has led to designation and classification of substantial river miles. The U.S. Forest Service manages 76 rivers under WSRA, amounting to over 4000 river miles, and the Bureau of Land Management covers almost 2000 river miles in 15 rivers. Over 150 rivers and over 10,000 river miles in all are within the WSRA system. Only recently, however, has the force of WSRA's management mandates been felt in ways that limit land use options and thus could be included as a potent component of ecosystem management policy. The cases in what is known as the "Oregon Trilogy" recognized WSRA's "protect and enhance" standard as an enforceable mandate. *See* Oregon Natural Desert Association v. Green, 953 F.Supp. 1133 (D.Or.1997); National Wildlife Federation v. Cosgriffe, 21 F.Supp.2d 1211 (D.Or.1998); Oregon Natural Desert Association v. Singleton, 47 F.Supp.2d 1182 (D.Or.1998). The following case provides an epilogue to the trilogy.

Oregon Natural Desert Association v. Singleton

United States District Court for the District of Oregon, 1999.
75 F.Supp.2d 1139.

■ REDDEN, DISTRICT JUDGE.

This is an action brought by environmental groups (collectively "ONDA"), against the Bureau of Land Management ("BLM") and three individuals. Oregon Cattlemen's Association appears as an intervenor-defendant. ONDA challenges the BLM's management of the Main, West Little, and North Fork Owyhee River corridors, alleging that the BLM failed to prepare an environmental impact statement ("EIS") analyzing the effect of cattle grazing on the area, as required by the National Environmental Policy Act ("NEPA"), 42 U.S.C. §§ 4321–4370a, and that its management plan violates the BLM's mandate under the Wild and Scenic Rivers Act ("WSRA"), 16 U.S.C. §§ 1271–1284.

* * *

Findings of Fact

Background

In 1984, Congress designated 120 miles of the Main Owyhee River as a federal wild and scenic river pursuant to the WSRA. In the Oregon Omnibus Wild and Scenic Rivers Act of 1988, Pub.L. 100–557, codified at 16 U.S.C. § 1274(a)(91), Congress added 57 miles of the West Little Owyhee and nine miles of the North Fork Owyhee to the national wild and scenic rivers system. Congress classified all three segments as "wild." A wild river area is defined under the WSRA as "free of impoundments and generally inaccessible except by trail, with watersheds or shorelines essentially primitive and waters unpolluted." 16 U.S.C. § 1273(b). The "wild" classification is the most restrictive of three possible classifications. Id.

Section 3 of the WSRA required the BLM to issue a "comprehensive management plan" to "provide for the protection of the river values" within three fiscal years after designation. 16 U.S.C. § 1274(d)(1). The WSRA requires that the plan "address resource protection, development of lands and facilities, user capacities, and other management practices necessary or desirable to achieve the purposes of this chapter." 16 U.S.C. § 1264(d)(1).

Conditions in the river corridor at plan implementation

In September 1991, the BLM issued a final management plan ("the Plan"). The Plan identified five outstandingly remarkable values ("ORVs") on the Main Owyhee: scenery, geology, recreation, wildlife, and cultural. The ORVs identified for the West Little and North Fork Owyhee included recreation, scenery and wildlife. The Plan did not designate botanical or fishery ORVs, but characterized vegetation as a "key component of the visual resource, important to watershed values, wildlife habitat, and a vital part of the natural setting for recreation."

* * *

Although cattle had been grazing the river corridor for many years at the time the Plan was written and implemented, the BLM recognized that in some of the river areas accessible to livestock, cattle grazing had created substantial negative effects. Cattle are grazed on 67 miles of the 186–mile river system, and the BLM found that 18 of these miles constituted "areas of livestock concern," i.e., showed noticeable negative effects created by grazing. The areas most affected by livestock grazing were trail crossings and "water gaps," the places where livestock come to the river to drink.

The Environmental Assessment ("EA") issued with the Plan noted that areas within at least seven of 11 grazing allotments and one trail area showed negative effects from livestock grazing, and that these negative effects had a direct impact on the scenic, recreational, and watershed ORVs of the Owyhee Rivers.

* * *

The BLM's range management objectives for the river corridor

Section 1281(a) of the WSRA specifies the management duties of the federal agency charged with managing the designated river:

Each component of the national wild and scenic rivers system shall be administered in such manner as to protect and enhance the values which caused it to be included in said system without, insofar as is consistent therewith, limiting other uses that do not substantially interfere with public use and enjoyment of these values. In such administration primary emphasis shall be given to protecting its esthetic, scenic, historic, archeologic, and scientific features. Management plans for any such component may establish varying degrees of intensity for its protection and development, based on the special attributes of the area.

16 U.S.C. § 1281(a).

The Plan promulgated by the BLM specified the following objectives for range management:

> Maintain or improve the vegetative cover of key species and the visual aspect of native perennial plants, within the soil and vegetative capabilities of ecological sites, in the corridor by 1999. Maintain proper utilization of key species. Minimize livestock impacts on vegetation and soils, within the river corridor, at water gaps/trail crossings, on uplands, and in riparian areas. Minimize livestock/recreation conflicts at water gaps/trail crossings by 1999.

The Plan's first management prescription for grazing was to "[i]nventory the river corridors to determine riparian areas and potentials." The inventory was necessary because the BLM did not at the time the Plan was written have baseline data on the level of grazing that had occurred at the time the rivers were designated or that was occurring at the time the Plan was being written.

However, Mr. Taylor [of BLM] testified at the evidentiary hearing on September 13, 1999, that this inventory has not been done, and will not be done unless and until a funding request is granted. If the funding becomes available, the inventory is scheduled to begin in 2001.

To achieve its goal of "maintaining or improving" vegetation, the Plan established three "utilization" standards: livestock could not consume more than 40% of the annual growth of "key grass species" in upland areas except for certain winter allotments, where 50% consumption was permitted, and livestock could not consume more than 30% of the annual growth of "current years leaders" for willows in riparian areas.

However, at the time the standards were set, the BLM had done no utilization studies for riparian areas. It appears from the Plan and other parts of the administrative record that the 30, 40 and 50% utilization standards represented the grazing levels in existence at the time the Plan was being written.

The Plan further provided that

> [h]erbaceous riparian vegetation will be managed to insure a properly functioning riparian system. Management may include restrictions on use levels, seasons or where feasible and compatible exclusionary fencing. Key sites will be monitored and use levels and/or management may be revised, on a case by case basis, through the allotment evaluation process. Any changes in use levels and or management must ensure plan objectives are being met. Water gaps and trail crossings will be managed so that vegetative cover does not decrease and, if possible, increases. Alternate sources of water, fencing and improved herding practices will be utilized where possible to reduce or eliminate livestock impacts at water gaps and trail crossings. Primary focus will be given to areas where camping/livestock conflicts exist.

Grazing impacts since promulgation of the Plan

The BLM asserts that since the Plan was promulgated, the overall impact of grazing on the river corridor has decreased, and four of the areas of concern have improved.

* * *

Mr. Taylor testified that [an] indicator of improved conditions in the river corridor was the BLM's recent finding of attainment of Properly Functioning Condition of Riparian Areas on 146.5 miles of the river areas (93%), with 6.8 miles (4.4%) functioning at risk. According to the BLM, "properly functioning riparian condition" means that adequate vegetation, landform, or large woody debris is present to:

1) dissipate stream energy associated with high waterflows, thereby reducing erosion and improving water quality;

2) filter sediment, capture bedload, and aid floodplain development;

3) improve flood-water retention and ground-water recharge;

4) develop root masses that stabilize streambanks against cutting action;

5) develop diverse ponding and channel characteristics to provide the habitat and the water depth, duration, and temperature necessary for fish production, waterfowl breeding, and other uses; and

6) support greater biodiversity.

Only 67 miles of the 186–mile river system are accessible to livestock. The evidence does not reveal how many of those 67 miles are encompassed within the 146.5 miles found to be properly functioning, how many are within the 39.5 miles still to be assessed, and how many are included in the 6.8 miles which are functioning at risk. (The areas of concern identified in 1993 comprise about 18 miles). The attainment of properly functioning condition in 146.5 miles, without more, does not provide the court with any basis for determining the condition of the areas of concern.

* * *

In August 1999, Katie Fite, a biologist who had been active in bringing conditions in Deary Pasture to the BLM's attention, wrote a letter to Mr. Taylor describing her observations of areas around the West Little Owyhee in July 1999, and enclosed photographs of some of those areas. Her observations included:

1. Trampled creek banks and signs of heavy grazing at Jackson Creek.

2. Severely damaged springs and seeps near the Anderson Crossing road, with livestock having completely "stripped the vegetation and pounded it into a mass of mud." Ms. Fite also noted that livestock were "wallowing well above their knees in the slime that passed for water here."

3. Salt located in several places near wet meadows and springs along the main road between Antelope Creek and Anderson Crossing, near Exchange Spring, and within 20 feet of a running spring.

4. Cattle defecating in the water, trampling stream banks, and grazing heavily upstream of Anderson Crossing.

Even the intervenor's expert witness, Gar Lorain, has opined that hot springs or other springs "could have unique [plant] species subject to fairly heavy grazing. These areas should be fenced to protect their unique characteristics." Declaration of Gar Lorain, Exhibit 2, p. 2. However, there is no indication that the BLM has responded to these observations with any reduction in the number of grazing animals or modification of grazing practices.

Conclusions of Law

The WSRA states the policy of the United States that certain designated rivers which possess "outstandingly remarkable scenic, recreational, geologic, fish and wildlife, historic, cultural or other similar values," shall be "preserved in free-flowing condition" and that the rivers and their immediate environments be "protected for the benefit and enjoyment of present and future generations." 16 U.S.C. § 1271.

Although the BLM asserts that its grazing management practices have generated improvements in the areas of concern first identified in 1993, the court concludes that the assertion is unsubstantiated by objective evidence except for the closure of Deary Pasture. Perhaps the most troubling evidence is Mr. Taylor's testimony that the numbers of animals and the seasons of use have remained completely unchanged since implementation of the Plan, except when grazing permits have been increased to exploit good water years.

* * *

The BLM's finding that most of the river corridor is in properly functioning riparian condition is not probative with respect to improvement in the areas of concern because the percentages provide no specific information about areas of concern. Other data gathered since 1993–the surveys of the West Little Owyhee, for example–consistently indicate heavy cattle grazing and erosion.

The Plan provided that restrictions on levels and seasons of use would be implemented where necessary to ensure that utilization standards were met, riparian vegetation was in a properly functioning condition, and livestock impacts on vegetation and soils within the river corridor, at water gaps and trail crossings were minimized so that vegetative cover would not decrease and, if possible, would increase. See, e.g., AR Tab 178, p. 30–31. None of this has been done. Mr. Taylor testified that the BLM has neither made changes to seasons of use nor reduced the number of AUMs permitted for any of the allotments since the Plan was implemented. In fact, Mr. Taylor admitted that the BLM has actually increased the number of AUMs in some allotments, because greater than anticipated rainfall had yielded more vegetation. The court is troubled by this indication that the BLM regards beneficial natural events as justifications for increased grazing, rather than as opportunities for recovery and enhancement of natural resources.

The Public Interest and the Balance of Equities

Factors favoring closure of the areas of concern

The WSRA charges the BLM with administering "each component of the national wild and scenic rivers system" in a manner which will "protect and enhance the values which caused it to be included in said system." 16 U.S.C. § 1281(a). In that administration, "primary emphasis shall be given to protecting its esthetic, scenic, historic, archeologic, and scientific features." Id.

The values and special attributes of each river segment are enumerated in the 1993 Plan. Recreation values on the Main Owyhee include rafting, drift boating, kayaking, hiking, photography, nature study, fishing, hunting, camping, and rockhounding. The West Little river segment offers very high quality primitive recreation experiences, including off-trail backpacking, swimming, hiking, wildlife viewing, and hunting. The North Fork Owyhee offers very high quality backpacking opportunities, early season expert level kayaking, hiking, hunting, camping, wildlife viewing and photography.

The scenic values of the Owyhee River include its dramatic landforms, whitewater, and slow-moving pools. The West Little Owyhee is characterized by canyons, flat sagebrush plateaus, and secluded pools confined by sheer rock walls; between the pools are reaches that flow as riffles or rapids during periods of high water, and become sandy or gravelly dry beaches in the dry, hot summer months. The North Fork of the Owyhee combines canyon bottoms overshadowed by steep canyon walls with flat high sagebrush desert.

Over 200 species of wildlife are found in the river canyons and the sagebrush desert on the rims. Birds include nesting raptors such as hawks, kestrel, falcon, and golden and bald eagles; game birds such as partridge, quail, mourning dove and sage grouse; waterfowl; and song birds. Mammal species include California bighorn sheep, mule deer, wild horses, pronghorn antelope, mountain lion, bobcat, coyote, badger, beaver, otter, muskrat, marmot, raccoon, porcupine and rabbit.

Several plant species within the canyonlands are classified as federal or state sensitive species or are on "watch lists." The preservation of all of these features is in the public interest.

Congress has classified the three river segments as "wild," the most restrictive of three WSRA classifications. A wild river area is defined under the WSRA as "free of impoundments and generally inaccessible except by trail, with watersheds or shorelines essentially primitive and waters unpolluted." 16 U.S.C. § 1273(b). The public interest also includes the public policy stated by Congress of preserving these rivers in as pristine and unpolluted a condition as possible.

It has been almost seven years since the BLM recognized that cattle grazing was creating noticeable negative effects on the rivers' values in some parts of the corridor. The BLM found that grazing conflicted with recreational values where livestock congregated, grazed and defecated around campsites; that the visual impact of livestock trailing and grazing affected scenic and recreation values; and that the ecological condition of upland and riparian areas was being degraded by livestock grazing, trampling and defecation. The BLM designated specific areas of concern in 1993, and stated its goal of managing those areas so as to maintain or improve the vegetative cover of key species and the visual aspect of native perennial plants, ensure proper utilization of key species, minimize livestock impact on vegetation and soils, and reduce livestock/recreation conflicts.

Although the BLM asserts that it has met these objectives, the evidence shows that, with the exception of the areas around Deary Pasture, grazing in the areas of concern has been neither reduced nor otherwise regulated. The same observations of heavy grazing, trampling, and bank erosion reported in 1993 have continued well into 1999. The persistent degradation of these areas, and the BLM's apparent inability to manage cattle grazing in a manner which would repair and restore the areas of concern has serious negative consequences for the recreational, scenic and ecological values of the designated rivers. These negative effects have existed for many years and there is no indication that they will change significantly in the future.

Economic impact on permittees of closing areas of concern

The grazing allotments within the river corridor enable private cattle and sheep ranchers to graze their livestock on publicly-owned land, typically at below-market rates. Although the parties dispute the dollars involved, it is uncontroverted that grazing permits constitute a public subsidy to livestock operators, sometimes a substantial one. The court notes that grazing privileges on publicly-owned land are not property rights, and the government is under no obligation either to provide or to continue them.

At present, the total AUMs for the 19 grazing allotments and 58 ranching operations in the river corridor are approximately 104,000. Closure of the areas of concern would represent a loss of 26,976 AUMs, or about 26% of the total in the river corridor.

Economic impact on Malheur County of closing areas of concern

According to the government's evidence, the total personal income in Malheur County is $491 million. According to a study by plaintiffs' expert witness Hans Radtke, an agricultural economist, the total income generated by the livestock industry in Malheur County is approximately $30 million a year—about 6% of the county's total personal income. The intervenor disputes this amount, arguing that agriculture and agriculture-related businesses generate about $92 million per year. Neither the government nor the intervenor has submitted evidence on the economic consequences to the affected permittees of closing the areas of concern. However, according to figures submitted by the plaintiffs, which neither the government nor the intervenor disputes, the maximum amount of personal income lost to Malheur County by eliminating 26,976 AUMs would be $692,204—approximately .023% of the county's total livestock income, using Radtke's $30 million figure. Because neither the government nor the intervenor has provided the court with evidence on prospective income loss to livestock-related businesses caused by the elimination of 26,976 AUMs, I make no finding on the economic impact to livestock-related businesses within Malheur County.

* * *

The public interest in requiring the BLM to implement the Congressional mandate contained in the WSRA is manifest, as is the public's interest in preserving and enhancing the extraordinary values of the Owyhee Rivers. The continued degradation of the areas of concern within the river corridor constitutes irreparable harm, and there are no legal remedies available to redress this harm. While a 25% reduction in subsidized grazing privileges will have an adverse economic effect on some of the individual permit holders, its overall effect on the county's economy is negligible.

For these reasons, the public interest and the balance of equities require the issuance of an injunction directing the BLM to exclude the areas of concern from any further grazing by domestic livestock.

Note and Questions

1. In deciding whether to enjoin BLM from issuing AUMs in the disputed areas, the court determined that while the reduction in subsidized grazing privileges would have an adverse economic effect on some of the individual permit holders, its overall effect on the county's economy would be negligible. What if the economic impact on the county would have been substantial? Does that matter under WSRA? Should it?

2. WSRA defines "outstanding resource values" as including not only "scenic, recreational, geologic, fish and wildlife, historic, [and] cultural" values, but also "other similar" values as well. 16 U.S.C. § 1271. In their guidelines for administration of the program, the Departments of Interior and Agriculture have clarified that ecological values are ORVs. *See* 47 Fed. Reg. 39,454, 39,455–57 (Sept. 7, 1982).

3. Given its limitation to free-flowing rivers, it should be no surprise that a principal objective of WSRA is to keep designated rivers in that condition. Section 7 of WSRA thus prohibits the Federal Energy Regulatory Commission, an agency within the Department of Energy that oversees nongovernmental hydroelectric projects and dams, from licensing any such project "on or directly affecting" a WSRA river. 16 U.S.C. § 1278(a). All federal agencies are prohibited from assisting any water project that would have a "direct and adverse effect on the values for which such river was established." *Id.*

4. As the Oregon trilogy demonstrates, WSRA's "protect and enhance" mandate goes well beyond limiting dams. Section 10 and 12 of WSRA outline the duties of federal river corridor management agencies and of federal agencies administering lands adjacent to the river corridor, respectively. Section 10 requires the federal land management agency with jurisdiction over the land through which the river flows to prepare the river management plan and constrain its land management actions pursuant to the protect and enhance mandate. The Departments of Interior and Agriculture interpret this section as codifying a "nondegradation and enhancement policy" that applies to all designated rivers regardless of classification. 47 Fed. Reg. 39,454, 39,458 (Sept. 7, 1982). Section 12 then extends the protect and enhance mandate, and thus the nondegradation and enhancement policy, to "any lands which include, border upon, or are adjacent to" the designated river. Federal agencies with jurisdiction over such lands, which may include agencies other than the river corridor management agency, need not prepare WSRA management plans, but must incorporate WSRA management purposes into all policies, regulations, contracts, and other actions that affect the designated river. *See* Wilderness Society v. Tyrrel, 918 F.2d 813 (9th Cir.1990). In short, WSRA preempts the multiple-use regime under which most federal land management agencies operate, vaulting preservation to a dominant-use status for designated rivers. For further background on these aspects of the WSRA program, see Charlton H. Bonham, *The Wild and Scenic Rivers Act and the Oregon Trilogy*, 21 PUBLIC LANDS & RESOURCES L. REV. 109 (2000); Sally K. Fairfax et al., *Federalism and the Wild and Scenic Rivers Act: Now You See It, Now You Don't*, 59 WASH. L. REV. 417 (1984); Peter M. K. Frost, *Protecting and Enhancing Wild and Scenic Rivers In the West*, 29 IDAHO L. REV. 313 (1992–93).

5. Why would a state seek designation of a Wild and Scenic River in its borders? The state must first so designate the river under state law, then it must petition the Secretary of the Interior for federal designation. If the river is designated, the state must manage nonfederal lands within the designation boundary consistent with WSRA's mandate at no cost to the federal government. So what's in it for the state? Indeed, several counties in California asked that question when, in June 1980, Governor Jerry Brown petitioned the Department of Interior to designate five river segments in northern California as WSRA rivers. The counties within which the river segments were located challenged the designations and managed, until the wee hours of the Carter Administration, to secure judicial injunctions against final approval by Secretary of Interior Cecil Andrus. The Ninth Circuit lifted the injunction in time for Andrus to sign the designations–his last official act in office–and the Ninth Circuit later rejected the merits of the counties' challenge to the final designations. *See* Del Norte County v. United States, 732 F.2d 1462 (9th Cir.1984). What do you suppose were the counties' reasons for challenging the designations, and why would the Governor have petitioned for designation despite the counties' opposition?

6. Only free-flowing rivers, of which few remain, can be designated for protection under the regulatory reach of the Wild and Scenic Rivers Act. In 1997, the Clinton Administration launched a non-regulatory program to coordinate protection and restoration of a limited number of "American Heritage Rivers" (AHR). *See* Exec. Order 13061, 62 Fed. Reg. 48,860 (Sept. 11, 1997). This American Heritage Rivers Initiative is designed to promote the environmental, economic, and cultural aspects of rivers and their river communities. In 1997, local communities, acting in coordination with their state, local, and tribal governments, nominated rivers for inclusion in the AHR program, and Vice President Gore announced the first batch of 14 AHRs in July 1998. A federal interagency task force then determined how existing environmental, economic, and cultural programs could best be refocused to benefit each AHR and its local community. Each AHR is assigned a federal employee to serve as "River Navigator" to help implement AHR management plans and act as a liaison to federal agencies. Federal agencies are responsible for matching AHR community needs with available federal resources within their respective jurisdictions, including pollution control, greenway and alternative transportation facilities, historic structure restoration, biological resources inventories, and similar efforts. Although critics of the program derided it as a step along the perilous road to federal regulation of private and local land uses, in fact the AHR Initiative remained entirely non-regulatory and has been continued thus far by the Bush Administration. For further background on the program, see Thomas Downs, *American Heritage Rivers Initiative: A Harbinger of Future White House Environmental Policy?*, 29 Envtl. L. Rep. (ELI) 10065 (1999), and for updated information, including a "state of the river" report for each AHR, see EPA's web site, American Heritage Rivers, www.epa.gov/rivers.

7. WSRA protects rivers that are free-flowing, and the AHR Initiative is non-regulatory in scope. Neither program, therefore, gets to the nub of the most controversial topic in riverine ecosystem management—dams. Dam building in the 20th century was practically a national pastime, and certainly a national policy. Over 75,000 public and private dams (over six feet high) have been erected on this nation's rivers, impeding over 600,000 miles of flowing water. As we have for the filling of wetlands, have we turned the corner on building dams? Most likely we have. There appears to be a broad consensus in federal and state policy that, as Bruce Babbitt quipped in the later years of his tenure as Secretary of Interior, we overdid it with dams. But the far thornier question is whether to remove existing dams to restore the free-flowing qualities of impounded waterways. Dams provide tremendous economic values to local and distant populations, including hydroelectric power, irrigation water, and recreation. But they fundamentally alter the ecosystem dynamics of the imprisoned waterway, and they can be as deadly as any predator to fish that are trapped behind them, or swept over spillways to their death, or pureed in power-generating turbine blades. Babbitt pushed the politics of dam removal front and center with his infamous "dam-busting tour" of 1998, and the Endangered Species Act listings of numerous runs of salmon in Pacific Northwest rivers, and of sturgeon in the Missouri River, have forced federal dam managers, principally the Bureau of Reclamation and the Army Corps of Engineers, to consider dam removal as a serious policy option. *See* Richard A. Lovett, *As Salmon Stage Disappearing Act, Dams May Too*, 284 SCIENCE 574 (1999). Few issues will more squarely test riverine ecosystem management policy than how national, state, and local interests deal with dam removal in the next decade. For a concise history of national policy on dams from its building phase to the current movement toward removal, see Christine A. Klein, *Dam Policy: The Emerging Paradigm of Restoration*, 31 Envtl. L. Rep. (ELI) 10486 (2001); Christine A. Klein, *On Dams and Democracy*, 78 OREGON L. REV. 641 (1999).

Note on Invasive Species in Freshwater Ecosystems

One of the more pernicious and potentially most intractable problems for freshwater ecosystems is the introduction of non-native species, also known as the problem of "invasive," "alien," or "exotic" species. While assigning the label of invasive to a species involves in some part a subjective determination by humans based on what species assembly we deem the "right" set of constituents, there is no question that human actions, intentional and unintended, have accelerated the rate of introductions of new species into unfamiliar ecosystems around the globe. Moreover, although species recruitment is one of the inevitable and dynamic consequences of natural ecosystem disturbances such as flood and fire, as well as of pure accident (e.g., birds blown off course), human intervention has clearly pushed that effect into the realm of being too much of a good thing. In short, alongside global climate change in terms of dimension and magnitude, or perhaps even ahead of it, the problem of invasive species is one with profound potential effects on global biodiversity. *See* Peter M. Vitousek et al., *Biological Invasions as Global Environmental Change*, AMERICAN SCIENTIST, Sept.-Oct 1996, at 468.

This phenomenon is present in all ecosystem types world-wide, but is particularly acute in freshwater ecosystem settings, and even more particularly in lakes and rivers. In the Great Lakes, for example, discharge of ballast water from international commercial vessels has introduced dozens of "hitchhiker" species into that environment. Not every introduced species succeeds in its new setting, but it takes only one to throw the existing dynamic equilibrium between species into disarray. In the Great Lakes, for example, ballast water introduced the zebra mussel, which has an amazing capacity to filter water (about 1 liter per day) and to reproduce. It has become the dominant mussel species in the Great Lakes ecosystem and, while it has contributed measurably to increased clarity of the water, that in itself has disrupted the food web in the entire system because light is reaching deeper into the water column. The zebra mussel has quickly moved beyond the Great Lakes and threatens many river systems. Indeed, freshwater mussels already were among the most rapidly vanishing species domestically, and the zebra mussel may push a good number to extinction.

The list of similar human-induced ecosystem disruptions is long and growing. For an extensive source of information on the zebra mussel and other invasive plant and animal aquatic species, see National Aquatic Nuisance Species Clearinghouse, http://www.aquaticinvaders.org.; and the University of Florida's Aquatic, Wetland, and Invasive Plant Information Retrieval System, http://aquat1.ifas.ufl.edu/database.html.

The legal response to invasive species has been hampered by the pervasive and invidious nature of the problem. International trade has opened countless pathways to species introductions, making it virtually impossible to cut off all possible avenues. Thus, we seldom know of the invasion before it has happened, and by then it is usually too late. It has proven extremely difficult to eradicate an invasive species from its conquered ecosystem without doing yet more harm to the ecosystem. So, what can law bring to the table? Some have argued that more aggressive regulation and enforcement of obvious pathways such as ballast water is called for. *See* Sandra B. Zellmer, *The Virtues of "Command and Control" Regulation: Barring Exotic Species from Aquatic Ecosystems*, 2000 U. Illinois L. Rev. 1233. That may be well and good, but will most likely only put a dent in the problem.

Our existing domestic law of invasive species is, therefore, understandably rather unfocused and ineffective. Early efforts included the Lacey Act of 1900, which authorized the Secretary of the Interior to control the importation of species deemed "injurious to human beings, to the interests of agriculture, horticulture, forestry, or the wildlife or the wildlife resources of the United States." 18 U.S.C. § 42. Much later, in the 1970s, Congress passed the Federal Noxious Weed Control Act of 1974, which authorized the Secretary of Agriculture to "regulate importation and interstate movement of listed noxious weed species," *see* 7 U.S.C. §§ 2801–2814, and President Carter in 1974 issued Executive Order 11987 to require federal agencies to "restrict the introduction of exotic species into the natural ecosystems." 42 Fed. Reg. 26,949 (1977). Still the problem grew.

At the core of the more recent responses is the Nonindigenous Aquatic Nuisance Prevention and Control Act of 1990 (NANPCA), as amended by the National Invasive Species Act (NISA). 16 U.S.C. §§ 4701–4751. NANPCA focused on the ballast water problem in the Great Lakes, and NISA expanded the scope nationally. The statute also established the Aquatic Nuisance Species Task Force to "identify the goals, priorities, and approaches for aquatic nuisance prevention." *Id.* § 4722. In an effort to coordinate these authorities and bring new focus to the invasive species battle, in 1999 President Clinton issued Executive Order 13112, which, among other things, established the Invasive Species Council to supervise the federal response to invasive species. *See* 64 Fed. Reg. 6183 (Feb. 3, 1999). The Order defines invasive species as "an alien species whose introduction does or is likely to cause economic or environmental harm of harm to human health." Federal agencies must review their actions regarding invasive species and must refrain from authorizing, funding, or carrying out actions that could promote introductions of invasive species.

On another front, environmental groups have failed to turn the Environmental Impact Statement (EIS) requirement of the National Environmental Policy Act into an invasive species weapon. In one case, for example, the court dismissed claims that the EIS covering the extension of an airport runway in Hawaii was deficient because it failed to consider invasive species impacts. The court found that the plaintiffs had not specified which invasive species were more likely to be introduced under the extension. *See* National Parks & Conservation Association v. U.S. Department of Treasury, 222 F.3d 677 (9th Cir.2000); *see also* Environmental Defense Fund v. Corps of Engineers, 348 F.Supp. 916 (N.D.Miss.1972) (EIS for waterway project not defective based on invasive species analysis).

Can you think of a more effective legal framework for preventing invasive species introductions? How about for managing ecosystems after an introduction has occurred and the new species has become dominant? Is this something law can really have any appreciable effect over? Alas, in the end we may, over generations, come to think of the zebra mussel as native to the Great Lakes. We most likely will not have a choice.

2.   WETLANDS

As previously described, wetland losses, while slowing dramatically in recent years, have been widespread in the history of this nation. The ecological values lost to the legacy of draining and filling are immeasurable, as are the habitat values that continue to be delivered from the 270 million acres of wetlands we still have in place. So, how should we manage those remaining resources?

A threshold question we inevitably confront along the way to reaching that management issue is where, exactly, are these wetlands, and which of them are subject to federal, state, or local regulatory authority? That question has befuddled federal policy for decades. Section 404 of the Clean Water Act establishes the primary federal program for answering those questions of boundary and application. Section 404 is jointly administered by the U.S. Army Corps of Engineers (Corps), with primary responsibility for implementing a permitting program regulating discharges of "fill" material into waters of the United States, including wetlands, and by the Environmental Protection Agency (EPA), which has authority to define environmental guidelines the Corps must follow. The materials in this section explain how these two agencies have dealt with the problem of defining where and how this program serves to define the core of our national wetland ecosystem management policy.

a.   BOUNDARIES

The term "wetland" implies the existence of "dryland" and thus a geographic limit to the ecosystem unit identified for management under Section 404 of the CWA. Identifying where wet ends and dry begins under Section 404, however, has not been as simple as it may seem. Indeed, wetness is not the exclusively dispositive factor. As the two cases that follow illustrate, questions of federalism and statutory construction have thrust legal issues to the forefront of what otherwise might be thought of as a straightforward bio-physical question.

United States v. Riverside Bayview Homes, Inc.

Supreme Court of the United States, 1985.
474 U.S. 121.

■ JUSTICE WHITE delivered the opinion of the Court.

This case presents the question whether the Clean Water Act (CWA), 33 U.S.C. § 1251 et seq., together with certain regulations promulgated under its authority by the Army Corps of Engineers, authorizes the Corps to require landowners to obtain permits from the Corps before discharging fill material into wetlands adjacent to navigable bodies of water and their tributaries.

I

The relevant provisions of the Clean Water Act originated in the Federal Water Pollution Control Act Amendments of 1972, 86 Stat. 816, and have remained essentially unchanged since that time. Under §§ 301 and 502 of the Act, 33 U.S.C. §§ 1311 and 1362, any discharge of dredged or fill materials into "navigable waters"–defined as the "waters of the United States"–is forbidden unless authorized by a permit issued by the Corps of Engineers pursuant to § 404, 33 U.S.C. § 1344. After initially construing the Act to cover only waters navigable in fact, in 1975 the Corps issued interim final regulations redefining "the waters of the United States" to include not only actually navigable waters but also tributaries of such waters, interstate waters and their tributaries, and nonnavigable intrastate waters whose use or misuse could affect interstate commerce. 40 Fed.Reg. 31320 (1975). More importantly for present purposes, the Corps construed the Act to cover all "freshwater wetlands" that were adjacent to other covered waters. A "freshwater wetland" was defined as an area that is "periodically inundated" and is "normally characterized by the preva-lence of vegetation that requires saturated soil conditions for growth and reproduction." 33 CFR § 209.120(d)(2)(h) (1976). In 1977, the Corps refined its definition of wetlands by eliminating the reference to periodic inundation and making other minor changes. The 1977 definition read as follows:

> The term "wetlands" means those areas that are inundated or saturat-ed by surface or ground water at a frequency and duration sufficient to support, and that under normal circumstances do support, a prevalence of vegetation typically adapted for life in saturated soil conditions. Wetlands generally include swamps, marshes, bogs and similar areas. 33 CFR § 323.2(c) (1978).

In 1982, the 1977 regulations were replaced by substantively identical regulations that remain in force today. See 33 CFR § 323.2 (1985).[2]

Respondent Riverside Bayview Homes, Inc. (hereafter respondent), owns 80 acres of low-lying, marshy land near the shores of Lake St. Clair in Macomb County, Michigan. In 1976, respondent began to place fill materials on its property as part of its preparations for construction of a housing development. The Corps of Engineers, believing that the property was an "adjacent wetland" under the 1975 regulation defining "waters of the United States," filed suit in the United States District Court for the Eastern District of Michigan, seeking to enjoin respondent from filling the property without the permission of the Corps.

The District Court held that the portion of respondent's property lying below 575.5 feet above sea level was a covered wetland and enjoined respondent from filling it without a permit. Respondent appealed, and the Court of Appeals remanded for consideration of the effect of the intervening 1977 amendments to the regulation. 615 F.2d 1363 (1980). On remand, the District Court again held the property to be a wetland subject to the Corps' permit authority. Civ. No. 77–70041 (May 10, 1981).

Respondent again appealed, and the Sixth Circuit reversed. 729 F.2d 391 (1984). The court construed the Corps' regulation to exclude from the category of adjacent wetlands—and hence from that of "waters of the United States"—wetlands that were not subject to flooding by adjacent navigable waters at a frequency sufficient to support the growth of aquatic vegetation. The court adopted this construction of the regulation because, in its view, a broader definition of wetlands might result in the taking of private property without just compensation. The court also expressed its doubt that Congress, in granting the Corps jurisdiction to regulate the filling of "navigable waters," intended to allow regulation of wetlands that were not the result of flooding by navigable waters. Under the court's reading of the regulation, respondent's property was not within the Corps' jurisdiction, because its semiaquatic characteristics were not the result of frequent flooding by the nearby navigable waters. Respondent was therefore free to fill the property without obtaining a permit.

We granted certiorari to consider the proper interpretation of the Corps' regulation defining "waters of the United States" and the scope of the Corps' jurisdiction under the Clean Water Act, both of which were called into question by the Sixth Circuit's ruling. We now reverse.

II

[The Court determined that a narrow reading of the Corps' regulation is not required in order to prevent it from causing a regulatory taking of property].

2. The regulations also cover certain wetlands not necessarily adjacent to other waters. See 33 CFR §§ 323.2(a)(2) and (3) (1985). These provisions are not now before us.

III

Purged of its spurious constitutional overtones, the question whether the regulation at issue requires respondent to obtain a permit before filling its property is an easy one. The regulation extends the Corps' authority under § 404 to all wetlands adjacent to navigable or interstate waters and their tributaries. Wetlands, in turn, are defined as lands that are "inundated or saturated by surface or ground water at a frequency and duration sufficient to support, and that under normal circumstances do support, a prevalence of vegetation typically adapted for life in saturated soil conditions." 33 CFR § 323.2(c) (1985). The plain language of the regulation refutes the Court of Appeals' conclusion that inundation or "frequent flooding" by the adjacent body of water is a sine qua non of a wetland under the regulation. Indeed, the regulation could hardly state more clearly that saturation by either surface or ground water is sufficient to bring an area within the category of wetlands, provided that the saturation is sufficient to and does support wetland vegetation.

The history of the regulation underscores the absence of any requirement of inundation. The interim final regulation that the current regulation replaced explicitly included a requirement of "periodi[c] inundation." 33 CFR § 209.120(d)(2)(h) (1976). In deleting the reference to "periodic inundation" from the regulation as finally promulgated, the Corps explained that it was repudiating the interpretation of that language "as requiring inundation over a record period of years." 42 Fed.Reg. 37128 (1977). In fashioning its own requirement of "frequent flooding" the Court of Appeals improperly reintroduced into the regulation precisely what the Corps had excised.

Without the nonexistent requirement of frequent flooding, the regulatory definition of adjacent wetlands covers the property here. The District Court found that respondent's property was "characterized by the presence of vegetation that requires saturated soil conditions for growth and reproduction," App. to Pet. for Cert. 24a, and that the source of the saturated soil conditions on the property was ground water. There is no plausible suggestion that these findings are clearly erroneous, and they plainly bring the property within the category of wetlands as defined by the current regulation. In addition, the court found that the wetland located on respondent's property was adjacent to a body of navigable water, since the area characterized by saturated soil conditions and wetland vegetation extended beyond the boundary of respondent's property to Black Creek, a navigable waterway. Again, the court's finding is not clearly erroneous. Together, these findings establish that respondent's property is a wetland adjacent to a navigable waterway. Hence, it is part of the "waters of the United States" as defined by 33 CFR § 323.2 (1985), and if the regulation itself is valid as a construction of the term "waters of the United States" as used in the Clean Water Act, a question which we now address, the property falls within the scope of the Corps' jurisdiction over "navigable waters" under § 404 of the Act.

IV

A

An agency's construction of a statute it is charged with enforcing is entitled to deference if it is reasonable and not in conflict with the expressed intent of Congress. Chemical Manufacturers Assn. v. Natural Resources Defense Council, Inc., 470 U.S. 116, 125 (1985); Chevron U.S.A. Inc. v. Natural Resources Defense Council, Inc., 467 U.S. 837 (1984). Accordingly, our review is limited to the question whether it is reasonable, in light of the language, policies, and legislative history of the Act for the Corps to exercise jurisdiction over wetlands adjacent to but not regularly flooded by rivers, streams, and other hydrographic features more conventionally identifiable as "waters."[8]

On a purely linguistic level, it may appear unreasonable to classify "lands," wet or otherwise, as "waters." Such a simplistic response, however, does justice neither to the problem faced by the Corps in defining the scope of its authority under § 404(a) nor to the realities of the problem of water pollution that the Clean Water Act was intended to combat. In determining the limits of its power to regulate discharges under the Act, the Corps must necessarily choose some point at which water ends and land begins. Our common experience tells us that this is often no easy task: the transition from water to solid ground is not necessarily or even typically an abrupt one. Rather, between open waters and dry land may lie shallows, marshes, mudflats, swamps, bogs—in short, a huge array of areas that are not wholly aquatic but nevertheless fall far short of being dry land. Where on this continuum to find the limit of "waters" is far from obvious.

Faced with such a problem of defining the bounds of its regulatory authority, an agency may appropriately look to the legislative history and underlying policies of its statutory grants of authority. Neither of these sources provides unambiguous guidance for the Corps in this case, but together they do support the reasonableness of the Corps' approach of defining adjacent wetlands as "waters" within the meaning of § 404(a). Section 404 originated as part of the Federal Water Pollution Control Act Amendments of 1972, which constituted a comprehensive legislative attempt "to restore and maintain the chemical, physical, and biological integrity of the Nation's waters." CWA § 101, 33 U.S.C. § 1251. This objective incorporated a broad, systemic view of the goal of maintaining and improving water quality: as the House Report on the legislation put it, "the word 'integrity' . . . refers to a condition in which the natural structure and function of ecosystems is [are] maintained." H.R.Rep. No. 92–911, p. 76 (1972). Protection of aquatic ecosystems, Congress recognized, demanded broad federal authority to control pollution, for "[w]ater moves in hydrologic cycles and it is essential that discharge of pollutants be controlled at

---

8. We are not called upon to address the question of the authority of the Corps to regulate discharges of fill material into wetlands that are not adjacent to bodies of open water, see 33 CFR §§ 323.2(a)(2) and (3) (1985), and we do not express any opinion on that question.

the source." S.Rep. No. 92–414, p. 77 (1972), U.S. Code Cong. & Admin. News 1972, pp. 3668, 3742.

In keeping with these views, Congress chose to define the waters covered by the Act broadly. Although the Act prohibits discharges into "navigable waters," see CWA §§ 301(a), 404(a), 502(12), 33 U.S.C. §§ 1311(a), 1344(a), 1362(12), the Act's definition of "navigable waters" as "the waters of the United States" makes it clear that the term "navigable" as used in the Act is of limited import. In adopting this definition of "navigable waters," Congress evidently intended to repudiate limits that had been placed on federal regulation by earlier water pollution control statutes and to exercise its powers under the Commerce Clause to regulate at least some waters that would not be deemed "navigable" under the classical understanding of that term. See S.Conf.Rep. No. 92–1236, p. 144 (1972); 118 Cong.Rec. 33756–33757 (1972) (statement of Rep. Dingell).

Of course, it is one thing to recognize that Congress intended to allow regulation of waters that might not satisfy traditional tests of navigability; it is another to assert that Congress intended to abandon traditional notions of "waters" and include in that term "wetlands" as well. Nonetheless, the evident breadth of congressional concern for protection of water quality and aquatic ecosystems suggests that it is reasonable for the Corps to interpret the term "waters" to encompass wetlands adjacent to waters as more conventionally defined. Following the lead of the Environmental Protection Agency, see 38 Fed.Reg. 10834 (1973), the Corps has determined that wetlands adjacent to navigable waters do as a general matter play a key role in protecting and enhancing water quality:

> The regulation of activities that cause water pollution cannot rely on ... artificial lines ... but must focus on all waters that together form the entire aquatic system. Water moves in hydrologic cycles, and the pollution of this part of the aquatic system, regardless of whether it is above or below an ordinary high water mark, or mean high tide line, will affect the water quality of the other waters within that aquatic system.

> For this reason, the landward limit of Federal jurisdiction under Section 404 must include any adjacent wetlands that form the border of or are in reasonable proximity to other waters of the United States, as these wetlands are part of this aquatic system. 42 Fed.Reg. 37128 (1977).

We cannot say that the Corps' conclusion that adjacent wetlands are inseparably bound up with the "waters" of the United States—based as it is on the Corps' and EPA's technical expertise—is unreasonable. In view of the breadth of federal regulatory authority contemplated by the Act itself and the inherent difficulties of defining precise bounds to regulable waters, the Corps' ecological judgment about the relationship between waters and their adjacent wetlands provides an adequate basis for a legal judgment that adjacent wetlands may be defined as waters under the Act.

This holds true even for wetlands that are not the result of flooding or permeation by water having its source in adjacent bodies of open water. The Corps has concluded that wetlands may affect the water quality of adjacent lakes, rivers, and streams even when the waters of those bodies do not actually inundate the wetlands. For example, wetlands that are not flooded by adjacent waters may still tend to drain into those waters. In such circumstances, the Corps has concluded that wetlands may serve to filter and purify water draining into adjacent bodies of water, see 33 CFR § 320.4(b)(2)(vii) (1985), and to slow the flow of surface runoff into lakes, rivers, and streams and thus prevent flooding and erosion, see §§ 320.4(b)(2)(iv) and (v). In addition, adjacent wetlands may "serve significant natural biological functions, including food chain production, general habitat, and nesting, spawning, rearing and resting sites for aquatic ... species." § 320.4(b)(2)(i). In short, the Corps has concluded that wetlands adjacent to lakes, rivers, streams, and other bodies of water may function as integral parts of the aquatic environment even when the moisture creating the wetlands does not find its source in the adjacent bodies of water. Again, we cannot say that the Corps' judgment on these matters is unreasonable, and we therefore conclude that a definition of "waters of the United States" encompassing all wetlands adjacent to other bodies of water over which the Corps has jurisdiction is a permissible interpretation of the Act. Because respondent's property is part of a wetland that actually abuts on a navigable waterway, respondent was required to have a permit in this case.

* * *

We are thus persuaded that the language, policies, and history of the Clean Water Act compel a finding that the Corps has acted reasonably in interpreting the Act to require permits for the discharge of fill material into wetlands adjacent to the "waters of the United States." The regulation in which the Corps has embodied this interpretation by its terms includes the wetlands on respondent's property within the class of waters that may not be filled without a permit; and, as we have seen, there is no reason to interpret the regulation more narrowly than its terms would indicate. Accordingly, the judgment of the Court of Appeals is Reversed.

Solid Waste Agency of Northern Cook County v. United States Army Corps of Engineers

Supreme Court of the United States, 2001.
531 U.S. 159.

■ CHIEF JUSTICE REHNQUIST delivered the opinion of the Court.

Section 404(a) of the Clean Water Act (CWA or Act), 86 Stat. 884, as amended, 33 U.S.C. § 1344(a), regulates the discharge of dredged or fill material into "navigable waters." The United States Army Corps of Engineers (Corps), has interpreted § 404(a) to confer federal authority over an abandoned sand and gravel pit in northern Illinois which provides habitat for migratory birds. We are asked to decide whether the provisions of § 404(a) may be fairly extended to these waters, and, if so, whether Congress could exercise such authority consistent with the Commerce Clause, U.S. Const., Art. I, § 8, cl. 3. We answer the first question in the negative and therefore do not reach the second.

Petitioner, the Solid Waste Agency of Northern Cook County (SWANCC), is a consortium of 23 suburban Chicago cities and villages that united in an effort to locate and develop a disposal site for baled nonhazardous solid waste. The Chicago Gravel Company informed the municipalities of the availability of a 533–acre parcel, bestriding the Illinois counties Cook and Kane, which had been the site of a sand and gravel pit mining operation for three decades up until about 1960. Long since abandoned, the old mining site eventually gave way to a successional stage forest, with its remnant excavation trenches evolving into a scattering of permanent and seasonal ponds of varying size (from under one-tenth of an acre to several acres) and depth (from several inches to several feet).

The municipalities decided to purchase the site for disposal of their baled nonhazardous solid waste. By law, SWANCC was required to file for various permits from Cook County and the State of Illinois before it could begin operation of its balefill project. In addition, because the operation called for the filling of some of the permanent and seasonal ponds, SWANCC contacted federal respondents (hereinafter respondents), including the Corps, to determine if a federal landfill permit was required under § 404(a) of the CWA, 33 U.S.C. § 1344(a).

Section 404(a) grants the Corps authority to issue permits "for the discharge of dredged or fill material into the navigable waters at specified disposal sites." Ibid. The term "navigable waters" is defined under the Act as "the waters of the United States, including the territorial seas." § 1362(7). The Corps has issued regulations defining the term "waters of the United States" to include "waters such as intrastate lakes, rivers, streams (including intermittent streams), mudflats, sandflats, wetlands, sloughs, prairie potholes, wet meadows, playa lakes, or natural ponds, the use, degradation or destruction of which could affect interstate or foreign commerce.... " 33 CFR § 328.3(a)(3) (1999).

In 1986, in an attempt to "clarify" the reach of its jurisdiction, the Corps stated that § 404(a) extends to intrastate waters:

a. Which are or would be used as habitat by birds protected by Migratory Bird Treaties; or

b. Which are or would be used as habitat by other migratory birds which cross state lines; or

c. Which are or would be used as habitat for endangered species; or

d. Used to irrigate crops sold in interstate commerce. 51 Fed.Reg. 41217.

This last promulgation has been dubbed the "Migratory Bird Rule."

The Corps initially concluded that it had no jurisdiction over the site because it contained no "wetlands," or areas which support "vegetation typically adapted for life in saturated soil conditions," 33 CFR § 328.3(b) (1999). However, after the Illinois Nature Preserves Commission informed the Corps that a number of migratory bird species had been observed at the site, the Corps reconsidered and ultimately asserted jurisdiction over the balefill site pursuant to subpart (b) of the "Migratory Bird Rule." The Corps found that approximately 121 bird species had been observed at the site, including several known to depend upon aquatic environments for a significant portion of their life requirements. Thus, on November 16, 1987, the Corps formally "determined that the seasonally ponded, abandoned gravel mining depressions located on the project site, while not wetlands, did qualify as 'waters of the United States' . . . based upon the following criteria: (1) the proposed site had been abandoned as a gravel mining operation; (2) the water areas and spoil piles had developed a natural character; and (3) the water areas are used as habitat by migratory bird [sic] which cross state lines."

During the application process, SWANCC made several proposals to mitigate the likely displacement of the migratory birds and to preserve a great blue heron rookery located on the site. Its balefill project ultimately received the necessary local and state approval. By 1993, SWANCC had received a special use planned development permit from the Cook County Board of Appeals, a landfill development permit from the Illinois Environmental Protection Agency, and approval from the Illinois Department of Conservation.

Despite SWANCC's securing the required water quality certification from the Illinois Environmental Protection Agency, the Corps refused to issue a § 404(a) permit. The Corps found that SWANCC had not established that its proposal was the "least environmentally damaging, most practicable alternative" for disposal of nonhazardous solid waste; that SWANCC's failure to set aside sufficient funds to remediate leaks posed an "unacceptable risk to the public's drinking water supply"; and that the impact of the project upon area-sensitive species was "unmitigatable since a landfill surface cannot be redeveloped into a forested habitat."

Petitioner filed suit under the Administrative Procedure Act, 5 U.S.C. § 701 et seq., in the Northern District of Illinois challenging both the Corps' jurisdiction over the site and the merits of its denial of the § 404(a) permit. The District Court granted summary judgment to respondents on the jurisdictional issue, and petitioner abandoned its challenge to the Corps' permit decision. On appeal to the Court of Appeals for the Seventh Circuit, petitioner renewed its attack on respondents' use of the "Migratory Bird Rule" to assert jurisdiction over the site. Petitioner argued that respondents had exceeded their statutory authority in interpreting the CWA to cover nonnavigable, isolated, intrastate waters based upon the presence of migratory birds and, in the alternative, that Congress lacked the power under the Commerce Clause to grant such regulatory jurisdiction.

The Court of Appeals began its analysis with the constitutional question, holding that Congress has the authority to regulate such waters based upon "the cumulative impact doctrine, under which a single activity that itself has no discernible effect on interstate commerce may still be regulated if the aggregate effect of that class of activity has a substantial impact on interstate commerce." 191 F.3d 845, 850 (C.A.7 1999). The aggregate effect of the "destruction of the natural habitat of migratory birds" on interstate commerce, the court held, was substantial because each year millions of Americans cross state lines and spend over a billion dollars to hunt and observe migratory birds. The Court of Appeals then turned to the regulatory question. The court held that the CWA reaches as many waters as the Commerce Clause allows and, given its earlier Commerce Clause ruling, it therefore followed that respondents' "Migratory Bird Rule" was a reasonable interpretation of the Act.

We granted certiorari, 529 U.S. 1129, 120 S.Ct. 2003, 146 L.Ed.2d 954 (2000), and now reverse.

Congress passed the CWA for the stated purpose of "restor[ing] and maintain[ing] the chemical, physical, and biological integrity of the Nation's waters." 33 U.S.C. § 1251(a). In so doing, Congress chose to "recognize, preserve, and protect the primary responsibilities and rights of States to prevent, reduce, and eliminate pollution, to plan the development and use (including restoration, preservation, and enhancement) of land and water resources, and to consult with the Administrator in the exercise of his authority under this chapter." § 1251(b). Relevant here, § 404(a) authorizes respondents to regulate the discharge of fill material into "navigable waters," 33 U.S.C. § 1344(a), which the statute defines as "the waters of the United States, including the territorial seas," § 1362(7). Respondents have interpreted these words to cover the abandoned gravel pit at issue here because it is used as habitat for migratory birds. We conclude that the "Migratory Bird Rule" is not fairly supported by the CWA.

This is not the first time we have been called upon to evaluate the meaning of § 404(a). In United States v. Riverside Bayview Homes, Inc., 474 U.S. 121 (1985), we held that the Corps had § 404(a) jurisdiction over wetlands that actually abutted on a navigable waterway. In so doing, we noted that the term "navigable" is of "limited import" and that Congress evidenced its intent to "regulate at least some waters that would not be deemed 'navigable' under the classical understanding of that term." Id., at 133, 106 S.Ct. 455. But our holding was based in large measure upon Congress' unequivocal acquiescence to, and approval of, the Corps' regulations interpreting the CWA to cover wetlands adjacent to navigable waters. See id., at 135–139, 106 S.Ct. 455. We found that Congress' concern for the protection of water quality and aquatic ecosystems indicated its intent to regulate wetlands "inseparably bound up with the 'waters' of the United States." Id., at 134, 106 S.Ct. 455.

It was the significant nexus between the wetlands and "navigable waters" that informed our reading of the CWA in *Riverside Bayview*

Homes. Indeed, we did not "express any opinion" on the "question of the authority of the Corps to regulate discharges of fill material into wetlands that are not adjacent to bodies of open water...." Id., at 131–132, n. 8, 106 S.Ct. 455. In order to rule for respondents here, we would have to hold that the jurisdiction of the Corps extends to ponds that are not adjacent to open water. But we conclude that the text of the statute will not allow this.

* * *

We ... decline respondents' invitation to take what they see as the next ineluctable step after *Riverside Bayview Homes*: holding that isolated ponds, some only seasonal, wholly located within two Illinois counties, fall under § 404(a)'s definition of "navigable waters" because they serve as habitat for migratory birds. As counsel for respondents conceded at oral argument, such a ruling would assume that "the use of the word navigable in the statute ... does not have any independent significance." We cannot agree that Congress' separate definitional use of the phrase "waters of the United States" constitutes a basis for reading the term "navigable waters" out of the statute. We said in *Riverside Bayview Homes* that the word "navigable" in the statute was of "limited effect" and went on to hold that § 404(a) extended to nonnavigable wetlands adjacent to open waters. But it is one thing to give a word limited effect and quite another to give it no effect whatever. The term "navigable" has at least the import of showing us what Congress had in mind as its authority for enacting the CWA: its traditional jurisdiction over waters that were or had been navigable in fact or which could reasonably be so made. See, e.g., United States v. Appalachian Elec. Power Co., 311 U.S. 377 (1940).

Respondents ... contend that, at the very least, it must be said that Congress did not address the precise question of § 404(a)'s scope with regard to nonnavigable, isolated, intrastate waters, and that, therefore, we should give deference to the "Migratory Bird Rule." See, e.g., Chevron U.S.A. Inc. v. Natural Resources Defense Council, Inc., 467 U.S. 837 (1984). We find § 404(a) to be clear, but even were we to agree with respondents, we would not extend *Chevron* deference here.

Where an administrative interpretation of a statute invokes the outer limits of Congress' power, we expect a clear indication that Congress intended that result. See Edward J. DeBartolo Corp. v. Florida Gulf Coast Building & Constr. Trades Council, 485 U.S. 568, 575 (1988). This requirement stems from our prudential desire not to needlessly reach constitutional issues and our assumption that Congress does not casually authorize administrative agencies to interpret a statute to push the limit of congressional authority. See ibid. This concern is heightened where the administrative interpretation alters the federal-state framework by permitting federal encroachment upon a traditional state power. See United States v. Bass, 404 U.S. 336, 349 (1971) ("[U]nless Congress conveys its purpose clearly, it will not be deemed to have significantly changed the federal-state balance"). Thus, "where an otherwise acceptable construction of a statute would raise serious constitutional problems, the Court will construe the statute to avoid such problems unless such construction is plainly contrary to the intent of Congress." DeBartolo, supra, at 575.

Twice in the past six years we have reaffirmed the proposition that the grant of authority to Congress under the Commerce Clause, though broad, is not unlimited. See United States v. Morrison, 529 U.S. 598 (2000); United States v. Lopez, 514 U.S. 549 (1995). Respondents argue that the "Migratory Bird Rule" falls within Congress' power to regulate intrastate activities that "substantially affect" interstate commerce. They note that the protection of migratory birds is a "national interest of very nearly the first magnitude," Missouri v. Holland, 252 U.S. 416, 435 (1920), and that, as the Court of Appeals found, millions of people spend over a billion dollars annually on recreational pursuits relating to migratory birds. These arguments raise significant constitutional questions. For example, we would have to evaluate the precise object or activity that, in the aggregate, substantially affects interstate commerce. This is not clear, for although the Corps has claimed jurisdiction over petitioner's land because it contains water areas used as habitat by migratory birds, respondents now, post litem motam, focus upon the fact that the regulated activity is petitioner's municipal landfill, which is "plainly of a commercial nature." But this is a far cry, indeed, from the "navigable waters" and "waters of the United States" to which the statute by its terms extends.

These are significant constitutional questions raised by respondents' application of their regulations, and yet we find nothing approaching a clear statement from Congress that it intended § 404(a) to reach an abandoned sand and gravel pit such as we have here. Permitting respondents to claim federal jurisdiction over ponds and mudflats falling within the "Migratory Bird Rule" would result in a significant impingement of the States' traditional and primary power over land and water use. See, e.g., Hess v. Port Authority Trans–Hudson Corporation, 513 U.S. 30, 44 (1994) ("[R]egulation of land use [is] a function traditionally performed by local governments"). Rather than expressing a desire to readjust the federal-state balance in this manner, Congress chose to "recognize, preserve, and protect the primary responsibilities and rights of States ... to plan the development and use ... of land and water resources...." 33 U.S.C. § 1251(b). We thus read the statute as written to avoid the significant constitutional and federalism questions raised by respondents' interpretation, and therefore reject the request for administrative deference.

We hold that 33 CFR § 328.3(a)(3) (1999), as clarified and applied to petitioner's balefill site pursuant to the "Migratory Bird Rule," 51 Fed.Reg. 41217 (1986), exceeds the authority granted to respondents under § 404(a) of the CWA. The judgment of the Court of Appeals for the Seventh Circuit is therefore Reversed.

■ JUSTICE STEVENS, with whom JUSTICE SOUTER, JUSTICE GINSBURG, and JUSTICE BREYER join, dissenting.

In 1969, the Cuyahoga River in Cleveland, Ohio, coated with a slick of industrial waste, caught fire. Congress responded to that dramatic event, and to others like it, by enacting the Federal Water Pollution Control Act

(FWPCA) Amendments of 1972, 86 Stat. 817, as amended 33 U.S.C. § 1251 et seq., commonly known as the Clean Water Act (Clean Water Act, CWA, or Act).

The Act proclaimed the ambitious goal of ending water pollution by 1985. § 1251(a). The Court's past interpretations of the CWA have been fully consistent with that goal. Although Congress' vision of zero pollution remains unfulfilled, its pursuit has unquestionably retarded the destruction of the aquatic environment. Our Nation's waters no longer burn. Today, however, the Court takes an unfortunate step that needlessly weakens our principal safeguard against toxic water.

It is fair to characterize the Clean Water Act as "watershed" legislation. The statute endorsed fundamental changes in both the purpose and the scope of federal regulation of the Nation's waters. In § 13 of the Rivers and Harbors Appropriation Act of 1899(RHA), 30 Stat. 1152, as amended, 33 U.S.C. § 407, Congress had assigned to the Army Corps of Engineers (Corps) the mission of regulating discharges into certain waters in order to protect their use as highways for the transportation of interstate and foreign commerce; the scope of the Corps' jurisdiction under the RHA accordingly extended only to waters that were "navigable." In the CWA, however, Congress broadened the Corps' mission to include the purpose of protecting the quality of our Nation's waters for esthetic, health, recreational, and environmental uses. The scope of its jurisdiction was therefore redefined to encompass all of "the waters of the United States, including the territorial seas." § 1362(7). That definition requires neither actual nor potential navigability.

The Court has previously held that the Corps' broadened jurisdiction under the CWA properly included an 80–acre parcel of low-lying marshy land that was not itself navigable, directly adjacent to navigable water, or even hydrologically connected to navigable water, but which was part of a larger area, characterized by poor drainage, that ultimately abutted a navigable creek. United States v. Riverside Bayview Homes, Inc., 474 U.S. 121 (1985). Our broad finding in *Riverside Bayview* that the 1977 Congress had acquiesced in the Corps' understanding of its jurisdiction applies equally to the 410–acre parcel at issue here. Moreover, once Congress crossed the legal watershed that separates navigable streams of commerce from marshes and inland lakes, there is no principled reason for limiting the statute's protection to those waters or wetlands that happen to lie near a navigable stream.

In its decision today, the Court draws a new jurisdictional line, one that invalidates the 1986 migratory bird regulation as well as the Corps' assertion of jurisdiction over all waters except for actually navigable waters, their tributaries, and wetlands adjacent to each. Its holding rests on two equally untenable premises: (1) that when Congress passed the 1972 CWA, it did not intend "to exert anything more than its commerce power over navigation," and (2) that in 1972 Congress drew the boundary defining the Corps' jurisdiction at the odd line on which the Court today settles.

Notes and Questions

1. What a difference 15 years can make! Acting far ahead if its time, the *Riverside Bayview* Court gave great deference to the Corps' "ecological judgment" about the "ecological integrity" of the wetland resource, but 15 years later, when deep in the midst of the ecosystem movement, the Court gave little deference at all to the Corps in *SWANCC*. The more pertinent *legal* question, of course, is whether the difference between "adjacent" and "isolated" wetlands is a meaningful difference under the Clean Water Act. The *Riverside Bayview* Court found it was reasonable, in the adjacent wetland context, that the Corps "concluded that wetlands adjacent to lakes, rivers, streams, and other bodies of water may function as integral parts of the aquatic environment even when the moisture creating the wetlands does not find its source in the adjacent bodies of water." In the isolated wetland context involved in *SWANCC*, by contrast, the majority opinion devotes no attention whatsoever to the ecological role of the wetland resource in deciding that the CWA does not reach that far. Is that because isolated wetlands do not "function as integral parts of the aquatic environment." Did Congress think that? Did the Corps? Did the Court? Or, even if isolated wetlands are integral parts of aquatic ecosystems, is the problem that they are not part of the legally-relevant aquatic ecosystem–the kind Congress had in mind in the CWA?

2. Indeed, perhaps the real question ought to be not whether *SWANCC* unjustifiably cuts off federal jurisdiction at the line between adjacent and isolated wetlands, but whether *Riverside Bayview* unjustifiably extends the line beyond surface waters to adjacent wetlands. Was *Riverside Bayview* correctly decided in this regard? Did Congress *really* mean to cover adjacent wetlands?

3. As important as isolated wetlands may be ecologically, isn't the *SWANCC* majority correct that the ecological processes that make isolated wetlands valuable are not the same as those that make adjacent wetlands valuable? The *Riverside Bayview* Court did, after all, focus specifically on hydrological process connections–specifically, the connections between adjacent wetlands and waters that are unquestionably within the scope of federal authority. Those connections do not exist in the case of isolated wetlands. Clearly, these distinctions are real *ecologically*. The majority and dissent in *SWANCC* simply differed over how relevant Congress thought that ecological distinction was *politically*.

4. Most Court-watchers agree the important message of *SWANCC* is to be found in the Court's constitutionally-based concern that the regulation of isolated wetlands under the Migratory Bird Rule would "claim federal jurisdiction over ponds and mudflats ... in a significant impingement of the States' traditional and primary power over land and water use." The Migratory Bird Rule has long been a target of such Commerce Clause challenges, with mixed results in the lower courts. *See* Hoffman Homes v. Administrator, 999 F.2d 256 (7th Cir.1993) (upholding the rule but finding insufficient evidence of the presence of migratory birds); Leslie Salt Co. v. United States, 55 F.3d 1388 (9th Cir.1995) (upholding the rule); United

States v. Wilson, 133 F.3d 251 (4th Cir.1997) (rejecting federal authority to regulate isolated wetlands that merely "could affect" interstate commerce). The setting thus was ripe for Supreme Court treatment of the issue after the Court's somewhat revolutionary Commerce Clause decision in United States v. Lopez, 514 U.S. 549 (1995), the first decision in over 60 years in which the court struck down federal legislation as exceeding Congress's authority to regulate interstate commerce. *See* Jonathan H. Adler, *Wetlands, Waterfowl, and the Menace of Mr. Wilson: Commerce Clause Jurisprudence and the Limits of Federal Wetland Regulation*, 29 ENVTL. LAW 1 (1999). The Court's *dicta* in *SWANCC* appears to have endorsed the view that there is, indeed, a limit to how far Congress can regulate to protect wetlands, and that isolated wetlands are too far afield. *See* Jonathan Adler, *The Ducks Stop Here? The Environmental Challenge to Federalism*, 9 SUPREME COURT ECONOMIC REV. 205 (2001). But how far do you think the Court believes this concern reaches?

Recall, for example, the endangered Delhi Sands Flower-loving Fly that introduces this text in Chapter 1. Its habitat certainly is as isolated as the waters in *SWANCC*, and the Endangered Species Act restrictions its listing imposed on surrounding government and commercial development surely appears to be a form of land use control. The split D.C. Circuit opinion rejecting the Commerce Clause challenges to those regulations, National Association of Home Builders v. Babbitt, 130 F.3d 1041 (D.C.Cir.1997) (reproduced in Chapter 6), preceded the Supreme Court's decision in *SWANCC*. Did the court of appeals get it wrong? If so, what does that mean for the Endangered Species Act, and for the implementation of ecosystem management law and policy in general? The converse question is also a puzzle: If private activities affecting isolated wetlands and the Delhi-sands Flower Loving Fly are within Congress' power to regulate, what isn't? Would there be any stopping federal ecosystem management law and policy from permeating every nook and cranny of land use decision making? Would *anything* be beyond the purview of federal ecosystem management?

5. The lower federal courts have been quick to apply *SWANCC* in a variety of circumstances with the effect of cutting off federal jurisdiction over isolated waters. *See, e.g.*, Rice v. Harken Exploration Co., 250 F.3d 264 (5th Cir.2001) (construing *SWANCC* to mean that the CWA applies only to a "body of water that is actually navigable or is adjacent to an open body of navigable water"); U.S. v. Newdunn, 195 F.Supp.2d 751 (E.D.Va.2002) (Corps had no jurisdiction over 38 acres of wetlands in the absence of proof of connection to navigable waters); U.S. v. Rapanos, 190 F.Supp.2d 1011 (E.D.Mich.2002) (dismissing criminal prosecution for unauthorized filling of wetlands because Corps did not prove area was navigable or an adjacent water). Meanwhile, EPA and the Corps continue to wrestle with how to interpret *SWANCC*. As of this writing, the agencies have yet to issue comprehensive guidance to field personnel on how to apply the CWA consistent with the Court's decision.

6. An implicit message of *SWANCC*, of course, is that states are free to regulate isolated wetlands, to fill in whatever gap in coverage the opinion opened. Many observers quickly reacted to *SWANCC*, however, as if political and economic constraints would prevent states from taking such action, and thus derided the decision as exposing isolated wetlands to rampant destruction. *See, e.g,* Jon Kusler, *The SWANCC Decision and the States–Fill in the Gaps or Declare Open Season?*, 23 National Wetlands Newsletter 9 (Mar.–Apr. 2001). But in fact, many states have responded to *SWANCC* by enacting or recommending the enactment of relatively aggressive regulatory programs to protect isolated wetlands now beyond the reach of the federal government. *See* Michael Gerhardt, *The Curious Flight of the Migratory Bird Rule*, 31 Envtl. L. Rep. (ELI) 10079 (2001); *New and Revised State Wetland Regulations Take Effect Across the County*, 23 National Wetlands Newsletter 18 (Sept.–Oct. 2001). Have the Court's critics treated it unfairly? Did they underestimate the political will of states to act in realms where the federal government may not? You can find descriptions of the state wetlands regulation programs that "fill the gap" through the Association of State Wetland Managers, www.aswm.org.

7.   Assuming states on balance do fill the gap *SWANCC* opened for isolated wetlands, are you comfortable with the new ecosystem management framework for wetlands overall, one where the federal government regulates wetlands from lakes and rivers to the outer edge of their adjacent wetlands, with or without concurrent state regulation, then states take over *exclusively* for isolated wetlands?

8.   Unlike federal public land units such as national forests (discussed in Chapter 8), where boundaries are sharply drawn lines on a map, ecosystem management for wetlands does not lend itself well to definition of the relevant ecosystem by discrete management unit. Congress chose to take a geographic unit approach in Section 404, but that approach inevitably leads to the kind of jurisdictional line drawing the Court grappled with in *Riverside Bayview* and *SWANCC*. Can you think of another way of defining "wetlands" for ecosystem management purposes that avoids this problem?

9.   *Riverside Bayview* and *SWANCC* involve the question of the *geographic* boundary of federal regulatory jurisdiction over wetlands. Another boundary is erected by the definition of *activities* subject to that jurisdiction. Even where an area indisputably is a surface water or an adjacent wetland subject to federal control, the activity taking place within the wetlands must fit the parameters of Section 404 for the regulatory program to attach. This activity-based limit on the Section 404 program has also proved significant and controversial. For example, in 1997 the D.C. Circuit held that mechanized drainage, excavation, and channelization of wetlands is not subject to Section 404 merely because of the "incidental fallback" of dirt and debris from the machine parts. *See* National Mining Association v. Army Corps of Engineers, 145 F.3d 1399 (D.C.Cir.1998). The Corps had previously taken the position, in the so-called "Tulloch Rule" (named after one of the cases leading to its development), that incidental fall back does trigger Section 404 and that, since it is virtually impossible to conduct mechanized activities in wetlands without some incidental fallback, such activities necessarily require permits. 58 Fed. Reg. 45,008 (Aug. 25, 1993).

After the D.C. Circuit opinion in *National Mining Association*, the Corps revised its rules to impose only a rebuttable presumption that such activities require permits. The presumption is removed if "project-specific evidence shows that the activity results in only incidental fallback." 66 Fed. Reg. 4550, 4575 (Jan. 17, 2001). New litigation has been initiated over that interpretation, and further confusion over the issue was added by the Ninth Circuit's subsequent decision that Section 404 covers "deep ripping," which is a form of plowing using extremely long plow prongs to open wetlands up to drainage. *See* Borden Ranch Partnership v. U.S. Army Corps of Engineers, 261 F.3d 810 (9th Cir.2001), cert. granted 122 U.S. 2355 (2002).

10. In *Riverside Bayview* the Court made short-shrift of the "spurious" argument that the requirement that a developer seek a Corps permit under CWA Section 404 before filling jurisdictional wetlands constitutes a taking of property. The landowner in that situation seeks a judicial remedy before exhausting available administrative remedies–the permit procedure–and thus, under settled doctrine, has no ripe takings case to present to a court. *See* Williamson County Regional Planning Commission v. Hamilton Bank, 473 U.S. 172 (1985). But what about the case where a permit has been sought and denied? Is that a so-called regulatory taking of property? There have been more such takings challenges lodged against Section 404 wetlands permit decisions than have been brought against any other federal environmental program. Early in the program's history those challenges failed routinely. *See* Deltona Corp. v. United States, 657 F.2d 1184 (Ct.Cl. 1981); Jentgen v. United States, 657 F.2d 1210 (Ct.Cl.1981). In the late 1980s, however, courts began issuing more landowner-friendly rulings, albeit based on rather extreme circumstances. *See* Florida Rock Industries v. United States, 791 F.2d 893 (Fed.Cir.1986); Loveladies Harbor, Inc. v. United States, 15 Cl.Ct. 381 (1988). Takings litigation under Section 404 then gained steam after the Supreme Court's decision in Lucas v. South Carolina Coastal Council, 505 U.S. 1003 (1992), ruling that newly-enacted land use regulations that both deprive the landowner of all economic value of the land and go beyond codifying pre-existing common law nuisance restrictions constitute takings per se–i.e., without regard to a balancing of public and private interests. Since then, takings challenges under Section 404 have mounted, though still with mixed results. *See* Good v. United States, 189 F.3d 1355 (Fed.Cir.1999) (no taking); Palm Beach Isles Associates v. United States, 231 F.3d 1354 (Fed.Cir.2000) (taking found). For a thorough discussion of the background and status the regulatory takings issue in the wetlands regulation context, see Robert Meltz, *Wetlands Regulation and the Law of Regulatory Takings*, 30 Envtl. L. Rep. (ELI) 10468 (2000); Robert Meltz, *Wetland Regulation and the Law of Property Rights "Takings,"* 23 NATIONAL WETLANDS NEWSLETTER (ELI) 1 (May–June 2001).

11. There are a number of federal programs relating to wetlands protection that will withstand Commerce Clause and Takings Clause scrutiny, even after *SWANCC*, because they are non-regulatory in scope and based on a source of federal authority other than the power to regulate interstate commerce, such as the spending power. For example, the so-called Swamp-buster Program withdraws agricultural subsidies from farmers that plant commodity crops (corn, wheat, soybeans, etc.) on lands converted from wetlands lands after 1985. 16 U.S.C. § 3821. The subsidy restriction does not apply to normal farming operations in wetlands, such as rice produc-tion, or to draining for non-crop uses such as cranberry farming or raising poultry. Because it involves withdrawal of subsidies to induce compliance with its goals, rather than regulation, the dicta in *SWANCC* regarding the limits of Congress's Commerce Clause authority should pose no threat to Swampbuster. *See* United States v. Dierckman, 201 F.3d 915 (7th Cir.2000) (finding Swampbuster is based on Congress's spending power, not its interstate commerce power). Several other nation-wide agriculture pro-grams, including the Wetland Reserve Program and the Conservation Reserve Program, are geared specifically to provide subsidies to qualifying farmers who retire wetlands from agricultural uses, and thus should also withstand scrutiny under the evolving Commerce Clause jurisprudence. While these programs have their own inherent limits, they have proven valuable for wetland conservation policy and are always at the forefront of Congress' Farm Bill deliberations. *See* Roger L. Pederson, *Farms and Wetlands Benefit from Farm Bill Conservation Measures*, 23 National Wetlands Newsletter 9 (Sept.–Oct. 2001). Other wetland programs depend for their authority on the power of the federal government to regulate itself. For example, the Fish and Wildlife Coordination Act requires that federal agencies, when considering whether to construct or finance water resources development projects, give equal consideration to wildlife conser-vation, including wetlands habitat impacts, as they give to other purposes of proposed projects. 16 U.S.C. § 661. For a description of these and over a dozen more federal programs that directly or indirectly establish wetlands conservation policy, see U.S. GAO, Wetlands Overview: Federal and State Policies, Legislation, and Programs, GAO/RCED–92–79FS (Nov. 1991).

b. APPLICATIONS

Riverside Bayview and *SWANCC* decide where Section 404 applies. When actions in areas within the federal side of that line meet the action-based parameters of Section 404–i.e., result in a nonexempt discharge of fill material into wetlands–they trigger the procedures and standards of the Section 404 permitting program. Although the Corps of Engineers is the lead federal agency for these purposes, as explained in more detail below several other federal agencies are involved in the process:

Corps of Engineers: makes site-specific wetland delineations; pro-mulgates nationwide and regional general permits; makes decisions on individual permits; takes enforcement actions; promulgates regulations to implement these programs.

EPA: issues standards under CWA section 404(b)(1) governing the environmental criteria for location of fill locations; can veto Corps permits based on environmental criteria.

U.S. Fish and Wildlife Service: reviews and comments on Corps permit applications pursuant to the Fish and Wildlife Coordination Act, 16 U.S.C. §§ 661–666c; may elevate disputes with the local Corps offices over permit issuance decisions to Corps Headquarters.

National Oceanic and Atmospheric Administration: reviews and comments on Corps permit applications and may also elevate local permit decisions.

The CWA provides several exemptions from permit requirements for activities in areas that otherwise might fit the parameters. For example, routine, ongoing farming operations are exempt from permitting, as are many maintenance activities for dikes, berms, dams, and bridges. *See* 33 U.S.C. § 1344(f)(1); 33 C.F.R. § 323.4. These exemptions do not allow new farming or construction in wetlands, however.

Activities requiring permits might qualify for one of the Corps' Nationwide Permits. These are so-called general permits, or permits by rule, that apply to specified low-impact activities with little or no application and review process. *See* 33 U.S.C. § 1344(c); 33 C.F.R. pt. 330. The Corps must find that activities covered by such permits generally will have minimal adverse impacts, separately and cumulatively. Some of the general permits require pre-discharge notifications to allow the Corps, on a case-by-case basis, to review impacts and deny the benefit of the general permit. The Corps may issue general permits on a regional and local basis as well.

Activities not qualifying for exemption or a general permit must undergo the Corps individual permit review process. *See* 33 C.F.R. pt. 325. An important facet of this program are the EPA's guidelines, issued pursuant to CWA Section 404(b)(1), governing where the Corps may specify sites for disposal of fill material. *See* 33 U.S.C. § 1344(b)(1). The criteria, which are binding on the Corps, include such factors as the effects on wildlife, on ecosystem diversity, productivity and stability, and on shorelines and beaches. *See* 40 C.F.R. pt. 230. EPA has used the guidelines to impose a "alternatives" analysis on Corps decisions, under which disposals in wetlands that are not associated with "water-dependent" activities are presumed to have financially practicable alternatives that are less environmentally damaging. *Id.* § 230.10(a).

For proposed disposals that meet EPA's environmental criteria and alternatives test, the Corps conducts a "public interest" analysis involving a wide variety of factors covering economic, environmental, and social interests. *See* 33 C.F.R. pt. 320. The Corps also submits the application to review by several other federal agencies through interagency consultation required under CWA Section 404(q), though recommendations derived from this process do not bind the Corps. Even the states get involved, as the Corps must obtain the state's certification that issuance of the permit will not cause a violation of state water quality standards developed pursuant to the CWA. *See* 33 U.S.C. § 1341(c). Lastly, although it has been used infrequently, EPA may veto Corps issuance of a permit if EPA concludes it would have "unacceptable adverse effect on . . . shellfish beds and fishery areas . . . wildlife, or recreational areas." 33 U.S.C. § 1344(c).

The premise of any permit program, of course, is that *some* of the regulated activity will be allowed. But under what conditions, and at what cost to the both the regulated entity and the affected ecosystem? As the following article describes, the Corps and EPA have grappled with these questions for many years under the Section 404 program, and have recently gravitated to an approach, known as "mitigation banking," that is designed to allow both sides of the equation to get most of what they need, if not all of what they want. See what you think of both the design and implementation of this effort to achieve a "win-win" ecosystem management policy.

J. B. Ruhl and R. Juge Gregg, Integrating Ecosystem Services Into Environmental Law: A Case Study of Wetlands Mitigation Banking

20 Stanford Environmental Law Journal 365 (2001).

I. Introduction

The federal wetland mitigation banking experience, a habitat trading program that has been in existence for over a decade, presents an opportunity for examining how ecosystem service values could be integrated into existing environmental law frameworks. In wetlands mitigation banking, a "bank" of wetlands habitat is created, restored, or preserved and then made available to developers of wetlands habitat who must "buy" habitat mitigation as a condition of federal government approval for development in wetland areas. Wetlands provide extensive and important services to human populations, including flood control and water quality improvement. If environmental law protects ecosystem services, evidence to that effect should exist in the structure and performance of the wetlands banking program. In particular, a program allowing what essentially amounts to trading of wetlands—exchanging acres destroyed in one location for acres created or improved elsewhere—ought to take into account the service values of the wetlands being traded. This article tests that hypothesis, exploring both the legal authority and the actual practice that exists in wetlands mitigation banking with respect to accounting for ecosystem service values that wetlands provide.

The genesis of wetlands mitigation banking was the revelation of widespread evidence that various forms of mitigation for wetlands losses used in the 1980s were not adequately protecting environmental values. During the 1990s, government and industry moved toward the banking program as a cornerstone of wetlands mitigation. This movement presented an opportunity to introduce greater emphasis on service values in the goals of mitigation. Indeed, there are now several specific regulatory provisions within the wetlands regulation and mitigation program that are particularly suited to incorporating ecosystem service values into the regulatory decision making process. Yet, while the existing legal framework of wetlands banking clearly accommodates integration of ecosystem service values with little or no changes to regulatory text, nothing in the regulations explicitly requires or encourages that approach generally. In short, the authority to integrate ecosystem service concerns into the wetlands mitigation banking program is implicit, but implementation in any broad, deliberate policy form remains only a latent potential.

* * *

II. LAW AND POLICY BACKGROUND

Section 311 of the Clean Water Act (CWA) prohibits "the discharge of any pollutant by any person," which, because of the way those terms are defined, also prohibits filling of wetlands. Nevertheless, section 404(a) of the statute authorizes the Secretary of the Army, through the Army Corps of Engineers (the Corps), to "issue permits for the discharge of dredged or fill material in the navigable waters of the United States at specified disposal sites." Pursuant to section 404(b)(1) of the CWA, the Environmental Protection Agency (EPA) must promulgate substantive permitting standards, known as the "404(b)(1) Guidelines," which the Corps must follow in administering the permit program. Thus, under the CWA, wetlands may be filled only if a permit is granted in accordance with the 404(b)(1) guidelines. These permits, known ubiquitously as "404 permits," "wetland permits," or "Corps permits," have become the cornerstone for federal protection of wetland resources. The permitting program, however, admits of many exceptions and nuances making it less than straightforward to determine whether a permit is required for a particular fill activity, and how to get one. Many routine land development activities do, however, require and receive a 404 permit. And along the way, permit applicants and the agencies often confront the issue of "mitigation" as one of the conditions the developer must satisfy in order to obtain the permit.

The Corps' guidelines for administering wetlands mitigation require it to review 404 permit applications using a preference "sequencing" approach. The first preference is to require the applicant to avoid filling wetland resources; the second preference is to require minimization of adverse impacts that cannot reasonably be avoided; and the least desirable preference is to require the developer to provide compensatory mitigation for those unavoidable adverse impacts that remain after all minimization measures have been exercised. The least desirable option, compensatory mitigation, is the basis for wetlands trading.

Both EPA and the Corps traditionally have preferred on-site to off-site locations for the compensatory mitigation activity, and have preferred in-kind mitigation to mitigation that uses a substantially different type of wetland. Regardless of location, EPA and the Corps prefer measures that restore prior wetland areas as the highest form of mitigation, followed by enhancement of low-quality wetlands, then creation of new wetlands, and, least favored of all, preservation of existing wetlands. To take an extreme example, if compensatory mitigation is deemed appropriate for a project involving fill of mangrove swamp wetlands in Florida, on-site restoration of an area of prior mangrove swamp wetlands would be a favored mitigation strategy, whereas off-site preservation of existing cranberry bog wetlands in Maine would be least favored.

Notwithstanding its official status as the least-favored alternative in the agencies' sequencing pecking order, compensatory mitigation has been used frequently in the 404 program. Compensatory mitigation frees up highly valued wetlands for more comprehensive and flexible development. While attractive for these purposes, the project-by-project compensatory mitigation approach has been widely regarded as having failed miserably in terms of environmental protection. Whether mitigation was accomplished onsite or near-site, the piecemeal approach complicated the Corps' ability to articulate mitigation performance standards, monitor success, and enforce conditions; not surprisingly, the success rate for this approach suffered as a result.

In light of these problems, during the late 1980s the Corps and EPA started shifting compensatory activities increasingly from on-site to off-site mitigation, thus opening the door to the wetlands mitigation banking technique. This approach, its proponents argued, would prove advantageous both in terms of economic efficiency and ecological integrity, aggregating small wetlands threatened by development into larger restored wetlands in a different location. It is defined generally as "a system in which the creation, enhancement, restoration, or preservation of wetlands is recognized by a regulatory agency as generating compensation credits allowing the future development of other wetland sites." In its most basic form, wetlands mitigation banking allows a developer to protect wetlands at one site in advance of development and then draw down the resulting bank of mitigation "credits" as development is implemented and wetlands at another site are filled. Indeed, the concept has progressed beyond this personal bank model. Today, large commercial and public wetlands banks, not tied to a particular development, sell mitigation piecemeal to third-party developers in need of compensatory mitigation.

* * *

III. THE POTENTIAL ROLE FOR ECOSYSTEM SERVICE VALUATION

As the preceding discussion suggests, there are three legal instruments that directly address wetlands mitigation banking. First, under the Clean Water Act section 404(b)(1), Congress delegated to EPA the responsibility for issuing regulations governing the location of wetlands fill-sites to ensure adequate environmental protection. These regulations are known as the "404(b)(1) Guidelines."[1] Second, in 1990, the United States Army Corps of Engineers (Corps) and the EPA signed a memorandum of agreement, known as the "Mitigation Guidance," clarifying the role each plays in wetlands mitigation under the 404(b)(1) Guidelines. And finally, as promised in the Mitigation Guidance, several agencies issued a guidance document in 1995 detailing operation of a wetlands mitigation bank, known as the "Banking Guidance."

This section discusses how these three instruments could be used to integrate consideration of ecosystem service values into wetland mitigation

---

1. [*Note*: These are found at 40 C.F.R. pt. 230.—Eds.]

banking decision making with little or no change to regulatory text. For now, however, the authority to do so remains only implicit, leaving any more comprehensive and deliberate integration of ecosystem services in the wetlands mitigation banking program a latent potential.

A. *404(b)(1) Guidelines*

The 404(b)(1) Guidelines provide extensive descriptions of wetlands values that the Corps should consider in assessing potential mitigation requirements. These guidelines are the reference point for both the Mitigation Guidance and the Banking Guidance, and provide clear regulatory authority to consider ecosystem service values such as those derived from the water purification and recreational opportunities that wetlands provide.

Subparts D through F of the Guidelines focus on the negative effects of disrupting wetlands, identifying the functions and values that may be lost due to the discharge of dredged or fill materials. Subpart F, entitled "Potential Effects on Human Use Characteristics," focuses exclusively on wetlands functions used for the benefit of humans; it thus deals most directly with values and functions that can be considered ecosystem services. Subpart F identifies five general human uses that are potentially impacted by wetlands development: (1) municipal and private water supplies; (2) recreational and commercial fisheries; (3) water-related recreation; (4) aesthetics; and (5) parks. For each category, the Guidelines chronicle specific impacts that such developments could have on wetlands values.

Subparts D through F thus acknowledge that certain human activities influence wetland functioning, and Subpart F explains how humans benefit from wetland functioning. The scope of Subpart F does not cover the full range of service values associated with wetlands; for example, it fails to address the value to humans of such functions as sedimentation control, nutrient cycling, flood control, and energy fixation. The tone and content of the section clearly indicates, though, that EPA's authority under section 404(b)(1) includes protection of wetland service values generally.

B. *The Mitigation Guidance*

Initially, the Corps and EPA "clashed over the proper role of mitigation in the ... permitting process." However, the Mitigation Guidance that the two agencies adopted in 1990 clarified the role of wetlands mitigation under the 404(b)(1) Guidelines. The Mitigation Guidance divides mitigation into three phases—avoidance, minimization, and compensatory mitigation—and required that those phases be conducted sequentially. Thus, the Corps "first makes a determination that potential impacts have been avoided to the maximum extent practicable ... ;" if there are any "remaining unavoidable impacts," the Corps is to mitigate them "to the extent appropriate and practicable by requiring steps to [2] minimize impacts, and ... [3] compensate for aquatic resource values." Mitigation banking is an option only if the third phase, compensatory mitigation, is reached. The Mitigation Guidance explicitly endorses mitigation banking as a form of compensatory mitigation and promised additional guidance on the subject. With respect to compensatory mitigation generally, the Mitigation Guidance requires that it be used "for unavoidable adverse impacts which remain after all appropriate and practicable minimization has been required;" expresses preferences for on-site mitigation and for wetlands restoration (as opposed to wetlands creation); and requires that "functional values" be examined. Thus, the Mitigation Guidance simply requires that functional value be examined and compensation provided—preferably on-site—for unavoidable adverse impacts.

The declaration of purpose is strong but the tactics to achieve it are not well-defined. Under the Mitigation Guidance, the Corps "*will strive* to achieve a goal of no overall net loss of values and functions." The methodologies used to determine whether this goal is being met are, however, only broadly described. The Mitigation Guidance simply advocates that "qualified professionals" tailor generally recognized assessment techniques to each site, constrained only by the requirement that they "consider" the ecological functions listed in the 401(b)(1) Guidelines.

* * *

According to the Mitigation Guidance, "mitigation should provide, at a minimum, one for one functional replacement (i.e., no net loss of values), with an adequate margin of safety." Nevertheless, because determining functional replacement may be difficult, "in the absence of more definitive information on the functions and values of specific wetlands sites, a minimum of one-to-one acreage replacement may be used as a reasonable surrogate for no net loss of functions and values." Thus, the Mitigation Guidance purports to require a margin of safety that ensures a one-to-one functional replacement, but if information is uncertain—as it usually will be—the Mitigation Guidance only requires a more easily quantified and non-functional one-to-one acreage replacement.

The Mitigation Guidance pays homage to the idea of "functions and values" in numerous instances. It commits the agencies to "strive to achieve a goal of no overall net loss of [wetlands] values and function" and purports to base "[t]he determination of what level of mitigation constitutes "appropriate" mitigation . . . solely on the values and functions of the aquatic resource that will be impacted." The Mitigation Guidance's attempts at quantitative valuation repeatedly focus on wetlands values and function, but it never defines these essential terms.

* * *

C. The Banking Guidance

In 1995, five United States agencies published the Banking Guidance, as promised in the Mitigation Guidance, in order to detail the use and operation of mitigation banks. The document's introduction declares that the "objective of a mitigation bank is to provide for *the replacement of the chemical, physical, and biological functions of wetlands and other aquatic resources which are lost* as a result of authorized impacts." This perspective is later broadened to acknowledge that "[t]he overall goal of a mitigation bank is to provide *economically efficient and flexible mitigation opportunities*, while fully compensating for wetland and other aquatic resource losses in a manner that contributes to the long-term ecological functioning of the watershed within which the bank is to be located." The Banking Guidance thus qualifies the goal of replacing ecological functioning by acknowledging economic realities.

The Banking Guidance describes the intricacies of creating a wetlands mitigation bank, but, like the Mitigation Guidance, is vague on what exactly is being "banked." Also like the Mitigation Guidance, the document relies heavily on the term "function." For example, site selection requires agencies to "give careful consideration to the ecological suitability of a site for achieving the goal and objectives of a bank, i.e., that it possess the physical, chemical and biological characteristics to support establishment of the desired aquatic resources and functions." Similarly, credit for wetland preservation is contingent upon the "functions" provided or augmented by the preserved land, and "credit may be given for the inclusion of upland areas occurring within a bank only to the degree that such features increase the overall ecological functioning of the bank."

Because the crediting and debiting procedure forms the heart of a wetlands mitigation bank, the determination of what will be counted as "currency" is crucial. The Banking Guidance focuses initially on the use of "aquatic functions" as its banking currency—a currency that is easily "exchanged" or translated into service values. But the Banking Guidance then follows the lead of the Mitigation Guidance and allows acreage to be a surrogate measure if functional assessment is impractical. The Banking Guidance then takes one more step back from its vision and allows any "appropriate functional assessment methodology . . . acceptable to all signatories" to be used to quantify credits. Once again, therefore, the official guidance provides an opportunity, but not a requirement, to rely on ecosystem indicators as the assessment methodology.

D. *Conclusion: Implicit Authority but No Explicit Requirement*

Each of the three cornerstone policies supporting wetlands mitigation banking supports integrating greater use of ecosystem services in the wetlands mitigation banking decision making process. Certainly nothing in the 404(b)(1) Guidelines, the Mitigation Guidance, or the Banking Guidance precludes the use of ecosystem service factors. The emphasis in each of the sources of authority on wetlands functions opens the door to more focused attention to the service values of those functions. Nevertheless, the functions emphasis falls short of explicit adoption of ecosystem services as a central or even relevant factor in wetlands mitigation banking decisions. At best, therefore, the current legal framework of wetlands mitigation banking establishes the implicit authority, but no explicit requirement, for the consideration of ecosystem services.

* * *

Given the current state of affairs, there is little promise for the integration of ecosystem service valuation methods into wetlands mitigation banking until methods of wetland assessment are significantly improved. In the absence of widely available, readily applied methods for calculating and comparing ecosystem service values of the wetlands being traded, the Corps will likely put constraints on trading markets to compensate for ecosystem function losses not recognized by acre-based methods. These constraints significantly undercut the market and information advantages ecosystem service valuation would impart to wetlands mitigation banking in general, thus further reducing any incentive to apply such methods.

Hence, unless some way is developed to capture the ecosystem service value of wetlands without costly, time-consuming, and complicated valuation methods—e.g., by measurement of readily determinable indicators of ecosystem service value—wetlands mitigation banking is likely to rely most heavily on acre-based and narrow function-based methods and highly regulated "markets" for trades. Nevertheless, if such assessment methods can be developed,...the authority to require their use is implicit in the existing legal framework of the section 404 program. By using these new assessment methods, the wetlands mitigation banking program would surely come closer to meeting its environmental protection objectives.

Notes and Questions

1. Based on the foregoing description, do you consider wetlands mitigation banking an ecosystem management policy? How would you articulate it as such? Try doing so using the three design components of ecosystem management policy articulated in Chapter 7:

	Currency	*Goals*	*Methods*
Wetlands Mitigation Banking			

2. The basic premise of the wetlands mitigation banking program is that wetlands values are fungible and transportable within certain limits, and thus can be traded back and forth between locations in defined geographic areas with no net loss to overall wetlands values. Commentators have questioned the premises of mitigation banking on two levels, not only for wetlands banking but for any habitat mitigation compensation program. First, as discussed in the preceding excerpt, so long as the measure of "value" remains bundled into relatively crude units such as wetland acres, the premise of fungibility seems suspect, thus placing the equivalency of trades beyond meaningful evaluation. Second, as the unit of valuation becomes more refined, thus making trade equivalence more measurable, the transportability of value becomes more questionable. For example, if very accurate measures of wetland ecosystem service values could be used, it would become apparent that wetland mitigation banking moves those values from one community to another community with virtually no public participation in the decision. Why, just because a developer and the mitigation banker agree, should wetlands policy endorse, indeed promote, such ecosystem service transfers? *See* James Salzman and J.B. Ruhl, *Currencies and the Commodification of Environmental Law*, 53 STAN. L. REV. 607 (2000).

3.   Recent scientific studies suggest that the failure of the fungibility premise is a profound defect in the wetlands mitigation banking program as implemented. In June 2001, the National Research Council, an arm of the National Academy of Sciences, concluded that compensatory wetlands mitigation is an abject failure. *See* NATIONAL ACADEMY OF SCIENCES/NATIONAL RESEARCH COUNCIL, COMPENSATING FOR WETLAND LOSSES UNDER THE CLEAN WATER ACT (June 2001). For one thing, the Corps does a poor job, according to the report, accounting for the acre-for-acre trades, meaning even on an acre-counting basis the program is losing "value." But more problematic is that acres of "enhanced" or "recreated" wetlands generally are not as functionally valuable as the acres that are lost to development. Fungibility, in other words, is not being met either on an acre basis or on a functions basis. *See also* R. Eugene Turner et al., *Count It by Acre or Function–Mitigation Adds Up to Net Loss of Wetlands*, NATIONAL WETLANDS NEWSLETTER, Nov.-Dec. 2001, at 5; CHESAPEAKE BAY FOUNDATION, MARYLAND NONTIDAL WETLAND MITIGATION: A PROGRESS REPORT (1997). After publication of the NRC report, the Corps of Engineers issued a defense of its program in the form of a guidance document, *see* U.S. Army Corps of Engineers, Regulatory Guidance Letter No. 01–1 (Oct. 31, 2001), which numerous environmental advocacy organizations immediately condemned as "arrogant" and evidence of an "anything goes approach," and which even some congressional and Bush Administration interests found alarmingly premature given the Corps' lack of coordination with EPA. *See* National Wildlife Federation et al., Press Release, Conservation Community Outraged by Army Corps of Engineers' Reversal on Wetlands Policy (Nov. 6, 2001); *EPA, NMFS Slam Corps RGL on Wetlands Mitigation*, ENDANGERED SPECIES AND WETLANDS REPORT, Mar. 2002, at 13; *Corps Releases Mitigation Guidance*, NATIONAL WETLANDS NEWSLETTER, Nov.–Dec. 2001, at 21; *Army Corps Urged by House Chairman to Withdraw Guidance, Seek EPA Comment*, 32 Env't Rep. (BNA) 2404 (2001).

4.   The second fundamental premise of wetland mitigation banking–transportability–also is under fire. One recent study of wetland banking in Florida found that wetland banking trades, even within the same watershed, have produced "a transfer of wetlands from highly urbanized, high-population density areas to more rural low-population areas." *See* Dennis M. King & Luke W. Herbert, *The Fungibility of Wetlands*, 19 NATIONAL WETLANDS NEWSLETTER, 10, 11 (Sept.–Oct. 1997). And other studies question the merits of the assumption that trading many small isolated wetlands for a large, contiguous wetland necessarily improves the ecological value of the wetlands. Many species actually are adapted to using complexes of many small, detached wetlands given greater variability of conditions, insurance against total loss in natural disasters, and source-sink population dynamics that may not operate in a contiguous wetland of equal total size. *See* Raymond D. Semlitsch, *Size Does Matter: The Value of Small Isolated Wetlands*, 22 National Wetlands Newsletter 5 (Jan.-Feb. 2000). So, moving wetlands around, even if total acreage and functions are held constant, may prove detrimental to some human populations as well as to plant and wildlife species.

5. For comprehensive overviews of wetland values and valuation methods, see Paul F. Scodari, Measuring the Benefits of Federal Wetland Programs (Envtl. L. Inst. 1997); Special Issue, *The Values of Wetlands; Landscape and Institutional Perspectives*, 35 Ecological Economics 1 (2000). For a thorough description of the history and operation of wetlands mitigation banking programs, see Environmental Law Inst., Wetland Mitigation Banking (1993); Royal C. Gardner, *Banking on Entrepreneurs: Wetlands, Mitigation Banking, and Takings*, 81 Iowa L. Rev. 527 (1996).

6. If wetlands could be valued with sufficient accuracy to support mitigation banking, why bother with banking at all? Couldn't we simply allow the person proposing to fill the wetlands to provide *monetary* compensatory mitigation? For descriptions and criticisms of such programs, known as "in-lieu" or "fee" mitigation, see U.S. General Accounting Office, GAO–01–325, Wetlands Protection: Assessments Needed to Determine Effectiveness of In-Lieu-Fee Mitigation (2001); Royal C. Gardner, *Money for Nothing? The Rise of Wetland Fee Mitigation*, 19 Va. Envtl. L. J. 1 (2000);

3. Watersheds

Wouldn't it be reasonable to expect a federal statute known as the Watershed Protection and Flood Prevention Act to lay out and implement a comprehensive national framework for, say, watershed protection and flood prevention? Alas, statutes often have grandiose names with very little substance to back it up. The federal statute that goes by that name, *see* 16 U.S.C. § 1001 et seq., could hardly be characterized as a watershed-based initiative. It starts by recognizing that "[e]rosion, floodwater, and sediment damages in the watersheds of the rivers and streams of the United States, causing loss of life and damage to property, constitute a menace to the national welfare." Yet the statute does little more than direct the Department of Agriculture to cooperate and cost-share with state and local governments "for the purpose of preventing such damages, of furthering the conservation, development, utilization, and disposal of water, and the conservation and utilization of land and thereby preserving, protecting, and improving the Nation's land and water resources and the quality of the environment."

It is reassuring that the environmental protection slipped in there at the end, though it is less than clear how "development, utilization, and disposal of water, and the . . . utilization of land" will serve that objective. Suffice it to say that there is no national comprehensive legislation defining and implementing a "watershed approach." But can a coherent national policy on the topic be pieced together? We turn to the following materials to see. They are organized into three components: (1) the basic overarching theme of watershed management; (2) the variety of approaches for implementing that theme; and (3) the experience with particular applications of watershed-level ecosystem management.

a.   THEME

William E. Taylor and Mark Gerath, The Watershed Protection Approach: Is the Promise About to Be Realized?

11 Natural Resources & Environment 16 (Fall 1996).

The concept of managing water resources and protecting water quality on a geographical or watershed basis has been evolving since the late 1800s. Only in the last few years, however, have water resource and protection organizations, both public and private, begun to focus seriously on a watershed approach to address remaining water quality problems. Until recently, water quality regulation and management focused on control of point source dischargers of pollutants. Technology and water quality-based effluent limitations on point sources, in conjunction with implementation of antibacksliding and antidegradation requirements, have resulted in a substantial reduction in point source pollutant loadings to waters and consequent upgrading of many receiving waters.

To achieve further cost-effective improvements in water quality, water resource managers recognize that they must now focus on a more comprehensive approach to water quality management, including continued control of point source discharges as well as control of nonpoint source discharges, preservation of habitat, and ground water protection and flow.

This article gives a brief history of watershed protection efforts and an analysis of why the watershed approach makes technical sense. * * *

Historical Development of the Watershed Protection Approach

For several decades, agencies such as the Army Corps of Engineers, the Soil Conservation Service (now known as the Natural Resources Conservation Service) and the Federal Energy Regulatory Commission (FERC) have funded and approved water resource projects designed to achieve multiple objectives which in turn required watershed-based planning and management. Early federal initiatives typically involved basinwide projects for flood control, municipal water supply, irrigation, hydroelectric power generation, recreation, and water quality improvement as part of a single project.

Section 3 of the 1917 Newland Act, 33 U.S.C. § 701 (1988) (repealed 1994), gave the Corps authority to undertake a comprehensive study of watersheds for flood control improvements. The Corps reports served as the basis for most river planning documents for the next several decades. Early basinwide plans tended to address water resource development rather than quality and focus on structural solutions rather than nonstructural pollution control-based planning. In the early 1960s, the focus of water resources management turned from a large basinwide economic development approach to a more regional development and water quality protection approach.

The Water Resources Planning Act of 1965 (WRPA), 42 U.S.C. § 1962 (1988 & Supp. V. 1993), evolved from several years of congressional review of river basin management plans. The WRPA recognized water pollution as a major national concern and attempted to coordinate federal programs to address both water quantity and quality. It did not result in significant change in national water policy, however, because Congress was unwilling to put regulatory teeth into the law or to cede any authority to local or regional basin entities.

The Federal Water Pollution Control Act of 1972, Pub. L. No. 92–500, 86 Stat. 816 (1972) (Clean Water Act or CWA), provided the United States Environmental Protection Agency (EPA) with authority and funding mechanisms specifically directed to watershed protection. The Clean Water Act's principal purpose was the restoration and maintenance of the "chemical, physical and biological integrity of the nation's waters." Section 102(a) of the Act directed EPA, in cooperation with other federal agencies, states and dischargers, to "prepare or develop comprehensive plans for preventing, reducing or eliminating pollution in navigable waters and ground waters." Section 102(c) and (d) provided federal grants for states to develop comprehensive water control plans consistent with the basin planning process under the Water Resources Planning Act. Section 208 of the CWA required states to develop "area wide waste treatment management plans" including land use-based pollution sources.

* * *

Despite all the available CWA statutory authority to begin watershed protection, implementation of these provisions has been limited because EPA and states initially focused on point source dischargers under section 402 of the CWA, the National Pollutant Discharge Elimination System (NPDES) permit program. The Water Quality Act of 1987, P.L. 100–4 (Feb. 4, 1987), added more specific planning-based approaches to the Clean Water Act, including section 319, which specifically required states to develop watershed-based approaches to polluted nonpoint source runoff. In addition, Section 320 established the national estuary program which adopts the planning and implementation of water quality management activities for an estuary's entire drainage area. Currently, there are twenty-eight approved national estuary projects.

Since the 1987 amendments to the Clean Water Act, states and federal agencies have begun to embrace watershed protection approaches seriously. A 1993 watershed management conference drew well over 1,000 participants from state and federal water resource agencies as well as other public and private organizations. The proceedings from that conference have served as a guide for many states implementing watershed protection approaches and for EPA's development of its watershed protection guidance. In a series of guidance documents published within the last year, EPA has finally begun to define and establish a framework for implementing a watershed protection approach.

* * *

What Is the Watershed Protection Approach and Does It Make Technical Sense?

Although each watershed will have different aquatic resources, pollutant loadings, land uses and regulatory programs, it is possible to define a watershed protection approach. In its latest guidance, *Watershed Approach Framework* (Final Draft May 15, 1996), EPA defined a watershed approach as a "coordinating framework for environmental management that focuses public and private sector efforts to address the highest priority problems within hydrologically-defined geographic areas, taking into consideration both ground and surface water flow." The goals are to protect and restore aquatic ecosystems and to protect human health. To be successful, the approach must include consideration of all environmental concerns, including protection of critical habitats, such as wetlands within the watershed, in addition to protection of surface and ground waters.

Much of the momentum behind the watershed approach derives from its very clear advantages for enhancing water quality. Water quality and quantity within a receiving water are affected by the sum of human activities and environmental processes in the hydrologic basin. It makes sense to coordinate water quality management in recognition of the sum of activities.... For example, the simple act of synchronizing wastewater discharge permits within a basin helps the permitting agency coordinate its data collection and modeling while more readily considering the combined impacts of the various dischargers and background loadings. The logical extension of this technique is the explicit consideration and management of nonpoint sources of pollutants and water consumption within the basin. The alternative approach of considering each pollutant source in isolation is not only less efficient, but also not suitable for assessing the full range of potential pollutant loads and necessary control measures.

While the watershed approach provides significant technical advantages, it can be very difficult to administer. The watershed approach is by its nature more complex than either traditional technology or water quality-based permitting. Full implementation of the watershed approach requires an understanding of: (1) the sources of each important pollutant throughout the basin; (2) the source and frequency of different rates of water flow through the system; and (3) how pollution control strategies are likely to affect the loadings and/or flows. For many pollutants (e.g., nutrients or short-lived toxics), it is desirable to understand the rate of pollutant loss from the system to avoid overestimation of loadings. The site-specific chemistry (e.g., bioavailability and toxicity) of the pollutant may bear consideration to develop effluent limitations appropriate for the specific watershed. Finally, it may be necessary to consider two critical flow periods: the low flow events traditionally examined in wasteload allocations at which point source effluents tend to dominate, and wet weather events during which nonpoint source pollutant loadings increase. Thus, permitting agencies face a data collection effort with significant spatial and temporal demands.* * *

A key element of most watershed protection strategies is the use of mathematical modeling to estimate water quality. A narrowly focused point source permitting approach often allowed the discharge to be considered in isolation, and background receiving water concentration was usually assumed rather than measured. The watershed approach puts all contaminant sources on an equal footing and requires a relatively sophisticated model to track all the loadings, dilutions, and attenuation mechanisms.* * *

Another complexity associated with the watershed approach is the issue of regulatory jurisdiction. The watershed approach, by its very nature, recognizes watershed boundaries but it must also acknowledge political ones. Many significant loadings (e.g., mercury from atmospheric deposition and nutrients from septic tanks) come from diffuse and remote (from a regulatory and geographical sense) sources. Controlling these types of sources will likely require new regulatory initiatives such as evaluation and regulation of air emissions for long-distance and long-term impacts. Conversely, many of the critical issues in watershed management are under local or regional control. For example, control of development, with simultaneous control of water consumption, runoff, and septic loadings, is generally local. To be successful, watershed management must motivate local officials and taxpayers to consider impacts across jurisdictional and geographical lines.

As discussed above, a key premise of the watershed approach is that nonpoint sources can and will be controlled on a basis similar to point source dischargers. This assumption is questionable. The agencies charged with regulating water quality through the NPDES program have little or no authority over many important nonpoint sources. This regulatory gap was the subject of a recent debate within EPA regarding trading of point source and nonpoint source pollutants. The EPA Office of Enforcement, recognizing EPA's tenuous ability to regulate many nonpoint sources, wants to hold the point source discharger responsible for the performance of the nonpoint source controls. On the other hand, the Office of Water, wishing to facilitate trading, wants to minimize the point source discharger's responsibility following the trade.

While pollutant trading is relatively well established under the Clean Air Act, it is more complicated under the watershed approach because the location and timing of the relevant loadings within the watershed greatly affect the results. It may be necessary, for example, to assure that the partners in a trade affect the water quality within the same river reaches. EPA is currently developing an effluent trading guidance document to clarify how such trades may be accomplished and credited.

Complexities such as jurisdictional questions and pollutant trading associated with watershed protection have led and will continue to lead to a proliferation of different approaches. Most states are rapidly moving to a watershed based approach and many hundreds of citizen groups are becoming directly involved in those efforts. EPA has published a number of watershed protection guidance documents in the last year and maintains a site on the World Wide Web to discuss the approaches different states are adopting. EPA has even established a Watershed Academy to educate state managers on watershed protection strategies.

The need and benefits of the watershed approach are becoming increasingly clear. Local, state and federal agencies are beginning to use any regulatory mandates available to them to further the watershed approach. Efforts are underway to reorganize regulatory agencies, to revamp the data collection and permitting processes, and to involve stakeholders to an unprecedented degree. In many cases, the state agencies and even the citizen watershed groups are out in front of EPA as the case studies below illustrate. Water pollution control agencies agree that the low-hanging fruit has been picked on the water pollution control tree. The watershed approach, despite its complexity, is a cost-effective means to provide the next highest level of water quality.

Notes and Questions

1. Taylor and Gerath adopt an EPA definition of "the watershed approach" that calls for a "coordinating framework for environmental management that focuses public and private sector efforts to address the highest priority problems within hydrologically-defined geographic areas, taking into consideration both ground and surface water flow." Putting aside, for the moment, the rather open-ended nature of that definition, many other commentators have agreed with the core premise of adopting the hydrologically-defined geographic area of a watershed as the unit of water resources ecosystem management policy. This approach, they posit, fairly closely ties the unit to the "problem-shed" within which the relevant socioeconomic and environmental dynamics play out. *See* Joe Cannon, *Choices and Institutions in Watershed Management*, 25 WM. & MARY ENVTL. L. & POLICY REV. 379 (2000); William Goldfarb, *Watershed Management: Slogan or Solution*, 21 BOSTON COLLEGE ENVTL. AFFAIRS L. REV. 483 (1994); A. Dan Tarlock, *Reconnecting Property Rights to Watersheds*, 25 WM. & MARY ENVTL. L. & POLICY REV. 69 (2000).

2. Notwithstanding their general endorsement of the watershed-based approach in theory, most of these same authors, and many like them, would also agree with Goldfarb's general assessment that "analysis of the concept's historical evolution and current applications discloses that it lacks specific descriptive or operational meaning. There are also significant theoretical problems in defining and applying watershed management." Goldfarb, *supra*, at 504. The devil, in other words, is in the details. For example, while there is general concurrence that a watershed-based approach is most likely the best, if not the necessary, approach to take for addressing nonpoint source pollution, few agree on how to do it. Basic design issues have to be hammered out before the watershed approach becomes more than a theory. Thus, in his epic exploration of the details of the watershed approach, Robert Adler identifies five core design issues for resolution, which can be summarized as follows:

- *scale*: What is the appropriate scale? Are programs that cover broadly defined regions more appropriate than those on a smaller scale? What is the right unit of watershed for watershed-based approaches, taking into account the tension that exists between appropriate ecological considerations and necessary political realities?

- *boundary*: What are appropriate boundaries? Should aquatic ecosystem restoration and protection programs be organized according to natural, rather than political boundaries. Should programs be based on watersheds? Ducksheds? Forestsheds? Bearsheds? How can we reconcile the overlapping nature of the physical boundaries of watersheds, the ecological boundaries that may be more open and complex, and existing political and decision making boundaries?

- *control*: Simply, who's in charge? Is the watershed approach a "bottom-up" or "top-down" political process? Is it a federal concern or a state concern? How does public participation enter the decision making process?

- *mission*: What is the fundamental mission of watershed programs? Will watershed management be primarily procedural, as in serving a coordination function, or substantive, as in imposing regulatory standards and constraints?

- *consistency*: How consistent need be, and can be, the various national and regional watershed approaches? To what extent does consistency limit flexibility to adapt to varying ecological factors?

See Robert W. Adler, *Addressing Barriers to Watershed Protection*, 25 Envtl. Law 973, 1088–1104 (1995); Robert W. Adler, *Model Watershed Protection*, 18 National Wetlands Newsletter 7, 8 (July–Aug. 1996). As you review the remaining materials in this section illustrating various approaches and applications of the watershed approach, consider how coherently they articulate answers to Adler's five factors.

b.   APPROACHES

Department of Agriculture, Department of Commerce, Department of Defense, Department of Energy, Department of the Interior, Environmental Protection Agency, Tennessee Valley Authority, Army Corps of Engineers, Unified Federal Policy for a Watershed Approach to Federal Land and Resource Management

65 Fed. Reg. 62566 (Oct. 8, 2000).

Background

More than 800 million acres of the Nation's land are managed by Federal agencies. These public lands contain significant physical and biological resources and are important to millions of Americans for multiple uses, such as drinking water, irrigation, transportation, recreation, and wildlife habitat. Federal land managers are responsible for protecting and restoring these resources.

The objective of the Federal Water Pollution Control Act of 1972, as amended, which is commonly referred to as the Clean Water Act, is to "restore and maintain the chemical, physical and biological integrity of the Nation's waters." Although Federal agencies are working to implement the applicable requirements of the Clean Water Act, further progress is needed both to prevent degradation of high quality waters and sensitive aquatic ecosystems and to accelerate the restoration of degraded water resources. This policy provides a foundation to help ensure that Federal land and resource management activities meet these goals and that the Federal government serves as a model for water quality stewardship.

* * *

Unified Federal Policy for a Watershed Approach to Federal Land and Resource Management

Introduction

Federal agencies manage large amounts of public lands throughout the country. To protect water quality and aquatic ecosystems on these public lands, Federal agencies have developed the following policy to reduce water pollution from Federal activities and foster a unified, watershed-based approach to Federal land and resource management. This policy is intended to accelerate Federal progress towards achieving the goals of the Clean Water Act (Federal Water Pollution Control Act of 1972, 33 U.S.C. § 1251 et seq.). This policy applies only to Federal lands and resources and does not affect water rights laws, procedures, or regulations. This policy does not supersede or otherwise affect existing State or Tribal authority under the Clean Water Act. The Federal agencies also acknowledge that, in international waters, the watershed approach is subject to the international treaties and agreements affecting those waters.

I. Policy Goals

We, the Federal agencies who have signed this policy, are committed to managing the Federal lands, resources, and facilities in our care as models of good stewardship and effective watershed management.

We recognize that State, Tribal, and local programs for watershed protection and improvement are currently underway and producing positive results. We also recognize the success of locally led, voluntary, watershed groups in planning and implementing water quality improvement actions. This policy seeks to build upon those existing efforts and expand cooperation among Federal, Tribal, State and local partners. This policy will enhance these programs by improving consistency among Federal agency watershed protection programs. We acknowledge that those Federal agencies without established programs will face an additional challenge to implement this policy and that the pace and level of implementation will vary by agency.

The following policy has two goals: (1) Use a watershed approach to prevent and reduce pollution of surface and ground waters resulting from Federal land and resource management activities; and (2) Accomplish this in a unified and cost-effective manner.

To develop a unified Federal policy that meets these two goals, we incorporated the following guiding principles:

A.   Use a consistent and scientific approach to manage Federal lands and resources and to assess, protect, and restore watersheds.

B.   Identify specific watersheds in which to focus our funding and personnel and accelerate improvements in water quality, aquatic habitat, and watershed conditions.

C.   Use the results of watershed assessments to guide planning and management activities in accordance with applicable authorities and procedures.

D.   Work closely with States, Tribes, local governments, private landowners, and stakeholders to implement this policy.

E.   Meet our Clean Water Act responsibility to comply with applicable Federal, State, Tribal, interstate, and local water quality requirements to the same extent as non-governmental entities.

F.   Take steps to help ensure that Federal land and resource management actions are consistent with applicable Federal, State, Tribal, and local government water quality management programs.

II.   Agency Objectives

To accomplish these policy goals, we propose to use existing funding, personnel, and authorities to pursue the following objectives. All agencies will implement this policy as individual agency laws, missions, funding, and fiscal and budgetary authorities permit.

A.   We will develop a science-based approach to watershed assessment for Federal lands. Watershed assessment information will become part of the basis for identifying management opportunities and priorities and for developing alternatives to protect or restore watersheds.

1.   We will develop consistent procedures for delineating, assessing, and classifying watersheds.* * *

2.   We will conduct assessments of watersheds that have significant Federal lands and resources.* * *

B.   We will use a watershed management approach when protecting and restoring watersheds.

1.   We will work collaboratively to identify priority watersheds.* * *

2.   Using existing legal authorities, we will develop a process and guidelines for identifying and designating waters or watersheds on

Federal lands that may have significant human health, public use, or aquatic ecosystem values and a need for special protection.

3. We will implement pollution prevention and controls, consistent with applicable legal authorities.* * *

4. We will improve watershed conditions through restoration and adaptive management. We will strive to work with States, Tribes, local governments, private landowners, and interested stakeholders to improve the condition of priority watersheds. Changes in management strategies and restoration efforts will focus on watersheds where Federal land and resource management activities can meaningfully influence surface and ground water quality and aquatic resources.

5. We will base watershed management on scientific principles and methods. We will use scientific information from research and management experience in designing and implementing watershed planning and management programs, and setting management goals (e.g., desired conditions). To expand current knowledge, we will collaborate to identify research needs and contribute to or sponsor research, as appropriate.

6. We will identify and incorporate watershed management goals into our planning, programs, and actions. We will periodically review and amend, as appropriate, policies and management plans for Federal lands and resources to meet goals for watershed protection and improvement. We will incorporate adaptive management principles into our programs. Our watershed goals will seek to minimize adverse water quality impacts due to ongoing and future management programs, minimize impairment of current or future uses, and restore watersheds where applicable State and Tribal water quality requirements under the Clean Water Act are not achieved due to activities occurring on Federal lands.

7. We will help Tribes and States develop science-based total maximum daily loads (TMDLs). We will assist and support State and Tribal efforts to develop and implement TMDLs in watersheds with significant Federal land and resource management activities. We will provide technical assistance, tools, and expertise. We will use TMDL results in watershed planning and subsequent resource management activities to meet applicable State and Tribal water quality requirements under the Clean Water Act.

* * *

Glossary of Terms

These definitions are intended only to help you understand the policy better, and do not change the meanings of terms defined by law or regulation.* * *

Adaptive management: A type of natural resource management in which decisions are made as part of an ongoing science-based process.

Adaptive management involves testing, monitoring, and evaluating applied strategies, and incorporating new knowledge into management approaches that are based on scientific findings and the needs of society. Results are used to modify management policy, strategies, and practices.

* * *

Watershed: A geographic area of land, water, and biota within the confines of a drainage divide. The total area above a given point of a water body that contributes flow to that point.

Watershed approach: A framework to guide watershed management that: (1) uses watershed assessments to determine existing and reference conditions; (2) incorporates assessment results into resource management planning; and (3) fosters collaboration with all landowners in the watershed. The framework considers both ground and surface water flow within a hydrologically defined geographical area.

Watershed assessment: An analysis and interpretation of the physical and landscape characteristics of a watershed using scientific principles to describe watershed conditions as they affect water quality and aquatic resources. Initial watershed assessments will be conducted using existing data, where available. Data gaps may suggest the collection of additional data.

Watershed condition: The state of the watershed based on physical and biogeochemical characteristics and processes (e.g., hydrologic, geomorphic, landscape, topographic, vegetative cover, and aquatic habitat), water flow characteristics and processes (e.g., volume and timing), and water quality characteristics and processes (e.g., chemical, physical, and biological), as it affects water quality and water resources.

U.S. EPA, Office of Water, Draft Framework for Watershed–Based Trading

(May 1996).

EXECUTIVE SUMMARY

Why Is EPA Publishing This Framework Now?

In response to President Clinton's *Reinventing Environmental Regulation* (March 1995), EPA is strongly promoting the use of watershed-based trading. Trading is an innovative way for water quality agencies and community stakeholders to develop common-sense, cost-effective solutions for water quality problems in their watersheds. Community stakeholders include states and water quality agencies, local governments, point source dischargers, contributors to nonpoint source pollution, citizen groups, other federal agencies, and the public at large. Trading can allow communities to grow and prosper while retaining their commitment to water quality.

The bulk of this framework discusses effluent trading in watersheds. Remaining sections discuss transactions that, while not technically fulfilling the definition of "effluent" trades, do involve the exchange of valued water quality or other ecological improvements between partners responding to market initiatives. This document therefore includes activities such as trades within a facility (intra-plant trading) and wetland mitigation banking.

Trading and Water Quality

Trading is not a retreat from Clean Water Act (CWA) goals. It can be a more efficient, market-driven approach to meet those goals. EPA supports only trades that meet existing CWA water quality requirements.

Similarly, support for trading does not represent any change in EPA's traditional enforcement responsibilities under the CWA. EPA encourages innovation in meeting water quality goals but will not depart from its enforcement and compliance responsibilities under the CWA. Trades that depend on fundamental chance in EPA's enforcement and compliance responsibilities will not be allowed.

EPA encourages trades that will result in desired pollution controls at appropriate locations and scales. Water quality standards must be met throughout watersheds. A buyer cannot arrange for reductions from a downstream discharger if violations of water quality standards would result. Generally, trades will shift additional load reductions to upstream sources. Thus, discharges will be reduced in the area between the sources.

Trading Provides Flexibility

Trading provides watershed managers with opportunities to facilitate implementing loading reductions in a way that maximizes water quality and ecological improvements. Managers can encourage trades that result in desired pollution controls, preferred reduction locations, and optimal scales for effective efforts.

Trading can fully use the flexibility of existing regulatory programs. The following examples illustrate this flexibility and demonstrate how trading can contribute to the cost-effectiveness of meeting water quality objectives.

* * *

Trading Encourages Environmental Benefits

Regardless of who trades and how, the common goal of trading is achieving water quality objectives, including water quality standards, more cost-effectively. Some communities will use trading to meet their waterbodies' designated uses at a lower cost than the cost without trading. Other communities will use trading to expand a waterbody's designated uses for the same amount they would have spent preserving fewer uses without trading. Communities can also use trading to maintain water quality in the face of proposed new discharges.

* * *

In particular, trading offers significant opportunities to expand non-point source pollution reductions beyond current levels. Point/nonpoint and nonpoint/nonpoint trading can facilitate nonpoint source reductions where they otherwise would not have occurred. In so doing, it can help address one of the sources of water pollution that is most persistent and difficult to reduce (economically, technically, and politically).

Beyond implementing trades, the process communities go through when they consider a trading option moves them toward more complete management approaches and more effective environmental protection. Identifying trading opportunities involves examining all pollution sources at once when evaluating technical and financial capabilities to achieve loading reductions. This brings regulated and unregulated sources together with other watershed stakeholders and engages them in a partnership to solve water quality problems.

* * *

Economic Benefits of Trading

One of the most immediately visible benefits of trading is the money some sources save while meeting pollution control responsibilities. Sources that "sell" loading reductions can also benefit financially and can invest proceeds in research and development, for example, or use them to offset other costs.

These economic benefits reach beyond dischargers to consumers and communities. Trading can keep municipal wastewater treatment or storm-water utility charges from increasing as quickly or by as much as they might without trading. Trading also can keep costs to consumers down as industry and business save on pollution control costs.

* * *

Who Might Trade?

Many sources or contributors to water pollution might consider trad-ing. Point source dischargers, nonpoint sources, and indirect dischargers may all participate in trades.

Point sources are direct dischargers that introduce pollutants into waters of the United States. Examples of point sources include [publicly owned treatment works (POTWs)], private wastewater treatment facilities, industrial dischargers, federal facilities that discharge pollutants. active and inactive mining operations, aquaculture operations, and municipal stormwater outfalls (generally communities with populations over 100,000). Point sources are regulated under the National Pollutant Discharge Elimi-nation System (NPDES) established under section 402 of the CWA. Many point source dischargers are required to comply with national discharge standards developed for industrial categories.

Indirect dischargers are industrial or commercial (i.e., nonresidential) dischargers that discharge pollutants to a POTW. Many indirect dischargers "pretreat" their wastewater prior to releasing effluent to POTW collection systems. Pretreatment includes pollution prevention and waste minimization practices, as well as on-site and off-site pollution control technology. Indirect dischargers are regulated under certain circumstances by POTWs according to CWA requirements. Many indirect dischargers also comply with national discharge standards developed for industrial categories.

Nonpoint sources are more diffuse, conveying pollution via erosion, runoff, and snowmelt to surface waters. Nonpoint sources also pollute groundwater via infiltration; this pollution can sometimes reach surface waters. Nonpoint sources include agriculture, silviculture, urban development, construction, land disposal, and modification of flow and channel structure. The CWA does not require federal controls for nonpoint sources. Instead, it requires that states, with EPA funding and technical support, develop and implement programs to control nonpoint sources.

Five Types of Trading in a Watershed Context

Generally, the term "trading" describes any agreement between parties contributing to water quality problems on the same waterbody that alters the allocation of pollutant reduction responsibilities among the sources. Such agreements also may include third parties, such as state agencies, local agencies, or brokerage entities. This framework groups trades into five categories:

1. **Point/Point Source Trading:** a point source(s) arranges for another point source(s) to undertake greater-than-required reductions in pollutant discharge in lieu of reducing its own level of pollutant discharge, beyond the minimum technology-based discharge standards, to achieve water quality objectives more cost-effectively.

2. **Intra-plant Trading:** a point source allocates pollutant discharges among its outfalls in a cost-effective manner, provided that the combined permitted discharge with trading is no greater than the combined permitted discharge without trading and discharge from each outfall complies with the requirements necessary to meet applicable water quality standards.

3. **Pretreatment Trading:** an indirect industrial source(s) that discharges to a POTW arranges for greater-than-required reductions in pollutant discharge by other indirect sources in lieu of upgrading its own pretreatment beyond the minimum technology-based discharge standards, to achieve water quality goals more cost-effectively.

4. **Point/Nonpoint Source Trading:** a point source(s) arranges for control of pollutants from nonpoint source(s) to undertake greater-than-required pollutant reductions in lieu of upgrading its own treatment beyond the minimum technology-based discharge standards, to achieve water quality objectives more cost effectively.

5. **Nonpoint/Nonpoint Source Trading:** a nonpoint source(s) arranges for more cost-effective control of other nonpoint sources in lieu

of installing or upgrading its own control or implement pollution prevention practices.

These categorizations are broad and might not reflect all possible trading combinations. As communities gain experience with trading and as EPA improves its understanding of the opportunities afforded by watershed-based decision making, the Agency will provide information about additional forms of trading.

Trading Arrangements

Trading arrangements can take many different forms. There are varying degrees of complexity related to the number of partners involved, the pollutant or reduction traded, and the form of the trade. Trading programs that involve point sources or indirect discharges require EPA's preapproval of trades.

Under trading arrangements, the total pollutant reduction must be the same or greater than what would be achieved if no trade occurred. A "buyer" and "seller" agree to a trade in which the buyer compensates the seller to reduce pollutant loads. Buyers purchase pollutant reductions at a lower cost than what they would spend to achieve the reductions themselves. Sellers provide pollutant reductions and may receive compensation.

Sources may negotiate trades bilaterally or may trade within the context of an organized program. Sources may negotiate prices or exchange rates for loading reductions themselves, or they may face those established by a market. A buyer and seller may be the only parties to trading, or third parties-public or private-may become involved.

* * *

Notes and Questions

1. Using Adler's proposed set of five factors that define watershed-based programs (see previous section), fill in this table to describe and compare the two programs outlined in the preceding agency policy statements:

	Unified Federal Land Management Policy	EPA Watershed Trading Policy
scale		
boundary		
control		
mission		
consistency		

Do either of the policies provide sufficient detail to allow you clearly to articulate how it approaches each of these design factors? If so, are watershed-based approaches really as hard to assemble as Adler and others suggest? If, on the other hand, you had difficulty expressing how the policies address these design issues, does that suggest the watershed-based approach as a general matter has limited utility? Could the policy statements be more clearly articulated? What language or other editing, *specifically*, would you recommend for doing so?

2.   Now that you've described the basic design parameters of the two programs, can you articulate them as ecosystem management policies? Try doing so using the three components of ecosystem management policy articulated in Chapter 7:

	Currency	*Goal*	*Method*
Unified Federal Land Management Policy			
EPA Watershed Trading Policy			

Again, do the policy statements allow you to describe the underlying ecosystem management frameworks with sufficient clarity? Do the ecosystem management frameworks of the two policies differ substantially?

3.   The Unified Federal Policy says that the signatory agencies will manage federal lands using a science-based watershed management approach. What does "science-based" mean, and what other approaches does it preclude? Who makes decisions using a "science-based" approach? Scientists? Is the public not a part of a "science-based" approach? Would you be comfortable leaving federal land management policy decisions exclusively to scientists? What if the scientists can't agree?

4.   The watershed-based trading program EPA has in mind is known as a "cap and trade" program, because government regulation of some form limits the available "commodity" being traded in order to create scarcity, and thus the incentive to trade. This is an inherent necessity in pollutant trading programs, for in the absence of a cap, pollutant dischargers would have no incentive to trade. It may sound straightforward, but as Richard E. Ayers has outlined, a number of difficult design questions confront efforts to institute such cap and trade programs. *See* Richard E. Ayers, *Expanding the Use of Environmental Trading Programs into New Areas of Environmental Regulation*, 18 PACE ENVTL. L. REV. 87,106–112 (2000). In summary, these are:

- *Determining the scope and geographical extent of the market.* Usually this means corresponding the trading "shed" to the resource unit in question, such as a watershed, but this can cause complications when the resource shed is very small (few traders) or large (lack of control over geographic distribution of trades).

- *Determining the geographic relationship between parties.* There may be areas within the resource shed that are of critical environmental value and thus off limits to trading. Or there may be directional issues within the resource shed, such as water flow, that influence who may trade what with whom. And, for a variety of reasons, trading could lead to "hot spots," areas where far more buying than selling of rights to pollute occurs, which could disproportionately expose the local human populations or ecological resources to adverse effects.

- *Determining the types of pollution sources permitted to be traded.* For example, if the concern in a watershed is nutrient pollution, which pollutants can be cross-traded, and between whom? There are many kinds and sources of nitrogen-based nutrients. As the trading unit becomes more universal, the scarcity of tradable units falls.

- *Determining a baseline.* A cap and trade program, to be effective, must ensure that the cap is lower than existing discharges. Determining this baseline of existing levels of emissions is an essential first step.

- *Determining the allocation of emission rights.* Unlike programs based on auctioning of emission rights or taxing use of rights, cap and trade programs require a mechanism of initially allocating the emission units being traded. This could be based on past emission levels of each trader, or past emission efficiency (e.g., units of pollutants emitted per unit of production).

- *Determining the method of quality assurance.* How will we know, as trading takes place, that emission rights that are "retired" or "moved" as part of a trade are actually retired or not emitted by both parties?

How well does EPA's watershed-based trading policy address these determinations? EPA maintains extensive information on the ongoing refinement of its trading policy at its website, *see www.epa.gov/OWOW/Watershed/trading.htm.* It also provides a site that consolidates water quality information from federal and state agencies into a graphic watershed-based tracking system accessible to the public. *See www.epa.gov/waters/.*

5. The Clinton Administration conceived of both the Clean Water Action Plan, from which the watershed-based trading policy was spawned, and the Unified Federal Policy late in its tenure. Since then, the Bush Administration has barely acknowledged their existence. Does this mean the Bush Administration is not interested in watershed-based approaches to freshwater ecosystem management? Were there aspects of the Clinton Administration policies that went too far toward promoting watershed-based management?

c. APPLICATIONS

William E. Taylor and Mark Gerath, The Watershed Protection Approach: Is the Promise About to Be Realized?

11 Natural Resources & Environment 16 (Fall 1996).

[The introductory portion of this article is excerpted *supra*]

Watershed Protection Case Studies

Four current watershed programs from around the country provide examples of innovative state approaches to watershed protection.

Case 1: State Agency Reorganizes and Citizen Group Is Empowered. The Charles River watershed in Massachusetts is highly developed. The river drains into Boston Harbor and has urban and residential development along its entire length. The river is also considered to be a important cultural, scenic, and recreational resource. Among the problems facing the river are the presence of fecal coliform bacteria at concentrations above the water quality criteria. In addition, there was widespread concern about the quantity of river flow as affected by groundwater and surface water withdrawals from the basin. Both of these problems clearly stem from development on which the traditional wastewater permitting process has little effect.

Statewide, these issues have contributed to the reorganization of the state water regulation divisions (e.g., water supply, wastewater permitting) around watershed boundaries as well as formation of basin teams with other state and federal agencies (e.g., U.S. Fish and Wildlife, EPA). Wastewater permitting has been coordinated in time to facilitate data collection and wasteload allocations. In addition, the Massachusetts Department of Environmental Protection has integrated its program of water withdrawal permitting into a program of watershed management. This program modification greatly facilitates data collection on critical stream flows as well as subjecting water withdrawals to evaluation relative to ecological issues and wasteload allocations. Multidisciplinary and multiagency teams within each basin now facilitate planning.

In the Charles River basin, the Charles River Watershed Association (CRWA), a nonprofit citizen group, has begun a study of the basin because of the problems of nonpoint source pollution and management of water quantity. In an ambitious five-year program, the organization will collect water quality samples, develop mathematical models of water quantity and quality, perform stakeholder outreach, and help to develop water quality management plans. The CRWA study uses local academic resources and citizen volunteers and is funded by government grants and the stakeholders.

* * *

Case 2: Pollutant Trading in North Carolina Watersheds. Nutrient loading to the Nuese and Tar–Pemlico watersheds in North Carolina has led to significant water quality impacts. The North Carolina environmental agency determined that the loadings to the watershed derive from both point sources (such as municipal treatment works) and nonpoint sources (e.g., animal wastes). Following the development of a TMDL for each system, the state established a program that effectively allows pollutant trading between the regulated entity, the dischargers, and an unregulated one, the nonpoint sources. The state sets targets for effluent nutrient concentrations for certain major NPDES-registered dischargers. If the discharger cannot meet those targets, the discharger pays into a state-administered fund to be used for the control of nutrients from nonpoint sources. The amount of the payment is proportional to the degree of noncompliance. This is a promising program that illustrates the importance of creativity in developing a management system in which the discharger gains flexibility, yet the agency retains requisite authority to further water quality goals. Of course, it is likely that the point sources will bear a heavier burden than the nonpoint sources but they may achieve "compliance" at a lower cost than in the absence of trading.

Case 3: Pollutant Trading of Toxic Metal in California. Copper loading from point sources and nonpoint sources to south San Francisco Bay has led to noncompliance with the water quality criteria. The Regional Water Quality Control Board oversaw the permitting of point sources. Following an inventory of copper sources and development of a TMDL, the Board found that copper loading to the basin should be reduced by 950 pounds per year. The major point sources were already well controlled with respect to copper. The various basin stakeholders (e.g., the POTWs, the municipalities, and environmental groups) and the Board concluded that control of nonpoint sources (notably sediment transport in stormwater) was likely to be the most cost-effective solution to the problem. Other, less obvious controls (e.g., curtailment of copper sulfate application for aquatic weed control and removal of copper from automobile brake pads) were also investigated. The resulting agreements on copper control measures were to be incorporated into a basin plan and the NPDES permits of the major point source dischargers.

Despite the planned trading of copper loading from the nonpoint to point sources, the major wastewater dischargers were still dissatisfied with their effluent limits. They appealed the basin plan to the State Water Quality Control Board, which overturned the plan. They based their challenge on the applicability of the water quality criterion as well as on the technical basis of the TMDL. This case study highlights the technical and regulatory complexity of the TMDL process and the high stakes involved in water quality management. Despite the failure of the basin management plan for copper, the major municipal wastewater dischargers have reduced their copper loading by 50 percent through better management of indirect dischargers and optimization of their treatment processes. Thus, increased awareness of the problem led to substantial water quality gains.

Case 4: Washington State Department of Ecology (DOE) Implements TMDLs. The State of Washington reorganized its wastewater permitting functions around watersheds by synchronizing permits and by establishing schedules for basin investigations and modeling and evaluation. DOE has developed TMDLs for a variety of pollutants including nutrients and toxics. The TMDL process led to the use of innovative methods for addressing some of the more difficult problems facing the watershed. For example, in some cases, development of a conservative water quality model has eliminated the requirement that an explicit reserve for uncertainty be included in the development of a TMDL. Thus, the reserve is implicit in the TMDL and the full, modeled TMDL can be distributed among the sources.

Washington DOE has an interesting approach to pollutant trading. Rather than organizing explicit trades, DOE modeled its watershed management on the Clean Lakes Program which presents the costs and benefits of a number of management schemes to the stakeholders and reaches a consensus on the best one. This approach is more likely to result in buy-in on the part of the regulated parties and has the potential to lead to cost-sharing.

Notes and Questions

1. How critical was it that a watershed-based approach be used in each of the case studies Taylor and Gerath recount? Can you think of a more effective approach that could have been employed to address the resource management issue that was posed in each case? EPA provides additional case studies of watershed-based initiatives at Watershed Success Stories, http://cleanwater.gov/success/.

2. The case studies Taylor and Gerath describe do not appear to have been motivated by mandatory legal authority. In many cases, collaborative efforts and policy momentum seem to have arisen spontaneously. Can law mandate such outcomes? How would you draft such a law?

Barton H. Thompson, Jr., Markets for Nature

25 Wm. & Mary Environmental Law and Policy Review 261 (2000).

I.   Introduction: Regulatory Markets versus Markets for Nature

The environment is a good in more than one sense of the word. The environment is beneficent to humanity, nurturing us, entertaining us, enlightening us, and providing us with the foundations of life—air, water, food, and a sustaining climate. To many, the environment reflects innate virtue as either God's handiwork or the aesthetic consequence of elegant physical laws. For these reasons, there is also human demand for protecting, sustaining, and enjoying the environment. The environment, in short, is also an economic "good." Although we are used to receiving for free many of the services and amenities provided by the environment, those services and amenities have value to us for which we would each be willing to pay some sum. "Natural" resources such as water, petroleum, and fish are already economic commodities, but as a consequence of their consumptive values. The values provided by a preserved nature raise the possibility that market systems also can support efforts to protect watersheds and other natural ecosystems. Markets for nature hold out the promise of a third rail, along with regulation and education, for preservation efforts.

* * *

III. Ecosystem Service Markets: Preserving Watersheds

* * *

Healthy ecosystems provide a variety of commercially valuable services that we take largely for granted because, for millennia, we have received the services free of charge. Such services include partial stabilization of climate, detoxification and decomposition of wastes, air purification, generation and renewal of soil and soil fertility, crop pollination, and pest control. Because many of these services are extremely valuable, efforts to preserve ecosystems may be able to capitalize on the value to bring in additional funding. The key is to find institutions or individuals who benefit from these ecosystem services and are willing to invest in their preservation.

* * *

In the twentieth century, a growing number of water suppliers turned to technological solutions such as filtration and disinfection to address water quality problems. Technological solutions, however, are often extremely expensive. Moreover, technology alone may not provide safe drinking water, as demonstrated by outbreaks of serious illnesses like cryptosporidiosis in cities that filter and disinfect their water supplies. Technological solutions also can have negative side effects in terms of both aesthetics (e.g., the taste of chlorine in drinking water) and health (e.g., the possible carcinogenicity of chlorine by-products). The abandonment of reservoirs because technological approaches could not keep pace with land development and thus pollution provides perhaps the most vivid illustration of the limitations of technological solutions and the critical importance of protecting watershed lands.

In theory, water suppliers can try to protect watershed lands through either regulation or the acquisition of property interests. In practice, many water suppliers have found acquisition more effective than regulation for several reasons. First, regulations are often difficult to enforce effectively, particularly where the watershed is in a political jurisdiction distant from the water supplier. Second, local communities typically fight significant regulatory efforts by distant water users.

* * *

A. Acquisitions Motivated by Watershed Services

Investment by water suppliers, cities, and others in the ecosystem services provided by riparian land is not hypothetical. In perhaps the best known example, New York City has chosen to invest over a quarter of a billion dollars in the acquisition and preservation of up to 350,000 acres of land in the Catskill watershed. Regulations under the federal Safe Drinking Water Act require water suppliers to filter their water unless the supplier can demonstrate that it has taken other steps, including protection of the watershed, that will adequately protect its customers from the risks of contamination. As noted earlier, filtration can be very expensive and is not always effective. New York City obtained an exemption from the filtration requirement by not only acquiring sensitive watershed lands, but funding watershed-based efforts to minimize pollution from farming and development, paying to improve sewage facilities in the watershed, and updating and extending its regulation of watershed activities to effectively police septic systems and other local sources of pollution. New York City estimates that the total cost of its entire watershed-protection program through 2010 will be about $1.5 billion, far less than the $4–8 billion cost of constructing a filtration plant (which would also entail annual operating costs of about $300 million).

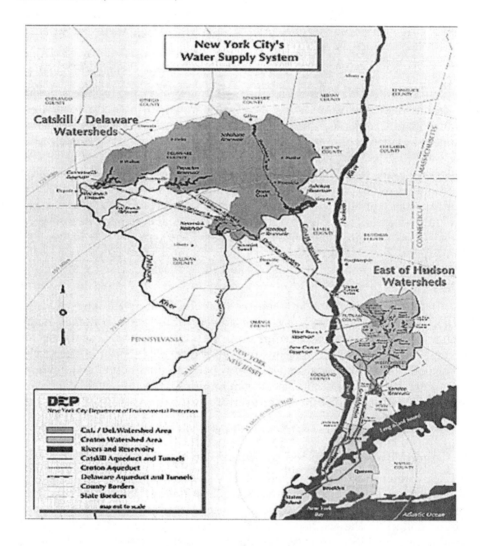

New York City's efforts to preserve watershed land are not unique. Water companies have long acquired land in their watersheds to protect the quality of their drinking water. On average, water companies in the United

States own only about two percent of the land in their watersheds. In some portions of the nation, however, water suppliers control a much higher percentage, including most of the remaining open space.

Renewed emphasis on drinking water quality, including the federal Environmental Protection Agency's filtration regulations, are today driving additional land acquisition. In the late 1990's, more than 140 cities were considering watershed conservation in an effort to ensure safe drinking water for their customers. In 1998, Seattle increased its ownership of land in the South Fork Tolt River watershed from about 30 to 70 percent through a land exchange with the Weyerhaeuser Company; previously Seattle had acquired close to 100 percent land ownership in the Cedar River watershed, its other major water source, through land exchanges with the federal government. Portland, Maine is actively purchasing land within 1,000 feet of its main reservoir and tributaries. Salt Lake City assesses its water customers a small additional monthly fee to pay for land preservation in the city's Provo River watershed. Charlotte–Mecklenburg Utilities, in North Carolina, uses a portion of its capital improvement budget each year to acquire watershed lands. In a joint interstate effort, New Jersey and New York have purchased over 17,000 acres of land in Sterling Forest, the watershed for almost a quarter of New Jersey's population. Both Syracuse, New York and Rochester, Minnesota have embarked on programs to protect the cities' water supplies by paying riparian farmers to establish buffer zones along key water bodies. Rather than protecting riparian land from any development, some water suppliers have used easements, leases, or other financial incentive programs to minimize the size of the footprint that local activities impose on the land.

Fewer examples exist of cities or others investing in watershed preservation for flood control purposes, but interest is growing in such "soft" solutions to flood damage. With funding from a voter-approved initiative, California's Napa County plans to spend $160 million to acquire 500 acres of flood plain; the county expects that the acquisition will significantly reduce flood damages, which totaled $500 million in the last four decades of the twentieth century. Local governments near Boston acquired rights to 8,000 acres of wetlands capable of holding some 50,000 acres of water during potential flood periods, rather than building a $100 million system of dams and levees or bearing continued flood damages averaging $17 million annually. Littleton, Colorado, acquired over 600 acres of land for both flood control and park purposes.

B. Potential Barriers to Capitalizing on Ecosystem Services

One should have a healthy dose of skepticism regarding how often water companies, local governments, and other entities will find it worthwhile to preserve watershed lands. A number of the situations where water suppliers have chosen to preserve watershed lands, for example, involve unique settings that are not likely to be widely duplicated. New York City, for example, was able to escape building a multi-billion dollar filtration plant not simply because it planned to acquire riparian property, but also because it promised to engage in extensive regulation, both mandatory and voluntary, of the watershed. New York City could do this because of its historic regulatory authority over activities in the watershed; many cities will lack this power and find it difficult to obtain such authority over the almost certain opposition of local governments. But the opportunity to enlist cities, water suppliers, and others in the preservation of watersheds is significant enough to justify examining how public policies either support or undermine incentives to invest in ecosystem services.

* * *

One approach ... would be the creation of a special watershed district empowered to tax individuals or entities benefitting from the watershed services and to use the revenues to acquire watershed lands or take other protective actions. Local communities have long used special districts as a means to overcome collective action problems in the provision of services with public good qualities. Farmers, for example, have formed irrigation districts to import and distribute water supplies and pest control districts to exterminate crop-threatening insects. Residents of flood-prone areas have formed flood control districts to finance dams and other engineering control measures. Local communities have formed school districts to provide public education. Each of these various districts is governed by a board of directors, generally elected by the affected population, which decides which measures to take and how to apportion the total cost among the local population, and then implements the measures and collects the necessary funds. The concept of a watershed services district would simply take this tried-and-true approach, which has been used to date primarily to finance engineering solutions to problems, and use it to ensure adequate provision of natural services.

Notes and Questions

1. Why was the New York City watershed management experience Thompson describes so successful? Consider the previous discussion of Professor Adler's five design parameters of watershed-based policy—scale, boundary, control, mission, and consistency. Were any of these poorly defined in the New York City case? For further background on the New York City drinking water watershed program, see Symposium, *The New York City Watershed in the 21st Century*, 22 FORDHAM ENVTL. L.J. 417 (2001); Michael C. Finnegan, *New York City's Watershed Agreement: A Lesson in Sharing Responsibility*, 14 PACE ENVTL. L. REV. 577 (1997).

2. Thompson suggests that a local "watershed services district" governance unit model would simply apply the tried-and-true approach taken in many other local management contexts. How simple would this be? Many local district governments, such as flood control districts, irrigation districts, and soil conservation districts, face the challenge of calculating a relevant policy metric. Irrigation districts distribute irrigation water and thus have to determine a price for units of water. How would a watershed services district specify the unit for determining how much different taxpayers within its jurisdiction benefit from the watershed services it manages. To the extent that actual watershed ecosystem service values are beyond monetary calculation, would acres of property suffice as a surrogate? Property values? How would you draft a state watershed services district enabling statute?

Pacific Coast Federation of Fishermen's Associations, Inc. v. National Marine Fisheries Service

United States Court of Appeals, Ninth Circuit, 2001.
253 F.3d 1137.

■ GOODWIN, CIRCUIT JUDGE:

Six environmental organizations sued the National Marine Fisheries Service ("NMFS") for declaratory and injunctive relief to challenge four biological opinions which had the effect of clearing the way for 23 proposed timber sales in the Umpqua River watershed in southwestern Oregon. The district court granted substantial relief and the defendant agency, together with intervening timber operators, appeal.

The Pacific Coast Federation of Fishermen's Associations, Inc. and five other organizations representing fishermen and environmental concerns are collectively referred to as "Pacific Coast." Their principal claim is that the "no jeopardy" opinions issued by NMFS filed in Seattle, where the agency has its regional headquarters, were arbitrary and inadequately supported by the "best available science" as required by the Endangered Species Act ("ESA"). At the heart of the controversy is the impact of proposed timber sales on the Umpqua River cutthroat trout and the Oregon Coast coho salmon. Douglas Timber Operators ("DTO") and the Northwest Forestry Association were allowed to enter the cases as defendant-intervenors. The cases have been consolidated for this appeal.

Pacific Coast alleged that NMFS acted arbitrarily and capriciously in reaching the conclusion that the proposed timber sales are not likely to jeopardize the continued existence of the listed species. The district court found that NMFS had acted arbitrarily and capriciously by assessing Aquatic Conservation Strategy ("ACS") compliance only at the watershed level, by failing to evaluate short-term degradations, and by failing to fully and sufficiently incorporate the watershed analysis consistently with the "best available science" requirements set by the ESA. The district court granted summary judgment in favor of Pacific Coast.

* * *

The NMFS issued four biological opinions stating that 23 timber sales in the Umpqua River Basin were not likely to jeopardize the continued existence of the Umpqua cutthroat trout and the Oregon Coast coho salmon. The proposed sales are within the range of the northern spotted owl, and therefore fall within the region covered by the Northwest Forest Plan ("NFP"). The United States Forest Service ("USFS") and the BLM adopted the NFP in 1994. The plan was designed to provide a comprehensive management program for 24.5 million acres of federal forest lands throughout the range of the spotted owl. See Seattle Audubon Society v. Lyons, 871 F.Supp. 1291, 1304 (W.D.Wash.1994), aff'd 80 F.3d 1401 (9th Cir.1996). One of the key components of the NFP is the ACS, a comprehensive plan designed to maintain and restore the ecological health of the waterways in the federal forests.

There are four components to the ACS: (1) key watersheds (the best aquatic habitat, or hydrologically important areas), (2) riparian reserves (buffer zones along streams, lakes, wetlands and mudslide risks), (3) watershed analysis (to document existing and desired watershed conditions), and (4) watershed restoration (a long-term program to restore aquatic ecosystems and watershed health). The ACS also has binding standards and guidelines that restrict certain activities within areas designated as riparian reserves or key watersheds. Additionally, ACS has nine objectives designed to maintain or restore properly functioning aquatic habitats.

* * *

Pacific Coast argued, and the district court agreed, that NMFS acted arbitrarily and capriciously by ... ignoring site-specific project effects and limiting its ACS compliance analysis to the watershed scale ...

* * *

WATERSHED SCALE ACS CONSISTENCY

In determining ACS consistency for the 23 timber projects challenged in this case, NMFS analyzed the projects' consistency with ACS at the watershed level. A watershed ... generally covers between 20 to 200 square miles of land. This equates to between 12,800 and 128,000 acres. The largest watershed considered with reference to projects at issue here is 350 square miles, or 224,000 acres. By contrast, a project site generally covers only a few sections (square miles) or fractions of sections. The NMFS conducts its analysis of the program by assessing the affects of any project level degradation on the entire watershed. Any degradation that cannot be measured at the watershed level is considered to be consistent with both ACS standards and objectives and therefore warrants a "no jeopardy" finding.

Pacific Coast contends that the watershed measure effectively masks all project level degradation. This argument raises two questions: (1) whether, because a 128 acre project represents only 1% to 0.1% of a watershed, any degradation would be perceptible at the watershed level; and (2) whether any effect was given to the cumulative degradation in an ACS.* * *

The NMFS contends that the proper level to evaluate ACS consistency is the watershed, because NFP and ACS are aimed at maintaining and restoring millions of acres of forest lands. Given that overall protection of forest and water resources is the concern of both NFP and ACS, it does not follow that NMFS is free to ignore site degradations because they are too small to affect the accomplishment of that goal at the watershed scale. For some purposes, the watershed scale may be correct, but NFP does not provide support for so limiting NMFS review. The purpose of ACS is to maintain and restore ecosystem health at watershed and landscape scales to protect habitat for fish and other riparian-dependent species and resources and restore currently degraded habitats. This general mission statement in NFP does not prevent project site degradation and does nothing to restore habitat over broad landscapes if it ignores the cumulative effect of individual projects on small tributaries within watersheds. The agency also must determine "how the proposed project or management action maintains the existing condition or moves it within the range of natural variability." The NMFS relies on this requirement to show that consistency will be attained at the watershed level. However, it is unclear whether NMFS performed an analysis of the cumulative effect of small degradations over a whole watershed. Pacific Coast asserts that NMFS did not consider cumulative effect. The NMFS had an opportunity to place in the record evidence demonstrating that it considered cumulative effect. We find nothing to show that it did. Appropriate analysis of ACS compliance is undertaken at both the watershed and project levels.

* * *

Its disregard of projects with a relatively small area of impact but that carried a high risk of degradation when multiplied by many projects and continued over a long time period is the major flaw in NMFS study. Without aggregation, the large spatial scale appears to be calculated to ignore the effects of individual sites and projects. Unless the effects of individual projects are aggregated to ensure that their cumulative effects are perceived and measured in future ESA consultations, it is difficult to have any confidence in a wide regional no-jeopardy opinion.... If the effects of individual projects are diluted to insignificance and not aggregated, then Pacific Coast is correct in asserting that NMFS's assessment of ACS consistency at the watershed level is tantamount to assuming that no project will ever lead to jeopardy of a listed species.

Notes and Questions

1. What was the problem with NMFS using a watershed-based approach to assess the impact of timber leases in *Pacific Coast Federation*? Did BLM use the wrong watersheds? Did it use watersheds that were too large, or too small? Should it not have used watersheds at all? Does the case suggest a flaw in the "watershed approach" generally?

2. The challenge for the agencies in *Pacific Coast Federation* was to devise a system for identifying when project-specific actions would jeopardize the continued existence of the protected species. This is simply one form of a problem that plagues environmental law—how to measure, account for, and respond to the cumulative effects of many independent actions, none of which has substantial effects in isolation. Using ecosystem-level units of impact assessment, such as watersheds, may make us more cognizant of the impact of cumulative effects on large scales, but, as *Pacific Coast Federation* suggests, doing so does not necessarily answer or simplify the challenge of knowing how to manage the many small actions that lead to those accumulated impacts. Indeed, the system the agencies adopted in *Pacific Coast Federation* avoided many management issues because it screened out most of the project-specific actions from closer evaluation by assessing impacts on the watershed only on a project-by-project basis–i.e., without evaluating cumulative effects.

By contrast, the very heart of environmental impact assessment under the National Environmental Policy Act (NEPA) is to evaluate cumulative effects. NEPA requires federal agencies, before they make a final decision on major actions they propose to carry out, fund, or authorize, to prepare a detailed statement of the environmental impact of the proposed action, known as an environmental impact statement (EIS). 42 U.S.C. § 4332(C). Regulations the Council of Environmental Quality (CEQ) has promulgated to implement NEPA, and which are binding on other federal agencies as they fulfill their NEPA duties, require agencies to consider the cumulative impacts of "connected actions" and "cumulative actions" in the same EIS. "Connected actions" are actions that "automatically trigger other actions which may require environmental impact statements," or "cannot or will not proceed unless other actions are taken previously or simultaneously," or "are independent parts of a larger action and depend on the larger action for their justification." 40 C.F.R. § 1508.25(a)(1). "Cumulative actions" are actions "which when viewed together with other proposed actions have cumulatively significant impacts." *Id.* § 1508.25(a)(2). And "cumulative impacts" are "the impact on the environment which results from the incremental impact of the action when added to other past, present, and reasonably foreseeable future actions regardless of what agency (Federal or non-Federal) or person undertakes such other actions." *Id.* § 1508.7. Thus, even when a proposed major federal action may have insignificant effects on the environment standing alone, NEPA would require assessment of the cumulative effects of the proposed action plus the effects of any connected or cumulative actions. Is this the kind of approach the *Pacific Coast Federation* court is demanding the agencies take? Is it the kind of approach that should be used in all "watershed-based" ecosystem management?

D. CASE STUDY: THE APALACHICOLA-CHATTAHOOCHEE-FLINT RIVER BASIN WATER WARS

The old adage, *whisky is for drinking and water is for fighting,* has long defined history in the American West. But in these days, it applies just as well east of the Mississippi.

Deep in the North Georgia hills, just a few hundred feet off the southernmost reaches of the Appalachian Trail, a small mountain brook marks the headwaters of the Chattahoochee River. As it meanders its way out of the Chattahoochee National Forest, through the quaint Bavarian-style town of Helen, the water soon empties into Lake Lanier, a huge reservoir north of Atlanta impounded in the 1940s by the U.S. Army Corps of Engineers' Buford Dam. Below the dam, cooled water spills out and works its way toward Atlanta, brushing by just north of that major southeastern city and then drifting westward toward Alabama. At West Point Lake Dam, the river veers more sharply southward and becomes the boundary between Alabama and Georgia. It passes by Columbus, Georgia on its east bank, then later the Alabama plantation town of Eufaula. At Sneads, Alabama, where Lake Seminole is impounded, it joins the Flint River, which has its origins near the south side of Atlanta, and crosses into Florida. There it becomes the Apalachicola River, a ribbon of water slicing across the Florida panhandle and emptying into the Gulf of Mexico at Apalachicola. This collection of rivers, over 750 river miles in all, makes up the Apalachicola–Flint–Chattahoochee River system, known by shorthand as "the ACF."

Apalachicola-Chattahoochee-Flint River Basin

The ACF is the lifeblood of one of the richest, most biologically diverse estuaries in the world—Apalachicola Bay. Life is so good in the Bay, its oysters grow faster than anywhere on earth (the Bay supplies 10 percent of the nation's oysters) and many species of fish found in the Gulf spend part their lives there. The Nature Conservancy lists the Bay as one of its hottest of biodiversity hotspots in the world. The Apalachicola River itself, plus its floodplain of over 180,000 acres, is home to one of the highest diversities of freshwater fish, amphibians, and crayfish in the nation.

Since the 1940s, however, biodiversity has had a tough competitor—humans, or, more specifically, dams. Under congressional mandate, the Corps of Engineers began "taming" the Chattahoochee with a series of major dams designed to impound water to meet a variety of human needs. The ready supply of water proved irresistible to residential and industrial development throughout the region. Population growth in the ACF basin boomed, concentrated in Atlanta. The area became one of the hottest regional economies in the nation. A hotspot of biodiversity and of economic vitality—the ACF had it all.

But trouble was on the horizon. A series of record droughts in the 1980s illustrated the limits of ACF water. In 1989, Georgia and the Corps proposed diverting more water from the Corps' impoundments to quench Atlanta's thirst. Georgia then applied to the Corps to add yet another major impoundment in the state—this one on the Tallapoosa River just five miles from where it crosses into Alabama. Alabama, fearing that less water flowing into the state and along its boundary with Georgia would mean less potential for its own economic growth, immediately brought litigation to halt both plans under a variety of federal laws, most prominently the National Environmental Policy Act. Florida, fearing less water emptying into Apalachicola Bay could damage the Bay ecosystem and the oyster and recreational fishing industries it supports, soon joined the fray.

Western states have resolved their many interstate water allocation battles in three ways: (1) litigation before the U.S. Supreme Court under its original jurisdiction; (2) congressional allocation; and (3) interstate compacts approved by Congress. Because water disputes of any substantial magnitude have been rare in the east, these methods have been seriously field tested in eastern settings only a few times, and not at all in recent history. But the ACF dispute was sizing up to be a biggie, with serious potential to head to the Supreme Court if the states could not agree. To avoid that high stakes proposition, in 1992 the three states entered into negotiations that led to a 1997 interstate compact to negotiate some more. *See* Pub. L. No. 105–104, 111 Stat. 2219 (1997) (unlike the only other eastern compact, the Delaware River Compact between Pennsylvania, Delaware, New York, New Jersey, and the federal government, the ACF compact includes the federal government as a participating, non-voting member, but it can exercise a veto power over what the states propose).

The negotiations since then have been protracted, focusing on each state's model of river flow conditions experienced under an array of climate and population projections. Unable to reach quick consensus, the states extended their self-imposed deadlines numerous times, hired respected mediators, and employed the best legal and technical experts money can buy, but to no avail. As of this writing, there was still no consensus proposal on the table.

Clearly, the politics of the ACF are as complex as is its ecosystem. The social and economic fabric of the ACF watershed illustrates why. Consider just a sample of who wants a piece of the ACF—that is, besides the ecosystem itself:

Atlanta Drinking and Industrial Water Suppliers: Located not far downstream of Lake Lanier, the Atlanta metropolitan area population has grown from 1 million in 1950 to over 4.0 million today. With groundwater supplying less than 2 percent of the water for Atlanta area residents and businesses, they have turned to Lake Lanier and the Chattahoochee. Overall, more than 2.5 million Atlanta area residents rely on water from the river or Lake Lanier for their domestic water supply. Local water supply authorities draw over 130 million gallons per day from the lake and over 275 million gallons per day from the river to supply residents and industry. To serve expected population and industrial growth, Atlanta planning officials would like to raise those figures to 297 and 408 million gallons per day, respectively, in the next 20 years. But over 65 percent of all that withdrawn water is returned to the river, albeit as treated wastewater, and the Atlanta area must maintain a healthy downstream flow in the river below the city–over 480 million gallons a day–to dilute the wastewater in order to meet Clean Water Act standards. So the impact of Atlanta on minimum daily water flow at the other end of the watershed is muted to a substantial degree, though, as you will see, many Florida interests don't see it that way.

Electric Power Utilities: Beginning in the 1930s, hydropower generated from dams has been an important source of electricity throughout the Southeastern United States. The Chattahoochee and Flint Rivers contribute substantially to this relatively inexpensive source of power. The U.S. Army Corps of Engineers operates several federal dams on the Chattahoochee, together supplying power to over 80,000 homes in the Southeast. The Georgia Power Company also operates dams on the Chattahoochee and Flint Rivers. Because they can come ''on line'' very quickly after water is diverted through their turbines, hydropower plants offer an extremely efficient way of meeting peak power demands, such as on hot summer afternoons. Fossil fuel plants take much longer to stoke up and wind down their turbines in response to demand fluctuations. But the hydropower turbines don't spin if the water doesn't flow, thus pitting the power plants in direct competition with upstream water diversions such as those going to Atlanta water supply companies, and with recreational interests that advocate more water storage in the reservoirs.

Shipping Industry. During the early 1800s, riverboat shipping traffic on the ACF system was a thriving industry, making the Port of Apalachicola one of the transportation centers of the Southeast. Cotton, timber, honey, and turpentine were the main goods shipped. But rail shipping proved a potent competitor, and riverboat shipping along the ACF river declined steadily in the latter part of the century and was negligible by the early 1900s. Improved dredging by the Corps, required by federal legislation in 1945, and the introduction of large barges that could compete more effectively against rail and truck shipping, returned some vitality to the ACF shipping industry in the mid–1900s. Fertilizer, coal, and agricultural products became the commodities of choice for shipping up and down the ACF. But the Corps needed to keep the river channel dredged and river levels sufficiently high in order to accommodate the deeper draft barges.

One means of doing so, which the Corps has used since the early 1990s when drought conditions became more common, is to release water from upstream reservoirs in slugs to raise the river levels at times when barge traffic is present. Still, even with these "navigation windows," the narrow ACF makes a poor venue for barges, allowing a tug to push only two at a time. Barge trips dropped from 204 in 1993 to 47 in 1999. Meanwhile, the Corps spends $4.5 million per year to dredge the channel and $2.5 million to operate the locks. That's right, over $7 million annually for 45 or so barge trips. You do the math! In addition, the dredging unquestionably damages the river ecosystem, and the pulses of water spill fish into floodplain pools that become isolated after the water quickly subsides, dooming the trapped fish. A May 2000 release to allow two barges to navigate up the river stranded thousands of fish in this manner. The slow-filling Lake Lanier is still recovering from the consequences of that release episode.

Farm Irrigation. The 26 southwest Georgia counties are dominated by agricultural economies, generating $1.6 billion in agricultural product revenue annually. These agricultural operations also used 325 million gallons of water per day, mostly for crop irrigation, and are projected to use 570 million gallons per day by 2050. Most of the irrigation water is drawn not from lakes or rivers, but rather from the Floridan Aquifer, a huge, highly productive limestone aquifer stretching from southern Georgia well into Florida. The relation between withdrawals from the aquifer and the chief surface water resource in the area, the Flint River, is not fully understood. But the impact of irrigation on the ACF system in general was sufficiently clear to prompt the Georgia legislature to pass legislation in 2000 authorizing payments to farmers who draw directly from the river to stop irrigating. Regulation of agricultural water conservation practices nonetheless remains a touchy issue in these parts.

Lake Recreation Economies. In addition to supplying residential and industrial water to urban Atlanta, Lake Lanier has become the city's playground. At 38,500 tree-rimmed surface acres, Lake Lanier is a boater's and retiree's heaven. Its shores are dotted with marinas, million dollar homes, resort hotels, and golf courses. Houseboats as long as 120 feet are not uncommon. Its recreational economy generates billions of dollars in revenue. All of this depends, however, on there being water in Lake Lanier. Yet the water level was 9 feet below normal in the summer of 2001, leaving boat docks high and dry, mangling boat propellers in never before seen shallows, and scarring the "lake appeal" of sprawling estate homes. If the dam gates were closed today and rainfall were to return to normal, it would take three months to return to normal levels. Since neither of those events seems likely to happen anytime soon, Lake Lanier recreational interests have to fight with all the other interests for every drop. The same is true of local economies built on the recreational amenities offered by the other major impoundments along the Chattahoochee. The question is whether these recreational economies, which depend on keeping water stored in reservoirs built to serve *the river*, are becoming the tails wagging the dog.

The Bay Fishing Industry. At the opposite end of the ACF watershed from Lake Lanier, 544 miles from the headwaters of the Chattahoochee, lies the Apalachicola Bay, home to the most productive oyster beds in the nation and the center of a highly productive estuary. A small but sustainable oyster and fishing industry has been based in Apalachicola for decades. But it is a far cry from the estates of Lake Lanier—most oyster harvesters and fishermen live week to week in fairly hard-scrabble circumstances. Their very livelihood depends on one thing above all else—water flowing out of the mouth of the Apalachicola River. But not just any flow. It has to be the right amount at the right time—the so-called natural flow regime. This demand is not a human invention. The life cycle of oysters and the value of the estuary as nursery habitat for shrimp, mullet, flounder, red snapper, and grouper depend on a fluctuating supply of freshwater throughout the year. By and large, that's all many of the Florida interests want from the ACF system–water at the end of the pipe the way nature intended it to be delivered. Franklin County, through which the Apalachicola River flows, has no aspirations of withdrawing water to launch another Atlanta. There is but one traffic light in the entire county! For those leading this modest lifestyle, however, whether what they think is a modest request can be met has become a major concern.

Any western water war presents an equally diverse set of interests. What makes the ACF truly a new type of water dispute–one not experienced even in the west, is that one of the states, in this case Florida, is in the mix primarily to advance an *ecological* dimension. To be sure, environmental factors have entered into the positions states have taken elsewhere in water allocation disputes, but primarily as a means of supporting the proposed economically-driven split of water quantity. Florida's ecologically-based position presents not only a quantity component, but also a qualitative factor in terms of the natural flow regime. This feature truly is new to the universe of interstate water disputes. As one observer has concluded:

> the "natural flow regime" approach to allocation proposed by Florida elevates environmental concerns to a new level in water quantity disputes. As a practical matter, the protection of Apalachicola Bay and its oysters represents a significant economic incentive for Florida's position, but the environmental elements are unmistakable. In any event, the water wars have made their way east, and they represent a new and complicated issue on the horizon of water law.

C. Grady Moore, *Water Wars: Interstate Water Allocation in the Southeast,* 14 Natural Resources & Env't 5 (1999).

Notes and Questions

1. Put yourself in the following circumstances, changing which ACF stakeholder has hired you, and consider what you would propose:

- You have been hired by all the stakeholders as a group to mediate a solution among them.

- You have been hired to negotiate in the mediation for one of the stakeholders.

- You have been hired by one of the stakeholders to draft federal and state legislation that best ensures its needs are met and to lobby for its adoption.

- You have been hired by one of the stakeholders to lobby the Corps of Engineers to convince it to act in your client's best interests.

- You are an aide to the governor of one of the states involved–Georgia, Alabama, or Florida–asked to outline the best overall position for the state to take in negotiations with the other states.

2. What is the relevant ecosystem for tackling any of the assignments outlined in the previous note? Is it the entire ACF watershed? Should particular stretches of the system—say, the segments between dams—be considered separately? Are farmers who draw from the Floridan aquifer in south Georgia in or out of the ecosystem?

3. Remember *Lamsilis subangulata*? Better known, to a few people at least, as shinyrayed pocketbook, it is one of the mussels found in the ACF system that has been listed as an endangered species under the Endangered Species Act (see its description in Chapter 2). Threats to the endangered mussel species living in the ACF include excessive nutrient levels from agriculture, siltation from land deforestation, toxic runoff from industry, loss of habitat to reservoir inundation, decreased water flow due to diversions, and competition from invasive species. None of the species is found in the Chattahoochee today. Which of the interests outlined above could make the best use of these species in advancing its objectives? How would you go about doing so?

4. If the three states fail to reach an agreement on water allocation in the ACF system, they could take their fight to the U.S. Supreme Court. If the dispute over ACF water were to reach the Supreme Court, the Court would decide the issue using the federal common law doctrine of equitable apportionment. The Constitution assigns the Supreme Court original jurisdiction over disputes between the states. U.S. Const art. III, sec. 2, cl. 2. Because the states have adopted a range of legal regimes for dealing with intrastate water rights, none of which takes into account impact on other states, the Court has developed federal common law principles for dividing the water when interstate disputes arise. *See* Hinderlider v. LaPlata River & Cherry Creek Ditch Co., 304 U.S. 92 (1938). The basic theme of the Court's approach is to divide the interstate water so as to balance benefits and injury with a sense of fairness to both states. This doctrine of equitable apportionment takes into account not only each state's respective water law, but also economic impacts, climate conditions, available water use conservation measures, and the overall impact of diversions on existing uses. The Court appoints a Special Master to engage in this multi-factored analysis, then reviews the Special Master's recommendations. The doctrine has long been employed in the west, *see, e.g.*, Kansas v. Colorado, 206 U.S. 46 (1907), and has occasionally been used to resolve disputes between eastern states, *see e.g.*, Connecticut v. Massachusetts, 282 U.S. 660 (1931). But no case has presented issues quite like those the ACF case would pose. Usually the Court is called upon to decree an annual amount or minimum flow to which each state is entitled. In the ACF case, however, Florida presumably would claim that, primarily for ecological reasons (albeit with incidental economic impacts), upstream states must deliver a particular "natural" flow regime that fluctuates throughout the year.

5. Although the Court's equitable apportionment jurisprudence certainly leaves room for incorporating ecological factors into the analysis, the precedents do not suggest how the Court would do so. The Court has, however, ruled that the doctrine applies not only to water, but to allocation of resources that run within interstate waters, such as anadromous fish. *See* Idaho v. Oregon, 462 U.S. 1017 (1983). And the Court has held that the doctrine imposes on states "an affirmative duty ... to take reasonable steps to conserve and even to augment natural resources within their borders for the benefit of other States." *Id.* at 1025. Yet, when downstream states claim injury from upstream diversions, the Court generally requires the downstream state to "prove by clear and convincing evidence some real and substantial injury or damage." *See id.* at 1027; Missouri v. Illinois, 200 U.S. 496, 521 (1906) (first holding that the doctrine requires a showing that the injury is of "serious magnitude clearly and fully proved"). How would you argue on behalf of Florida that this standard has been met?

6. If the Supreme Court were asked to divide ACF water between the states, to what extent would the concept of ecosystem services (developed more fully in Chapter 7) be useful in the Court's determination? After all, isn't that why Florida wants ACF water—for the services it provides to Apalachicola Bay? If the Court can equitably apportion a river's water and the fish that travel in it, couldn't the Court also equitably apportion the ecosystem services the water delivers? And as a general matter, should the doctrine of equitable apportionment be expanded to allow, or even to require, that interstate river disputes be resolved according to an *ecologically* equitable apportionment? For a suggestion of such an approach for international water apportionment, see A. Dan Tarlock, *Safeguarding International River Ecosystems in Times of Scarcity*, 3 U. DENVER WATER L. REV. 231 (2000).

7. For an excellent background on the ACF dispute, including a comparison of the three principal solution options—congressional legislative allocation, interstate compact allocation, and Supreme Court equitable apportionment—see Dustin S. Stephenson, *The Tri-State Compact: Falling Waters and Fading Opportunities*, 16 JOURNAL OF LAND USE AND ENVTL. L. 83 (2000). An excellent series of articles on the situation by environmental journalist Bruce Ritchie appeared in the *Tallahassee Democrat*, Nov. 4–12, 2001.

CHAPTER 11

COASTAL AND MARINE ECOSYSTEMS

Chapter Outline
A. Biodiversity in Coastal and Marine Ecosystems
B. Primer on the Law of Coastal and Marine Ecosystems
    1. Historical Background
    2. Current Scene
C. Biodiversity Management in Coastal and Marine Ecosystems
    1. Beaches and Coastal Lands
        a. State Land Management Frameworks
        b. Federal Agency Consistency
    Note on the Public Trust Doctrine
    2. Estuaries
        a. The National Estuary Program
        b. The Chesapeake Bay Agreement
    Note on the Bay–Delta Agreement
    Note on Hypoxia in Coastal Waters
    3. Living Resources
        a. Fisheries Management
            1. Regulating Fish Harvests
            2. Protecting Fish Habitat
    Note on the Marine Mammals Protection Act
        b. Marine Protected Areas
D. Case Study: The Plight of the White Marlin

A. BIODIVERSITY IN COASTAL AND MARINE ECOSYSTEMS

Coastal and marine ecosystems, which cover over 70 percent of the earth's surface, pose the irony of being both highly productive in terms of goods and services provided *to* humans, and highly sensitive to the assaults caused *by* humans. Both conditions are the result of the sheer complexity of the matrix of ecological factors we call the coastal zone—the beaches, bays, estuaries, tidelands, reefs, and other features that form the transition between terrestrial ecosystems and the open ocean. The complex ecosystem dynamics of the coastal zone involve immense inputs of water, nutrients, and sediments from terrestrial and freshwater sources, plus cold, nutrient-rich sea water from upwelling currents that run along the continental shelves. The result is an ecological mixing bowl that, combined with the open ocean, accounts by some estimates for over two-thirds of global ecological goods and services. Yet, because the mixing bowl depends on continued inputs from land and sea, human activities in those regions have profound effects on the coastal zone recipe. By all measures humans have altered that recipe dramatically, the only question being whether we have also done so irreversibly.

Anyone who has walked along a shoreline will appreciate the sheer diversity of life and ecological processes found where land meets sea. Estimates are that this thin ribbon of land and water accounts for over one-third of the earth's ecological goods and services, making it by far the most productive ecosystem type per acre. The coastal zones of the United States (excluding Alaska) include 34,000 square miles of estuaries and almost 60,000 miles of ocean shoreline. Estuaries, where freshwater from rivers and streams meets salty seawater, are vitally important to birds, mammals, fish, shellfish and other wildlife. They support habitat for 75 percent of our nation's commercial fish stocks and 75 percent of the listed threatened and endangered birds and mammals. In particular, the wetlands and other land areas ringing the aquatic component of estuaries help to filter pollution and sediment from upland runoff, stabilize the shoreline, provide flood control, and support highly productive plant life and other habitat. Indeed, in terms of net primary production per acre, coastal mangrove forests are perhaps the most productive ecosystem on the planet. Other important systems in the coastal zone include seagrass and algae beds, kelp forests, tidal marshes, and coral reefs.

By comparison, the open ocean is not nearly as productive as the coastal zone ecosystems, but it is far from the void it was once thought to be. United States jurisdiction extends to more than 2.2 million nautical square miles of ocean. Ocean waters over the U.S. continental shelf have always played an important role in fishery harvests. The deep ocean has been far less explored, but is increasingly understood to support a considerable array of pelagic (living in the vertical water column) and benthic (living on the ocean floor) marine life. The ocean itself is also crucially important to overall global health, serving as a sort of planetary biological pump regulating numerous ecological conditions. It is a vast sink for global carbon and nutrients, and its deep currents modulate global weather patterns.

As productive and important to global functioning as it is, the complex of coastal and marine ecosystems is nonetheless under substantial assault by human land-based and water-based activities. Many people fail to appreciate the connections between their land-based actions and the health of coastal and marine ecosystems, and most also are largely unaware of the impacts of water-based industries on coastal and marine resources.

Coastal areas are among the most popular places to live and locate industry in the United States. The coastal zone, defined for demographic purposes as all areas within 50 miles of shoreline, constitutes 17% of the U.S. land area, but is inhabited by more than 53% of the nation's population—over 140 million people. Coastal populations continue to grow, a trend that could result in three-quarters of the U.S. population's living in the coastal zone by 2020. This continuing pressure to develop the coastal zone has resulted in significant conversion of habitat, particularly of coastal wetlands and seagrass beds. But the impacts of land-based development on coastal ecosystems go far beyond direct habitat losses. In its recent comprehensive report on the issue, *Protecting the Oceans from Land-based Activi-*

ties, the United Nations Environment Program explains that by altering runoff flow regimes and coastal land barrier effects, coastal development accelerates erosion of shorelines. Also, urban land uses in the coastal zone, as well as agricultural land uses there and far inland, contribute to runoff of sediment, nutrient, and other pollutants into coastal ecosystems, and direct discharge of treated sewage and industrial effluent visibly impairs coastal waters. For example, the U.S. Environmental Protection Agency's *1994 National Water Quality Inventory* rates nutrients and bacteria as the primary pollutants that are impairing estuaries, with urban runoff, sewage plants, agriculture, and industry discharges, in that order, as the leading sources. Even land-based air pollution, given the ocean's efficiency at absorbing atmospheric chemicals, is increasingly becoming a concern in the marine environment.

Water-based industries have taken their toll on coastal and marine ecosystems as well. The shipping industry discharges and spills millions of tons of oil per year, with obvious and immediate impacts on marine birds, mammals, and fish. The fishing industry, which experienced massive capitalization and technological advances in the 1970s and 1980s, is at risk of fishing itself out of existence. Of 200 marine stocks of fish harvested worldwide, the U.S. National Oceanographic and Atmospheric Administration (NOAA) considers at least a third to be below estimated long term sustainable levels, and a recent survey of the American Fisheries Society identifies 82 North American marine fish species considered at risk of extinction. As one might expect, these are comprised largely of high trophic level species that grow slowly, mature late in life, and reproduce less frequently—generally, the fish humans like to eat for food and prize as recreational trophy catches.

So how bad is it? Is it all doom and gloom? Can the impacts be reversed? The fact is that we don't know. A recent study, the first ever to comprehensively examine the effects of human actions on coastal ecosystems since the rise of modern *homo sapiens*, suggests that long time lags exist between cause and effect and that the effects usually are larger than ever imagined. *See* Jeremy B. C. Jackson et al., *Historical Overfishing and the Recent Collapse of Coastal Ecosystems*, 293 SCIENCE 629 (2001). The researchers conclude that historical overfishing exceeds pollution, habitat loss, and climate change as the leading factor in this story. As overfishing directly removes members of complex food chains, cascade effects through the food chains are substantial but may take decades or longer to play out. Long after a species at a particular trophic level is decimated, others may perish simply as a result of the new food chain dynamic. There is also mounting evidence, however, that marine species are suffering increased disease as the stresses of pollution and ocean warming decrease host resistance and improve conditions for opportunistic parasites and other disease vectors. *See* C.D. Harvell et al., *Emerging Marine Diseases— Climate Links and Anthropogenic Factors*, 285 SCIENCE 1505 (1999). No one knows whether these and other effects are preordained to become worse given what has happened up to now, or whether they could be reversed by action, perhaps drastic action, taken immediately.

Indeed, some may argue the situation is not that grim. Although many of our coastal waters are threatened by the assaults mentioned above, many remain in good condition. Many local economies and industries depend on healthy coastal and marine ecosystems, and thus have an incentive to protect them.

Perhaps the policy measures discussed in the remainder of this chapter already have turned the tide of degradation. There is evidence in both directions. In a recent report, EPA described the ecological and environmental conditions of the nation's coastal waters, showing a mixed bag. *See* USEPA, Office of Water, Clean Water Action Plan: National Coastal Condition Report (2002). EPA summarized its findings as follows:

> Existing data show that the overall condition of the U.S. coastal waters as fair to poor, varying from region to region and that 44% of estuarine areas in the U.S. are impaired for human use or aquatic life use. To determine the overall condition of the Nation's estuaries, EPA measured seven coastal condition indicators, including water clarity, dissolved oxygen, sediments, benthos, fish contamination, coastal wetlands loss, and eutrophication. These indicators were rated in estuaries in each region of the country (northeastern, southeastern, Gulf of Mexico, west coast, and Great Lakes regions). The condition of each resource was rated as good, fair, or poor. The indicators were combined to describe the overall coastal condition for each of the region.

> The northeastern estuaries, Gulf of Mexico and the Great Lakes are in fair to poor ecological condition, while southeastern and west coast estuaries are in fair ecological condition. Water clarity is good in west coast and northeastern estuaries, but fair in the Gulf of Mexico, southeastern estuaries, and the Great Lakes. Dissolved oxygen conditions are generally good and sediment contaminant conditions are generally poor throughout the estuaries and Great Lakes of the United States. Eutrophication in coastal waters is increasing throughout much of the United States and results in poor eutrophic conditions in the Gulf of Mexico, west coast and northeastern estuaries and in fair to good conditions in the remaining estuaries of the continental United States.

> Living resources are in fair condition in estuaries throughout the United States, although small changes in water quality could cause this condition to worsen and result in a poor rating. Living resources in the Great Lakes, northeastern estuaries, Gulf of Mexico and the west coast are currently in poor condition. Contaminant concentrations in fish tissues are low throughout the estuarine waters of the United States with exceptions in selected northeastern estuaries, Gulf of Mexico estuaries and the Great Lakes. Fish consumption advisories exist throughout the Gulf of Mexico and northeastern coastal areas, although these advisories largely pertain to offshore species (e.g., king mackerel).

USEPA, Office of Water, National Coastal Condition Report Fact Sheet (2002), *available at* http://epa.gov/owow/oceans/nccr/nccrfs.html. EPA provided the following "score card" showing the bottom line for each coastal region of the United States:

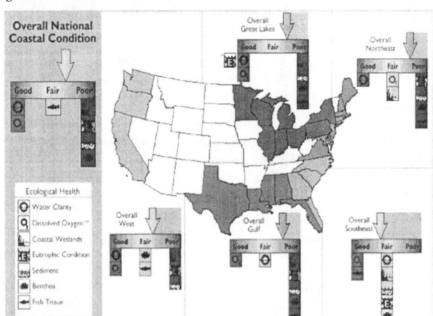

* No indicator data available
** Does not include the hypoxic zone in offshore Gulf of Mexico waters.

Notes and Questions

1. What is your "take away" message from EPA's conclusions? With many estuarine resources in "good" condition for several factors, how concerned should we be about the need for yet more ecosystem management law for coastal and marine settings? How much should we be willing to invest to restore the other factors to good condition?

2. Global warming could have many impacts on fish and other coastal and marine species. Large-scale increases in the heat content of the world's oceans have been observed to occur over the last 45 years, which, according to some studies, can be closely associated with anthropogenically-induced carbon emissions. *See* Tim P. Barnett, *Detection of Anthropogenic Climate Change in the World's Oceans*, 292 SCIENCE 270 (2001). There is historical evidence that ocean temperature variations attributable to natural climate change have affected fishery populations, suggesting that rapid temperature changes due to carbon emissions could do the same. *See* Bruce P. Finney et al., *Impacts of Climatic Change and Fishing on Pacific Salmon Abundance Over the Past 300 Years*, 290 SCIENCE 795 (2000). According to the U.S. EPA, some bodies of water may become too warm for the fish that currently inhabit those areas; but warmer temperatures may also enable fish in cold ocean waters to grow more rapidly. Global warming may also change the chemical composition of the water that fish inhabit: the amount of oxygen in the water may decline, while pollution and salinity levels may increase. Loss of wetlands could diminish habitat and alter the availability of food for some fish species. Higher water temperatures may have the most important implications for inland fisheries, but wetland loss, salinity changes, and higher temperatures are all likely to affect finfish and shellfish in the coastal zone as well. The most vulnerable species are those that either reproduce in coastal wetlands, spend their entire lifetimes in an estuary, or both. Scientists generally expect fish on the high seas to be less affected by global warming than coastal and inland fisheries.

3. Sea level is rising more rapidly along the U.S. coast than worldwide. Studies by EPA and others have estimated that along the Gulf and Atlantic coasts, a one foot (30 cm) rise in sea level is likely by 2050 and could occur as soon as 2025. In the next century, a two foot rise is most likely, but a four foot rise is possible. Sea level will probably continue to rise for several centuries, even if global temperatures were to stop rising a few decades hence. Rising sea level inundates wetlands and other low-lying lands, erodes beaches, intensifies flooding, and increases the salinity of rivers, bays, and groundwater tables. Coastal marshes and swamps are particularly vulnerable to rising sea level because they are mostly within a few feet of sea level. As the sea rises, the outer boundary of these wetlands will erode, and new wetlands will form inland as previously dry areas are flooded by the higher water levels. As EPA points out, however, the amount of newly created wetlands will be much smaller than the area of wetlands that are lost, because the amount of dryland within a few feet above the wetlands is much less than the area of wetlands that would be lost if sea level rises a few feet. Moreover, developed areas are and more often will be protected with bulkheads, dikes, and other structures that will keep new wetlands from forming inland. Many states are taking these prospects quite seriously and beginning now to prepare for management of rising sea levels and diminishing coastal wetlands using many of the authorities covered in this chapter. *See* Proceedings, Rising Tides, Eroding Shores: The Legal and Policy Implications of Sea Level Rise and Coastal Erosion (Univ. Md. Law School, Apr. 20, 2001).

4. Increased human-induced ocean noise is also beginning to show up as a disturbance in coastal and marine environments. Seismic surveys, the whine of motors, test explosives blasts, Navy sonar systems, and a host of other intrusions have made a cacophony out of some coastal and ocean waters. Marine mammal researchers are increasingly concerned that these unnatural noises may disrupt the hearing of whales and other marine mammals, *see* David Malakoff, *A Roaring Debate Over Ocean Noise*, 291 SCIENCE 576 (2001), and litigation has been initiated to challenge some noise-producing activities for alleged failure adequately to take those effects into account. *See, e.g.*, Hawaii County Green Party v. Clinton, 124 F.Supp.2d 1173 (D.Haw.2000) (dismissing suit challenging Navy development and deployment of low-frequency active sonar defense system).

5. Ecosystem management policies in coastal and marine settings have generally followed one of two approaches: the Integrated Coastal Management (ICM) approach and the Large Marine Ecosystem Management

(LME) approach. *See* Roger B. Griffis and Katharine W. Kimball, *Ecosystem Approaches to Coastal and Ocean Stewardship*, 6 ECOLOGICAL APPLICATIONS 708 (1996). ICM techniques focus on coordinating the policy and governance process for a defined eco-region along issue-specific lines. LME, the more science-driven of the two approaches, defines large-scale hydrographic regimes within which management decisions are focused on thematic rather than species-specific objectives. The two methods are not mutually exclusive; indeed, combined they recognize what others have stated as a fundamental reality of ecosystem management—that it is both a political and a scientific endeavor.

6.   We've organized this chapter into subcategories of ecosystems such as beaches, coastal waters, estuaries, and bays. These are, at best, rough geographic divisions that bear only limited ecological validity. The field of oceanography is, essentially, geography for marine ecologists, and at its more advanced levels includes ecological divisions far more detailed than we undertake here. For a lucid account of current thinking in the field, see ALAN LONGHURST, ECOLOGICAL GEOGRAPHY OF THE SEA (1998).

7.   For a survey of the various threats to coastal and marine ecosystems written in the context of law and policy development, see W. M. von Zharen, *Ocean Ecosystem Stewardship*, 23 WM. & MARY ENVTL. LAW AND POLICY FORUM 1, 2–27 (1998). For scientific and socioeconomic discussions see Anne Platt McGinn, *Safeguarding the Health of Oceans*, Worldwatch Paper No. 145 (Worldwatch Inst. 1999); U.S. DEPARTMENT OF THE INTERIOR, U.S. GEOLOGICAL SURVEY, STATUS AND TRENDS OF THE NATION'S BIOLOGICAL RESOURCES, *available at* http://biology.usgs.gov/; U.S. DEPARTMENT OF THE INTERIOR, NATIONAL BIOLOGICAL SERVICE, OUR LIVING RESOURCES 259–284 (1995); NATIONAL ACADEMY OF SCIENCES/NATIONAL RESEARCH COUNCIL, PRIORITIES FOR COASTAL ECOSYSTEM SCIENCE (1994). An excellent, if sobering, survey of case studies and research on rising marine species morbidity and mass mortality events is available in Bruce McKay and Kieran Mulvaney, *A Review of Major Marine Ecological Disturbances*, 18 ENDANGERED SPECIES UPDATE 14 (Jan.–Feb. 2001).

B.   PRIMER ON THE LAW OF COASTAL AND MARINE ECOSYSTEMS

Like the law of the other ecosystems treated in this part of the text, the law of coastal and marine ecosystems evolved initially as an adjunct to the development of property rights and in response to the demands of resource extraction industries. Only recently has anything approaching an ecosystem management model surfaced in this body of laws, and it has done so far less coherently than for the terrestrial and freshwater ecosystems studied in other chapters.

1.   HISTORICAL BACKGROUND

The law of coastal and marine ecosystems has unfolded against a context of property and resource ownership that is more complicated than for terrestrial environments. The general framework in force in the United States today, subject to nuances and exceptions not studied here, is as follows: the private owner of land in the coastal zone owns seaward only to the mean high tide water line; the states control the waters (and lands below them) from that point seaward for three miles; federal jurisdiction begins at that point and extends outward another nine miles to the edge of the internationally-recognized 12–mile limit of the territorial sea, and then still farther, albeit for more limited purposes, to the edge of the 200–mile Exclusive Economic Zone. It has not always been so, however, as international law long held that national jurisdiction ended at a three-mile territorial sea limit. All ocean resources beyond that point were a common pool resource for which open access by all nations was the default rule, subject to modification by international law. The federal government thus was largely excluded from a meaningful role in coastal and marine ecosystem management law—the states dominated along the coastal zone, including the nation's then 3–mile territorial sea, and beyond that only international law could regulate anyone's actions.

That traditional set of affairs was rocked in the 1970s by several developments. Domestically, of course, generic environmental laws such as the Clean Water Act, Clean Air Act, and Endangered Species Act increased federal power everywhere, including the coastal zone. But another new federal law came on line during that period to address coastal issues in particular. The Coastal Zone Management Act of 1972 (CZMA), 16 U.S.C. § 1451–65, established a cooperative venture between the federal and coastal state governments whereby states that agree to regulate coastal development, pursuant to nationally-prescribed goals, receive federal funding and a federal commitment not to act inconsistent with the state's coastal management program. The federal government thereby entered into the coastal arena without upsetting the dominant role states had historically enjoyed.

Just as the CZMA was beginning its tenure, the culmination of movement in the international law forum led to enactment of the other principal federal marine ecosystem legislation, the Magnuson–Stevens Fishery Conservation and Management Act of 1976 (Magnuson–Stevens Act), 16 U.S.C. § 1801–1882. President Truman dealt the first blow to the traditional international rule of open access when he declared in 1945 that the United States would exercise regulatory control over "conservation zones in those areas of the high seas contiguous to the coast of the United States wherein fishing activities have been or in the future may be developed and maintained on a substantial scale." Presidential Proclamation No. 2667, 10 Fed. Reg. 12303 (1945). In 1964 Congress enacted the Bartlett Act, Pub. L. No. 88–308, 78 Stat. 194 (1964) (repealed 1977), to prohibit foreign vessel fishing in the United States' territorial sea. The Contiguous Fisheries Zone Act, Pub. L. No. 88–308, 78 Stat. 194 (1964) (repealed 1977), followed in 1966 to formally adopt a 9–mile wide "contiguous zone" beginning at the outer limit of the states' 3–mile territorial sea, but the law established no governance framework for the area beyond state waters. Most other nations made similar or more extensive claims, but by the 1970s most nations were still dissatisfied with the international law of fisheries management generally. Because of the lack of enforcement provisions in most of the bilateral and multilateral international fishing agreements that had been hammered out, open access remained the *de facto* rule of law. Work thus began in the United Nations 1974 Law of the Sea Conference to develop a uniform rule of a 12–mile territorial sea and a 200–mile fishery conservation zone.

Displeased with the Conference's progress on this issue, in 1976 the United States enacted the Magnuson–Stevens Act, unilaterally declaring exclusive fisheries jurisdiction in the 200–mile fishery conservation zone and certain portions of the continental shelf beyond that, and establishing a dominant federal role for governance in the new jurisdiction. Several years later the Law of the Sea Conference settled on the respective 12–mile and 200–mile limits, and President Reagan renamed the fisheries zone the Exclusive Economic Zone (EEZ). Thus the present jurisdictional configuration was not in place, with all the terminology applicable today, until 1982. This explains the rather late start the federal government got in managing coastal and marine fishery resources.

Somewhat like the CZMA in its reliance on shared local and federal roles, the Magnuson–Stevens Act establishes eight Regional Fishery Management Councils to design and implement fisheries management plans that meet nationally-prescribed goals. Also like the CZMA, the Magnuson–Stevens Act did not intrude significantly on the traditional powers of the states to regulate fishing in their coastal waters. Only in extremely limited circumstances may federal regulation preempt state authority.

Since their original enactment, both the CZMA and the Magnuson–Stevens Act have received significant amendments the effect of which, arguably, has been to move them closer to the ecosystem management approach. The CZMA was amended in 1990 by the Coastal Zone Act Reauthorization Amendments (CZARA). Pub. L. No. 101–508, tit. VI, subtit. C, 104 Stat. 1388 (1990). Section 6217 of CZARA added a special provision to the CZMA dealing with coastal nonpoint source pollution. *See* 16 U.S.C. § 1455b. The Section 6217 program, as it is often called, requires states with coastal management programs to also develop a coastal nonpoint pollution control program to retain federal funding. CZARA thus moved the CZMA beyond its original focus on development in coastal areas, to recognizing the important indirect effects of land-based activities on the coastal zone. Similarly, the Magnuson–Stevens Act was amended most recently and most substantially through the Sustainable Fisheries Act of 1996 (SFA). Pub. L. No. 104–297, 110 Stat. 3559 (1996). The SFA amendments expanded federal authority to identify, respond to, and regulate overfishing, required fishery management plans for the first time to identify "essential fish habitat,"and added to the national criteria by which fishery plans are measured. The amendments thus recognized that fishery management will usually involve not only regulation of harvests, but also conservation of habitat. After their amendments, therefore, both the CZMA and the Magnuson–Stevens Act incorporated the more holistic approach characteristic of ecosystem management. The details of both laws, as amended, as well as a host of related laws and initiatives, are covered in later sections of this chapter.

2.   CURRENT SCENE

Although they may provide a foundation, neither the CZMA nor the Magnuson–Stevens Act can fairly be described as having had comprehensive ecosystem management foremost in mind at their inception. The subsequent amendments to both laws added tidbits of ecosystem focus, but still, by 1990 we lacked a comprehensive national ecosystem management policy for coastal and marine ecosystems.

Responding to this gap, and in concert with its overall push to reorient environmental policy toward ecosystem management, the Clinton Administration initiated several ecosystem management initiatives for the coastal and marine settings. The EPA took its watershed approach, developed initially for freshwater ecosystems, seaward in its *Coastal Watershed Protection Strategy. See* U.S. EPA, *Estuaries and Your Coastal Watershed,* www.epa.gov/owow/oceans. The U.S. National Oceanographic and Atmospheric Administration (NOAA) teamed up with a group of conservation organizations known as Restore America's Estuaries to develop a *National Strategy to Restore Estuarine Habitat,* a collaborative effort of federal, state, scientific, and community leaders to develop estuary restoration methods and goals. *See* Steve Emmett–Mattox, *Restoration at the Edge—A National Strategy for America's Estuaries,* 23 NATIONAL WETLANDS NEWSLETTER 7 (July–Aug. 2001). Indeed, this effort actually has resulted in something tangible—in 2000 the Estuary Restoration Act was enacted to coordinate estuary restoration plans and to fund $275 million in federal money toward estuary restoration over the next 5 years. *See* 33 U.S.C. § 2901; Pub. L. No. 106–457. NOAA also recently developed final regulations, excerpted later in this chapter, to guide Regional Fishery Management Councils in the identification of essential fish habitat under the Sustainable Fisheries Act amendments to the Magnuson–Stevens Act. *See* 67 Fed. Reg. 2343 (Jan. 17, 2002) (amending 50 C.F.R. part 600). Under authority of an Executive Order President Clinton issued in 2000, NOAA also is directing a coordinated program to assemble and manage Marine Protected Areas–sanctuaries and other preserves designed to replicate in the marine ecosystem what wilderness and other preserve designations accomplish on land. Finally, EPA and NOAA together have overseen review and approval of the first submitted state Coastal Nonpoint Pollution Control Programs, as required by section 6217 of the CZARA (discussed above).

These developments, many of which are covered in more detail later in this chapter, represent the increased focus coastal and marine ecosystems have received in the past decade. It remains to be seen, however, where these developments will lead. None alone constitutes a comprehensive national ecosystem management policy for coastal and marine ecosystems, and any might be redirected by the Bush Administration, the current Congress, or their respective successors.

Notes and Questions

1. We have a U.S. Forest Service to oversee comprehensive management of public national forests (see chapter 8) and a Bureau of Land Management to oversee comprehensive management of federal public range lands (see chapter 9). The closest federal equivalent for coasts and oceans is the National Oceanographic and Atmospheric Administration (NOAA), a division of the Department of Commerce. NOAA's strategic plan assigns responsibility for much of its research and conservation of the ocean's living marine resources to the National Marine Fisheries Service (NMFS), also known as NOAA Fisheries. NMFS implements six statutes pertinent to coastal and marine ecosystem management: the Magnuson–Stevens Fishery Conservation and Management Act, which regulates fisheries within the U.S. Exclusive Economic Zone (see this chapter); the Endangered Species Act, which protects threatened or endangered species (see Part II); the Marine Mammal Protection Act, which regulates the taking or harassment of marine mammals (see this chapter); the Lacey Act, which prohibits fishery transactions that violate state, federal, American tribal, or foreign laws; the Fish and Wildlife Coordination Act, which authorizes NMFS to collect fisheries data and to advise other agencies on environmental decisions affecting living marine resources; and the Agricultural Marketing Act, which authorizes a voluntary seafood inspection program. NMFS implements these programs principally through several divisions: the Sustainable Fisheries Division, the Protected Resources Division, and the Habitat Conservation Division.

Elsewhere within NOAA, the National Ocean Service's Office of Ocean and Coastal Resource Management (OCRM) implements two additional laws that are integral to coastal and marine ecosystem management, both of which are covered in this chapter: the Coastal Zone Protection Act and the Marine Protection, Research, and Sanctuaries Act. Through this authority OCRM administers 21 Estuarine National Research Preserves and 12 National Marine Sanctuaries, and is the federal representative in 29 state coastal management plans covering over 99 percent of the nation's coastlines.

Overall, therefore, NOAA's combined responsibilities bring it close to being the coastal and marine equivalent of the major federal land management agencies. A significant difference is that NOAA does not have direct jurisdiction over the mineral resources found in the EEZ, that being the responsibility of the Department of the Interior's Minerals Management Service under authority of the Outer Continental Shelf Lands Act, 43 U.S.C. § 1331–1356.

2. State ownership of and jurisdiction over the coastal zone was long assumed a right of statehood for the original states as successors to the King's interests in tidelands and the adjacent seas, and for later admitted states based on the doctrine of equal footing, which provides that states admitted into the Union after adoption of the Constitution are entitled to the same rights as the original states in their tidal waters and submerged lands. *See* Pollard's Lessee v. Hagan, 44 U.S. 212 (1845). It was widely assumed that this sovereignty extended not only to the states' inland submerged lands, but also to their claimed three-mile band of coastal waters. State constitutions and statutes treated this arrangement as uncontroversial well into the 20th century, as did the federal government and, it appeared, the United States Supreme Court. This seemingly settled state of affairs was thrown asunder in United States v. California, 332 U.S. 19 (1947), and United States v. Texas, 339 U.S. 707 (1950), wherein the Court held that it was in fact the federal government that had established claim to the three-mile territorial sea and thus had paramount rights to the entire continental shelf, including the minerals in the seabeds beneath what previously had been assumed to be these states' three-mile territorial seas. The states, in other words, had no title or interest in the territorial sea or its submerged lands off their respective coasts after all. This ruling, understandably, caused great controversy. After some congressional jockeying with President Truman over terms, the United States ceded these rights back to the states under the Submerged Lands Act of 1953, while retaining its navigation rights and the rights to regulate within its constitutional scope of power over interstate commerce, navigation, national defense, and international affairs. See 43 U.S.C. §§ 1301–1315. Then, to make the extent of federal power and jurisdiction clear, Congress adopted the Outer Continental Shelf Lands Act, mentioned in the previous note, to establish federal ownership and management of the submerged lands seaward of the three-mile state coastal jurisdiction line. As a nuance to this strange jurisdictional history, the Submerged Lands Act allowed states to make claims beyond the three-mile line based on historical claims. The Supreme Court later determined that the historic boundaries of Texas (based on its claims as an independent nation) and Florida (based on congressional approval of its constitution) along the Gulf of Mexico extended three marine leagues (about 10 miles) from the coast. See United States v. Florida, 363 U.S. 21 (1960); United States v. Louisiana, 363 U.S. 1 (1960). The final turn of events was the successful suit the United States brought in 1975 against the Atlantic Coast states to extend to those states the reasoning of United States v. California., i.e., that their rights to territorial seas flowed not as a right of statehood, but from the federal government's cession of its paramount rights under the Submerged Lands Act. See United States v. Maine, 420 U.S. 515 (1975).

3.   As messy as the history of international, national, and state boundary lines has been, the boundary between private and public land ownership along the coast can also be just as complicated. As a general rule, the states, either by original statehood or the equal footing doctrine, have title to lands subject to the ebb and flow of the tides on the basis that they are not capable of private development and that their general uses, such as navigation and fishing, are public in nature. See Shively v. Bowlby, 152 U.S. 1 (1894). And, as a general rule, most states follow the federal rule that the state owns these lands seaward of the mean high tide of all high tides. See Borax Consolidated v. City of Los Angeles, 296 U.S. 10 (1935). But if tidelands were conveyed into private hands by the United States prior to statehood, or were conveyed by other nations prior to acquisition by the United States through cession or treaty, the private landowner may have a claim seaward of the high tide line, though such grants are narrowly construed. *See* The Rebeckah, 1 C. Rob. 227. And some states take the position that they do not own tidelands that are not under waters navigable in fact. *See, e.g.,* Lee v. Williams, 711 So.2d 57 (Fla.5th Dist.Ct.App. 1998) (Florida includes only tidelands under navigable waters as sovereign lands). As for attempts by states to convey tidelands and other submerged lands they own into private hands, see the discussion of the Public Trust Doctrine later in this chapter.

4.   Some states have implemented coastal conservation measures going far beyond anything required in federal law. Indeed, South Carolina took such aggressive measures to protect beaches that it wound up in the U.S. Supreme Court. In 1988, the state enacted its Beachfront Management Act, which had the direct effect of barring a beach lot owner, who had purchased his lots before the new law was enacted, from erecting any permanent structures on his property. At the time, the legislation provided for no waivers or permits. A landowner alleged that the state had taken his property without just compensation in violation of the Fifth Amendment. The Supreme Court held that, because the state had denied all economically or beneficial use of all the land in question, it had effected a regulatory taking unless, in the state courts' view, the new law functionally did no more than codify the state's background common law nuisance restrictions on the land. *See* Lucas v. South Carolina Coastal Council, 505 U.S. 1003 (1992). For a discussion of South Carolina's beach conservation efforts before and since the *Lucas* case, see Douglas T. Kendall, *Preserving South Carolina's Beaches: The Role of Local Planning in Managing Growth in Coastal South Carolina*, 1 SOUTH CAROLINA ENVTL. L.J. 61 (2000).

5.   Coastal states also have a long history of regulating recreational and commercial fishing in their waters. Ironically, this is what recently landed Massachusetts in hot water under the Endangered Species Act. The state established a licensing program for commercial gillnets and lobster pots. The state's licensees, however, fished in waters also occupied by right whales, an endangered species protected under the ESA. Because the use of the licensed gear was associated with whale deaths, a federal appellate court found that the state could be held liable for taking the whales in violation of the ESA because its licenses were a "but for" causative factor in the illegal takes. *See* Strahan v. Coxe, 127 F.3d 155 (1st Cir.1997). What incentive does such vicarious liability present to states considering whether to regulate fishing in their waters?

6.   International law has also entered into the domestic law development scene in recent years. In June 1992, at the United Nations Conference on Environment and Development (UNCED) in Rio de Janeiro, participating nations adopted the so-called Rio Declaration to formally endorse the concept of sustainable development and to commit to formulating national plans of action for achieving it. *See* Rio Declaration on Environment and Development, UNCED, U.N. Doc. A/CONF.151/Rev. 1, 31 I.L.M. 874 (1992). UNCED's blueprint for what national sustainable development plans should contain is known as the Agenda 21 document, which covers an array of topics and provides guiding principles of sustainable development. See UNCED Agenda 21, U.N. Doc. A/CONF.151.26 (1992). Chapter 17 of Agenda 21 addresses ocean resources, recognizing that "[t]he marine environment–including the oceans and all seas and adjacent coastal areas– forms an integrated whole that is an essential component of the global life-support system and a positive asset that present opportunities for sustainable development." *Id*. § 17.1. The goal of all nations, according to Agenda 21, thus should be to devise "new approaches to marine and coastal areas management and development, at the national, subregional, regional, and global levels, . . . that are integrated in content and are precautionary and anticipatory in ambit." Many of the domestic U.S. laws and initiatives discussed in this chapter have been adopted or amended since Agenda 21 was promulgated. As you review them, consider how these programs live up to the expressed Agenda 21 goal. For a comprehensive assessment in that regard, notable not only for its breadth of programmatic coverage but also its depth of analysis of each program, see Robin Kundis Craig, *Sustaining the Unknown Seas: Changes in U.S. Ocean Policy and Regulations Since Rio '92*, 32 Envtl. L. Rep. (Envtl. L. Inst.) 10190 (2002). *See also A Review of Developments in U.S. Ocean and Coastal Law 1994–1996*, 2 Ocean and Coastal L.J. 457 (1997).

7. This chapter focuses primarily on the handful of federal laws and initiatives that can be cobbled together into an ecosystem management framework for coastal and marine ecosystems. For a comprehensive survey of the numerous other federal and state laws that apply to some degree or another to coastal and marine ecosystem issues, see W.M. von Zharen, *Ocean Ecosystem Stewardship*, 23 Wm. & Mary Envtl. Law and Policy Forum 1, 2–27 (1998).

C. Biodiversity Management in Coastal and Marine Ecosystems

Because of its wide geographic spread, the United States enjoys a vast diversity of marine ecosystems in five major regions: Northeast, Southeast, Alaska, Pacific coast, and western Pacific oceanic. While each of these regions presents familiar components of marine ecosystems—beaches, tide-lands, estuaries, continental shelf waters, and open ocean—each is distinct given the broad array of climate and geography our nation spans. There may be a common set of problems, including beach erosion, nonpoint source pollution, and overfishing, but solutions will vary by location. The topic of coastal and marine ecosystem management thus is a large umbrella under which many laws and polices will fit.

We have chosen a rather obvious, geographically-focused approach for organizing the materials you will study—start at the coastline and work out to the open seas. By and large, this also happens to be approach the applicable legal framework takes. It is, of course, a flawed approach, as the coastal and marine environment comprises one of the most open set of ecosystem dynamics imaginable. Indeed, even what happens well inland of the coastline, in terrestrial and freshwater ecosystems, can have a profound influence on coastal and marine systems. Hence, as you review the following materials, consider how the legal framework (and thus future editions of this text) might be reconceived to more effectively achieve a coherent policy of coastal and marine ecosystem management.

1. BEACHES AND COASTAL LANDS

The Coastal Zone Management Act (CZMA), 16 U.S.C. §§ 1451–1465, is the nation's primary foundation for beach and coastal area conservation. The CZMA authorizes the Department of Commerce to administer a federal grant program to encourage coastal states to develop and implement coastal zone management programs for the purpose of protecting, developing, and enhancing coastal zone resources, which include wetlands, flood plains, estuaries, beaches, dunes, barrier islands, coral reefs, and fish and wildlife and their habitat. The coastal states include any bordering an ocean or the Gulf of Mexico, Long Island Sound, or the Great Lakes. The coastal zone within these states includes coastal waters and adjacent shorelands in proximity to the shoreline. It includes islands, transitional and intertidal areas, salt marshes, wetlands, and beaches. *Id.* § 1453(1). The objectives of the grant program and related CZMA provisions are to improve the management of the coastal zone resources within those states, which necessarily includes managing private and public land development actions in the coastal zone.

Two features set the CZMA apart from many other federal natural resource management laws. First, it relies heavily on states to implement national policy through state-designed land management decision making frameworks. Second, it obligates federal agencies to implement their respective actions in a manner consistent with state coastal management programs. The result is a form of ecosystem management that is quite decentralized but which reaches a broad array of actors and actions.

a. STATE LAND MANAGEMENT FRAMEWORKS

One of the congressional findings supporting the CZMA was that

The key to more effective protection and use of the land and water resources of the coastal zone is to encourage the states to exercise their full authority over the lands and waters in the coastal zone by assisting the states, in cooperation with Federal and local governments and other vitally affected interests, in developing land and water use programs for the coastal zone, including unified policies, criteria, standards, methods, and processes for dealing with land and water use decisions of more than local significance.

16 U.S.C. § 1451(i). The CZMA thus outlines a national policy on coastal zone management, but allows states to devise their own plans for fulfilling those goals and to use state law to implement them. To receive federal

approval, which triggers the requirement of federal agency consistency (discussed in the next section), a state coastal management program must describe "permissible land uses and water uses within the coastal zone" and "the means by which the state proposes to exert control over the land uses and water uses ... including a list of relevant State constitutional provisions, laws, regulations, and judicial decisions." *Id.* § 1455(d)(2). This legal framework must provide for "adequate consideration of the national interest involved in planning for, and managing the coastal zone, including the siting of facilities ... which are of greater than local significance," *id.* § 1455(d)(8), and it must include "procedures whereby specific areas may be designated for the purpose of preserving or restoring them for their conservation, recreational, ecological, historical, or esthetic values." *Id.* § 1455(d)(9).

So long as the national goals are satisfactorily addressed, the states have considerable latitude in the design of their land use management frameworks. The CZMA lays out three general schemes from which the states can choose: (1) local implementation of the state plan and state-promulgated standards, subject to state review; (2) direct state regulation; and (3) state review of state and local decisions for consistency with the state plan. The intensity of land use regulation can vary under any of these approaches, and some states go well beyond the minimum necessary scope of regulation to implement the CZMA's national goals, albeit others do not. The following case provides an example of a state coastal management program that takes the more aggressive approach.

Kirkorowicz v. California Coastal Commission,

Court of Appeal, Fourth District, Division 1, California, 2000.
100 Cal.Rptr.2d 124.

FACTUAL AND PROCEDURAL BACKGROUND

The Kirkorowiczes' 21.47 acre property consists of three separate legal parcels. It is irregularly shaped, zoned rural/residential permitting one single-family residential unit per parcel and private stables by right, and located in the Olivenhain community of the City of Encinitas (City). This periodically flooded property lies north of San Elijo Lagoon Preserve within the 100–year floodplain of the Escondido Creek, which forms its eastern boundary. Consistent with the Encinitas General Plan Policies, which encourage equestrian activities in the floodplain, the Kirkorowiczes currently board horses on their property using corrals and fences.

In 1994, the Kirkorowiczes applied to City for a [City Development Permit (CDP)] to board a maximum of 42 horses and to construct an approximately 1,728 square foot enclosed stable, a storage area for supplies and manure, a driveway and a car/horse trailer turnaround area. The proposal included a 27,000 square foot building pad to be created by placing 8,700 cubic yards of fill on the site. City hired biologist Vincent N. Scheidt to evaluate the permit application, to survey the biological resources and to determine whether the proposal would impact wetlands. In a series of reports between October 1995 and July 1996, Scheidt determined that parts of the project were proposed to be developed in wetlands; concluded the project would have a direct, minor impact on marginal wetland habitat areas; identified the wetlands plant species found onsite; mapped the wetlands onsite; and concluded the project would result in a direct loss of .44 acre of jurisdictional wetlands, but that the potentially significant loss was mitigable. On behalf of the Kirkorowiczes, John W. Brown, senior biologist with Dudek & Associates, visited the site in March 1996 and concluded that using United States Army Corps of Engineers' standards for determining jurisdictional wetlands, "wetland hydrology is absent from this portion [the locus of the proposed fill] of the site." However, the Kirkorowiczes' habitat mitigation plan conceded that project implementation would permanently impact .44 acre of very low-quality wetland. Given the historical use of the property as grazing land, the wetlands resource was described as "degraded" in the environmental documentation.

The original project site plan proposed two driveways. Due to the horizontal curve along Manchester Avenue and the obstruction of visibility due to the dirt embankment on Manchester Avenue across from the project site, the original driveway design did not meet City standards for safe stopping sight distance. Consequently, the project was revised to place a single driveway at the southerly end of the building pad that met sight-line requirements for safe ingress from and egress to Manchester Avenue. During City review, the Kirkorowiczes agreed to reduce the proposed wetlands fill and move the pad area closer to the northern boundary to minimize possible wetlands and visual impacts. Later, before the Commission, they further agreed to reduce the area of wetlands impact to .35 acre.

On January 9, 1997, the City Planning Commission (CPC) approved the project, concluding the project would not impose any significant environmental impact that would not be reduced to a level below significant by the required mitigation measures established by the EIA. The San Elijo Lagoon Conservancy, a local preservation group, appealed the CPC's decision to the City Council. On May 14, the City Council approved the project and the issuance of a CDP for construction of the 1,728 square foot stable facility to board up to 39 horses, involving placement of approximately 8,700 cubic yards of fill in approximately .44 acre of wetlands within the 100–year floodplain. On June 2, the matter was appealed to the Commission. On April 8, 1998, by a vote of nine to zero, the Commission followed its staff's recommendation and denied the project as inconsistent with City's certified [Local Comprehensive Plan (LCP)] regarding floodplain development (Land Use Policy 8.2 of the Land Use Plan (LUP)) and protection of wetlands (LUP Resource Management Element Policy 10.6). Specifically, the staff concluded the proposed fill of wetlands to accommodate vehicle access and a turnaround is not permitted under the LCP and that other alternatives to provide safe access to the site and avoid filling the wetlands had not been adequately explored.

The Kirkorowiczes' petition for a writ of administrative mandamus challenged the Commission's decision on several grounds. However, by the time of the hearing on May 7, 1999, the parties agreed that in light of this court's decision in Bolsa Chica Land Trust v. Superior Court (1999) 71 Cal.App.4th 493, 83 Cal.Rptr.2d 850, the Kirkorowiczes' intended construction of a stable under an implied historic use exception was no longer permitted in the wetlands and thus their argument focused on whether their project would affect jurisdictional wetlands. In granting the petition, the trial court declared:

> In this matter, it appears to the Court that the Coastal Commission conducted its proceedings operating under the presumption that the subject property was a protected wetland. However, after having reviewed the administrative record, the Court finds that there is not substantial evidence to support a finding that the Kirkorowicz' property is a protected wetland as contemplated by the Coastal Act, § 30121 and the City's Local Coastal Program Policy 10.6. Accordingly, absent substantial evidence to support the conclusion that the property involves a wetland, the decision denying the CDP was erroneous.

On July 19, 1999, the judgment issuing the peremptory writ was filed, directing the Commission to set aside its April 8, 1998 decision denying the CDP, to rehear the matter and to determine whether there are protected wetlands on the Kirkorowiczes' property. The Commission timely appealed.

THE STANDARD OF REVIEW

Because this matter came to the trial court on a petition for a writ of mandate under Code of Civil Procedure section 1094.5, that court was required to determine whether substantial evidence supported the Commission's findings and whether those findings supported its decision. * * * In determining whether substantial evidence supports the Commission's decision, we look to the "whole" administrative record and consider all relevant evidence, including that evidence which detracts from the decision. Although this task involves some weighing to fairly estimate the worth of the evidence, that limited weighing does not constitute independent review where the court substitutes its own findings and inferences for that of the Commission. Rather, it is for the Commission to weigh the preponderance of conflicting evidence, as we may reverse its decision only if, based on the evidence before it, a reasonable person could not have reached the conclusion reached by it.

THE RECORD CONTAINS SUBSTANTIAL EVIDENCE THAT WETLANDS EXIST ON THE KIRKOROWICZES' PROPOSED PROJECT SITE

Emphasizing wetlands are areas within the coastal zone that may be covered periodically or permanently with shallow water and are characterized by hydric soils and/or hydrophytes, the Commission contends the administrative record contains substantial evidence there are wetlands on the Kirkorowiczes' property. It asserts the area in which the Kirkorowiczes intend to put their driveway and turnaround is covered periodically with shallow water, is marked by the presence of hydrophytes, and was identi-

fied by City's consulting biologist as wetlands. In response, the Kirkorowiczes echo their plea below that in light of our decision in Bolsa Chica Land
Trust v. Superior Court, supra, "absolutely prohibiting" the Commission
from approving any encroachment into wetlands, except for the enumerated uses permitted by section 30233, it is incumbent upon the Commission
to take a much closer look at what, in fact, constitutes protected wetlands
and thus their case requires reevaluation. They assert that the occasionally
wet area on their property is not remotely of the character the Coastal Act
was intended to protect. Rather, they contend the Act was intended to
protect areas predominantly much wetter than the area in dispute. They
argue the periodic coverage of an area with water is not determinative of
whether property constitutes a wetland; the wet areas on their property are
not part of a marine environment; the mere presence of some hydrophytes
is not determinative; and there is no substantial evidence of wetlands
hydrology or hydric soils on their property.

The Coastal Act defines "wetland" in section 30121 as "lands within
the coastal zone which may be covered periodically or permanently with
shallow water and include saltwater marshes, freshwater marshes, open or
closed brackish water marshes, swamps, mudflats, and fens." Although the
Commission relies on the cited definition to identify "wetlands," by necessity it has expanded the definition, given the highly variable environmental
conditions along the California coast rendering some wetlands not readily
identifiable by simple means. In its expanded definition, the Commission
includes [in guidelines] those lands where "saturation with water is the
dominant factor determining the nature of soil development and the types
of plant and animal communities living in the soil and on its surface."
Consequently, the Commission's regulations further describe wetlands as
"land where the water table is at, near, or above the land surface long
enough to promote the formation of hydric soils or to support the growth of
hydrophytes ..." (Cal.Code Regs., tit. 14, § 13577(b)(1)), thus recognizing
"the presence or absence of hydrophites [sic] and hydric soils are the most
useful, but not exclusive, criteria to determine whether a particular parcel
is a 'wetland.' " The Commission explains:

> "channels, lakes, reservoirs, bays, estuaries, lagoons, marshes, and the
> lands underlying and adjoining such waters, whether permanently or
> intermittently submerged, to the extent that such waters and lands
> support and contain significant fish, wildlife, recreational, aesthetic, or
> scientific resources." (§ 5812, subd. (a).)

* * *

Our review of the administrative record convinces us that substantial
evidence supports the Commission's finding that protected wetlands exist
on the Kirkorowiczes' property. Guided by the foregoing broad definition of
wetlands, substantial evidence shows the proposed site requiring fill is
periodically covered with shallow water and is replete with hydrophytes,
and that it specifically has been identified by City's consulting biologist,
Scheidt, as wetlands. Preliminarily, located within the 100–year floodplain
of the Escondido Creek, the testimonial and photographic evidence shows

the development area is periodically covered with shallow water, as even moderate rain causes the area to become submerged and flooded. In fact, since 1982, the Kirkorowiczes' property has been flooded to an extreme depth on an average of two to three times a year, and on countless other occasions has been significantly impacted by average rainfall regardless whether occurring during a wet or dry season. Another professional biologist confirmed this "obvious" inundation lasting for some time each year. Secondly, the record is replete with evidence of the presence of hydrophytes in the development area. The development site involves two broadly overlapping plant communities: ruderal, non-native, mostly upland vegetation associated with intensive agricultural or equestrian site usage, and very low-quality, but mostly native, floodplain fringe wetlands vegetation in areas which are subject to longer periods of hydration during the rainy season. Consulting biologist Scheidt inventoried the flora and fauna he observed and classified the species based on their primary habitat. Of the 42 species of flora he recorded, 18 were wetland species. Four of these species are included in the Commission's representative list of wetland species in its Interpretive Guideline. A third-party professional biologist who conducted a field site inspection concluded the site to be a high-quality functional wetland habitat dominated by these wetland species. Scheidt concluded that a portion of the property proposed for development supported areas of jurisdictional wetlands, relying primarily on the presence or absence of indicator wetland plant species. He noted: "All wetland areas, even those in a heavily disturbed state, are considered significant biological resources in so far as they have a potential to buffer adjacent, higher quality areas. In this case, much higher quality wetland habitat is present to the south beyond the limits of the proposed site development area."

* * *

Emphasizing their property constitutes marginal degraded wetlands at best and is not designated as an environmentally sensitive habitat area (ESHA), the Kirkorowiczes argue it is of no environmental consequence and thus not worthy of protection. To the contrary, the Coastal Act by its definition of wetland (§ 30121) does not distinguish between wetlands according to their quality. Indeed, section 30233 limits development in all wetlands without reference to their quality. This is so because of the dramatic loss of over 90 percent of historical wetlands in California and their critical function in the ecosystem. * * * Simply stated, in determining whether a wetland is protected under the Coastal Act and the LCP, the quality of the wetland is essentially legally irrelevant. As City's biologist Scheidt explained, "[a]ll wetland areas, even those in a heavily disturbed state, are considered significant biological resources in so far as they have a potential to buffer adjacent, higher quality areas. In this case, much higher quality wetland habitat is present to the south beyond the limits of the proposed site development area." The logic of this argument is apparent, for the failure to preserve and protect degraded or disturbed wetlands buffering adjacent higher quality wetlands will inevitably jeopardize, compromise and eventually erode the latter.

* * *

The judgment is reversed. The Commission is entitled to costs on appeal.

Notes and Questions

1.   The Kirkorowiczes argued that, based on their poor quality, "the wet areas on their property are not part of a marine environment." What marine environment did they have in mind, of which these wet areas were not a part, and of what environment were they a part? What if they had shown that the areas, while often wet, had no hydrologic connection to coastal waters? Under the CZMA, the coastal zone extends inland from the shoreline "only to the extent necessary to control shorelands, the uses of which have a direct and significant impact on the coastal waters, and to control those geographic areas which are likely to be affected by or vulnerable to sea level rise." 16 U.S.C. § 1453(1). Did the Kirkorowiczes' land fit that standard?

2.   As our appreciation of the connectedness of ecosystems advances, defining the inland limits of the coastal zone according to the "extent necessary to control shorelands" standard could lead to some interesting results. Hawaii's coastal zone, for example, has long included the entire state, albeit for rather obvious reasons. Most states use either a defined distance (e.g., in California, the 1000–yard wide strip adjoining the coast), or a politically-defined boundary (e.g., in North Carolina, all land in a county bounded by the coast). Where would an *ecologically*-defined line be drawn in your state?

3.   Even accepting that the regulated areas of the Kirkorowiczes' property were properly within the state's defined coastal zone *and* a part of the marine environment, does it make sense to construct a coastal ecosystem management program in which, as the court put it, "the quality of the wetland is essentially legally irrelevant." In other words, should biological quality make *no* difference? It undoubtedly is true, as the state argued, that "wetland areas … in a heavily disturbed state have a potential to buffer adjacent, higher quality areas," but does that make them "significant biological resources" worthy of the same level of protection as the higher quality areas?

4.   Of the 35 states and territories eligible for CZMA participation, many embraced the CZMA quickly and moved toward rapid adoption of coastal programs for federal approval, with 25 reaching that point by 1980. In the "recalcitrant" states, however, issues arose about the wisdom and desirability of adding yet an additional layer of land use control, particularly one motivated by federal initiative. Indeed, Florida went so far as to enact a law, known as the "no-nothing-new bill," requiring that any CZMA program the state environmental agency submitted for federal approval be based exclusively on existing state laws and regulations. This precipitated a rather long and bruising battle in the early 1980s between the Florida legislature, the Governor's office, and the federal NOAA to hash out a plan all parties could accept. The federal government saw Florida, with a coastline second in magnitude only to Alaska, as a vital member in the CZMA club. The Florida legislature wanted no new regulations. The end result was a compromise in which existing law and policy were reconfigured, along with nonstatutory initiatives, to create a plan the federal government could respectably declare a step forward, but which did not step forward with new regulatory teeth. *See* Daniel O. O'Connell, *Florida's Struggle for Approval Under the Coastal Zone Management Act*, 25 Natural Resources J. 61 (1985). Some states were even more resistant to the idea of joining the CZMA club than Florida–Texas, Georgia, and Ohio did not enter the program until after 1995. Currently, 95,331 national shoreline miles (99.9%) are managed by the program. Of the remaining 108 miles, 45 lie within Indiana, which is in the process of program development, and the rest within Illinois, which is not participating. A description of the national program and the state programs is available at http://www.ocrm.nos.noaa.gov/czm/.

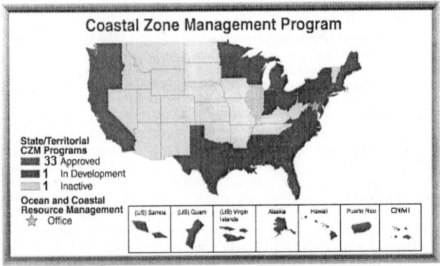

Coastal Zone Management Program

State/Territorial CZM Programs
33 Approved
1 In Development
1 Inactive

Ocean and Coastal Resource Management
☆ Office

(US) Samoa | (US) Guam | (US) Virgin Islands | Alaska | Hawaii | Puerto Rico | CNMI

National Ocean Service 2002

5. As previously noted, Section 6217 of the Coastal Zone Act Reauthorization Amendments (CZARA), Pub. L. 101–508, tit. VI, subtit. C, 104 Stat. 1388 (1990), added a special provision to the CZMA dealing with coastal nonpoint source pollution. *See* 16 U.S.C. § 1455b. The Section 6217 program, as it is often called, requires states and territories with coastal zone management programs to also develop a Coastal Nonpoint Pollution Control Program to retain federal funding under the CZMA. In its CNPCP, a state or territory is to describe how it will implement nonpoint source pollution "management measures" that conform to criteria NOAA and the EPA have specified in a series of guidances for six categories of *inland* land uses: agriculture, forestry, urbanization, marinas and boating, hydromodification, and wetland and riparian areas. *See* U.S. EPA, Guidance Specifying Management Measures for Sources of Nonpoint Pollution in Coastal Waters (1993).

Rather than focusing on water quality, the statute and the agencies' specified management measures focus primarily on control technology and specification of "best management" practices, *see* 16 U.S.C. § 1455b(g)(5), and states may adopt the management measures that are appropriate given local physical and political conditions. This approach was intended to avoid difficult issues of causation–i.e., linking specific nonpoint sources to specific water quality impacts–and to avoid creating yet another unwieldy and expensive water quality improvement program. Yet, as they were in response to the original CZMA coastal plan process (see previous note), many states resented the new planning and management step as both and unfunded mandate and yet another layer of federally-induced regulation of local land use, in this case pushing yet further inland in focus.

Repeated threats by coastal states to leave the CZMA program and attacks on the program in Congress, while unavailing, eventually led EPA and NOAA to appease the states by adding greater flexibility to the program at the administrative level. By July 1996, all 29 states with approved CZMA coastal zone programs had submitted CNPCP proposals and the process of negotiation with the federal agencies was in full swing. Yet as of this writing, now over ten years after CZARA was enacted, only eight plans have received final and full approval from NOAA and EPA. For more information on the Section 6217 program, including the relevant implementing regulations and interpretive guidances, see U.S. EPA, *Coastal Nonpoint Source Pollution Control Program*, http://www.epa.gov/OWOW/NPS/coastnps.html, and NOAA, *Coastal Nonpoint Pollution Control Program*, http://www.ocrm.nos.noaa.gov/czm/6217/. For a critique of Section 6217, see Andrew Solomon, *Section 6217 of the Coastal Zone Act Reauthorization Amendments of 1990: Is there Any Point?*, 31 ENVTL. LAW 151 (2001). Solomon argues that the voluntary nature of the CZMA generally, plus the rather faint threat of withdrawing CZMA funding for failure to compile a CNPCP, has positioned states to exercise leverage over EPA and NOAA to force them to make the CNPCP too flexible and undemanding.

b. FEDERAL AGENCY CONSISTENCY

Why would a state, even if it wishes to manage its coastal resources comprehensively, bother with subjecting its state coastal management program (CMP) to federal review and approval under the CZMA? The answer, obviously, is that the CZMA dangles some additional carrots. First, federal grants are available for implementation of an approved coastal management program. 16 U.S.C. § 1455(a). More importantly, and the subject of the following case, is that approval of a state CMP triggers the requirement that federal agencies ensure that activities they carry out, fund, or authorize in the coastal zone are "carried out in a manner which is consistent to the maximum extent practicable with the enforceable policies of approved state management programs." *Id.* § 1456(c)(2). After working with state CMP managers and making any appropriate changes to the proposed action, federal agencies and applicants provide a consistency statement to the CMP, along with supporting information. A state CMP reviews the federal action to determine if the proposed action will be consistent with the CMP. These federal consistency reviews are the responsibility of the lead state CMP agency. As the following case demonstrates, this state consistency review power can significantly influence federal agency action.

California Coastal Commission v. United States

United States District Court, S.D. California, 1998.
5 F.Supp.2d 1106.

■ Miller, District Judge.

Plaintiff California Coastal Commission seeks a preliminary injunction against defendants United States of America, Department of the Navy, and Secretary of the Navy enjoining the disposal of dredged material from the San Diego Bay previously designated for coastal beach replenishment. This dredging of the bay is part of a homeporting project by which the Navy will base a Nimitz class aircraft carrier. Defendants oppose the motion for a preliminary injunction. After careful consideration of all the pleadings, parties' arguments and applicable law, the court rules as follows:

BACKGROUND

This case deals with the fundamental federal policy of conforming a federal coastal project to meet the dictates, to "the maximum extent possible," of state coastal management plans. This federal policy is codified in federal legislation known as the Coastal Zone Management Act (CZMA) 16 U.S.C. §§ 1451, 1456(c)(1).

Plaintiff California Coastal Commission (Commission) is the state agency responsible for review of federal agency projects for consistency with the federally approved California Coastal Management Program (CMP). The Commission reviewed and approved the Homeporting project of defendants United States of America, Department of the Navy, and Secretary of the Navy (collectively, the Navy) which included the dredging of portions of the San Diego Bay and the use of dredged sandy material for beach replenishment along certain San Diego coastal communities. In 1995, the Navy submitted Consistency Determination (CD) 95–95 which discussed the specifics of the dredging and disposal of the sandy material. Specifically, CD 95–95 called for the deposit of approximately 7.9 million cubic yards of material to Imperial Beach, Mission Beach, Del Mar and Oceanside to replenish areas affected by erosion. Additionally, 2 million cubic yards of other material not suitable for replenishment was to be disposed of in the ocean itself at site LA–5 approximately 4.5 miles off the coast of Point Loma. The remaining material, unsuitable for ocean disposal, would be confined to a new wharf structure at NASNI.

On November 16, 1995 the Commission concurred with CD 95–95 and the Navy commenced its dredging project in September, 1997. Shortly thereafter, live ordnance and munitions were discovered in the dredged material deposited on the beach.

In October, 1997 the Navy requested that the Commission concur with modifications to the project which would permit the disposal of 2.5 million cubic yards of dredged material earlier designated for beach replenishment at the LA–5 site. The Navy contends it requested the modifications in order to continue dredging while a long term solution was found. According to the Navy, an interruption of the dredging would result in excessive dredging expenses and a possible delay in the Homeporting project. On October 17, 1997 the Navy submitted a new CD (CD–140–97) which proposed that all remaining sediment be dumped into the ocean at LA–5 and that some inner channel materials be used for beach replenishment. The Navy asserts that CD 140–97 called for the use of a 3 inch ordnance grate to screen out larger ordnance in the outer channel. The Navy recognized however, that it did not know the exact size of the ordnance in the outer channel and the possibility existed that some of the ordnance was too small to be sifted through the grate. Thus, the Navy could not guarantee that all ordnance would be removed through its grating system proposed in CD 140–97.

* * *

The Commission objected to CD 140–97 stating that the amended project was not consistent with the requirement of the Coastal Zone Management Act (CZMA) that a project conform to a state coastal management plan to the maximum extent possible, that alternatives were available which would permit the Navy to complete the dredging as originally planned, and that the Navy had failed to document the cost of alternatives.

* * *

On November 19, 1997 the Navy sought and received a permit modification from the U.S. Army Corps of Engineers (Corps) which authorized the Navy to dispose of the remaining materials at LA–5. This modification was issued pursuant to § 404 of the Clean Water Act 33 U.S.C. § 1344(CWA) which gives the Corps authority to regulate the Navy's dredging and disposal operations for the project. The Corps approved this modification without Commission concurrence which the Navy contends thereby became unnecessary. On November 19, 1997 the Navy sent a letter to the Commission indicating that the Navy intended to continue dredging and disposal of previously designated beach replenishment at the LA–5 site without the Commission's concurrence. The Navy also indicated that it planned to "fully investigate beach nourishment options for placement of sand from the Homeporting project in coordination with the California Coastal Commission."

The Commission now moves for a preliminary injunction enjoining the Navy from further dredging and disposal of beach replenishment until the alternatives outlined in the Harris Report, CD 161–97 and other reports generated by the Commission are explored. The Commission submits the

Navy is in violation of the CZMA as it has not demonstrated that the disposal of all material at the LA–5 site is consistent to the maximum extent practicable with the state's CMP under the CZMA. Further the Commission argues that the Navy has never demonstrated, as required by state and federal law, that alternatives to ocean dumping or other mitigation measures are unfeasible or impracticable. The Commission believes that injunctive relief is appropriate as the public will suffer irreparable injury from the continued dredging and disposal of beach replenishment which would otherwise be irretrievably lost. The Commission submits it is likely to ultimately succeed on the merits of its claim.

The Navy opposes the motion stating that the provisions of the CMP are not applicable as the ordnance laden material is not suitable for beach replenishment, that consistency with the CMP does not require the Navy to violate other applicable federal or state laws (in this case § 404 of the CWA) and that the discovery of ordnance was an unforeseeable event which, under the CZMA, allows the Navy to deviate from the CMP. Additionally, the Navy argues that a preliminary injunction would impose a great hardship on the Navy and that the Commission has failed to establish the likelihood of success or that the balance of hardships tips in its favor.

DISCUSSION

Federal agencies seeking to engage in project activity in a coastal zone must comply with the requirements of the CZMA. 16 U.S.C. § 1456(c)(1), (2). Section 307(c)(1)(A) of the CZMA states, in pertinent part:

> Each federal agency activity within or outside the coastal zone that affects any land or water use or natural resource of the coastal zone shall be carried out in a manner which is consistent to the maximum extent practicable with the enforceable policies of approved state management programs.

A federal agency is required to submit a "consistency determination" (as previously discussed "CD") to the state no later than 90 days before the proposed activity indicating that the federal activity would likely affect the coastal zone. 15 C.F.R. § 930.34.

The California Coastal Act (CCA) addresses the federal activity in this case. Under § 30233(a) of the CCA:

> [t]he diking, filling, or dredging of open coastal waters ... shall be permitted where there is no feasible less environmentally damaging alternative, and where feasible mitigation measures have been provided to minimize adverse environmental effects.

Section 30233(b) of the CCA further provides that:

> [d]redging and spoils disposal shall be planned and carried out to avoid significant disruption to marine and wildlife habitats and water circulation. Dredge spoils suitable for beach replenishment should be transported for such purposes to appropriate beaches or into suitable longshore current systems.

There is no private right of action under the CZMA itself. City and County of San Francisco v. United States, 443 F.Supp. 1116, 1127 (N.D.Cal.1977), aff'd, 615 F.2d 498 (9th Cir.1980). Judicial review of a federal agency action under the CZMA is obtained through the Administrative Procedure Act (APA). 5 U.S.C. §§ 701–706. However, in a case such as this where Congress has provided in the CZMA more than one method in achieving the Act's purpose of protecting the nation's coastal zones, the principles of equitable discretion should be applied. Weinberger v. Romero–Barcelo, 456 U.S. 305, 315–318 (1982). Accord Friends of the Earth v. U.S. Navy, 841 F.2d 927, 934–935 (9th Cir.1988).

INJUNCTIVE RELIEF

The threshold issue in the analysis of whether the Commission should be afforded equitable injunctive relief is what is the standard this court should employ for judicial review. The Navy urges this court to apply a deferential standard by which the Navy's determination to dump dredged materials previous designated for beach replenishment must be upheld unless it was "arbitrary, capricious, an abuse of discretion, or otherwise not in accordance with the law." 5 U.S.C. § 706(2)(A). The Navy argues application of this narrow standard must result in affirmance of its action under the following rationale: The Navy, in effect, has the exclusive right to determine whether the sediments in question are "suitable for beach replenishment," the sediments as they exist in their pre-dredge state are not suitable for beach replenishment purposes, within the meaning of Cal.Pub.Res.C § 30233, because of the presence of ordnance and munitions, and, therefore, disposal of these materials into deep ocean water complies with the CZMA and CCA. The Navy then concludes that because there is sufficient evidence (existence of ordnance and munitions) to support its conclusion, this court must defer.

Initially, as mentioned above, the deferential standard of review should not apply in this case. The Ninth Circuit has recognized that traditional equitable discretion should be applied in determining whether the Navy has complied with the requirement of the CZMA that a homeporting project conform to the relevant state coastal management plan. See Friends of the Earth v. U.S. Navy, supra. 841 F.2d 927 (9th Cir.1988). The CZMA was enacted by Congress to clearly encourage the wise use of coastal resources through adoption of state coastal plans. Traditional judicial review subserves that stated legislative intent.

* * *

Preliminary injunctive relief is available if the party meets one of two tests: (1) a combination of probable success and the possibility of irreparable harm, or (2) the party raises serious questions and the balance of hardship tips in its favor. Arcamuzi v. Continental Air Lines, Inc., 819 F.2d 935, 937 (9th Cir.1987). "These two formulations represent two points on a sliding scale in which the required degree of irreparable harm increases as the probability of success decreases." Id. Under both formulations, however, the party must demonstrate a "fair chance of success on the merits" and a "significant threat of irreparable injury." Id.

Likelihood of Success on the Merits

The Commission argues that the Navy cannot show, as required under the CCA, that no less environmentally damaging alternative exists or that feasible mitigation measures have been provided to minimize adverse environmental effects. Specifically, the Navy has submitted only CD 140–97 for consideration, has withdrawn CD 161–97, and has never submitted other analyses, including the Harris Report. Thus, the Navy's position that no feasible alternative exists other than to waste a valuable resource is predicated upon an incomplete factual record and unilateral determinations made by the Navy without the benefit of Commission input. On this record, the Navy has not shown that the dredging and disposal is consistent to the maximum extent practicable with the enforceable policies of approved state management programs in violation of the CZMA.

The Navy's position that it considered a reasonable range of alternatives as required under § 30233(a) of the CCA which it addressed in CD 140–97 and CD 161–97 simply does not answer the question. The Navy acknowledges in its pleadings and oral argument it "remains willing to negotiate a reasonable solution" to offshore dumping. This position presupposes there may indeed be a feasible alternative to wasting this valuable beach replenishment resource. Until pending alternatives have at least been considered by both parties, it is illogical to conclude that offshore dumping is consistent with the CCA to the maximum extent possible. Finally, this portion of the analysis does not depend on whether the discovery of ordnance was an unforeseen event which warrants deviation from the CCA. As long as a reasonable alternative to dumping may be found with further expeditious study by the parties this factor is not material.

The Navy's contention that it has submitted feasible, less environmentally damaging alternatives and has provided certain measures to mitigate the adverse environmental effects all in compliance with the CCA and CZMA, is not borne out by the record. Specifically, the Navy's alternatives are contained in CD 161–97 and other analyses [including one known as the Harris Report] which have either been withdrawn from consideration by the Navy or never submitted in final form to the Commission. The Navy cannot meritoriously argue that these alternatives contained in the CD 161–97 or the subsequent analyses are properly before the Commission at this time and therefore in compliance with the CCA or CZMA. Therefore, as the Navy has failed to demonstrate that it has complied with the requirements of the CZMA and CCA and has failed to allege an acceptable exemption from these requirements, the court finds that on the present record the Commission would likely succeed on the merits of its case against the Navy for disposing of beach replenishment materials off the coast of California in a manner inconsistent with the federal and state law.

Irreparable Harm and the Balancing of Hardships

* * *

Legitimate considerations of irreparable harm and hardship balance in favor of the Commission. One or more viable alternatives to ocean dumping of a valuable natural resource may presently exist and be quickly identified through further expeditious study and good faith negotiation by the parties. A reasonable additional period of time should be afforded for that contingency. Any offshore dumping of this resource during this period of study represents an irretrievable loss which such study and negotiation could prove to be an unnecessary and costly waste.

Any excess dredging fees to be paid by the Navy, as well as any short term delay in the completion of dredging operations for this homeporting project are more than counterbalanced by the need to allow an additional period of expedited study and negotiation by the parties during which offshore dumping operations cease. Thus, a preliminary injunction is granted enjoining the Navy from disposing at LA–5 or any other offshore dumping site dredging material previously designated for beach replenishment purposes. This preliminary injunction is conditioned upon the Commission's expeditious study of proposed alternatives to offshore dumping, including those set forth in the Harris Report, and the good faith of the parties to negotiate a resolution which is the stated goal of both sides. The court reserves jurisdiction to modify or dissolve this preliminary injunction upon shortened notice.

Notes and Questions

1. What is the Navy supposed to do with the sand? Would *you* like to walk along a beach that had been "renourished" with the Navy's sand? Of course, all the court held is that the Navy had not yet met its burden of proving no less damaging alternative to ocean dumping exists. But how far would the Navy have to go to prove this? For example, what if, at great expense, the Navy could clean the sand of munitions? Would the CZMA require that it do so?

2. The court explains that the CZMA consistency determination requires a finding that "that no less environmentally damaging alternative exists or that feasible mitigation measures have been provided to minimize adverse environmental effects." The Commission and the court, however, focused not on the ecological impact of disposal of the sand on the *ocean* ecosystem, but on the loss of the dredged sand itself, which both portrayed as a valuable *beach* ecology resource that should not be wasted. The Commission and the court thus appear to take as a given that using the sand for what is known as beach "restoration" or "nourishment" is ecologically superior to dumping it in the ocean. Even putting aside the presence of munitions in the sand, is it clear that beach nourishment is such a good thing ecologically, not to mention economically?

As for cost, beach nourishment projects can carry hefty price tags. The most expensive to date involved nourishing the beaches from Sandy Hook to Barnegat Inlet in New Jersey, to the tune of $1.16 billion, but dozens of local projects have topped $150 million in cost, much of it federally subsidized. *See Surf's Up. So Is Costly Bid to Shield "Sand Castles,"* USA Today, Apr. 9, 2001, at 12A. On the other hand, over the past 50 years federal beach nourishment expenditures have run only about $15 million per year, which is far less than is spent of federal dollars to subsidize activities with far less economic impact than the tourism and recreational pursuits that our beaches and coasts support.

Ecologically, however, beach nourishment can disrupt subaerial zone habitat by altering temperature and compaction conditions, and can alter wave and runoff conditions important to adjacent shallow subtidal habitat. Of particular concern, for example, is the effect it has on turtle eggs as nests are buried deeper and compacted, as well as on the ability of the female turtles to reach areas of the beach above tidal influence to lay the eggs. *See* U.S. Fish and Wildlife Service, Life History and Environmental Requirements of Loggerhead Turtles (1988); National Academy of Sciences, Decline of the Sea Turtles: Cause and Prevention (1990); Sarah L. Minton et al., *The Effect of Beach Nourishment with Aragonite Versus Silicate Sand on Beach Temperature and Loggerhead Sea Turtle Nesting Success*, 13 J. Coastal Resources 904 (1997).

For an extensive study of both the economy and the ecology of beach nourishment, searching for ways to balance the economic benefits and the ecological consequences, see National Academy of Sciences, Beach Nourishment and Protection (1995), *available at* http://books.nap.edu/books/0309052904/html/index.html. The Academy concluded that beach nourishment is a viable shore protection technique and means for restoring lost recreational assets, but that it must be carefully designed, implemented, and monitored through multidisciplinary methods.

3. One significant limitation on the state consistency review power flows from the CZMA exclusion from the coastal zone of any "lands the use of which is by law subject solely to the discretion of or which is held in trust by the Federal Government." 16 U.S.C. § 1453(1). The Supreme Court has held, however, that states may nonetheless exercise direct environmental regulation over non-federal activities occurring on federal lands. *See* California Coastal Commission v. Granite Rock Company, 480 U.S. 572 (1987). In *Granite Rock*, the state imposed environmental conditions on mining operations the company proposed to carry out on national forest lands pursuant to Forest Service regulations. The company argued that the exclusion of federal lands in the CZMA from the coastal zone preempted the state from imposing the environmental conditions, but the Court held that "even if all federal lands are excluded from the CZMA definition of coastal zone the CZMA does not automatically preempt all state regulation of activities on federal land." *Id.* at 593. Hence, although the state could not use CZMA consistency review to limit the mining activities, it could use its police power authority directly to impose environmental conditions on the company's mining operations.

4. An approach similar to the CZMA's consistency step, but perhaps delivering more punch, is used in the Coastal Barrier Resources Act of 1982 (CBRA), 16 U.S.C. §§ 3501–3510, as substantially amended by the Coastal Barrier Improvement Act of 1990, Pub. L. 101–591 (1990), and administered by the Department of the Interior. CBRA establishes the Coastal Barrier Resources System (CBRS), consisting of undeveloped coastal barriers–natural formations that protect landward aquatic habitats from the direct forces of waves, tides, and wind–and other identified coastal areas. Included in the classification of coastal barriers are the following:

Bay Barriers—barriers that enclose a pond or marsh by connecting two headlands

Tombolos—sand or gravel beaches that connect islands to each other or to the mainland

Barrier Spits—barriers which are attached to the mainland at one end and extend into the open water

Barrier Islands—barriers completely detached from the mainland

Dune or Beach Barriers—wide sandy beaches with dunes or hills

Fringing Mangroves—bands of mangrove trees along tropical/subtropical shores.

The CBRA recognizes that these features are unique landscapes that serve as a protective barrier against the forces of wind and tidal actions caused by coastal storms. In addition, coastal barriers provide a protective habitat for a variety of aquatic species. Thus, Congress declared three important goals to be met through the statute: (1) minimize loss of human life by discouraging development in high-risk areas; (2) reduce wasteful expenditure of Federal resources; and (3) protect the natural resources associated with coastal barriers.

Congress initially defined the CBRS as a collection of specific undeveloped coastal barriers along the Atlantic and Gulf of Mexico coasts, later expanding it to include additional areas along the Atlantic and Gulf coasts, the Great Lakes, Puerto Rico, and the U.S. Virgin Islands. Over 1,200 miles of shoreline, comprising approximately 1.3 million acres, make up the 560 units included today within the CBRS.

Unlike the CZMA, inclusion of a unit in the system is initiated by a federal decision, though the Interior Department consults with states and will consider voluntary state inclusion of state or local lands into the system. Under the statute, with limited exceptions no new federal expenditures or direct or indirect financial assistance may be used for any purpose within the system, including construction, improvement, or purchase of structures, roads, and boat facilities, and the stabilization of shoreline areas. Thus, for example, federal flood insurance under the National Flood Insurance Program is not available for structures located in the CBRS. While this approach does not induce state-led management of coastal resources, the prevalence of federal funding of land use activities in coastal areas (e.g., flood insurance and beach renourishment) suggests that the withdrawal of such funding in CBRS areas could have significant effects on land use.

5. Another distinction between the CZMA consistency review procedure and the approach the CBRA uses is in the nature and source of the authority involved. In the CBRA, the federal government has restricted itself—the state is not involved. In the CZMA, by contrast, the federal government has empowered the states to restrict federal agency decisions. Is that constitutional? Can Congress authorize the states to trump the supremacy of federal law and authority? Or does the CZMA merely incorporate state programs as federal law and authorize the states to implement consistency review as a matter of federal law?

Note on the Public Trust Doctrine

Virtually all of the ecosystem management materials covered in this text are based in federal, state, or local legislative initiatives. Does the common law have any role to play in ecosystem management? Generally, the answer has been no. Although the rare case has used doctrines of nuisance law to address broad environmental injuries, *see, e.g.*, Georgia v. Tennessee Copper Co., 240 U.S. 650 (1916) (using common law public nuisance law to address air pollution); Reserve Mining Co. v. United States, 514 F.2d 492 (8th Cir.1975) (using common law nuisance to address water pollution), these are, for the most part, limited applications of common law doctrine to extreme cases of pollution. Nothing approaching a judicial doctrine of ecosystem management emerges from them.

One common law principle that many scholars have argued could defy this trend is known as the Public Trust Doctrine. The name is impressive, suggesting great possibilities. Indeed, in his landmark 1970 article, inspiring many since then to envision a Public Trust Doctrine motivating broad goals of natural resources conservation, Professor Joseph Sax outlined an ambitious agenda for just those purposes. *See* Joseph L. Sax, *The Public Trust Doctrine In Natural Resource Law: Effective Judicial Intervention*, 68 MICH. L. REV. 471 (1970). Sax argued that "[o]f all the concepts known to American law, only the public trust doctrine seems to have the breadth and substantive content which might make it useful as a tool of general application for citizens seeking to develop a comprehensive legal approach to resource management problems." *Id.* at 474. Thirty years and hundreds of law review articles later, however, this vision remains largely unfulfilled hope. Why?

The Public Trust Doctrine traces its roots to the Institutes of Justinian in Roman Law, which declared that there are three things common to all people: (1) air; (2) running water; and (3) the sea and its shores. Along with the Romans, this principle invaded England and became part of its common law, which the states imported with minor variations after the American Revolution. While the British version held that tidelands were held by the King for the benefit of all English subjects, the American version replaced the crown with the states, and the courts became the doctrine's chief enforcer.

The scope of the trust imposed by the Public Trust Doctrine can be thought of in several dimensions. First, it has a geographic reach that must be defined. In the American version, this has generally meant all lands subject to the ebb and flow of the tide, and all waters navigable in fact, such as rivers, lakes, ponds, and streams. Next, the uses that the trust protects and prohibits must be defined. In American jurisprudence, fishing, commerce, and navigation are core protected uses, with other uses such as boating, swimming, anchoring, and general recreation being recognized as well in most states. Uses inconsistent with those protected values may be prohibited–that is, even if the state wishes to facilitate such incompatible uses, it may be restrained from doing so. Finally, the Public Trust Doctrine carries with it restrictions on the alienation of public trust lands to private interests when to do so would undermine the protected public uses. Clearly, these are dimensions in which ecosystem management operates as well, so the thought of linking the Public Trust Doctrine with ecosystem management is by no means far-fetched.

A series of nineteenth century U.S. Supreme Court cases, focused principally on the scope of property rights associated with statehood, breathed apparent life into these parameters of the Public Trust Doctrine. First, in Martin v. Waddell's Lessee, 41 U.S. (16 Pet.) 367 (1842), the Court applied the doctrine in a case involving resolution of title to tidelands and tidal rivers. Next, in The Daniel Ball, 77 U.S. (10 Wall.) 557 (1870), the Court held that "[t]hose rivers must be regarded as public navigable rivers in law which are navigable in fact." But, as Professor Sax described it, the lodestar case of the Public Trust Doctrine, at least for purposes of thinking about it as a tool of resource conservation, came in the Court's 1892 opinion in Illinois Central Railroad Co. v. Illinois, 146 U.S. 387 (1892). The Court held that Illinois could not sell fee interests in the land under Chicago Harbor to private developers because

> the State holds the title to the lands under the navigable waters. . . . It is a title held in trust for the people of the State that they can enjoy the navigation of the waters, carry on commerce over them, and have liberty of fishing therein freed from the obstruction or interference of private parties.

Id. at 452. Almost 100 years later, the Court reiterated the principle using similar terms in Phillips Petroleum Co. v. Mississippi, 484 U.S. 469, 476 (1988) ("our cases firmly establish that the States, upon entering the Union, were given ownership over all the lands beneath the waters subject to the tides' influence"). Yet, that about sums up the Public Trust Doctrine as far as the U.S. Supreme Court is concerned–the states may not alienate fee title in tidelands, shores, and other public trust lands in violation of the Public Trust Doctrine. Suffice it to say that the Court has not championed Professor Sax's vision of doing more with the doctrine, limiting its jurisprudence largely to questions of who owns what, and much less so to the federalism question of what a state may do with public trust lands acquired as a matter of statehood.

The *Phillips Petroleum* decision did remind us, however, that "[i]t has been long established that the individual states have the authority to define the limits of the lands held in public trust." *Id.* at 475. As it is fundamentally a state law doctrine, therefore, many state courts have opined on the scope of the Public Trust Doctrine as well, some with a vigor not found in the U.S. Supreme Court jurisprudence. One famous case from California, regarding the diversion of water from Mono Lake, ruled that "[t]he state has an affirmative duty to take the public trust into account in the planning and allocation of water resources, and to protect public trust uses whenever feasible." National Audubon Society v. Superior Court of Alpine County, 658 P.2d 709 (Cal.1983). This and other state cases like it, however, are mindful of the "publicness" of public trust lands, emphasizing uses such as navigation, fishing, and recreation, and not necessarily preservation or even active conservation. Even in the Mono Lake case, for example, the court held that in exercising the public trust "the state must bear in mind its duty as trustee to consider the effect of the taking [of water] on the public trust, and to preserve, so far as consistent with the public interest, the *uses* protected by the trust." *Id.*

Even in the states, therefore, the Public Trust Doctrine has had its chief impact as an arbiter of property rights and, thereby, as a tool mainly to facilitate public access to and use of tidelands and beaches. It has by no means been transformed into a judicial ecosystem management program in any state. It is true that an occasional state case suggests an ecologically-oriented purpose to the doctrine. Perhaps the most noted case in this regard is from Wisconsin, in which the court found that the doctrine required that wetland areas be limited to uses consistent with natural conditions. *See* Just v. Marinette County, 201 N.W.2d 761 (Wis.1972). Several more recent cases are variations on that theme. *See e.g.*, Aspen Wilderness Workshop, Inc. v. Colorado Water Conservation Board, 901 P.2d 1251 (Colo.1995) (state must avoid injury to creek ecology from ski resort's water appropriation request); Selkirk–Priest Basin Association v. Idaho ex rel. Andrus, 899 P.2d 949 (Idaho 1995) (doctrine allows environmental group standing to challenge timber sales on ground that sedimentation could injure fish spawning grounds); Vander Bloemen v. Wisconsin Department of Natural Resources, 551 N.W.2d 869 (Wis.App.1996) (doctrine extends to protection of lakeside ecology). By and large, however, the state courts have declined to mobilize Professor Sax's vision of the Public Trust Doctrine as a means of effective and broad judicial intervention in resource management policy. There is, simply put, no ecosystem management duty to be found in the judiciary's version of the Public Trust Doctrine.

By contrast, since Professor Sax's seminal work, many environmental law scholars have charted and claimed all sorts of ecosystem management goals for the Public Trust Doctrine. Of course, doing so requires that one or more of the doctrine's parameters be expanded beyond present judicial interpretations. So, for example, courts could extend the geographic scope to encompass regulation of private lands adjacent to public trust lands, or they could add ecosystem dynamics to the protected "uses." The doctrine might even be transformed from its current status as a restriction on state power to alienate public trust lands or to allow incompatible uses, to one imposing an affirmative *duty* of ecosystem management. For examples of these and other academically posited stretchings of the doctrine, see JACK H. ARCHER ET AL., THE PUBLIC TRUST DOCTRINE AND THE MANAGEMENT OF AMERICA'S COASTS (1994); Jack H. Archer and Terrance W. Stone, *The Interaction of the Public Trust and the "Takings" Doctrines: Protecting Wetlands and Critical Coastal Areas*, 20 VERMONT L. REV. 81 (1995); Robin Kundis Craig, *Mobil Oil Exploration, Environmental Protection, and Contract Repudiation: It's Time to Recognize the Public Trust in the Outer Continental Shelf*, 30 Envtl. L. Rep. (Envtl. L. Inst.) 11104 (2000); Ralph W. Johnson and William C. Galloway, *Can the Public Trust Doctrine Prevent Extinctions?*, in BIODIVERSITY AND THE LAW 157 (William J. Snape III ed., 1996).

What explains the chasm between the judicial and the academic visions of the Public Trust Doctrine? What keeps academics coming back to the Public Trust Doctrine, asking ever more of it, but repels judges from taking it farther? One rather obvious possibility is that, not long after Professor Sax suggested how its latent power could be tapped, the legislative revolution of the 1970s unfolded to bring one after the other of comprehensive environmental laws into being. In short, who needs the Public Trust Doctrine? By comparison to the targeted legislative agenda that brought on line the Clean Water Act, Coastal Zone Management Act, National Forest Management Act, Endangered Species Act, and other resource management laws spawned in that era and which remain the workhorses of ecosystem management today, the Public Trust Doctrine seems, like many common law doctrines, hopelessly open-ended, amorphous, and unwieldy. Perhaps, in addition to seeing no critical need to go down the road Sax mapped, courts see trouble ahead were they tempted to start the journey.

Indeed, in a countercurrent to the Saxian vision, Professor Richard Lazarus and a few other academics with impeccable "green" credentials have argued that the Public Trust Doctrine, if shaped as Sax wanted, could actually be antithetical to proactive and innovative environmental and resource management. *See* Richard J. Lazarus, *Changing Conceptions of Property and Sovereignty in Natural Resources: Questioning the Public Trust Doctrine*, 71 IOWA L. REV. 631 (1986); Richard Delgado, *Our Better Natures: A Revisionist View of Joseph Sax's Public Trust Theory of Environmental Protection, and Some Dark Thoughts on the Possibility of Law Reform*, 44 VAND. L. REV. 1209 (1991). For one thing, they argue, it places too much reliance on a judiciary that is not always in tune with what the academic vision of the Public Trust Doctrine appears to want have happen. The growth of the police power state, its authorities grown and channeled since World War II into a huge administrative law apparatus, seem a far better prospect for carrying out an environmentalist agenda. And, they point out, at its core the Public Trust Doctrine is about property rights in the form of public rights to *use* the environment of public trust lands. Like any trust, the purpose of the public trust lands is not merely to preserve their corpus, but to put them to public use. The administrative law version of ecosystem management is potentially more flexible–its course charted by congressional will–and thus may not lead to a regime biased toward use versus preservation (though, for the most part, "multiple use" doctrine is the rule for public lands managed outside of the Public Trust Doctrine).

In any event, whatever it is, and whatever the wisdom of the Saxian vision of it, the Public Trust Doctrine is not a font of ecosystem management law as it stands today. True, that could change, but any movement in that regard does not appear on the horizon. Yet, hope for Public Trust Doctrine's place in the larger picture of resource management will not seem to go away. It is, as Professor Carol Rose put it, an "arresting phrase" with tremendous rhetorical power to remind us of why we might think of something like ecosystem management. *See* Carol M. Rose, *Joseph Sax and the Idea of the Public Trust*, 25 Ecology L.Q. 351 (1998). It is also, as Professor Charles Wilkinson observed, decidedly enduring and pan-cultural, owing its roots to Roman law laid down centuries ago and coming to us via British common law. *See* Charles F. Wilkinson, *The Headwaters of the Public Trust: Some Thoughts on the Source and Scope of the Traditional Doctrine*, 19 Envtl. Law 425 (1989). We think of this note, therefore, as a placeholder for potential expansion of the topic in future editions of this text, though we expect over time it will more likely become a candidate for editing.

2. Estuaries

Estuaries, also known as bays, lagoons, harbors, inlets, or sounds, are partially enclosed bodies of water formed where freshwater from rivers and streams flows into the ocean, mixing with the salty sea water. The defining feature of an estuary is the mixing of fresh and salt water. Estuaries and the lands surrounding them are places of transition from land to sea, and from fresh to salt water. Although influenced by the tides, estuaries are protected from the full force of ocean waves, winds, and storms by the dunes, reefs, barrier islands, and other features that define an estuary's seaward boundary.

Biologically, the tidal, sheltered waters of estuaries support unique communities of plants and animals, specially adapted for life at the margin of the sea. Because many different habitat types are found in and around estuarine environments, they are among the most diverse, complex, and productive on earth, creating more organic matter each year than comparably-sized areas of forest, grassland, or agricultural land.

With over 50 percent of the nation's population living within 50 miles of the sea, it is no wonder that estuaries are also of tremendous economic value. Tourism, fisheries, and other commercial activities thrive on the wealth of natural resources estuaries supply. Because estuaries offer protected coastal waters, they also are used extensively as harbors and ports vital for shipping, transportation, and industry. The EPA recently quantified some of these economic values:

Estuaries provide habitat for more than 75% of America's commercial fish catch, and for 80–90% of the recreational fish catch. Estuarine-dependent fisheries are among the most valuable within regions and across the nation, worth more than $1.9 billion in 1990, excluding Alaska.

Nationwide, commercial and recreational fishing, boating, tourism, and other coastal industries provide more than 28 million jobs. Commercial shipping alone employed more than 50,000 people as of January, 1997.

There are 25,500 recreational facilities along the U.S. coasts–almost 44,000 square miles of outdoor public recreation areas. The average American spends 10 recreational days on the coast each year. In 1993 more than 180 million Americans visited ocean and bay beaches–nearly 70% of the U.S. population. Coastal recreation and tourism generate $8 to $12 billion annually.

In just one estuarine system–Massachusetts and Cape Cod Bays–commercial and recreational fishing generate about $240 million per year. In that same estuary, tourism and beach-going generate $1.5 billion per year, and shipping and marinas generate $1.86 billion per year

See USEPA, Office of Water and Watersheds, *About Estuaries, available at* http://www.epa.gov/owow/estuaries/about1.htm.

Notwithstanding the important ecosystem values estuaries deliver to humans, human-induced impairment of estuarine resources is particularly acute in many areas of the nation. On the list of sources of impairment: nutrients from agriculture, sewage treatment, septic tanks, and residential yards; pathogens from released sewage, agricultural manure, medical waste, boat waste, and urban runoff; habitat alteration from transportation, residential, commercial, and agricultural development; and toxins from industrial pollution, urban runoff, and agricultural pesticides; and introduced species. In its comprehensive study of these conditions in states bordering the Gulf of Mexico, the EPA found the most troubling trend to be declining wetland habitat, with Gulf states having lost 41 to 54 percent of historic estuarine wetlands during the last 200 years. *See* USEPA, THE ECOLOGICAL CONDITION OF ESTUARIES IN THE GULF OF MEXICO (July 1999), *available at* http://www.epa.gov/ged/gulf.htm. Other conditions are varied throughout the states, with EPA issuing a "report card" on 11 ecological indicators for each showing some indicators in the "good-no problem" category for most states, while the score for others in some states was as low as "poor-severe problem." Texas and Louisiana showed particularly serious concerns with nutrients; Alabama and Mississippi rated good to fair on most indicators; Florida was representative of the Gulf as a whole.

The law of estuaries is as complex as are their ecological and economic features. Estuaries are, not surprisingly, included within the definition of coastal waters under the Coastal Zone Management Act and thus are fully within the scope of state coastal management programs as described in the previous section of this chapter. *See* 16 U.S.C. § 1453(3) (referring to "sounds, bays, lagoons, bayous, ponds, *and estuaries*"). Ecosystem management for estuarine resources is also embodied in a more focused manner in the Clean Water Act's National Estuary Program (NEP) and in a special program devoted to the nation's largest single estuary, the Chesapeake Bay.

a. THE NATIONAL ESTUARY PROGRAM

Like the Coastal Zone Management Act, covered in the previous section, the National Estuary Program (NEP) targets a broad range of issues going far beyond just water quality, and it engages state and local governments and communities in the land and resource management process. The following article provides an excellent overview of the NEP as it operates today.

Matthew W. Bowden, An Overview of the National Estuary Program

11 Natural Resources & Environment 35 (Fall 1996).

After more than three decades of environmental regulation aimed at specific media and individual activities, a trend is developing toward a broader, more holistic approach to environmental management. Today's buzzword and the first guiding principle in the U.S. Environmental Protection Agency's (EPA) new Five Year Strategic Plan is "ecosystem protection." Watersheds are one of the most widely accepted types of ecosystems, and a little-known watershed management program called the National Estuary Program (NEP) is at the forefront of EPA's recent focus on ecosystem protection.

The addition of section 320 to the Clean Water Act (CWA), 33 U.S.C. § 1330, in 1987 established the NEP. . . . Rather than taking the traditional command-and-control regulatory approach, the NEP seeks to involve all affected stakeholders in the process of identifying an estuary's problems and developing management measures to solve those problems. . . . [T]he NEP's final work product is not enforceable regulation. Rather, the ultimate goal of the NEP is to prepare a comprehensive management plan for estuaries selected for the program. Although such plans are not binding regulation, they represent an influential joint statement of public policy by federal and state governments and affected stakeholders on how the estuary should be managed. As such, management plans can foster changes in existing regulation and development of new approaches to protecting estuarine resources.

Nomination and Funding of Estuary Programs

Under section 320 of the CWA, the governor of a state may nominate an estuary within the state to the NEP by submitting an application to EPA. If the nomination is accepted, EPA convenes a "Management Conference" to begin assessing the condition of the estuary. After identifying the estuary's problems, the Management Conference begins work on a management plan. EPA provides 75 percent of the funding for the Management Conference, while state and local governments or other nonfederal sources supply the remaining 25 percent.

* * *

Management Conference Participation and Organization

Once an estuary is accepted in the NEP, EPA and the state governor select members of a Management Conference to carry out the program. Section 320 of the CWA provides that the Management Conference must include representatives of all states located in the estuarine zone, agencies and local governments having jurisdiction over any significant part of an estuary, affected industries, educational institutions, and the general public. As a general rule, most Management Conferences consist of four committees: a Policy Committee, a Management Committee, a Science/Technical Advisory Committee, and a Citizen Advisory Committee.

The Policy Committee is the governing body of the Management Conference. Members of this committee normally include high-level representatives from EPA, key state agencies, and local governments. The Policy Committee may also include representatives from business, industry, and environmental groups. The Policy Committee sets goals, objectives and priorities and provides overall direction for the conference. It usually selects members of the other three committees and must ultimately approve a management plan for the estuary.

* * *

The Citizen Advisory Committee serves as the main voice of the stakeholders in the estuary and its watershed. Representatives of business and industry, associations, environmental and civic groups, farming and fishing groups, educators, and other affected and interested citizens on this committee communicate the concerns of the public and the regulated community to the Management Conference. The Citizen Advisory Committee also helps inform the public about the activities and work product of the Management Conference.

EPA promotes the NEP as a model of collective decisionmaking among various stakeholders with competing interests. The Management Conference is designed to foster collaborative decision making and consensus building around conflicting interests. Ideally, the Management Conference structure encourages open discussion and compromise that produce widespread support for the actions needed to restore and protect the estuary.

* * *

Duties of the Management Conference

Section 320(b) of the CWA directs the Management Conference to perform the following:

- assess trends in the estuary's water quality, natural resources, and uses;

- identify environmental problems by collecting and analyzing relevant data;

- determine relationships between pollutant loadings to changes in water quality and natural resources;

- develop a Comprehensive Conservation and Management Plan to restore and maintain the estuary;

- develop plans for coordinated implementation of the management plan by federal, state, and local regulatory agencies;

- monitor effectiveness of the management plan; and

- review all federal assistance programs and development projects to determine their consistency with the management plan.

These responsibilities reveal that the NEP is primarily a study and planning program. However, the Management Conference will address impacts on the estuary from a wide variety of human activities that occur across a large geographic area.

* * *

[T]he Management Conference must identify the environmental problems throughout this area and seek to link these problems with their probable causes. EPA defines this as the "characterization" process. This process involves analysis of existing information and knowledge on the condition of the estuary, its problems, the probable causes of these problems, and any apparent trends. Existing sources of information on estuaries are usually quite extensive, including research initiatives and EPA or state environmental databases, research on fisheries and coastal habitats from the National Oceanic and Atmospheric Administration or the state wildlife agency, and research and data from other federal and state agencies, academic institutions, and the private sector.

* * *

Comprehensive Conservation and Management Plan

After completing the characterization process, the Management Conference will begin work on a Comprehensive Conservation Management Plan (CCMP) for the estuary. CCMPs are not direct, enforceable regulation. This important detail prompted one witness testifying before a House subcommittee to describe the NEP as a program which "may encourage those states with good intentions to gum away. The teeth, however, are lacking." HOUSE MERCHANT MARINE AND FISHERIES COMMITTEE, COASTAL WATERS IN JEOPARDY, H.R. DOC. NO. 38, 101st Cong., 1st Sess. at 27 (1989) (quoting testimony of Oliver A. Houck, Professor of Law, Tulane University on September 28, 1988).

While this may be true to some extent, when properly developed, CCMPs represent a blueprint for restoring and protecting an estuary. They contain numerous recommendations for regulatory changes based on available scientific data. Consequently, CCMPs can serve as an important catalyst for changing a host of existing regulations and adopting new regulations necessary to accomplish its goals. If implemented, these recommendations would have a significant impact on the various stakeholders who live and operate in the watershed.

EPA guidance requires that CCMPs contain five main components. First and foremost, a CCMP must contain proposed solutions for the priority problems identified by the Management Conference. Although some CCMPs contain innovative solutions to identified problems, most proposed solutions consist of recommendations for more stringent and comprehensive controls and enforcement in existing regulatory regimes.

* * *

The second part of a CCMP is a plan for implementing its specific recommendations. Accordingly, CCMPs include strategies for obtaining and maintaining support in the public and private sectors for recommended initiatives. These strategies generally involve coordinating and focusing various government agencies on the goals of the CCMP and developing political support in the private sector and regulated communities to accomplish these goals.

The third element of a CCMP is a plan for financing the various recommended corrective actions. Invariably, solutions proposed by CCMPs are extremely expensive and implementation costs can run into the tens of millions of dollars. Financing can come from various sources. Federal funding is available from the CWA's State Revolving Fund and nonpoint source program along with several other EPA programs. However, state and local sources will incur most of the costs for implementing CCMP recommendations. This often requires tax increases or imposition of costs on the public, property owners, businesses and industries, and other citizens directly affected by increased regulation.

* * *

The fourth component of a CCMP is a plan for monitoring the effectiveness of implementation of proposed solutions. Monitoring is needed to determine whether management actions are having their intended impact and whether they should be modified in any way. Monitoring also helps assess the success of the CCMP.

The fifth and last part of a CCMP is the federal consistency review. The Management Conference must review all federal financial assistance programs and federal development projects to determine their consistency with the CCMP. If federal programs or projects conflict with the CCMP, they may require modification. Consequently, in addition to fostering numerous regulatory changes, CCMPs can have a significant impact on federal activities within the study area.

Within 120 days after completion of the CCMP and after providing for public review and comment, EPA must approve the plan if it satisfies the requirements of the NEP and if the state governor concurs. Section 320 of the CWA states that, "upon approval of a [CCMP], such plan *shall be implemented*." 33 U.S.C. § 1330(f)(2) (emphasis added). Although there are no authorities construing the ultimate effect of this mandate, the language is quite strong and implies, at a minimum, that federal agencies must do everything in their power to implement a CCMP. This would also include making any necessary changes in federal programs and projects to ensure that federal activities in the study area are consistent with the CCMP.

The NEP is a watershed-based management approach to protecting our nation's estuaries. Acceptance into the NEP leads to a comprehensive evaluation of all the problems in a given estuary and the development of strategies to address the most serious problems threatening the long-term health of the estuary. Public participation in this effort is one of the conceptual cornerstones of the NEP, although meaningful participation by the public may not be fully realized in all cases. Management plans created under the NEP do not have the force of law, but they can and do motivate federal, state and local regulators into action.

Notes and Questions

1. Through its CCMP device, the NEP takes the same approach as is taken under the Coastal Zone Management Act through Coastal Management Program, described in the previous section of this chapter. The article describes the CCMP as a blueprint for management of the estuary's ecological and economic resources, but notes the potential force of the federal action "consistency" review step in putting that blueprint into action. As the materials on the Coastal Zone Management Act explore, a consistency review step can indeed alter federal agency actions. But is it strong enough to serve as the primary ecosystem management instrument for estuaries? Or, as Professor Houck put it in his congressional testimony cited in the article, is the NEP, even with the consistency review step, all gums and no teeth?

2. The management challenge facing the NEP is compounded by the rather open ecological features of an estuary. As the transition from land to sea and freshwater to marine water, what jurisdictional boundary makes the most ecological sense for the NEP? Clearly, what happens in an estuary will depend in large part on what happens in the rivers leading to it, which leads in turn to the upland watersheds of those rivers. *See* USEPA, *Coastal Watersheds*, http://www.epa.gov/owow/oceans/factsheets/fact1.html (summarizing the impacts of land based activities on coastal watersheds). As shown in the figure below, the watersheds of the estuaries that are in the NEP cover a lot of ground, with many extending through several states. Is the NEP, as described above, equipped to manage on that scale of ecosystem dynamics? *See* Randall B. Wilburn, *Mediate Water Quality Problems on a Watershed Basis? Done That, What's Next?*, 31 ST. BAR OF TEXAS ENVTL. L.J. 151, 155–161 (2001). Will the coastal nonpoint pollution program, added to the Coastal Zone Management Act through section 6217 of the Coastal Zone Act Reauthorization Amendments (described in the previous section of this chapter), provide scope and depth of regulation where the NEP might not?

Watersheds of National Estuaries

Shaded areas show the watersheds for the 11 National Estuaries.

USEPA

4. What goes up must come down. To further complicate matters, consider that each estuary's inland watershed shown in the previous figure is influenced by an "airshed" which, as shown in the next figure, can cover a much larger area. While not as well-defined as a watershed, airsheds represent prevailing atmospheric patterns that circulate and carry particulate and gaseous matter within the roughly-defined airshed boundary. They are defined more formally as the area responsible for emitting 75 percent of the air pollution reaching a body of water. Because different pollutants behave differently in the atmosphere, the airshed for different pollutants will vary. Air pollutants such as oxidized nitrogen thus may be emitted far outside the boundary of a watershed but within the airshed that covers the watershed. This and other pollutants thus may settle within the watershed and, if they dissolve in water or bind with soils that erode into water, may eventually enter the estuary. The nitrogen added through this process of atmospheric deposition acts as a nutrient within the estuarine ecosystems, and, like all good things, there can be too much of nutrients (see the discussion of "hypoxia" later in this chapter). Indeed, studies show that well over 20 percent of the total nitrogen entering some estuarine watersheds comes through atmospheric deposition. Other pollutants that can enter and severely affect estuaries through atmospheric deposition include mercury, sulfur gases, and heavy metals. For more on this problem and ongoing regulatory responses taking place through EPA's Atmospheric Deposition Initiative, see U.S. EPA, *What are the Major Effects of Common Atmospheric Pollutants on Water Quality, Ecosystems, and Human Health,* *available at* http://www.epa.gov/owow/oceans/airdep.

5. Section 315 of the Coastal Zone Management Act, 16 U.S.C. § 1461, establishes the National Estuarine Research Reserve System (NEERS). As stated in the program's implementing regulations, 15 C.F.R. § 921.1, the NEERS has the following mission:

> the establishment and management, through Federal-state cooperation, of a national system of estuarine research reserves representative of the various regions and estuarine types in the United States. National Estuarine Research Reserves are established to provide opportunities for long-term research, education, and interpretation.

The NERRS was established by the CZMA to help address the problem of current and potential degradation of coastal resources brought about by increasing and competing demands for these resources. Prior to establishment of the NERRS, the management of estuarine resources was inadequate, and scientific understanding of estuarine processes necessary for improving management was increasing slowly and without national coordination. There were no ready mechanisms to detect trends in estuarine conditions, or to provide information on these trends, the overall significance of estuaries, and possible solutions to the growing problems. The NERRS is portrayed as one part of the solution for maintaining healthy coastal resources. NERRS research, education, and resource stewardship programs are tools that can help fill gaps in knowledge, and guide decisionmaking so that our estuaries can sustain multiple uses over the long term. Accordingly, the goals of NEERS are to:

- Ensure a stable environment for research through long-term protection of National Estuarine Research Reserve resources;

- Address coastal management issues identified as significant through coordinated estuarine research within the System;

- Enhance public awareness and understanding of estuarine areas and provide suitable opportunities for public education and interpretation;

- Promote Federal, state, public and private use of one or more Reserves within the System when such entities conduct estuarine research; and

- Conduct and coordinate estuarine research within the System, gathering and making available information necessary for improved understanding and management of estuarine areas.

15 C.F.R. § 921.1(b). Currently, the NEERS includes 25 reserves distributed in 22 states and covering a total of 1 million acres. For more information on the NEERS program, see http://www.ocrm.nos.noaa.gov/nerr/.

6. One of the surprises of the 106th Congress was the enactment of the Estuaries and Clean Waters Act of 2000. Pub. L. No. 106–457, 114 Stat. 1957 (2000). The legislation is nonregulatory in character, but does: (1) establish an estuary restoration program under the authority of the U.S. Army Corps of Engineers, which has primary authority for permitting activities in wetlands (see chapter 10); (2) reauthorize the NEP and the Chesapeake Bay program (see *infra*). With respect to estuary restoration, the legislation establishes the Estuary Habitat Research Council, made up of five federal agencies, and requires that it develop a national strategy for restoring 1 million acres of estuarine habitat over the next 10 years through a combination of federal, state, and local funding. The law defines estuary restoration as:

- the reestablishment of chemical, physical, hydrologic, and biological features and components associated with an estuary;

- the cleanup of pollution for the benefit of estuary habitat;

- the control of nonnative and invasive species;

- the reintroduction of a species native to the estuary

- the construction of reefs to promote fish and shellfish production and to provide estuary habitat for living resources

Clearly, this new initiative is designed to strike at the factors identified in the beginning of this section as principal threats to estuarine resources in our nation. The legislation authorizes $275 in federal funding over fives years to do so, yet, as noted, provides no new regulatory authorities.

b. THE CHESAPEAKE BAY

The nation's largest estuary, the Chesapeake Bay, is not part of the National Estuary Program, but rather is managed through a multi-jurisdictional cooperative agreement between the federal government and several states that is administered as the Chesapeake Bay Program. Indeed, the Chesapeake Bay Program, initiated in 1983 through the first Chesapeake Bay Agreement, was the model for the National Estuary Program. The major difference is that the Chesapeake Bay Program is a multi-party agreement in which the member states have agreed to adopt watershed level regulation of land use and environmental quality to meet the ecological performance goals set for the estuary. The following materials describe how this agreement came into being, its evolution to the current program agreement adopted in 2000, and the ecological impact the program has had.

The Chesapeake Bay watershed encompasses 66,387 square miles. Its year 2000 population is estimated to be 15,594,241. Major populated places include: Baltimore, MD; Washington, DC; Norfolk, VA; Richmond, VA; Arlington, VA.

Chesapeake Bay Program (2002).

Harry R. Hughes and Thomas W. Burke, Jr., The Cleanup of the Nation's Largest Estuary: A Test of Political Will

11 Natural Resources & Environment 30 (Fall 1996).

The Chesapeake Bay's largest tributary, the Susquehanna River, rises in Cooperstown, New York, a sleepy little village more famous for its Baseball Hall of Fame than for being the headwaters of North America's largest estuary. At Havre de Grace, Maryland, 240 miles south of Cooperstown, the Susquehanna is released from its banks and blossoms into the Chesapeake Bay, famous for other joys of summer—blue crabs, rockfish, sailboats, sunshine, and Silver Queen corn. Two hundred miles south of Havre de Grace, past the ports of Baltimore, Cambridge, Newport News, and Norfolk, are the lighthouses of Cape Charles and Cape Henry, Virginia, the twin towers that mark the Bay's end and the beginning of the open Atlantic Ocean.

Over 64,000 square miles of land between Cooperstown, New York, and Cape Charles, Virginia, drain into the Chesapeake Bay through her tributaries. This estuary, where fresh water from these vast river networks and salt water from the Atlantic mix, boasts over 2,700 different species of plants and animals, more than any other similar area on the North American continent. Rivers like the Susquehanna, which drains central Pennsylvania, and the Potomac, whose hundreds of tributaries drain parts of Maryland, Virginia, West Virginia, and Pennsylvania, bring a mixture of good and bad to the Bay. In Maryland, smaller rivers like the Patuxent, Patapasco, Choptank, Chester, Sassafras, and Severn also carry fresh water down to the Bay, but like the larger rivers, they too transport other "cargo"—nitrogen, phosphorous, sediment, and toxic chemicals. In Virginia, the Rappahannock, James, and Elizabeth Rivers empty similar loads into the Bay.

The British call the Thames River "liquid history." America too has its liquid history, and much of it runs into the Chesapeake Bay. Antietam Creek, Bull Run, the Shenandoah and Appomattox Rivers conjure images of blue and grey warriors, struggling to assert their definition of freedom. The canals that link the Chesapeake with the Delaware River and the Ohio River speak to us of an even earlier time, when "way out West" meant places just over the Appalachians. St. Mary's City (Maryland's first capitol), Annapolis (the nation's first capitol) and Richmond (the Confederate States' only capitol) all owe their existence to "their" rivers, the Potomac, Severn, and James.

Today, 13 million people in six states and the District of Columbia call the Chesapeake's watershed home. That number will climb to 15 million by the year 2020. Two major ports, wide-ranging basic and high-value manufacturing, hi-tech, bio-tech and information processing businesses are a part of the Bay area's economy. The U.S. Naval Academy and scores of other colleges and universities, as well as the "business" of government are headquartered in this region. A highly developed tourism industry depends on the Bay while grain, chickens, fruit, vegetables, tobacco, dairy cattle and beef make agriculture the number one economic enterprise of the region and its largest single land use. The economy and use of the Bay, therefore, is as diverse as its biology.

Looking Back

The Chesapeake Bay of the eighteenth and nineteenth centuries was a prolific food source with abundant oysters, fish, and waterfowl beyond imagination. The Bay was also an avenue of commerce. The Baltimore clipper and paddle-wheel steamboat opened many rivers, streams, and creeks to trade and commerce. Little towns, isolated from one another and hard to reach by road or rail, became connected at the wharf. Baltimore became the hub of the watershed, drawing in seafood, crops, and raw materials, and sending out finished goods and, as the second busiest port of entry after New York City, immigrants.

But the industrialization of the region brought a new use for the Bay—garbage disposal. From foundries, factories, and farms, waste material was dumped in the Bay "free of charge." We built factories beside the rivers and streams for fuel-free power and then used the same waterways as ready-made sewers without concern for the long-term health of the resource. The Bay has suffered at our hands. By 1850, over 50 percent of Maryland forests had been cut down. The deep-water ports of Bladensburg, Port Tobacco, and Joppatown were closed to navigation by 1830 due to siltation and runoff. Where ocean-going ships once loaded cargo for England and European ports, only canoes and dories could navigate by the end of the War of 1812. The Bay yielded oyster harvests in excess of 20 million bushels a year in the 1880s, but only a mere fraction (approximately 1 percent) of the historic harvest is available today. Overfishing and impoundments which block spawning runs have reduced the shad population in the Bay to 10 percent of its historic level.

By the 1940s and 1950s, the cumulative impact of historic abuses and the surging effect of the population explosion in war-driven Washington, D.C., Baltimore and Newport News, finally began to overwhelm the Bay.

In 1972, Hurricane Agnes did heavy damage to bay grasses, and they did not recover. Scientists began to realize that poor water quality inhibited bay grass recovery and growth. Large areas of the Bay were suffering from anoxia (no dissolved oxygen) or hypoxia (little dissolved oxygen) and toxics were a persistent problem, especially in major industrial areas—Baltimore Harbor, the waters around Norfolk, and the Anacostia River in the District of Columbia.

A new environmental group, the Chesapeake Bay Foundation, began to sound the alarm and rally people with the cry, "Save the Bay." At the same time, the Alliance for the Chesapeake dedicated itself to bringing together government, business and environmentalists in a cooperative spirit to solve common problems.

Maryland's U.S. Senator Charles "Mac" Mathias called on the U.S. Environmental Protection Agency to study the Bay to determine the causes of its decline and offer suggestions for the Bay's recovery. That study, funded in 1976 and completed in 1983, provided the scientific foundation for action, including several scientific breakthroughs. The first and most important discovery was that excess nutrients (phosphorus and nitrogen) were significantly responsible for the water quality decline. These nutrients (from sewage treatment plant outfalls, farm fields and feed lots and suburban lawns and construction sites) over-enrich algae in the Bay, which then multiplies so fast and thick that it blocks sunlight from bay grasses, killing them. Bay grasses are the "nursery" areas for small fish and crabs and a critical food source for many types of waterfowl. To compound the problem, when this algae dies, it uses up the available oxygen in the water as it decomposes, seriously limiting the water's ability to sustain aquatic life or even to assimilate pollution.

Another technical breakthrough involved modeling the way the Bay works, at first with a scaled-down physical model that spread out over two acres, and later with Cray supercomputers. With the help of the computers, the models aided in determining the flow dynamics of the Bay and helped pinpoint where nutrients were coming from and how land use affects water quality in the Bay.

It is important to understand that the Chesapeake does not "drain" out into the Atlantic. For the most part, what goes in, stays in. Further, the ratio of watershed land to water—how many acres of land drain into so many acres of water—is almost 17 to 1 for the Chesapeake Bay. For comparison, the Great Lakes ratio is only 7 to 1 and the Mediterranean Sea ratio is 1 to 1. In the Chesapeake watershed, actions on land have profound effects in the water.

But inactions on land can also profoundly affect the water. When preliminary EPA study findings began to be reported, it was clear that we could not wait or let this report sit on a shelf, gathering dust. We had to take action. We began by working with citizen groups, city and county governments, and neighboring states. Both Governor Thornburg of Pennsylvania and Governor Dalton of Virginia overcame longstanding "institutional" and geographic biases against a coalition where, "someone might tell them what to do." Additionally, Pennsylvania had to confront the "it doesn't affect me" syndrome to form the beginnings of a cooperative partnership. Broad grassroots support also was building, and a new coalition of federal agencies, state and local governments, and scientists was named the "Chesapeake Bay Program." Its mission was to understand the reasons behind the Bay's decline and to begin to implement recovery measures.

By 1983, our early work paid off and the working coalition of federal agencies, state and local governments, and citizens had progressed to the point where a formal declaration, in the form of a "Chesapeake Bay Agreement," was signed by the leaders of the six interests most involved with the Bay—the U.S. EPA, signing on behalf of all federal agencies; the State of Maryland; the Commonwealths of Virginia and Pennsylvania; the District of Columbia; and the Chesapeake Bay Commission, an alliance of state legislators from Maryland, Virginia, and Pennsylvania.

This agreement, although short, was a huge step forward in the restoration of the Bay. It brought together the principal leaders in the watershed, committed them to action, and created an organized structure dedicated to the systematic and scientific analysis of the Bay's problems and the refinement of a science-based, consensus-driven plan for the Chesapeake's cleanup.

The 1987 Chesapeake Bay Agreement

In 1987, a second Chesapeake Bay Agreement was signed by the same six parties who signed the original compact. This agreement, longer and more specific, set out goals, objectives, and commitments in six areas:

1. Living Resources: Providing for restoration and protection of the Bay's living resources, their habitats and ecological relationships was the goal established in the first area of this new agreement. The agreement declared the Bay's living resources to be the ultimate measure of the restoration's progress. The health and abundance of benthos, fish, shellfish, plants, and waterfowl would signal the cleanup's success.

2. Water Quality: To "[r]educe and control point and non-point sources of pollution to attain the water quality condition necessary to support the Living Resources of the Bay," was the objective in the second area. But how clean should the Bay be to support these living resources? And which living resources? Before any decisions could be made, scientists and Bay managers first had to determine the water quality levels necessary to support all the varied forms of life in the Bay. This investigation is ongoing.

3. Population Growth and Development: "Plan for and manage the adverse environmental effects of human population growth and land development in the Chesapeake Bay watershed." Growth devours forest and farm land, wetlands and shoreline as well as local governments' budgets to build new infrastructure. But a strong economy finances environmental action. Balancing population growth and economic development with environmental concerns and the restoration of the Chesapeake Bay is a difficult task.

4. Public Information, Education and Participation: "Promote general understanding among citizens ... the problems facing the Bay and policies and programs designed to help it. Foster individual responsibility and stewardship of the Bay's resources." Action follows consensus. Public understanding and public participation allow people to make informed decisions and provide the political support for tough choices.

5. Public Access: "Promote increased opportunities for public appreciation and enjoyment of the Bay and its tributaries." If people cannot use the Bay, why should they support its restoration? With 95 percent of the Bay's shoreline privately owned, encouraging use and the concept of personal stewardship is a key to success.

6. Governance: "Support and enhance the present comprehensive efforts and perpetuation of commitments necessary to ensure long-term results." Laws and regulations are necessary to effect the Bay cleanup, but continuity of management and steady commitment also are important.

Politically, the 1987 Chesapeake Bay Agreement put the responsibility for success directly on the elected leaders of the Bay states. The Bay cleanup was now "officially" part of executive and legislative agendas. The agreement also provided a yardstick for measuring success. With over fifty separate commitments, watchdog groups and concerned citizens could measure success and levels of dedication to Bay restoration by the number of commitments fulfilled.

For Bay managers and scientists, the 1987 agreement provided goals. The most important was the 40 percent reduction in *controllable* nitrogen and phosphorus loadings to the Bay by the year 2000. This commitment precipitated significant action at the state and local levels, in the agriculture industry, and in the business arena, including the construction and upgrade of wastewater treatment plants, and new ways for farmers to plow, manage herds, control manure, and harvest crops. The agreement also affected changes in growth management regulations and even in the process for building highways and housing developments.

The agreement called for management plans for dozens of species of fish and shellfish (crabs, oysters, rockfish, white and yellow perch, bluefish, American shad), wildfowl and tidal and nontidal wetlands. The signatories agreed to eliminate blockages to spawning fish trying to swim upstream. To date, the Bay states have invested millions of dollars in fish ladders, fish elevators, and removal of stream blockages, thereby freeing hundreds of miles of stream and river for migration of shad, rockfish, and perch. Projects like the fish "elevator" at the Conowingo Dam on the Susquehanna River, and the destruction of small dams and creation of fish ladders on the James River are helping many species make a comeback.

Finally, the 1987 Bay Agreement created an extensive state and federal infrastructure dedicated to the Chesapeake's restoration and operated on a consensus basis. All of the signatories have committed significant state resources. Maryland, for instance, has spent an average of $80 million a year on direct Chesapeake Bay activities, augmented by $20 million a year in federal support since 1987. Local governments and business have shouldered their share of the responsibility by implementing new regulations and ordinances, adhering to higher permit levels and by spending money directly on Bay-saving construction. The Bay cleanup has even become an international model for multijurisdictional cooperation and success. Visitors from China, Japan, Russia, Europe, South America, and Africa make pilgrimages to the Bay to discover the recipe for the Bay cleanup's success. They are interested in finding out how three states, the District of Columbia, over fifty federal agencies, and more than 2,000 local governments develop and cooperatively implement a broad-based array of programs without laws mandating coordination or cooperation.

* * *

What must be done for the Bay now seems clear:

- Assume individual responsibility for the Bay. We all must understand that what we do, whether at home or in the workplace, affects the environment and waters like the Chesapeake Bay. In the Bay watershed, helping people recognize and adopt an ethic of personal stewardship is at the top of the "how to save the Bay" list.

- Reach the nitrogen and phosphorus reduction goals as close to the year 2000 as possible. Government actions, legislation, personal habits, and business practices all should be geared toward cutting the input of nutrients to local waterways and the Bay.

- Manage growth to protect the economy and preserve the ecology. A healthy economy and a healthy Bay are not mutually exclusive. Law and regulation should not be crafted at the expense of either.

A few years ago, there was a Chesapeake Bay education campaign that used a large arrowhead, pointing to the drain of a bathroom sink, the tailpipe of a car, the edge of a farm field, the smokestack of a power plant or factory, a shower drain, suburban lawn, boat holding tank, storm drain, dam, suburban development or city. It said, "The Bay Starts Here!" The idea, of course, was simple. The Bay, and the Bay cleanup, starts with us. It begins at our doorsteps, in our homes and offices and where we play or relax, whether that is on the water, on the golf links or in our own backyards.

In the final analysis, the challenge to preserve and protect the Chesapeake is not a test of how clever our scientists are or how smart sewage treatment plant engineers can be. It is, rather, a test of our political will, a measure of our strength of commitment. We know what the problems are. We know what needs to be done to fix them. The test is whether we will, not whether we can.

Notes and Questions

1. The governance structure of the Chesapeake Bay Agreement is quite unusual, and may explain both why it appears to be effective and why it may be difficult to generalize for ecosystem management. As Professor Jon Cannon has postulated, multi-jurisdictional collaborative resource management programs such as this present three types of transaction costs: the costs of gathering information relevant to the management goals (information costs); the costs of negotiating, monitoring, and enforcing the agreement (coordination costs); and the costs of controlling strategic behavior by participants, such as free-riding, that may impeded cooperation (strategic costs). *See* Jon Cannon, *Choices and Institutions in Watershed Management*, 25 Wm. & Mary Envtl. L. & Policy Rev. 379, 394 (2000). The Chesapeake Bay Program has achieved a rather complex, decentralized, but nonetheless efficient structure for controlling these costs. The ultimate decision making body for the Agreement is the Executive Council, which consists of the Governors of Virginia, Maryland, and Pennsylvania, the Administrator of the USEPA, the Mayor of the District of Columbia, and the Chair of the Chesapeake Bay Commission. The Executive Council draws on the work of many advisory committees, but most of its decisions are implemented through the Chesapeake Bay Commission, an appointed advisory body composed of state legislators, agency heads, and citizen representatives. The Commission works to implement the Executive Council's decisions through legislative and administrative initiatives. Cannon argues that this structure works because management of the Bay involves a relatively small number of jurisdictions with a relatively high number of shared concerns and goals given the configuration of the states and the physical characteristics of the Bay. In particular, typical upstream versus downstream issues often posed in watershed contexts are not as acute in the Bay. The Chesapeake Bay Program, therefore, may suggest a governance ideal that cannot efficiently be transported to other resource management contexts because of physical and ecological constraints.

2. As the foregoing history illustrates, over time a primary concern of the Chesapeake Bay Program has become controlling the nitrogen and phosphorous inputs into the estuary. The 1987 agreement had expressed the goal of 40 percent reduction of nitrogen and phosphorous entering the mainstem of the Bay. The agreement also stated that the parties would "reduce and control point and nonpoint sources of pollution to attain the water quality condition necessary to support the living resources of the Chesapeake Bay." Based on modeling and monitoring in the years that followed, the parties to the agreement determined that the program had paid insufficient attention to the quality of the Bay's *tributaries*, where most spawning grounds and essential habitat are found, and needed to intensify its focus on nonpoint sources of nutrients, particularly from agriculture. In 1992, therefore, the parties amended the agreement to require development of tributary-specific strategies aimed primarily at nonpoint sources of nitrogen and phosphorous. *See* Chesapeake Bay Agreement: 1992 Amendments (Aug. 12, 1992).

3. Some conditions in the Chesapeake Bay began to improve after the 1992 amendments focused efforts on tributaries (see previous note), but many did not. Blue crab populations continued declining, bay grass acreage increased slightly, and oyster populations rebounded only marginally. Most importantly, the nitrogen and phosphorous reduction goals, while seemingly within reach, were not met, and rising population and urbanization suggested that more concerted efforts would be required not only to gain on the remaining increment, but to avoid losing ground. On June 28, 2000, therefore, the program parties adopted a new and more far-reaching agreement, known as Chesapeake 2000, with the effect of greatly increasing the program's attention to comprehensive management of the watershed lands. Chesapeake 2000 outlines 93 commitments detailing protection and restoration goals in 5 main categories: (1) living resource protection and restoration; (2) vital habitat protection and restoration; (3) water quality protection and restoration; (4) sound land use control; and (5) stewardship and community engagement. Among the commitments are to:

- increase the number of native oysters tenfold by 2010
- restore blue crabs by establishing harvest targets
- reduce the loss rate of farmland and forest land to urban development by 30 percent by 2012
- permanently preserve from development 20 percent of the lands in the watershed by 2010

The focus on preserving forested land added an important new dimension to the program, recognizing the watershed-based link between terrestrial ecosystems and coastal estuary ecosystems. This focus appears to be working. Riparian forest buffers along streambanks have increased substantially already, to 118,000 miles–about 60 percent of all streambank miles in the watershed. Over 6.7 million acres of land in the watershed are preserved from development, just 1.1 million acres short of the 20 percent goal. Gradually, therefore, the Chesapeake Bay Program has evolved from its inception in 1983, when the principal focus was the Bay, to including tributaries and, later still, the watershed lands as co-equal partners in the management challenge. For these details and other information on the Chesapeake Bay Program's history, status, and plans, see http://www.chesapeakebay.net.

4. The latest news on the Chesapeake Bay Program involves three agreements the participants entered in December 2001 to mitigate the effects of stormwater runoff. One agreement commits the Program to developing and demonstrating innovative stormwater control methods and technologies that local governments and private landowners can employ. Another agreement commits several Program participants to working to achieve 50 restoration targets for one of the major watersheds in the system, the Anacostia River watershed. The final agreement, called "Building for the Bay," enlists several cooperating organizations, including the National Association of Home Builders, to help the Program promote voluntary adoption of 22 principles that reduce the environmental effects of residential and commercial development "through employment of site design practices which restore natural hydrology, retain rainwater onsite, reduce reliance on constructed conveyances, and otherwise prevent the erosion and scouring of natural streambeds." *See* Builders for the Bay Program (Dec. 2001), *available at* www.chesapeakebay.net/press.htm.

Note on the Bay–Delta Plan

The West Coast version of the Chesapeake Bay Program, in terms of scope and ambition, is the San Francisco Bay/Sacramento–San Joaquin Delta Estuary Plan (Bay–Delta Plan). By contrast, however, the Bay–Delta Plan management framework is structurally far more complex than is the case for the Chesapeake Bay Program. The Bay–Delta Estuary is one of the largest ecosystems for fish and wildlife habitat and production in the United States, and is the largest on the West Coast. Efforts to manage it comprehensively began in 1960s with a series of decisions by the California State Water Rights Board, predecessor to the current State Water Resources Control Board (SWRCB), affecting water rights in the federal Central Valley Project (CVP) and the State Water Projects (SWP), both of which influence the Bay–Delta. The state also enacted the Porter–Cologne Water Quality Control Act of 1969, Cal. Water Code §§ 13000 et seq., which established the SWRCB and a group of Regional Water Quality Control Boards whose charge was to develop water quality control plans for their respective regions of the state. These plans are to define beneficial uses of water rights, water quality objectives, and a program of implementation.

During the 1970s and 1980s, the state and federal governments became increasingly concerned that the impaired flow conditions and increasing salinity of water in the Bay–Delta basin were degrading the ecosystem. In

1994, the Governor's Water Policy Council and the federal government's Federal Ecosystem Directorate, known collectively as CALFED, entered into a framework agreement to focus on coordination of water quality standards, water project operations, and ecosystem resources. To implement the plan, USEPA began work on water quality standards for the Bay–Delta Estuary under the Clean Water Act. *See* 59 Fed. Reg. 813 (Jan. 1994) (proposed standards); 60 Fed. Reg. 4664 (Jan. 1995) (final standards). The SWRCB adopted a water quality control plan to implement the standards in May 1995. *See* SWRCB Resolution No. 95–24 (May 22, 1995).

The four primary objectives of the Bay–Delta Plan are the overall objectives for each of the key program areas it created: water quality, ecosystem quality, water supply, and vulnerability of Delta functions. Secondary objectives within each of these areas tie back to the primary objectives:

- Provide good water quality for all beneficial uses;

- Improve and increase aquatic and terrestrial habitats and improve ecological functions in the Bay–Delta to support sustainable populations of diverse and valuable plant and animal species;

- Reduce the mismatch between Bay–Delta water supplies and current and projected beneficial uses dependent on the Bay–Delta system;

- Reduce the risk to land use and associated economic activities, water supply, infrastructure, and the ecosystem from catastrophic breaching of Delta levees.

To implement the Bay–Delta Plan, a clear definition of the problems had to be addressed and a range of solution alternatives developed. Second, to comply with the the National Environmental Policy Act (NEPA) and the California Environmental Quality Act (CEQA), a program level or first-tier environmental impact statement (EIS) and environmental impact report

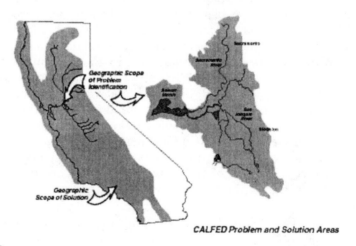

CALFED Problem and Solution Areas

CALFED 1999

(EIR) had to be prepared to identify impacts associated with the various alternatives. Then, as specific projects would be contemplated within the Bay–Delta Plan area, project-level or second-tier EIS/EIR documents are prepared for each element of the selected solution.

Unlike the Chesapeake Bay Program, the scope of CALFED's work is within one state, but within that more limited scope almost two dozen federal and state agencies have some regulatory or management responsibility for some aspect of the Bay–Delta, and all have a seat on the CALFED Policy Group. Each agency assigns staff to CALFED; however, CALFED is not itself an agency. Rather, CALFED, in essence, is a forum through which the agencies coordinate and evaluate their actions. Each agency implements elements of the Bay–Delta Plan that are consistent with and within the scope of its respective statutory responsibility. Notwithstanding this cumbersome structure, within five years of the Bay–Delta Plan approval the CALFED participants had achieved remarkable progress toward implementation.

As CALFED has described the five-year period, see CALFED Bay–Delta Program, http://calfed.ca.gov, in the first phase of implementation CALFED members developed a range of alternatives, consisting of hundreds of actions. The Program conducted meetings and workshops to obtain public input, prepared a Notice of Intent and Notice of Preparation pursuant to NEPA and CEQA, and held public scoping sessions to determine the focus and content of the EIS/EIR. The first phase concluded in September 1996 with the development of a range of alternatives for achieving long-term solutions to the problems of the Bay–Delta estuary.

During Phase II, the Program conducted a comprehensive program-wide environmental review process. A draft programmatic EIS/EIR and interim Phase II Report identifying three draft alternatives, each outlining many different specific projects and actions, and program plans was released on March 16, 1998. The release of the documents was followed by a 105–day public comment period. On June 25, 1999, CALFED again released a draft programmatic EIS/EIR followed by a 90 day comment period. The final programmatic EIS/EIR was released July 21, 2000, followed on August 28, 2000, by the Record of Decision (ROD) and Implementation Memorandum of Understanding (MOU) between the CALFED participants. *See* http://calfed.ca.gov/adobe_pdf/rod.

The ROD completed Phase II, and the MOU began Phase III–implementation of the preferred alternative. The first seven years of Phase III are referred to as Stage 1, and are designed to lay the foundation for the following years. Site-specific, detailed environmental review will occur during this phase prior to the implementation of each proposed action. Implementation of the CALFED Bay–Delta Plan is expected to take 30 years. Projects cover areas such as water supply reliability, ecosystem restoration and watershed management, water quality controls, levee systems, scientific research, and basic program management. At some point along the way, it is envisioned that CALFED may actually be transformed, by concurrent federal and state legislation, into some form of joint governance mechanism closer in framework to the Chesapeake Bay Program.

The question, as with the Chesapeake Bay Program, will be how to balance state and federal sovereignty with the joint project's mission in such a way as to control information, coordination, and strategic costs.

Note on Hypoxia in Coastal Waters

The increasing focus on nonpoint source pollution in the National Estuary Program, the Chesapeake Bay Program, and the Bay–Delta Plan stems from the alarming phenomenon of hypoxia conditions that is becoming more frequent and widespread in estuarine settings. Hypoxia means an absence of oxygen reaching living tissues. In coastal waters, it is characterized by low levels of dissolved oxygen, so that not enough oxygen is available to support fish and other aquatic species. Hypoxia results from too much of a good thing—nutrients. Nutrients, obviously, are necessary to aquatic environments at some level. But if nutrients entering the system exceed the capacity of the food web to assimilate them, the accelerated production of organic matter, manifested in the massive algae blooms that induce hypoxia, begins to choke the system with its own waste—the process known as eutrophication. The increased primary productivity leads to increased flux of organic matter to the bottom, causing bottom water hypoxia, altered energy flow, and stresses to fisheries resources. Particularly vulnerable are benthic (bottom level) species such as crabs and lobsters that are unable to migrate quickly enough to escape the toxic conditions.

Dead crab following acute hypoxia conditions
in the Gulf of Mexico (USEPA 2002)

While hypoxia occurs in the Chesapeake Bay, Bay–Delta, and other of our nation's estuaries, nowhere in the United States is the hypoxia problem more of a concern than where the Mississippi River empties into the Gulf of Mexico. There the hypoxic zone has reached a size larger than the state of New Jersey—an 8000 square mile area running from the Mississippi Delta to the Texas–Louisiana border. It is most acute there during the summer months, when the weather is warm and the Gulf is especially calm, because of a lack of any "churning" of the layers of fresh and salt water. When this stratification of the water column occurs, the lower saltwater layer becomes cut off from the resupply of oxygen from fresh surface waters. This is when the rich nutrient content of the Mississippi River water becomes a liability to the system. The nutrients encourage the growth of algae, and as other aquatic organisms feast on the algae blooms that generate large amounts of fecal matter. This abundance of organic waste sinks to the saltier depths where it decomposes, using what remains of the available oxygen and creating a hypoxic zone. This condition can persist well into autumn, until storms and other high winds begin to churn the water layers and bring much-needed oxygen back to the bottom.

Some level of hypoxia events is natural in many estuarine ecosystems as annual floods flush high levels of naturally-occurring nutrients into the marine system. The largest recorded hypoxic zone in the Gulf at the time, for example, was after the massive floods on the Mississippi in 1993—it neared 8,000 square miles, an area as large as the entire Chesapeake Bay. But increasingly, hypoxia conditions are more persistent, occur more frequently, and affect larger areas even in relatively dry years. The 1993 record has been surpassed in years since without massive floods—in 2001 the record of over 8000 square miles was set. Some reasons include stream channelization, which moves nutrient flow faster downriver, and the loss of riparian buffers that impede flow of nutrients from land into streams. But the core of the problem is simply too much nutrient loading—about 40 percent higher today than the average loads experienced during 1955–1970.

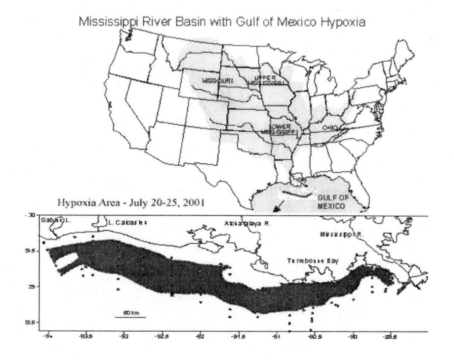

The excess supply of nutrients that accounts for the surge in hypoxia conditions may come from a wide range of sources: runoff from developed land, atmospheric deposition, soil erosion, and agricultural fertilizers. Sewage and industrial discharges can also contribute nutrients. Of all these causes, agriculture has been implicated as both the most significant contributor of excess nutrients and the most difficult problem to solve. *See* Dan Ferber, *Keeping the Stygian Waters at Bay*, 291 SCIENCE 968 (2001).

In recognition of the growing and alarming nature of the hypoxia problem, Congress in late 1998 passed the Harmful Algal Bloom and Hypoxia Research and Control Act of 1998. *See* Pub. L. No. 105–383, Title VI., § 604. During the fall of 1997, the Clinton Administration had formed the multi-agency Mississippi River/Gulf of Mexico Watershed Nutrient Task Force to research Gulf hypoxia in conjunction with the White House Office of Science and Technology Policy's Committee on Environment and Natural Resources. The 1998 law required the Task Force to submit a scientific assessment of Gulf hypoxia and a plan for reducing, controlling, and mitigating its effects. After voluminous comments on a draft plan (see 65 Fed. Reg. 42690 (July 2000)), in January 2001, in one of its last official acts, the Clinton Administration released the Task Force's report. *See* ACTION PLAN FOR REDUCING, MITIGATING, AND CONTROLLING HYPOXIA IN THE NORTHERN GULF OF MEXICO (Jan. 2000), *available at* http://www.epa.gov/msbasin/actionplan.htm.

The Action Plan focuses on two primary approaches: (1) reduce nitrogen loads to streams and rivers in the Mississippi River Basin, and (2) restore and enhance denitrification and nitrogen retention within the Basin. These prescriptions seem both obvious and easier said than done. The magnitude of the management challenge Gulf hypoxia poses is staggering. The Basin covers 1.2 million square miles. The river pours 612,000 cubic feet per second into the Gulf of Mexico. Over 12 million people live just in the 125 counties that directly border the Mississippi River itself. In other words, geographically, physically, and demographically, this is a monster-sized problem.

Accordingly, the Action Plan's finding and recommendations are rather sweeping in scope. The Task Force endorsed an "adaptive management" approach like the framework outlined in Chapter 7 of this text, and it identified what it believes will be the important indicators of implementation and results, such as riparian acres preserved. But overall, the tone is visionary far more than it is detailed. Representative are the following remarks:

> There are no simple solutions that will reduce hypoxia in the Gulf. An optimal approach would take advantage of the full range of possible actions to reduce nutrient loads and increase nitrogen retention and denitrification. This should proceed within a framework that encourages adaptive management and accomplishes this in a cost-effective manner. While reduction of nitrogen is the principal focus of this framework, many of the actions needed to reduce nitrogen loads will complement and enhance existing efforts to restore water quality

throughout the basin. With additional assistance, this national effort to reduce Gulf hypoxia will be implemented within the existing array of State and Federal laws, programs, and private initiatives.

The tools provided by the Clean Water Act, and the programs established under the last several Farm Bills, the Coastal Wetlands Planning, Protection, and Restoration Act, and Water Resources Development Acts, are critical to implementing this plan. Because nutrient overenrichment is a widespread problem, these existing national programs and initiatives incorporate specific elements intended to reduce nutrient loadings to surface waters and to foster restoration of natural habitats capable of removing nutrients from waters. They include the following:

- encouraging nonpoint source pollutant reductions under the Clean Water Act, the Farm Bill, Coastal Zone Amendments and Reauthorization Act, and State cost-sharing programs;

- implementation of the Environmental Quality Incentives Program (EQIP) to assist grain and livestock producers in reducing excessive nutrients' movement to water resources;

- implementation of the Conservation Reserve Program, Wetlands Reserve Program, Corps of Engineers Environmental Restoration Programs, and Agricultural Extension Education Programs to promote restoration and enhancement of natural systems for nitrogen retention and denitrification;

- implementation of nutrient management through State and Tribal efforts to implement watershed-based approaches to water quality management, including monitoring and assessing waters, adoption of water quality standards, which include nutrient criteria, developing total maximum daily loads (TMDLs), and implementing point source controls for nutrients through the National Pollutant Discharge Elimination System (NPDES);

- promoting public-private partnerships to restore buffers;

- promoting cost-effective flood control alternatives and implementing projects under the Coastal Wetlands Planning, Protection, and Restoration Act that result in nitrogen removal from the Mississippi and Atchafalaya Rivers;

- supporting actions by non-water quality State and Tribal agencies, private landowners, and agricultural and other industries to reduce nitrogen loadings to the basin; and

- providing voluntary incentives for nitrogen reductions from point and nonpoint sources.

Action Plan at 9–10.

Nothing too specific to be found there, and not much has happened in the Bush administration to add flesh to these bones since the Clinton Administration published the Action Plan. The "What's New" page on the Task Force's web site has few entries since January 2002. *See* http://www.epa.gov/msbasin/newhtm. In the only major action, the Task Force, with new members, held its 8th meeting in early February 2002, the major agenda item of which was to "familiarize new Task Force members with Action Plan." *See* http://www.epa.gov/msbasin/agenda20702.htm. Moreover, as described in Chapter 10 of this text, many of the "tools" the Action Plan says are already provided by the Clean Water Act and other programs to manage hypoxia have, for the most part, already proven to be ineffective for controlling nonpoint source runoff, particularly from agriculture. For more on that problem, see Sarah White, *Gulf Hypoxia: Can A Legal Remedy Breathe Life Into The Oxygen Depleted Waters?*, 5 DRAKE J. OF AGRICULTURAL LAW 519 (2000). Overall, Gulf hypoxia, like many of the problems examined in this text, remains a known and accelerating ecological dilemma with no concrete solution in sight.

This is not the case elsewhere, however, as some other nations have confronted hypoxia with more deliberate and forceful measures. In Denmark, for example, a massive lobster die-off in 1986 in the Kattegat straight heightened public awareness and demands for action. Denmark enacted tighter restrictions on farmers, making them account for all manure and commercial fertilizer applications, and forced wastewater treatment plants to reduce phosphorous emissions. The government also began an aggressive program of buying and restoring riparian lands and paying farmers to plant winter wheat to soak up soil nitrogen. The combination of approaches appears to be working, as phosphorous loads have decreased by 80 percent and algae growth appears to have been stunted. *See* John S. MacNeil, *Off Denmark, A Drawn–Out War Against Hypoxia*, 291 SCIENCE 969 (2001). Could these measures be replicated in the United States? Why haven't they been? What measures, with any reasonable chance of success *and* of being implemented, would you recommend for the United States?

3. LIVING RESOURCES

As our focus moves from the coastline outward to the continental shelf and open seas, ecosystem management concerns also shift from an emphasis on the impacts of land-based activities to one emphasizing the importance of water-based activities. The activity of primary concern is the harvesting of living resources, more commonly known as fishing. To be sure, other activities on the continental shelf, such as oil and mineral exploration and pollution from vessels, have had significant marine resource impacts, but *no* activity by any stretch of the imagination has had as profound an impact on the ocean ecosystem as has fishing. Ecosystem management in that setting has relied primarily on the two techniques covered in this section–management of fisheries and protection of marine areas through preserves.

a. FISHERIES MANAGEMENT

It was, at one time, unthinkable that we could exhaust the ocean's vast living resources. Today, the question is whether we are on an irreversible path toward their virtual depletion. And there appears to be no end to our voracious appetite for fish and other marine food sources. Humans consumed 71.7 million tons of fish, crustaceans, and molluscs in 1990, and 93.8 million tons in 1997–a one-third increase in 7 years. This rise in consumption, mirrored by a rise in harvesting production, might suggest that the ocean's resources are indeed limitless, but that is far from the truth. Increasingly, fishing operations (and thus human consumption) are depleting "higher" food chain species and relying more and more on "lower" species. And, as fishing is more often than not indiscriminate in terms of target species, a tremendous amount of unwanted species, known as "bycatch," is simply discarded, usually dead, back into the ocean. The result is that today estimates are that at least 50 major world fish stocks are "fully exploited," meaning they can withstand no increase in harvests and remain sustainable, and at least 25 percent are overfished or depleted. *See* U.N. Food and Agriculture Organization, State of the World Fisheries and Aquaculture–SOPHIA 2000 (2001), *available at* www.fao.org. The bottom line is that the fishing industry, if it intends to remain economically viable, must face significant self-or externally-imposed regulation. The following excerpt provides a concise explanation of how matters have evolved from the state of bounty to one of scarcity.

United Nations Food and Agriculture Organization, Code of Conduct for Responsible Fisheries

(October 31, 1995).

Preface

From ancient times, fishing has been a major source of food for humanity and a provider of employment and economic benefits to those engaged in this activity. The wealth of aquatic resources was assumed to be an unlimited gift of nature. However, with increased knowledge and the dynamic development of fisheries after the second world war, this myth has faded in face of the realization that aquatic resources, although renewable, are not infinite and need to be properly managed, if their contribution to the nutritional, economic and social well-being of the growing world's population is to be sustained.

The widespread introduction in the mid-seventies of exclusive economic zones (EEZs) and the adoption in 1982, after long deliberations, of the United Nations Convention on the Law of the Sea provided a new framework for the better management of marine resources. The new legal regime of the ocean gave coastal States rights and responsibilities for the management and use of fishery resources within their EEZs which embrace some 90 percent of the world's marine fisheries. Such extended national jurisdiction was a necessary but insufficient step toward the efficient management and sustainable development of fisheries. Many coastal States continued to face serious challenges as, lacking experience and financial and physical resources, they sought to extract greater benefits from the fisheries within their EEZs.

In recent years, world fisheries have become a market-driven, dynamically developing sector of the food industry and coastal States have striven to take advantage of their new opportunities by investing in modern fishing fleets and processing factories in response to growing international demand for fish and fishery products. By the late 1980s it became clear, however, that fisheries resources could no longer sustain such rapid and often uncontrolled exploitation and development, and that new approaches to fisheries management embracing conservation and environmental considerations were urgently needed. The situation was aggravated by the realization that unregulated fisheries on the high seas, in some cases involving straddling and highly migratory fish species, which occur within and outside EEZs, were becoming a matter of increasing concern.

[substantive provisions of the Code are reproduced later in this section]

Notes and Questions

1. For thorough accounts of how the trends discussed in the Code of Conduct preamble have evolved in the United States, and how legal and policy frameworks have been constructed in response, see MICHAEL L. WEBER, FROM ABUNDANCE TO SCARCITY: A HISTORY OF U.S. MARINE FISHERIES POLICY (2002); SUSAN HANNA ET AL., FISHING GROUNDS 23–35 (2000); MICHAEL J. BEAN & MELANIE J. ROWLAND, THE EVOLUTION OF NATIONAL WILDLIFE LAW 148–151 (1997); Eldon V.C. Greenberg, *Ocean Fisheries*, in ENVIRONMENTAL LAW: FROM RESOURCES TO RECOVERY 260–274 (Celia Campbell–Mohn et al. eds, 1993).

2. Overfishing can have profound effects on ecosystem dynamics well beyond the depletion of target and bycatch species populations, yet often these impacts are indirect and delayed in nature. For example, researchers are piecing together an explanation for the collapse of kelp forests off the Alaskan coast that begins with the impacts of whaling practices from 50 years ago. After World War II, whaling in Pacific waters steadily decimated the great whale species populations. As a result, Orcas, or killer whales, had to "fish down" the food chain, turning to other species as prey. Their initial target was sea lions. As sea lions faced the twin threat of Orcas, as well as human overfishing of their own prey fish species, their populations also began to dwindle. Orcas next turned to sea otters to fill the gap. As sea otters took the brunt, their food of choice, sea urchins, began to explode in numbers. And what is the sea urchin's staple? Kelp. Thus, the effects of whaling over 50 years ago set in motion an ecological reverberation effect that only recently has manifested itself in depleted kelp forests. *See* Jay Withgott, *A Whale of a Chain Reaction*, 295 SCIENCE 1457 (2002). Does this potential for complex, time-delayed ecological cascades suggest any particular approach to fisheries management? Consider as you review the following materials whether any of the legal frameworks covered could prevent this story from being repeated for other ecological contexts of the coastal and marine environment.

1. Regulating Fish Harvests

The United States, obviously, falls squarely within the challenge the U.N. Food and Agriculture Organization outlined in its 1995 Code of Conduct. At the core of our response is the Magnuson–Stevens Fishery Conservation and Management Act (Magnuson–Stevens Act), 16 U.S.C. §§ 1801–1882, as amended by the Sustainable Fisheries Act (SFA), Pub. L. No. 104–297, 110 Stat. 3559 (1996), which provides the primary statutory framework for the protection and management of the nation's marine fishery resources. It establishes eight Regional Fishery Management Councils (Councils), each of which has the authority and responsibility to govern conservation and management of the fisheries under its geographical jurisdiction. *See* 16 U.S.C. § 1852. The Councils perform this function by developing and implementing fishery management plans (FMPs) and amendments thereto. After a Council develops an FMP, the National Marine Fisheries Service (NMFS) and/or the National Ocean and Atmospheric Administration (NOAA), acting on behalf of the Secretary of Commerce, evaluate the FMP and determine whether it complies with the Magnuson–Stevens Act and other applicable law. The Secretary of Commerce may then approve, disapprove, or partially approve the FMP. *See id.* § 1854.

The approval of an FMP requires several steps: (1) an initial review of the FMP to ensure its consistency with the Magnuson–Stevens Act and other applicable law; (2) the publishing of the FMP in the Federal Register, followed by the commencement of a 60–day public comment period; and (3) the approval, disapproval, or partial approval of the FMP within 30 days of the end of the comment period. The Secretary may not adopt an FMP recommended by a Council if that FMP violates any of the ten "National Standards" for FMPs established by the Magnuson–Stevens Act, as amended. These are described as follows:

> Any fishery management plan prepared, and any regulation promulgated to implement any such plan, pursuant to this title shall be consistent with the following national standards for fishery conservation and management:
>
> (1) Conservation and management measures shall prevent overfishing while achieving, on a continuing basis, the optimum yield from each fishery for the United States fishing industry.
>
> (2) Conservation and management measures shall be based upon the best scientific information available.
>
> (3) To the extent practicable, an individual stock of fish shall be managed as a unit throughout its range, and interrelated stocks of fish shall be managed as a unit or in close coordination.
>
> (4) Conservation and management measures shall not discriminate between residents of different States. If it becomes necessary to allocate or assign fishing privileges among various United States fishermen, such allocation shall be (A) fair and equitable to all such fishermen; (B) reasonably calculated to promote conservation; and (C)

carried out in such manner that no particular individual, corporation, or other entity acquires an excessive share of such privileges.

(5) Conservation and management measures shall, where practicable, consider efficiency in the utilization of fishery resources; except that no such measure shall have economic allocation as its sole purpose.

(6) Conservation and management measures shall take into account and allow for variations among, and contingencies in, fisheries, fishery resources, and catches.

(7) Conservation and management measures shall, where practicable, minimize costs and avoid unnecessary duplication.

(8) Conservation and management measures shall, consistent with the conservation requirements of this Act (including the prevention of overfishing and rebuilding of overfished stocks), take into account the importance of fishery resources to fishing communities in order to (A) provide for the sustained participation of such communities, and (B) to the extent practicable, minimize adverse economic impacts on such communities.

(9) Conservation and management measures shall, to the extent practicable, (A) minimize bycatch and (B) to the extent bycatch cannot be avoided, minimize the mortality of such bycatch.

(10) Conservation and management measures shall, to the extent practicable, promote the safety of human life at sea.

16 U.S.C. §§ 1853(a)(1)-(10). In addition, the Secretary may propose federal management measures if such measures become necessary and the relevant Council failed to propose an appropriate measure within a reasonable amount of time. *See id.* at § 1854(c)(1). After the Secretary approves an FMP, the relevant Council implements it through a "framework adjustment." *See* 50 C.F.R. § 648.90.

In 1996, Congress enacted the SFA in order to further protect fishery stocks, many of which had become severely depleted by the mid–1980s. The SFA strengthened the Magnuson–Stevens Act by requiring NMFS and the Councils (1) to prevent overfishing and rebuild depleted fish populations, and (2) to report, assess, and minimize bycatch. The statute defines "overfishing" as the "rate or level of fishing mortality that jeopardizes the capacity of a fishery to produce the maximum sustainable yield on a continuing basis." 16 U.S.C. § 1802(29). The SFA regulates fishing mortality rates (F) with reference to this maximum sustainable yield (MSY), which is defined by regulation as "the largest long-term average catch or yield that can be taken from a stock or stock complex under prevailing ecological and environmental conditions." 50 C.F.R. § 600.310(c)(1)(i). Hence, as one legislator put it, the SFA "essentially says we can only catch that portion of the fish that represents interest. This is called the maximum sustainable yield. Without touching the principal fish; that being the critical population necessary to replenish the stock year after year, we will continue to have fish." 141 Cong. Rec. H10232 (daily ed. Oct. 18, 1995) (statement of Rep.

Wayne Gilchrest, House Subcommittee on Fisheries Conservation, Wildlife and Oceans).

In addition, however, the SFA also requires that all fishery management plans rebuild depleted fish populations within a period "as short as possible," but "not to exceed ten years." 16 U.S.C. § 1854(e)(4)(A). The SFA thus not only prohibited all fishing at levels exceeding MSY, but also required fish populations to be rebuilt to a biomass level (B) that "allows the fishery to produce MSY on a continuing basis." *See id.* §§ 1802(28)–(29). The SFA imposed an October 11, 1998 deadline on all Councils to develop the necessary rules and regulations to comply with this and other provisions of the statute, and imposes a statutory duty on NMFS to ensure that the Councils do so. *See id.* § 1854(c)(1).

The following materials assess the Magnuson–Stevens Act from three perspectives. First, we present a recent case illustrating the type of disputes that arise under the Act and how the courts approach them. Second, we provide a critique of the legislation from an economist and legal scholar. Finally, we present the central substantive provisions of the FAO Code of Conduct to allow evaluation of the extent to which the United States has fulfilled the Code's vision.

Natural Resources Defense Council, Inc. v. Daley

United States Court of Appeals, District of Columbia Circuit, 2000.
209 F.3d 747.

■ Harry T. Edwards, Chief Judge:

Paralichthys dentatus, or summer flounder, a commercially valuable species of flounder, dwells off the Atlantic coast and are harvested primarily between May and October from North Carolina to Maine. The summer flounder fishery is an "overfished" fishery, in the process of recovering from severe depletion prevalent during the late 1980s and early 1990s. The Secretary of Commerce, advised by the National Marine Fisheries Service ("the Service"), the principal appellee in this case, annually sets a fishing quota limiting each year's summer flounder catch, pursuant to the Magnuson-Stevens Fishery Conservation and Management Act ("the Fishery Act"), 16 U.S.C. §§ 1801–1883 (1994 & Supp. IV 1998). This case involves appellants' challenge to the Service's quota for the 1999 summer flounder harvest.

Before the District Court, appellants alleged that the 1999 quota did not provide sufficient assurance that it would meet the conservation goals of the Fishery Act and attendant regulations. Appellants also claimed that the Service's conclusion that the quota had no significant environmental impact was based on an inadequate environmental assessment, thereby violating the National Environmental Policy Act ("NEPA"). On cross-motions for summary judgment, the District Court granted judgment in favor of appellees. See Natural Resources Defense Council, Inc. v. Daley, 62 F.Supp.2d 102 (D.D.C.1999).

We reverse the District Court and remand the case to the Service for further proceedings consistent with this opinion. The 1999 quota, when adopted, had a documented 18% likelihood of meeting the statute's conservation goals. We hold that, under the Fishery Act, the disputed quota is insufficient to meet Congress' mandate to the Service to prevent overfishing and to assure that specific conservation goals are met. We also hold that the Service's proposal to supplement the quota with other purportedly protective measures does not satisfactorily ameliorate the quota's glaring deficiencies. Because of our disposition on these grounds, we have no need to reach appellants' NEPA claims.

I. BACKGROUND

A. Regulatory Background

The Fishery Act was enacted to establish a federal-regional partnership to manage fishery resources. Under the statute, there are eight Regional Fishery Management Councils "to exercise sound judgment in the stewardship of fishery resources." 16 U.S.C. §§ 1801(b)(5), 1852(a) (Supp. IV 1998). Management Councils propose and monitor fishery management plans "which will achieve and maintain, on a continuing basis, the optimum yield from each fishery." Id. § 1801(b)(4) (1994). Management Councils submit management plans to the Secretary of Commerce (functionally the Service), who may then adopt them through notice and comment rulemaking. See id. § 1854(a) (Supp. IV 1998). An "optimum yield" under the statute is defined as the "maximum sustainable yield from the fishery." Id. § 1802(28)(B) (Supp. IV 1998). If a fishery is "overfished," the management plan must "provide[ ] for rebuilding to a level consistent with" the maximum sustainable yield. Id. § 1802(28)(C). A fishery is "overfished" if the rate of fishing mortality "jeopardizes the capacity of a fishery to produce the maximum sustainable yield on a continuing basis." Id. § 1802(29).

The Service defines overfishing and optimum yield according to the fishing mortality rate ("F"). F represents that part of a fish species' total mortality rate that is attributable to harvesting by humans, whether through capture or discard. Fish are "discarded" for many reasons, including, for example, when they are the wrong species, undersized, or not valuable enough. Values for F can range anywhere from 0 to over 2, and only indirectly represent the amount of fish captured by industry. For instance, an F of 1.4 means that about 20% of all summer flounder that are alive at year 1 will be alive at year 2. There is a specific F, termed "F submax," that is defined as that fishing mortality rate that will maximize the harvest of a single class of fish over its entire life span. Overfishing is fishing in excess of F submax. Therefore, the basic goal of a management plan is to achieve F submax, thereby preventing overfishing and assuring optimum yield.

B. The Summer Flounder Fishing Quota

From a commercial standpoint, the summer flounder is one of the most important species of flounder in the United States. All parties agree that the summer flounder fishery is "overfished" and has been for some time. The Mid–Atlantic Fishery Management Council ("MAFMC"), covering New York, New Jersey, Delaware, Pennsylvania, Maryland, Virginia, and North Carolina, developed the original summer flounder management plan with the assistance of two other regional Management Councils and the Atlantic States Marine Fisheries Commission ("the Commission"), a consortium of 15 coastal states and the District of Columbia. The Service approved the original management plan in 1988; however, the Service has amended the plan several times. At the time relevant to the instant case, the plan was designed to achieve a fishing mortality rate equal to F submax by 1998.

Pursuant to the management plan, the Service must set a quota each year fixing the total weight of summer flounder that may be harvested by commercial and recreational fishers. This quota is referred to as the "total allowable landings" for the year, or "TAL." The Service allocates 60% of the TAL to commercial fisheries and 40% of the quota to recreational fisheries, and states receive allocations based upon their share of the summer flounder fishery. States may subdivide their allocated commercial quota between "incidental" and "directed" catch. Directed fisheries intentionally harvest summer flounder. Fishers who catch juvenile flounder, or who are part of the directed fishery for another species and catch summer flounder unintentionally, have harvested incidental catch.

The TAL must meet several requirements. It must be consistent with the 10 national standards of fishery conservation and management set out in the Fishery Act. See 16 U.S.C. § 1851(a)(1)-(10) (1994 & Supp. IV 1998). Most relevant to the instant case, the quota must embody conservation measures that "shall prevent overfishing while achieving, on a continuing basis, the optimum yield from each fishery for the United States fishing industry." Id. § 1851(a)(1) (1994). The quota must also be "consistent with" the fishery management plan. See id. § 1854(b)(1). Finally, under the applicable regulations, the Regional Administrator of the Service must annually adopt a final rule "implement[ing] the measures necessary to assure that the applicable specified F will not be exceeded." 50 C.F.R. § 648.100(c) (1999). The "applicable specified F" is also referred to as the "target F."

There is a relatively direct relationship between the TAL and the likelihood of achieving the target F. In general, the higher the TAL, the less likely a plan is to achieve the target F. In other words, the lower the target F, the lower the TAL must be to attain the target F. The basic dispute between the parties concerns whether the 1999 TAL provides a sufficient guarantee that the target F for summer flounder will be achieved.

For 1999, the summer flounder fishery management plan mandated a target F equivalent to F submax, which was 0.24. The Summer Flounder Monitoring Committee, a MAFMC committee, had recommended a TAL of 14.645 million pounds, while MAFMC had recommended a TAL of 20.20 million pounds. The Service rejected MAFMC's recommendation as "unacceptably risk-prone" for several reasons: (1) it had an "unacceptably low probability" of 3% of achieving the target F; (2) it had a 50% probability of achieving an F of 0.36, which was "significantly higher" than the target F; (3) the proposal relied on unpredictable data; and (4) MAFMC had "yet to specify a harvest level that has achieved the annual target F." Fisheries of the Northeastern United States; Summer Flounder, Scup, and Black Sea Bass Fisheries, 63 Fed.Reg. 56,135, 56,136 (1998) (to be codified at 50 C.F.R. pt. 648) (proposed Oct. 21, 1998) ("Proposed TAL"). The Service also rejected the Summer Flounder Monitoring Committee's recommendation of a 14.645 million pound TAL. Although the Committee's recommendation had a 50% chance of achieving the target F, the Service rejected the proposal without any meaningful explanation.

On October 21, 1998, the Service proposed a TAL of 18.52 million pounds. See id. All parties agree that, at most, the Service's proposal afforded only an 18% likelihood of achieving the target F. The Service also proposed an incidental catch restriction "to address discards in this fishery that should further reduce the overall mortality." Id. This measure provided that, within the commercial fishery, 32.7% of the allocated quota be committed to incidental catch. In the end, then, the Service proposed a TAL of 7.41 million pounds for recreational harvest, 7.47 million pounds for directed commercial harvesting, and 3.64 million pounds for incidental commercial catch, for a total of 18.52 million pounds. See id. The Service also considered recent changes in minimum mesh size. On this point, the Service noted that, while MAFMC felt that the "recently adopted mesh provision requiring 5.5 inch" mesh throughout the net would "substantially reduce discard and discard mortality," the alleged benefits of mesh had yet to be verified by anyone. Id.

Between the time of proposal of the 1999 TAL and its adoption, the Service concluded that it did not have the authority to impose any incidental catch restrictions on the states. Therefore, the Service merely recommended that the states adopt the incidental catch proposal, making the proposal entirely voluntary. The Commission, the body representing 15 coastal states and the District of Columbia, also declined to command the states to adopt the proposal. According to an advisor to the Service's Assistant Administrator for Fisheries, this development "result[ed] in an unknown but probably substantial reduction in the likelihood that [MAFMC's] rebuilding schedule will be achieved," and he therefore recommended that the Service adopt the Summer Flounder Monitoring Committee's recommended 14.645 million pound TAL.

The Service rejected this recommendation and, on December 31, 1998, issued the final TAL, adopting its initial proposal. The Service acknowledged that the Summer Flounder Monitoring Committee's recommended quota had a 50% chance of achieving the target F, while the Service's TAL had only an 18% chance of achieving the target F. See Fisheries of the Northeastern United States; Summer Flounder, Scup, and Black Sea Bass Fisheries, 63 Fed.Reg. 72,203, 72,203–04 (1998) (codified at 50 C.F.R. pt. 648) ("Final TAL"). The Service also recognized that the incidental catch provisions were entirely voluntary. See id. at 72,204. The Service simply recommended that states adopt the additional incidental catch provisions "[t]o improve the probability of achieving the target [F]." Id. Nowhere did the Service analyze the effect on fishing mortality of shifting from a mandatory to a voluntary incidental catch provision.

The Service responded to comments that the TAL did not sufficiently assure achievement of the target F by stating that: (1) the TAL had a higher probability of meeting the target F than MAFMC's 20.2 million pound recommendation; and (2) the incidental catch recommendations "would improve the likelihood that the target fishing mortality rate would be attained." Id. at 72,206. In response to other comments, the Service suggested that the 5.5 inch minimum mesh provision might ameliorate other mortality concerns, but acknowledged that the requirement had not been in effect long enough to determine its efficacy. See id. at 72,208.

Appellants filed suit in District Court on January 29, 1999, seeking, inter alia, (1) a declaratory judgment that defendants violated the Fishery Act, the Administrative Procedure Act ("APA"), and NEPA, and (2) remand to the agency to impose a new summer flounder TAL. The District Court upheld the Service's adoption of the 18.52 million pound TAL, deferring to the agency under Chevron U.S.A. Inc. v. Natural Resources Defense Council, Inc., 467 U.S. 837, 104 S.Ct. 2778, 81 L.Ed.2d 694 (1984). The District Court first determined that §§ 1851 (a)(1) and (a)(8) in the Fishery Act evinced competing interests between advancing conservation and minimizing adverse economic effects and that Congress offered no insight as to how to balance these concerns. See Natural Resources Defense Council, 62 F.Supp.2d at 106–07. In addition, the trial court found that the Fishery Act expressed no clear intent as to the particular level of certainty a TAL must guarantee to be consistent with 16 U.S.C. § 1851(a)(1). See id. at 107. Given these perceived ambiguities, the District Court deferred to the Service pursuant to Chevron Step Two. This appeal followed.

II. ANALYSIS

As we recently held in Associated Builders & Contractors, Inc. v. Herman, 166 F.3d 1248 (D.C.Cir.1999), [i]n a case like the instant one, in which the District Court reviewed an agency action under the [APA], we review the administrative action directly. See Troy Corp. v. Browner, 120 F.3d 277, 281 (D.C.Cir.1997); Gas Appliance Mfrs. v. Department of Energy, 998 F.2d 1041, 1045 (D.C.Cir.1993). In other words, we accord no particular deference to the judgment of the District Court. See Gas Appliance Mfrs., 998 F.2d at 1045. Rather, on an independent review of the record, we will uphold [the agency's] decision unless we find it to be "arbitrary, capricious, an abuse of discretion, or otherwise not in accordance with law." 5 U.S.C. § 706(2)(A) (1994). Id. at 1254.

As for the Service's disputed interpretations of the Fishery Act, we are guided by the Supreme Court's seminal decision in Chevron U.S.A., Inc., [467 U.S. at 837], [which] governs review of agency interpretation of a statute which the agency administers. Under the first step of Chevron, the reviewing court "must first exhaust the 'traditional tools of statutory construction' to determine whether Congress has spoken to the precise question at issue." Natural Resources Defense Council, Inc. v. Browner, 57 F.3d 1122, 1125 (D.C.Cir.1995) (quoting *Chevron*, 467 U.S. at 843 n. 9, 104 S.Ct. 2778). * * * If, however, "the statute is silent or ambiguous with respect to the specific issue," Chevron, 467 U.S. at 843, Congress has not spoken clearly, and a permissible agency interpretation of the statute merits judicial deference. Id. Bell Atlantic Telephone Companies v. FCC, 131 F.3d 1044, 1047 (D.C.Cir.1997). Although agencies are entitled to deferential review under *Chevron* Step Two, our judicial function is neither rote nor meaningless: [W]e will defer to [an agency's] interpretation[ ] if [it is] reasonable and consistent with the statutory purpose and legislative history. * * * This case presents a situation in which the Service's quota for the 1999 summer flounder harvest so completely diverges from any realistic meaning of the Fishery Act that it cannot survive scrutiny under *Chevron* Step Two.

As an initial matter, we reject the District Court's suggestion that there is a conflict between the Fishery Act's expressed commitments to conservation and to mitigating adverse economic impacts. Compare 16 U.S.C. § 1851(a)(1) (directing agency to "prevent overfishing" and ensure "the optimum yield from each fishery"); with id. § 1851(a)(8) (directing agency to "minimize adverse economic impacts" on fishing communities). The Government concedes, and we agree, that, under the Fishery Act, the Service must give priority to conservation measures. It is only when two different plans achieve similar conservation measures that the Service takes into consideration adverse economic consequences. This is confirmed both by the statute's plain language and the regulations issued pursuant to the statute. See id. § 1851(a)(8) (requiring fishery management plans, "consistent with the conservation requirements of this chapter," to take into account the effect of management plans on fishing communities); 50 C.F.R. § 600.345(b)(1) (1999) ("[W]here two alternatives achieve similar conservation goals, the alternative that ... minimizes the adverse impacts on [fishing] communities would be the preferred alternative.").

The real issue in this case is whether the 1999 TAL satisfied the conservation goals of the Fishery Act, the management plan, and the Service's regulations. In considering this question, it is important to recall that the Service operates under constraints from three different sources. First, the statute requires the Service to act both to "prevent overfishing" and to attain "optimum yield." 16 U.S.C. § 1851(a)(1). Overfishing is commonly understood as fishing that results in an F in excess of F submax. Since F submax for 1999 was equivalent to 0.24, this constraint required the Service to issue regulations to prevent F from exceeding 0.24. Second, any quota must be "consistent with" the fishery management plan adopted by the Service. See id. § 1854(b)(1). In this case the fishery management plan called for an F of 0.24. Therefore, the quota had be to "consistent with" achieving that F. Third, the Service is required to adopt a quota "necessary to assure that the applicable specified F will not be exceeded." 50 C.F.R. § 648.100(c). The "applicable specified F" for 1999 was F submax, or 0.24.

All of these constraints, then, collapse into an inquiry as to whether the Service's quota was "consistent with" and at the level "necessary to assure" the achievement of an F of 0.24, and whether it reasonably could be expected to "prevent" an F greater than 0.24. In other words, the question is whether the quota, as approved, sufficiently ensured that it would achieve an F of 0.24. Appellants argue that the quota violates applicable standards under both *Chevron* Step One and *Chevron* Step Two. Because we find appellants' *Chevron* Step Two arguments convincing, we have no need to reach their alternative argument that the Service violated NEPA by relying on an inadequate environmental assessment in promulgating the final rule.

Appellants' *Chevron* Step One "plain meaning" argument is virtually indistinguishable from their *Chevron* Step Two reasonableness argument. Appellants acknowledge that the statutory terms "assure," "prevent," and "consistent with" do not mandate a precise quota figure. However, appellants contend that a TAL with only an 18% likelihood of achieving the target F is so inherently unreasonable that it defies the plain meaning of the statute. This is an appealing argument on the facts of this case, because, as we explain below, the Service's action is largely incomprehensible when one considers the principal purposes of the Fishery Act. Nonetheless, we still view this case as governed by *Chevron* Step Two. The statute does not prescribe a precise quota figure, so there is no plain meaning on this point. Rather, we must look to see whether the agency's disputed action reflects a reasonable and permissible construction of the statute. In light of what the statute does require, short of a specific quota figure, it is clear here that the Service's position fails the test of *Chevron* Step Two.

The 1999 quota is unreasonable, plain and simple. Government counsel conceded at oral argument that, to meet its statutory and regulatory mandate, the Service must have a "fairly high level of confidence that the quota it recommends will not result in an F greater than [the target F]." Fishermen's Dock Coop., Inc. v. Brown, 75 F.3d 164, 169–70 (4th Cir.1996). We agree. We also hold that, at the very least, this means that "to assure" the achievement of the target F, to "prevent overfishing," and to "be consistent with" the fishery management plan, the TAL must have had at least a 50% chance of attaining an F of 0.24. This is not a surprising result, because in related contexts, the Service has articulated precisely this standard. See National Marine Fisheries Service, Final Fishery Management Plan for Atlantic Tunas, Swordfish and Sharks, Vol. I, at 288, reprinted in J.A. 382 (April 1999) (concluding that the Service should choose management measures that have "at least a 50–percent confidence in target reference points," and when choosing between two alternatives with a greater than 50% probability, should choose the higher "unless there are strong reasons to do otherwise").

The disputed 1999 TAL had at most an 18% likelihood of achieving the target F. Viewed differently, it had at least an 82% chance of resulting in an F greater than the target F. Only in Superman Comics' Bizarro world, where reality is turned upside down, could the Service reasonably conclude that a measure that is at least four times as likely to fail as to succeed offers a "fairly high level of confidence."

* * *

As we noted at the outset of this opinion, the Service's quota for the 1999 summer flounder harvest so completely "diverges from any realistic meaning" of the Fishery Act that it cannot survive scrutiny under *Chevron* Step Two. See GTE Serv. Corp., 205 F.3d at 421. The Service resists this result by suggesting that we owe deference to the agency's "scientific" judgments. See Br. for Appellees at 33. While this may be so, we do not hear cases merely to rubber stamp agency actions. To play that role would be "tantamount to abdicating the judiciary's responsibility under the Administrative Procedure Act." A.L. Pharma, Inc. v. Shalala, 62 F.3d 1484, 1491 (D.C.Cir.1995). The Service cannot rely on "reminders that its scientific determinations are entitled to deference" in the absence of reasoned analysis "to 'cogently explain' " why its additional recommended measures satisfied the Fishery Act's requirements. Id. at 1492 (quoting Motor Vehicle Mfrs. Ass'n, Inc. v. State Farm Mut. Auto. Ins. Co., 463 U.S. 29, 48, 103 S.Ct. 2856, 77 L.Ed.2d 443 (1983)). Indeed, we can divine no scientific judgment upon which the Service concluded that its measures would satisfy its statutory mandate.

Here, the adopted quota guaranteed only an 18% probability of achieving the principal conservation goal of the summer flounder fishery management plan. The Service offered neither analysis nor data to support its claim that the two additional measures aside from the quota would increase that assurance beyond the at-least–50% likelihood required by statute and regulation.

Notes and Questions

1. As the *NRDC v. Daley* case makes abundantly clear, fisheries biology, as a branch of applied ecology, has focused on a quest for the perfect fish harvest optimization "algorithm," and the result has been a regulatory structure that is expressed through highly technical, quasi-mathematical concepts such as F, F submax, and TAL. As the ecologist P.A. Larkin suggested several decades ago, this approach opens the door to the fundamental question: What shall be optimized? *See* P.A. Larkin, *Fisheries Management–An Essay for Ecologists*, 9 ANNUAL REVIEW OF ECOLOGY AND SYSTEMATICS 57 (1978). As Larkin explained then, traditional fisheries biologists sought to optimize maximum sustained yield. Later, attention turned to maximum sustained economic return, and then, with the Magnuson–Stevens Act, to an optimum sustained yield approach. But the expansion of EEZs to 200–miles invited many countries to pursue "variations of long-standing policies that place fisheries matters in a much broader context of economic and political optimization. The field is wide open for modeling extravaganzas which may expose the consequences of virtually any proposed course of action; and the recipes for action that emerge will be the logical consequences of the predisposing assumptions." *Id.* at 69. Is there any evidence in *NRDC v. Daley* that the "optimization" that had taken place for the summer flounder was more political than biological in focus? Put another way, how is that, as the court put it, we wound up with a management regime for the summer flounder straight out of Superman Comics' Bizarro world?

2. The "modeling extravaganzas" of which P.A. Larkin wrote suggest a science-policy interface that depends on confidence that (1) the models are sound; (2) the scientific information used in the models is reliable; and (3) the policy decision makers actually follow the model results. For example, as the *NRDC v. Daley* court described it, that case "collapse[d] into an inquiry as to whether the Service's quota was 'consistent with' and at the level 'necessary to assure' the achievement of an F of 0.24, and whether it reasonably could be expected to 'prevent' an F greater than 0.24." How confident are you that we can reliably determine what F is? How confident are you that, if we know F for a species, NMFS and the Regional Councils will implement quotas that will achieve F? The Stanford Fisheries Project is examining these questions based on a survey of 25 years of historical records from 11 U.S. fisheries. *See The Use of Science in U.S. Fisheries*, http://fisheries.stanford.edu. Thus far the researchers have found that the fishery quotas Regional Councils set often fall above or below the range recommended by the scientists advising the Councils. Is this evidence of "political optimization" or simply lack of understanding?

3. Those political and policy issues aside, the basic legal grist of the Magnuson–Stevens Act boils down to issues like that presented in *NRDC v. Daley*. At the very least, therefore, F, TAL, and their related acronyms do provide fact-based inquiries for the law to apply. The *NRDC v. Daley* court explained how these legal hooks operate on three levels:

> First, the statute requires the Service to act both to "prevent overfishing" and to attain "optimum yield." 16 U.S.C. § 1851(a)(1). Overfishing is commonly understood as fishing that results in an F in excess of F submax. Second, any quota must be "consistent with" the fishery management plan adopted by the Service. See id. § 1854(b)(1). In this case the fishery management plan called for an F of 0.24. Third, the Service is required to adopt a quota "necessary to assure that the applicable specified F will not be exceeded." 50 C.F.R. § 648.100(c). The "applicable specified F" for 1999 was F submax, or 0.24.

As Professor Oliver Houck has argued, without these legal hooks there may be little hope of producing anything but political optimization in domestic fisheries management. *See* Oliver A. Houck, *On the Law of Biodiversity and Ecosystem Management*, 81 Minn. L. Rev. 869, 949–53 (1997). Does *NRDC v. Daley* suggest to you that there is an adequate legal backstop to prevent fisheries management from falling into that trap?

4. What do you think of the court's interpretation in *NRDC v. Daley* that "under the Fishery Act, the Service must give priority to conservation measures. It is only when two different plans achieve similar conservation measures that the Service takes into consideration adverse economic consequences." NMFS has interpreted the statute this way. *See* 50 C.F.R.

§ 600.345(b). Do you agree with the court's and the agency's interpretation of the statute? Do you agree with their interpretation as a matter of policy in general? If this is true, couldn't the agency ensure that it will never have to take into account adverse economic consequences simply by characterizing different management plans as achieving dissimilar conservation measures? Several courts have grappled with these issues. One found, quite to the contrary of *NRDC v. Daley*, that NMFS failed adequately to evaluate the economic impact of summer flounder quotas on North Carolina fishing operations, characterizing the quota as "a buzzsaw to mow down whole fishing communities in order to save some fish." North Carolina Fisheries Ass'n v. Daley, 27 F.Supp.2d 650, 667 (E.D.Va.1998). A thorough summary of these and other cases on the topic is found in Kristen M. Fletcher, *When Economic Analysis and Conservation Clash: Challenges to Economic Analysis in Fisheries Management*, 31 Envtl. L. Rep. (ELI) 1168 (2001).

Shi-Ling Hsu and James E. Wilen, Ecosystem Management and the 1996 Sustainable Fisheries Act

24 Ecology Law Quarterly 799 (1997).

The record of fisheries conservation in the U.S. since the [Magnuson-Stevens Fisheries Conservation and Management Act (FCMA)] is checkered. In some regions, councils have been able to insulate the scientific determination of optimum yield from the intensely acrimonious decisions about which groups using what kinds of gear should catch the fish. In other regions, however, councils have not overcome industry pressures to increase harvest targets, sometimes justified by directives that weaken the mandate to set clear biological targets with vague admonitions to consider economic or social factors. In some regions, managers have been unable even to gather rudimentary data necessary to begin managing fisheries, such as landings records.

This ... highlights one of the most significant failures of fisheries conservation under the FCMA: the failure to contain entry and capacity growth in fisheries. As economists have warned for forty years, fishing capacity in open access fisheries tends to be directly proportional to their profitability. If no authority regulates the fish stocks, too many vessels end up chasing too few fish, which can lead to a complete collapse of the fishery. Regulators may be able to maintain a healthy biomass by employing management measures like shortened seasons to limit harvests to sustainable levels. However, because such measures encourage overcapitalization, as in the halibut case where the whole fleet geared up to fish a five day season, they create significant economic waste. The original FCMA avoided addressing the inevitability of overcapitalization and left this responsibility to the regional councils, where political pressures to leave fisheries open to all entrants were most strongly expressed. As a result of this, most of the fisheries in the U.S. are vastly overcapitalized. Short seasons govern even those in good shape biologically, leading to relatively poorer quality fish, and encouraging fishing with methods that emphasize volume and induce bycatch, discarding, and other forms of waste.

THE 1996 SUSTAINABLE FISHERIES ACT

The 1996 Sustainable Fisheries Act (the "SFA"), which amended the 1976 Magnuson Act, reflects another step in fine tuning the fisheries regulatory apparatus established under the original act. At the same time, the SFA, like the original act, is clearly the outcome of political logrolling, tradeoffs, and compromises among disparate groups with different agendas. The policy impacts, achieved by a negotiation process that included input from conservation organizations and fishing industry representatives, are by no means inconsequential, but they should not be taken for a systemic overhaul of fisheries management.

Stamps of conservation groups are imprinted upon numerous provisions of the SFA. Perhaps the most significant change the SFA effected is the removal of some discretion regarding "overfished" fisheries. If the Secretary of Commerce (the "Secretary") determines that a fishery is "overfished," she is required to immediately notify the appropriate regional fishery management council and give the council one year to develop a fishery management plan that ends overfishing and rebuilds the stock of fish. If the regional council fails to develop a plan within one year, the Secretary is required to prepare a plan within nine months. The plan to end overfishing must do so within a time frame that is "as short as possible, taking into account the status and biology of any overfished stock of fish . . . and the interaction of the overfished stock of fish within the marine ecosystem," but must generally be accomplished in less than ten years. The regional council developing the fishery management plan is still responsible for specifying "objective and measurable criteria for identifying when the fishery . . . is overfished." The SFA also establishes new requirements regarding bycatch—fish that fishermen catch incidentally when fishing for another species. Fishery management plans under the SFA must be consistent not only with the seven national standards described above, but also with the three additional ones specified by the SFA, including one that requires plans to minimize bycatch or minimize the mortality from bycatch. Further, the first national standard, mandating that management achieve the "optimum yield" from each fishery, previously defined as "maximum sustainable yield . . . as modified by any relevant economic, social, or ecological factor," has been altered to allow only that maximum yield be reduced by any such relevant factors. This prevents councils from raising allowable harvests in response to local pressure for larger allocations.

* * *

THE OVERCAPITALIZATION PROBLEM

While the SFA contains some notable conservation advances, it does little to address the most fundamental cause of overfishing and waste—the chronic overcapitalization of fishing industries. The technological resourcefulness of fishermen has historically made a mockery of the most stringent and carefully crafted command and control regulations aimed at reducing fishing effort. For example, reductions in season lengths have encouraged fishermen to build bigger, faster vessels with more short-term harvesting capacity, necessitating further reductions in season lengths. Limitations or restrictions on gear types or capacity (e.g., net size regulations) have invited substitution of other inputs that partially thwart the regulations' original purpose, leading to further attempts by regulators to contain fishing technology's impact on overall harvest levels. Even in cases where limited entry programs have been instituted to freeze capacity and prevent further entry by new boats, there has been a need for additional measures to control capacity as fishermen have continued to increase individual vessel capacity on existing boats. In fact, measures that constrain fishing capital growth by fiat focus only on the symptom of the problem and not on the cause, which is the open access nature of the resource.

A few fisheries have adopted a measure that attacks the fundamental property rights problem: individual fishermen quotas (IFQs). IFQs grant rights to harvest a given percentage of the biologically determined total annual allowable catch. They are, in effect, a property right to the potential harvest. As a result, they change the incentives fishermen face in a radical way. Under IFQs, a fisherman does not need to build a bigger boat to outfish his competitors before regulators close the season; he may fish whenever it is efficient to do so during the season. While IFQs have only been adopted in three fisheries in the U.S., they have been adopted in over fifty fisheries worldwide.

The impacts of IFQs on fisheries are a remarkable counterpoint to the status quo in traditional fisheries managed by closed seasons and gear restrictions. IFQs have reversed the race to overcapitalize, because they encourage fishermen to downsize and adopt fishing practices more suitable to producing higher valued products year-round. Reduced overcapitalization offers many benefits. First, the product itself improves, as fish formerly frozen because of short seasons are available fresh throughout the year. Second, fishermen begin to act as stakeholders of the resource, since detrimental actions more clearly impact their own potential revenues. This is an important byproduct since fishermen and regulators tend to view most modern regulated fisheries as an adversarial struggle. Finally, in fisheries where there were formerly significant amounts of bycatch and discards, fishermen reduce waste, particularly if bycatch is included in their quota.

Notes and Questions

1. When Individual Fishing Quotas (IFQs) are made transferrable, they are referred to as Individual Transferrable Quotas (ITQs). The point of IFQs, as the authors of the article explain, is to reduce the motivation each individual fishing operation has to race to maximize its individual harvest share of a fishery-wide quota. The IFQ relieves the pressure to invest in technology to compete in the race. The point of transforming IFQs into ITQs, then, is to allow a market in IFQs that will direct the quotas to the most efficient fishing operations. The result is that the holders of ITQs should be the fishing operations that have most efficiently balanced investment in technology and intensity of fishing against the yield. Yet, as elegant as the economic theory sounds, not everyone is enamored of IFQs or ITQs. One commentator argues that they are subject to three structural flaws. First, they require an initial allocation of individual quotas that becomes entrenched economically in the form of reliance, and the quotas are thus politically difficult to change later in the light of better scientific evidence about appropriate harvest levels. Second, they impose a heavy information burden in order to set the initial quotas at reliable levels. Finally, they produce the possibility of "hot spots" as quotas may be exercised, particularly under ITQ systems, in concentrated areas and times. *See* Sharon R. Siegel, *Applying the Habitat Conservation Model to Fisheries Management: A Proposal for a Modified Fisheries Planning Requirement*, 25 COLUM. J. ENVTL. LAW 141, 151–54 (2000). But aren't these three problems evident, perhaps far more than the authors of article suggest, in conventional fishery-wide quota systems? How can ecosystem management policies be crafted that avoid or mitigate these three phenomena?

2. Would IFQs or ITQs have solved the summer flounder overfishing problem dealt with in *NRDC v. Daley*? How would you set up an IFQ or ITQ program to address the overfished flounder stocks?

3. The IFQ and ITQ instruments focus on the relationship between the success of fisheries management and the economic structure of property rights in the fisheries themselves. Open access regimes, in which exclusivity of fishing rights does not exist or is poorly enforced, seem inevitably to lead to counter-productive "racing" and over-capitalization behavior. Far more effectively than fishery-wide quotas, IFQs and ITQs provide the right economic incentives to their holders to conserve fishing effort and capital. But what of the Native American fisheries? Did they not prosper despite a lack of poorly defined private property rights? This conventional lore may in fact be at odds with the historical record. For example, evidence suggests that the tribes of the Pacific Northwest maintained thriving salmon fisheries for thousands of years prior to European settlement by relying on a complex, but strictly enforced, system of exclusive tribal rights to different salmon streams. In essence, they developed tribal FQs and tribal TQs long before any economic theory supported the case for doing so. *See* D. Bruce Johnsen, *Customary Law, Scientific Knowledge, and Fisheries Management among Northwest Coast Tribes*, 10 N.Y.U. ENVTL. L.J. 1 (2001). Does this suggest that economists such as Liu and Wilen are on the right track when it comes to modern fisheries management? For an additional perspective on overcapitalization in the fishing industry and the role ITQs could play, see Robert J. McManus, *America's Saltwater Fisheries: So Few Fish, So Many Fihshermen*, 9 NATURAL RESOURCES & ENV'T 13 (Spring 1995).

United Nations Food and Agriculture Organization, Code of Conduct for Responsible Fisheries

(October 31, 1995).

Preface

[the opening paragraphs of the Preface are reproduced earlier in the chapter]

The Committee on Fisheries (COFI) at its Nineteenth Session in March 1991 called for the development of new concepts which would lead to responsible, sustained fisheries. Subsequently, the International Conference on Responsible Fishing, held in 1992 in Cancún (Mexico) further requested FAO to prepare an international Code of Conduct to address these concerns. The outcome of this Conference, particularly the Declaration of Cancún, was an important contribution to the 1992 United Nations Conference on Environment and Development (UNCED), in particular its Agenda 21. Subsequently, the United Nations Conference on Straddling Fish Stocks and Highly Migratory Fish Stocks was convened, to which FAO provided important technical back-up. In November 1993, the Agreement to Promote Compliance with International Conservation and Management Measures by Fishing Vessels on the High Seas was adopted at the Twenty-seventh Session of the FAO Conference (Annex 1).

Noting these and other important developments in world fisheries, the FAO Governing Bodies recommended the formulation of a global Code of Conduct for Responsible Fisheries which would be consistent with these instruments and, in a non-mandatory manner, establish principles and standards applicable to the conservation, management and development of all fisheries. The Code, which was unanimously adopted on 31 October 1995 by the FAO Conference, provides a necessary framework for national and international efforts to ensure sustainable exploitation of aquatic living resources in harmony with the environment.

* * *

This Code sets out principles and international standards of behaviour for responsible practices with a view to ensuring the effective conservation, management and development of living aquatic resources, with due respect for the ecosystem and biodiversity. The Code recognises the nutritional, economic, social, environmental and cultural importance of fisheries, and the interests of all those concerned with the fishery sector. The Code takes into account the biological characteristics of the resources and their environment and the interests of consumers and other users. States and all those involved in fisheries are encouraged to apply the Code and give effect to it.

Article 1—Nature and Scope of the Code

1.1 This Code is voluntary. However, certain parts of it are based on relevant rules of international law, including those reflected in the United Nations Convention on the Law of the Sea of 10 December 1982.* * *1.3 The Code provides principles and standards applicable to the conservation, management and development of all fisheries. It also covers the capture, processing and trade of fish and fishery products, fishing operations, aquaculture, fisheries research and the integration of fisheries into coastal area management. * * *

Article 2—Objectives of the Code

The objectives of the Code are to:

a.   establish principles, in accordance with the relevant rules of international law, for responsible fishing and fisheries activities, taking into account all their relevant biological, technological, economic, social, environmental and commercial aspects;

b.   establish principles and criteria for the elaboration and implementation of national policies for responsible conservation of fisheries resources and fisheries management and development;

c.   serve as an instrument of reference to help States to establish or to improve the legal and institutional framework required for the exercise of responsible fisheries and in the formulation and implementation of appropriate measures;

d.   provide guidance which may be used where appropriate in the formulation and implementation of international agreements and other legal instruments, both binding and voluntary;

e.   facilitate and promote technical, financial and other cooperation in conservation of fisheries resources and fisheries management and development;

f.   promote the contribution of fisheries to food security and food quality, giving priority to the nutritional needs of local communities;

g.   promote protection of living aquatic resources and their environments and coastal areas;

h.   promote the trade of fish and fishery products in conformity with relevant international rules and avoid the use of measures that constitute hidden barriers to such trade;

i.   promote research on fisheries as well as on associated ecosystems and relevant environmental factors; and

j.   provide standards of conduct for all persons involved in the fisheries sector

* * *

Article 6—General Principles

6.1   States and users of living aquatic resources should conserve aquatic ecosystems. The right to fish carries with it the obligation to do so in a responsible manner so as to ensure effective conservation and management of the living aquatic resources.

6.2   Fisheries management should promote the maintenance of the quality, diversity and availability of fishery resources in sufficient quantities for present and future generations in the context of food security, poverty alleviation and sustainable development. Management measures should not only ensure the conservation of target species but also of species belonging to the same ecosystem or associated with or dependent upon the target species.

6.3   States should prevent overfishing and excess fishing capacity and should implement management measures to ensure that fishing effort is commensurate with the productive capacity of the fishery resources and their sustainable utilization. States should take measures to rehabilitate populations as far as possible and when appropriate.

6.4   Conservation and management decisions for fisheries should be based on the best scientific evidence available, also taking into account traditional knowledge of the resources and their habitat, as well as relevant environmental, economic and social factors. . . .

6.5   States and subregional and regional fisheries management organizations should apply a precautionary approach widely to conservation, management and exploitation of living aquatic resources in order to protect them and preserve the aquatic environment, taking account of the best scientific evidence available. The absence of adequate scientific information should not be used as a reason for postponing or failing to take measures to conserve target species, associated or dependent species and non-target species and their environment.

* * *

6.8   All critical fisheries habitats in marine and fresh water ecosystems, such as wetlands, mangroves, reefs, lagoons, nursery and spawning areas, should be protected and rehabilitated as far as possible and where necessary. Particular effort should be made to protect such habitats from destruction, degradation, pollution and other significant impacts resulting from human activities that threaten the health and viability of the fishery resources.

6.9   States should ensure that their fisheries' interests, including the need for conservation of the resources, are taken into account in the multiple uses of the coastal zone and are integrated into coastal area management, planning and development.

* * *

6.19   States should consider aquaculture, including culture-based fisheries, as a means to promote diversification of income and diet. In so doing, States should ensure that resources are used responsibly and adverse impacts on the environment and on local communities are minimized.

Article 7—Fisheries Management

7.1   General

7.1.1 States and all those engaged in fisheries management should, through an appropriate policy, legal and institutional framework, adopt measures for the long-term conservation and sustainable use of fisheries resources. Conservation and management measures, whether at local, national, subregional or regional levels, should be based on the best scientific evidence available and be designed to ensure the long-term sustainability of fishery resources at levels which promote the objective of their optimum utilization and maintain their availability for present and future generations; short term considerations should not compromise these objectives.
* * *

7.1.3 For transboundary fish stocks, straddling fish stocks, highly migratory fish stocks and high seas fish stocks, where these are exploited by two or more States, the States concerned, including the relevant coastal States in the case of straddling and highly migratory stocks, should cooperate to ensure effective conservation and management of the resources. This should be achieved, where appropriate, through the establishment of a bilateral, subregional or regional fisheries organization or arrangement.

* * *

7.5   Precautionary approach

7.5.1   States should apply the precautionary approach widely to conservation, management and exploitation of living aquatic resources in order to protect them and preserve the aquatic environment. The absence of adequate scientific information should not be used as a reason for postponing or failing to take conservation and management measures. 7.5.2 In implementing the precautionary approach, States should take into account, inter alia, uncertainties relating to the size and productivity of the stocks, reference points, stock condition in relation to such reference points, levels and distribution of fishing mortality and the impact of fishing activities, including discards, on non-target and associated or dependent species, as well as environmental and socio-economic conditions. 7.5.3 States and subregional or regional fisheries management organizations and arrangements should, on the basis of the best scientific evidence available, inter alia, determine:

a.   stock specific target reference points, and, at the same time, the action to be taken if they are exceeded; and

b.   stock-specific limit reference points, and, at the same time, the action to be taken if they are exceeded; when a limit reference point is approached, measures should be taken to ensure that it will not be exceeded.

7.5.4   In the case of new or exploratory fisheries, States should adopt as soon as possible cautious conservation and management measures, including, inter alia, catch limits and effort limits. Such measures should remain in force until there are sufficient data to allow assessment of the impact of the fisheries on the long-term sustainability of the stocks, whereupon conservation and management measures based on that assessment should be implemented. The latter measures should, if appropriate, allow for the gradual development of the fisheries.

Notes and Questions

1.   The Code of Conduct certainly says all the right things when it comes to fisheries management. But will it just be another in the long line of international environmental policy documents that said all the right things and, because of deficient international enforcement provisions, have had no meaningful impact? Perhaps it is too soon to tell. The FAO points to numerous bilateral, subregional, and regional actions inspired by the Code of Conduct. *See* FAO Fisheries Department, *Code of Conduct for Responsible Fisheries*, http://www.fao.org/fi/agreem/codecond/codecond.asp. These too, however, may suffer from lack of adequate enforcement mechanisms or political will. Ultimately, the real test may be whether the Code of Conduct shapes *domestic* law reform. How well does U.S. fisheries law fare when tested against the Code of Conduct principles? What specific legal reforms would you institute to bring our law closer to the international ideal?

2. Article 7.5 of the Code of Conduct endorses state adoption of what has come to be known as the Precautionary Principle, which has become a bedrock of international environmental law and of many domestic legal regimes as well, most notably that of Germany. At its core, the Precautionary Principle counsels that scientific uncertainty should not be used to justify taking economically appealing actions with potentially adverse environmental consequences. Stated more proactively, scientific uncertainty ought not preclude regulation of economic activity if it is feared (but not necessarily known) that the environmental consequences could be significant. In more concrete terms, the Precautionary Principle suggests using cost-benefit analysis cautiously, as it is subject to ends-driven manipulation, and emphasizing interdisciplinary, long-term, risk averse approaches to decision making. Beyond this, little meat fills out the bones of the Precautionary Principle. It is found as almost a premise of many international policies such as the Code of Conduct. But with no realistic proposition of enforcement in that realm, it remains largely a principle and not a mandate at the international level. Many scientists and legal scholars, therefore, are working feverishly to provide more definition to the Precautionary Principle in a variety of domestic law contexts where enforcement can be more directly achieved. *See* PERSPECTIVES ON THE PRECAUTIONARY PRINCIPLE (Ronnie Harding and Elizabeth Fisher eds., 1999); PROTECTING PUBLIC HEALTH & THE ENVIRONMENT: IMPLEMENTING THE PRECAUTIONARY PRINCIPLE (Carolyn Raffensperger and Joel Tickner eds., 1999).

Still, is the Precautionary Principle nonetheless comprised of too amorphous and potentially unstable a set of precepts to put forth as the foundation of domestic fisheries management law? The Code of Conduct, for example, leaves many open-ended questions unanswered in its call for member states to

> adopt as soon as possible cautious conservation and management measures, including, inter alia, catch limits and effort limits. Such measures should remain in force until there are sufficient data to allow assessment of the impact of the fisheries on the long-term sustainability of the stocks, whereupon conservation and management measures based on that assessment should be implemented. The latter measures should, if appropriate, allow for the gradual development of the fisheries.

How cautious is cautious enough? What is sufficient data? When is long-term sustainability achieved? How gradually should emerging fisheries be developed to be sufficiently precautionary? Some commentators contend that U.S. fisheries law does not go far enough explicitly to implement the Precautionary Principle, yet they offer little concrete description of legal text that would do so and answer questions like these. *See, e.g.*, Michele Territo, *The Precautionary Principle in Marine Fisheries Conservation and the U.S. Sustainable Fisheries Act of 1996*, 24 VERMONT L. REV. 1351 (2000). Do you believe our domestic law inadequately reflects the Precautionary Principle? What *specific* legal provisions would you draft to improve our precautionary performance?

3.   Article 6.19 of the Code of Conduct promotes use of aquaculture, which entails the controlled propagation of aquatic species for commercial, recreational, or public uses. Aquaculture already is a big business in the United States, accounting for almost $1 billion on production revenue. The epicenter of the industry is in the southeastern states, which account for two-third of production revenue, and the principal product is food fish, primarily catfish, trout, and salmon. *See* U.S. DEPARTMENT OF AGRICULTURE, NATIONAL AGRICULTURAL STATISTICS SERVICE, 1998 CENSUS OF AGRICULTURE, *available at* http://www.nass.gov/census/census97/aquaculture.

Isn't aquaculture the perfect answer to overfishing of wild species? The problem is that aquaculture presents its own set of environmental impacts, some quite alarming in potential. One concern is that fish may be farmed in areas where they are not native species, meaning that escaped eggs or fry have the potential to invade and disrupt local ecological dynamics. *See* Rosamond L. Naylor et al., *Aquaculture–A Gateway for Exotic Species*, 294 SCIENCE 1655 (2001). Even where they are local, farmed fish may have been genetically altered, such that intermingling with native species can alter population gene resources. Also, like terrestrial farms, fish farms take up space, often displacing valuable natural coastal habitat. And, also like some terrestrial farms, fish farming concentrates animal waste and animal disease and discharges them into local ecosystems. But, ironically, one of the primary concerns with aquaculture is that many fish farms use fish feed made from processed wild fish. We are, in other words, zealously fishing one species in the wild to feed another in captivity. *See* REBECCA GOLDBERG ET AL., PEW OCEANS COMMISSION, MARINE AQUACULTURE IN THE UNITED STATES: ENVIRONMENTAL IMPACTS AND POLICY OPTIONS (2001). Does the Code of Conduct adequately warn of and balance these adverse consequences in its promotion of aquaculture?

4.   How different are the three approaches to fisheries management covered in this section: the species-wide harvest limits applied under the conventional Fisheries Management Plan framework; the market-based approach of tradable Individual Fishing Quotas; and the U.N.'s Code of Conduct principles? Using the framework for evaluation of ecosystem management programs developed in Chapter 7 of this text, compare the three approaches:

	Currency	*Goals*	*Methods*
Conventional Fishery Management Plans			
Tradable Individual Fishing Quotas			
U.N. Code of Conduct			

2. Protecting Fish Habitat

Prior to the Sustainable Fisheries Act of 1996 (SFA), the primary focus of fishery management under the Magnuson–Stevens Act had been on regulating fish harvests. The quality of fish habitat, however, is undeniably a significant factor in fishery harvest sustainability. Thus, the SFA introduced a new component to federal regulation of fisheries management–the Essential Fish Habitat (EFH) program.

The statutory provisions for EFH are quite streamlined, establishing two core programs supported by a relatively tame regulatory authority:

16 U.S.C. § 1855

(b) FISH HABITAT.—

(1) (A) The Secretary shall, within 6 months of the date of enactment of the Sustainable Fisheries Act, establish by regulation guidelines to assist the Councils in the description and identification of essential fish habitat in fishery management plans (including adverse impacts on such habitat) and in the consideration of actions to ensure the conservation and enhancement of such habitat.

* * *

(2) Each Federal agency shall consult with the Secretary with respect to any action authorized, funded, or undertaken, or proposed to be authorized, funded, or undertaken, by such agency that may adversely affect any essential fish habitat identified under this Act.

* * *

(4) (A) If the Secretary receives information from a Council or Federal or State agency or determines from other sources that an action authorized, funded, or undertaken, or proposed to be authorized, funded, or undertaken, by any State or Federal agency would adversely affect any essential fish habitat identified under this Act, the Secretary

shall recommend to such agency measures that can be taken by such agency to conserve such habitat.

The National Marine Fisheries Service (NMFS) has promulgated a series of comprehensive regulations to implement these EFH provisions. The agency adopted a proposed rule in early 1997, *see* 62 Fed. Reg. 19723 (Apr. 23, 1997), and an interim final rule late that year, *see* 62 Fed. Reg. 66531 (Dec. 19, 1997), all the while soliciting and integrating public comment into a final rule making. The agency's regulatory efforts culminated in January 2002 with promulgation of a comprehensive final rule, excerpts of which are reproduced below. Under the final NMFS regulations, the basic scheme for EFH follows the SFA's two-staged process of EFH identification followed by federal agency consultation that leads to, at most, voluntary recommendations for EFH management.

First, Regional Fishery Management Councils (Councils) must identify in their Fishery Management Plans (FMPs) the EFH for each life stage of each managed species in the fishery management unit. Councils must identify as EFH those habitats that are necessary to the species for spawning, breeding, feeding, or growth to maturity. Councils must describe EFH in text and must provide maps of the geographic locations of EFH or the geographic boundaries within which EFH for each species and life stage is found. Councils also can identify EFH that is especially important ecologically or particularly vulnerable to degradation as "habitat areas of particular concern" (HAPC) to help provide additional focus for conservation efforts. HAPC status, however, carries with it no additional substantive protections or procedural safeguards. Councils must evaluate the potential adverse effects of fishing activities on EFH and must include in FMPs management measures that minimize adverse effects to the extent practicable. Councils must identify other nonfishing activities that may adversely affect EFH and recommend actions to reduce or eliminate these effects.

Once EFH is identified, NMFS and the Councils must implement the EFH coordination, consultation, and recommendation requirements of the Magnuson–Stevens Act. NMFS will make available descriptions and maps of EFH to promote EFH conservation and enhancement. The regulations encourage Federal agencies to use existing environmental review procedures to fulfill the requirement to consult with NMFS on actions they fund, carry out, or authorize that may adversely affect EFH, and the regulations contain procedures for abbreviated or expanded consultation in cases where no other environmental review process is available. Consultations may conducted at a programmatic and/or project-specific level. In cases where adverse effects from a type of actions will be minimal, both individually and cumulatively, a General Concurrence procedure further simplifies the consultation requirements. The regulations encourage coordination between NMFS and the Councils in the development of nonbinding EFH Conservation Recommendations to federal or state agencies for actions that would adversely affect EFH. Federal agencies must respond in writing within 30 days of receiving EFH Conservation Recommendations from NMFS. If the action agency's decision is inconsistent with the NMFS's EFH Conservation Recommendations, the agency must explain its reasoning and NMFS may request further review of the decision.

NMFS's rule making was not without controversy. Although the EFH program is not nearly as imposing as the framework the Endangered Species Act establishes for protection of species and their habitat, some fishing industry interests were concerned that NMFS may go too far in regulation, while many conservation groups were concerned that NMFS may fail to give EFH any teeth at all. Some of the more interesting and controversial provisions of the rule are presented in the following excerpt.

Department of Commerce, National Marine Fisheries Service, Magnuson–Stevens Act Provisions; Essential Fish Habitat (EFH)

67 Fed. Reg. 2343 (Jan. 17, 2002).

SUMMARY: NMFS issues this final rule to revise the regulations implementing the essential fish habitat (EFH) provisions of the Magnuson–Stevens Fishery Conservation and Management Act (Magnuson–Stevens Act). This rule establishes guidelines to assist the Regional Fishery Management Councils (Councils) and the Secretary of Commerce (Secretary) in the description and identification of EFH in fishery management plans (FMPs), the identification of adverse effects to EFH, and the identification of actions required to conserve and enhance EFH. These regulations also detail procedures the Secretary (acting through NMFS), other Federal agencies, and the Councils will use to coordinate, consult, or provide recommendations on Federal and state actions that may adversely affect EFH. The intended effect of the rule is to promote the protection, conservation, and enhancement of EFH

* * *

NMFS amends 50 CFR part 600 as follows:

§ 600.10   Definitions

Essential fish habitat (EFH) means those waters and substrate necessary to fish for spawning, breeding, feeding, or growth to maturity. . . . "necessary" means the habitat required to support a sustainable fishery and the managed species' contribution to a healthy ecosystem;

Subpart J–Essential Fish Habitat (EFH)

* * *

§ 6000.810   Definitions and word usage

* * *

Habitat areas of particular concern means those areas of EFH identified pursuant to § 600.812(a)(8). Healthy ecosystem means an ecosystem where ecological productive capacity is maintained, diversity of the flora and fauna is preserved, and the ecosystem retains the ability to regulate itself. Such an ecosystem should be similar to comparable, undisturbed ecosystems with regard to standing crop, productivity, nutrient dynamics, trophic structure, species richness, stability, resilience, contamination levels, and the frequency of diseased organisms.

* * *

§ 600.815   Contents of Fishery Management Plans

(a) Mandatory contents-

(1) Description and identification of EFH-

(i)Overview. FMPs must describe and identify EFH in text that clearly states the habitats are habitat types determined to be EFH for each life stage of the managed species. FMPs should explain the physical, biological, and chemical characteristics of EFH and, if known, how these characteristics influence the use of EFH by the species/life stage. FMPs must identify the specific geographic location or extent of habitats described as EFH. FMPs must include maps of the geographic locations of EFH or the geographic boundaries within which EFH for each species and life stage is found.

* * *

(iv) EFH determination

* * *

(C) If a species is overfished and habitat loss or degradation may be contributing to the species being identified as overfished, all habitats currently used by the species may be considered essential in addition to certain historic habitats that are necessary to support rebuilding the fishery and for which restoration is technologically and economically feasible. Once the fishery is no longer considered overfished, the EFH identification should be reviewed and amended, if appropriate.

(D)Areas described as EFH will normally be greater than or equal to aquatic areas that have been identified as "critical habitat" for any managed species listed as threatened or endangered under the Endangered Species Act.

(E)Ecological relationships among species and between the species and their habitat require, where possible, that an ecosystem approach be used in determining the EFH of a managed species. EFH must be designated for each managed species, but where appropriate, may be designated for assemblages of species or life stages that have similar habitat needs and requirements. If grouping species or using species assemblages for the purpose of designation EFH, FMPs must include a justification and scientific rationale. The extent of the EFH should be based on the judgment of the Secretary and the appropriate Council(s) regarding the quantity and quality of habitat that are necessary to maintain a sustainable fishery and the managed species' contribution to a healthy ecosystem.

* * *

(2) *Fishing activities that may adversely affect EFH*

* * *

(ii) *Minimizing adverse effects.* Each FMP must minimize to the extent practicable adverse effects from fishing on EFH, including EFH designated under other Federal FMPs. Councils must act to prevent, mitigate, or minimize any adverse effects from fishing, to the extent practicable, if there is evidence that a fishing activity adversely affects EFH in a manner that is more than minimal and not temporary in nature....

(iii) *Practicability.* In determining whether it is practicable to minimize an adverse effect from fishing, Councils should consider the nature and extent of the adverse effect on EFH and the long and short-term costs and benefits of potential management measures to EFH, associated fisheries, and the nation, consistent with national standard 7. In determining whether management measures are practicable, Councils are not required to perform formal cost/benefit analysis.

(iv) *Options for managing adverse effects from fishing.* Fishery management options may include, but are not limited to:

(A) *Fishing equipment restrictions....*

(B) *Time/area closures....*

(C) *Harvest limits....*

* * *

(4) *Non-fishing related activities that may adversely affect EFH.* FMPs must identify activities other than fishing that may adversely affect EFH. Broad categories of such activities include, but are not limited to: dredging, filling, excavation, mining, impoundment, discharge, water diversions, thermal additions, actions that contribute to non-point source pollution and sedimentation, introduction of potentially hazardous materials, introduction of exotic species, and the conversion of aquatic habitat that may eliminate, diminish, or disrupt the functions of EFH. For each activity, the FMP should describe known and potential adverse effects to EFH.

* * *

(7) *Prey species.* Loss of prey may be an adverse effect on EFH and managed species because the presence of prey makes waters and substrate function as feeding habitat, and the definition of EFH includes waters and substrate necessary to fish for feeding. Therefore, actions that reduce the availability of major prey species, either through direct harm or capture, or through adverse impacts to the prey species, may be considered adverse effects on EFH if such actions reduce the quality of EFH. FMPs should list the major prey species for the species in the fishery management unit and

discuss the location of prey species' habitat. Adverse effects on prey species and their habitats may result from fishing and non-fishing activities.

(8) *Identification of habitat areas of particular concern.* FMPs should identify specific types or areas of habitat within EFH as habitat areas of particular concern based on one or more of the following considerations:

(i) The importance of the ecological function provided by the habitat.

(ii)The extent to which the habitat is sensitive to human-induced environmental degradation.

(iii)Whether, and to what extent, development activities are, or will be, stressing the habitual type.

(iv)The rarity of the habitat type.

Subpart K–EFH Coordination, Consultation, and Recommendations

§ 600.905   Purpose, scope, and NMFS/Council cooperation.

(a) *Purpose.* These procedures address the coordination, consultation, an recommendation requirements of sections 305(b)(1)(D) and 305(b)(2–4) of the Magnuson–Stevens Act. The purpose of these procedures is to promote the protection of EFH in the review of Federal and state actions that may adversely affect EFH.

(b) *Scope.* Section 305(b)(1)(D) of the Magnuson–Stevens Act requires the Secretary to coordinate with, and provide information to, other Federal agencies regarding the conservation and enhancement of EFH. Section 305(b)(2) requires all Federal agencies to consult with the Secretary on all actions or proposed actions authorized, funded, or undertaken by the agency that may adversely affect EFH. Sections 305(b)(3) and (4) direct the Secretary and the Councils to provide comments and EFH Conservation Recommendations to Federal or state agencies on actions that affect EFH. Such recommendations may include measures to avoid, minimize, mitigate, or otherwise offset adverse effects on EFH resulting from actions authorized, funded, or undertaken by that agency. Section 305(b)(4)(B) requires Federal agencies to respond in writing to such comments.

Notes and Questions

1.   NMFS has clearly adopted a watershed-based approach to EFH management and conservation. Consider how it handled the charge that in so doing it acted outside its statutory authority:

Comments on Using an Ecosystem or Watershed Approach to Resource Management

Comment A: A number of commenters representing non-fishing interest state that the Magnuson–Stevens Act does not authorize a risk-averse or ecosystem approach to EFH. These commenters thought that the focus should be limited to fish species and not ecosystem principles.

Response A: NMFS provided a detailed response to this comment in the preamble to the interim final rule at 62 FR 66532–66533, and the response remains the same. In summary, the Magnuson–Stevens Act provides authority for the link between EFH and the managed species' contribution to a healthy ecosystem in a number of places. Ecosystem concepts are common in the statutory definitions of "fishery resources," "conservation and management," and "optimum." The fact that the Magnuson–Stevens Act directs the Councils to address the degradation and loss of EFH from both fishing and non-fishing activities through conservation and enhancement measures further reflects support for the ecosystem-based management of marine and anadromous fisheries. Ecosystem management encourages sustainable resource use and recognized the uncertainties inherent in management and the need to make risk-averse decisions. This regulation embraces those concepts and urges Councils to seek environmental sustainability in fishery management, within the current statutorily prescribed fishery management framework (i.e., management by FMPs).

67 Fed. Reg. 2349.

This response resembles the approach other resource management agencies have taken when responding to similar "ultra vires" claims lodged against their adoption of ecosystem management principles. Because few federal laws explicitly authorize ecosystem management as a policy model or goal, agencies are left to wedging it into existing laws through open-ended statutory terms that are sufficiently flexible to encompass ecosystem management principles. A statutorily-specified resource "sustainability" goal, in particular, is amenable as administratively-devised justifications for ecosystem management. Should agencies be allowed to graft ecosystem management onto existing laws that clearly were passed without explicit congressional intent endorsing its underlying principles? Can an agency interpret sustainability to mean one thing at one time (e.g., sustained economic yield) and later to mean quite another thing (e.g., sustained ecosystem dynamics) with no change in the statute? Of course, NMFS can point to the Sustainable Fisheries Act as a legislative endorsement of some change in approach, but was it an endorsement of ecosystem management? For more on this theme, see Lee Banaka and Dennis Nixon, *Essential Fish Habitat and Coastal Zone Management: Business As Usual Under the Magnuson–Stevens Act?*, 30 GOLDEN GATE U. L. REV. 969 (2000).

2. Many fisheries biologists argue that conventional fisheries management policy for too long ignored the obvious connection between habitat and fishery. The dogma of catch limits and time/place limits does absolutely nothing to advance habitat conservation, yet, according to some research, investment in habitat conservation, particularly of coastal and inland wetlands, could yield greater fishery performance results at less cost. *See* Special Focus Issue: *Wetland Dependent Fisheries*, 22 NATIONAL WETLANDS NEWSLETTER 1 (Nov.–Dec. 2000). Will the EFH program take advantage of that connection? *See* 67 Fed. Reg. 2347 (affirming EFH can include wetland areas).

3. As fruitful as wetland conservation may be to fisheries performance, haven't we seen time and again that land-based upland activities can profoundly degrade coastal and estuarine resources? EFH, however, is defined as including only waters and substrate, *see* 16 U.S.C. § 1802(10), thus precluding designation of upland areas as EFH. By contrast, designation of critical habitat for species listed under the Endangered Species Act can include areas outside the species' occupied range. *See* 16 U.S.C 1532(5)(A). Why did Congress not take the same approach for EFH? For comparison of the EFH and ESA programs on this and other scores, see Kristen M. Fletcher and Sharonne E. O'Shea, *Essential Fish Habitat: Does Calling it Essential Make it So?*, 30 Envtl. Law 1 (2000). Notwithstanding that EFH cannot include upland areas, can activities on upland areas that affect EFH areas be subjected to EFH consultation procedures? How would you make the argument under the statute and regulations that such an approach could be taken?

4. One of the more controversial aspects of the EFH program is the focus NMFS has put on non-fishing activities. Fishing may deplete fish, but generally without affecting the physical or biological components of habitat. Yet fishing techniques can have dramatic impacts on fish habitat in some contexts. In particular, there is mounting evidence that trawling, which drags large nets across the ocean floor, has scraped large areas clean of all functional habitat. *See* David Malakoff, *Papers Posit Grave Impact of Trawling*, 282 Science 2168 (1988). NMFS has recognized that trawling's "mobile fishing gear" can reduce the structural complexity of fish habitat, making it more difficult for fish to hide from their predators and interfering with benthic (bottom) ecosystem processes. *See* NMFS, Essential Fish Habitat: FAQs (Jan. 2002).

Note on the Marine Mammals Protection Act

The focus of the preceding materials has been on fish species–the biologically-relevant jurisdictional limit of the Magnuson–Stevens Act. But as debate over fishery management emerged as a serious legislative initiative in the 1970s, support also built for an initiative directed at conservation of marine mammals. Many scientists and conservationists portrayed marine mammals as not only ecologically important, but also ethically entitled to protection given their apparent intelligence and social behavior. Although a patchwork of federal and state laws provided some level of protection to a limited array of marine mammals, this approach proved to be neither coordinated nor effective. A number of highly publicized stories–such as the slaughter of harp seal pups in Canada, the hunting of several whale species to near extinction, and the killing of dolphins and porpoises as bycatch from industrialized tuna fishing operations–brought the problem into sharp focus. By the early 1970s, the calls had grown quite loud in support of broader federal protective measures for all marine mammals. The Marine Mammal Protection Act of 1972 (MMPA), 16 USC §§ 1361—1421h, is designed to provide such comprehensive levels of protection to this other valuable form of living marine resources.

The primary goal of the MMPA is to "maintain the health and stability of the marine ecosystem," which the statute links to a marine mammal's "optimum sustainable population keeping in mind the carrying capacity of the habitat." 16 U.S.C. § 1361(6). At the core of the statute, therefore, is this notion of an optimal sustainable population, defined as "the number of animals which will result in the maximum productivity of the population or the species, keeping in mind the carrying capacity of the habitat and the health of the ecosystem of which they form a constituent element." *Id.* § 1362(9). To achieve this ideal for marine mammal species, the original enactment established a blanket moratorium on "take" of marine mammals, *id.* § 1372 (a), defining take, much as Congress did later in the Endangered Species Act, to include such acts as killing, capturing, harming, and harassing. *Id.* § 1362(12). This moratorium, however, excluded certain activities, the most significant of which was "incidental taking" of marine mammals in the course of commercial fishing operations–i.e., bycatch of marine mammals. *Id.* § 1371(a)(2). Congress added this exemption in response to concerns that after years of little or no protection, imposing the blanket taking moratorium on commercial fishing for all marine mammal species could seriously disrupt the industry. Thus, instead of being subject to the blanket moratorium, commercial fishing operations were required to obtain permits from the Department of Commerce, acting through the National Marine Fisheries Service (NMFS) in consultation with the independent Marine Mammal Commission, to authorize such bycatch takes.

NMFS implemented the commercial fishing incidental take program through permits designed to allow takes only of "nondepleted" marine mammal stocks. The initial goal of the permit program, moreover, was to reach "zero mortality" through the adoption of practicable fishing techniques. In practice, however, the permit program was rather laxly administered, and no clear statutory standard allowed conservationists to invoke rigorous judicial review. Amendments in 1981 and 1984 attempted to reach compromises, focusing in particular on the tuna-dolphin problem, but in general the framework remained unsatisfying to both fishing and conservation interests through the 1980s.

This uneasy state of affairs was rocked in 1988 when, in Kokechik Fishermen's Association v. Secretary of Commerce, 839 F.2d 795 (D.C.Cir. 1988), the D.C. Circuit held that NMFS had improperly implemented the incidental take permit system because it had failed to ensure that the authorized commercial fishing operations would not also result in takings of depleted stocks or of stocks whose status had not been determined. The case had been brought by Alaskan subsistence salmon fishing interests challenging NMFS's issuance of a general incidental take permit to Japanese operations in the Pacific Northwest, which had authorized take of over 6000 Dall's porpoises over a three-year period. But the plaintiffs got more than they bargained for. Given how indiscriminate commercial fishing is in terms of ability to screen out particular bycatch species, other fisheries in U.S. waters that had the benefit of general or "small scale" incidental take permits were in the same boat, so to speak, as the Japanese salmon fishing operations. Strict application of *Kokechik* would have required in practical terms either shutting down most domestic commercial fishing or abandoning the MMPA.

To avoid having to make that choice, and to buy time to design an alternative, Congress in 1988 established a five-year exemption from the incidental take permit requirement for vessels that allowed NMFS to station observers aboard to study more closely the impacts of fishing on marine mammals. 16 U.S.C. § 1383a. Along with the Regional Fisheries Management Councils and the Marine Mammal Commission, NMFS was required to develop a proposal based on its study that would more effectively manage the interaction between marine mammals and commercial fishing. The agencies submitted their proposal on January 1, 1992, which fishing and conservation interests alike criticized as based on insufficient evidence. The fishing and conservation interests crafted a joint counterproposal, which NMFS in turn criticized as administratively impracticable. The battle eventually led to the adoption of sweeping amendments to the MMPA in 1994. *See* Pub. L. No. 103–238, 108 Stat. 532 (1994).

One of the key changes the amendments made was to replace the incidental take permit requirement with a program for establishing "Take Reduction Plans" and related regulatory authorities. Under this approach, NMFS must prepare and periodically revise stock assessment reports for all marine mammals that reside in U.S. waters, including population estimates, productivity rates, mortality rates, geographic range, and estimated "potential biological removal" (PBR)–the maximum number of animals, excluding natural mortality, that the stock can withstand losing without affecting its ability to reach and maintain an optimal sustainable population level. 16 U.S.C. § 1386. Any stock for which human-caused mortality exceeds the PBR, as well as any stock designated as depleted under the MMPA or as endangered or threatened under the Endangered Species Act, is designated a "strategic stock." *Id.* § 1362(19). If NMFS determines that a fishery's operations frequently or occasionally cause mortality or serious injury to a strategic stock, the relevant fishing and conservation interests must work with NMFS to develop a Take Reduction Plan. *Id.* § 1387(f). The plan must have as its short-term goal to reduce mortality to PBR levels, and as its five-year goal to reduce fishing impacts to "insignificant levels approaching a zero mortality and serious injury rate." *Id.* § 1387(f)(2). NMFS is authorized in the plans to adopt time, place, gear, and technique restrictions on fishing operations in order to achieve PBR. If the fishery complies with the plan, it has complied with the MMPA incidental take prohibition.

As some commentators have observed, the 1994 amendments moved the MMPA away from the original "presumption of sacredness," under which all marine mammals received blanket protection from take in the absence of an incidental take permit, to a regime closer to the sustainable yield approach of the Magnuson–Stevens Act (and many other public resource management laws discussed in other chapters of this text). *See* George A. Chmael II et al., *The 1994 Amendments to the Marine Mammal Protection Act*, 9 Natural Resources & Env't 18 (Spring 1995). Other commentators have noted the apparently intentional ambiguities Congress has sprinkled into the statute that vex attempts to pin down congressional intent. For example, the statute states that "[w]henever consistent with" the primary objective of maintaining the health and stability of the marine ecosystem, "it should be the goal to obtain an optimum sustainable population," yet it is not clear under any provision of the statute when achieving optimum sustainable population would be inconsistent with the health and stability of the marine ecosystem. *See* MICHAEL J. BEAN AND MELANIE J. ROWLAND, THE EVOLUTION OF NATIONAL WILDLIFE LAW 135–36 (3d ed. 1997).

Also, although the greater focus the 1994 amendments brought to stock assessment may suggest endorsement of an ecosystem management approach to marine mammal habitat and ecosystem components–after all, that is the statute's expressed purpose–the statute never has required any formal ecosystem management framework. The statute repeatedly defines terms, such as optimum sustainable population, with the suggestion that it is "keeping in mind the carrying capacity of the habitat," but that is where habitat is kept–in mind only. The 1994 amendments also introduced several ecosystem study and monitoring programs, but no authority to manage habitat or other ecosystem components. *See* Susan C. Alker, *The Marine Mammal Protection Act: Refocusing the Approach to Conservation*, 44 UCLA L. REV. 527 (1996). Overall, then, the MMPA stands as yet another example of how identifying the "law" of ecosystem management is at present really an exercise in piecing together bits of legislation into incomplete, but sometimes effective, programs of regulation and conservation.

b. MARINE PROTECTED AREAS

One ecosystem management approach we have seen used frequently in terrestrial settings is preservation of the ecological status quo, at least as nearly as possible, through creation of refuges, wilderness areas, and other land reserves. The Marine Protection, Research, and Sanctuaries Act of 1972 (MPRSA), 16 U.S.C. §§ 1431–1447(f), provides the analogous authority for coastal and marine ecosystems. The MPRSA authorizes the Secretary of Commerce, acting through the National Ocean Service (NOS), to "designate any discrete area of the marine environment as a national marine sanctuary and promulgate regulations implementing the designation." *Id.* § 1433(a). The factors NOS must use in determining sanctuary status include "the area's natural resource and ecological qualities, including its contribution to biological productivity, maintenance of ecosystem structure, . . . and the biogeographic representation of the site." *Id.* § 1433(b)(1)(A). Extensive consultation and notice requirements attach to the sanctuary designation process, requiring NOS to notify affected federal, state, and local entities, prepare an environmental assessment, and complete public notice and comment. *See id.* § 1434.

Once designated, national marine sanctuaries are generally managed for limited multiple uses such as recreation, education, commercial fishing, and shipping, though resource uses such as fishing are subject to regulation. *Id.* § 1431. Regulations thus vary by sanctuary, but, overall, NOS must facilitate public and private uses of the resources within sanctuaries "to the extent compatible with the primary objective of resource protection." *Id.* § 1431(b)(2). Also, federal agencies must consult with NOS regarding actions they undertake, inside or outside the sanctuary, that are likely to "destroy, cause the loss of, or injure any sanctuary resource." *Id.* § 1434(d)(1)(A). NOS can recommend "reasonable and prudent alternatives" to the action, and the action agency must justify any departure from them through a written statement. *Id.* § 1434(d).

The most potentially controversial exercise of conservation authority in sanctuaries involves the power NOS has to restrict commercial fishing. But NOS is limited in this respect by a cumbersome process and difficult statutory burdens. Regional Fishing Councils established under the Magnuson–Stevens Act (see this chapter, *supra*) are entitled to decide whether fishing regulations are needed in a sanctuary and, if so, to draft them. A Council's decision "shall be accepted and issued as proposed regulations by the Secretary unless the Secretary finds that the Council's action fails to fulfill the purposes and policies of this chapter and the goals and objectives of the proposed designation." *Id.* § 1434(a)(5). That burden, plus the agency's obligation to facilitate sanctuary resource use, make it unlikely that NOS will employ its regulatory authority in a way that restricts commercial fishery access beyond the Councils' wishes. Indeed, most litigation over NOS regulations has involved restrictions on recreational pursuits such as diving and personal watercraft, with courts generally upholding the restrictions as within the agency's authority. *See* Craft v. National Park Service, 34 F.3d 918 (9th Cir.1994) (diving); Personal Watercraft Industry Association v. Department of Commerce, 48 F.3d 540 (D.C.Cir.1995) (small craft use); United States v. Fisher, 22 F.3d 262 (11th Cir.1994) (salvage collectors).

Through its Office of the National Marine Sanctuaries, NOS now administers over a dozen national marine sanctuaries located throughout the Pacific and Atlantic coasts, the Gulf of Mexico, Hawaii, and other United States territories. They range in size from the one-quarter square mile Fagatele Bay sanctuary in American Samoa, to the 5,300 square-mile Monteray Bay sanctuary on the California coast. The total protected area is nearly 18,000 square miles. NOS has issued consolidated regulations for the sanctuary system, see 15 C.F.R. part 922, and descriptions of the program and the various sanctuaries are available at the NOS website, National Marine Sanctuaries, http://www.sanctuaries.nos.noaa.gov.

Although many observers laud its objectives, few rave about the National Marine Sanctuary program's accomplishments. The designation process has proven cumbersome and politically handicapped, and meaningful conservation is not guaranteed for a designated sanctuary in any event. *See* MICHAEL J. BEAN AND MELANIE J. ROWLAND, THE EVOLUTION OF NATIONAL WILDLIFE LAW 338–40 (3rd ed. 1997). Hence, in order to expand on the ideal of marine sanctuaries embodied in the National Marine Sanctuaries program, in May 2000 President Clinton issued the following Executive Order to coordinate a system "Marine Protected Areas" (MPA) that would include the sanctuaries and any other marine area protected under related federal laws. The term "marine protected area" has been in use for over two decades, and the practice of establishing marine protected areas has been around for centuries. A marine protected area has come to mean different things to different people, based primarily on the level of protection provided by the MPA. Some see MPAs as sheltered or reserved areas where little, if any, use or human disturbance should be permitted. Others see them as specially managed areas designed to enhance multiple uses such as recreation, fishing, and mineral exploration. Consider how President Clinton defined the term and outlined its management implications in his 2000 Executive Order:

Executive Order 13158, Marine Protected Areas

65 Fed. Reg. 34909 (May 26, 2000).

By the authority vested in me as President by the Constitution and the laws of the United States of America and in furtherance of the purposes of the National Marine Sanctuaries Act (16 U.S.C. 1431 et seq.), National Wildlife Refuge System Administration Act of 1966 (16 U.S.C. 668dd-ee), National Park Service Organic Act (16 U.S.C. 1 et seq.), National Historic Preservation Act (16 U.S.C. 470 et seq.), Wilderness Act (16 U.S.C. 1131 et seq.), Magnuson–Stevens Fishery Conservation and Management Act (16 U.S.C. 1801 et seq.), Coastal Zone Management Act (16 U.S.C. 1451 et seq.), Endangered Species Act of 1973 (16 U.S.C. 1531 et seq.), Marine Mammal Protection Act (16 U.S.C. 1362 et seq.), Clean Water Act of 1977 (33 U.S.C. 1251 et seq.), National Environmental Policy Act, as amended (42 U.S.C. 4321 et seq.), Outer Continental Shelf Lands Act (42 U.S.C. 1331 et seq.), and other pertinent statutes, it is ordered as follows:

Section 1. Purpose. This Executive Order will help protect the significant natural and cultural resources within the marine environment for the benefit of present and future generations by strengthening and expanding the Nation's system of marine protected areas (MPAs). An expanded and strengthened comprehensive system of marine protected areas throughout the marine environment would enhance the conservation of our Nation's natural and cultural marine heritage and the ecologically and economically sustainable use of the marine environment for future generations. To this end, the purpose of this order is to, consistent with domestic and international law: (a) strengthen the management, protection, and conservation of existing marine protected areas and establish new or expanded MPAs; (b) develop a scientifically based, comprehensive national system of MPAs representing diverse U.S. marine ecosystems, and the Nation's natural and cultural resources; and (c) avoid causing harm to MPAs through federally conducted, approved, or funded activities.

Sec. 2. Definitions. For the purposes of this order: (a) "Marine protected area" means any area of the marine environment that has been reserved by Federal, State, territorial, tribal, or local laws or regulations to provide lasting protection for part or all of the natural and cultural resources therein.

(b) "Marine environment" means those areas of coastal and ocean waters, the Great Lakes and their connecting waters, and submerged lands thereunder, over which the United States exercises jurisdiction, consistent with international law.

(c) The term "United States" includes the several States, the District of Columbia, the Commonwealth of Puerto Rico, the Virgin Islands of the United States, American Samoa, Guam, and the Commonwealth of the Northern Mariana Islands.

Sec. 3. MPA Establishment, Protection, and Management. Each Federal agency whose authorities provide for the establishment or management of MPAs shall take appropriate actions to enhance or expand protection of existing MPAs and establish or recommend, as appropriate, new MPAs. Agencies implementing this section shall consult with the agencies identified in subsection 4(a) of this order, consistent with existing requirements.

Sec. 4. National System of MPAs. (a) To the extent permitted by law and subject to the availability of appropriations, the Department of Commerce and the Department of the Interior, in consultation with the Department of Defense, the Department of State, the United States Agency for International Development, the Department of Transportation, the Environmental Protection Agency, the National Science Foundation, and other pertinent Federal agencies shall develop a national system of MPAs. They shall coordinate and share information, tools, and strategies, and provide guidance to enable and encourage the use of the following in the exercise of each agency's respective authorities to further enhance and expand protection of existing MPAs and to establish or recommend new MPAs, as appropriate:

(1) science-based identification and prioritization of natural and cultural resources for additional protection;

(2) integrated assessments of ecological linkages among MPAs, including ecological reserves in which consumptive uses of resources are prohibited, to provide synergistic benefits;

(3) a biological assessment of the minimum area where consumptive uses would be prohibited that is necessary to preserve representative habitats in different geographic areas of the marine environment;

(4) an assessment of threats and gaps in levels of protection currently afforded to natural and cultural resources, as appropriate;

(5) practical, science-based criteria and protocols for monitoring and evaluating the effectiveness of MPAs;

(6) identification of emerging threats and user conflicts affecting MPAs and appropriate, practical, and equitable management solutions, including effective enforcement strategies, to eliminate or reduce such threats and conflicts;

(7) assessment of the economic effects of the preferred management solutions; and

(8) identification of opportunities to improve linkages with, and technical assistance to, international marine protected area programs.

* * *

(f) To better protect beaches, coasts, and the marine environment from pollution, the Environmental Protection Agency (EPA), relying upon existing Clean Water Act authorities, shall expeditiously propose new science-based regulations, as necessary, to ensure appropriate levels of protection for the marine environment. Such regulations may include the identification of areas that warrant additional pollution protections and the enhancement of marine water quality standards. The EPA shall consult with the Federal agencies identified in subsection 4(a) of this order, States, territories, tribes, and the public in the development of such new regulations.

Sec. 5. Agency Responsibilities. Each Federal agency whose actions affect the natural or cultural resources that are protected by an MPA shall

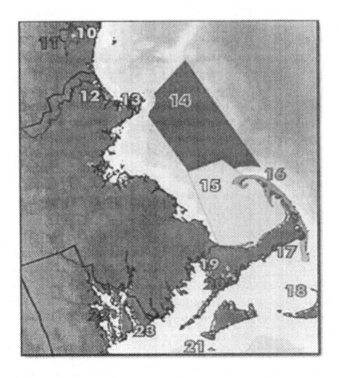

Marine Protected Areas off the Cape Cod coast: (10) Great Bay National Wildlife Refuge (11) Great B Estuarine Research Reserve (12) Parker River National Wildlife Refuge (13) Thacher Island National W (14) Gerry E. Studds/Stellwagen Bank National Marine Sanctuary (15) Cape Cod Bay Northern Right Whal Habitat (16) Cape Cod National Seashore (17) Monomoy National Wildlife Refuge (18) Nantucket Nation Refuge (19) Mashpee National Wildlife Refuge (20) Waquoit Bay National Estuarine Research Reserve (2 Long Island National Wildlife Refuge (22) Narragansett Bay National Estuarine Research Reserve (23) National Wildlife Refuge (Marine Protected Areas Initiative, 2002)

identify such actions. To the extent permitted by law and to the maximum extent practicable, each Federal agency, in taking such actions, shall avoid harm to the natural and cultural resources that are protected by an MPA. In implementing this section, each Federal agency shall refer to the MPAs identified under subsection 4(d) of this order.

Sec. 6. Accountability. Each Federal agency that is required to take actions under this order shall prepare and make public annually a concise description of actions taken by it in the previous year to implement the order, including a description of written comments by any person or organization stating that the agency has not complied with this order and a response to such comments by the agency.

* * *

Notes and Questions

1. Do MPAs work? It is, after all, rather difficult to contain fish through lines drawn on a map of the ocean. There is mounting evidence, however, that marine preserves do serve as sinks for recovering fish stocks. In Florida waters, for example, sport anglers have landed a disproportionate number of world and state record trophy fish from waters adjacent to a 60–square mile area near Cape Canaveral that was closed to access in 1962 for security reasons. Some researches believe the closed area—the functional equivalent of a reserve—provided sheltered nursery habitat for the surrounding areas. Similar results are being recorded near other marine preserve areas around the world. *See* Callum M. Roberts et al., *Effects of Marine Preserves on Adjacent Fisheries*, 294 Science 1920 (2001). But there is substantial disagreement over the reliability of these findings. Many other researchers and interested parties attribute the purported effects of the Florida and other "no take" areas to other causes, such as overall increases in regulation, unusually superior habitat regimes, ecological change, and experimental bias. *See Marine Reserves and Fisheries Management*, 295 Science 1233 (2002). The scientists, in other words, don't agree. What should ecosystem management policy do in that case? For further thoughts on this question in light of the uncertainty about the effectiveness of MPAs and the controversial nature of MPA management, see National Academy of Sciences/National Research Council, Marine Protected Areas: Tools for Sustaining Ocean Ecosystems (2001).

2. Not all MPAs prohibit human activities as comprehensively as does the Florida research site involved in the reserve productivity study discussed in the previous note. As in terrestrial public lands settings, the management norm for most MPAs is closer to a multiple use mandate rather than an ideal of wilderness. How effective is a multiple-use model, particularly one that allows commercial and recreational fishing, in meeting the objective of biodiversity conservation in MPA areas?

Indeed, the nation's first MPA provides an example of the range of possibilities that are available under the MPA approach and the limitations associated with different approaches. About 70 miles west of Key West in Florida lie the Dry Tortugas—seven small coral and sand islands Ponce de Leon charted in 1513 and named after their dry, yet turtle-abundant, environs. They were well known in the 17th and 18th centuries as a haven for pirates attacking merchant ships in the Gulf of Mexico. Well into the 20th century, their remote location and inhospitable (for humans) environment had left them largely pristine, home to more than 400 species of reef fish, lush corals, seabird rookeries, and crystal clear waters. In 1935, they became the Dry Tortugas National Monument, and in 1990 most of the surrounding waters were included in the Florida Keys National Marine Sanctuary. In 1992, the entire area also was included in the Dry Tortugas National Park, within which commercial fishing was banned.

Yet, notwithstanding its long history of MPA status and inclusion in overlapping national sanctuary and park programs, the Dry Tortugas ecosystem faced numerous stresses, most notably from recreational fishing. Modern fishing and boating technology has made the Dry Tortugas well within reach of recreational boating and sport fishing from the Keys and Florida's lower Gulf Coast. For example, the National Park had 18,000 visitors in 1984, and over 84,000 in 1999. The sizes of individual fish being taken from the area were falling substantially below healthy species averages, suggesting a path toward overfished status. With the threat of fishing primarily in mind, Congress allowed the Sanctuary managers, in the Florida Keys National Marine Sanctuary and Protection Act of 1990, to use "zoning" as a means of designating limited uses for different areas. In July 1997, the Sanctuary instituted a comprehensive network of zones, including a series of 23 "no-take" areas covering only one percent of the Sanctuary area, but over 65 percent of the shallow reef habitat. A much larger no-take preserve had been proposed in 1994, but was dropped from the final plan in response to heated opposition from commercial and recreational fishing operations, divers, and other recreational interests used to moving about and using the Sanctuary with relatively little impediment.

The Sanctuary managers regrouped, however, and in 1998 formed a multi-stakeholder planning initiative, known as Tortugas 2000, to study the resurrection of the large ecological reserve concept. In May 1999, the working group approved a proposal that represented consensus between fishing and conservation interests. The plan called for a 151–square nautical mile Tortugas Ecological Reserve split into two sections: Tortugas North, which protects extensive coral resources and in which diving and snorkeling using mooring buoys (no anchoring) will be allowed by permit; and Tortugas South, which protects important fish resources and in which only research is allowed. Florida approved the inclusion of state waters in the reserve in April 2001, and in July 2001 the National Ocean Service issued its formal approval. Later that month, the Department of the Interior carved out a no-take Research Natural Area from a portion of the national park adjacent to the sanctuary, pushing the total no-take reserve area to almost 200 square nautical miles. Upon full implementation, the Tortugas Ecological Reserve will provide the largest permanent no-take MPA in the nation and the second largest in the world. For more information on this history and the current implementation progress status, see http://www.fknms.nos.gov/tortugas.

3. What do you take from the fact that fishery resources in the Dry Tortugas continued to decline for decades after the initial designation of the area as an MPA in 1935, primarily from recreational fishing? Does this suggest that the MPA approach must make frequent and significant use of permanent no-take reserves in order to be truly effective as an ecosystem management instrument? How should the process of identification of the need for and location of no-take areas be carried out, and who should be involved?

4. The Bush Administration recently announced, through the Secretary of Commerce, that it has decided to continue federal development of the MPA program:

> The Administration has decided to retain Executive Order 13158 on marine protected areas. America must strive to harmonize commercial and recreational activity with conservation. We can do both.

> This Administration is committed to improving conservation and research in order to preserve our great marine heritage. It is a national treasure. It must be protected and dutifully maintained.

> At the Department of Commerce alone, the President's budget included $3 million in first time funding to support marine protected area activities consistent with existing law. If approved by Congress, these dollars can help us better manage this critical effort.

> I also plan to appoint a Marine Protected Area Advisory Committee comprised of key experts and stakeholders. The membership will include academic, state and local, non-governmental and commercial interests. The process will be open and will draw on America's great reservoir of experience and expertise.

> Past MPA designations like the Dry Tortugas in the Florida Keys were successful because they followed a well-planned process and secured grassroots support. The Dry Tortugas MPA offers a model for the years ahead.

> Conservation can be balanced with commercial and recreational activity. It is our stewardship responsibility. We will work with the Department of Interior, the Environmental Protection Agency and other federal agencies to safeguard our valuable coastal and ocean resources for the tomorrows in which we all will live.

Statement by Secretary of Commerce Donald L. Evans Regarding Executive Order 13158, Marine Protected Areas (June 4, 2001).

5. In cooperation with the Department of the Interior and working closely with other organizations, the Department of Commerce, through the National Oceanic and Atmospheric Administration (NOAA), has created the Marine Protected Areas Center (MPA Center) to coordinate the effort to implement the Executive Order, develop a framework for a national system of MPAs, and provide Federal, State, territorial, tribal, and local governments with the information, technologies, and strategies to support the system. *See* U.S. Departments of Commerce and the Interior, *The National MPA Initiative*, available at http://mpa.gov.

6. In the implementation of the Executive Order, the Departments of Commerce (through NOAA) and the Interior have identified two parallel tracks to focus the national initiative:

> *Network Design:* Evaluating the adequacy of existing levels of protection for marine resources and recommending new MPAs, and/or strengthening existing MPAs to establish a comprehensive and representative network.

> *Science-based Management:* Using science (both natural and social) to develop objective information, technical tools, and management strategies needed to support a national MPA network.

Is it clear where these tracks focus the MPA program? Is the Executive Order clear as to the objectives, authorities, and criteria this MPA initiative should employ? Not everyone believes it is. One vocal critic of the MPA program is the Recreational Fishing Alliance (RFA), which lobbies intensely on behalf of fishery conservation, and particularly in opposition to the effects of commercial fishing, but also objects to protective measures that restrict what it contends are the relatively minor impacts of recreational fishing. The RFA argues as follows with respect to MPAs:

> Designed correctly, MPAs can be useful for fishery conservation/management purposes as a part of a fishery management plan and could be implemented with the endorsement of the recreational fishing community if they accommodate the following:

> (A) There must be a clear identification of the conservation problem. Traditional management practices (gear restrictions, quotas, bag limits, closed seasons etc.) have been evaluated and do not provide sufficient conservation and management remedies to the affected stocks of fish;

> (B) The proposal for a specifically-identified MPA must include measurable criteria to determine the conservation benefit to the affected stocks of fish and contain economic impact information on how the proposed actions would affect fishermen;

> (C) The proposal also should allow for other types of recreational fishing, such as trolling for pelagic species, that would not have an impact on demersal stocks of concern, as an example.

> (D) Any closed areas within a MPA should be established with a sunset provision. On that date-certain, the zones will automatically reopen unless there is scientific proof that the closure should remain in effect and those findings are communicated to the public through a process integrating substantial public review and comment;

> (E) The plan provides a timetable for periodic review of the continued need for any closed area at least once every three years and an estimated time-line for removing the closure;

(F) The closed area is no larger than that which is supported by the best available scientific information;

(G) The fishery management measures are part of a fishery management plan as required by the Magnuson–Stevens Act as amended by the proposed Freedom to Fish Act.

See Recreational Fishing Alliance, RFA Position Paper on MPAs, *available at* http://www.savefish.com/mpaprop.html. Using the framework for ecosystem management policy analysis developed in Chapter 7 of this book, compare the MPA ecosystem management frameworks as described in the Executive Order and MPA Initiative versus the criteria outlined in the RFA proposal:

	Currency	**Goals**	**Methods**
Executive Order			
RFA Proposal			

7. Protection from overfishing is the purpose usually associated with an MPA, but other water-based actions can prove difficult to manage without using an MPA approach. In the case of the Florida manatee, for example, the chief threat to the species' survival is boating. The species is cold-intolerant, relatively slow swimming, and spends much of its time near the surface foraging on marine vegetation. These traits put it right in the path of one of Floridians' favorite pastimes—boating. Over 800,000 vessels are registered in the State of Florida, with an additional 400,000 vessels registered in other states regularly using Florida waters. Not surprisingly, the leading non-natural cause of manatee deaths today is collision with boats and boat propellers, primarily in seven Florida counties. Although enforcement of "speed zones" can help alleviate fatal boat collisions, federal and state agencies have limited resources to patrol all of Florida's waters. Thus, Florida agencies and the U.S. Fish and Wildlife Service, exercising its authority under the Endangered Species Act and Marine Mammals Protection Act, have established a network of "manatee refuges," where boating is highly regulated, and "manatee sanctuaries," where all waterborne activity is regulated. *See, e.g.,* USFWS, Manatee Protection Areas in Florida, 67 Fed. Reg. 680 (Jan. 7, 2002) (designating two new manatee refuge areas totaling more than 1500 surface acres). Should boating in Florida's waters be more severely restricted to protect the manatee? Should the total number of registered boats be capped? Should large areas of coastal waters be posted as completely off limits to boating?

8. The states are becoming more aggressive in the use of MPAs. In California, for example, recent legislation, the Marine Life Protection Act (MLPA), requires that the Department of Fish and Game develop a plan for establishing networks of marine protected areas in California waters to protect habitats and preserve ecosystem integrity, among other things. *See*

California Fish & Game Code, §§ 2850—2863. Sponsored by the Natural Resources Defense Council, the bill was supported by conservation, diving, scientific and educational groups. The purpose of the MLPA is to improve the array of MPAs existing in California waters through the adoption of a Marine Life Protection Program and a comprehensive MPA Master Plan. The MLPA states that "marine life reserves" (defined as no-take areas) are essential elements of an MPA system because they "protect habitat and ecosystems, conserve biological diversity, provide a sanctuary for fish and other sea life, enhance recreational and educational opportunities, provide a reference point against which scientists can measure changes elsewhere in the marine environment, and may help rebuild depleted fisheries." The Master Plan requires that recommendations be made for a preferred alternative network of MPAs with "an improved marine life reserve component." The MLPA further states that "it is necessary to modify the existing collection of MPAs to ensure that they are designed and managed according to clear, conservation-based goals and guidelines that take full advantage of the multiple benefits that can be derived from the establishment of marine life reserves." The six goals are:

> To protect the natural diversity and abundance of marine life, and the structure, function, and integrity of marine ecosystems.

> To help sustain, conserve, and protect marine life populations, including those of economic value, and rebuild those that are depleted.

> To improve recreational, educational, and study opportunities provided by marine ecosystems that are subject to minimal human disturbance, and to manage these uses in a manner consistent with protecting biodiversity.

> To protect marine natural heritage, including protection of representative and unique marine life habitats in California waters for their intrinsic value.

> To ensure that California's MPAs have clearly defined objectives, effective management measures, and adequate enforcement, and are based on sound scientific guidelines.

> To ensure that the state's MPAs are designed and managed, to the extent possible, as a network.

Id. § 2853(b). These goals were used as guiding principles in the development of the process used by a Master Plan Team to formulate recommendations of networks of MPAs. Does this approach strike you as fundamentally different from the federal MPA approach outlined in the Executive Order?

9. If you had a limited budget to place 10 new MPAs into being, where would you put them? Would it make sense to place them in areas of highest biodiversity? Of highest threat to human-induced degradation? Of lowest cost? There is an extensive effort in "bio-geography" underway to make the case that "hot spots" of biodiversity are where we should put our money for the most bang from the buck. *See* Callum M. Roberts, *Marine Biodiversity Hotspots and Conservation Priorities for Tropical Reefs*, 295 SCIENCE

1280 (2002). The argument is that by focusing financial and regulatory effort on making these areas no-take MPAs, we yield the greatest protection of the greatest concentration of biodiversity at the most efficient level of effort. Given the harsh unlikelihood that the entire marine environment will ever attain no-take MPA status, isn't this a sensible second-best alternative for those interested in maximizing the impact of MPAs?

10.   All told, only about one percent of the coastal and ocean waters within U.S. jurisdiction qualify as an MPA, and of that only ten percent receives the highest level of protection—a no-take prohibition of all fishing and other resource extraction activities. Is this tool of ecosystem management, used frequently in terrestrial settings, simply underused in the coastal and marine ecosystem context? Why hasn't it been used more?

D.   CASE STUDY: THE WHITE MARLIN

On of the most problematic management issues in marine and coastal settings involves what are known as highly migratory fish (HMS), which include tuna species, marlin, oceanic sharks, sailfishes, and swordfish. See 16 U.S.C. § 1802(2) (provision of the Magnuson–Stevens Act defining HMS). These fish roam the oceans with no respect for political boundaries. They are top-level predators in the marine food chain. They are also highly prized by humans for food and as sporting trophies. In this case study we examine the challenge domestic ecosystem management policy faces in conserving these species, an issue that was rocked to the core by the following recent development:

Department of Commerce, National Marine Fisheries Service (NMFS), 90–Day Finding for a Petition to List Atlantic White Marlin

66 Fed. Reg. 65676 (Dec. 20, 2001).

On September 4, 2001, NMFS received a petition from the Biodiversity Legal Foundation and James R. Chambers requesting NMFS to list the Atlantic white marlin (*Tetrapturus albidus*) as threatened or endangered throughout its range, and to designate critical habitat under the ESA. The petition contained a detailed description of the species, including the present legal status; taxonomy and physical appearance; ecological and fisheries importance; distribution; physical and biological characteristics of its habitat and ecosystem relationships; population status and trends; and factors contributing to the population's decline. Potential threats identified in the petition include: (1) overutilization for commercial purposes; (2) inadequacy of existing regulatory mechanisms; (3) predation; and (4) other natural or man-made factors affecting the species' continued existence. The petitioners also included information regarding how the species would benefit from being listed under the ESA, cited references and provided appendices in support of the petition.

Under the ESA, a listing determination can address a species, subspecies, or a distinct population segment (DPS) of a species (16 U.S.C. 1532 (16)). The petitioners requested that NMFS list Atlantic white marlin throughout its entire range. They are found in warm waters throughout tropical and temperate portions of the Atlantic Ocean and its adjacent seas (Caribbean, Mediterranean and Gulf of Mexico). A highly migratory pelagic species, they are found predominantly in the open ocean over deep water, near the surface in the vicinity of major ocean currents where their prey is concentrated. Their food resources include small fishes and invertebrates such as squid that can be swallowed whole.

The petitioners provided a detailed narrative justification for their petitioned action, describing past and present numbers and distribution of Atlantic white marlin. Information regarding its status was provided for the entire range of the species. The petition was accompanied by appropriate supporting documentation, including the most recent stock assessment for this species.

In 1997, the Atlantic white marlin was listed as overfished under the Magnuson–Stevens Fishery Conservation and Management Act (16 U.S.C. 1801 et seq.). In April 1999, NMFS published Amendment 1 to the Atlantic Billfish Fishery Management Plan, which included rebuilding programs and measures to reduce bycatch and bycatch mortality for Atlantic billfish, including white marlin. The International Commission for the Conservation of Atlantic Tunas (ICCAT), responsible for management of tunas and tuna-like fishes of the Atlantic Ocean also considers the Atlantic white marlin to be overfished. Several binding recommendations have been adopted by ICCAT over the last few years to reduce landings and improve data and monitoring. The most recent recommendation in November 2000 included a two-phase rebuilding plan involving further landing reductions and the development of more rebuilding measures after the next stock assessments in 2002.

The petitioners assert that existing protection for Atlantic white marlin at both the national and international level is inadequate to conserve the species or prevent its slide to extinction. The population's decline has been documented thoroughly by ICCAT's scientific advisors, the Standing Committee for Research and Statistics (SCRS). According to the petitioners, the primary cause of the Atlantic white marlin decline is due to bycatch in the international swordfish and tuna fisheries. The most recent stock assessment conducted in July of 2000 (SCRS/00/23) indicates that by the end of 1999: (1) the total Atlantic stock biomass had declined to less than 15 percent of its maximum sustainable yield level; (2) fishing mortality was estimated to be at least seven times higher than the sustainable level; (3) overfishing has taken place for over three decades; and (4) the stock is less productive than previously estimated, with a maximum sustainable yield smaller than 1,300 metric tons. The population's abundance was last at its long-term sustainable level in 1980. Reduction in prey species availability may also be a threat to the species, with two of its important prey species,

Atlantic bluefish and squid, listed as overfished under the Magnuson–Stevens Act (16 U.S.C. 1801 et seq.).

Petition Finding

Based on the above information and the criteria specified in 50 CFR 424.14 (b)(2), NMFS finds that the petitioner presents substantial scientific and commercial information indicating that a listing of Atlantic white marlin may be warranted. Under section 4 (b)(3)(A) of the ESA, this finding requires that NMFS commence a status review on Atlantic white marlin. NMFS is now initiating this review. Within one year of the receipt of the petition (by September 3, 2002), a finding will be made as to whether listing the Atlantic population of the white marlin as threatened or endangered is warranted, as required by section 4 (b)(3)(B) of the ESA. If warranted, NMFS will publish a proposed rule and take public comment before developing and publishing a final rule.

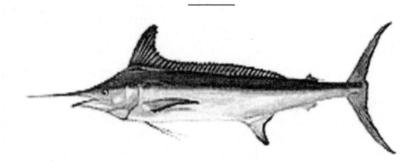

The White Marlin

The factors leading NMFS to find that listing of the white marlin as an endangered species "may be warranted" reveal the complexity of coastal and marine ecosystems, and thus the challenge of any effort to manage them. The white marlin is a long-lived, top-level predator fish that lives in the open ocean and travels long distances following schools of mackerel, tuna, dolphin fish, and squid. It is one of the lions of the ocean; yet, its behavioral characteristics have spelled doom for its continued survival.

The marlin's problem can be summarized as follows: marlins like to eat tuna; humans also like to eat tuna; when humans hunt tuna, marlins are the collateral damage. The method of choice for humans to hunt tuna is the pelagic longline gear rig. The gear consists of a mainline, often miles long, suspended in the water column by floats and from which up to several thousand baited hooks are attached on leaders called gangions. The lines remain in the water for a day or so, and then they are reeled in and the caught tuna are removed from the hooks one-by-one. This method, however, is not especially discriminating–in addition to tuna, longline fishing also hooks sea birds, marine mammals, and billfish such as marlins. This so-called "bycatch" usually is not retained, either for economic or regulatory reasons, but it is nonetheless usually very much dead.

Commercial longline fishing bycatch is without question the leading cause of white marlin mortality. U.S commercial fishing operations killed 40.8 metric tons of white marlin as bycatch in 2001, mostly from the longline fishery bycatch (by contrast, recreational fishing landed 0.2 metric tons). Decades of this and other fishing pressures on white marlins throughout Atlantic waters have decimated the population. Things have gotten even worse since the data NMFS used in its listing petition finding. By the beginning of 2000, the biomass of white marlin had been driven to an estimated 13 percent of the sustainable level (MSY, as described earlier in this chapter) and it continues to decline. The sustainable fishing level rose to roughly 8 times the MSY level by 2000. As a result, the white marlin population is now close to, if not beyond, the point of recruitment failure–the population level below which there are too few adult breeders to replace the population even under the best of circumstances. If the rate of decline continues, this point will soon be passed irreversibly and the marlin will become extinct. Some estimate this will occur in less than five years unless something is done.

But what can be done? More to the point, what does ecosystem management have to offer as a domestic policy instrument to help avoid white marlin extinction? In fact, the United States has already taken some decisive action. As noted in the petition finding, in 1997 NMFS declared the white marlin an overfished species under the Magnuson–Stevens Act (described earlier in this chapter). NMFS has instituted limited "time/area" closures of areas to longline fishing in a few areas of domestic waters, *see* NMFS, Atlantic Highly Migratory Species–Pelagic Longline Fishing, 65 Fed. Reg. 47213 (Sept. 1, 2000) (codified at 50 C.F.R. pt. 635), and it has prohibited commercial harvests of white marlin in all U.S. waters. Bycatch, however, is not considered commercial harvest. Moreover, marlin, like other HMS fish, regularly travel outside U.S. waters, where only international agreements can restrict fishing harvests.

As NMFS notes, in international waters the International Commission for the Conservation of Atlantic Tunas (ICCAT) is responsible for management of tunas and tuna-like fishes of the Atlantic Ocean and adjacent seas. The organization was established in 1969, at a Conference of Plenipotentiaries, which prepared and adopted the International Convention for the Conservation of Atlantic Tunas in Rio de Janeiro, Brazil, in 1966. (For a history of ICCAT and related international regimes for HMS, see Karen L. Smith, *Highly Migratory Fish Species: Can International and Domestic Law Save the North Atlantic Swordfish?*, 21 NEW ENG. L. REV. 10 (1999).) The Convention established exclusive jurisdiction in ICCAT as the only fisheries organization that can undertake the range of work required for the study and management of tunas and tuna-like fishes in the Atlantic, which include about 30 species of direct concern. Such studies include research on biometry, ecology, and oceanography, with a principal focus on the effects of fishing on stock abundance. The Commission's work requires the collection and analysis of statistical information relative to current conditions and trends of the fishery resources in the Convention area. The Commission also undertakes work in the compilation of data for other fish species that are caught during tuna fishing (bycatch) in the Convention area, and which are not investigated by another international fishery organization–e.g., the white marlin.

Not everyone is pleased with ICCAT's work–it is known in some circles as the "International Commission to Catch All Tunas." Yet ICCAT also considers the white marlin to be overfished and has adopted some restrictions on landings. ICCAT has not, however, banned commercial harvests in international waters. Tuna fishing operations from many other nations thus are free keep their white marlin and other "bycatch" for commercial use, or to discard it. It is difficult, therefore, for U.S. domestic regulation to manage the fishing-based mortality rate for the white marlin and other HMS fish.

In addition to the commercial longline tuna fishing industry, another major player in the white marlin's future is the recreational fishing industry, represented by organizations such as The Billfish Foundation and the Recreational Fishing Alliance. Few fish are more prized among sport

A prize-winning white marlin.

fishermen than the white marlin. At the 23rd Annual White Marlin Open held in 1996, the world's largest billfishing tournament with over 235 boats entering, the winner landed a 73.5 pound marlin and for that won a prize of $401,140. *See World Fishing Firsts at White Marlin Open*, www.beach-net.com/wmopen.html. The competition in 2000 had total prize money of over $1.4 million, with the white marlin category again receiving the highest proportion of money.

Overall, recreational billfish anglers spend more than $2 billion annually on their sport. The NMFS estimates that, whereas the total economic activity generated by commercial fishing is roughly $27 billion annually, the recreational fishing industry generates over $100 billion of economic impact. *See* NMFS, 2001 STOCK ASSESSMENT AND FISHERY EVALUATION FOR ATLANTIC HIGHLY MIGRATORY SPECIES (2001). Needless to say, therefore, the recreational fishing industry is keenly interested in how NMFS treats the white marlin listing petition. *See, e.g.*, The Billfish Foundation, *Endangered Species Listing–Wrong Tool*, SALTWATER SPORTSMAN, Dec. 2001, at 57.

Indeed, listing of the white marlin as endangered under the ESA would be a drastic approach to management of the species in domestic waters. As the materials in Part II of this text explain, bycatch of white marlin would be considered "incidental take" of the species for which a permit would be required. Recreational fishing for white marlin would be flatly prohibited, and recreational fishing with the incidental potential to land a marlin would be regulated.

But what are the alternatives? Is there any option short of ESA listing and rigorous enforcement of the incidental take prohibition that will reverse the decline of the white marlin, and does any option (even ESA listing) really solve the problem that, as an HMS, U.S. law cannot truly manage the white marlin's ecosystem? Consider how these other approaches fare under those criteria:

1. Ban longline tuna fishing at all times in all domestic waters, but leave recreational and other commercial tuna harvesting methods alone.

2. Develop longline fishing bycatch reduction methods and enforce them through stiff fines and by posting NMFS observers or video monitoring systems (VMS) aboard longline tuna fishing vessels. (For more on VMS methods see Blue Water Fisherman's Association v. Minetta, 122 F.Supp.2d 150 (D.D.C.2000), and NMFS, Notice of Reconsideration of VMS Requirements, 66 Fed. Reg. 1907 (Jan. 10, 2001).)

3. Avoid direct banning or regulation of longline tuna fishing as in (1) and (2), but impose a significant per fish white marlin "bycatch fee" that would be enforced through counts made by onboard observers or VMS.

4. Buy out the domestic longline tuna fishing fleet and ban any new longline tuna fishing vessels from operating within domestic waters. (For an example of a buyout proposal see The Atlantic Highly Migratory Species Act of 2001, H.R. 1367, 107th Cong. (2001).)

5. Establish a chain of white marlin Marine Protected Area preserves within domestic waters, within which no commercial or recreational fishing for tuna or billfish is allowed.

6. Establish a voluntary tuna fishing certification program for domestic and foreign vessels that use verified bycatch reduction methods, and institute a massive public education campaign centering around "Marvin the Marlin" in an effort to convince consumers to eat only "marlin friendly" tuna–i.e., tuna processed only from certified vessels' harvests.

7. Levy a tax on the tuna processing industry or on retail sales of tuna, and use the revenue to finance a white marlin recovery effort that would include any or all of options (2) through (6).

What would the petitioner, Mr. Chambers, have to say about each of these proposals? How about the commercial fishing industry? How about the recreational fishing industry? Sport fishing boat manufacturers? The tuna processing industry? Consumers?

Of course, another interested party is NMFS, which is in the unusual position of potentially managing the white marlin through two different statutory programs administered through two different divisions of the agency. The NMFS Highly Migratory Species Division, which is part of the NMFS Office of Sustainable Fisheries, administers the Magnuson–Stevens Act for the white marlin and other HMS. *See* http://www.nmfs.noaa.gov/sfa/hmspg.html. The decision whether to grant the ESA listing petition, however, is under review by the Endangered Species Division of the NMFS Office of Protected Resources. *See* http://www.nmfs.noaa.gov/prot_res/PR3/PR3overview.html. Assume you are the NMFS HMS Division employee tasked with managing the white marlin. How would you react to the petition and what action would you take to influence the outcome? Assume you are the NMFS ESA Division employee tasked with reviewing the petition. How would you interact with the employees of the HMS Division?

Considering that this is a chapter on ecosystem management, does it also strike you as odd that none of the proposals outlined above involves management of an *ecosystem*? In other words, they all involve efforts to reduce the direct take of white marlins. Is there any way to address the plight of the white marlin through habitat management? Isn't that the fundamental challenge for recovering highly migratory fish species–i.e., that it's not about the habitat? So, is the white marlin really a problem in ecosystem management? On the other hand, as a top-level predator in its domain, the white marlin surely is an important facet of the relevant ecosystem, and thus its demise can be expected to alter the dynamics thereof.

It seems an odd twist of fate that a fish worth over $400,000 as a trophy prize for recreational fishermen is being driven to extinction because it is scrap to tuna fishing operations. If there is a way to avoid what appears increasingly to be an inevitable demise of the white marlin, it had better be implemented soon. Any ideas?

CHAPTER 12

EXTREME ECOSYSTEMS

Chapter Outline

A. Fragile Ecosystems
    1. Deserts
    2. Islands
    3. Coral Reefs
B. Human–Dominated Ecosystems
    1. Urban America
        a. Impact Assessment Methods
        b. Management Frameworks
    2. Agricultural Lands
    3. Recreational Areas

An implicit assumption of the conventional ecosystem management regimes studied thus far in this part of the text is that, with some degree of carefully planned management, it will be possible to sustain a reasonable balance of sorts between human use and ecosystem dynamics. This assumption, controversial even in the settings already studied, becomes increasingly tenuous as we move toward either of two extremes in ecosystem conditions. At one extreme are ecosystems so fragile that anything beyond a de minimis human presence may do substantial harm. Even ecosystem management itself, to the extent it implies monitoring and manipulating ecosystem components and processes, may impose intolerable damage to these ecosystems. At the other extreme are ecosystems already so dominated by human presence that the question of harm to the ecosystem dynamics that existed prior to human intrusion is moot. Indeed, in these cases a new set of ecosystem dynamics has evolved, one that cannot be described in any detail except as an anthropogenic consequence. While polar opposites in terms of ecosystem dynamics, in both cases the central question for these extreme ecosystems is the role of humans.

In many cases either or both of the extreme ecosystem types may be found in free-standing conditions (e.g., the North Pole; New York City), or embedded within larger ecosystems managed under conventional legal regimes (e.g., coral reefs in a Marine Protected Area; ski areas in a national forest). In either setting a threshold management issue is whether to craft specialized laws and policies to respond to the qualities that make the resource an extreme case. This is not an easy question, for the risk of over-specializing may be to create an inefficient patchwork of legal regimes across the landscape, whereas relying on the generalized mandates of conventional ecosystem management authorities could prove ineffective in extreme and particularized settings.

Under either policy approach, another significant challenge is deciding whether the difficulty of achieving some balance between people and ecosystem is worth it. An easy approach for fragile ecosystems would be to simply exclude humans altogether, and the easy approach for human-dominated ecosystems would be to throw in the towel and write off the ecosystem values as a lost cause. That may not be a politically viable official policy, but it may be the only financially viable approach in many settings.

Consider, for example, the images of the Gulf of Mexico beach at Destin, Florida captured on the next page. With its miles of sugar-white sand beaches, seemingly endless supply of beach-front rentals, and loads of restaurants and entertainment venues for people of all ages, Destin is one of the most popular beach resort destinations in the nation. (One of your authors is a frequent visitor; the other, alas, is stranded a thousand miles from the nearest ocean!) But all those people pose a serious threat to one of the most fragile of ecosystems—coastal dunes. The dunes are a vital coastal resource, protecting inland areas from flooding during storm surges and providing prime habitat for coastal vegetation and wildlife. Critical to dune dynamics, however, are the coastal grasses and vines that stabilize the sands. But this vegetative anchor is extremely sensitive to disturbance. Few coastal wildlife species can impose the same weight and force that a human would, even a small child, when frolicking about in the dunes. On the other hand, Destin wouldn't be Destin without frolicking on the beach. And every linear foot of "no humans" dune area is a serious sacrifice for Destin's economy. So the question is, where do the humans get to frolic and where do the dunes get to stay?

The top picture in the series illustrates Destin's answer to that question. The result of municipal and state land use decisions has been to create a sharp division between "humans dominate" and "dunes dominate" zones. The sign advising us to "protect dune areas" marks the boundary between the two opposite extremes. To the west (lower left picture) lies the human playground of condos and beaches raked clean every morning by motorized beach patrols. Save for the thin slivers of "dune areas" fronting some condominiums, there is no pretense in this zone of trying to keep a fully functioning dune ecosystem in the picture. By contrast, to the east lies Henderson Beach State Recreation Area (lower right picture), where signs every 50 feet or so remind us not to venture into the dunes, even for a leisurely stroll. Visitors to the recreation area are shunted on long, narrow boardwalks across the dunes to reach the shoreline, where only there can they stay. Glancing to the east and west while standing at the dividing sign, as one of your authors has done many times, provides an almost surreal contrast, one that immediately drives home the challenges of managing the extreme ecosystems.

In this chapter we survey a selection of fragile and human-dominated ecosystems and the policies we use to manage them. In neither case is our treatment exhaustive of the possibilities, for they are as varied and as specialized as the many different ecosystem extremes that exist. Rather, we have chosen what we believe are the most representative and frequently encountered examples of each extreme ecosystem type and of its associated policy instruments.

The sign in the top picture marks the boundary between Destin's public beaches and Henderson Beach State Recreation Area. Beachfront condominiums and restaurants (lower left picture) extend for miles to the west before another protected area can be found. The protected dunes of the recreation area (lower right picture) extend about a mile to the east before another stretch of developed beachfront appears. Is there any way to introduce more human presence into dune ecosystems without decimating the dunes? Is there any way to introduce more dunes into the developed beachfront without decimating its economic potential?

A. FRAGILE ECOSYSTEMS

In Lucas v. South Carolina Coastal Council, 505 U.S. 1003 (1992), Justice Scalia's plurality opinion crafted a *"per se"* category of regulatory takings of property for instances where regulation enacted after a property owner takes title deprives the land of all economically beneficial use. In such cases, he explained, a taking has occurred as a matter of law and the only remaining question is what compensation is due. An exception to this categorical treatment, however, exists when "background principles" of state nuisance and property law—i.e., the common law in place when title was taken—would have in any event restricted the property owner from engaging in the use subsequently restricted by regulation. Commenting on this doctrine in his concurring opinion, Justice Kennedy took issue with the narrow quality of Justice Scalia's so-called "nuisance exception" to the *per se* takings rule:

> The common law of nuisance is too narrow a confine for the exercise of regulatory power in a complex and interdependent society. The State should not be prevented from enacting new regulatory initiatives in response to changing conditions, and courts must consider all reasonable expectations whatever their source. The Takings Clause does not require a static body of state property law; it protects private expectations to protect private investment. I agree with the Court that nuisance prevention accords with most common expectations of property owners who face regulation, but I do not believe this can be the sole source of state authority to impose severe restrictions. Coastal property may present such unique concerns for a fragile land system that the State can go further in regulating its development and use than the common law of nuisance might otherwise permit.

505 U.S. at 825.

What "fragile land systems" might qualify under Justice Kennedy's approach for exemption from the *per se* regulatory takings category, and what form might the state's "severe restrictions" take? As with each of the ecosystem regimes studied thus far, answers to these and related questions often hinge on the ownership status of the land in question—public or private—since the government has more latitude to restrict land use on its own lands. But the approaches taken on public lands face a demand for public use that does not exist in the private land context, where the right to exclude is paramount. Thus, in either ownership regime, imposing "severe restrictions" on behalf of protecting "fragile land systems" is bound to cause conflict. Some examples of varying levels of restriction imposed in different fragile ecosystem settings illustrate these challenges.

1. Desert Ecosystems

Although they teem deceptively with abundant biodiversity, deserts are not especially resilient ecosystems. Tank tracks from World War II training exercises remain clearly visible in parts of the Mojave desert, and the devastation of the Gulf War on Kuwait's desert lands, despoiled by over 250 million gallons of oil that Iraqi troops intentionally released (20 times more than the *Exxon Valdez* spill), will take over $1 billion and several decades to restore to some semblance of functional health. *See* Ben Shouse, *Kuwait Unveils Plan to Treat Festering Desert Wound*, 293 Science 1410 (2001). And restoration is a term that can be used only loosely in desert settings. In his studies of western ghost towns, for example, Robert Webb of the U.S.

Geological Survey has found that while vegetative communities have been established over a century after boom towns turned bust, the species composition is significantly different in many areas. Hardy, long-lived species such as creosote have been replaced by fast-breeding, short-lived plants like cheese-bush. *See* Kathryn Brown, *Ghost Towns Tell Tales of Ecological Boom and Bust*, 290 SCIENCE 35 (2000).

In a modern day version of this story, consider how the court in the following opinion describes and addresses the effects of off-road vehicles (ORV) on a portion of California desert. As ORV use became increasingly popular, both President Nixon and President Carter acted to control their presence on federal public lands through Executive Orders. *See* Exec. Order 11644, 37 Fed. Reg. 2877 (Feb. 8, 1972); Exec. Order 11989, 42 Fed. Reg. 26959 (May 25, 1977). Congress also addressed the issue of ORV use on Bureau of Land Management lands in the Federal Land Planning and Management Act. The case involves the manner in which the agency implemented these authorities in a particularly sensitive desert area.

Sierra Club v. Clark

United States Court of Appeals for the Ninth Circuit, 1985.
22 ERC 1748, 15 Envtl. L. Rep. 20,319

■ POOLE, CIRCUIT JUDGE:

Plaintiffs Sierra Club, Desert Protective Council and California Native Plant Society ("Sierra Club") filed this action seeking judicial review under the Administrative Procedure Act, 5 U.S.C. § 706(1), of the failure of defendants Secretary of the Interior, Director of the Bureau of Land Management ("BLM"), and California State Director of BLM ("Secretary") to close Dove Springs Canyon to off road vehicle ("ORV") use. Sierra Club appeals from the district court's denial of their motion for summary judgment, and the grant of the Secretary's cross-motion for summary judgment. We affirm.

FACTS

Dove Springs Canyon is located in the California Desert Conservation Area ("Desert Area"), established in 1976, 43 U.S.C. § 1781, under the Federal Land Policy Management Act ("the Act"), 43 U.S.C. § 1701 et seq. The Desert Area covers approximately 25 million acres in southeastern California, approximately 12.1 million of which are administered by the BLM. Dove Springs Canyon is comprised of approximately 5500 acres; 3000 acres are designated "open" for unrestricted use of ORVs.

Dove Springs Canyon possesses abundant and diverse flora and fauna. Over 250 species of plants, 24 species of reptiles, and 30 species of birds are found there. It also offers good habitat for the Mojave ground squirrel, the desert kit fox, and the burrowing owl. Because the rich and varied biota is unusual for an area of such low elevation in the Mojave Desert, the Canyon was once frequented by birdwatchers and naturalists, as well as hikers and fossil hunters.

Recreational ORV usage of Dove Springs Canyon began in 1965 and became progressively heavier in the ensuing years. By 1971, the Canyon was being used intensively by ORV enthusiasts. It became especially popular because the site's diverse terrain, coupled with relatively easy access, provides outstanding hill-climbing opportunities. By 1979, up to 200 vehicles used the Canyon on a typical weekend; over 500 vehicles used it on a holiday weekend. In 1973, the BLM adopted its Interim Critical Management Program for Recreational Vehicle Use on the California Desert ("Interim Program") which designated Dove Springs Canyon as an ORV Open Area, permitting recreational vehicle travel in the area without restriction.

Extensive ORV usage has been accompanied by severe environmental damage in the form of major surface erosion, soil compaction, and heavy loss of vegetation. The visual aesthetics have markedly declined. The character of the Canyon has been so severely altered that the Canyon is now used almost exclusively for ORV activities.

In July of 1980 Sierra Club petitioned the Secretary of the Interior to close Dove Springs Canyon to ORV use under the authority of Executive Order No. 11644, as amended by Executive Order No. 11989, and 43 C.F.R. § 8341.2 because of "substantial adverse effects" on the vegetation, soil and wildlife in the Canyon. The Secretary responded that the matter would be addressed in the California Desert Conservation Plan and Final Environmental Impact Statement ("the Final Plan").

The Final Plan approved by the Secretary in December 1980 maintained unrestricted ORV use in Dove Springs of 3000 of the 5500 acres. Sierra Club filed this action on January 6, 1981, alleging that the Secretary's failure to close Dove Springs violated Executive Order No. 11644, as amended by Executive Order No. 11989, and 43 C.F.R. § 8341.2; 43 U.S.C. § 1732(b), which requires the Secretary to prevent "unnecessary or undue degradation of the lands;" and 43 U.S.C. §§ 1781(b) and (d), which require the Secretary to maintain and conserve resources of the Desert Area under principles of "multiple use and sustained yield." Sierra Club sought declaratory relief and a writ of mandate compelling closure.

* * *

ANALYSIS

The district court ruled that the plaintiffs' complaint was an attack upon the Canyon's initial designation as an "ORV freeplay area" in the Final Plan, and refused to address plaintiffs' contention that the Executive Orders and the Regulation required closure of the area after the Final Plan was adopted.... The district court also ruled that the Secretary's and BLM's exercise of discretion under the Act in designating the Canyon as open mooted the plaintiffs' claim. The plaintiffs in the district court and on appeal contend, however, that the closure standard contained in the Executive Orders and the Regulation applies independently of the designation process. The plain meaning of the provisions supports their view.

The Regulation provides:

> Notwithstanding the consultation provisions of § 8342.2(a), where the authorized officer determines that off-road vehicles are causing or will cause considerable adverse effects ... the authorized officer shall immediately close the areas or trails affected.... Such closures will not prevent designation ..., but these lands shall not be opened to the type(s) of off-road vehicle to which it was closed unless the authorized officer determines that the adverse effects have been eliminated and measures implemented to prevent recurrence.

43 C.F.R. § 8341.2(a). This provision creates a separate duty to close without regard to the designation process; it does not automatically become inoperative once the Secretary exercises his discretion to designate the land.

The district court erred in its analysis and conclusion that the Secretary's designation mooted the Sierra Club's claims. Nevertheless, we do not reverse the district court on account of this error because we must affirm if the record fairly presents any basis for affirmance. On appeal from a grant of summary judgment, we review the record de novo, using the same standard as the district court under Fed.R.Civ.P. 56(c). Because we have decided that the closure standard of the Executive Orders and the Regulation applies independently of the designation of the land as open under the Act, the issue before us is whether the damage to Dove Springs Canyon amounts to "considerable adverse effects" which require the Canyon's closure. The parties agree that there is no genuine issue as to the extent of the damage to the Canyon, and therefore resolution of this issue depends upon whether the Secretary's interpretation of this phrase or that of the Sierra Club is to control.

Traditionally, an agency's interpretation of its own regulation is entitled to a high degree of deference if it is not unreasonable. The Secretary interprets "considerable adverse effect" to require determining what is "considerable" in the context of the Desert Area as a whole, not merely on a parcel-by-parcel basis. The Secretary contends such a broad interpretation is necessary and is consistent with 43 U.S.C. § 1781(a)(4) which expresses a congressional judgment that ORV use is to be permitted "where appropriate."

* * *

Sierra Club argues that ... the Secretary's interpretation should not be adopted because it is unreasonable. Sierra Club insists that the sacrifice of any area to permanent resource damage is not justified under the multiple use management mandate of 43 U.S.C. § 1702(c) that requires multiple use "without permanent impairment of the productivity of the land and the quality of the environment." In further support of its position Sierra Club adverts to the requirement in the Act that the Secretary prevent "unnecessary and undue degradation" of the public lands, 43 U.S.C. § 1732(b). In addition, Sierra Club contends, when Congress established the Desert Area it intended the Secretary to fashion a multiple use and sustained yield management plan "to conserve [the California desert] resources for future generations, and to provide present and future use and enjoyment, particularly outdoor recreational uses, including the use, where appropriate, of off-road recreational vehicles." 43 U.S.C. § 1781(a)(4). Sierra Club argues that it is unreasonable for the Secretary to find ORV use "appropriate" when that use violates principles of sustained yield, substantially impairs productivity of renewable resources and is inconsistent with maintenance of environmental quality.

We can appreciate the earnestness and force of Sierra Club's position, and if we could write on a clean slate, would prefer a view which would disallow the virtual sacrifice of a priceless natural area in order to accommodate a special recreational activity. But we are not free to ignore the mandate which Congress wrote into the Act. Sierra Club's interpretation of the regulation would inevitably result in the total prohibition of ORV use because it is doubtful that any discrete area could withstand unrestricted ORV use without considerable adverse effects. However appealing might be such a resolution of the environmental dilemma, Congress has found that ORV use, damaging as it may be, is to be provided "where appropriate." It left determination of appropriateness largely up to the Secretary in an area of sharp conflict. If there is to be a change it must come by way of Congressional reconsideration. The Secretary's interpretation that this legislative determination calls for accommodation of ORV usage in the administrative plan, we must conclude, is not unreasonable and we are constrained to let it stand.

* * *

Under the California Desert Conservation Area Plan, approximately 4 percent (485,000 acres) of the total acreage is now open to unrestricted ORV use. Dove Springs itself constitutes only 0.025 percent of BLM administered lands in the Desert Area. Although all parties recognize that the environmental impact of ORV use at Dove Springs is severe, the Secretary's determination that these effects were not "considerable" in the context of the Desert Area as a whole is not arbitrary, capricious, or an abuse of the broad discretion committed to him by an obliging Congress.

Notes and Questions

1. *Sierra Club v. Clark* illustrates the difficulty of managing particularly fragile areas that lie within larger public land management units, especially when the amorphous "multiple use" mandate applies. Dove Springs Canyon, at 5500 acres, was a minuscule fraction of the Desert Area's 25 million acres. It provided, however, a concentration of biodiversity that was particularly susceptible to damage from a concentration of ORV usage. The BLM strategy for implementing the multiple use mandate seems to have been to concentrate ORV use into contained areas such as the Canyon and write them off from a conservation perspective, leaving the remaining areas free for different management objectives. While this spelled doom for Dove Springs Canyon, was it an appropriate strategy for management of the

Desert Area as a whole? Assuming that some ORV use must be allowed in the Desert Area—after all, Congress decreed as much—how would you have managed the distribution of ORV areas? Should they be allowed at moderate levels everywhere, or at intensive levels in special areas only? Should areas of intensive use be periodically rotated throughout the Desert Area, so as to allow recovery in used areas over time?

2.   The court's reasoning was that BLM was within its discretion to define "considerable adverse effects" such that the numerator was the 3000 acres of Dove Springs Canyon open to ORV use, but the denominator was the entire Desert Area's 25 million acres. Does that make sense? Will any area under, say, 10,000 acres that is opened to concentrated ORV use ever affect the entire Desert Area so substantially as to trigger the "considerable adverse effects" standard under this approach? On the other hand, where would ORV use ever be allowed if the test were to focus only on the area where ORV "freeplay" use is concentrated? That approach might lead to decisions that spread ORV use out over larger areas, so as to make the effects more diffuse and thus less likely to be "considerable." Which outcome would you prefer? Do you have an alternative approach to recommend?

3.   The Dove Springs Canyon area fell prey to Congress's attempt to balance ORV use and desert resource conservation in a large, defined public land area within which no distinctions were made for fragile ecosystems. Dove Springs was, in other words, nothing special within the California Desert Conservation Area. As the following statutory excerpt illustrates, however, Congress has identified some desert resources for special treatment, setting them apart from surrounding public land resources.

Mojave National Preserve

16 U.S.C.A. § 410aaa–41

§ 410aaa–41. Findings

The Congress hereby finds that—

(1) Death Valley and Joshua Tree National Parks, as established by this subchapter, protect unique and superlative desert resources, but do not embrace the particular ecosystems and transitional desert type found in the Mojave Desert area lying between them on public lands now afforded only impermanent administrative designation as a national scenic area;

(2) the Mojave Desert area possesses outstanding natural, cultural, historical, and recreational values meriting statutory designation and recognition as a unit of the National Park System;

(3) the Mojave Desert area should be afforded full recognition and statutory protection as a national preserve;

(4) the wilderness within the Mojave Desert should receive maximum statutory protection by designation pursuant to the Wilderness Act [16 U.S.C.A. § 1131 et seq.]; and

(5) the Mojave Desert area provides an outstanding opportunity to develop services, programs, accommodations and facilities to ensure the use and enjoyment of the area by individuals with disabilities, consistent with section 504 of the Rehabilitation Act of 1973 [29 U.S.C.A. § 794], Public Law 101–336, the Americans With Disabilities Act of 1990 (42 U.S.C. 12101) [42 U.S.C.A. § 12101 et seq.], and other appropriate laws and regulations.

§ 410aaa–42. Establishment

There is hereby established the Mojave National Preserve, comprising approximately one million four hundred nineteen thousand eight hundred acres, as generally depicted on a map entitled "Mojave National Park Boundary—Proposed", dated May 17, 1994, which shall be on file and available for inspection in the appropriate offices of the Director of the National Park Service, Department of the Interior.

§ 410aaa–47. Withdrawal

Subject to valid existing rights, all Federal lands within the preserve are hereby withdrawn from all forms of entry, appropriation, or disposal under the public land laws; from location, entry, and patent under the United States mining laws; and from disposition under all laws pertaining to mineral and geothermal leasing, and mineral materials, and all amendments thereto.

§ 410aaa–50. Grazing

(a) The privilege of grazing domestic livestock on lands within the preserve shall continue to be exercised at no more than the current level, subject to applicable laws and National Park Service regulations.

(b) If a person holding a grazing permit referred to in subsection (a) of this section informs the Secretary that such permittee is willing to convey to the United States any base property with respect to which such permit was issued and to which such permittee holds title, the Secretary shall make the acquisition of such base property a priority as compared with the acquisition of other lands within the preserve, provided agreement can be reached concerning the terms and conditions of such acquisition. Any such base property which is located outside the preserve and acquired as a priority pursuant to this section shall be managed by the Federal agency responsible for the majority of the adjacent lands in accordance with the laws applicable to such adjacent lands.

§ 410aaa–52. Preparation of management plan

Within three years after October 31, 1994, the Secretary shall submit to the Committee on Energy and Natural Resources of the United States Senate and the Committee on Natural Resources of the United States House of Representatives a detailed and comprehensive management plan for the preserve. Such plan shall place emphasis on historical and cultural sites and ecological and wilderness values within the boundaries of the preserve. . . .

Notes and Questions

1. Withdrawals of land from larger public land multiple use areas into specialized single or dominant use units may provide a means of isolating particularly fragile areas and subjecting them to different management protocols. Of course, doing so tends to concentrate the multiple uses allowed outside the special unit into a smaller area, thus intensifying the competition between advocates of the different uses.

2. In the Mojave National Preserve example, Congress allowed domestic livestock grazing to remain as a permitted use in the specially protected area. Why? If you read through Chapter 9 of this text, you will learn that grazing privileges on federal public lands are not property rights, and thus Congress could extinguish them with no fear of facing property takings claims. So why did Congress designate a special Mojave National Preserve area and allow grazing to remain?

2.   ISLAND ECOSYSTEMS

Destin, Florida, which is featured in the opening of this chapter, is located on a barrier island, a thin sliver of sand separated from the mainland by a wide bay and marsh system. Barrier islands like it rim the Texas, Florida, and Alabama coasts on the Gulf of Mexico as well the Atlantic coasts of New Jersey, Delaware, Maryland, North Carolina, and South Carolina. And Destin is not alone in being a popular spot. One study found that in 100 coastal counties, towns located on barrier islands approved construction of 54,000 homes in 1998, worth over $6.5 billion in home construction value alone. *See Boom on the Beach*, USA TODAY, July 27, 2000, at 4A. Home lot prices for choice oceanfront real estate regularly exceed $1 million. Commercial real estate prices are soaring as well (the two-bedroom condominium unit one of your authors frequents in Destin, one of 30 units in the building, is available for $450,000!).

What will all these people mean for the barrier island ecosystems? Islands, for obvious reasons, present unusual ecosystem dynamics. Isolated from the mainland and one another, their assemblage of species often depends on chance and accident, such as how finches reached the Galapagos Islands. This makes island ecosystems particularly vulnerable to disruption from natural disaster and invasive species, both of which have been exacerbated by human impact. Development in island habitat reduces ecosystem resilience to natural flood and storm events and exposes island species to higher risks of extirpation. *See* Thomas Brooks and Michael Leonard Smith, *Caribbean Catastrophes*, 294 SCIENCE 1469 (2001). Non-native species introduced deliberately or unintentionally by human action can come to dominate an island ecosystem food web within decades. *See* Erik Stokstad, *Parasitic Wasps Invade Hawaiian Ecosystem*, 293 SCIENCE 1241 (2001). Beyond these dramatic potential effects, the sheer presence of so much humanity in island settings poses tremendous growth management problems. In the Florida Keys, for example, sewage effluent poses a significant threat to island resources. Sydney T. Bacchus, *Knowledge of*

Groundwater Responses, 18 ENDANGERED SPECIES UPDATE 79 (2001). And Hawaii, our nation's island state, leads all others in listed endangered species.

Do islands need special protection from these development pressures? Should development be banned in some circumstances, or is the staggering economic potential of island development sufficient to justify turning their ecosystems from fragile to human-dominated? Consider how one coastal state has addressed the problem in the following case, which illustrates how hard-fought ecosystem management battles may be and how important the meaning of a word can be in the outcome.

Estate of Edgar E. Sims, Jr. v. Department of Environmental Protection

Office of Administrative Law, State of New Jersey, 1994.
95 N.J.A.R.2d (EPE) 6.

INITIAL DECISION AND FINAL AGENCY DECISION

KANE, ALJ:

STATEMENT OF THE CASE AND PROCEDURAL HISTORY

Pursuant to N.J.A.C. 7:7E–3.21, the respondent classified property owned by the petitioner as a "bay island". The petitioner disagreed with this determination claiming instead that the property known as Rum Point, is part of the barrier island of the City of Brigantine, located immediately north of Atlantic City, New Jersey. On May 3, 1993, this matter was transmitted to the Office of Administrative Law (OAL) to be heard as a contested case pursuant to N.J.S.A. 52:14B–1 to–15 and N.J.S.A. 52:14F–1 to–13. The hearing was conducted at the Atlantic City OAL, 1201 Bacharach Boulevard, Atlantic City, New Jersey on April 25, 26 and May 2, 3, 11 and 17, 1994. The record was held open until June 29, 1994 in order to permit the filing of post-hearing briefs. An initial decision was rendered on August 5, 1994.

ISSUES

1. Whether evidence submitted in support of the summary statement to N.J.A.C. 7:7E–3.21, should be excluded from the consideration of what constitutes a bay island.

2. Whether the respondent properly classified Rum Point as a bay island pursuant to the criteria set forth at N.J.A.C. 7:7E–3.21.

INTRODUCTION

Over 22,000 years ago the coast line of New Jersey began 75 miles east of its current location. As the ice age concluded and the ice packs receded northward and began to melt, the ocean levels began to rise over 400 feet which had the effect of causing the coast line to recede westwardly.

As the ocean, over thousands of years, continued to consume the land mass, barrier islands and bay islands were formed and destroyed. These dramatic geologic changes have, within historic time, resulted in the formation of the barrier island and back bay system known as the New Jersey coast.

Located in Atlantic County, the barrier islands of Brigantine and Absecon Island were formed. Absecon Island contains New Jersey's gambling mecca, Atlantic City, while Brigantine Island lies to the north immediately across Absecon Channel. The dramatic geologic changes which formed Brigantine and Absecon Islands also resulted in the formation of Rum Point, which is located in the Absecon Channel between these two barrier islands.

Absecon Island and Brigantine Island are connected by State Route 87 which leaves Absecon Island in the vicinity of Harrah's Hotel Casino and Trump Castle Hotel Casino, proceeds northward across Absecon Channel, and first makes land fall on the island of Rum Point. Petitioner claims that, geologically, Rum Point is part of the Brigantine barrier island system while the respondent has classified Rum Point as a bay island pursuant to N.J.A.C. 7:7E–3.21.

Whether Rum Point is classified as a bay island or part of the barrier island of Brigantine significantly impacts on the type, location, and extent of development which would be permitted on Rum Point.

LOCATION OF RUM POINT

Rum Point is located at the confluence of St. George's Thorofare and Absecon Channel, one mile west of where Absecon Channel meets the Atlantic Ocean, between the barrier islands of Absecon and Brigantine. Rum Point is one of the stepping stone islands utilized by the causeway which links Absecon Island with Brigantine Island. State Route 87 leaves Absecon Island in the vicinity of Harrah's Hotel Casino and Trump Castle Hotel Casino, commonly known as the Marina District. The causeway proceeds over Absecon Channel making land fall on Rum Point.

Petitioner's property consists of less than eight acres of undeveloped uplands surrounded on three sides by a large expanse of coastal wetlands, a small beach, and tidal waterways. The fourth side abuts the causeway which acts as a boundary line between the portion of Rum Point owned by the petitioner and the remainder of the island owned by Resorts International Hotel Casino. The entire island is known as Rum Point; however, for the purposes of this opinion, Rum Point will refer to that tract of land owned by the petitioner.

Prior to 1926, access to Brigantine Island was gained through a ferry between Absecon Island and Brigantine Island in the same location as the current causeway or Brigantine Bridge, also named the Haneman Memorial Bridge. Between 1926 and 1927, when the bridge was built between the two barrier islands, a causeway was built on fill across Rum Point, Boot Island

and onto Brigantine Island. This causeway had the effect of linking the islands of Rum Point and Boot Island with the barrier island.

* * *

With the elimination of the ferry service in 1926–1927, and the completion of the causeway between the two barrier islands, Brigantine began to grow rapidly in both commercial and residential development. Within the last 25 years, the lack of available developable land on Brigantine Island subsequently increased the pressure to develop uplands on Boot Island and a narrow band of uplands adjacent to the causeway. This area quickly filled with single family residences and small condominium complexes.

The causeway transversing Rum Point is undeveloped due largely to the unavailability of uplands. Tidal waters and some coastal wetlands abut the filled area which serves as the basis for the causeway. Large pieces of concrete rubble have been placed adjacent to the causeway in several areas facing St. George's Thorofare in order to reinforce the causeway and prevent soil erosion caused by wave action. To this date, the Rum Point property has remained undeveloped.

FINDINGS OF FACT AND CONCLUSIONS OF LAW

Since purchasing the property prior to his death in 1986, Edgar Sims, Jr., and since that point, his estate, have had numerous meetings and discussions with the respondent concerning the classification of Rum Point as a bay island. Finally, by letter dated October 30, 1992 DEP confirmed its classification of Rum Point as a bay island which in turn triggered the within appeal.

The conclusion that Rum Point was a bay island was based upon materials submitted by the petitioner and the definition of bay island set forth at N.J.A.C. 7:7E–3.21 including the rationale statement accompanying the bay island rule, including the USGS map of the site. The October 30, 1992 letter stated specifically:

Rum Point remains today an undeveloped parcel of land surrounded by tidal waters with the exception of the right-of-way of Route 87 and its accompanying underpass. The property is adjacent to highly sensitive areas. Extensive wetlands surrounds the uplands portion of the site which serves as a year round haven for numerous animal species on a year round and seasonal basis. An exceptionally good shellfish habitat is found throughout the Absecon inlet, St. George's and Panama Thorofares and the Department has taken great care to protect and foster this valuable natural resource.

Rum Point is also subject to storm inundation and tidal flooding and classified by the Federal Emergency Management Act as a "V"-Zone. Route 87 is the only storm evacuation route for residents of the City of Brigantine. Any development which would contribute greater numbers of vehicles to the already strained evacuation capacity of the road system is unacceptable.

The definition of a Bay Island set forth in N.J.A.C. 7:7E–3.21 states:

> Bay Island are islands or filled areas surrounded by tidal waters, wetlands, beaches or dunes, lying between the mainland and barrier islands.

Petitioner contends that the DEP overstepped its bounds when it classified Rum Point as a bay island because it utilized factors and criteria not contained within the plain language of N.J.A.C. 7:7E–3.21. Respondent counters that in addition to the language set forth in N.J.A.C. 7:7E–3.21, its decision to classify Rum Point as a bay island was augmented by a consideration of the factors set forth in the rule's summary statement which accompanied the rule at 22 N.J. Register 1188(a).

This statement details certain factors concerning environmental impact when classifying a tract of land. This rule rationale statement states:

> New Jersey Bay Islands are for the most part inaccessible and undeveloped. Many of these islands are former wetlands where upland areas have been created by past filling, particularly with dredge spoils. Many are suitable for future dredge spoil disposal. They are adjacent to areas with high environmental sensitivity, particularly wetlands, intertidal flats, tidal waterways, shellfish beds, and endangered and threatened wildlife habitats. Development of the islands would pose a great threat to these natural resources and habitat. The majority of, if not all, bay islands are valuable wildlife habitats or have the potential to become habitat through the implementation of management techniques. Their value, in part, stems from their isolation from human activity as compared to the intense development and beach usage of oceanfront barrier islands. For example, sandy areas are used by beach nesting birds such as least tern, black skimmer, and piping plover, and vegetated areas are used by colonial nesting birds such as heron and non-colonial birds such as the marsh hawk. Bay islands are also subject to flooding and by virtue of their location function as bridges between the mainland and barrier islands. If developed, these islands would pose added storm evacuation problems. They are usually distant from public services, and therefore unsuitable for development.

In summary, bay islands have historically remained undeveloped and their importance for the maintenance of wildlife habitat requires that development be limited to only areas that are already developed.

* * *

[Finding that the island in question met the criteria prescribed in the regulation and its rationale statement, the Administrative Law Judge initially held for the Department, and on administrative appeal of that ruling the Department issued the following final decision]

FINAL AGENCY DECISION

■ SHINN, COMMISSIONER:

* * *

Rum Point meets the regulatory definition of a bay island. First, the ALJ found, "an examination of the most recent topographical maps demonstrates that Rum Point is located between the barrier island of Brigantine and the mainland." He further found that "the construction of the causeway linking Rum Point to the barrier island did not alter its essential characteristic and nature as that of a bay island which is part of a flood delta complex." Although petitioner argues in one of its exceptions that Rum Point is located between two barrier islands and not between a barrier island and the mainland, it does not follow that because Rum Point lies between two barrier islands (Absecon and Brigantine), it does not also lie between a barrier island and the mainland. This circumstance satisfies one condition of the definition set forth in N.J.A.C. 7:7E–3.21(a). Rum Point is also surrounded by tidal water, wetlands, beaches and dunes, except where the constructed causeway links it to Boot Island. Thus, the second condition of the definition is met since that definition expressly provided at the time of decision that the presence of a causeway is to be discounted.

Notes and Questions

1. New Jersey's approach seems to have been to let development dominate on barrier islands and conservation dominate on bay islands. As the case illustrates, this approach makes the status of a particular island all important and, apparently, worth fighting over. But why the sharp distinction? Is it because bay islands are inherently more ecologically valuable than barrier islands? Nothing in the opinion suggests that is the case. Is it because barrier islands are inherently more valuable for development than bay islands? Mr. Sims apparently found bay island development economically viable. So why establish an ecosystem management regime that depends so crucially on this distinction?

2. One of the classic experiments in ecology took place in the 1970s when two ecologists who helped spearhead the discipline of island biogeography, Daniel Simberloff and E.O. Wilson, fumigated several islands in the Florida Keys in an effort to relate species diversity to geographic area. Several ecologists had been exploring the concept that species diversity increases in a function relationship with increases in geographic area and independent of habitat diversity within the area. Thus, it was thought, islands may experience dramatic changes in species composition after a hurricane sweeps prior occupants off the surface, but the total number of species that reappear will be the same as before, and the equilibrium level of species will be a function of the size of the island, not the habitat composition on the island. If this "size matters" theory were true, it could help guide not just island ecosystem management, but the design of any sort of terrestrial or aquatic preserve. By wiping out all the species on an island, Simberloff and Wilson could follow the emergence of a new species assemblage and test the species-area hypothesis. Indeed, their experiment turned out just as the theory predicted, and many ecologists since then have worked to test and refine the theory. For an account of the experiment and its impact on island ecology, see CHARLES C. MANN AND MARK L. PLUMMER, NOAH'S CHOICE: THE FUTURE OF ENDANGERED SPECIES 53–81 (1996).

3. Mountains are the "sky islands" of terrestrial settings, presenting many of the same ecosystem dynamics as do their aquatic cousins—isolation, pronounced exposure to extreme natural events, and delicate balances of biodiversity. This feature is seldom as pronounced as it is in the "archipelago" of mountain ranges running from southern Arizona and New Mexico into the Mexican states of Sonora and Chihuahua. Research has found that these mountains, while containing discrete biodiversity complexes, also interact through the migration of species using riparian habitat travel corridors between the ridge tops. As these riparian corridors have become increasingly fragmented from urban, agricultural, and recreational development, however, the biota distribution in the various mountains has shown increasing levels of mountain-specific endemism. Like aquatic islands, in other words, the mountain sky islands are profoundly influenced by what happens between them, in the larger landscape context, and thus preserving just the mountains is not sufficient to conserve their ecosystem dynamics. *See* Peter Warshall, *Southwestern Sky Island Ecosystems*, in NATIONAL BIOLOGICAL SERVICE, OUR LIVING RESOURCES 318 (1995). Is island ecosystem management, whether of the aquatic or terrestrial variety, largely a question of managing what takes place elsewhere?

3.   CORAL ECOSYSTEMS

The world's coral reefs cover only about 113,720 square miles—less than 0.1 percent of the globe's ocean area—but are second only to tropical rain forests in plant and animal diversity. Coral reefs house over one-fourth of all marine life. Even more so than rain forests, however, coral reefs are extremely sensitive to environmental disturbance. Even slight variations in temperature, light, salinity, oxygen, and nutrient load can fundamentally alter coral reef dynamics. Thus, unlike rain forests, coral reefs are small islands of fragility sprinkled throughout the huge, dominant ocean ecosystem. Variations in conditions that the ocean as a whole may shrug off thus can have devastating consequences for ecosystems. *See* U.S. EPA, Office of Water and Watersheds, *Coral Reefs and Your Coastal Watershed*, http://www.epa.gov/owow/oceans/factshhets/fact4.html.

Indeed, it is estimated that over 25 percent of the world's coral reefs are severely damaged. About half of those losses are attributed to direct and indirect human pressures, such as coastal development, waste dumping, oil spills, nonpoint source pollution, overfishing, and coral mining. The remainder is attributable to temperature-induced "bleaching," which results when the algae that inhabit the coral polyps die or are expelled due to stress. *See* Craig Quirolo, *Coral Disease and Monitoring in the Florida Keys*, 19 ENDANGERED SPECIES UPDATE 15 (2002).

The Florida coral reef tract, which is the third largest in the world after the Great Barrier Reef in Australia and the Belize reef, is by no means immune to these problems. The same is true of the United States' other reef systems, off of the Virgin Islands and the Hawaiian Islands. But managing coral reefs independently of the surrounding ocean waters is, of course, impossible. Coral is stationary. Given this constraint, how should ecosystem management of our nation's coral reefs proceed? In addition to the Coral Reef Conservation and Restoration Partnership Act of 2000, discussed in Chapter 2, consider whether the following measures have any promise in that regard:

Coral Reef Protection

Executive Order 13089.
63 Fed. Reg. 32701 (June 16, 1998).

By the authority vested in me as President by the Constitution and the laws of the United States of America and in furtherance of the purposes of the Clean Water Act of 1977, as amended (33 U.S.C. 1251, et seq.), Coastal Zone Management Act (16 U.S.C. 1451, et seq.), Magnuson–Stevens Fishery Conservation and Management Act (16 U.S.C. 1801, et seq.), National Environmental Policy Act of 1969, as amended (42 U.S.C. 4321, et seq.), National Marine Sanctuaries Act, (16 U.S.C. 1431, et seq.), National Park Service Organic Act (16 U.S.C. 1, et seq.), National Wildlife Refuge System Administration Act (16 U.S.C. 668dd-ee), and other pertinent statutes, to preserve and protect the biodiversity, health, heritage, and social and economic value of U.S. coral reef ecosystems and the marine environment, it is hereby ordered as follows:

Section 1. Definitions.

(a) "U.S. coral reef ecosystems" means those species, habitats, and other natural resources associated with coral reefs in all maritime areas and zones subject to the jurisdiction or control of the United States (e.g., Federal, State, territorial, or commonwealth waters), including reef systems in the south Atlantic, Caribbean, Gulf of Mexico, and Pacific Ocean.

(b) "U.S. Coral Reef Initiative" is an existing partnership between Federal agencies and State, territorial, commonwealth, and local governments, nongovernmental organizations, and commercial interests to design and implement additional management, education, monitoring, research, and restoration efforts to conserve coral reef ecosystems for the use and enjoyment of future generations. The existing U.S. Islands Coral Reef Initiative strategy covers approximately 95 percent of U.S. coral reef ecosystems and is a key element of the overall U.S. Coral Reef Initiative.

* * *

Sec. 2. Policy.

(a) All Federal agencies whose actions may affect U.S. coral reef ecosystems shall: (a) identify their actions that may affect U.S. coral reef ecosystems; (b) utilize their programs and authorities to protect and enhance the conditions of such ecosystems; and (c) to the extent permitted by law, ensure that any actions they authorize, fund, or carry out will not degrade the conditions of such ecosystems.

* * *

Sec. 3. Federal Agency Responsibilities. In furtherance of section 2 of this order, Federal agencies whose actions affect U.S. coral reef ecosystems, shall, subject to the availability of appropriations, provide for implementation of measures needed to research, monitor, manage, and restore affected ecosystems, including, but not limited to, measures reducing impacts from pollution, sedimentation, and fishing. To the extent not inconsistent with statutory responsibilities and procedures, these measures shall be developed in cooperation with the U.S. Coral Reef Task Force and fishery management councils and in consultation with affected States, territorial, commonwealth, tribal, and local government agencies, nongovernmental organizations, the scientific community, and commercial interests.

Sec. 4. U.S. Coral Reef Task Force. The Secretary of the Interior and the Secretary of Commerce, through the Administrator of the National Oceanic and Atmospheric Administration, shall co-chair a U.S. Coral Reef Task Force ("Task Force"), whose members shall include, but not be limited to, the Administrator of the Environmental Protection Agency, the Attorney General, the Secretary of the Interior, the Secretary of Agriculture, the Secretary of Commerce, the Secretary of Defense, the Secretary of State, the Secretary of Transportation, the Director of the National Science Foundation, the Administrator of the Agency for International Development, and the Administrator of the National Aeronautics and Space Administration. The Task Force shall oversee implementation of the policy and Federal agency responsibilities set forth in this order, and shall guide and support activities under the U.S. Coral Reef Initiative ("CRI"). All Federal agencies whose actions may affect U.S. coral reef ecosystems shall review their participation in the CRI and the strategies developed under it, including strategies and plans of State, territorial, commonwealth, and local governments, and, to the extent feasible, shall enhance Federal participation and support of such strategies and plans. The Task Force shall work in cooperation with State, territorial, commonwealth, and local government agencies, nongovernmental organizations, the scientific community, and commercial interests.

Sec. 5. Duties of the U.S. Coral Reef Task Force.

(a) Coral Reef Mapping and Monitoring. The Task Force, in cooperation with State, territory, commonwealth, and local government partners, shall coordinate a comprehensive program to map and monitor U.S. coral reefs.* * *

(b) Research. The Task Force shall develop and implement, with the scientific community, research aimed at identifying the major causes and consequences of degradation of coral reef ecosystems. This research shall include fundamental scientific research to provide a sound framework for the restoration and conservation of coral reef ecosystems worldwide.* * *

(c) Conservation, Mitigation, and Restoration. The Task Force, in cooperation with State, territorial, commonwealth, and local government agencies, nongovernmental organizations, the scientific community and commercial interests, shall develop, recommend, and seek or secure implementation of measures necessary to reduce and mitigate coral reef ecosystem degradation and to restore damaged coral reefs. These measures shall include solutions to problems such as land-based sources of water pollution, sedimentation, detrimental alteration of salinity or temperature, overfishing, over-use, collection of coral reef species, and direct destruction caused by activities such as recreational and commercial vessel traffic and treasure salvage. In developing these measures, the Task Force shall review existing legislation to determine whether additional legislation is necessary to complement the policy objectives of this order and shall recommend such legislation if appropriate.* * *

Sec. 6. This order does not create any right or benefit, substantive or procedural, enforceable in law or equity by a party against the United States, its agencies, its officers, or any person.

WILLIAM J. CLINTON

President Clinton later took a more specific and direct approach to protect a particular reef system under the Antiquities Act of 1906, which states in virtually its entirety as follows:

> The President of the United States is authorized, in his discretion, to declare by public proclamation historic landmarks, historic and prehistoric structures, and other objects of historic or scientific interest that are situated upon the lands owned or controlled by the Government of the United States to be national monuments, and may reserve as a part thereof parcels of land, the limits of which in all cases shall be confined to the smallest area compatible with the proper care and management of the objects to be protected. When such objects are situated upon a tract covered by a bona fide unperfected claim or held in private ownership, the tract, or so much thereof as may be necessary for the proper care and management of the object, may be relinquished to the Government, and the Secretary of the Interior is authorized to accept the relinquishment of such tracts in behalf of the Government of the United States.

16 U.S.C. 431. By characterizing coral reefs in the U.S. Virgin Islands as being of "scientific interest," President Clinton issued the following justification for designating the reefs as a national monument:

Establishment of the Virgin Islands Coral Reef National Monument

Presidential Proclamation 7399.
66 Fed. Reg. 7364 (January 22, 2001).

By the President of the United States of America

A Proclamation

The Virgin Islands Coral Reef National Monument, in the submerged lands off the island of St. John in the U.S. Virgin Islands, contains all the elements of a Caribbean tropical marine ecosystem. This designation furthers the protection of the scientific objects included in the Virgin Islands National Park, created in 1956 and expanded in 1962. The biological communities of the monument live in a fragile, interdependent relationship and include habitats essential for sustaining and enhancing the tropical marine ecosystem: mangroves, sea grass beds, coral reefs, octocoral hardbottom, sand communities, shallow mud and fine sediment habitat, and algal plains. The fishery habitats, deeper coral reefs, octocoral hardbottom, and algal plains of the monument are all objects of scientific interest and essential to the long-term sustenance of the tropical marine ecosystem.

* * *

As part of this important ecosystem, the monument contains biological objects including several threatened and endangered species, which forage, breed, nest, rest, or calve in the waters. Humpback whales, pilot whales, four species of dolphins, brown pelicans, roseate terns, least terns, and the hawksbill, leatherback, and green sea turtles all use portions of the monument. Countless species of reef fish, invertebrates, and plants utilize these submerged lands during their lives, and over 25 species of sea birds feed in the waters. Between the nearshore nursery habitats and the shelf edge spawning sites in the monument are habitats that play essential roles during specific developmental stages of reef-associated species, including spawning migrations of many reef fish species and crustaceans.

The submerged monument lands within Hurricane Hole include the most extensive and well-developed mangrove habitat on St. John. The Hurricane Hole area is an important nursery area for reef associated fish and invertebrates, instrumental in maintaining water quality by filtering and trapping sediment and debris in fresh water runoff from the fast land, and essential to the overall functioning and productivity of regional fisheries. Numerous coral reef-associated species, including the spiny lobster, queen conch, and Nassau grouper, transform from planktonic larvae to bottom-dwelling juveniles in the shallow nearshore habitats of Hurricane Hole. As they mature, they move offshore and take up residence in the deeper coral patch reefs, octocoral hardbottom, and algal plains of the submerged monument lands to the south and north of St. John. The monument lands south of St. John are predominantly deep algal plains with scattered areas of raised hard bottom. The algal plains include communities of mostly red and calcareous algae with canopies as much as half a meter high. The raised hard bottom is sparsely colonized with corals, sponges, gorgonians, and other invertebrates, thus providing shelter for lobster, groupers, and snappers as well as spawning sites for some reef fish species. These algal plains and raised hard bottom areas link the shallow water reef, sea grass, and mangrove communities with the deep water shelf and shelf edge communities of fish and invertebrates.

Section 2 of the Act of June 8, 1906 (34 Stat. 225, 16 U.S.C. 431), authorizes the President, in his discretion, to declare by public proclamation historic landmarks, historic and prehistoric structures, and other objects of historic or scientific interest that are situated upon the lands owned or controlled by the Government of the United States to be national monuments, and to reserve as a part thereof parcels of land, the limits of which in all cases shall be confined to the smallest area compatible with the proper care and management of the objects to be protected.

WHEREAS it appears that it would be in the public interest to reserve such lands as a national monument to be known as the Virgin Islands Coral Reef National Monument:

NOW, THEREFORE, I, WILLIAM J. CLINTON, President of the United States of America, by the authority vested in me by section 2 of the Act of June 8, 1906 (34 Stat. 225, 16 U.S.C. 431), do proclaim that there are hereby set apart and reserved as the Virgin Islands Coral Reef National Monument, for the purpose of protecting the objects identified above, all lands and interests in lands owned or controlled by the United States within the boundaries of the area described on the map entitled "Virgin Islands Coral Reef National Monument" attached to and forming a part of this proclamation. The Federal land and interests in land reserved consist of approximately 12,708 marine acres, which is the smallest area compatible with the proper care and management of the objects to be protected.

All Federal lands and interests in lands within the boundaries of this monument are hereby appropriated and withdrawn from all forms of entry, location, selection, sale, or leasing or other disposition under the public land laws, including but not limited to withdrawal from location, entry, and patent under the mining laws, and from disposition under all laws relating to mineral and geothermal leasing, other than by exchange that furthers the protective purposes of the monument. For the purpose of protecting the objects identified above, the Secretary shall prohibit all boat anchoring, except for emergency or authorized administrative purposes.

For the purposes of protecting the objects identified above, the Secretary shall prohibit all extractive uses, except that the Secretary may issue permits for bait fishing at Hurricane Hole and for blue runner (hard nose) line fishing in the area south of St. John, to the extent that such fishing is consistent with the protection of the objects identified in this proclamation.

Lands and interests in lands within the monument not owned or controlled by the United States shall be reserved as a part of the monument upon acquisition of title or control thereto by the United States.

The Secretary of the Interior shall manage the monument through the National Park Service, pursuant to applicable legal authorities, to implement the purposes of this proclamation. The National Park Service will manage the monument in a manner consistent with international law.

The Secretary of the Interior shall prepare a management plan, including the management of vessels in the monument, within 3 years, which addresses any further specific actions necessary to protect the objects identified in this proclamation.

The establishment of this monument is subject to valid existing rights. Nothing in this proclamation shall be deemed to revoke any existing withdrawal, reservation, or appropriation; however, the national monument shall be the dominant reservation.

Warning is hereby given to all unauthorized persons not to appropriate, injure, destroy, or remove any feature of this monument and not to locate or settle upon any of the lands thereof.

IN WITNESS WHEREOF, I have hereunto set my hand this seventeenth day of January, in the year of our Lord two thousand one, and of the Independence of the United States of America the two hundred and twenty-fifth.

WILLIAM J. CLINTON

Notes and Questions

1. The Coral Reef Task Force approach represents an effort to identify a fragile ecosystem type and apply concerted conservation measures to it wherever it occurs. A similar effort has been initiated at the international level through the International Coral Reef Initiative, which now has over 90 member countries. *See* http://www.environnement.gouv.fr/icri/index.html. The designation of the Virgin Islands Coral Reef National Monument represents an effort to identify a discrete example of the fragile ecosystem type and direct focused conservation measures toward it. President Clinton established a similar coral preserve when he designated a 200–kilometer-long necklace of islands northwest of Hawaii as the Northwestern Hawaiian Islands Coral Reef Ecosystem Reserve. *See* Exec. Order 13178, 65 Fed. Reg. 76903 (Dec. 7, 2000). Through both approaches direct injury internal to the coral reefs from anchoring, fishing, and mineral extraction can be addressed, but in neither case can the conservation regime insulate the coral reef from degradation of surrounding marine resources from pollution, temperature warming, fishing, and other external forces of injury. How much promise do these approaches, separately or combined, thus hold for coral reef conservation?

2. To drive the previous point home, the greatest single threat to coral reefs likely is global warming, something neither the Coral Reef Task Force nor special protection status can competently address. Coral is white. The color in corals is supplied by microscopic algae that inhabit the coral in a symbiotic relationship. Warm water temperatures can stress the corals, causing them to expel the algae and thus lose their color and in many cases die out. This phenomenon, known as "bleaching," is being recorded as unprecedented levels worldwide, and is considered a more prominent threat to coral than is pollution or any other source of degradation. *See Warmer Waters More Deadly to Coral Reefs than Pollution*, 290 SCIENCE 682 (2000). Then there is the problem of dust, from Africa. That's right, researchers believe that great dust storms in Africa spread dust westward across the Atlantic Ocean to be deposited over coral rich waters in the Caribbean Sea. The dusts introduce pathogens and nutrients into surrounding waters, apparently damaging corals even in remote, otherwise undisturbed areas. See John C. Ryan, *Dust in the Wind: Fallout from Africa May Be Killing Coral Reefs an Ocean Away*, WORLD-WATCH, Jan.-Feb. 2002, at 32. Obviously, these are not the kind of injuries that can be controlled by designating a coral reef for special protection.

3.   The work of the Coral Reef Task Force is documented at http://coralreef.gov. For an excellent background and assessment of the Coral Reef Task Force, see Robin Kundis Craig, *The Coral Reef Task Force: Protecting the Environment Through Executive Order*, 30 Envt'l L. Rep. (ELI) 10343 (2000). The United States is represented in the International Coral Reef Initiative through the Department of Commerce's National Oceanic and Atmospheric Administration. *See NOAA's Coral Health and Monitoring Program*, http://coral.aoml.noaa.gov/icri.

4.   The Antiquities Act has more frequently been used to designate special protection sites in terrestrial ecosystems. Indeed, presidential decrees under the Antiquities Act were responsible for designation of significant protected areas within the Grand Canyon, Glacier Bay, Death Valley, and Carlsbad Caverns, all of which enjoyed bipartisan support. But President Clinton caused considerable controversy when, acting on a strategy then Secretary of Interior Bruce Babbitt devised, late in his tenure he designated vast areas of federal public lands in western states as national monuments. (He designated the Virgin Islands reef monument on his last day in office.) Because the Antiquities Act allows the Executive to do so without Congress's involvement, and then to restrict land uses within the designated area, this was seen by many in Congress, the state and local governments, and the private landowner community as an end run way of injecting enhanced conservation measures into public lands otherwise subject to the "multiple use" management mandate, but by many others as the only viable way of establishing ecosystem management regimes in those lands given Republican control of Congress. We never said ecosystem management isn't political! For background on the Clinton Administration's Antiquities Act initiative, see Christine A. Klein, *Preserving Monumental Landscapes Under the Antiquities Act*, 87 CORNELL L. REV. __ (forthcoming 2002); Justin James Quigley, *Grand-Staircase Escalante National Monument: Preservation or Politics*, 19 J. LAND, RESOURCES, AND ENVT'L L. 55 (1999); Sanjay Ranchod, *The Clinton National Monuments, Protecting Ecosystems With the Antiquities Act*, 25 HARV. ENVT'L L. REV. 535 (2001).

B.   HUMAN-DOMINATED ECOSYSTEMS

The biologist Stephen Palumbi recently summed up what is increasingly the reality for ecosystem management policy development: "Human impact on the global biosphere now controls many major facets of ecosystem function." Stephen R. Palumbi, *Humans as the World's Greatest*

Evolutionary Force, 293 Science 1786 (2001). This naked truth is hard for many environmental advocates to accept. For them, it suggests that we ought to throw in the towel. For many others, however, it is nothing less than a call to arms for ecosystem management. In this section we consider the consequences of human domination in several ecosystem settings, and explore some ways of addressing their future.

1. Urban America

Urban America hardly needs an introduction. It seems to be everywhere, and growing incessantly in what is known derisively as "sprawl." But in fact the urban, built-up environment accounts for a small fraction—well under 10 percent—of our nation's land mass; it's just that it is such a visible and fast-changing domain that makes it so front and center in the environmental policy debate. Also, urbanization has tended to be selective in terms of type of setting, favoring coasts, lakes, mountains, and other areas where, it so happens, so much of ecosystem management policy seems to be focused. Urbanization thus has become a central theme in ecosystem management circles.

Far from being ecological wastelands, however, there is life in the city. A new breed of urban ecologists are finding exceedingly diverse and complex web of life in distinctly urbanized locations. Metropolitan Phoenix, for example, has over 75 species of bees, 200 species of birds, and hundreds of insect species documented within its area housing 2.8 million people. As one astute ecologist concluded, Phoenix is not moving or shrinking, so we ought to consider ways of designing urban development there and elsewhere to foster urban ecosystem richness. *See* Keith Kloor, *A Surprising Tale of Life in the City*, 286 Science 663 (1999). On the other hand, ought not "sprawl" be prevented from further intruding into what is now relatively undisturbed habitat? Is there a way of knowing where urban "ecosystems" should flourish and where they should be contained? In the following materials, we explore these issues by examining frameworks for assessing and managing the impact of urban development on ecosystems.

a. IMPACT ASSESSMENT METHODS

Mark Lorenzo, Sizing Up Sprawl
9 Wild Earth 72 (Fall 1999).

The threats to ecological integrity and biodiversity resulting from the production of goods, ranging from resource extraction through factory processing and packaging are well recognized. Less well recognized is the destructive potential of consumption in the form of expansive lifestyles. Land conversion and development for new housing, recreation and retailing demanded by growing populations with rising aspirations, increasingly takes place in farm fields, forests and wetlands at the interface of suburbia and wildlands. This form of sprawling, auto-dependent development, or "sprawl" is subjecting ecosystems and species to death by a thousand cuts.

The destructive environmental impacts of sprawl-type development are many and varied. There are biological impacts, such as the fragmentation of habitat and associated biodiversity losses, environmental impacts such as degraded air and water, and spiritual impacts such as loss of a unique sense of place, diminished cultural connections to the land, and reduced opportunities for wild experiences. We must recognize the immense, and in many ways irreplaceable, value of our natural environment to better comprehend what is lost from extensive development in relatively natural areas.

Advocates for wildlands protection and conservation have new allies in an unlikely profession—economics. Practitioners of ecological economics do what some say is impossible or even immoral, yet others regard as increasingly necessary. Ecological economists try to estimate the dollar value of ecological processes and entities. Quantifying the economic value of wildlands, for example, is certainly a methodological challenge which requires an interdisciplinary understanding and synthesis of both ecology and economics.

However, many objections to this approach seem based either on ignorance of common statistical methods of valuation, or on a moral distaste for assigning a dollar value to something as "priceless" as an old-growth forest. Yet, "priceless" resources are routinely destroyed in favor of resource extraction or development for which substantial economic value is claimed. Ecological economists believe it may often prove more useful to attempt to fully "price" the ecological goods and services at risk, than to be surprised and outraged when "priceless" translates effectively to "$0.00" in the political calculus of land-use policy.

The values and benefits of the ecological systems that are the necessary foundation for all life on this planet can be roughly categorized and partly quantified in biological and economic terms. We must be careful, though, to remember that the whole is far greater than the sum of all the parts. The art of assessing particular damages to well-defined parts of ecosystems has made substantial progress, especially where impacts on human health and well-being are concerned. However, the state of the art for biological assessment and economic valuation at landscape scales and for complex interactive systems is in a much more formative stage.

In particular, assessing and valuing the categories of greatest public importance, such as ecosystem services and biological diversity overall, remain the premier challenges for the emerging science of ecological economics. There are synergistic, threshold and future effects that are especially difficult to assess rigorously, but which are increasingly apparent across North America's landscape, and prominent in public discussions of the sprawl phenomenon. Simply because it is a formidable task to assess and/or value some of the necessary factors does not in any way diminish their importance.

Indeed, early work has found that when the total public value of ecosystems and biological diversity are estimated using the best available methods, their value in situ far exceeds the more typically measured values of consumption or "development" of the resource(s) in question. This finding justifies much more caution about further incursions into remaining natural areas or wildlands.

* * *

Assessing the Environmental Effects of Sprawl

The tools and methods of ecological economics and conservation biology together can provide important basis for comprehensive assessments of the environmental effects of sprawl. Ecological economics assesses the human value of ecosystems and changes in ecosystems. This often includes previously well-defined economic categories such as consumptive or use values, and can also include non-consumptive values including existence, legacy, and option values. Ecological economics today especially focuses on the value to humans of natural "ecosystem services," across many categories, such as the water collection and purification services performed by forested watersheds, and the potential costs to replace such.

Conservation biology assesses the biological value of ecosystems and changes in ecosystems. This discipline is distinguished from ecology per se by urgent concern with the rates of loss of biological diversity—a factor of both wildlife species diversity and abundance. Practitioners develop solutions to reduce and reverse the unnaturally high current rates of species extinctions, which can take the form of biological reserve strategies. Conservation biology developed from studies of island biogeography that found biodiversity to be highly associated with the absolute size of "undisturbed" ecosystems, and their relative levels of fragmentation, congruity and connectivity. For example, the "area effect" relating habitat size to biodiversity shows that a 90% reduction in habitat area may result in a 50% loss of all species formerly present.

Sprawl-type development especially destroys quality wildlife habitat and fragments natural ecosystems, due to extensive construction activity, water diversion, paving and creation of road networks in relatively natural areas. Thus, many ecosystem services become less productive, ecological integrity is diminished, and biodiversity is lost from those areas previously having the greatest relative ecological value. Conservation biology can offer causal mechanisms linking development, disturbances and biological impacts, and ecological economics can provide methods for quantifying the actual or probable losses in economic terms.

What then are all the relevant factors in a comprehensive framework to assess the environmental effects of sprawl? Three broad categories of potential impacts must be included: 1) human health impacts; 2) ecosystem and biodiversity impacts; and, 3) recreational, aesthetic, spiritual and future impacts (see below). Each of these broad areas involves multiple factors, including typical environmental impact categories as well as the many dynamic processes of ecological systems. While these lists of factors are extensive, the categorical distinctions are somewhat arbitrary and important factors are no doubt still absent. Eventually the scientific tools will evolve to describe, assess and quantify more of these at various scales, but can we afford to wait?

* * *

A Holistic Framework for Assessing the Environmental Impacts of Sprawl

[A] holistic framework for assessing the environmental impacts of sprawl at various ecological scales can be devised. Both the quality of the development and the quality of the land to be developed must be involved in a comprehensive assessment of sprawl's environmental impacts. All these factors must be considered to fully address the question of sprawl's environmental impacts, i.e. relative to our environment, what is threatened or lost when we build into and upon relatively natural areas? And more broadly; how are we disorganizing natural systems in the process of organizing ourselves?

These questions can be framed more rigorously as a formula for analysis, involving all the potential assessment factors listed previously by categories, as follows:

A Formula to Evaluate the Environmental Impacts of Sprawl

"SPRAWL" = A Function of (I–M) x E

(Impacts–Mitigation)(Ecological Value)

"IMPACTS" = A Function of (C1) + (C2) + (C3)

(Category 1: Human Health Impacts of Sprawl) + (Category 2: Impacts on Ecosystem Goods, Services and Biodiversity from Sprawl) + (Category 3: Aesthetic, Recreational, Spiritual and Future Impacts from Sprawl)

"MITIGATION" = A Function of (M1) + (M2) + (M3)

(Mitigation of Category 1 impacts, for example through pollution prevention and control, vehicle miles traveled reduced by fees or access to alternative transport, etc ...) + (Mitigation of Category 2 impacts, i.e. through habitat protection measures, runoff management, etc ...) + (Mitigation of Category 3 impacts, i.e through greenways, buried power lines, etc ...)

"ECOLOGICAL–VALUE" = A Function of (B) + (EGS) + (NA)

(Biodiversity)(Ecosystem Goods and Services)(Natural Aesthetics)

As application of this formula reveals, the term sprawl is generally used where both the environmental impacts and ecological values are relatively high, for example large developments in rural areas. In a continuum of land development that ranges from relatively low environmental impacts on less ecologically valuable sites, to relatively higher impacts on more ecologically valuable sites, "sprawl" refers much more often to the latter. Due to the nature of the land involved in sprawl development, and even with mitigation efforts, ecological integrity has become increasingly damaged across landscapes, resulting in the diminished biodiversity.

Of course, no technical analysis can put a definitive dollar value upon the environmental impacts of sprawl for a landscape or a single site. This is because considerable interpersonal differences in values must be taken into account for numerous factors. One person's invaluable forested landscape, appears to another as a timber supply, or another as a prime second home site.

However, for purposes of decisions about land use, it is essential to understand that public, non-consumptive value is an additive function, whereas private consumptive value is subtractive relative to the fundamental value of natural systems, including biodiversity. Development of sites in agricultural and forested landscapes may provide profits to few but unquestionably diminishes great value for many. Increased application of ecological economics can help show the full costs of sprawl, and reinforce the case for wildlands protection and restoration.

Notes and Questions

1. What is "sprawl" and why is it occurring? Assume that we could accurately apply Lorenzo's formula to weigh and compare the ecological cost of different patterns of urban land development. What pattern would we choose? Would that pattern materialize as a function of unrestricted market dynamics, or would it require regulation?

2. Is sprawl bad? Who says? If it is so bad, why is it happening? Are the people who are moving out into the previously rural areas being deceived by developers eager to sell homes, or duped by local politicians greedy for tax dollars, or guided to ruin by some invisible force against their will? Indeed, the public distaste for sprawl seems matched only by its appetite for it. And many commentators come to sprawl's defense, questioning the calls for regulation Lorenzo and others have made. For example, Gregg Easterbrook reveals the irony of the anti-sprawl movement with his observation that "the ideal restaurant would have terrific food, moderate prices, and would be unpopular, so lines would never inconvenience diners. Legislatures could make restaurants less crowded by, say, mandating that some tables be kept vacant even when customers are queued. Those already seated would surely benefit. But others would stew over being denied service, while business and jobs would be lost." Gregg Easterbrook, *The Case for Sprawl*, NEW REPUBLIC, Mar. 15, 1999. And Nicole Garnett has noted that voters frequently approve local sprawl initiatives but defeat statewide ones, suggesting that the reason is that voters realize that they cannot escape the restrictions of state growth controls nearly as easily as they can local rules. People want just the right amount of sprawl, in other words, and want to retain local control over exactly how much that is. *See* Nicole Stelle Garnett, *Trouble Preserving Paradise?*, 87 CORNELL L. REV. 158 (2001). At bottom, then, sprawl occurs because people value the many personal benefits of living in previously rural areas, even as they recognize the adverse environmental consequences of doing so. Clearly, that is a tradeoff the anti-sprawl movement deems unwise, but who is to decide?

3. Critics of sprawl suggest that the alternative, referred to hopefully as "smart growth," should be a tighter urban envelope facilitated through polices favoring urban redevelopment, infill (using unused urban pockets), and denser development patterns. *See* Patrick Gallagher, *The Environmental, Social, and Cultural Impacts of Sprawl*, 15 NATURAL RESOURCES & ENVIRONMENT 219 (2001); John W. Frece, *Smart Growth: Prioritizing State Investments*, 15 NATURAL RESOURCES & ENV'T 236 (2001). Does Lorenzo's formula favor that pattern? In other words, is it necessarily the case that a denser, more contained pattern of urban development is a desirable ecosystem management policy? Conservation biologists appear to assume that is the case, but as one researcher has pointed out, we actually know very little about urban ecosystems. Most of the foundations of ecosystem management are built on conservation biologists' studies of landscapes that are relatively undisturbed by human presence. Do we know enough about urban ecosystems to choose the optimum urban development pattern? *See* James R. Miller and Richard J. Hobbs, *Conservation Where People Live and Work*, 16 CONSERVATION BIOLOGY 330 (2002).

4. Obviously, the ecological values to which Lorenzo refers would be estimated at a social scale. Like any piece of land, however, different individuals will assign different values to undisturbed habitat. If the ecological value of areas under development pressure is as high as Lorenzo suggests, why don't those who value it so much in its undeveloped state purchase it? Well, they do. Private, non-profit land trusts directed at securing ecosystem resources in and around urban land uses have exploded in number, finances, and acquisitions in recent years. Over 1,200 such entities exist, boasting a total of over 1 million members. As described in Chapter 3, large land conservation organizations such as the Nature Conservancy, Ducks Unlimited, and the Trust for Public Land are national in scope and have large paid staffs. By contrast, most land trusts are staffed by volunteers and local in focus. Operating primarily by acquisition of fee simple title and conservation easements, these smaller land trusts have conserved over 5 million acres of habitat in the United States. *See* Jean Hocker, *Land Trusts: Key Elements in the Struggle Against Sprawl*, 15 NATURAL RESOURCES & ENV'T 244 (2001). Land trusts benefit from the generous tax consequences land donors receive for land contribution, suggesting that tax policy can play an important role in ecosystem management. *See* Nancy A. McLaughlin, *The Role of Land Trusts in Biodiversity Conservation on Private Lands*, 38 IDAHO L. REV. 453 (2002). For more information on land trusts, see the Land Trust Alliance's home page at http://www.lta.org.

b. MANAGEMENT FRAMEWORKS

While not explicitly applying Lorenzo's formula, many urban settings provide case studies for the question of where and how urbanization should be allowed to dominate. These settings fall into two familiar categories—those in which the Endangered Species Act is the dominant legal force and those in which it is not. In the following materials we provide an example from each group.

The first case study involves a collaborative planning process underway in Southeast Orlando, which is situated between two regionally significant systems: the Econlockhatchee River (The Econ) and Boggy Creek. The area includes portions of two major drainage basins (Boggy Creek and Lake Hart), a connected system of lakes and small water bodies, high concentrations of wetlands, and a great diversity of plants and wildlife, many of which are protected by the City ordinance as well as Federal and State regulations. Though much of this habitat forms contiguous corridors, some areas have been altered by roadways, agricultural conversion, ditching, and cattle grazing. The area is also in the path of central Florida's voracious appetite for sprawling growth fueled by the presence of Disneyworld, other recreational attractions, and an emerging technology economy.

A diverse group of citizens believed that an opportunity exists in the Southeast Orlando area to create a permanently protected ecological system that is both regionally significant and maintains the integrity of on-site drainage and wildlife corridors. Envisioned as a Primary Conservation Network, or PCN, this area can also become a "mitigation bank" that allows smaller wetland areas outside the network to be transferred to areas of increased importance and viability. As the Primary Conservation Network becomes the mitigation receiving zone, gaps in the system could be recaptured, helping to reinforce the overall integrity of this ecology. Careful siting of trails, parks, and ponds would also allow the PCN to serve as an integrated community amenity. Details of the vision and its implementation follow:

City Of Orlando, Planning And Development Department, Southeast Orlando Sector Plan Vision Statement

http://www.cityoforlando.net/planning/deptpage/sesp/sesp.htm (2000).

Introduction

The Southeast Orlando Sector Plan is one of the largest urban planning and development projects ever undertaken by the City of Orlando. The area covered by the Plan consists of more than 19,300 acres and is within a 10 to 20 minute driving distance of Downtown Orlando, many of the region's entertainment attractions, as well as other regional job and education centers. The Plan area is located directly adjacent to the ever-expanding Orlando International Airport, which is already the 16th busiest airport in the United States and the 25th busiest airport in the world. In addition, the State of Florida has authorized the construction of a high speed rail system linking Tampa, Orlando, and Miami, with the hub of the system being the Orlando International Airport. With the Southeast Orlando Sector Plan, the City is proactively formulating a sustainable development strategy that builds and strengthens the livability of the entire community.

The City of Orlando has identified the Southeast Orlando Sector Plan area as a Future Growth Center with the Orlando International Airport as the primary economic and employment generator. In the near future, the Greater Orlando Aviation Authority plans to construct a fourth runway, expand terminal facilities, build new on-site roadways, pursue regional rail transit linkages, and actively market airport-related office and industrial development on the airport property. In addition, this growth will generate a significant need for convenient housing to serve airport employees. The City's projections indicate a potential for over 13,300 residential units, 2.1 million square feet of retail space, 3.3 million square feet of office space, 1,950 hotel rooms, 4.7 million square feet of industrial space, and 600,000 square feet of civic/government space within the Southeast Plan area by the year 2020. Again, by 2020, the Plan area could house more than 28,000 residents.

In order to build and sustain a viable community, development must feature a mixture of land uses which allow for increased accessibility, diversity, and opportunities for social interaction within the context of an integrated amenity framework. Utilizing the neighborhood as the basic community building unit, the center of residential neighborhoods will be defined by public space and activated by locally-oriented civic and commercial facilities. Employment, shopping and services will be concentrated in town, village, and neighborhood centers that are compact and walkable. The City will also be utilizing design concepts that provide a strong connection between nature and the built environment.

In order to accomplish this vision, the City of Orlando has entered into a unique partnership with the southeast area property owners, the Greater Orlando Aviation Authority, Orlando Utilities Commission, representatives of other local, regional, and state agencies affected by or having permitting jurisdiction of aspects of the project, and representatives of interest groups concerned with building successful communities. The partnership is responsible for master planning and the technical design of infrastructure, developing and implementing a financing strategy for the early provision of public facilities such as schools, the preparation of the urban design/land use plan, along with development standards and an administrative process to implement the Plan.

What Orlando Wishes to Accomplish in the Southeast Orlando Sector Plan Area—Building Community

Identity and community are often lost in the faceless growth of many suburban areas. Each developer works on a separate time-line to build and market their property. Public facilities and civic spaces are often placed on the left-over land, if they're included at all. The result is a series of sterile and unconnected subdivisions, rather than a cohesive community of people and places. The large size of the Southeast Orlando Sector Plan area, as well as the coordinated effort underway to plan for the area, provide an opportunity to create a unique image and character that is immediately identifiable to visitors and residents. Like some older communities (Winter Park or Downtown Orlando), Southeast Orlando will be immediately identifiable through its urban design, particularly in its streetscapes, the trees planted along major boulevards, and the scale of local streets.

A sense of community will be nurtured through the new town's physical structure. The area will be identified by the pattern of residential neighborhoods that focus on town, village and neighborhood centers; by the design of homes and commercial buildings; and by the proximity to nature. Each residential neighborhood will be scaled to the pedestrian, making casual interactions possible. Schools and parks will be focal points for neighborhood activities, rather than anonymous institutions within large no man's lands. And commercial districts will integrate public facilities and spaces, creating a civic atmosphere typical of more traditional downtowns. The design and landscape improvements to Narcoossee Road will be an extremely important first step in developing an identity for the community.

Balanced Land Uses

At buildout, the Southeast Plan area could very well be a mid-size town of 50,000 to 60,000 people. A full array of land uses, services, amenities and activities are being planned to fill the needs of the ultimate population. The Southeast Plan provides for a coordinated approach to transportation, employment, shopping and services, by concentrating such uses in places that are compact and walkable. The concept of balanced land uses will create a community which is more diverse and accessible, leading to greater opportunities for social interaction and growth. The new neighborhoods will include a richer mix of building types than can be found in conventional suburban neighborhoods—from sideyard houses, semi-detached houses, cottages, secondary units, courtyard apartments, and mid-rise apartments to shopfronts and offices with apartments above. Think of Park Avenue in Winter Park or Edgewater Drive in College Park.

* * *

Preservation of Natural Systems

The Southeast Orlando Sector Plan area presents the City and the development community with significant environmental opportunities. The Southeast Plan area is situated between two regionally significant systems; the Econlockhatchee River (the Econ) and Boggy Creek. The site itself includes portions of two major drainage basins (Boggy Creek and Lake Hart), a connected system of lakes and small water bodies, high concentrations of wetlands, and a great diversity of plants and wildlife, many of which are protected by the City's Growth Management Plan and Federal and State regulations. Though much of this habitat forms contiguous corridors, some areas have been altered by agricultural conversion, ditching, and cattle grazing.

An opportunity exists in the Plan area to create a permanently protected ecological system that is both regionally significant and maintains the integrity of on-site drainage and wildlife corridors. Sensitive site planning will ensure that natural habitats are protected and natural features become an integral part of the community through a designated Primary Conservation Network. These resources will be treated as key amenities, rather than as edges to developments. Public access will be permitted while important natural features and sensitive habitats are preserved. Pedestrian and bicycle paths and trails will be constructed along creek, canal or wetland edge systems, thus serving a dual function of allowing public access to open space and providing paths to destinations along the edges of linear parks. Major public facilities, such as schools, parks, and recreation centers will be linked by these open space/bicycle and pedestrian trail systems. In the Southeast Plan area, there is an opportunity for open spaces to shape and enhance neighborhoods, to provide a scenic resource from roads, and to serve as permanent wildlife corridors.

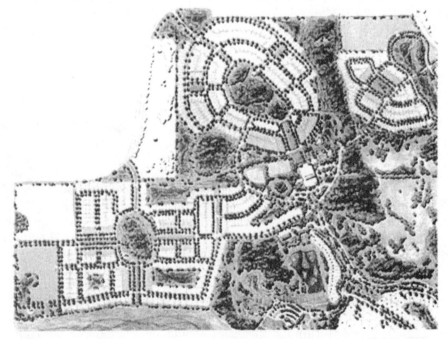

Depiction of what future land use in the Southeast Orlando area will look like under the Southeast Orlando Sector Plan. How would this pattern of land use fare under Lorenzo's sprawl impact formula (Sprawl = (Impacts - Mitigation) x Ecological Value)?

The Southeast Orlando Sector Plan adopted the following provisions to implement the Primary Conservation Network concept:

City of Orlando, Planning and Development Department, Southeast Orlando Sector Plan
(May 10, 1999).

Primary Conservation Network and Ecological Systems

Design standards and criteria have been established for the development proposed to occur within Southeast Orlando in order to comply with existing regulatory guidelines and restrictions while protecting the ecological integrity of the natural resources within the study area. These standards are not meant to restrict landowners developmental potential, but to guide and direct development in an ecological sensitive direction. The Primary Conservation Network (PCN), as envisioned, would protect wetland communities and habitat for numerous common and protected wildlife species while allowing passive recreation uses such as pedestrian and bike trails. The following standards are broken down by those required under normal regulatory review processes (primarily applicable to wetland/habitat areas outside the defined PCN); and those applicable to the entire planning area. These standards should provide for a more functional natural environment within the proposed development as well as provide opportunities to enhance and preserve natural communities and existing wildlife corridors.

* * *

Storm Drainage

a. Storm Drainage Requirements. All future development in the Southeast Orlando planning area shall be required to discharge stormwater at rates not to exceed historic runoff rates and volumes. Stormwater detention and water quality facilities will be required for all development within the Southeast Orlando planning area, as determined during the review process for each development. The specifications and standards of the Southeast Orlando Stormwater Master Plan, when developed, shall be followed to the greatest extent feasible.* * *

b. Integration with Existing Storm Drainage Systems. Existing drainageways and wetlands should be maintained or enhanced in a natural state to the greatest extent feasible. In lower-density areas, drainage systems should recharge on-site groundwater by using swales and surface systems, rather than concrete-lined or underground storm drains. All urban runoff should be treated on-site with biological retention and filtration areas.

c. Joint Use Stormwater/Open Space Opportunities. the location, function, and design of all stormwater facilities should be coordinated with open space and park areas, in order to provide for joint use opportunities, wherever possible.

d. Interconnected System. Where possible, greenways with trails should line riparian corridors and storm drainageways connecting to destinations such as schools, parks, and Neighborhood Centers. Coordinate an open lands system among property owners to use land efficiently and retain wildlife movement corridors.

Additional Development Guidelines

a.   Roads crossing wetland systems should be fitted with oversized culverts where feasible to facilitate and maintain wildlife corridors.

b.   Upland buffers, preservation areas, and wetland systems should be maintained so as to prevent invasion by nuisance and/or exotic species listed.

c.   Recreation opportunities within or adjacent to the PCN should be limited to passive uses such as biking or hiking trails or other educational opportunities. Golf courses can provide valuable linkages in the overall PCN, but should not be eligible for PCN credits except where active vegetation has been retained.

d.   Every attempt should be made to mitigate for impacts to wetlands and listed wildlife species such as gopher tortoise within the study area through preservation and/or enhancement of habitat.

e.   Retaining existing native vegetation and the use of native drought-resistant plants in both residential, commercial, and common use areas is encouraged.

f.   Minimize additional roads crossing and encroachments across/into the PCN.

g.   Encourage the placement of stormwater management ponds, utility facilities, and other non-residential land uses adjacent to the defined PCN.

h.   While not a requirement, the City and developers should attempt to maintain a 500 foot minimum width for environmentally sensitive lands to allow wildlife movement.

i.   Design surface water management systems to discharge pre-treated stormwaters to preserved wetlands in such a way as to maintain and/or enhance their current hydrology.

j.   Reduce and/or eliminate fencing as a means to delineate property ownership's wherever practicable.

k.   Create an area-wide signage program designating PCN boundaries, alerting drivers at critical intersections of roads and the PCN, and educating residents within Southeast Orlando of the value, functions, and restrictions within the PCN.

l.   Encourage a domesticated animal control program including free or reduced price sterilization, community sponsored humane society, and other programs to reduce displacement and harm to existing animal species.

Notes and Questions

1.   Orlando's Primary Conservation Network concept is designed to bring about the pattern of development shown in the picture reproduced above.

Would Lorenzo approve? Is this the kind of development pattern you think his formula for estimating the ecological costs of sprawl would favor or reject?

2. How sharp are the PCN's teeth? The measures listed in the PCN materials are chock full of qualifications such as "where feasible," "should be," and "where practicable." Will this make fruition of the PCN's vision harder to accomplish? Should Orlando have *mandated* all of the design criteria specified in the PCN plan? If not, what will motivate developers to conform to the PCN program, and what will empower the public to demand that they do so?

3. As far as large urban ecosystem planning efforts go, the Orlando plan is unusual in that it was *not* motivated by concerns over compliance with the Endangered Species Act. Indeed, the ESA is increasingly becoming *the* land use control mechanism in urban settings. *See* Nancy Perkins Spyke, *Charm in the City: Thoughts on Urban Ecosystem Management*, 16 J. LAND USE & ENVTL. LAW 153 (2001). The next case study in urban ecosystem management involves another regional planning effort, but this one, besides being on the opposite side of the nation, was carried out under the specter of Endangered Species Act regulatory controls. Consider whether the design of the urban planning effort was influenced substantially by that presence.

National Wildlife Federation v. Babbitt

United States District Court, Eastern District of California, 2000.
128 F.Supp.2d 1274.

■ LEVI, DISTRICT JUDGE.

Plaintiffs challenge the United States Fish and Wildlife Service's issuance of an incidental take permit to allow development in the Natomas Basin, a 53,000 acre tract of largely undeveloped land stretching to the North of the City of Sacramento. The Natomas Basin contains habitat of the Giant Garter Snake, a threatened species under the federal Endangered Species Act, and the Swainson's hawk, a threatened species under the California Endangered Species Act. The parties now bring cross-motions for summary judgment.

I.   Background and Procedural History

A.   The Natomas Basin

The Natomas Basin ("Basin") is a low-lying region of predominately agricultural lands in the Sacramento Valley consisting of approximately 53,000 acres. The Basin is part of a larger flood plain known as the American Basin, and is situated at the northern end of the City of Sacramento ("City"). Approximately 22% of the area of the Basin is within the City limits, with the remainder to the north, in the jurisdiction of Sacramento and Sutter Counties. Because the area was subject until recently to frequent flooding, the Basin has remained relatively immune from development despite its proximity to a growing metropolitan region.

The Basin provides habitat or potential habitat for a number of species listed as endangered or threatened under federal or state law. Of the 11,387 acres of the Basin that lie within the City, approximately 30% has been developed, while roughly 55% is in crop land and 15% is vacant or in its natural state. The land containing natural vegetation is primarily located along irrigation canals, drainage ditches, pastures, and uncultivated fields.

In 1986, heavy spring rains caused significant flood damage in the Sacramento area. In response, the Army Corps of Engineers ("Corps") undertook a study of proposals to improve flood control measures. This study culminated in the Corps' issuance, in 1991, of the "American River Watershed Investigation Feasibility Report" ("Feasibility Report"). The Feasibility Report proposed to provide 200 year flood protection through construction of the Auburn Dam and a series of levee improvements.

The United States Fish and Wildlife Service ("the Service") reacted to the Corps' proposal with concern, noting the importance of the Basin to waterfowl using the Pacific Flyway for migration and to certain native species listed as endangered or threatened under federal or state law. The Service issued a report in 1991, the "American River Watershed Investigation, Natomas Area," ("1991 Report"), that examined the "indirect impacts" from development likely to result from flood control in the Basin. The 1991 Report considered a 41,000 acre subarea of the Basin that provided upland or wetland habitat and concluded that nearly the entire area—39,200 acres—would be developed if a 200 year flood control plan were implemented. The 1991 Report found that the biological effects of this degree of urbanization in the Basin would be a dramatic loss of wildlife habitat.

The 1991 Report contained an extensive discussion of potential means of mitigating for the urbanization that would follow upon flood control. The 1991 Report recommended that an area totaling 17,650 acres in the Natomas Area be acquired and managed as a wetland/upland complex, to offset the expected loss of 22,717 acres of such habitat. Excluding acquisition costs, the 1991 Report anticipated the nonrecurring cost for development of acquired lands into habitat at approximately $171,675,000—over $9700 per acre. In addition, the Report estimated annual management costs of $8,825,000, or $500 per acre.

In 1991, the Sacramento Area Flood Control Authority ("SAFCA") began the process of applying for an incidental take permit ("ITP") under § 10 of the Endangered Species Act ("ESA"), 16 U.S.C. §§ 1531 et seq.
* * *

In 1993, the Service listed the Giant Garter Snake ("GGS") as a threatened species under the ESA. The listing notice identified 13 distinct populations of the GGS, of which the American Basin population was one of the largest. See 58 Fed.Reg. 54053 (Oct. 20, 1993). The Natomas Basin population is a subpopulation of the American Basin population of the GGS. After the listing of the GGS, "interest renewed ... in developing a habitat conservation plan." In January 1994, the Natomas Basin Habitat Conservation Plan Working Group ("Working Group") was formed, and began development of a habitat conservation plan ("Plan," or "HCP"),[2] as required to qualify for an ITP. The Working Group was comprised of "representatives of land owners of a large proportion of the affected area."

* * *

B.   The Final HCP

The Natomas Basin HCP is intended "to promote biological conservation along with economic development and the continuation of agriculture within the Natomas Basin." The HCP lists 26 species that are "potentially subject to take," and which are to "be included in the state and federal permits issued in accordance with the Plan." The proposed permit authorizes incidental take resulting from urban development, as well as any incidental take that may occur through rice-farming or result from management of the Plan's reserve lands. The HCP was developed as a regional conservation plan for the entire Natomas Basin, and was intended for use in connection with ITP applications for each of the municipalities and water companies with interests in the Basin . . .

The Plan is administered by the Natomas Basin Conservancy ("NBC") which has the responsibility to establish and oversee "a concerted Basin-wide program for acquiring and managing mitigation lands on behalf of the permittees. Specifically, the NBC will be responsible for collecting and managing mitigation fees required by the City and Counties, for using the fees to establish mitigation lands, and for managing the mitigation lands for the benefit of the covered species." The Plan provides for a Technical Advisory Committee ("TAC"), comprised of representatives from the Service, the California Department of Fish and Game, and any permittee, as well as outside experts, "to advise the NBC in implementing" the HCP.

The Plan calls upon the NBC to assemble connected 400 acre blocks of reserve lands—with one block of at least 2500 acres—for the benefit of the Giant Garter Snake and to protect Swainson's hawk habitat and nesting areas. The HCP states that "to the maximum extent practicable, the [Natomas Basin] HCP will ensure that habitat acquisition will be provided in advance of habitat conversion resulting from urban development in the Natomas Basin." Funding for land acquisition, however, is derived from the collection of mitigation fees for development. Thus, with regard to the phasing of land acquisition, the HCP actually requires only that, after an initial acquisition of 400 acres, which is to be made "as soon as possible,"

---

2. Section 10 of the ESA allows the Secretary to issue an incidental take permit ("ITP"), authorizing its holder to take some members of protected species when such taking is incidental to carrying out an otherwise lawful activity. See 16 U.S.C. § 1539(a). To obtain an ITP, an applicant must develop and submit an HCP, which specifies (1) the likely impact from the proposed takings; (2) the steps the applicant will take to minimize and mitigate such impacts and the funding available for such mitigation; (3) alternative actions considered, and the reasons for not selecting them; and (4) such other measures as the Secretary may require as necessary or appropriate for the purposes of the plan. See 16 U.S.C. § 1539(a)(2)(A); see also Part II.A, infra.

"no more than one year shall elapse between receipt of a fee and expenditure of that fee in the purchase or other acquisition of mitigation land."

The Plan is based on certain key principles and assumptions. First, the Plan assumes that only 17,500 acres of Basin land will be developed over the 50 year life of the permit, and that a substantial proportion of the undeveloped land will remain in agriculture, particularly rice, which is believed to have unique value as habitat for the GGS. The Plan's conclusion that a ratio of .5 acres of reserve lands for each 1 acre of developed land will ensure the biological needs of the protected species is based on the assumption that a considerable portion of the undeveloped and agricultural lands in the Basin will remain undeveloped, thereby augmenting the habitat value of the reserve lands. Second, the Plan pursues a regional approach to conservation. Whereas without the Plan, individual landowners could pursue separate permit applications, or develop their land without securing an ITP, the HCP is intended to provide a consolidated approach under which resources may be pooled and conservation lands may be purchased throughout the Basin. Third, the HCP treats all Basin lands as fungible, as equally valuable habitat. Thus, the HCP requires developers to "mitigate" for the anticipated take of individuals or habitat by payment of a fee for each acre developed. Rather than differentiating among lands according to their value as habitat for protected species, the HCP requires all landowners within the Permit area to pay a mitigation fee for developing their land, regardless of whether any particular parcel has or lacks habitat value. Depending on one's point of view, this uniform treatment is either a strength or a weakness of the Plan. It is a strength because mitigation fees are to be collected on all acreage and are used "to set aside 0.5 acres of habitat land for each 1.0 acres of gross development that occurs in the Basin."[7] It is a potential weakness because the Plan does not attempt to identify, prior to intensified development under the ITP, particular parcels for acquisition as reserves, based upon the importance of those parcels as habitat, but simply specifies acquisition criteria, and leaves specific reserve acquisition to the future decisionmaking of the NBC.

Finally, the Plan is based upon what it calls "adaptive management." The Plan recognizes that the current state of knowledge as to the conservation needs of protected species is imperfect, and that its assumptions as to the amount, location, and pace of development in the Basin and as to the adequacy of the mitigation fee to accommodate increased expenses may prove inaccurate. The Plan addresses these uncertainties through its "adaptive management" provisions, which permit the Plan's conservation strategy to be adjusted based on new information. The HCP's conservation

---

7. The initial base fee was set at $2240 per acre developed, of which $1829 was allocated for acquisition of land, $142 for restoration, enhancement and monitoring, $150 for administration of operation and monitoring, $75 for an operations and monitoring endowment, and $44 for administration of fee collection. Because these figures represent the amount collected for each acre developed, the amount available for each acre of mitigation land is twice the given figures. For example, the HCP assumes that land can be acquired for $1829 x 2, or $3658 per acre, and sets aside $142 x 2, or $284 per acre for restoration, enhancement, and monitoring.

program can be modified under the adaptive management provisions if: (1) new information results from ongoing research on the GGS or other covered species; (2) recovery strategies under Fish and Wildlife Service recovery plans for the GGS or the Swainson's hawk differ from the measures contemplated by the HCP; (3) certain of the HCP's mitigation measures are shown through monitoring to require modification; or (4) the HCP's required minimum block sizes for reserve lands are shown to require revision. The Plan anticipates that the NBC will make discretionary decisions in future years based upon new information. The NBC will decide, for example, which lands to purchase, depending on a variety of future considerations difficult now to predict, and whether to change the mix of in and out of Basin reserve lands and agricultural as opposed to marsh reserve lands.

* * *

IV.   Endangered Species Act Claims

The nub of plaintiffs' challenge to the HCP concerns its strategy of adaptive management. Plaintiffs object that in the face of incomplete information as to a number of important issues—including the conservation needs of the covered species, the likely location and pace of development in the Basin, and the preferred location and availability of reserves—the Plan does not undertake studies to develop better information, but simply creates a structure and describes a process for reaching and adjusting decisions in the future based on developing information. Plaintiffs contend that many of the Service's findings are arbitrary because of the uncertainty inherent in the HCP's deferred decisionmaking scheme.* * *

With miscellaneous exceptions, plaintiffs' ESA claims fall into two categories. First, plaintiffs challenge as arbitrary the Service's findings as to the adequacy of the Plan's provisions, particularly those related to funding and mitigation; second, plaintiffs contend that the Service's findings regarding the biological effects of the Plan on covered species are arbitrary and capricious. As explained below, under the APA's deferential standard of review, the Service's findings largely pass muster with respect to the Plan as a whole; however, with respect to the City's Permit, the Service's findings do not. Many of the provisions of the HCP are based on the assumption that all of the land-use agencies with jurisdiction over parts of the Basin will become permittees. Similarly, the Service's findings are plainly geared toward the regional nature of the HCP, and do not adequately reckon with the local nature of the Permit or analyze what would happen if the City's lands were developed under the HCP, while the lands outside the City limits were developed piecemeal, by individual landowners, outside the HCP and the protections provided by the HCP. The importance of this point is most obvious with respect to the funding mechanism, which relies on future fee increases to fund current increases in land acquisition costs. Because reserve land acquisition lags behind the development that funds it, the funding mechanism must play catch up, passing increased costs on to the next developer. The Plan can cope with increased expenses for mitigation, monitoring, and the like, only so long as there exists a ready supply of land to be developed under the HCP. The biological findings, too, are based on inadequate consideration of the tension between the regional Plan and the local Permit. The record contains no particularized analysis of the importance of the City's lands as habitat, and no consideration of how the species will fare if the City's lands are developed under the Plan and some or all of the remainder of the Basin is fully developed outside the HCP. Similarly, although the 1997 Biological Opinion concedes that large blocks of reserve lands may not be possible if only the City is a permittee, there is no analysis of the consequences—particularly for the GGS—of abandoning the goal of large connected blocks of reserve lands.

Notes and Questions

1. The central flaw the court identifies is not that the Natomas Basin plan fails to satisfy the criteria of the Endangered Species Act, but that its manner of implementation rests on the faulty premise that Sacramento could begin to issue development approvals under the plan without the other local government jurisdictions in the Natomas Basin having sought to enter the conservation program. Obviously, Sacramento can't force other jurisdictions to enter into the plan, but since the plan is based on Basin-wide habitat assumptions, the court ruled that Sacramento cannot act as if those jurisdictions have entered the plan. This result illustrates the difficulty and importance of intergovernmental cooperation at the local level when ecosystem management goals are not embedded in federal or state law. *See* James W. Spensley, *Using Intergovernmental Agreements to Manage Growth*, 15 NATURAL RESOURCES & ENV'T 240 (2001).

2. Compare the Orlando and Natomas Basin urban ecosystem management plans. Which rests on greater scientific foundations? Which provides greater accountability for planning and development decisions? Which provides greater opportunity for public input? Which includes greater provisions for monitoring and assessment? Which is more likely to produce the intended pattern of development? Which intended pattern of development provides the better ecosystem management program?

3. In both cases, the conservation plan will exact an economic cost on development. The Natomas Basin plan pegged the cost at $2240 per acre of developed land. The Orlando plan costs are unspecified. In either case the costs will be passed on the end consumer, such as the residential home buyer. How would you feel if, upon purchasing a home for, say, $100,000, you learned that $2500 of that cost was devoted to habitat conservation taking place 30 miles away? Have you contributed that much to habitat conservation efforts in cash donations this year? Over the past 5 years? In your lifetime?

4. For that matter, is it equitable to place the cost of urban ecosystem management on *new* development? The Natomas Basin plan relied on a fee on new development to fund its $170 million land acquisition and $8 million annual operations costs. But new development in the Basin did not cause the area's endangered species problem. Past development did that. Existing developed land also would benefit not only from the ecological values of the newly protected lands, but also from the market impact of increased scarcity of developable land and costs of development. If you believe it is appropriate to charge $2240 for each acre of new development in the Basin, ought not owners of acres of *existing* developed land be assessed an equal fee? How successful would a candidate for local office be running on such a platform? Are there other ways of allocating some of the cost of urban land development to existing land uses or some surrogate?

2.   AGRICULTURAL LANDS

> For farmers and ranchers, for people who make a living on the land, every day is Earth Day. There's no better stewards of the land than people who rely on the productivity of the land. And we can work with our farmers and ranchers to help improve the environment.

President Bush offered this perspective on farming in America when he signed the 2002 Farm Bill on May 13, 2002. Our nation's farm policy has for decades rested, and continues to rest, on this bedrock premise—that farmers are the "best stewards of the land" because they depend on the land's productivity for their livelihood. But consider what farming means to the ecosystems within and around the farmed lands. First, any pre-farm undisturbed habitat is removed. About 900 million acres of the United States—45 percent of the nation's land mass—have experienced this transformation and remain in agricultural uses today. Next, while the number of environmentally conscious farmers is strong and growing, in most cases farming tills soil, irrigates, fertilizes, applies pesticides, plants monoculture crops or grazes domesticated livestock, removes the commodity product and starts the process over again. Finally, farming practices generally rely on the ability to move excess fertilizer, animal waste, and pesticides off of the farm land, and onto surrounding lands and waters.

The end result is that farming American-style leads to massive soil erosion, depletion of surface and underground water supplies, pollution of air and water with nutrients, pesticides, and sediments, and depletion of soil resources. A few facts and figures drive this picture home:

- 930 million acres of habitat have been converted to farming uses
- 25 percent of all cropland has become highly erodible
- 2 billion tons of soil are eroded annually from farms by wind and water
- 331 million tons of eroded farm soils empty each year into the Gulf of Mexico alone
- 55 million acres of cropland are irrigated
- 48 million acres of cropland have become saline, most due to irrigation
- 750 million pounds of pesticides are released annually into surface and ground waters

- farms produce 200 times as much animal waste as the nation's human waste
- Maryland's 300 million chickens produce 720 million pounds of waste annually
- farm runoff releases 1.16 million tons of phosphorous into the nation's waters each year
- farm runoff releases 4.65 million tons of nitrogen into the nation's waters each year
- ammonia from hog waste releases 179 million pounds of nitrogen into the atmosphere each year in North Carolina alone

See J.B. Ruhl, *Farms, Their Environmental Harms, and Environmental Law*, 27 ECOLOGY L.Q. 263 (2000). Not surprisingly, the EPA recently determined that runoff from agriculture is the leading cause of impairment of lakes, rivers, and wetlands in our nation's waters. *See* OFFICE OF WATER, U.S. EPA, NATIONAL WATER QUALITY INVENTORY 1994 REPORT TO CONGRESS (1994).

The policy response to this unusual form of ecological "stewardship" has not been what one might imagine. Unlike other polluting industries, most of which have been target of command-and-control style federal and state regulation since the 1970s, farming has enjoyed extensive exemptions from environmental regulation. The policy for most industries is that the polluter pays—that is, the polluter must assume the costs of internalizing the social costs of the polluting activity. Instead, the environmental policy for agriculture has been to pay farmers to avoid causing environmental harms. The following article provides an excellent history and overview of how that irony came about and the possible future directions in which it is headed.

Christopher R. Kelley and James A. Lodoen, Federal Farm Program Conservation Initiatives: Past, Present, and Future

9 Natural Resources & Env't 17 (Winter 1995).

In testimony before Congress during the 1990 farm bill debate, the U.S. Department of Agriculture's Assistant Secretary for Economics, Dr. Bruce Gardner, rhetorically asked "Why do we need a conservation title in the farm bill?" His response captured the long-standing ambivalence over conservation's role in federal farm policy:

The purpose [of a conservation title] is to maintain the environment for farmers and the rest of us and to maintain the productive capacity of agriculture for future generations. At the same time we have to keep in mind the overall objective of farm policy, to promote an economically healthy agriculture and an internationally competitive one, and moreover, we have to achieve these objectives at the minimum possible cost to the taxpayer.

Conservation Issues and Agricultural Practices and Oversight on the Forestry Title of the 1990 Farm Bill: Hearings Before the Subcomm. on Conservation and Forestry of the Senate Comm. on Agriculture, Nutrition, and Forestry, 101st Cong., 1st & 2d Sess. 469 (pt. XI) (1990). As Dr. Gardner recognized, the "overall objective" of federal farm policy is to improve the agricultural economy. From its New Deal beginnings, farm policy has been primarily designed to stabilize commodity prices and to support farm income. Stated bluntly, "[federal] farm policy should be taken for what it is, namely, industrial policy with some economic benefits for farmers and their industrial partners." WILLIAM P. BROWNE, SACRED COWS AND HOT POTATOES: AGRARIAN MYTHS IN AGRICULTURAL POLICY 35 (1992).

Federal farm policy has been ambivalent toward conservation initiatives. On the one hand, sound conservation practices can produce short-and long-term economic benefits. On the other hand, the natural resources capital on which agriculture is based has not been the primary concern of those who have guided federal farm policy for the last sixty years. Instead, farmers' wealth relative to the wealth of the nonfarm population has mattered most. Nevertheless, conservation and the federal farm programs have been linked in varying degrees throughout the programs' history, although this linkage has not been inspired by an unadulterated desire to conserve natural resources. Rather, conservation initiatives have largely served to garner political support or otherwise to assist in achieving a given program's primary economic objectives.

If the past is prologue, the economic well-being of the agricultural sector will continue to be the predominant concern of federal farm policy for the foreseeable future. Conservation, however, is likely to assume a greater role for at least two reasons. First, agriculture is coming under greater environmental scrutiny. Examination of the farm programs' impact on the environment is giving credence to the contention that "[m]any farmers in these programs manage their farms to maximize present and future program benefits, sometimes at the expense of environmental quality." NAT'L RESEARCH COUNCIL, ALTERNATIVE AGRICULTURE 10 (1989) (ALTERNATIVE AGRICULTURE). As a result, Congress can be expected to continue to respond to concerns about the environmental impact of farm policy. Second, budgetary constraints and international competition will accelerate the trend toward market-oriented policies, thus diminishing the programs' traditional role in supply management. At the same time, the economic consequences of ending income transfers to farmers, particularly the decline in land values that may result, are likely to discourage an abrupt halt to income transfers. Paying farmers for adopting conservation practices is likely to expand, perhaps ultimately replacing other current mechanisms for supporting farm income, albeit with reduced fiscal outlays.

Conservation's past and present role in the federal commodity programs—the flagships of federal farm policy—and the future of conservation initiatives are examined here. In the process, one fact emerges as inescapable—just as farming alters the natural environment, federal farm programs are not environmentally neutral. They produce both harm and good, with the final accounting being a matter of perspective and debate. The debate's scale is not insignificant for there is considerable support for the proposition that the farm programs have done more harm than good to the environment.

Inauspicious Beginnings

The earliest significant linkage of conservation with federal farm policy was short-lived. As the nation entered the Great Depression, the farm sector was suffering from low income, surplus crops, and eroded cropland. Farming employed about 20 percent of the civilian labor force and accounted for almost 8 percent of the gross national product. Improving the farm economy, therefore, was viewed as an important first step in restoring the nation's economy. *See* Kristen Allen & Barbara J. Elliott, *The Current Debate and Economic Rationale for U.S. Agricultural Policy in* U.S. AGRICULTURE IN A GLOBAL SETTING: AN AGENDA FOR THE FUTURE 9, 17–18 (M. Ann Tutwiler ed., 1988). Prior to enactment of the first New Deal farm legislation, national land-use planning and the retirement of submarginal lands from crop production was debated as a way to improve the farm economy. Conservation, however, was not the only goal of those urging land-use planning—"a fervent desire to avoid imposing controls on individual farmers' operations also prompted its promotion." Harold F. Breimyer, *Agricultural Philosophies and Policies in the New Deal*, 68 MINN. L. REV. 333, 338 (1983).

In the Agricultural Adjustment Act of 1933, the New Deal's first farm legislation, production controls prevailed over land-use planning as the mechanism for improving commodity prices. The use of conservation as a tool to promote economic interests had not been forgotten, however. When the 1933 Act's production controls and funding mechanism were declared unconstitutional in *United States v. Butler*, 297 U.S. 1 (1936), Congress responded less than two months later with the Soil Conservation and Domestic Allotment Act of 1936.

The 1936 Act paid participating farmers to plant "soil-building" grasses and legumes, instead of "soil-depleting" commodities such as corn and cotton. In so doing, it inaugurated the idea of making "green payments" to farmers, a concept still in currency today. Nevertheless, the Act's fundamental economic objectives were revealed by the classification of wheat, then in surplus, as "soil depleting" despite its soil-conserving characteristics. Moreover, although soil conservation improved, the legislation did little to stabilize commodity production and improve farm income. Consequently, Congress once again turned to production controls and enacted the Agricultural Adjustment Act of 1938. After surviving constitutional challenges in *Mulford v. Smith*, 307 U.S. 38 (1939), and *Wickard v. Filburn*, 317 U.S. 111 (1942), the 1938 Act became the cornerstone of today's commodity programs.

Originally intended as temporary measures, the commodity programs begun by the Agricultural Adjustment Act of 1938 have survived for six decades. During these years, they have weathered considerable criticism.

While some of this criticism reflects a fundamental disagreement with government intervention in the farm economy, other criticism has been more focused. For example, commodity traders and handlers have challenged the trade-distorting effects of government-supported prices and the loss of production through acreage reduction programs. Consumers and taxpayers have questioned the programs' marketplace and government costs. Farmers have criticized the programs' failure to improve prices and the increasing complexity of the program rules. Finally, environmentalists have asserted that the basic mechanisms used to support commodity prices and farm income have had negative environmental consequences.

Because some environmentalists have indicted the basic mechanisms used by the programs, an understanding of this criticism requires an understanding of the incentives and disincentives the programs create. In turn, this dictates examining what the federal commodity programs do.

The Role of the Federal Commodity Programs

The commodity programs have two primary functions—price support and income support. Commodity prices are supported through nonrecourse loans. The loans are made post-production, usually for a nine-month term. The loan rate established for each price-supported commodity effectively sets the commodity's market price floor because producers can forfeit the crop securing the loan to the government in lieu of repaying the loan. In addition to establishing a market price floor, nonrecourse loans provide individual farmers with capital at a time when market prices typically are at their lowest, thus providing a direct incentive for participation. Nonrecourse loans also tend to spread marketings throughout the year. *See generally* J.W. Looney, *The Changing Focus of Government Regulation of Agriculture in the United States*, 44 MERCER L. REV. 763, 781–89 (1993).

Nonrecourse loans are dwarfed in economic significance by deficiency payments, the primary form of farm income support. Deficiency payments are direct income transfers to participating producers of feed grains (corn, sorghum, oats, and barley), wheat, rice, and upland cotton. 7 U.S.C. §§ 1444f(c), 1445b–2(c), 1441–2(c), 1444–2(c). Deficiency payments use the concept of "target prices," a per-unit sum deemed by Congress to cover the commodity's cost of production. Payments are based on the difference between the commodity's "target price" and the commodity's averaged market price or the commodity's loan rate, whichever difference is less. If the averaged market price exceeds the target price, no deficiency payments are made.

Commodity programs are annual programs, and program enrollment is voluntary. To participate in a commodity's deficiency program, farmers may have to set aside land, idling it from production. Under this "production adjustment" requirement, known as the acreage reduction program (ARP), the idled land must be dedicated to a conservation use, also known as the acreage conservation reserve.

The annual decision to participate is usually based on economic considerations. When an ARP is in effect, the decision turns on whether the projected market price will justify setting aside the required acreage. In most years for most program crops, the overwhelming majority of eligible acreage is enrolled. Because enrollment reduces financial risks, program participation is often required by lenders.

The deficiency payment rate for each commodity is a per-unit rate. Payments are made on "program production," a calculation that involves multiplying a farm's "permitted acres" for a commodity by that farm's yield for the commodity. Prior to 1985, program yields were subject to adjustment based on actual production. Average yields were frozen in 1985, however, and have not been adjusted since.

A farm's permitted acres for a commodity are the crop acreage base minus the ARP acreage if an ARP is in effect. Only the production on a farm's "crop acreage base" for each particular commodity is eligible for deficiency payments. Crop acreage bases are calculated on a five-year moving average. If, for example, a farmer planted less corn in a particular year, the farm's corn acreage base would be reduced in subsequent years. Recapturing lost base takes time because a producer cannot increase a program crop's acreage base above its existing level without losing that year's eligibility for payments on all program crops.

* * *

Commodity Programs' Effect on Conservation

Federal farm policy is marked by contradictions and unintended consequences. Some of the contradictions are obvious. For example, "[s]et asides are designed to control supplies and increase market prices, but price supports induce farmers to increase supply, and by so doing, to depress market prices." Robbin A. Shoemaker, A Model of Participation in U.S. Farm Programs 14 (U.S. Dep't of Agric., Econ. Res. Serv., Tech. Bull. No. 1819, Aug. 1993). The contradictions also extend to other agricultural sectors. Policies that increase feed grain prices, for example, disfavor livestock farmers who feed grain to their livestock. The environment, in particular, is buffeted by contradictory incentives and disincentives, producing results that were probably never intended.

The most fundamental criticism made by environmentalists against the commodity programs is that the availability of deficiency payments encourages farmers to plant program crops in continuous cycles to preserve crops' acreage bases. The environmental consequences of this incentive have been summarized as follows:

> The farm programs support crops that tend to require high agrichemical inputs and are associated with high rates of soil erosion. Other less-erosive and less-agrichemical-dependent crops receive little government support. The programs reward farmers for specializing in program crops year after year, resulting in further soil depletion and pest problems, which in turn lead to a greater need for agrichemical inputs. The programs tend to discourage farmers from planting other crops and from using more diversified crop rotations.

U.S. GEN. ACCOUNTING OFFICE, ALTERNATIVE AGRICULTURE: FEDERAL INCENTIVES AND FARMERS' OPTIONS 3 (Pub. No. PEMD–90–12, Feb. 1990).

* * *

In addition to reducing the incentives for engaging in monocultural production, recent program rules have created disincentives to converting highly erodible land and wetland to cropland. Also, highly erodible land in production must be farmed in accordance with an approved conservation plan beginning in 1995. The highly erodible land and wetland conservation provisions are commonly known as "sodbuster" and "swampbuster," respectively. Of the two, the swampbuster provisions have elicited greater attention.

Under the swampbuster provisions, a person is ineligible for any benefits administered through the U.S. Department of Agriculture (USDA) if

[1]   [t]he person produces an agricultural commodity on wetland that was converted after December 23, 1985; or

[2]   [a]fter November 28, 1990, the person converts a wetland by draining, dredging, filling, leveling or other means for the purpose, or to have the effect, of making the production of an agricultural commodity possible.

16 U.S.C. § 3821(a), (b). The first of these two "triggers" was enacted in the Food Security Act of 1985 (FSA); the second in the Food, Agriculture, Conservation, and Trade Act of 1990 (FACTA). Under the FSA, a wetland could be converted with impunity (but for a possible Clean Water Act § 404 violation) so long as an agricultural commodity was not produced. FACTA closed that "loophole" by making the triggering event the conversion of the wetland. Also, under the FSA, benefits were denied only for the crop year in which the agricultural commodity was produced. 16 U.S.C. § 3821(a)(1)– (3). Violations of FACTA result in ineligibility for benefits for the crop year in which the conversion occurred and for all subsequent crop years until the converted wetland is restored. 16 U.S.C. § 3821(b).

Wetlands may still be farmed under natural conditions, such as a drought, provided the action of the person farming the wetland "does not permanently alter or destroy natural wetland characteristics." 16 U.S.C. § 3822(b)(1)(D). This means that no action can be taken to increase the effects on the water regime beyond that which existed on such land before December 23, 1985, unless it is determined that the effect on remaining wetland values would be minimal. 7 C.F.R. § 12.33(a).

There are numerous exemptions from ineligibility under the swampbuster provisions. 16 U.S.C. § 3822(b). In addition to these exemptions, the loss of USDA benefits may be avoided by restoration of a wetland that had been converted prior to December 23, 1985. This "mitigation through restoration of another converted wetland" provision is available if the wetland that was converted was either frequently cropped or converted between December 23, 1985, and November 28, 1990. 16 U.S.C. § 3822(f).

* * *

For some environmentalists the disincentives imposed by the sodbuster and swampbuster provisions are not sufficient to make farm programs a net gain for the environment for they only add "another layer of regulation without altering the more fundamental incentives imbedded in the structure of agricultural policies." Paul Faeth, Paying the Farm Bill: U.S. Agricultural Policy and the Transition to Sustainable Agriculture 3 (1991). Others bemoan the fact that the provisions apply only to program participants, while objecting to the government's near total nonenforcement of the swampbuster provisions. Anthony N. Turrini, *Swampbuster: A Report from the Front*, 24 Ind. L. Rev. 1507, 1511 (1991) ("[f]ortunately for wetlands ... most producers participate in federal farm programs."). Still others find in the conservation provisions' combination of directives and exemptions a continuation of the congressional "ambivalence in creating an aggressive environmental program for agriculture...." Linda A. Malone, *Reflections on the Jeffersonian Ideal of an Agrarian Democracy and the Emergence of an Agricultural and Environmental Ethic in the 1990 Farm Bill*, 12 Stan. Env'tl L. Rev. 3, 46 (1993).

The Future

Although they are not a part of the basic commodity programs, the Conservation Reserve Program (CRP) and the Wetland Reserve Program (WRP) may provide a rough guide for the direction of federal farm policy. Under the CRP, farmers are paid to take highly erodible land out of production for an extended period, usually ten years. 16 U.S.C. §§ 3831–3836. Under the WRP, wetlands are removed from production through permanent or thirty-year easements. 16 U.S.C. §§ 3837–3837f. Both programs have had their critics, but the basic concept of paying farmers to further environmental goals generally has been well-received by farmers and environmentalists.

Paying farmers to behave in an environmentally responsible manner offers an alternative to command-and-control regulations based on the "polluter pays" principle. The concept, often described as "green payments," has been suggested as a substitute for deficiency payments. As such, income transfers to farmers would cease to be linked to production as they are today. Instead, the payments would be "coupled" to conservation practices. *See* Tim Osburn, *U.S. Conservation Policy—What's Ahead?*, Agric. Outlook, Nov. 1993, at 36.

While attractive to many farmers and environmentalists, green payments present a number of political problems. The most fundamental problem is that the favored recipients of green payments are not likely to be the same farmers favored by current income transfers. Because current income transfers are coupled with production, larger producers are favored. In addition, current programs favor producers in the Midwest, particularly in the upper and lower Great Plains. Green payments, on the other hand, are likely to be parceled out where the environmental problems are most acute and nearest to population centers, resulting in a shifting of payments to both coasts and to major watersheds.

In addition, green payments do not present the ideal mechanism for redirecting payments to the small family farmer whose maintenance is often the rallying cry and justification for income transfers to agriculture. If, for example, green payments are "means tested" so that they are received by only those farmers who are less financially secure relative to other farmers and the nonfarm population, the environmental benefits may be diluted. To the extent environmentalists deplore the environmental harms resulting from large-scale, industrial agriculture, an argument could be made that the financial incentives should be directed toward reforming the behavior of those who farm the most land, thus putting the smaller farmer at a disadvantage.

Finally, there is no certainty that green payments would result in sufficient income to satisfy farmers. An idea attracting considerable attention in the 1995 farm bill debate is a program that would guarantee a farmer's per-acre return for a given crop at a prescribed level. *See* Joy Harwood, *Streamlining Farm Policy: The Revenue Guarantee Approach*, AGRIC. OUTLOOK, Apr. 1994, at 24. The notion that farmers' revenue should be guaranteed serves as a reminder that farm policy is economic policy.

In all likelihood, federal farm policy will remain an inconsistent and contradictory collection of economic incentives and disincentives. At the same time, conservation initiatives will expand, but only gradually. Recalling that price and income support has formed the core of federal farm policy for six decades, the redirection of that policy will more likely be evolutionary rather than revolutionary.

––––––

As the preceding article predicted, in the 2002 Farm Bill Congress continued the Conservation Reserve Program and Wetlands Preserve Program, but also enhanced additional "green payment" programs oriented toward providing incentives for actual farming and habitat enhancement practices rather than merely paying for land set asides. Prominent examples of this approach are the Environmental Quality Incentives Program and the Wetlands Habitat Incentive Program.

Farm Security and Rural Investment Act of 2002

H.R.2646 (2002).

SEC. 2301. ENVIRONMENTAL QUALITY INCENTIVES PROGRAM.

Chapter 4 of subtitle D of title XII of the Food Security Act of 1985 (16 U.S.C. 3839aa et seq.) is amended to read as follows:

SEC. 1240. PURPOSES.

The purposes of the environmental quality incentives program established by this chapter are to promote agricultural production and environmental quality as compatible goals, and to optimize environmental benefits, by—

(1) assisting producers in complying with local, State, and national regulatory requirements concerning—

(A) soil, water, and air quality;

(B) wildlife habitat; and

(C) surface and ground water conservation;

(2) avoiding, to the maximum extent practicable, the need for resource and regulatory programs by assisting producers in protecting soil, water, air, and related natural resources and meeting environmental quality criteria established by Federal, State, tribal, and local agencies;

(3) providing flexible assistance to producers to install and maintain conservation practices that enhance soil, water, related natural resources (including grazing land and wetland), and wildlife while sustaining production of food and fiber;

(4) assisting producers to make beneficial, cost effective changes to cropping systems, grazing management, nutrient management associated with livestock, pest or irrigation management, or other practices on agricultural land; and

(5) consolidating and streamlining conservation planning and regulatory compliance processes to reduce administrative burdens on producers and the cost of achieving environmental goals.

SEC. 1240A. DEFINITIONS.

In this chapter:

* * *

(2) ELIGIBLE LAND-

(A) IN GENERAL—The term "eligible land" means land on which agricultural commodities or livestock are produced.* * *

(3) LAND MANAGEMENT PRACTICE—The term "land management practice" means a site-specific nutrient or manure management, integrated pest management, irrigation management, tillage or residue management, grazing management, air quality management, or other land management practice carried out on eligible land that the Secretary determines is needed to protect from degradation, in the most cost-effective manner, water, soil, or related resources.

* * *

(6) STRUCTURAL PRACTICE—The term "structural practice" means—

(A) the establishment on eligible land of a site-specific animal waste management facility, terrace, grassed waterway, contour grass strip, filterstrip, tailwater pit, permanent wildlife habitat, constructed wetland, or other structural practice that the Secretary determines is needed to protect, in the most cost effective manner, water, soil, or related resources from degradation* * *

SEC. 1240B. ESTABLISHMENT AND ADMINISTRATION OF ENVIRONMENTAL QUALITY INCENTIVES PROGRAM.

(a) ESTABLISHMENT—

(1) IN GENERAL—During each of the 2002 through 2007 fiscal years, the Secretary shall provide cost-share payments and incentive payments to producers that enter into contracts with the Secretary under the program.

(2) ELIGIBLE PRACTICES—With respect to practices implemented under this chapter—

(A) a producer that implements a structural practice in accordance with this chapter shall be eligible to receive cost-share payments; and

(B) a producer that implements a land management practice, or develops a comprehensive nutrient management plan, in accordance with this chapter shall be eligible to receive incentive payments.

* * *

(e) INCENTIVE PAYMENTS—

(1) IN GENERAL—The Secretary shall make incentive payments in an amount and at a rate determined by the Secretary to be necessary to encourage a producer to perform 1 or more land management practices.* * *

SEC. 1240D. DUTIES OF PRODUCERS.

To receive technical assistance, cost-share payments, or incentive payments under the program, a producer shall agree—

(1) to implement an environmental quality incentives program plan (including a comprehensive nutrient management plan, if applicable) that describes conservation and environmental purposes to be achieved through 1 or more practices that are approved by the Secretary;

(2) not to conduct any practices on the farm or ranch that would tend to defeat the purposes of the program;

(3) on the violation of a term or condition of the contract at anytime the producer has control of the land—

(A) if the Secretary determines that the violation warrants termination of the contract—

(i) to forfeit all rights to receive payments under the contract; and

(ii) to refund to the Secretary all or a portion of the payments received by the owner or operator under the contract, including any interest on the payments, as determined by the Secretary; or

(B) if the Secretary determines that the violation does not warrant termination of the contract, to refund to the Secretary, or accept adjustments to, the payments provided to the owner or operator, as the Secretary determines to be appropriate;

SEC. 1240E. ENVIRONMENTAL QUALITY INCENTIVES PROGRAM PLAN.

(a) IN GENERAL—To be eligible to receive cost-share payments or incentive payments under the program, a producer shall submit to the Secretary for approval a plan of operations that—

(1) specifies practices covered under the program;

(2) includes such terms and conditions as the Secretary considers necessary to carry out the program, including a description of the purposes to be met by the implementation of the plan; and

(3) in the case of a confined livestock feeding operation, provides for development and implementation of a comprehensive nutrient management plan, if applicable.

(b) AVOIDANCE OF DUPLICATION—The Secretary shall, to the maximum extent practicable, eliminate duplication of planning activities under the program under this chapter and comparable conservation programs.

SEC. 1240H. CONSERVATION INNOVATION GRANTS.

(a) IN GENERAL—The Secretary may pay the cost of competitive grants that are intended to stimulate innovative approaches to leveraging Federal investment in environmental enhancement and protection, in conjunction with agricultural production, through the program.

* * *

SEC. 1240I. GROUND AND SURFACE WATER CONSERVATION.

(a) ESTABLISHMENT—In carrying out the program under this chapter, subject to subsection (b), the Secretary shall promote ground and surface water conservation by providing cost-share payments, incentive payments, and loans to producers to carry out eligible water conservation activities with respect to the agricultural operations of producers* * *

SEC. 1240N. WILDLIFE HABITAT INCENTIVE PROGRAM.

(a) IN GENERAL—The Secretary, in consultation with the State technical committees established under section 1261, shall establish within the Natural Resources Conservation Service a program to be known as the wildlife habitat incentive program (referred to in this section as the 'program').

(b) COST–SHARE PAYMENTS—

(1) IN GENERAL—Under the program, the Secretary shall make cost-share payments to landowners to develop—

(A) upland wildlife habitat;

(B) wetland wildlife habitat;

(C) habitat for threatened and endangered species;

(D) fish habitat; and

(E) other types of wildlife habitat approved by the Secretary.

(2) INCREASED COST SHARE FOR LONG–TERM AGREE-MENTS—

(A) IN GENERAL—In a case in which the Secretary enters into an agreement or contract to protect and restore plant and animal habitat that has a term of at least 15 years, the Secretary may provide cost-share payments in addition to amounts provided under paragraph (1).

Notes and Questions

1. There are over 35 million acres of agricultural land enrolled in the Conservation Reserve Program, most of which are planted in grasses and thus offer significant wildlife habitat resources. But studies have shown that farmers will exit the CRP program on expiration of their contracts if crop prices make land more valuable in production. The CRP is, after all, primarily designed to provide another form of economic support to the agricultural community. So what is the CRP–ecosystem management, economic management, or both? *See* Arthur W. Allen, *Agricultural Ecosystems*, in NATIONAL BIOLOGICAL SERVICE, OUR LIVING RESOURCES 423 (1995).

2. If programs like the CRP, WRP, and EQIP make sense to you as a tool of ecosystem management, should they be expanded into other economic sectors? Should we pay residential subdivision developers, golf courses, and ski areas to set aside more land for habitat? Of course, we don't do that as a general rule, and it is unlikely we ever will, but why not?

3. RECREATIONAL AREAS

Americans love the outdoors. Indeed, we may be loving it to death. Golfing, skiing, fishing, scuba diving, camping, rafting, mountain biking, hiking, jet skiing, horseback riding, hunting, snowmobiling, and other outdoor recreations expose their enthusiasts to the splendor of nature, no doubt committing millions of people to conservation values, but at what cost to the ecosystems within which these activities take place? Recall, for example, the devastation offroad vehicle recreation use caused the desert canyon in the *Sierra Club v. Clark* case covered in this chapter's section on desert ecosystems. The court there noted that extensive recreational usage had caused severe environmental damage in the form of major surface erosion, soil compaction, and heavy loss of vegetation. Is this kind of intense recreation a serious challenge to ecosystem management?

That question has been posed with increasing frequency on public lands, where the gradual withdrawal of public forests, lakes, grasslands, and other resources from commercial extractive uses has, ironically, opened the door to different kind of conflict. Past conflicts, many of which are recounted in earlier chapters in this text, focused on the ecological cost of logging, mining, and other traditional industries that dominated on public lands under the "multiple use" policy. For both economic and political reasons, however, those uses are in sharp decline, leading to increased commercial opportunity in the form of recreation and tourism. For example, the Forest Service estimates that of the national forests system's $130 billion annual contribution to the national economy, over $100 billion is attributable to recreation in the system's 23,000 recreation sites, 133,000 miles of trails, and 380,000 miles of roads.

But recreation is not ecologically cost-free. Mountain bikes rut out trails; snowmobile engines emit pollutants; and horses pollute in obvious ways. One or two of each a day may not be a problem along a mountain trail, but recreational usage on public lands has skyrocketed in the past decade, with no end in sight. It is not clear that the host ecosystems can withstand larger and larger onslaughts of tourists and recreationists without suffering injury to fundamental ecosystem dynamics. Indeed, even the personal experience of "being in the woods" has changed. The solitary hiker and canoeist is now hardly visible among the waves of mountain bikes and traffic jams of rubber rafts found in many public parks. To the traditionalist, fun in the wild has been trampled by wild fun.

Thus, as Professor Jan Laitos and Thomas Carr have observed, "the looming conflict in public land use will be between two former allies— recreation and preservation interests. Such a conflict is particularly likely to arise between low-impact, human-powered recreational users (preservationists) and high-impact, motorized recreational users (recreationists)." Jan G. Laitos and Thomas A. Carr, *The Transformation on Public Lands*, 26 Ecology L.Q. 140 (1999). Indeed, this conflict has already reached pitched-battle proportions in numerous settings. Perhaps nowhere is this more true than in the case of snowmobiles in Yellowstone National Park in Wyoming.

Yellowstone is one of the crown jewels in the National Park System's 376 units, which cover more than 83 million acres. There was no real system of national parks in the United States until August 25, 1916, when President Woodrow Wilson signed the Organic Act creating the National Park Service. Established as an agency of the Department of the Interior, the Park Service was responsible for protecting the 40 national parks and monuments then in existence. The Organic Act instructed the Park Service to

promote and regulate the use of [national parks] by such means and measures as conform to the fundamental purpose of the said parks, . . . which purpose is to conserve the scenery and the natural and historic objects and the wild life therein and to provide for the enjoyment of the same in such manner and by such means as will leave them unimpaired for the enjoyment of future generations.

16 U.S.C. § 1. As other national parks were established, primarily in the western states, most were added to Park Service jurisdiction, but many other natural park areas were established and administered as separate units by the War Department and the Department of Agriculture's Forest Service. No single agency provided unified management of the varied federal parklands until 1933, when all the units were transferred to Park Service jurisdiction.

In 1970, Congress declared in the General Authorities Act that all units of the system have equal legal standing in a national system. Areas of the National Park System, the act states,

> though distinct in character, are united through their inter-related purposes and resources into one national park system as cumulative expressions of a single national heritage; that, individually and collectively, these areas derive increased national dignity and recognition of their superb environmental quality through their inclusion jointly with each other in one national park system preserved and managed for the benefit and inspiration of all people of the United States.

16 U.S.C. § 1a–1. Within this system—now a true national system of federal parks—the Park Service uses 19 separate designations for classifying units, such as lakeshores, seashores, and battlefields.

Established on March 1, 1872, Yellowstone was the first national park and was included as one of the original parks under the Park Service's jurisdiction in the Organic Act. Indeed, Yellowstone is the first and oldest national park in the world. Preserved within its boundaries are Old Faithful Geyser and over 10,000 hot springs and geysers, the majority of the planet's total. These geothermal wonders are evidence of one of the world's largest active volcanoes, the caldera of which spans almost half of the park. Home of the grizzly bear, wolf, and free-ranging herds of bison and elk, the park's 2.2 million acres form the core of the Greater Yellowstone Ecosystem, one of the largest intact temperate zone ecosystems remaining on the planet. Yellowstone was designated an International Biosphere Reserve on October 26, 1976, and a World Heritage Site on September 8, 1978.

Yellowstone is also a popular spot. Over 2.7 million recreation visits were made to the park in 2001 (the record is over 3.1 million in 1992). The park's annual budget is over $27 million. On average about 140,000 winter visits are made to the park, most prominently for snowmobiling. Take it from your authors, snowmobiling is *a lot* of fun. But besides being speedy and noisy, snowmobiles also pollute heavily. The two-stroke engines used on snowmobiles are woefully inefficient, burning less than half of the fuel and dumping the rest into the environment. A single snowmobile emits more than 100 times as much pollution as a car in the same amount of running time. Given these characteristics, it is no wonder that the growing concentration of snowmobiling in Yellowstone led eventually to an air pollution problem. In 2002, for example, the Park Service provided employees in certain areas of the park with respirators to combat poor air quality attributable to snowmobiles. The Park Service estimates that snowmobiles produce up to 68 percent of Yellowstone's carbon monoxide pollution and 90 percent of the hydrocarbon emissions. *See* NATIONAL PARK SERVICE, AIR QUALITY CONCERNS RELATED TO SNOWMOBILE USAGE IN NATIONAL PARKS (2000).

So the question has been posed, should snowmobiles be allowed in Yellowstone and other national parks? Environmental protection interests argue that mass transit snow coaches can deliver people into the park's winter wonderland efficiently and with far less ecological impact. Snowmobile interests defend their "right" of personal access and argue that technological improvements in snowmobile performance can solve the problem. This conflict festered through the 1990s and into this decade, with policy flip-flops and rhetorical assaults characteristic of the many other ecosystem management issues studied in this text. As it prepared to make a final decision by winter of 2002, the Park Service summarized the history of the conflict in the following publication.

Department of the Interior, National Park Service, Special Regulations; Areas of the National Park System.

67 Fed. Reg. 15145 (Mar. 29, 2002).

In 1990 a Winter Use Plan was completed for Yellowstone National Park (YNP), Grand Teton National Park (GTNP), and the John D. Rockefeller, Jr., Memorial Parkway (the Parkway). In 1994 the National Park Service (NPS) and U. S. Forest Service (USFS) staff began work on a coordinated interagency report on Winter Visitor Use Management. This effort was in response to an earlier than expected increase in winter use. The 1990 Winter Use Plan projected 143,000 visitors for the year 2000. Winter visitors to YNP and GTNP in 1992–93 exceeded this estimate. Total visitors to YNP and GTNP in that year were, respectively, 142,744 and 128,159.

In 1994 the Greater Yellowstone Coordinating Committee (GYCC) composed of National Park Service Superintendents and National Forest Supervisors within the Greater Yellowstone Area (GYA), recognized the trend toward increasing winter use and identified concerns related to that use. The GYCC chartered an interagency study team to collect information relative to these concerns and perform an analysis of winter use in the GYA. This analysis, Winter Visitor Use Management: a Multi-agency Assessment, was drafted in 1997 and approved by the GYCC for final publication in 1999. The assessment identifies desired conditions for the GYA, current areas of conflict, issues and concerns, and possible ways to address them. The final document considered and incorporated many comments from the general public, interest groups, and local and state governments surrounding public lands in the GYA.

In May 1997, the Fund for Animals filed suit against the National Park Service (NPS). The suit alleged that the NPS had failed to conduct adequate analysis under the National Environmental Policy Act (NEPA) when developing its winter use plan for the areas, failed to consult with the U. S. Fish and Wildlife Service on the effects of winter use on threatened and endangered species, and failed to evaluate the effects of trail grooming on wildlife and other park resources. In October 1997, the Department of Interior (DOI) and the plaintiffs reached a settlement agreement. Under the agreement, the NPS agreed, in part, to prepare an environmental impact statement (EIS) for new winter use plans for the parks and the parkway. This settlement provision was satisfied with publication and distribution of the final environmental impact statement (FEIS) on October 10, 2000. A record of decision (ROD) was signed by Intermountain Regional Director Karen Wade on November 22, 2000 and subsequently distributed to interested and affected parties. The ROD selected FEIS Alternative G, which eliminates both snowmobile and snowplane use from the parks by the winter of 2003–2004, and provides access via an NPS-managed, mass-transit snowcoach system. The decision was based on a finding that existing snowmobile and snowplane use impairs park resources and values, thus violating the statutory mandate of the NPS.

Implementing aspects of this decision relating to designation of routes available for over-snow motorized access required a rule change for each park unit in question. Following publication of a proposed rule and the subsequent public comment period, a final rule was published in the *Federal Register* on January 22, 2001. The rule became effective on April 22, 2001. Full implementation of the plan and the rule changes do not occur until the winter of 2003–2004.

The Secretary of the Interior and others in the Department of the Interior and the National Park Service were named as defendants in a lawsuit brought by the International Snowmobile Manufacturers Association and several groups and individuals. The State of Wyoming intervened on behalf of the plaintiffs. The lawsuit asked for the decision, as reflected in the ROD and final rule, to be set aside. The lawsuit alleged that NPS failed to give legally mandated consideration to all of the alternatives, made political decisions outside the public process and contradictory to evidence and data, failed to give the public appropriate notice and participation, failed to adequately consider and use the proposals and expertise of the Cooperating Agencies, failed to properly interpret and implement the Parks' purpose, discriminated against disabled visitors, and improperly adopted implementing regulations. A settlement was reached on June 29, 2001, and, through its terms, NPS is acting as lead agency to prepare a Supplemental Environmental Impact Statement (SEIS). In accordance with the settlement, the SEIS will incorporate "any significant new or additional information or data submitted with respect to a winter use plan." Additionally, the NPS will consider new information and data submitted regarding new snowmobile technologies. A Notice of Intent to prepare a Supplemental Environmental Impact Statement was published in the *Federal Register* on July 27, 2001 (FR 39197).

As a term of the settlement, the State of Wyoming was designated as a "cooperating agency" for the development of the Supplemental EIS. Subsequent to the settlement, all other agencies that signed cooperating agency agreements during the earlier EIS process agreed to be cooperating agencies for the SEIS. These agencies are: the U. S. Forest Service, the States of Montana and Idaho, Fremont County, Idaho, Gallatin and Park Counties in Montana, and Park and Teton Counties in Wyoming. In addition, the Environmental Protection Agency (EPA) was invited to be a new cooperating agency in this effort.

The NPS determined that the preparation of a Supplemental EIS will further the purposes of the National Environmental Policy Act, which includes soliciting more public comments on the earlier decision and alternatives and considering new information not available at the time of the earlier decision. The purpose of this rule is to postpone the implementation of existing snowmobile regulations in Yellowstone National Park, the John D. Rockefeller, Jr., Memorial Parkway, and Grand Teton National Park for one year because the NPS has not had sufficient time to plan for and implement the NPS-managed, mass-transit, snowcoach-only system outlined in the existing Winter Use Plan and to complete the Supplemental EIS.

Notes and Questions

1. The Park Service released its Supplemental EIS in February 2002, with comments received through May 2002 and a final decision expected by January 2003. In the SEIS, the agency outlined three general alternatives:

> Alternative 1—Implement the current rule allowing access only by snowcoaches, phasing out snowmobiles over a period of several years.

> Alternative 2—Allow non-guided snowmobile access, but place a daily cap on usage and phase in new emission and noise standards.

> Alternative 3—Allow only guided snowmobile tour access, using snowmobiles equipped with the best available technology for emissions and noise, and encourage visitor use of snowcoaches.

Are any of these alternatives likely to satisfy all the interests weighing in on the question? Which alternative would you recommend the agency adopt, and why?

2. Two other hot-button "recreation vs. environment" controversies on federal public lands involve the regulation of "personal water craft," *see, e.g.*, 66 Fed. Reg. 35661 (July 6, 2001) (Bureau of Land Management restrictions on use in designated area), and management of ski areas, many of which are licensed to operate in national forests, *see* James Briggs, *Ski Resorts and National Forests: Rethinking Forest Service Management Practices for Recreational Use*, 28 B.C. ENVT'L AFFAIRS L. REV. 79 (2000). In chapter 8 we explore the increased emphasis on ecosystem management in national forests, which undoubtedly would influence the location, expansion, and operation of ski areas. In 1994 a federal ecosystem management task force recommended that the Park Service also adopt ecosystem management as its guiding policy principle. *See* NATIONAL PARK SERVICE, ECOSYSTEM MANAGEMENT WORKING GROUP OF THE RESOURCE STEWARDSHIP TEAM, ECOSYSTEM MANAGEMENT IN THE NATIONAL PARK SERVICE (1994). Would the "nonimpairment" mandate of the Organic Act provide a foundation for doing so? *See* William J. Lockhart, *New Nonimpairment Policy Projected for the National Park System*, 30 Envt'l L. Rep. (ELI) 10704 (2000).

PART IV

GLOBAL BIODIVERSITY

CHAPTER **13** Domestic Laws

CHAPTER **14** International Law

Most of our consideration of biodiversity so far has been limited to the United States, a country blessed with a remarkable array of ecosystems of species. Of course, it is hardly the only nation to enjoy those benefits. Biodiversity thrives throughout the world. Africa is home to elephants, lions, the plains of the Serengeti, and the abundant biodiversity that is native to Madagascar. Asia hosts pandas, tigers, and the dense forests of Thailand and Vietnam. South America is best known for the plants, birds, and wildlife that live in the Amazonian rainforest. Australia is home to many unique creatures—such as kangaroos, Tasmanian devils, and koalas—and to the ecosystem of the Great Barrier Reef. Much of Europe's original biodiversity has disappeared, but it is still home to many notable species and ecosystems. Even Antarctica contains a fragile yet dramatic ecosystem featuring phytoplankton, krill, marine mammals, and most of the world's fresh water. And the world's oceans are teeming with life ranging from tiny plants to whales.

Much of this global biodiversity is threatened, for the same reasons that biodiversity is threatened within the United States. Pollution, disease, predation, exotic species, and many other causes play a role in the disappearance of native ecosystems and species around the world. But the leading threat to biodiversity is habitat destruction. Indeed, habitat destruction is a greater threat in Asia, Africa, and South America than it is in the United States because of the combination of the rapidly growing human population and rising standard of living of those areas. The consequence is that much land that was once simply habitat for all types of wildlife and plants is now being sought for human residences, agriculture, and commercial development.

Generally, there are three ways of using the law to protect global biodiversity. First, each country could apply its own laws to protect the biodiversity within its borders. Second, a country could try to apply its own environmental laws to actions affecting biodiversity elsewhere around the world. Third, international law could be developed to create similar legal standards applicable to all countries. Consider the advantages and disadvantages of each approach as we examine the protection of global biodiversity in the following two chapters.

CHAPTER 13

DOMESTIC LAWS

Chapter Outline
A. Reliance on Each Country's Own Domestic Laws
   1. China
   2. Australia
   3. Latin America and the Caribbean
B. Extraterritorial Application of Domestic Environmental Laws

A. RELIANCE UPON EACH COUNTRY'S OWN DOMESTIC LAWS

Every nation has laws protecting biodiversity. Some laws seek to manage activities occurring in particular ecosystems, such as forests, coastal areas, or wetlands. Other laws regulate the time, place, and manner of hunting, fishing, logging, and the commercial exploitation of biodiversity. A smaller number of nations have laws specifically protecting endangered or threatened species or ecosystems. And a small, but growing, number of laws protect biodiversity as a whole.

The specific content of the domestic laws concerning biodiversity varies greatly from country to country. Poland prohibits the destruction of ant-hills in forests. The Bangladesh Fish Act provides seasonal protection to fish and their habitats. Germany requires ten percent of its land to be designated as interconnecting habitat conservation areas to protect biodiversity. Vietnam protects its forests by requiring the use of wood substitutes. Iceland allows only one unwieldy twelve-foot long tool, called a hafur, to be used to capture puffins. South Africa prohibits development that disturbs ecosystems or leads to a loss of biodiversity. Many nations have established nature preserves to protect their notable ecosystems and species from outside threats. Alas, the existence of such laws does not necessarily guarantee the actual preservation of a nation's biodiversity. Enforcement is a serious problem for many national and local governments. Financial resources and technical expertise limit the effective management of ecosystems.

So can individual countries enact and enforce adequate laws to protect the biodiversity that exists within their borders? Are some countries more capable of protecting their biodiversity than other countries? When should the international community, or other nations, take steps to protect the species, ecosystems, and genetic diversity within another country that is unable to accomplish such protection itself? Consider the answers to those questions in light of the following case studies from China, Australia, Latin America, and the Caribbean.

1.   CHINA

China is a vast, varied nation that hosts an incredible range of ecosystems and species. Most famously, it is the only home of the giant panda, the symbol of many efforts to protect biodiversity throughout the world today. But China is also the home for more than 1,250,000,000 people. China's rapid economic growth further challenges efforts to preserve ecosystems, species, and genetic resources. China's efforts to address the needs of its biodiversity are reflected in two documents prepared by the Chinese government with the assistance of the United Nations Environment Programme (UNEP): CHINA: BIODIVERSITY CONSERVATION ACTION PLAN (1994), and CHINA'S BIODIVERSITY: A COUNTRY STUDY (1997), *available at* <http://english.zhb.gov.cn/biodiv/state_dimp_en/country_study.html>. These documents provide the best overview of the status of and efforts to protect biodiversity within China, and they are complemented by countless additional materials, only some of which are cited in the following account.

China's biodiversity

"China's biodiversity ranks eighth in the world and first in the northern hemisphere." CHINA'S AGENDA 21: WHITE PAPER ON CHINA'S POPULATION, ENVIRONMENT, AND DEVELOPMENT IN THE 21ST CENTURY 171 (1994). Over 100,000 species of animals and nearly 33,000 plant species exist in 460 different types of ecosystems. Those ecosystems include forests, grasslands, deserts, wetlands, seas and coastal areas, and agricultural ecosystems. China hosts 212 different types of bamboo forests alone. China also has an unusual number of ancient and relic species because of its protection from historic geologic events such as the movement of glaciers. Such species and ecosystem diversity is complemented by an unsurpassed collection of genetic diversity. "The richness of China's cultivated plants and domestic animals are incomparable in the world. Not only did many plants and animals on which human survival depend originate in China, but it also retains large numbers of their wild prototypes and relatives." China's National Report on Implementation of the Convention on Biological Diversity: The Richness and Uniqueness of China's Biodiversity, *available at* <http://english.zhb.gov.cn/biodiv/state_con_en/china_biodiv_den.htm >.

The biodiversity of China has encountered countless threats for thousands of years, including the cultivation of more and more land for agriculture and the consequences of numerous wars. During the Great Leap Forward of 1958 to 1960, Mao Zedong targeted the "Four Pests": rats, sparrows, flies, and mosquitoes. The attack on sparrows enlisted schoolchildren to knock down nests and to beat gongs so that the sparrows could not find a place to rest. Only after sparrows were virtually eliminated throughout China did the country's leaders recognize the value of the birds in controlling insects. China faces many of the same threats as biodiversity in other countries, with the notable addition of the country's notorious air pollution. Habitat loss is the biggest threat to biodiversity in China. As in many other countries, rapid economic development and continued population growth exert relentless pressure on previously undeveloped areas that offered habitat to a diversity of wildlife and plants. Overgrazing of rangelands, erosion, and the adverse effects of tourism and mining further compromise the condition of ecosystems and species throughout China.

Forests have suffered an especially devastating toll throughout China. For years, China encouraged the wholesale destruction of forests for their timber—which was the country's primary fuel until coal recently replaced it—or simply the removal of trees to facilitate agricultural crops. Trees were cut indiscriminately in a planned effort to generate revenue for local education, health and infrastructure needs. As one villager remembered:

> When I was a child, there were jackals and foxes in the woods, but after the big trees were cut to fuel furnaces during the [Great Leap Forward], there wasn't even a rabbit. New trees grew, but then it was time to "learn from Dazhai." In fact, we didn't need terraces in our area, because the population was sparse. But our per-*mu* production was considered low. So we had to cut the trees. Whoever cut the most got the most political points, and the most gain.

JUDITH SHAPIRO, MAO'S WAR AGAINST NATURE: POLITICS AND THE ENVIRONMENT IN REVOLUTIONARY CHINA (2001). Another writer recalled when he was sent to Fuyuan County in northeastern China "charged with 'opening the wilderness' to convert the land to farmland and with making preparations for war with the Soviet Union:"

> During my five years in the county, beginning in 1969, when I arrived, the old forest was almost completely cut down and some animals were wiped out. One was an amphibious muskrat. During my second year, two hunters and a dog stayed with us. Every day they came back with a huge bag of animals that they were going to sell for two or three *yuan* each. The Army Corps only gave us 36 *yuan* a month, so that was a lot of money to us. So we all went out to hunt the muskrats, and after a few years, they were pretty much all gone. They were easy to hunt because they would stand up and cry out. I caught three. The disappearance of the muskrats affected the whole ecosystem. They eat fish; foxes eat them. It was a change in the whole food chain. Serious harm was done. But their fur was good.

Id. at 167. Fires and pests further degraded forest ecosystems. The result was that forest cover in the lush provinces of southwest China declined from 30% of the land in 1950 to 13% by 1999. The loss of forests, in turn, caused deadly flooding along the Yangtze River and devastated the natural ecosystems and the species within them. Tigers, for example, "stalk their prey from the cover and the shadow provided by forests. The relationship is pretty simple: no forests, no tigers." ROBERT B. MARKS, TIGERS, RICE, SILK, AND SILT: ENVIRONMENT AND ECONOMY IN LATE IMPERIAL SOUTH CHINA 323 (1998). Forests continue to disappear at an alarming rate, with the remaining forests often broken into smaller, fragmented areas. *See, e.g.*, THE ROOT CAUSES OF BIODIVERSITY LOSS 153–182 (Alexander Wood, Pamela Stedman–Edwards & Johanna Mang eds.) (2000) (chapter describing the loss of biodiversity in the forested areas of Deqin County in northern Yunnan Province and Pingwu County in northern Sichuan Province).

Other types of ecosystems confront similar threats. Overgrazing, farming, and plagues of rodents have caused the grassland steppes that account for one-third of China's total area to lose up to half of their grass yields in the past twenty years. Over seven million hectares of wetlands were reclaimed during the past thirty years. Once known as "the province with a thousand lakes," Hubei Province now has only 326 lakes and rivers left. Lime mining and handicraft production by local residents have damaged eighty percent of the coral reefs along the coast of Hainan Island. The overall result is that "continued destruction and deterioration of ecosystems has now become one of the most serious environmental problems in China." CHINA: BIODIVERSITY CONSERVATION ACTION PLAN, *supra*, at 10.

Biodiversity is also threatened by the direct exploitation of many species. "Plants are cut for fuel, building materials, food and medicine. Birds, mammals, reptiles, fish and many invertebrates are hunted and fished virtually everywhere they are available." *Id.* at 13. Commercial trade in wildlife is another serious threat. China is the world's largest exporter and a leading user of endangered species. Enforcement becomes even more difficult because of the huge demand for products derived from endangered species. Traditional Chinese medicine uses tiger bones (for arthritis and rheumatism), rhino horns (for fevers), and bear gall bladders. Nearly every tiger part is used as a tonic, an aphrodisiac, gourmet delicacies or some other purpose. Chinese pharmaceutical factories use 1,400 pounds of rhino horns annually, the product of about 650 rhinos. Panda pelts sell for as much as $10,000 and tiger bones are priced at $500 per pound, and a rhino horn can earn as much as $45,000. Villagers can earn ten years income from one tiger.

These pressures are evidenced in the placement of three native Chinese species among the World Wildlife Fund's list of the top ten most endangered species in the world. The giant panda is the most famous of those three species. Only one thousand pandas are left in the wild, and their numbers are still declining, albeit at a reduced rate. The threats to their survival include the loss of bamboo and habitat, a relatively small number of young pandas, genetic inbreeding, inability to survive in captivity, and poaching. The second species–the black rhinoceros–has suffered a 95% drop in population since 1970 so that only 2,000 are alive today. The third species–the Indo–Chinese tiger–is the most endangered. Estimates of the number of Indo–Chinese tigers alive in the wild range from 50 to 500, and with two of the four native Chinese tiger species already extinct, many fear that this tiger could disappear by the end of the century. The disappearance of native species is obvious in other ways as well. The town of "Wild Yak Gully now has no wild yaks; Wild Horse Sands, no wild horses," and the Town of Moose and the Town of Gazelle have no moose or gazelles. *Id.* at 15. Other notable Chinese species that are endangered include the Yangtze alligator, the white flag dolphin, the crested ibis, and certain Mongolian horses.

China's genetic diversity has suffered as well. "As habitats become fragmented and the total numbers of individuals within a species reduced, the genetic base becomes increasingly narrow. There is less and less opportunity to avoid inbreeding and to maintain genetic diversity.... In the case of domestic species, the problem is aggravated by intentional selective breeding for a narrow set of desired characteristics." *Id.* at 12. Genetic diversity among plants has dropped as rapid economic development along the coast and agricultural innovations have led to the disappearance of many traditional strains of native plants and crops.

China's efforts to protect its biodiversity

China's primary response to the threat to its biodiversity has been the creation of nature reserves. The Dinghushan National Natural Reserve was the first such reserve, established in 1956 in Guangdong Province to protect the subtropical evergreen forests and accompanying rare plants and animals. By 1999, 1,146 reserves covered 8.8% of China's land, with plans to expand the system to 1,200 reserves covering 10% of the land by 2010. Fourteen of the 124 national nature reserves are specifically identified as biosphere reserves. Thirteen of those reserves were for pandas, with plans for at least a dozen more panda reserves and corridors to connect the growing number of scattered reserves. One such reserve covers 45,000 square kilometers and protects 60 endangered animals and 300 rare plants. By contrast, efforts to establish a tiger reserve have failed to date because of the huge amount of land required by wild tigers, the lack of acceptable sites, and the ignorance about the precise needs of tigers. Forest ecosystems are well represented in the nature reserves. Wetland and coastal ecosystems have been included in reserves since the 1970's, while the creation of reserves for grassland and desert ecosystems are a new priority for the government.

The nature reserves, however, do not solve all of the problems faced by China's biodiversity. Most reserves are simply no hunting zones, not affirmative wildlife management areas. For example, over 15,000 people live in ninety villages within Xishuangbanna Nature Reserve in southwestern Yunnan Province, where "they engage in agriculture, forestry, animal production, fisheries, and small-scale retailing and commercial activities." CLEM TISDELL, BIODIVERSITY, CONSERVATION AND SUSTAINABLE DEVELOPMENT: PRINCIPLES AND PRACTICES WITH ASIAN EXAMPLES 147–48 (1999). More generally, "[s]ome engineering projects go on even in the core areas of nature reserves. In other reserves or scenic spots, tourism is promoted to develop the local economy, and while tourism can assist conservation when it is carried out properly, the prospects for quick profits may lead to abuses of the natural systems and species which the reserves protect." CHINA: BIODIVERSITY CONSERVATION ACTION PLAN, *supra*, at 21. Additionally, "illegal hunting and poaching of endangered animal and plant species occurs frequently" in reserves. *Id.* There is no general law regulating the operation of nature reserves. Management difficulties and inadequate funding also threaten many reserves. Reserve administrators and employees are often untrained to protect the species in their care. Most reserves do not even possess a list of species that live there.

The Chinese government is aware of these shortcomings, though, and it has charted an ambitious program to improve the effectiveness of nature reserves in protecting the country's biodiversity. Proposed actions include restrictions on free access to sensitive reserves, better pay and living conditions for reserve personnel (including allowances for families to live in nearby cities), efforts to "improve relations with local people and find ways for them to make a living without depleting the natural resources," and the establishment of new nature reserves "in regions with urgent need of biodiversity conservation," such as the coral reefs of Dongshan Island and seven proposed reserves to conserve wild rice, soybeans, and other agricultural crops. *Id.* at 36–40.

The nature reserves are joined by zoos, botanical gardens, and scientific study institutes. China's 28 zoological gardens and 143 zoological exhibition sites contain more than 600 species of animals. Over 13,000 species of plants are contained in more than 100 botanical gardens. Over one thousand scientists work together through the Chinese Research Network of Ecosystems to study and monitor ecosystem diversity. Genetic diversity is protected by "the world's largest resource bank of different varieties of crops, a number of gene and cell banks and 25 germ-plasm nurseries, which hold a total of 350 thousand specimens of germ-plasm for various species of trees and crops." CHINA'S AGENDA 21, *supra*, at 173.

Educational campaigns serve as another primary feature of China's efforts to protect its biodiversity. China has traditionally relied on exhortational campaigns to change people's conduct. China's biodiversity conservation action plan begins with an emphasis on the need "[t]o enhance the nation's awareness of the critical importance of our biodiversity and its conservation is our urgent task of the highest priority." CHINA: BIODIVERSITY CONSERVATION ACTION PLAN, *supra*, at ii. Such an educational focus appears in China's Agenda 21 plan, which calls for media teaching about biodiversity, the promotion of public events such as Earth Day and Bird Loving Week, and the use of a traveling Panda Exhibition. China also held a "National Program for Environmental Education and Publicity" that drew upon the resources of such organizations as the government's environmental departments, the Ministry of Broadcasting and Television, and the Chinese Communist Youth League. One recent program to protect the five thousand remaining *grus nigricollis*–a rare type of crane–is designed to "make the youth conscious of animal protection before they become poachers." Such efforts have helped convince 99% of the Chinese people that environmental pollution and ecological destruction are at least "fairly serious" issues. In particular, anyone who harms a panda must face "the censure of an angry public."

Yet all agree that more environmental education needs to be done. The greatest problem exists in rural areas where people ask why wild animals can no longer survive on their own, and where menus proclaiming "Rare Wild Animals Are Served" still appear in restaurants and hotels. The demand for the products of endangered species remains high. Years of teaching traditional Chinese medicine and delicacies is hard to reverse.

How do you convince a billion people to take aspirin instead of rhino horn pills? "Many Chinese still believe that wildlife species are endowed with magical powers capable of curing a myriad of ills, and are angered by pressure from countries such as the United States to ban the sale of endangered species." Daniel C.K. Chow, *Recognizing the Environmental Costs of the Recognition Problem: The Advantages of Taiwan's Direct Participation in International Environmental Law Treaties*, 14 STAN. ENVTL. L.J. 256, 299 (1995). Likewise, many still see tigers as pests, just as many ranchers fear the introduction of wolves and bears into the western United States. More generally, "[b]iodiversity conservation is a new technical term for many officials in the governments at all levels and for citizens who are lacking basic knowledge on biodiversity conservation." CHINA: BIODIVERSITY CONSERVATION ACTION PLAN, *supra*, at 33.

The biodiversity conservation action plan reveals a keen understanding of the importance of gaining public support for the task at hand:

> In general, people want government policies that do not require them to change their lifestyles, provide material benefits and development, and provide benefits today that will be paid for later. Politics to conserve biodiversity would be the opposite, requiring fundamental changes in people's relationship with the environment, restricting access to resources, foregoing material benefits, and paying today for abstract future benefits. Unless the public is convinced of the value of conserving biodiversity, and the government changes its policies accordingly, the chance of saving biodiversity is small.

Id. at 60. Thus the Chinese government seeks to help the media better publicize the importance of biodiversity conservation, "[w]ork with local theater groups to write and perform plays with a biodiversity message," and teach students of all ages about biodiversity in the nation's schools. *Id.* at 60–62.

Neither China's emphasis on nature reserves nor its use of educational campaigns actually regulates any conduct that threatens biodiversity. The development of Chinese wildlife law mirrors the development of Chinese environmental law (and indeed Chinese law) generally. Interest in the environment and interest in law both lagged until the 1970's, so not surprisingly, there was little Chinese environmental law. The People's Congress approved the Law on Environmental Protection—the first general Chinese environmental statute—in 1978. Article 15 of that law prohibits hunting and exploitation of rare wildlife. Then, in 1982, several provisions regarding environmental protection were added to China's constitution. Article 9 provides for state ownership of natural resources, ensures state protection of natural resources, and prohibits appropriation or damage of natural resources. Article 26 provides that "the State protects and improves the living environment and the ecological environment, prevents and remedies pollution and other public hazards." By 1994, China had enacted twelve national statutes, twenty national administrative regulations, over six hundred local laws and regulations, and three hundred other norms regulating the environment.

Chinese biodiversity law has developed in much the same fashion. The Forestry Law prohibits the hunting of animals in protected areas. The Water Law provides that the government "shall protect water resources and adopt effective measures to preserve natural flora, plant trees and grow grass, conserve water sources, prevent and control soil erosion and improve the ecological environment." The Grassland Law directs the government to protect grassland ecosystems, vegetation, and rare plants, and it prohibits harmful reclamation and construction activities. Genetic diversity is protected by Plant Variety Regulations, though China may need to strengthen those regulations to comply with the International Convention for the Protection of New Varieties of Plants (UPOV). *See* Lester Ross & Libin Zhang, *Agricultural Development and Intellectual Property Protection for Plant Varieties: China Joins the UPOV*, 17 PAC. RIM. L. J. 226 (1999).

The newest Chinese law seeks to abate the transformation of once fertile grassland ecosystems into lifeless deserts. Nomadic herders have lived in the grasslands of what is now Inner Mongolia for countless generations, but the 1950's brought a wave of Chinese immigrants adding more livestock and seeking to cultivate the naturally arid land bordering the Gobi Desert. Today, expanding desertification claims 2,500 square kilometers at a cost of $6.5 billion to China's economy each year. The effects of the dust have been seen as far away as Colorado, where particulate concentrations rose above permissible levels in April 2001 after the jet stream carried the dust all the way from China. In March 2002, another dust storm dumped 30,000 tons of dirt on Beijing, even as billboards around the city trumpeted the "Green Olympics" to be held there in 2008. The resulting international publicity prompted local television newscasters to affirm the government's resolve to "outwit" the dust storms. The first law to try to match wits with the dust was enacted by the National People's Congress (NPC) in August 2001. The law against desertification:

- States that land occupants have a duty not only to prevent desertification but also to restore areas that have already become desert;
- Promises unspecified preferential policies, tax breaks, subsidies and technical support to offset the cost of this unfunded mandate;
- Creates a new class of protected areas off-limits to development and calls for farmers and herders to be removed from those areas; and
- Authorizes local governments to grant land-use rights of up to 70 years to desertified areas if the landholder promises to undertake restoration efforts.

China Adopts Law to Control Desertification: A November 2001 Report from U.S. Embassy Beijing, *available at* <http://www.usembassy-china.org.cn/english/sandt/desertification_law.htm>. As Qu Geping, the chair of the NPC Environment and Resources Committee, explained, the anti-desertification law was designed to prevent the frequent dust storms that have sounded "a warning bell from nature."

Endangered wildlife is also protected by Chinese law. The Ministry of Forestry established the first list of Rare and Precious Species of China in 1969. A 1983 State Council circular orders that "[a]ll economic activities that affect the breeding and survival of endangered wildlife in their main nesting area should be banned." A general wildlife protection law was enacted in 1989. As of 1990, the List of State's Mainly Protected Wild Animals contained 96 animals; killing any species on that list is prohibited. Over a hundred other species appear on the list of animals to be protected by provincial and local governments; those governments are beginning to enact their own wildlife laws. Most recently, China tightened its wildlife laws in 1993, bowing to pressure from the United States and international environmental groups.

The existence of such laws is one thing; their actual implementation is another. To be sure, there are examples of very stringent enforcement of wildlife laws in China. The government recently executed numerous poachers for killing endangered pandas. Five poachers were sentenced to death in 1994 after they killed sixteen elephants in a nature reserve in Yunnan province and then engaged in a fierce gun battle with police. In 1995, nineteen hotels and restaurants on Hainan Island were closed and fined $34,000 for serving bear's paw, monkey brains and other wildlife. China has promised to step up such efforts to punish those who kill endangered species for financial gain. China has also acted to prohibit patented medicines from containing ingredients taken from endangered species. *See* John Copeland Nagle; *Why Chinese Wildlife Disappears as CITES Spreads*, 9 GEORGETOWN INT'L ENVTL. L. REV. 435 (1997). But the Chinese government admits its failure to adequately enforce the existing laws protecting biodiversity:

> While many laws and regulations intended to protect biodiversity exist, in practice they are often not enforced or enforced strictly, or when the violators are apprehended, the court system treats them very leniently. As a result, illegal hunting and collection of endangered animal and plant species is very widespread, and disputes arise continuously between management of nature reserves and local residents, hindering biodiversity conservation efforts.

CHINA: BIODIVERSITY CONSERVATION ACTION PLAN, *supra*, at 32.

A final part of China's biodiversity strategy is its active participation in international efforts to protect biodiversity. In 1980, China joined the Convention on International Trade in Endangered Species (CITES, which is discussed below at page 873). In 1992, it signed the Ramsar Convention for the protection of wetlands. That year also saw China become one of the first nations to ratify the Convention on Biological Diversity that was negotiated in Rio de Janeiro (discussed below at page 883). China then launched a "China Biodiversity Conservation Plan" in 1994, and it discussed the measures needed to protect biodiversity in its white paper documenting China's efforts to further its Agenda 21 environmental commitments. The Agenda 21 strategy states that "[t]he policy for biodiversity conservation in China is 'laying equal stress on both the development and utilization and the conservation and protection of natural resources' and 'he who develops, conserves; he who utilizes, compensates; he who destroys, restores.' " CHINA'S AGENDA 21, *supra*, at 171–72.

But critics question China's resolve to end its trade in endangered species. China resisted international calls for the destruction of existing rhino horn stocks. It declined to become a member of the Global Tiger Forum established by twelve Asian countries in 1994 to protect endangered tigers throughout Asia. It advanced a proposal that would create a farm to raise tigers in order to satisfy the demand for tiger parts, though that idea was withdrawn after environmentalists objected. Its limited efforts to stop that trade have subjected China to international criticism. For example, in 1993 the United States and other countries threatened to sanction China for failing to control the trade in tiger and rhino parts. That the United States decided not to penalize China was viewed as an exercise in diplomacy unrelated to China's actual progress in enforcing the treaty. China's efforts to protect its ecosystems suffer from similar limitations on resources and political will. As one observer writes, China's solid national biodiversity policy "has made very little difference to the peoples of southwest China, where many of the reserves lack staff, funds, infrastructure, or a management plan. The international conservation community has focused on the panda at the expense of other species." John Studley, *Environmental Degradation in China*, China Review (Spring 1999), *available at* <http://www.gbcc.org.uk/studley12.htm>.

Notes and Questions

1.   The Chinese government's biodiversity conservation action plan states that "the overall objective of biodiversity in China is to set in place as soon as possible measures for avoiding further damage, and, over the long term, for mitigating or reversing the damage already done." China: Biodiversity Conservation Action Plan, *supra*, at 35. It then identifies several specific objectives designed to achieve that goal: improve the basic research on biodiversity in China, improve the national network of nature reserves and other protected areas, conserve wild species that are significant for biodiversity, conserve genetic resources related to crops and domestic livestock, promote in-site conservation outside nature reserves, establish a nationwide information and monitoring network for biodiversity conservation, and coordinate biodiversity and sustainable development. *Id.* at 35–47. Are these objectives attainable? Are there other goals that China should pursue in its effort to preserve its biodiversity?

2.   Can China protect its biodiversity by itself? Could it do so if it received adequate funding and technical assistance from developed nations? How should other nations or the international community encourage or pressure China to protect its biodiversity?

2.   Australia

In Re Fund for Animals LTD

Administrative Appeals Tribunal, 1986.
9 A.L.D. 622.

JUDGMENTS: Gallop J (Presidential Member), R A Balmford (Senior Member) and D B Williams (Member)

(1) This is an application under para 80(1)(a) of the Wildlife Protection (Regulation of Exports and Imports) Act 1982 (the Wildlife Act) for the review of a decision by the Minister of State for Arts, Heritage and Environment declaring, pursuant to sub-s 10(1) of that Act, a management programme entitled "Kangaroo Conservation and Management in Queensland" to be an approved management programme for the purposes of that Act. . . .

(8) The significance of the approval of a management programme by the Minister pursuant to s10 of the Wildlife Act derives from para 21(b) and sub-para 31(c)(iv) of that Act which, omitting irrelevant parts, read as follows:

21 A person shall not, otherwise than in accordance with a permit or an authority, export— . . .

(b) a specimen that is, or is derived from, a native Australian animal . . .

31 . . . the Minister shall not grant a permit to export a specimen that is, or is derived from, a native Australian animal . . . unless the Designated Authority has advised him that he is satisfied— . . .

(c) where the specimen is an animal specimen other than a live animal, that– . . .

(iv) the specimen is, or is derived from, an animal specimen that was taken in accordance with an approved management programme. . . .

Historical background

. . . (33) There are three major components of the relationship between people and kangaroos in Australia. First, kangaroos are a major pest for the farming community. Second, a substantial kangaroo harvesting industry produces meat and skins for both domestic and export markets. Third, not only in Australia, but in other parts of the world, kangaroos are seen as a significant, and in symbolic terms perhaps the most significant, native Australian animal. As to this, it can be assumed that no-one actually wishes to see any of the species concerned become extinct; and there are certainly people who abhor the killing of any kangaroos at all. Most Australians would probably have feelings lying somewhere between these two extremes.

(34) The three aims of pest control, harvesting and conservation must somehow be reconciled. As to this, the Minister said at a public meeting in Charleville, Queensland, on 18 June 1985:

I am the Minister for the Environment with the responsibility to protect the various species of kangaroos in sufficient numbers throughout their various habitats in Australia.

The government as a whole also has a responsibility to ensure that Australia's primary producers do not have their properties so over-run with kangaroos that their livelihood and Australian prosperity is threatened. It is often difficult to strike a delicate balance between the two objectives. The purpose of the national management plan is to try to ensure that both of these objectives are met.

The kangaroo industry which includes shooters, pet food manufac-turers and the skin and fur traders is seen by the government as a tool in that management plan. We do not and I repeat do not see the industry as an end in itself. That may not please some of you but I would be less than honest if I suggested that I or the government was implementing the management plan to ensure the prosperity of the kangaroo industry.

As the Minister administering the Wildlife Act, the Minister's direct responsibility in respect of kangaroos is, as he says, their protection: but given that the government of which he is a member accepts a responsibility to control kangaroos as a pest, he must, in deciding to approve a manage-ment programme, take account of both responsibilities. It should, however, be noted that the sole object of the Wildlife Act, made very clear in its long title and in s3 is the protection and conservation of Australian wildlife, not the control of pests.

(35) Queensland, like other Australian States, paid bounties on kanga-roo scalps in the nineteenth and early twentieth centuries. Kangaroos were seen, and are still seen, as competing with domestic stock for water and for pasture, damaging fences and destroying crops. In the 40 years prior to 1917, bounties were paid by the Queensland Government on some 26 million scalps. A substantial kangaroo-harvesting industry, based on the sale of skins, then developed. In the late 1950s, the decline of the rabbit population led wholesale butchers, pet food manufacturers and cold-store operators to turn their attention to kangaroo meat. Export and domestic markets were established. However, for lack of quality control, the export market for human consumption soon collapsed, and most kangaroo meat now taken is used in Australia for pet food.

(36) In the 1960s, opposition to the commercial exploitation of kanga-roos developed both in Australia and overseas, particularly in the United States. The Customs (Prohibited Exports) Regulations made under the Customs Act 1901 had for many years prohibited the export of kangaroo products without the prior consent of the Minister for Customs. That consent had been given consistently so as to permit the operation of a substantial export industry. In the year 1960–61, for example, over two and a half million kilograms of kangaroo meat and over half a million skins were exported from Australia. In January, 1973, the then Minister for Customs announced that, from 1 April 1973, he would not permit the export of kangaroo products until advised by the Minister for Environment and Conservation that such exports were not likely to cause the species of kangaroo from which they were derived to become endangered.

(37) Fauna protection groups in the United States had expressed concern that kangaroos might be threatened with extinction as a result of overharvesting. Their efforts led in 1973 to the banning by the United States authorities, with effect from 30 December 1974, of the importation of products from red, eastern grey or western grey kangaroos....

(39) After considering the report of the working party, the Minister for the Environment and Conservation indicated that he would advise the Minister for Customs and Excise to allow the export of kangaroo products when certain requirements were met: namely, that he had received from each State wishing to export kangaroo products, a programme for the management of the kangaroo species involved; that he was satisfied that the programme was being effectively implemented; and that he had agreement from each State that an upper limit to harvesting (i.e., a "culling quota") would be fixed annually, following consultation between relevant State and Commonwealth authorities, and recommended to the Minister for Customs and Excise by the Minister for the Environment and Conservation.

(40) In due course, State management programmes were approved; and exports were permitted, from 21 August 1975 so far as Queensland was concerned, under the Customs (Prohibited Exports) Regulations as before. Negotiations between governments had led to a situation where the United States was prepared to accept imports of kangaroo products from a State if that State's management programme had been certified by the Commonwealth Government as complying with prescribed standards. This was done as the several State management programmes were approved. Exports to the United States then resumed, and have continued, although the matter has not been entirely resolved: it appears from the Minister's speech in June 1985 that at that date the red, eastern grey and western grey kangaroos were still listed as threatened species under the United States Endangered Species Act 1973, and there was no evidence before the Tribunal which would suggest that the situation had changed at the time of hearing. . . .

(43) The Wildlife Act was passed by parliament in 1982 and came into operation on 1 May 1984. The Regulations were made on 4 April 1984. Amendments to the Customs (Prohibited Export) Regulations with effect from 1 May 1984 removed from the Regulations all references, direct or indirect, to, *inter alia*, kangaroo products. The Commonwealth's control over the export of kangaroo products thus passed from the Minister for Customs to the Minister for the Environment and was made formally dependent upon that Minister's approval of management programmes . . .

[The court then held that the 1985 management program for kangaroos in Queensland failed to satisfy the Wildlife Act's requirement for such a management program because "the document as a whole gives the impression of something cobbled together in order to meet a series of requests by the Commonwealth, rather than a carefully drafted and consistent plan for the management of kangaroos in Queensland," and because of "significant discrepancies between the management programme as approved and published"].

PART III—THE SUBSTANCE OF THE MANAGEMENT PROGRAMME

The species

(108) Having reached our decision for the reasons stated, we nevertheless consider that in view of the time and effort devoted at the hearing by

both parties to the substantive merits of the management programme, it is appropriate that we should examine, at least in part, the issues which the parties wished to, and did, debate before us. The five species which are dealt with in the management programme under review are the red kangaroo *Macropus rufus*, the eastern grey kangaroo *M giganteus*, the western grey kangaroo *M fuliginosus*, the *wallaroo M robustus* and the whiptail wallaby *parry*.... All except the whiptail wallaby, the range of which extends only into a small part of northern New South Wales, are found over extensive areas of mainland Australia outside Queensland....

(109) The grey kangaroo was considered to be one species until as recently as 1972 but is now recognized as constituting two distinct species.... The presence of the Western Grey in southern Queensland was recorded in 1982 based on unpublished evidence, but as a "negligible component of the fauna." A more formal record was published in 1984 describing the distribution of the species over an area of 100,000 square kilometres. The authors suggest that "the geographic variation in relative density reported here for *fuliginosus* need not be much different from what it was a century ago." The report states that the two species are difficult to distinguish from the air, though easily distinguished when close at hand. In the field they can be identified separately at up to 100 metres in good light (300 metres with binoculars).

The Regulation

(110) To recapitulate, s10 of the Wildlife Act provides that "the Minister shall not declare a management programme to be an approved management programme unless he is satisfied of certain matters in relation to the programme." Those matters are prescribed by reg 5(1) of the regulations, and the relevant matters for the purposes of this application for review are:

(a) that there is available to the Designated Authority sufficient information concerning the biology of each species subject to the management programme, and the role of that species in the ecosystems in which it occurs, to enable the Designated Authority to evaluate a management programme for that species;

(b) that

(i) [as to imports]

(ii) in relation to permitting the export of specimens taken, or specimens derived from specimens that have been taken, in accordance with the management programme—discussions have been held by the Designated Authority with the relevant body or bodies having powers or duties under any law of the Commonwealth, a State or a Territory for the protection, conservation or management of animals or plants, or of both, subject to the management programme;

(c) after receiving and considering advice from the Designated Authority—that the management programme contains measures to ensure that the taking in the wild, under that management programme, of any specimen—

(i) will not be detrimental to the survival of the species or sub-species to which that specimen belongs; and

(ii) will be carried out at minimal risk to the continuing role of that species or sub-species in the ecosystems in which it occurs and so as to maintain the species or sub-species in a manner that is not likely to cause irreversible changes to, or long term deleterious effects on, the species, sub-species or its habitat; and

(d) after receiving and considering advice from the Designated Authority—that the management programme provides for adequate periodic monitoring and assessment of the effects of the taking of specimens under that management programme on the species or sub-species to which those specimens belong, their habitat and such other species or sub-species as are specified in writing by the designated Authority as likely to be affected by that taking.

(111) We are satisfied, on the evidence before us, that "discussions" within the undemanding meaning of sub-para (b)(ii) have taken place, and we do not propose to consider that aspect of the matter further.

Initial submissions of the applicant

(112) The applicant's principal submission was that the management programme should be set aside on the grounds that the Minister was not, or should not have been, satisfied as to the matters prescribed by reg 5(1). Specifically the applicant submitted ... that the management programme was deficient in the following respects:

(a) There was insufficient information available in terms of reg 5(1)(a) concerning the biology of the whiptail wallaby and its role in the ecosystem in which it occurs.

(b) The inclusion of the western grey kangaroo with the eastern grey kangaroo in the same quota allowed uninformed exploitation of the former species, which was inconsistent with the requirements of reg 5(1)(c)(i) relating to the survival of the species.

(c) There were not adequate measures to ensure the survival of the species concerned and their continuing role in the ecosystem in terms of reg 5(1)(c)(ii), particularly in respect of the eastern grey kangaroo and the wallaroo as records do not provide evidence that it is possible to sustain a harvest at the recommended levels; and

(d) There was insufficient provision for monitoring and assessment of the effects of the culling in respect of any of the species, contrary to the requirement of reg 5(1)(d). . . .

Information as to the whiptail wallaby

(114) As to the applicant's submission numbered (a) above, it was not in issue that less is known about the whiptail wallaby than about the other species subject to the management programme. Comparatively few published papers deal with any aspect of its biology, or its role in the ecosystem in which it occurs. Doctor Southwell has experience of counting whiptails and has a paper on that work ready to submit for publication. It became

apparent from Dr. Kirkpatrick's evidence that officers of the QNPWS were developing information on the ageing and sexing of this species sufficient to enable them to determine the age and sex of dead animals, and the information is being prepared for publication. The ability to age and sex the individuals which are killed is fundamental to any monitoring of a culling programme. It was clear from the evidence, including that of Dr. Southwell, who has extensive experience of the species in its habitat, that the stronghold of the whiptail wallaby is in steep, rocky country accessible to humans only on foot; and that in that country the animals are present in large numbers. It was also clear from the evidence of Dr. Kirkpatrick and of Mr. Houen of the Queensland Graingrowers Association, that whiptails are a major pest to farmers with properties adjacent to that habitat. Accordingly they have been shot as pests in large numbers for many years. . . .

(116) Regulation 5(1)(a) requires that there be "available to the Designated Authority sufficient information . . . to enable the Designated Authority to *evaluate* a management programme" for the species. Whatever "sufficient information" may mean in that context, it does not mean "sufficient information to enable the *preparation* of a management programme" for the species. Assume a species which has been described from a single individual, and which is known to inhabit the Great Victoria Desert, but as to which nothing else at all is known. A management programme, which provides for culling, is proposed for that species. The Director, as Designated Authority, may well, in evaluating that management programme, take into account the insufficiency of the information available as to the biology of the species concerned and its role in the ecosystem in which it occurs, and decide that, because of that insufficiency, no management programme which provides for the culling of that species should be approved. Almost no information is available about the species: but he is still able to *evaluate* the management programme. He has, in effect, decided as a result of his evaluation, that there is insufficient information available to enable the *preparation* of such a management programme. But that is not the question directly posed by the requirement in reg 5(1)(a).

(117) On the basis of the material contained in the preceding paragraphs, the Tribunal finds that there is sufficient information available to the Designated Authority concerning the biology of the whiptail wallaby and its role in the ecosystem in which it occurs to enable him to evaluate a management programme for that species as required by reg 5(1)(a).

Inclusion of the two grey kangaroos in the same quota

(118) We turn now to consider the applicant's submission numbered (b) in para (112), *supra*. Figure 3 [omitted here] indicates the area of Queensland in which the two species of grey kangaroo have been confirmed, comparatively recently, to be co-existing. . . . It was clear from the evidence that the two species are difficult to distinguish in the field. It is likely that the proportion of western to eastern greys in the overlap zone is approximately 1 to 12. Both species are widespread outside Queensland. There was no evidence to suggest that the population of the western grey kangaroo which is found in Queensland is thought likely to constitute a

separate sub-species of the western grey. Doctor Denny's evidence was that the biology of the species in Queensland is almost unknown, but he agreed under cross-examination that he had no reason to suppose that it differed from the biology of the species elsewhere in Australia. There was some evidence that shooters discriminated against western greys, despite the difficulties of identification; it was also suggested in evidence that the dictates of fashion could in the future result in shooters preferring western greys.

(119) The Tribunal noted with considerable concern that the culling quota for grey kangaroos as appearing in the management programme as approved in October 1985, like that appearing in the 1984 programme, did not separate the quotas for eastern and western greys, even though these are different and separate species. The reasons were stated in exhibits and oral evidence to be partly historical, because the separate identity of the two species was confirmed relatively recently, and partly practical, because of the difficulty of identifying the two species separately in the field. The western grey, in the zone where the two species co-exist, is present in considerably smaller numbers than the eastern grey. The Tribunal considered that a management programme in which the quotas for two species were expressly combined was not capable of being evaluated by the Designated Authority in such a way as to enable the Designated Authority to give to the Minister, in respect of the western grey, the advice required to be given by reg 5(1)(c) and (d). Those provisions are concerned with the survival of species as such; and the combination of two species under one culling quota ignores the overriding importance given to that matter by the Regulations.

(120) In this context s25 of the Wildlife Act.... is of considerable significance; the Regulations have clearly been enacted with a view to assisting in the implementation of that provision. The legislation is concerned with species, and there is no escape from a responsibility to identify separate species, despite the difficulties set out in para (109), *supra*

(121) On this point we accept the submission of Mr. Prineas, for the applicant, that the Regulations must be read as relating to the species in respect of the area in which the management programme is to operate. If that were not so, a species with a range extending across several States might well become extinct as a result of each State's implementing a management programme prepared on the assumption that the species was secure in the other States.

(122) Accordingly, the Tribunal finds that, in view of the combination of the two grey kangaroo species in one quota in the text of the management programme, the management programme does not contain measures to ensure that the taking in the wild, under the management programme, of any specimen of the western grey kangaroo will accord with the requirements of reg 5(1)(c).

[The court next held that the Queensland plan also violated the Wildlife Act's provisions requiring the enforcement of separate culling quotas for each species and requiring ongoing scientific monitoring and assessment.]

Regulation 5(1) generally

(127) As the hearing progressed, the Tribunal became more and more convinced that the terms of the Wildlife Act and of the Regulations needed to be interpreted in a manner which allows for a degree of reasonableness as to the extent and depth of knowledge as a prerequisite for a management programme. Regulation 5 rests on the Minister being "satisfied" about a number of different matters, and this leaves a broad discretion in his hands.

(128) Perfect knowledge is an unattainable ideal and it is a matter of fine judgment based on experience and professional insight to enable a Minister to be satisfied that the Designated Authority has sufficient knowledge to evaluate a management programme. Similarly it must remain a matter of judgment to assess what measures will not be detrimental to the survival of the species or may be carried out at minimum risk to its role in the ecosystem. Clearly, for example, there is more information available about the eastern grey kangaroo in Queensland than about the western grey, but this alone does not establish that inadequate information exists for the western grey. Furthermore, experienced scientists and wildlife managers can make use of information about the same species in different environments, or different species in similar environments, which, when properly interpreted, is a useful adjunct to or component of knowledge about the species in the environment in question. Better that than nothing, granted the need to keep collecting data, not all of which is published.

(129) The state of knowledge needs to be interpreted within a time sequence which takes account of the administrative experience, research programme and observations or field experience (including the observations and experience of landholders) which is available in respect of the species in question. In the particular case of kangaroos, it could be argued that information about measures taken with respect to dingoes, including the effects of seasons and of control measures, are significant factors in preserving the role of the kangaroo species in the ecosystem, or even of preserving the species themselves. Yet relatively little was made of this significant factor by any of the parties in this case, although it was referred to by Mr. Houen of the Queensland Graingrowers Association. Similarly, little attention was paid to economic aspects of the management programme in terms of costs of control measures including harvesting, damage to crops and improvements and the effects of variations in the prices of meat and of skins and their acceptability in domestic and overseas markets. Yet it could be validly argued, as it may well be in the future, that knowledge of each of these components is essential as a basis for monitoring the management programme. The information which is available needs to be assessed in the context of the time and the stage of development of knowledge by management authorities responsible for the administration of legislation . . . .

(132) Thus in the interpretation of the Wildlife Act, as a new piece of legislation involving co-operation between the Commonwealth and the States, there is an element of timeliness by which the stage of development of knowledge needs to be assessed. At any one time the level of knowledge about different species, and about different aspects of each individual species, varies significantly and only time itself will enable a more balanced understanding of the total characteristics of each species to be assessed. The management programme on the one hand needs to rest on a degree of uncertainty and imperfect knowledge, and on the other hand, point the way to those gaps in knowledge which can be filled with best effect in improving the management programme. In this respect, the Tribunal considers the mere existence of the Wildlife Act and of the management system conducted under reg 5 gives a degree of choice of action and options which may have been unwise before these arrangements were made, simply because the Act enables controls and corrections to be invoked before irreparable or irreversible change occurs. . . .

Notes and Questions

1. What is the purpose of a management program for kangaroos? As revised in 1990, the National Plan of Management of Kangaroos states three objectives:

- To maintain populations of kangaroos over their natural ranges;

- To contain the deleterious effects of kangaroos on other land management practices; and

- Where possible, to manage kangaroo species as a renewable natural resource providing the conservation of the species is not compromised.

Or, as two Queensland zoologists explain, "[m]anagement of wildlife populations can be divided into three areas: pest control, conservation, and sustained-yield harvesting." They continue, with considerable understatement, that "kangaroo management in Australia has been made difficult by conflicting objectives in addition to divergent community attitudes to kangaroos. . . . Kangaroos are harvested throughout Australia as both a renewable resource and a pest. While the aims for any management program are ultimately value judgements, if they are ambiguous or at least unclear then the outcome or appropriateness of the subsequent management action cannot be judged." TONY POPLE & GORDON GRIGG, COMMERCIAL HARVESTING OF KANGAROOS IN AUSTRALIA (1999), *available at* <http://www.ea.gov.au/biodiversity/trade-use/wild-harvest/kangaroo/harvesting/index.html>. What kind of management plan would best achieve each of those goals?

2. In 1995, the United States Fish & Wildlife Service removed the red kangaroo, the western grey kangaroo, and populations of the eastern grey kangaroo from the list of threatened species under the Endangered Species

Act. The FWS made special note of the success of Queensland in adopting a kangaroo management plan:

> The legislation protecting and conserving nature in Queensland is the Faunal Conservation Act 1974, which has been replaced by the Nature Conservation Act 1992. The new Queensland Act has been implemented for kangaroos, replacing the existing legislation. The Nature Conservation Act 1992 creates classes of protected areas; designates classes of wildlife; and provides for development of conservation plans to protect, use, and manage protected areas, critical habitats, and classes of wildlife. The Queensland kangaroo management program describes how the activities of shooters and dealers are regulated, how the size and/or composition of the population is to be monitored, the harvest regulations and checks to prevent illegal harvest or over-harvest, and other measures to ensure the conservation of the species. The approval of Queensland's kangaroo management program by the Commonwealth Government indicates an assurance that commercialism will not threaten the survival of kangaroo populations throughout their range.

Removal of Three Kangaroos From the List of Endangered and Threatened Wildlife, 60 Fed. Reg. 12887, 12902 (1995). The FWS also noted that the actual quota of kangaroos that can be culled must be approved by both state and federal authorities after receiving "public input from the rural community, pastoralists, graziers, shooters, dealers, the Department of Primary Industries, conservation groups, and politicians whose constituents are impacted by kangaroos." *Id.* at 12900.

3.   Does the killing of kangaroos for meat, leather, and other commercial purposes help or hinder the animals? The ecosystem? The Australian ambassador to the United States insisted that "[r]egrettable as commercial culling might be, it has proved to be the most humane available means of controlling agricultural and pastoral areas." F. Rawdon Dalyrmple, *Australia's Efforts for the Kangaroo*, CHRISTIAN SCIENCE MONITOR, Feb. 23, 1989, at 20. More recently, a South Australian state government report called for increased culling of koalas because they were displacing native species and ecosystem of Kangaroo Island. The report also criticized the management plan for koalas because of the plan's preference for sterilization and translocation instead of culling. Such "high-cost management options," the report explained, "are driven by socio-economic and tourism needs rather than sound ecological management and conservation principles." Catherine Hockley, *Advisory Panel Urges Kangaroo Island Plan; 20,000*, THE ADVERTISER, Oct. 31, 20001. Proponents of culling also insist that the alternative is poisoning and other less humane actions taken by individual ranchers. Nonetheless, the culling of kangaroos troubled the court. As it explained:

> The expression "culling quota" is generally used throughout the material before the Tribunal, and we have accordingly adopted that usage. We would, however, record our view that the process referred to is one of "killing" or "harvesting" rather than culling. The verb "to cull" implies, in our view, a degree of selectivity in the taking of the animals

which is absent from the process of killing hundreds of thousands of kangaroos and the manner in which that killing is performed, which appears from the evidence to be chiefly by spotlight shooting at night. . . . The word "cull" is perhaps intended to carry less emotional content than "kill"; it is, however, certainly less accurate. "Harvest", at least in the context of the industry, can be said to be reasonably accurate. . . .

In re Fund for Animals Ltd., para. 17.

4. The 1982 Wildlife Act is the leading federal statute regulating the commercial exploitation of Australian wildlife. Environment Australia, the federal agency charged with administering the law, explains that the Act "seeks to ensure that wildlife exports and imports take place within a controlled framework that meets the long term needs of all key interest groups. For this reason wildlife trade is regulated to include management and conservation measures that help to ensure that demands are met sustainably."   <http://www.ea.gov.au/biodiversity/trade-use/legislation/index.html>. Note that while the Act regulates commercial trade, it does not contain the kind of prohibition on certain habitat modification that has been controversial in the American Endangered Species Act.

The most recent Australian statute addressing biodiversity is the Environment Protection and Biodiversity Conservation Act 1999 (EPBC Act). The EPBC Act replaced five federal statutes, including the law protecting endangered species. It mandates government assessment and approval of any actions that are likely to have a significant impact on a matter of national significance. The six matters of national significance are the critical habitat of a listed threatened species or ecological community, any of Australia's 53 wetlands of international importance that are protected by the Ramsar Convention, any of the country's fourteen World Heritage properties (such as the Great Barrier Reef and Kakadu National Park), listed migratory species, Commonwealth marine areas, and nuclear actions. Note especially the concept of a "threatened ecological community," which does not have a counterpart under the American ESA. To date, 27 such communities have been listed. For example, the "Eastern Suburbs Banksia Scrub of the Sydney Region" is a plant community that includes tree, shrub and heath species growing in nutrient-poor sand deposits that has been reduced to a few remnants in the eastern and southeastern suburbs of Sydney because of such activities as development, invasion by exotic plants, grazing by horses and rabbits and erosion from vehicle and pedestrian use. The EPBC Act affords additional protections to any listed threatened species and ecological communities. Recovery plans must be implemented for such species and communities, while wildlife conservation plans must be prepared for conservation dependent species, listed migratory species, marine species, and cetaceans. Finally, a "threat abatement plan" must be prepared in response to any listed "key threatening process" that threatens the survival, abundance, or evolutionary development of a native species or ecological community. For more information about the EPBC,   *see*   Environment   Australia's   web   site   at

<http://www.ea.gov.au/epbc/about/index.html#>; Richard B. Stewart, *A New Generation of Environmental Regulation?*, 29 CAPITAL UNIV. L. REV. 21, 160–62 (2001). For detailed accounts of Australian environmental law generally, *see* Robert F. Blomquist, *Protecting Nature "Down Under": An American Law Professor's View of Australia's Implementation of the Convention on Biological Diversity—Laws, Policies, Programs, Institutions and Plans, 1992–2000*, 9 DICK. J. ENV. L. POL. 227 (2000); Kenneth M. Murchison, *Environmental Law in the United States: A Comparative Overview*, B.C. ENVTL. AFF. L. REV. 503 (1995).

5. Kangaroos are only one of the famous examples of Australia's biodiversity. Species such as koalas, Tasmanian devils, emus, and the presumably extinct Tasmanian tiger are known to animal lovers around the world. Many of Australia's ecosystems are equally famous. The Great Barrier Reef contains 350 types of coral and more than 1,500 species of fish, and it is the only living organism visible from space. The Australian outback covers 85% of the country with a hot, arid environment that hosts few people but a surprising group of species, including countless insects, emu, and the desert death adder. The diverse array of ecosystems that exist within the Australian continent also includes the tropical jungles of Kakadu National Park in northern Australia, the rugged Tasmanian wilderness, the alpine settings of the Snowy River and the Snowy Mountains, and the rainforests along the Queensland coast.

The staggering range of ecosystems and species in Australia is reminiscent of the abundant biodiversity in the United States. Both nations, moreover, enjoy stable economic and political systems. What could each country learn from the other about protecting biodiversity?

3. LATIN AMERICA AND THE CARIBBEAN

Lila Katz De Barrera–Hernandez & Alastair R. Lucas, Environmental Law In Latin America and the Caribbean: Overview and Assessment

12 GEORGETOWN INTERNATIONAL ENVIRONMENTAL LAW REVIEW 207, 219–224 (1999).

Notwithstanding the efforts made by Latin America and Caribbean (LAC) countries to enact environmental laws and to implement appropriate environmental regulation, significant problems remain. In several countries, development, particularly resource development, has encountered serious delays and increased costs as a result of uncertainty and lack of consistency in environmental requirements.... This article is based on a study undertaken for the Latin American Energy Organization (OLADE) that covers twenty-six nations of diverse natural, cultural, and economic backgrounds. Countries included are: Argentina, Barbados, Bolivia, Brazil, Chile, Colombia, Costa Rica, Cuba, Dominican Republic, Ecuador, El Salvador, Grenada, Guatemala, Guyana, Haiti, Honduras, Jamaica, Mexico, Nicaragua, Panama, Paraguay, Peru, Suriname, Trinidad & Tobago, Venezuela, and Uruguay. Their varied geography, location, and resource base translates into different patterns of development and use of resources. However, all twenty-six countries share an equal need for growth and a renewed political will to enhance the region's standard of living while taking the necessary steps to protect and manage the environment.

Beyond the pressing need for better living conditions, the LAC countries share other common and recurring problems of particular importance from an environmental point of view. These include the following:

- lack of or old and inefficient infrastructure;

- heavy concentration and migration of population to urban centers;

- centralization of decision making in places often removed from the directly affected areas;

- weak and fractured institutions;

- lack of public awareness of environmental dimensions and consequences of human actions and decisions; and

- lack of human and financial resources to develop and implement adequate environmental management structures at a country wide level.

The problems listed above, should be kept in mind while reading the following overview. . . .

3. Fractured Approach to Protection

Besides protected areas legislation, most [Latin America and Caribbean] LAC legislation deals with individual resources in isolation. The main areas that are the subject of this type of treatment are forests, water, soil, and–in the case of Caribbean and Central American countries–beaches and coastal areas. In a few cases, biodiversity protection may provide a broader, more integral basis for protection. In addition, some particularly sensitive ecosystems may be the subject of special protection; these include mangroves in the Dominican Republic and Venezuela, and wetlands in Costa Rica.

Provisions for the protection of individual resources may be either scattered throughout the environmental and sectoral legislation, or may be embodied in resource specific laws concerning forest, water, and soil conservation. Some of the resource-specific laws in force today may have differing policy origins, including providing for the protection and adequate supply of potable and irrigation water, soil conservation for agriculture, or the protection of the tourism industry. Although these laws are a far cry from the implementation of any form of integral ecosystem protection, they may nevertheless affect development activities.

In order to illustrate the difference between the approach taken in most countries in the region and a more integrated ecosystem-based approach, the following is an overview of the treatment given to some of the main resources dealt with by LAC law.

a.   Forests

This type of legislation generally deals with the resource in three different ways: (1) as a commercial renewable resource; (2) as a soil or watershed protection tool; and (3) as a value in itself. In all three cases, permits may be required for tree cutting and some prohibitions may apply to felling of protected species or in especially vulnerable terrain. Sustainability is generally the guiding criteria for permits under the first category, and land or water protection is the criteria for second category permits. Beyond the legislation on protected natural areas, the protection of forests *per se* may provide one of the strongest expressions of ecosystem protection. As forest laws come closer to the protection of ecosystem integrity, the criteria on exclusions and environmental assessment criteria required for permits and authorizations to conduct activities in such areas becomes stricter. A good example may be found in the Brazilian Forests Code, which provides for a classification of forests in relation to their use and regulates their management accordingly.

b.   Water

Laws and regulations dealing with water generally revolve around ensuring that adequate quality and quantities are available for human consumption. For example, Colombian law gives priority to the use of water for human consumption and provides that all other uses will be prohibited when works might endanger a community. Water legislation also includes discharge and quality standards, allocation of rights and permits, zoning and other restrictions to protect water sources and navigation, and, in some cases, basin management provisions. Generally, these provisions deal with water quite independently from the protection of the natural ecosystem.

c.   Land

Rather than finding land use and soil conservation provisions as part of ecosystem protection schemes, these norms may be found scattered throughout environmental and energy statutes as well as in planning and resource conservation legislation. For instance, in Peru, preservation, conservation, and sustainable use of agricultural land around metropolitan Lima and other cities with over 200,000 inhabitants is declared to be of social and national interest. Changes in use must be declared by law. In Cuba, the use of land entails the obligation to conserve or improve its fertility through use of adequate techniques, production methods, procedures and resources regardless of the purpose for its use. In a similar fashion, protection of beaches and coastal areas may generally be linked to the protection and promotion of the tourism industry. As a result, these provisions are generally directed at protecting the natural scenery through imposing restrictions on coastal uses, works or activities, or preventing littering and coastal contamination.

The region seems to favor individual resource-management legislation rather than the protection of ecosystems in an integrated manner. From a developer's point of view, this highlights the need to know the area and the different kinds of activities conducted in the proposed location in order to be able to assess the competing interests. Developers should scan a broad range of legal instruments to ensure compliance, and to prevent or mitigate conflicts with other uses.

Although the ecosystem impact of development activities in the region may be addressed through the mostly mandatory requirement for environmental licensing and EIAs, generally, it is important to note that only in a few countries, such as Colombia, Ecuador, Guyana, and Mexico, does environmental impact assessment legislation make specific reference to ecosystem protection. Even in those countries, references to ecosystems are generally limited to an EIA requirement that sensitive ecosystems be identified in EIA studies for mitigation purposes.

However, it may be worth noting that despite the generally negligible place taken by the concept of protection of ecosystems *per se* in the LAC environmental legislation, at least two countries, Mexico and Honduras, include ecosystems among the public goods that deserve official protection under criminal law by specifically including harm or damage to ecosystems as a punishable offense.

Notwithstanding the above, integral protection of ecosystems may result from the implementation of protected areas legislation, including laws on national parks and similar legislation. The issue of protected areas is discussed below.

4.   Preservation and Protection of Areas and Species

The LAC region is endowed with the world's richest reserves of biological resources. Protection of biodiversity, genetic resources, and areas of special environmental value are familiar concepts and concerns in the region—where increasing sustainable exploitation of the countries' natural resources (including biodiversity resources) is seen as a convenient route towards much needed economic development. Adherence to basic principles of sustainability and the countries' international commitments may thus be linked to the widespread enactment of framework legislation on protected natural areas–and, to a lesser extent, provisions on protection of species. However, in most cases these laws do not go beyond basic enabling statutes and their implementation is generally deferred, with demarcation and management plans to be completed at a later date. Moreover, provisions making room for exceptions to the application of the basic protection regimes are not uncommon, rendering them particularly vulnerable to competing development interests. It should also be noted that area and species protection are not necessarily intertwined, as they are in U.S. legislation.

Following a pattern that has been highlighted above as endemic to LAC environmental law, protected areas and species statutes generally belong to the youngest generation of environmental legislation, added to or superimposed over older protection instruments. As a result, while this type of legislation may empower the authorities to submit an entire area to a special management and authorization regime, these schemes may be difficult to reconcile with pre-existing ones such as systems of national parks or protected forests and with individual resource-protection laws. In some cases, pre-existing schemes may have been included in–or absorbed by–the different categories of protected areas. Newly created frameworks, such as basin management systems or indigenous reserves, may also interact and/or interfere with the regulation and management of species and areas.

A good example of how legislative overlap may impact on a project is provided by the Chilean legislation. Although that country's environmental statute takes the one window approach to environmental licensing, it may be difficult to reconcile that approach with what is provided under Chile's Protected Wildlife Areas statute. Both statutes legislate similar, even seemingly parallel, licensing processes under different authorities. Whereas the latter statute designates the Minister of Agriculture as the authority in charge of approving concessions in protected areas, the environmental statute and its regulations award overlapping jurisdiction to the National/Regional Environment Commission. The Protected Wildlife Areas Law requires preparation of an EIA study, which is equally mandatory for this type of project under the environmental statute. Similar overlapping provisions may be found in relation to the assignment of control and enforcement powers.

Although the existence of legislation on protected natural areas is quite widespread within LAC, the methods for on-the-ground definition of the protected areas, and the different categories of land within, as well as for the elaboration of the environmental management and monitoring plans, have been, for the most part, deferred to future regulatory and administrative activity. In general, no further guidelines are given beyond the broad definition of categories or uses contained in the basic laws. The lack of guideline criteria aggravates a generalized lack of baseline information on the countries' existing resources and on the prevailing environmental conditions, as well as on the environmental impacts of economic activities and practices on the region's environment.

Treatment of pre-existing rights within newly created areas is another issue that deserves attention. Although some statutes specify that exercise of rights that may have existed before the delimitation and creation of protected natural areas shall be subject to the incumbent legislation or, alternatively, that private rights shall be "respected"—it is also true that, in general, no further clarification is provided. In some cases expropriation may be anticipated as an alternative to the subsistence of private rights.

Finally, another issue confronting the adequate protection of areas is the practical problem arising at the implementation stage where lack of material (i.e., financial and technological) and human resources hinder effective management and control, especially in remote and inaccessible areas. Furthermore, penalizing technically illegal activities undertaken for subsistence purposes, such as slash-and-burn farming, wood-cutting, and hunting is highly unpopular and may have little practical effect where no realistic alternatives to these practices are available.

Notes and Questions

1.   The Latin American and Caribbean region is home to some of the most remarkable biodiversity in the world. Indeed, five of the ten so-called ecological mega-diversity countries are located within the region: Brazil, Colombia, Ecuador, Mexico, and Peru. The Amazon basin is teeming with birds, plants, reptiles, fish, and millions of insect species. Ten percent of the world's plant and animal species can be found in Colombia alone. Forty percent of all of the species native to tropical forests live within Latin American and the Caribbean. But the protection of that biodiversity presents significant challenges in a region that faces the institutional and economic constraints noted by Barrera–Hernandez and Lucas. How can each nation best confront those challenges?

2.   Costa Rica is famous for its biodiversity, attracting ecotourists to its lush rainforests and to its tropical beaches. Costa Rica has also adopted the most thorough law in the region to protect biodiversity. Its Biodiversity Law, enacted in 1998, contains a number of significant provisions:

- "The state will exercise total and exclusive sovereignty over the components of biodiversity."–§ 2

- "The biochemical and genetic properties of biodiversity, wild or domesticated, belong to the public domain."–§ 6

- A National Commission for the Management of Biodiversity is charged with developing and implementing national policies concerning biodiversity.–§ 14

- "The plans or the authorization for the use of mineral resources, land, flora, fauna, water or other natural resources as well as the location of human settlements and industrial or agricultural developments ... will consider in particular ... the sustainable conservation and use of, especially when it concerns plans or permits which affect biodiversity of the wild protected areas."–§ 52

- "Wild protected areas, besides those of the State, can be municipal, mixed or private property."–§ 60

- Access to the components of genetic diversity and the protection of intellectual property rights are regulated.–§§ 62–85

- "Biological education should be integrated in the educational plans of all anticipated levels, to achieve understanding of the value of biodiversity and the way in which it plays a part in the life and aspirations of every human being."–§ 86

- "At the discretion of the Technical Office of the Commission, an environmental impact assessment of proposed policies shall be requested when it considers that they could affect biodiversity."–§ 92

- "Every person will be authorized to act in administrative or jurisdictional headquarters for the defence and protection of biodiversity."–§ 105

Is there anything missing? In other words, what else would you include in an ideal domestic statute to protect biodiversity?

Note, however, that the existence of such a statute hardly guarantees the preservation of biodiversity. As Barrera–Hernandez and Lucas explain later in their article, "one of the main problems of the region is the weak enforcement of environmental legislation." They add that "weak enforcement seems to be caused more by lack of human and financial resources and institutional capacity than by lack of adequate legislation." Barrera–Hernandez & Lucas, *supra*, at 230–31. Enforcement is further challenged by confusing and redundant laws. The overlap between Chile's Protected Wildlife Areas statute and its general environmental statute that is described by Barrera–Hernandez and Lucas may even underestimate those challenges. Chile itself admits that while it "has literally hundreds of environmental laws on the books," many of them "were obsolete, unenforceable, contradictory or duplicative." Embassy of Chile–USA, Environmental Policy in Chile, *available at* <http:www.chile-usa.org/documents/political/envpol.htm>.

3.   Section 105 of Costa Rica's Biodiversity Law operates as an expansive citizen suit provision. Public participation in decisionmaking is less well-established elsewhere in the region. As Barrera–Hernandez and Lucas observe, "With few exceptions, participation is generally limited to project approval processes. However, there are some instances of indirect or 'institutionalized' participation, i.e., through the incorporation of [Environmental Nongovernmental Organization] or communities' representatives in advisory bodies.... Two features that are quite common throughout the region are the lack of provisions for intervenor funding and regulatory allowance for the application of alternative dispute resolution (ADR) mechanisms." Barrera–Hernandez & Lucas, *supra*, at 233–34. What is the most effective way in which Latin American or Caribbean residents could influence their nation's biodiversity policies?

4.   Nature preserves are the hallmark of the biodiversity strategies of most Latin American and Caribbean countries. The extent of those preserves varies greatly, as do the activities that are permitted within them. Chile, for example, relies upon a National System of Protected Wildlife Areas (SNASPE) that includes 30 national parks, 40 national reserves, and 11 national monuments that cover 18% of Chile's land. The protection afforded by Chilean law to wildlife on that land depends upon the specific type of area. National parks and national monuments are large and small, respectively, areas containing important native biodiversity where the extraction of natural resources is prohibited. National reserves are designed for both the conservation of biodiversity and the development of economic resources.

Chile is not the only country to struggle with the kinds of activities that are consistent with the designation of a nature reserve. Dan Tarlock has noted that "[m]odern theories of biodiversity protection seek to go beyond merely preserving and 'fencing off' large areas from development because the strategy is often not a *sufficient* biodiversity protection strategy and is not consistent with the norm of sustainable development." He describes "the shift from separatist to integrationist conservation thinking" in Brazil's nature reserve policy. Beginning in 1974, "extensive large reserves (in and out of the Amazon region) were created at least on paper," with an emphasis "on the creation of reserves unconnected to indigenous populations, who were perceived as a threat to the Amazon's biodiversity." By the end of the 1980's, "after intense pressure from the forest dwellers, extractive reserves were established in Brazil with the help of the World Bank and other international conservation organizations." Tarlock concludes that "[t]he long-term success of the reserves depends on the establishment of a collective property entitlement in those who depend on the reserve, but this will be difficult because there are many individual conflicting groups claiming these areas." A. Dan Tarlock, *Exclusive Sovereignty Versus Sustainable Development of a Shared Resource: The Dilemma of Latin American Rainforest Management*, 32 TEXAS INT'L L. J. 37, 58–59 (1997).

Would you agree that "protected areas are extremely important for the protection of biodiversity, yet requiring them to carry the burden for biodiversity conservation is a recipe for ecological and social failure?" PARKS IN PERIL: PEOPLE, POLITICS, AND PROTECTED AREAS 2 (Katrina Brandon, Kent H. Redford & Steven E. Sanderson eds. 1998). That was the premise of the report on the extensive case studies of nine protected areas within the Latin American and Caribbean region conducted by Parks in Peril, a division of the Nature Conservancy dedicated to protecting biodiversity within the region. The collective lessons of those case studies included (1) the importance of how and why the park was created in determining its future uses, management, and conflicts; (2) the significance of zoning as "a key park management tool" that "allows for different kinds of uses in different areas;" (3) uncertainty concerning the legal and customary rights to use the land, water, trees, wildlife and other natural resources within a park and immediately outside a park's boundaries; (4) a wide range of conflicts involving parks and a similar range of management responses; (5) the vulnerability of the parks to outside threats such as logging, mineral production, roads, colonization, agriculture, and tourism; (6) and "the lack of commitment or political will or power to promote environmental objectives" that made the national policy framework "arguably the single most important element within the case studies for the long-term survival of the sites." Katrina Brandon, *Comparing Cases: A Review of Findings*, in PARKS IN PERIL, *supra*, at 376–414.

5. Which nations have demonstrated the greatest ability to protect the biodiversity that lives within their borders? Certain Latin American or Caribbean countries? China? Australia? The United States? Is any nation capable of protecting its biodiversity without assistance or compulsion from other countries or the larger international community? What significance, if any, should be attached to the ratification of the Convention on Biological Diversity (to be discussed in much more detail in chapter 11) by China, Australia, and nearly every Latin American and Caribbean country–but not the United States?

B.   EXTRATERRITORIAL APPLICATION OF DOMESTIC ENVIRONMENTAL LAWS

If a country is unable or unwilling to protect the biodiversity that lives within its borders, can another country adopt such protections itself? A country's laws usually only apply to activities that occur in the country's territory or to the country's own citizens, but there are many exceptions to that general rule. A country may also decide to subsidize or otherwise encourage the steps needed to protect biodiversity around the world. The choice of carrots–*e.g.*, providing financial assistance–or sticks –*e.g.*, regulating or sanctioning harmful conduct–often determines the popularity of a country's unilateral efforts to protect biodiversity outside its own borders. The more fundamental question is whether, or in what circumstances, it is appropriate for a country to use its laws to influence the management of ecosystems and species in other parts of the world.

Asian Elephant Conservation Act of 1997

House Report 105–266 Part 1.
United States House of Representatives Committee on Resources, 1997.
105th Congress, 1st Session.

The Committee on Resources, to whom was referred the bill (H.R. 1787) to assist in the conservation of Asian elephants by supporting and providing financial resources for the conservation programs of nations within the range of Asian elephants and projects of persons with demonstrated expertise in the conservation of Asian elephants, having considered the same, report favorably thereon with an amendment and recommend that the bill as amended do pass. . . .

PURPOSE OF THE BILL

The purpose of H.R. 1787 is to create an Asian Elephant Conservation Fund and to authorize the Congress to appropriate up to $5 million per year to this Fund for each of the next five fiscal years to finance various conservation projects.

BACKGROUND AND NEED FOR LEGISLATION

In Asia, the relationship between man and elephant dates back almost 5,000 years when elephants were first captured and trained for use in religious ceremonies, war, and as draft animals. In fact, ancient Hindu scriptures frequently refer to elephants, the elephant-headed god Ganesh is revered throughout India, and the white elephant has special religious significance for Buddhists throughout Asia. In Chinese culture, elephants have played a special role in folklore, games, medicine and pageantry.

Asian elephants have also been used in forestry operations for many years. Today, wild elephants are still captured and trained for use in logging operations in Burma. Elsewhere throughout their range, domestic elephants are used for ceremonial, tourism and transportation purposes. These activities provide an important source of income to numerous local communities.

Sadly, the Asian elephant is now in grave danger and unless steps are immediately taken by the international community, the Asian elephant will largely disappear from most of its historical range. To date, the Asian elephant has been declared endangered and placed on the U.S. Endangered Species Act list, on the Red List of Mammals by the International Union for Conservation of Nature and Natural Resources (IUCN)-World Conservation Union, and on Appendix I of the Convention on International Trade in Endangered Species of Wild Fauna and Flora. Placement on Appendix I prohibits all commercial trade in Asian elephant products on a worldwide basis.

Despite these efforts, the population of Asian elephants living in the wild has dramatically fallen to about 40,000 animals, which is less than 10 percent of its elephant cousin (Loxodonta Africana) living in Africa. These wild populations are located in 13 countries in South and Southeast Asia. The largest population of 20,000 Asian elephants, or 50 percent of the total, resides in India; the smallest population of 50 animals is located in Nepal. What is equally distressing is the fact that there are only about 14 populations of 1,000 or more individuals in a contiguous area. Seven of these populations are found in Burma and India. In simple terms, this means that such drastic population fragmentation increases the likelihood of geographic extinctions and greatly reduces the long-term viability of the species. In addition, it has been estimated that there are about 16,000 domesticated elephants.

There are a number of important reasons why there has been a severe decline in the number of Asian elephants. The primary reason is the loss of habitat. All Asian elephants need a shady or forest environment, and this habitat is disappearing rapidly throughout Asia. Due to their sheer size and social structure, elephants need large areas to survive. Since Asian elephants inhabit some of the most densely populated areas of the world, forest clearance for homes and large-scale agricultural crops have resulted in a dramatic loss of thousands of acres of their habitat. In essence, elephants and man are in direct competition for the same resources.

Second, while poaching for ivory has not been an overriding reason for its decline, Asian elephants of both sexes are increasingly being poached for bones, hide, meat and teeth. Hide is used for bags and shoes in China and Thailand, and bones, teeth and other body parts are used in traditional Chinese medicine to cure various ailments. In fact, this type of poaching even threatens domestic elephants that are allowed to free-range in various forests.

Third, Asian elephants are still captured in the wild for domestication. In Burma, the country with the highest demand for work elephants, adult elephants are captured and trained for use in the timber industry. Regrettably, capture operations inevitably result in some mortalities and it does adversely affect the genetic pool of elephants living in the wild.

Finally, conflicts between elephants and people are increasing at an escalating rate. This is a direct result of the dramatic loss of forest habitat and the ensuing competition for the remaining resources. Every year, thousands of acres of agricultural crops are destroyed by elephants looking for food. In many cases, elephants encounter people where they were not found previously, thereby leading to the destruction of human lives and homes.

In countries where governments are concerned with this ever-increasing problem, measures taken are drastic and very expensive. For instance, in Malaysia, there was large-scale shooting of crop-raiding elephants in the late 1960s and, more recently, the construction of electric fences and translocation of problem elephants to protected areas. Other countries like Indonesia are taking short-term measures by capturing large numbers for domestication. However, they have found no long-term use for these domesticated elephants because there has been no traditional relationship between people and working elephants. In countries like Cambodia and Vietnam where no immediate solutions are provided by governmental authorities because of lack of financial resources, people are increasingly taking the law into their own hands and shooting the elephant offenders.

It is also important to understand that effective Asian elephant conservation and management efforts will have a positive effect on other species that reside in the same habitat. In the case of the Asian elephant, these include: the Asiatic wild dog, Clouded leopard, gaur, Great Pied hornbill, Hoolock gibbon, kouprey, Lion-tailed macaque, Malayan sun bear, peacock pheasant, rhinoceros and tiger. It is essential to the survival of these species that the Asian elephant not be allowed to disappear from this planet.

Finally, unlike the African elephant, there is no sport-hunting of Asian elephants and no large stockpiles of Asian elephant ivory in government warehouses in Asia or Southeast Asian countries.

COMMITTEE ACTION

... On Thursday, July 31, 1997, the Subcommittee on Fisheries Conservation, Wildlife and Oceans held a legislative hearing on H.R. 1787, the Asian Elephant Conservation Act of 1997. The Subcommittee heard testimony from Mr. Marshall P. Jones, Assistant Director for International Affairs, United States Fish and Wildlife Service, Department of the Interior; Dr. Terry Maple, President and Chief Executive Officer, Zoo Atlanta; Dr. Eric Dinerstein, Chief Scientist and Director, World Wildlife Fund; the Honorable Andy Ireland, Senior Vice President, Feld Entertainment, Inc.; Dr. Raman Sukumar, Center for Ecological Sciences, Indian Institute of Science; Dr. Mary C. Pearl, Executive Director, Wildlife Preservation Trust International; and Dr. Michael Stowe, Research Associate, Smithsonian Institution. All witnesses testified in strong support of the bill.

Mr. Marshall Jones of the Interior Department testified that:

On behalf of the Administration, the Service fully supports the enactment of this legislation and congratulates the Congress on its foresight in recognizing and addressing the plight of the Asian elephant. Asian elephants need active protection and management of their habitat, resolution of the deleterious conflicts with humans over land uses, better law enforcement activities to protect against poaching, reduction of captures from the wild, and better care and humane treatment of the remaining captive populations. They also need the restoration of the harmonious relationship that previously existed with humans through community education and awareness activities. It is indeed timely that this Subcommittee is now considering H.R. 1787 (which) acknowledges the problems of forest habitat reduction and fragmentation, conflicts with humans, poaching and other serious issues affecting the Asian elephant. The Act addresses the need to encourage and assist initiatives of regional and national agencies and organizations whose activities directly or indirectly promote the conservation of Asian elephants and their habitat, and it provides for the establishment of an Asian Elephant Conservation Fund, authorized to receive donations and appropriated funds. While many range governments have demonstrated a commitment towards conservation, the lack of international support for their efforts has been a serious impediment. . . .

CONGRESSIONAL BUDGET OFFICE COST ESTIMATE

. . . Summary: H.R. 1787 would establish a new fund to support the conservation of Asian elephants. The bill would direct the Secretary of the Interior to use amounts in the new fund to finance eligible conservation efforts, which may include specific projects such as research and education as well as ongoing activities such as law enforcement. For this purpose, the bill would authorize appropriations to the fund of $5 million for each of fiscal years 1998 through 2002. Also, the Secretary would be authorized to accept and use donated funds without further appropriation.

Assuming appropriation of the authorized amounts, CBO estimates that implementing H.R. 1787 would result in additional discretionary spending of about $10 million over the 1998–2002 period (with the remainder of the authorized $25 million estimated to be spent after 2002). The legislation would affect direct spending and receipts by allowing the Secretary to accept and spend donations; therefore, pay-as-you-go procedures would apply. Any such transactions, however, would involve minor, offsetting amounts. H.R. 1787 does not contain any intergovernmental or private-sector mandates as defined in the Unfunded Mandates Reform Act of 1995 (UMRA), and would have no impact on the budgets of state, local, or tribal governments.

Notes and Questions

1. Why is the plight of the African elephant a problem for the United States Congress? How does the committee report's description of the need for the legislation explain why it is the United States that should enact such legislation?

2. Congress appropriated almost three million dollars for the African Elephant Conservation Fund between 1998 and 2000. That money was used to support 51 projects in 18 countries, and it was also used to gain matching and in-kind support for those projects. The projects funded by the Fund include:

- Research on the movements of forest elephants in the Nouabalé-Ndoki National Park in the Republic of the Congo.

- Training for field personnel managing elephant populations in Kakum National Park in Ghana and Marahoue National Park in Côte d'Ivoirie.

- Genetic research on African elephants that will provide wildlife managers with data to monitor genetic erosion and inbreeding.

- The study and management of conflicts between people and elephants in the Muzarabani Rural District in Zimbabwe.

The U.S. Fish & Wildlife Service proclaims that "[s]ignificant improvement of the conservation status of African elephants is now evident in some places where the Fund has worked with on-the-ground partners." U.S. Fish & Wildlife Service, African Elephant Conservation Fund (Mar. 2001), *available at* <http://international.fws.gov/pdf/afecffs.pdf>.

3. The African Elephant Conservation Act is just one of a number of U.S. statutes designed to assist the protection of biodiversity overseas. The Neotropical Migratory Bird Conservation Act, Pub. L. No. 106–247, 114 Stat. 593 (2000), is the most recent such statute. The three purposes of the act are "(1) to perpetuate healthy populations of neotropical migratory birds; (2) to assist in the conservation of neotropical migratory birds by supporting conservation initiatives in the United States, Latin America, and the Caribbean; and (3) to provide financial resources and to foster international cooperation for those initiatives." *Id.* § 3. The act seeks to achieve these goals by funding conservation projects, authorizing federal appropriations and private donations to the Neotropical Migratory Bird Conservation Account, sharing information and encouraging meetings among people and organizations involved in protecting the birds, and developing appropriate international agreements. *Id.* §§ 5–7, 9. Which activities should receive the Secretary of the Interior's highest priority in implementing the new act?

Determination of Threatened Status for the Koala

United States Fish and Wildlife Service, 2000.
65 Fed. Reg. 26762.

Background

The koala (*Phascolarctos cinereus*) is an arboreal mammal found only in Australia. It has a compact body, large head and nose, large and furry ears, powerful limbs, and no significant tail. Mature koalas weigh from 4–15 kilograms (10–35 pounds), with larger animals in southern Australia. The koala is a marsupial, more closely related to kangaroos and possums than to true bears and other placental mammals. Koalas carry their young in a pouch for about 6 months. They occur in the forests and woodlands of central and eastern Queensland, eastern New South Wales, Victoria, and southeastern South Australia.

In a petition dated May 3, 1994, which we received on May 5, 1994, Australians for Animals (AFA) (in Australia) and the Fund for Animals (FFA) (in the United States) requested that the koala be classified as endangered in New South Wales and Victoria, and as threatened in Queensland. About 40 organizations in the United States and Australia were named as supporting the petition. The document included extensive data indicating that the koala has declined dramatically since European settlement of Australia began about 200 years ago and has lost more than half of its natural habitat because of human activity. Once numbering in the millions, the koala was intensively hunted for its fur up through the 1920s. It is totally dependent for food and shelter on certain types of trees within forests and woodlands. The destruction or degradation of this habitat would reduce the viability of populations, even if the animals were otherwise protected.

On October 4, 1994, we announced a 90–day finding that the petition presented substantial information indicating that the requested action may be warranted. That notice also initiated a status review of the koala. On February 15, 1995, we reopened the comment period on the status review until April 1, 1995. We sent a telegram to the U.S. embassy in Australia, asking that appropriate authorities be notified and asked to comment. We also presented the review directly to numerous concerned organizations and authorities. Of the approximately 400 responses received, the great majority were brief messages in support of listing, but several responses were from persons or organizations providing substantive comments based on firsthand knowledge of the situation.

On September 22, 1998, we proposed the koala as threatened throughout its range, and we sought public comments. We received over 3,000 responses: The vast majority were cards with a printed message endorsing the comments of the International Wildlife Coalition and supporting threatened status for the koala, but personal letters also expressed support for listing the species. We also received letters with substantive comments on the proposal from persons with direct knowledge of koala biology; many of those comments came from persons or groups who had offered opinions and information on earlier notices. We also sought information from scientists on a number of outstanding issues.

What Were the Comments of Those Who Opposed the Proposed Listing?

All of the Australian Federal and State authorities that commented on the proposal opposed it. They were joined by three other respondents, including two who represented zoological associations in Australia and the United States.

Dr. Colin Griffiths, Director of National Parks and Wildlife, submitted comments for Environment Australia, the agency responsible for koala policy on the national level. He stated that the Australian Government continues to object to our proposal to list the koala as a threatened species under U.S. law. Noting that, under the Endangered Species Protection Act 1992 (ESPA) no trade in koalas or koala products is permitted, Dr. Griffiths said "we have yet to see any explanation of how the listing of the koala in the United States would contribute to koala conservation." The submission also stated that the Endangered Species Scientific Subcommittee established under the ESPA has evaluated nominations of the koala both under "species that are endangered" and "species that are vulnerable." In each instance, the subcommittee concluded that the koala did not meet the criteria for listing at a national level.

We fully understand the view of the Australian Government on the status of a species that is native only within its boundaries, particularly where only an occasional zoo acquisition leaves the country. However, our Endangered Species Act (ESA) is international in scope, and we are compelled by law to evaluate petitions of species beyond U.S. boundaries.

Dr. Griffiths made the point that the Australian Government has taken a number of steps in koala conservation since the listing proposal came to us in 1994. A scientific advisory board has reported to the Minister of Environment that the species is relatively abundant and widespread nationally and not likely to become endangered within the next 25 years. In 1998, the legislation of the Commonwealth and the States protecting koalas was integrated into the National Koala Conservation Strategy. The Strategy was developed by the Australian and New Zealand Environment and Conservation Council and was included with the comments submitted by Environment Australia.

Finally, the submission made the objection raised by several others on the listing proposal: Australians particularly object to a rule in which we classify the species as threatened throughout its range rather than assess whether the koala warrants this classification in each State. While the ESA does not allow us to differentiate vertebrate populations solely on state or provincial boundaries (whereas we can on national boundaries), it does allow us to make these distinctions when significant biological differences exist between the populations. The issue that predominates is whether the three subspecies that have been described for koalas represent distinct vertebrate population segments.

Mr. Allan Holmes, Director of National Parks and Wildlife for the Department of Environment, Heritage and Aboriginal Affairs of South Australia, also made the point that the status of the koala varies regionally, and it is not considered nationally endangered or vulnerable. Koalas in South Australia are protected under the National Parks and Wildlife Act 1972 and are listed as rare under Schedule 9. In providing a history of koala management in the State, Mr. Holmes maintained that the classification as rare is misleading as the koala population in South Australia was at the western edge of its range even prior to European settlement. By 1930, the koala was considered extinct in South Australia, and, as a consequence, a population was established on Kangaroo Island and subsequently at other sites on the mainland. Koala habitat is patchy in South Australia, largely due to forest fragmentation caused by 150 years of agricultural development. Koalas introduced to these patches have established populations and have frequently exceeded carrying-capacity with consequent damage to food trees. The letter affirmed the commitment of the Government of South Australia to ensuring that koalas are conserved in the State and that they are managed in such a way that will sustain them and their habitat. Mr. Holmes concluded that the current situation in South Australia with local overpopulation and genetic founder effects illustrates that the threats to koalas are different across Australia and that a single classification may not best serve conservation efforts for the species.

Mr. Michael Taylor, Secretary of the Department of Natural Resources and Environment for the State of Victoria, said that the status of the species has continuously improved from the 1920s when it was probably endangered, to its current status as a widespread and common species. The koala is protected wildlife under the provisions of Victoria's Wildlife Act 1975, which protects all indigenous terrestrial vertebrates, and the Flora and Fauna Guarantee Act 1998, which seeks to insure that species not only survive but retain their evolutionary potential in the wild. Under the provisions of that law, any person or group can nominate a species for listing, and it will be assessed by an independent Scientific Advisory Committee. Victoria's submission noted that, while 359 taxa have been nominated, the koala has not been one of them. Moreover, the government of Victoria has subjected all of its native vertebrates to the World Conservation Union criteria, and, while over 200 taxa were listed as threatened at some level, the koala did not meet the criteria. . . .

Mr. Taylor presented the specific actions that Victoria has taken in recent years to protect koalas and their habitat. Victoria's Biodiversity Strategy calls for a reversal in the decline of native vegetation with a goal of no net loss by 2001. The Planning and Environment Act of 1987 includes the objective to assist the protection of biodiversity, and the Land for Wildlife Program provides mechanisms to conserve areas of environmental significance. The view of the Department of Natural Resources and Environment is that Victoria has a strong viable koala population in the wild, and thus listing would only divert attention from the species that are under threat.

Mr. Brian Gilligan, Director General of the New South Wales National Parks and Wildlife Service, wrote that the population there is intermediate in physical size between the larger southern koalas in Victoria and South Australia and the smaller northern koalas in Queensland. The population in New South Wales was decimated by hunting until it was estimated to contain only 1,000 koalas by 1920. Researchers believed the population had recovered to 5,000–10,000 koalas by the 1970s. The koala was listed as vulnerable under the New South Wales Endangered Fauna Act 1991 and more recently has the protection of threatened species and the Threatened Species Conservation (TSC) Act 1995, which replaced the earlier law. Because the koala is an ecological specialist, it is vulnerable to local extinctions. The letter details several steps that New South Wales has taken to help koala recovery in the State. Under the State Environmental Planning Policy 1995, a detailed habitat assessment is required before approving development of greater than 1 hectare in local government areas where koalas are known to exist. As required of any vulnerable species, the TSC Act requires the National Parks and Wildlife Service to prepare a recovery plan within 10 years. Also, the New South Wales government has begun creating forest reserves under the Regional Forest Agreements (RFAs). The State government has reserved 600,000 hectares so far, and, by their assessment, a large proportion of this land is koala habitat.

Mr. Greg Gordon of Queensland National Parks and Wildlife Service qualified his earlier comments in the proposed rule, that koalas could become vulnerable in the future. "I would see this as a long-term possibility only, as a result of continuing land clearing, assuming clearing is unchecked. It is difficult to put a time frame on this but I would think it would be many decades away, e.g. 50–100 years." Gordon wrote that the main problem is that most koala sites have poor habitat protection as they occur on privately managed land, which may be at risk of partial or total clearing at some time in the future. He added that in Queensland conservation measures for private lands are being developed, and more effective habitat protection is likely to be available in the medium term.

Mr. Mark S. Canty submitted a letter opposing the proposal. He contrasted the national system of "Landcare" groups that have been forming in Australia, with the RFAs being set up by the government with the goal of preserving 15 percent of forest types that existed in Australia prior to 1750. Mr. Canty said that the result of these preservation targets has been an increase in areas being cleared by landholders to avert government decrees, and he expressed his concern that listing the koala would have the same negative impact, with landholders not reporting koala sightings for fear of being told how to manage their property. Mr. Canty expressed the view that agriculture and housing developments represent a greater threat to koalas than forestry practices. We fully understand this viewpoint, and we are aware that even the perception of imposed solutions stimulated by those living far from the effected land can have a counterproductive effect. Nothing in this listing in any way limits or directs specific measures in Australia for the benefit of koala conservation, on either the State or the Federal level.

Ms. Christine Hopkins, Executive Director of the Australian Regional Association of Zoological Parks and Aquaria (ARAZPA), provided valuable information related to the koala from the international to the state level. The summary of status and legislation was developed by the Monotreme & Marsupial Taxon Advisory Group. Convener Gary Stator said that the Taxon Advisory Group could see no basis to list the species as endangered, and Ms. Hopkins said the ARAZPA could find no evidence in support of listing the species as threatened.

Senior officials at the American Zoological Association (AZA) have modified the position stated in the previous submission of the AZA. Ms. Kristin Vehrs, Dr. Michael Hutchins, and Mr. Robert Howarth maintain that the data provided fail to meet the listing criteria under the Act, specifically that the species is threatened throughout its entire range. While acknowledging that certain koala populations in New South Wales and Queensland continue to be threatened, studies conducted in Victoria and South Australia suggest that the koala has begun to reestablish itself there. AZA stated that while some areas may meet the habitat loss criterion for listing, none currently meet the overutilization criterion in this instance. They conclude that no commercial exploitation occurs, and the few koalas going to zoos for research and educational display do so under permits with conditions that are highly restrictive. AZA notes that while habitat loss has been extensive, the Commonwealth and each State have their own management plans to reverse that trend. We concur with the AZA comments that koalas do not face the same magnitude of threats throughout Australia. The criteria for a threatened species, however, is one that is likely to become endangered throughout all or a significant portion of its range.

What Were the Substantive Comments of Those Who Favored Listing the Koala as Threatened?

Ms. Valerie Thompson, North American Koala Population Manager for the AZA, expressed support for listing the koala as threatened. She based her view on field expeditions mapping koala habitat in conjunction with the Australia Koala Foundation. She also submitted letters from other AZA member institutions, responses to a packet of information on the listing that she had sent out as an Executive Committee member of the Marsupial and Monotreme Taxon Advisory Group. She concluded the AZA did not have a consensus on the koala listing and included letters from institutions and scientists in which nine favored listing, four opposed, and two abstained. The letters included with Ms. Thompson's submission reflected divergent views of the status of koala within the zoo community in the United States, as was evident from the submissions of the scientists in Australia. To list a species, we must determine it meets the criteria based on information from scientists surveying koalas and their habitat.

Mr. Michael Kennedy, Director of Humane Society International (HSI), reiterated support for the listing of the koala as threatened. He stated that habitat clearance, particularly in the States of New South Wales and Queensland, is the greatest threat to koala survival. HSI reviewed the legislative actions taken since the previous comment period. Nominations were submitted under the national ESPA 1992 to list the koala as "vulnerable" and "endangered" by different conservation groups; both of these nominations were denied, though some of the scientists evaluating the proposals favored them. In New South Wales, where four koala populations were nominated as "endangered" under the New South Wales TSC Act,

1995, HSI noted that only one of the nominations was successful. In 1996, the Australian Government published the first National State of the Environment Australia. The document concluded that the "greatest pressures on biodiversity come from demands on natural resources by increasing populations of humans, their affluence and their technology.* * * Habitat modification, has been and remains, the most significant cause of loss of biodiversity." The HSI letter stated that the Endangered Species Scientific Subcommittee (ESSS) recommended that vegetation clearance be recognized as a key threatening process as nominated by HSI. The Federal Minister for the Environment rejected the ESSS recommendation on legal but not biological grounds.

Ms. Deborah Tarbart, Executive Director of the Australia Koala Foundation (AKF), provided additional information on behalf of the foundation supporting the listing of the koala. The AKF has been actively adding areas to the Koala Habitat Atlas, and three of those areas were included as appendices with the submission. They demonstrate that a small percentage of primary koala habitat remains in particular areas that are associated with koalas. The AKF believes that overpopulation of koalas in some areas of Victoria and South Australia misdirects the debate, as they are atypical populations in isolated habitats....

Ms. Julie Zyzniewski, President of the Koala Council in Queensland, wrote that, while the State and local governments have adopted some measures to stabilize the population in southeast Queensland, habitat destruction in the rest of the State and elsewhere in Australia had worsened. The Koala Council therefore strongly supports listing in the belief that it will provide moral support for community-based organizations such as the Koala Council.

Ms. Donna Hart and Dr. Ron Orenstein of the International Wildlife Coalition, based in the United States and Canada, reiterated their support of the listing. They maintained that the decline in eucalyptus-dominated woodland in southeastern Australia continues, and the policies of the many Australian jurisdictions appear to be aimed at accelerating this decline rather than halting it. As this is not true of all areas, IWC would favor a State-by-State listing.

Dr. Frank N. Carrick of the University of Queensland makes several points in support of the listing proposal. Queensland is the only State where the koala can be "considered to approach a natural condition in terms of number, distributional range and genetic and demographic integrity." The State also has one of the world's highest rates of clearing of native vegetation. Moreover, the riparian or coastal and lower altitude forests favored by koalas are the forests most extensively destroyed and fragmented for agriculture, grazing, intensive forestry, and residential development. The high-density koala population in southeastern Queensland-which Dr. Carrick sees as having a vital role in the survival of the species over evolutionary time-is the area of fastest human population growth in Australia. As for the ability of government regulation to reverse these trends, Dr. Carrick expressed the view that the Queensland Nature Conservation

Act has inherent deficiencies that have resulted in the downgrading of the classification of the koala from "permanently protected" to the "common fauna" category.

We concur that the State with the most robust koala population in Australia also has the population at most serious risk. While we recognize that the Queensland government has enacted a State Planning Policy (SPP1/95) to control land allocation processes that are threatening koala populations, it will take years of monitoring to determine whether the Policy has been effective and the trend has been reversed in Queensland.

Dr. Tony Norton, Royal Melbourne Institute of Technology, commented primarily on the forestry assessments that have been undertaken since the proposed listing. These assessments will serve as a basis for setting new guidelines for land allocation, forest management, and forestry sawlog and woodchip quotas over the next 20 years. Dr. Norton found that none of the assessments that have been completed so far have delivered their intended goals of a world class forest conservation reserve system or world class forest management practices and concludes that the habitat of the koala in the wild is endangered. He therefore reasserted his support for the listing to force Australian governments to meet both national and international commitments from the preservation of the country's biodiversity.

Mr. Robert Bertram of the South East Forests Conservation Council provided thorough documentation of the demise of the koala population in one part of New South Wales. At present 39 percent of the high-quality koala habitat in the area is reserved in National Parks, and resource agreements prevent reducing the intensity of current logging operations in the remainder of the quality habitat. Claiming that the government has demonstrated disregard for the known science and the precautionary principle in making land-use decisions, the Council gave its view that the situation of the koala in the region and across New South Wales on public land is uncertain at best.

Mr. D.J. Schubert writing on behalf of the original petitioners expressed frustration with the delay in publishing the proposed rule from the petition submitted in May 1994. The AFA and FFA contend that conditions have only declined further since their earlier comments and that the koala now merits endangered status throughout its range. They concur with other comments that most of the habitat destruction is the result of timber, agriculture, mining, and development. Most of the clearing of eucalypt forests is for the export woodchip markets. The submission also points out that the Australian Government has redefined the forest to include woodlands, plantations, and other areas not regarded as native forest. The effect has been to increase the amount of land considered forest in Australia from 41 to 157 million hectares. The AFA–FFA submission documents the development of the RFAs in Victoria, where the process has proceeded faster than in other States, and maintains that the new assessment provides "virtually no benefit" for the koala and its habitat. Given the specificity of the food and habitat requirements of the koala, inclusion of additional areas as RFAs may give an artificially high estimate of the land area that constitutes potential koala habitat.

Why Should We Consider the Koala, a Species That Is Not Native to the United States and That Is Only Rarely Imported To Be Displayed in Zoos, for Listing Under the ESA?

This question is one that people asked in letters from the Government of Australia as well as the States within the country. As the koala does not naturally cross national boundaries and is not in legal international commercial trade, why should we take the considerable time to consider the species as threatened?

The ESA is not restricted to species native to the United States, or those subject to international trade. The Act considers national boundaries, but makes that consideration secondary to the concern for the survival of species. The Act obligates us to make a determination in response to a petition.

As for the priority of such foreign species, with so many other important priorities in international wildlife conservation, we have proceeded deliberately with the listing process, sometimes to the dismay of the petitioners. We have found that, during listing consideration, with its requirements for public comment and consideration of those comments in developing a final decision, sometimes important strides have been made by the countries in the conservation measures that have been developed or enforced. In such cases, the ESA provides an important conservation benefit.

What Is the Status of the Koala in Regard to the Five ESA Listing Factors?

* * *

D.   The Inadequacy of Existing Regulatory Mechanisms

Although State laws generally protect the koala from direct taking and commercial utilization, much of the petitioners' argument is based on a lack of regulatory mechanisms that adequately protect the habitat of the species. Although a significant portion of the koala's remaining habitat is on government land, such ownership does not preclude logging and other modification. Researchers have particular concern that deforestation for the woodchip market is proceeding without proper assessment of environmental impacts. Even if such impacts were taken into account, the petitioners argue the welfare of the koala would not be given adequate attention because the species is not listed pursuant to Australia's ESPA. We can look at the situation of the koala in each State to determine the adequacy of the current regulations.

Though the koalas of Queensland are the smallest in size, the State has the largest koala population, and the most remaining koala habitat of the States. Queensland also has one of the highest rates of clearing of native vegetation. Under the National Forestry Policy, the rate of clearfell-ing continues to be high on private lands. According to the 1996 assessment of the Australian and New Zealand Environment and Conservation Re-search Council, the koala population is stable in some areas, thinly scat-tered in many others, and in steep decline in some coastal areas. A consensus exists that the population overall is declining at different rates depending largely on the degree of development. The situation is particu-larly critical in southeast Queensland, where urbanization threatens the still substantial koala population. Despite legislation that includes the Nature Conservation Act 1992 and the State Planning Policy 1995, the major threat is poor habitat protection for most of the koala population.

In New South Wales, koalas were once abundant throughout the eastern half of the State and driven to near extirpation by the 1920s. The State government estimates that the population recovered to 5,000–10,000 by the 1970s, with the largest and most secure population in the northwest part of the State. The State government also is concerned that continued habitat fragmentation could lead to local extinctions. For that reason, the koala was listed as a vulnerable species under the NSW Endangered Fauna (Interim Protection) Act, 1991. When that law was replaced by the Threat-ened Species Conservation Act, 1995, the koala continued to be designated as vulnerable by the independent Scientific Committee created with the new legislation. The New South Wales Scientific Committee recently decid-ed that the Hawks Nest and Tea Gardens koalas meet the criteria of an endangered population.

Koalas are native to the Australia Capital Territory, although they were very rare by 1901. Currently the population is small and likely the descendants of several introductions from Victoria. Almost all of the koalas in Victoria represent the success of reintroduction efforts, as the species was extirpated in the State by the early 1900s, with the exception of three remnant populations. Koalas were introduced to Phillip and French Islands by the 1890s, and it is from translocations of these populations, which began to overcrowd their island habitats, that the present population largely descends. . . .

Land use practices vary enormously in different States, and they are currently undergoing evaluation and change in many jurisdictions. We conclude that the inadequacy of present regulations over a significant portion of the species' range is a factor in designating the koala as threatened. . . .

What Are the Available Conservation Measures as a Result of This Listing?

Although habitat loss was a crucial factor in the determination that the koala is threatened, specific critical habitat is not being proposed, as its designation is not applicable to foreign species.

Conservation measures provided to species listed as endangered or threatened under the ESA include recognition, international cooperation, recovery actions, requirements for Federal protection, and prohibitions against certain activities. Recognition through listing encourages conservation measures by Federal, international, and private agencies, groups, and individuals.

Section 7(a) of the Act, as amended, and as implemented by regulations at 50 CFR part 402, requires Federal agencies to evaluate their actions that are to be conducted within the United States or on the high seas with respect to any species that is proposed or listed as endangered or threatened and with respect to its proposed or designated critical habitat (if any). Section 7(a)(2) requires Federal agencies to ensure that activities they authorize, fund, or carry out are not likely to jeopardize the continued existence of a listed species or to destroy or adversely modify its critical habitat. If a proposed Federal action may affect a listed species, the responsible Federal agency must enter into formal consultation with the Service. We are not aware of such actions with respect to the species covered by this proposal, except as may apply to importation permit procedures.

Section 8(a) of the Act authorizes the provision of limited financial assistance for the development and management of programs that the Secretary of the Interior determines to be necessary or useful for the conservation of endangered and threatened species in foreign countries. Sections 8(b) and 8(c) of the Act authorize the Secretary to encourage conservation programs for foreign endangered and threatened species and to provide assistance for such programs in the form of personnel and the training of personnel.

Section 9 of the Act, and implementing regulations found at 50 CFR 17.21 and 17.31, set forth a series of general prohibitions and exceptions that apply to all threatened wildlife. These prohibitions, in part, make it illegal for any person subject to the jurisdiction of the United States to take, import or export, ship in interstate commerce in the course of commercial activity, or sell or offer for sale in interstate or foreign commerce any threatened wildlife. It also is illegal to possess, sell, deliver, transport, or ship any such wildlife that has been taken in violation of the Act. Certain exceptions apply to agents of the Service and State conservation agencies.

We may issue permits to carry out otherwise prohibited activities involving endangered and threatened wildlife under certain circumstances. Regulations governing permits are codified at 50 CFR 17.22, 17.23, and 17.32. Permits are available for scientific purposes, to enhance propagation or survival, or for incidental take in connection with otherwise lawful activities. These permits must also be consistent with the purposes and policy of the Act as required by Section 10(d). For threatened species, we may also issue permits for zoological exhibition, educational purposes, or special purposes consistent with the purposes of the Act.

Our policy ... is to identify to the maximum extent practicable at the time a species is listed those activities that would or would not constitute a violation of section 9 of the Act. The intent of this policy is to increase public awareness of the effects of this listing on proposed or ongoing activities involving the species. Importations into and exportations from the

United States, and interstate and foreign commerce, of koalas (including tissues, parts, and products) from New South Wales and Queensland without a threatened species permit would be prohibited. Koalas removed from the wild or born in captivity prior to the date the species is listed under the Act would be considered "pre-Act" and would not require permits unless they enter commerce. When a specimen is sold or offered for sale, it loses its pre-Act status. Currently, 10 zoological institutions in the United States hold koalas. . . .

Notes and Questions

1. Why is the plight of the koala the concern of the United States? Is the United States more or less justified in legislating to protect the koala than it is to protect African elephants? Should the United States defer to the wishes of the Australian government?

2. Would you answer the questions in note 1 differently if the kangaroo lived in Vietnam, or in another developing country that enjoys less wealth and a less established legal system than Australia?

Defenders of Wildlife v. Lujan

United States Court of Appeals for the Eighth Circuit, 1990.
911 F.2d 117.

■ John R. Gibson, Circuit Judge.

[In 1978, the Fish and Wildlife Service and the National Marine Fisheries Service promulgated a joint regulation providing that the consultation duties imposed by section 7 of the Endangered Species Act apply to the actions of federal agencies in foreign countries. The agencies changed their position in 1986 and issued a new regulation limiting the scope of the consultation duties to federal agency actions within the United States or on the high seas. Environmentalists sued to overturn the 1986 regulation and to require, for example, the Agency for International Development to engage in a section 7 consultation before funding development projects that could jeopardize the habitat of endangered crocodiles, elephants and leopards in Africa.]

. . . It cannot be denied that Congress has chosen expansive language which admits to no exceptions. Reduced to its simplest form, the statute clearly states that each federal agency must consult with the Secretary regarding any action to insure that such action is not likely to jeopardize the existence of any endangered species. We recognize, however, that the use of all-inclusive language in this particular section of the Act is not determinative of the issue. We must search the Act further for clear expression of congressional intent.

The Supreme Court extensively discussed the Act's ambitious purpose in *Tennessee Valley Authority v. Hill*. "The plain intent of Congress in enacting this statute was to halt and reverse the trend toward species extinction, whatever the cost. This is reflected not only in the stated policies of the Act, but in literally every section of the statute." The Court described the Act as "the most comprehensive legislation for the preservation of endangered species ever enacted by any nation."

In the Act, Congress declared that "the United States has pledged itself as a sovereign state in the international community to conserve to the extent practicable the various species of fish or wildlife and plants facing extinction." The Act lists various international agreements which guide this pledge. Congress also committed itself to meeting the international commitments of the United States to existing conservation programs. The Act further declares one of its purposes is to take the appropriate steps to achieve the purposes of the international treaties and conventions just mentioned.

The Act defines "endangered species" broadly and without geographic limitations. Furthermore, the Act sets out a detailed procedure for determining whether a species is endangered. This section states that the Secretary shall determine whether a species is endangered or threatened after taking into account "those efforts, if any, being made by any State or foreign nation ... to protect such species." The Secretary is instructed to give consideration to species which have been designated as requiring protection from unrestricted commerce by any foreign nation, or pursuant to any international agreement, and species identified as in danger of extinction by any State agency or by any agency of a foreign nation. Moreover, the Secretary is required to give actual notice to and invite comment from each foreign nation in which species proposed for listing as endangered are found.

The Secretary is instructed to publish a list of all species found to be threatened. Defenders asserts, and the Secretary does not contest, that "[a]s of May 1989, of 1,046 species listed as endangered or threatened, 507 were species whose range is outside the United States. In addition, there are 71 listed species whose range includes both United States and foreign territory." The listing process does not distinguish between domestic and foreign species.

The Act contains a section entitled "International Cooperation" which declares that the United States' commitment to worldwide protection of endangered species will be backed by financial assistance, personnel assignments, investigations, and by encouraging foreign nations to develop their own conservation programs. While the Secretary argues that this section and section 1538, dealing with imports and exports of wildlife, embody Congress' complete response to the international problem of endangered species, we are persuaded that this provision cannot be so neatly excised from the larger statutory scheme. Rather, we believe that the Act, viewed as a whole, clearly demonstrates congressional commitment to worldwide conservation efforts. To limit the consultation duty in a manner which protects only domestic endangered species runs contrary to such a commitment.

Based upon the foregoing examination of the Act as a whole, we are convinced that congressional intent can be gleaned from the plain language of the Act. Accordingly, we owe no deference to the Secretary's construction of the Act. *See Chevron*, 467 U.S. at 842–43. Furthermore, "[t]he judiciary is the final authority on issues of statutory construction and must reject administrative constructions which are contrary to clear congressional intent."

We believe that the answer to the extraterritorial issue can be found in the plain words of the statute. Our examination of the statute's legislative history, however, also reinforces our conclusion.

The original Environmental Species Act was enacted in 1973. Soon thereafter, the Secretary initiated a rulemaking process in order to implement the Act. In regard to the consultation requirement at issue here, the Secretary solicited comment from various agencies. Several agencies, including the Army Corps of Engineers, the State Department, and the Defense Department, expressed opposition to extraterritorial application. The Council on Environmental Quality, the Interior Department Solicitor's Office, and the General Counsel's Office of the National Oceanic and Atmospheric Administration, however, took the position that the consultation duty extended to foreign countries. After considering the extensive commentary, the Secretary concluded that Congress intended the duty to extend beyond the United States, and published a final rule on January 4, 1978, providing that: Section 7 . . . requires every Federal agency to insure that its activities or programs in the United States, upon the high seas, and in foreign countries, will not jeopardize the continued existence of a listed species. At that time, the Secretary justified the extraterritorial application by stressing the Act's broad, inclusive language; its legislative history; and its policy implications.

After these regulations were issued, Congress amended the consultation section of the Act to reflect its present version. The amendment was essentially a reorganization to allow additions to the rest of the section. The conference report to these 1978 amendments indicates that no substantive changes were intended. . . . In light of the fact that the "existing law" at the time of the 1978 amendments included the prior regulation requiring consultation on foreign projects, we believe that the above language provides strong evidence of the conference committee's tacit approval of the prior regulation. . . .

Despite this evidence of congressional intent, in 1983, the Secretary issued a notice of proposed rulemaking to revise the regulation. The proposed regulation eliminated the need for consultation on foreign projects and defined "action" to exclude foreign activities. The Secretary attributed its radical shift on extraterritorial application to "the apparent domestic orientation of the consultation and exemption processes resulting from the [1978] Amendments, and because of the potential for interference with the sovereignty of foreign nations."

We are compelled to reject this justification. We recognize that "[a]n administrative agency is not disqualified from changing its mind," and that "substantial deference is nonetheless appropriate if there appears to be have been good reason for the change." In this situation, however, the reasons offered for the change fall far short when examined in the context of the Act's language and legislative history previously discussed.

The Secretary places great emphasis upon the Act's treatment of the critical habitat clause, as support for its position. According to the Secretary, Congress could not have intended that the critical habitat provisions apply only to domestic projects while the consultation requirement extends to foreign projects. We are not persuaded. The Act reveals an intent to separately address the concerns raised by critical habitats and endangered species. The designation of critical habitat is governed by different procedures and standards than the listing of endangered species. Furthermore, we observe that the Secretary was not troubled by this alleged inconsistency when it promulgated its earlier regulation permitting differing geographic scopes of the two concerns. The evidence reveals that the consultation requirement and the critical habitat designation have been viewed as severable as to their geographical scope.

The Secretary claims that the domestic orientation of the consultation requirement is shown by the exemption provision added by the 1978 amendments. Specifically, the Secretary points out that exemptions are granted only if "the action is of regional or national significance," and require the weighing of public interests, which would be a gross intrusion upon the sovereignty of foreign nations. Again, we are unpersuaded. The exemption clauses provide that "the Governor of the State in which an agency action will occur, if any, . . . may apply to the Secretary for an exemption." This language, when considered with the substantive and persuasive evidence previously discussed, leads us to conclude that the exemption provisions do not limit the consultation requirement geographically. The Secretary also identifies other provisions of the Act which purportedly limit the consultation duty. We have carefully considered these arguments and believe that they do not compel a different result here. They merit no further discussion.

To support its construction of the Act, the Secretary relies heavily upon the canon of statutory construction that statutes are presumed to have domestic scope only. To overcome the presumption that the statute was not intended to have extraterritorial application, there must be clear expression of such congressional intent. We are convinced that evidence of such intent is found both in the words of the Act and in its legislative history as previously set forth. This evidence leaves us with the belief that Congress intended for the consultation obligation to extend to all agency actions affecting endangered species, whether within the United States or abroad.

The Secretary also expresses concerns about the impact on foreign relations stemming from extraterritorial application of the consultation duty. It urges that such a construction would be viewed as an intrusion upon the sovereign right of foreign nations to strike their own balance between development of natural resources and protection of endangered species. We note initially that the Act is directed at the actions of federal agencies, and not at the actions of sovereign nations. Congress may decide that its concern for foreign relations outweighs its concern for foreign wildlife; we, however, will not make such a decision on its behalf.

Notes and Questions

1. The Supreme Court reversed the Eighth Circuit because the environmental plaintiffs lacked standing to raise the extraterritoriality issue. *See* Lujan v. Defenders of Wildlife, 504 U.S. 555 (1992). Justice Stevens would have held that the plaintiffs did have standing, but he disagreed with the Eighth Circuit on the merits and would have held that the ESA does not apply extraterritorially. *Id.* at 585–89 (Stevens, J., concurring in the judgment). The issue remains unresolved by the Supreme Court.

2. Judge Gibson acknowledged the rule that statutes are presumed *not* to apply extraterritorially unless Congress clearly says so. Does the evidence cited by the court overcome that presumption? Justice Stevens emphasized the presumption against extraterritoriality in concluding that the ESA's consultation duty does not apply to projects conducted in foreign countries. *See Lujan*, 504 U.S. at 585–89. Is that presumption justified? What is its purpose? Jonathan Turley advocates the contrary presumption "that, unless expressly limited, Congress intends statutes to apply extraterritorially." Jonathan Turley, *"When in Rome": Multinational Misconduct and the Presumption Against Extraterritoriality*, 84 Nw. U.L. Rev. 598, 655–60 (1990). What is the justification for *that* presumption? Why have any presumption?

3. Similar questions surround the extraterritorial effect of other federal statutes affecting biodiversity. In United States v. Mitchell, 553 F.2d 996 (5th Cir.1977), the defendant U.S. citizen was charged with violating the Marine Mammal Protection Act (MMPA) by capturing dolphins in the Bahamas and taking them to an aquarium in England. The Fifth Circuit reversed Mitchell's conviction because the MMPA did not apply to his actions. The court explained:

> When Congress considers environmental legislation, it presumably recognizes the authority of other sovereigns to protect and exploit their own resources. Other states may strike balances of interests that differ substantially from those struck by Congress. The traditional method of resolving such differences in the international community is through negotiation and agreement rather than the imposition of one particular choice by a state imposing its law extraterritorially.

Id. at 1002.

Numerous federal courts have grappled with the extraterritorial application of the National Environmental Policy Act (NEPA), which requires an environmental impact statement (EIS) of any federal government project that will have a significant effect on the environment. In Environmental Defense Fund, Inc. v. Massey, 986 F.2d 528 (D.C.Cir.1993), the court held that NEPA obligated the National Science Foundation to prepare an EIS before incinerating food wastes in Antarctica. Other NEPA cases have reached differing results. *See* NRDC v. Nuclear Regulatory Comm'n, 647 F.2d 1345 (D.C.Cir.1981) (holding NEPA does not apply extraterritorially to NRC export licensing decisions); Sierra Club v. Adams, 578 F.2d 389 (D.C.Cir.1978) (assuming that NEPA applied to the construction of a highway in South America where the United States had 2/3 of the financial responsibility and controlled construction); NEPA Coalition of Japan v. Aspin, 837 F.Supp. 466 (D.D.C.1993) (rejecting the application of NEPA to the actions at American military installations in Japan); Greenpeace USA v. Stone, 748 F.Supp. 749 (D.Haw.1990) (holding that NEPA does not apply extraterritorially to the transport of U.S. chemical weapons from West Germany to Johnston Atoll for destruction there), *appeal dismissed*, 924 F.2d 175 (9th Cir.1991) (case moot because transport completed); People of Enewetak v. Laird, 353 F.Supp. 811 (D.Haw.1973) (holding that NEPA applies to U.S. trust territories in Pacific).

4. *Should* the ESA apply extraterritorially? What are the costs of using U.S. environmental laws to regulate environmental activities in other countries? Are they outweighed by the environmental benefits? Principle 12 of the Rio Declaration on Environment and Development states that "[u]nilateral actions to deal with environmental challenges outside the jurisdiction of the importing country should be avoided. Environmental measures addressing transboundary or global environmental problems should, as far as possible, be based on international consensus." In a leading discussion of the issue, the editors of the Harvard Law Review proposed a test stating that U.S. environmental statutes should not be applied extraterritorially unless the country in which the activity occurs does not have adequate environmental laws itself. *See Developments in the Law: International Environmental Law*, 104 Harv. L. Rev. 1484, 1609–38 (1991). How do those tests differ? Do any of the countries discussed earlier in this chapter–Australia, China, or the Latin American and Caribbean nations–have "inadequate" environmental laws so that the extraterritorial application of the laws of other countries is justified?

United States–Import Prohibition of Certain Shrimp and Shrimp Products

World Trade Organization Appellate Body, 1998
WT/DS58/AB/R; (98–3899); AB–1998–4
1998 WTO DS LEXIS 13 (1998)

[In 1987, the United States issued regulations pursuant to the ESA that required shrimp fishers to use Turtle Excluder Devices (TEDs) in areas where many endangered sea turtles are killed when caught in shrimp trawls. Without such devices, sea turtles are captured as shrimp nets drag through the water, and the turtles drown if they cannot escape the nets. TEDs are designed to guide the turtles out of the net while keeping the shrimp within the net. Section 609 of a general congressional appropriations statute enacted in 1989 banned the importation of shrimp from any nation that did not require its shrimp fishers to use the TEDs mandated by the ESA. At first, the State Department interpreted the amendment to apply only to those nations with shrimping fleets in the Gulf of Mexico, the Caribbean Sea, and the western Atlantic Ocean. The Earth Island Institute, an environmental group based in San Francisco, successfully challenged the limited application of the amendment in federal court, and in May 1996 the United States imposed a shrimp embargo on 40 nations that failed to comply with the TED requirements. In response, India, Malaysia, and Pakistan filed a complaint with the World Trade Organization (WTO) contending that the American shrimp importation ban violated international trade agreements.]

113.   Article XX of the [General Agreement on Tariffs and Trade] GATT 1994 reads, in its relevant parts:

Article XX *General Exceptions*

Subject to the requirement that such measures are not applied in a manner which would constitute a means of arbitrary or unjustifiable discrimination between countries where the same conditions prevail, or a disguised restriction on international trade, nothing in this Agreement shall be construed to prevent the adoption or enforcement by any Member of measures:

... (*b*) necessary to protect human, animal or plant life or health;

... (*g*) relating to the conservation of exhaustible natural resources if such measures are made effective in conjunction with restrictions on domestic production or consumption;

B.   *Article XX(g): Provisional Justification of Section 609*

125.   In claiming justification for its measure, the United States primarily invokes Article XX(g). Justification under Article XX(b) is claimed only in the alternative; that is, the United States suggests that we should look at Article XX(b) only if we find that Section 609 does not fall within the ambit of Article XX(g). We proceed, therefore, to the first tier of the analysis of Section 609 and to our consideration of whether it may be characterized as provisionally justified under the terms of Article XX(g)....

1.   "Exhaustible Natural Resources"....

128.   ... Textually, Article XX(g) is *not* limited to the conservation of "mineral" or "non-living" natural resources. The complainants' principal argument is rooted in the notion that "living" natural resources are "renewable" and therefore cannot be "exhaustible" natural resources. We do not believe that "exhaustible" natural resources and "renewable" natural resources are mutually exclusive. One lesson that modern biological sciences teach us is that living species, though in principle, capable of reproduction and, in that sense, "renewable", are in certain circumstances indeed susceptible of depletion, exhaustion and extinction, frequently because of human activities. Living resources are just as "finite" as petroleum, iron ore and other non-living resources....

132.   We turn next to the issue of whether the living natural resources sought to be conserved by the measure are "exhaustible" under Article XX(g). That this element is present in respect of the five species of sea turtles here involved appears to be conceded by all the participants and third participants in this case. The exhaustibility of sea turtles would in fact have been very difficult to controvert since all of the seven recognized species of sea turtles are today listed in Appendix 1 of the Convention on International Trade in Endangered Species of Wild Fauna and Flora ("CITES"). The list in Appendix 1 includes "all species *threatened with extinction* which are or may be affected by trade."

133.   Finally, we observe that sea turtles are highly migratory animals, passing in and out of waters subject to the rights of jurisdiction of various coastal states and the high seas.... The sea turtle species here at stake, i.e., covered by Section 609, are all known to occur in waters over which the United States exercises jurisdiction. Of course, it is not claimed that *all* populations of these species migrate to, or traverse, at one time or another, waters subject to United States jurisdiction. Neither the appellant nor any of the appellees claims any rights of exclusive ownership over the sea turtles, at least not while they are swimming freely in their natural habitat—the oceans. We do not pass upon the question of whether there is an implied jurisdictional limitation in Article XX(g), and if so, the nature or extent of that limitation. We note only that in the specific circumstances of the case before us, there is a sufficient nexus between the migratory and endangered marine populations involved and the United States for purposes of Article XX(g).

134.   For all the foregoing reasons, we find that the sea turtles here involved constitute "exhaustible natural resources" for purposes of Article XX(g) of the GATT 1994.

2.   "Relating to the Conservation of [Exhaustible Natural Resources]"

135.   Article XX(g) requires that the measure sought to be justified be one which "relates to" the conservation of exhaustible natural resources. [The appellate body concluded that section 609(b)(1) related to the conservation of exhaustible natural resources because that section "is not a simple, blanket prohibition of the importation of shrimp imposed without regard to the consequences (or lack thereof) of the mode of harvesting employed upon the incidental capture and mortality of sea turtles."]

3.   *"If Such Measures are Made Effective in conjunction with Restrictions on Domestic Production or Consumption"....*

144.   We earlier noted that Section 609, enacted in 1989, addresses the mode of harvesting of imported shrimp only. However, two years earlier, in 1987, the United States issued regulations pursuant to the Endangered Species Act requiring all United States shrimp trawl vessels to use approved TEDs, or to restrict the duration of tow-times, in specified areas where there was significant incidental mortality of sea turtles in shrimp trawls. These regulations became fully effective in 1990 and were later modified. They now require United States shrimp trawlers to use

approved TEDs "in areas and at times when there is a likelihood of intercepting sea turtles," with certain limited exceptions. Penalties for violation of the Endangered Species Act, or the regulations issued thereunder, include civil and criminal sanctions. The United States government currently relies on monetary sanctions and civil penalties for enforcement. The government has the ability to seize shrimp catch from trawl vessels fishing in United States waters and has done so in cases of egregious violations. We believe that, in principle, Section 609 is an even-handed measure.

145. Accordingly, we hold that Section 609 is a measure made effective in conjunction with the restrictions on domestic harvesting of shrimp, as required by Article XX(g).

C. *The Introductory Clauses of Article XX: Characterizing Section 609 under the Chapeau's Standards*

146. As noted earlier, the United States invokes Article XX(b) only if and to the extent that we hold that Section 609 falls outside the scope of Article XX(g). Having found that Section 609 does come within the terms of Article XX(g), it is not, therefore, necessary to analyze the measure in terms of Article XX(b).

147. Although provisionally justified under Article XX(g), Section 609, if it is ultimately to be justified as an exception under Article XX, must also satisfy the requirements of the introductory clauses—the "chapeau"—of Article XX, that is,

Article XX *General Exceptions*

Subject to the requirement that such measures are *not applied in a manner which would constitute a means of arbitrary or unjustifiable discrimination between countries where the same conditions prevail*, or *a disguised restriction on international trade*, nothing in this Agreement shall be construed to prevent the adoption or enforcement by any Member of measures: (emphasis added)

We turn, hence, to the task of appraising Section 609, and specifically the manner in which it is applied under the chapeau of Article XX; that is, to the second part of the two-tier analysis required under Article XX....

2. *"Unjustifiable Discrimination"*

161. We scrutinize first whether Section 609 has been applied in a manner constituting "unjustifiable discrimination between countries where the same conditions prevail." Perhaps the most conspicuous flaw in this measure's application relates to its intended and actual coercive effect on the specific policy decisions made by foreign governments, Members of the WTO. Section 609, in its application, is, in effect, an economic embargo which requires *all other exporting Members*, if they wish to exercise their GATT rights, to adopt *essentially the same* policy (together with an approved enforcement program) as that applied to, and enforced on, United States domestic shrimp trawlers. As enacted by the Congress of the United States, the *statutory* provisions of Section 609(b)(2)(A) and (B) do not, in

themselves, *require* that other WTO Members adopt *essentially the same* policies and enforcement practices as the United States. Viewed alone, the statute appears to permit a degree of discretion or flexibility in how the standards for determining comparability might be applied, in practice, to other countries.[158] However, any flexibility that may have been intended by Congress when it enacted the statutory provision has been effectively eliminated in the implementation of that policy through the 1996 Guidelines promulgated by the Department of State and through the practice of the administrators in making certification determinations. . . .

166. Another aspect of the application of Section 609 that bears heavily in any appraisal of justifiable or unjustifiable discrimination is the failure of the United States to engage the appellees, as well as other Members exporting shrimp to the United States, in serious, across-the-board negotiations with the objective of concluding bilateral or multilateral agreements for the protection and conservation of sea turtles, before enforcing the import prohibition against the shrimp exports of those other Members. . . .

168. Second, the protection and conservation of highly migratory species of sea turtles, that is, the very policy objective of the measure, demands concerted and cooperative efforts on the part of the many countries whose waters are traversed in the course of recurrent sea turtle migrations. The need for, and the appropriateness of, such efforts have been recognized in the WTO itself as well as in a significant number of other international instruments and declarations. As stated earlier, the Decision on Trade and Environment, which provided for the establishment of the CTE and set out its terms of reference, refers to both the Rio Declaration on Environment and Development and Agenda 21. Of particular relevance is Principle 12 of the Rio Declaration on Environment and Development, which states, in part:

> Unilateral actions to deal with environmental challenges outside the jurisdiction of the importing country should be avoided. *Environmental measures addressing transboundary or global environmental problems should, as far as possible, be based on international consensus.* (emphasis added). . . .

Moreover, we note that Article 5 of the Convention on Biological Diversity states:

> . . . each contracting party shall, as far as possible and as appropriate, cooperate with other contracting parties directly or, where appropriate, through competent international organizations, in respect of areas

---

158. Pursuant to Section 609(b)(2), a harvesting nation may be certified, and thus exempted from the import ban, if:

(A) the government of the harvesting nation has provided documentary evidence of the adoption of a program governing the incidental taking of such sea turtles in the course of such harvesting that is comparable to that of the United States; and

(B) the average rate of that incidental taking by vessels of the harvesting nation is comparable to the average rate of incidental taking of sea turtles by United States vessels in the course of such harvesting . . .

beyond national jurisdiction and on other matters of mutual interest, for the conservation and sustainable use of biological diversity....

169.   Third, the United States did negotiate and conclude one regional international agreement for the protection and conservation of sea turtles: The Inter–American Convention. This Convention was opened for signature on 1 December 1996 and has been signed by five countries, in addition to the United States, and four of these countries are currently certified under Section 609. This Convention has not yet been ratified by any of its signatories. The Inter–American Convention provides that each party shall take "appropriate and necessary measures" for the protection, conservation and recovery of sea turtle populations and their habitats within such party's land territory and in maritime areas with respect to which it exercises sovereign rights or jurisdiction. Such measures include, notably,

> the reduction, to the greatest extent practicable, of the incidental capture, retention, harm or mortality of sea turtles in the course of fishing activities, through the appropriate regulation of such activities, as well as the development, improvement and use of appropriate gear, devices or techniques, including the use of turtle excluder devices (TEDs) pursuant to the provisions of Annex III [of the Convention]....

172.   Clearly, the United States negotiated seriously with some, but not with other Members (including the appellees), that export shrimp to the United States. The effect is plainly discriminatory and, in our view, unjustifiable. The unjustifiable nature of this discrimination emerges clearly when we consider the cumulative effects of the failure of the United States to pursue negotiations for establishing consensual means of protection and conservation of the living marine resources here involved, notwithstanding the explicit statutory direction in Section 609 itself to initiate negotiations as soon as possible for the development of bilateral and multilateral agreements. The principal consequence of this failure may be seen in the resulting unilateralism evident in the application of Section 609. As we have emphasized earlier, the policies relating to the necessity for use of particular kinds of TEDs in various maritime areas, and the operating details of these policies, are all shaped by the Department of State, without the participation of the exporting Members. The system and processes of certification are established and administered by the United States agencies alone. The decision-making involved in the grant, denial or withdrawal of certification to the exporting Members, is, accordingly, also unilateral. The unilateral character of the application of Section 609 heightens the disruptive and discriminatory influence of the import prohibition and underscores its unjustifiability....

175.   Differing treatment of different countries desiring certification is also observable in the differences in the levels of effort made by the United States in transferring the required TED technology to specific countries. Far greater efforts to transfer that technology successfully were made to certain exporting countries—basically the fourteen wider Caribbean/western Atlantic countries cited earlier—than to other exporting countries,

including the appellees. The level of these efforts is probably related to the length of the "phase-in" periods granted—the longer the "phase-in" period, the higher the possible level of efforts at technology transfer. Because compliance with the requirements of certification realistically assumes successful TED technology transfer, low or merely nominal efforts at achieving that transfer will, in all probability, result in fewer countries being able to satisfy the certification requirements under Section 609, within the very limited "phase-in" periods allowed them.

176. When the foregoing differences in the means of application of Section 609 to various shrimp exporting countries are considered in their cumulative effect, we find, and so hold, that those differences in treatment constitute "unjustifiable discrimination" between exporting countries desiring certification in order to gain access to the United States shrimp market within the meaning of the chapeau of Article XX.

3. *"Arbitrary Discrimination"*

177. We next consider whether Section 609 has been applied in a manner constituting "arbitrary discrimination between countries where the same conditions prevail." We have already observed that Section 609, in its application, imposes a single, rigid and unbending requirement that countries applying for certification under Section 609(b)(2)(A) and (B) adopt a comprehensive regulatory program that is essentially the same as the United States' program, without inquiring into the appropriateness of that program for the conditions prevailing in the exporting countries. Furthermore, there is little or no flexibility in how officials make the determination for certification pursuant to these provisions. In our view, this rigidity and inflexibility also constitute "arbitrary discrimination" within the meaning of the chapeau....

185. In reaching these conclusions, we wish to underscore what we have *not* decided in this appeal. We have *not* decided that the protection and preservation of the environment is of no significance to the Members of the WTO. Clearly, it is. We have *not* decided that the sovereign nations that are Members of the WTO cannot adopt effective measures to protect endangered species, such as sea turtles. Clearly, they can and should. And we have *not* decided that sovereign states should not act together bilaterally, plurilaterally or multilaterally, either within the WTO or in other international fora, to protect endangered species or to otherwise protect the environment. Clearly, they should and do.

186. What we *have* decided in this appeal is simply this: although the measure of the United States in dispute in this appeal serves an environmental objective that is recognized as legitimate under paragraph (g) of Article XX of the GATT 1994, this measure has been applied by the United States in a manner which constitutes arbitrary and unjustifiable discrimination between Members of the WTO, contrary to the requirements of the chapeau of Article XX. For all of the specific reasons outlined in this Report, this measure does not qualify for the exemption that Article XX of the GATT 1994 affords to measures which serve certain recognized, legitimate environmental purposes but which, at the same time, are not applied

in a manner that constitutes a means of arbitrary or unjustifiable discrimination between countries where the same conditions prevail or a disguised restriction on international trade. As we emphasized in *United States— Gasoline*, WTO Members are free to adopt their own policies aimed at protecting the environment as long as, in so doing, they fulfill their obligations and respect the rights of other Members under the *WTO Agreement*. . . .

Notes and Questions

1. Should the United States ban the importation of shrimp from nations that do not require TEDs? The general issue of the proper relationship between free trade and environmental protection has generated enormous controversy and an enormous literature. The arguments on both sides are nicely summarized in David Hunter, James Salzman & Durwood Zaelke, International Environmental Law and Policy 1127–39 (2d ed. 2002); and Michael M. Weinstein & Steve Charnovitz, *The Greening of the WTO*, Foreign Affairs, Nov./Dec. 2001, at 147.

2. Why did the WTO conclude that the American ban was illegal? Prior to the establishment of the WTO in 1995, a GATT dispute panel had been less accepting of the effects of American environmental regulation on international trade. The leading dispute involved trade sanctions imposed by the United States pursuant to the Marine Mammal Protection Act banning the importation of tuna from nations that failed to protect dolphins from the effects of tuna fishing. In 1991, a GATT panel held that the import ban violated GATT. *See* General Agreement on Tariffs and Trade: Dispute Panel Settlement Panel Report on United States Restrictions on Imports of Tuna, 30 I.L.M. 1594 (1991). Two years later, another GATT panel reached a similar conclusion concerning the secondary embargo imposed upon intermediary nations who could not prove that any of their shipments to the United States contained tuna that had been caught by nations that failed to take dolphin-safe fishing. *See* General Agreement on Tariffs and Trade: Dispute Settlement Panel Report on United States Restrictions on Import of Tuna, 33 I.L.M. 839 (1994). While the dispute panel reports never took effect, they provoked a hostile response from environmentalists around the world.

3. After the WTO appellate panel issued its decision in the shrimp and sea turtle dispute, the United States agreed to import shrimp from nations that do not require TEDs when it is proved that the shrimp came from boats that used TEDs. In October 2001, a WTO appellate body upheld a renewed United States ban on the importation of shrimp from any nations who could not show that the shrimp were caught by boats using TEDs. The appellate body found that the United States remedied its previous unfair discrimination. The appellate body noted that the United States had made good faith efforts to negotiate a sea turtle conservation agreement with the Indian Ocean and southeast Asian nations affected by the law, and that it had helped those nations with technical advice to adopt fishing methods that were safe for the endangered sea turtles. *See* United States–Import Prohibition of Certain Shrimp and Shrimp Products, Recourse to Article 21.5 by Malaysia (WT/DS58/RW) (Oct. 23, 2001). To the Malaysian press, that decision "signals yet another victory for powerful developed nations to practise double standards and selective protectionism, in the name of conservation and the environment.... Like all developing countries, [Malaysia] has first to have the resources to care for the forests, oceans, and atmosphere. Denying the country a legitimate source of income [from its small wild shrimp industry] is only going to make it that much tougher, regardless of its willingness, to protect the environment; which of course is not quite the rich countries' real objective." *Turning Turtle on the Environment*, BUSINESS TIMES (MALAYSIA), Oct. 25, 2001, at 6. How would you respond to Malaysia's complaints?

4. Would the WTO allow the United States to apply section 7 of its Endangered Species Act to conduct occurring in other countries? Could the United States enforce ESA section 9 to a take of endangered species that occurs overseas?

CHAPTER 14

INTERNATIONAL LAW

Chapter Outline

The international character of biodiversity leads many to champion the role of international law in protecting it. Species that live in more than one country, and ecosystems that transcend national borders, are obvious candidates for legal actions that apply to every concerned nation. Public international law governs the actions of the governments of countries around the world; only rarely does it govern the activities of private organizations or individuals. Public international law includes treaties and other agreements between two or more nations, and the customary international law norms that govern every nation regardless of its consent. Treaties are the dominant mode of protecting biodiversity, though the types of treaties are almost as diverse as the plants and wildlife they seek to protect. The examples range from three countries that have agreed to protect one lake–*e.g.*, the Agreement on the Protection of Lake Constance Against Pollution, signed by Austria, Germany, and Switzerland in 1992–to the 182 parties to the Convention on Biological Diversity who have agreed to protect the full range of biodiversity throughout the world. But treaties, like contracts, only bind those who agree to them. The conspicuous unwillingness of some countries to approve prominent agreements respecting biodiversity–such as the failure of the United States to ratify the Convention on Biological Diversity–demonstrates the limits of treaties as a means of protecting biodiversity. Customary international law, consisting of those legal principles that have become so prevalent to be deemed law, purports to regulate the actions of the governments whether or not the country has agreed to such legal principles. It thus avoids the consent problem posed by treaties, but it presents its own challenges of defining what constitutes such customary international law and determining what consequences attach to a violation of it. Efforts to craft a customary international law solution to the problems of biodiversity have yet to succeed.

International environmental law generally, and the international law protecting biodiversity, have received vastly increased attention in recent years. While some wildlife treaties date back a century or more, the agreements of the past thirty years have attracted more parties and applied to more species than ever before. The international community came together in Stockholm in 1972, and in Rio de Janeiro in 1992, to grapple with the role of international law in protecting the environment and biodiversity. The accompanying treatises, journals, web sites, and other resources are far too numerous to fully engage in this chapter. Some of the best sources of information about the role of international law in protecting biodiversity include the Convention on Biological Diversity's website's at <http://www.biodiv.org/>; LAKSHAM D. GURUSWAMY ET AL., INTERNATIONAL ENVIRONMENTAL LAW AND WORLD ORDER: A PROBLEM-ORIENTED COURSEBOOK 792–953 (2d ed. 1999); and DAVID HUNTER, JAMES SALZMAN & DURWOOD ZAELKE, INTERNATIONAL ENVIRONMENTAL LAW AND POLICY 909–1124 (2d ed. 2002).

A.   TREATIES

Statement of Dr. William T. Hornaday, Investigation of the Fur–Seal Fisheries: Hearing Before the Senate Committee on Conservation of National Resources on Bill (S. 7242) Entitled "An Act to Protect the Seal Fisheries of Alaska, and for Other Purposes"

61st Congress, 2d Session (1910).

Doctor HORNADAY. I fancy that you have time, gentlemen, to consider only the facts that impinge directly upon the subject of the present hearing. It is my sincere belief that the fate of the fur seal today lies in the hands of this committee, the Secretary of Commerce and Labor, the Secretary of State, the Commissioner of Fisheries, and the President of the United States, and if that small group of men is not instrumental in securing immediately the measures that are absolutely necessary to the preservation of that industry, then I believe that it will be wiped out entirely in the very near future.

First, one word in regard to my own status. By profession I am a zoologist, but when I am at home I draw as much salary for being a practical business man as for my zoological work. My interest in this question is not alone that of a zoologist, nor is it a sentimental interest. As a business man and patriotic citizen I desire to see a very valuable industry saved to the Government and to the people of the United States.

I represent the Camp–Fire Club of America, which is a New on York organization of lawyers, doctors, businessmen, and others, all of whom are sportsmen and lovers of nature. One of the chief objects of the club is the preservation of wild life and forests. Our New York organization is allied with six other clubs, scattered all the way from Jamestown, N.Y., to Los Angeles, Cal. Our own club in New York contains about 350 members. A committee consisting of thirteen members, all of whom, with the exception of myself, are lawyers, has been formed to promote the better protection of wild life. It is called the committee on game protective legislation and preserves, and of it I have the honor to be chairman. One of the first acts of that committee was to consider the case of the fur seal. My expenses here are paid by a special subscription in the club for the work of this committee. I do not come here to represent New York Zoological Society. That organization has not elected to take an active part in the fur-seal matter, for the reason that it has so many interests in other directions.

THE CURSE OF "PELAGIC SEALING"

The present trouble lies first in the fact that the number of fur seals have diminished to a very low point, and secondly, that the situation is entangled with the affairs of other nations, particularly Canada and Japan. The evil at this moment arises from what is call "pelagic sealing," but I shall be careful not to dwell on the details of that, because I know that you are already familiar with it. "Pelagic sealing" means killing seals at sea; and in killing seals at sea all ages and sexes are killed, indiscriminately, and of all that are killed, fully one-half are lost. . . .

Owing to the lack of a treaty with Japan, the Japanese seal-hunters have the right to kill seals to within 3 miles of the shores of [the Pribilof Islands of Alaska]. They always have had that right, and during the past fifteen years have exercised it with merciless vigor and persistence. You will remember that about two years ago a party of Japanese landed on one of the Pribilof Islands and actually began to kill the seals on our own soil.

. . . The Canadians may kill fur seals anywhere outside of [the Canadian dead line], which is 60 miles from the shore of the islands, but Americans may not kill seals at sea anywhere.

Now, it is a well attested fact that every year, in the fall, the seals leave these islands and make the most wonderful migration that is made by any aquatic species in the world. It covers nine months in the year, and reaches about 2,000 miles. The months indicated here represent the seasonal progress of the herd. This region forms part of the hunting grounds of the pelagic sealers of Canada, who make their headquarters in Vancouver and Victoria. Until very recently, at least, the pelagic fleet contained about thirty vessels, manned by about twelve hundred men, using about three hundred boats. The sealers know in general the locality of the fur seals at different periods of the year. The migration route has been determined by the logs of the various pelagic sealing vessels that have taken fur seals at various times of the year throughout that great course. The greatest killing in done around the islands and close along the Pacific coast from San Francisco up to Prince William Sound. Between 1883 and 1897 a total of 304,713 skins of seals killed at sea were marketed, and undoubtedly an equal number were lost. In 1895 there were 56,291 seals killed at sea that were secured.

During the breeding season, or from July to October, the mother seals are obliged to leave the islands to go off shore from sixty to a hundred miles for food. The moment the mother seals leave the Japanese 3–mile dead line, the Japanese sealers are after them; and a mother seal, in order to get her food with which to suckle her young, is obliged to run the gauntlet of these Japanese vessels. The destruction of a mother seal means the starvation of the offspring on shore; and uncounted thousands of young seals have perished on our islands from that cause.

In 1895 a member of Congress, Mr. J. B. Crowley, assisted in counting about 30,000 young fur seals that starved to death on the breeding ground because their mothers had been killed while in quest of food.

Now, let us come down to the necessities of the case. It has been my pleasure to study both the life history and the political history of the fur seal during the past thirty years. I have accumulated during that time a great many documents and publications on the subject, and facts derived from men who are familiar with the fur seals and the islands from personal observation. Up to this date several of my personal friends have been advising the Government as experts, and although I have witnessed this awful and wasteful slaughter of the fur seal herd, I have resolutely held my peace until a few weeks ago. Now, however, as the total destruction of this industry is imminent and the fur seal is fast following the American bison, I feel that it is my duty as an American citizen to speak out.

To come directly to the point, from a careful study of the present situation and the past history that has led up to it, especially the Russian history of the fur-seal industry, I am convinced that we must do three things to save our fur-seal herd, as follows:

(1) We must immediately forbid the execution of a new lease for the killing of fur seals on these islands, for reasons that I will presently set forth;

(2) We must secure treaties with Canada, Russia, Japan, and Mexico which will put a stop to pelagic sealing; and

(3) We must declare a closed season for ten years.

TREATIES REGARDING PELAGIC SEALING.

I have been convinced all along that if a new lease is now made, to take the place of that which expires next April, it will greatly complicate the negotiations of our State Department, not only with Canada, but Japan and Russia as well. We believe that the Japanese and Russians are willing to enter into an agreement with us for the suppression of pelagic sealing as soon as we come to satisfactory terms with Canada. Last year our State Department endeavored to negotiate a seal treaty with Japan, but with no result.

Now, what is the state of affairs with Canada? During the past three years the State Department has submitted several propositions to Canada. I do not wish to say much regarding the work of the State Department, but there are some things that really must be said. I understand that certain propositions were submitted by Senator Root, then Secretary of State, and that they have all been rejected. The matter remains as much open and unsettled—and irritating—today as it was four years ago. Now, here is an important fact bearing on this subject:

In 1905 Mr. John Hay, who was then Secretary of State, prepared, with the assistance of Senator Dillingham, the basis of an agreement with the Canadian government that if ratified, would have settled this whole matter and put it on a good foundation for many years to come. That agreement, a copy of which I hold in my hand, proposed that Canada should have a compensation for the suppression of pelagic sealing by her people. The Canadians are not willing to give up their pelagic sealing privileges for nothing, for the simple reason that the Canadians have made much money out of it. Now, Mr. Hay and Sir Mortimer Durand agreed that it was right for Canada to have compensation—in the form of a small percentage of the net annual income from the killing lease—20 per cent of whatever the amount might be, and also the privilege of being represented by a commissioner on our islands. Well, just as this was on the point of being ratified into a treaty Secretary Hay died and the whole thing fell to the ground. Mr. Elliott exercised every form of persuasion and insistence that he knew of to induce this Government, through Congress, through cabinet officers and the President, to renew it along those lines, but I am told by Mr. Elliott that it never was done.

At the Paris tribunal, in 1893, we made a treaty with England and Canada, by which the Canadians were estopped from pelagic sealing at the 60–mile limit. We have no absolute treaty with Japan and Russia, and therefore the terms of international law hold good; that is to say beyond the 3–mile limit. We can not protect our fur seals from either Japanese or Russians. Canada is still bound by the Paris award to the 60–mile limit.

A CLEAR FIELD FOR TREATY MAKING

That opinion is extremely important at this time. It will take a year to make these treaties; and it may take even longer. I insist upon it, as representing three hundred and fifty good citizens who are interested in this subject, that the making of any new killing lease at this time would complicate our negotiations with foreign governments, and it might easily be the means of entirely defeating the purpose of our Secretary of State. We believe that the imperative need at the present hour is that this committee shall secure from Congress such action which will enable the Secretary of Commerce and Labor to abandon all thought of now executing a new lease, and at the same time, if possible, provide for a ten-year close season.

There is one other matter that I desire to mention, because it is certain to be brought forward by those who desire the execution of a new slaughtering lease. The law provides that the Secretary of Commerce and Labor shall have the power to restrict, or in other words, to place a limit upon the number of seals that may be killed under the lease, according to his discretion. It is claimed by some of the gentlemen who do not agree with us in the matter, that it is not necessary to have a new act passed, or any new regulation of any kind adopted on the theory that the Secretary of Commerce and Labor has the power to regulate killing operations upon our islands and suspend all killing if he chooses. I do not think that, as professional and business men, this committee is likely to lose sight of this fact: With a new lease in existence, and with thousands of dollars invested in the sealing industry, it will require a Secretary of Commerce and Labor with a tremendous amount of courage to go to the length of stopping the killing privilege entirely, even though he should think it ever so necessary.

I do not think that it would be right for this Government to expect any Secretary to assume the entire responsibility of virtually abrogating an existing lease where the holders of the lease have large vested rights, and assuredly would put in large-sized claims for damages.

THE LOSS TO THE UNITED STATES

We got last year (1909) $10.22 per skin for 14,336 skins, making our revenue about $150,000, whereas the protection of the islands with five revenue cutters and other charges on account of the fur-seal industry cost the Government about $340,000 for the year, thus entailing a loss to the Government of about $190,000 for 1909. Ever since 1895 we have had to maintain up there a fleet of from four to five vessels to protect the seal herds from total annihilation. You will remember that on one occasion a marauding force of Japanese seal killers landed on St. George Island and eight Japanese were killed by our guards. With the assistance of Mr. Elliott, I have ascertained from official records the cost of maintaining that patrol fleet, and our other expenses in connection with the fur-seal industry. The diagram I now submit, drawn exclusively from official, records shows that up to 1891, when the present lease was executed, we had received a net revenue of $5,738,724; that since that time, or from 1891 up to this time, the net loss for the twenty years has been $2,247,544; and that during that time the fur-seal herd has decreased from 4,700,000 to whatever it is now, which may reasonably be estimated at 60,000. As stated by Commissioner Bowers in his report, our loss on seals killed at sea easily runs up into more millions than we can reckon up. The worst of it all is that the situation today is so critical that nothing save quick and drastic action can save the industry from complete annihilation.

Now, gentlemen I think that is all that I need to offer on this occasion. In conclusion, I wish to repeat what I said before that the fur-seal industry rests in the hands of this committee, the Secretary of Commerce and Labor, the United States Commissioner of Fisheries, the Secretary of State, and the President of the United States.

I have read various articles in Canadian papers, voicing the sentiments of the Canadian government, to the effect that Canada is not willing to make any new treaty except on a basis of compensation and joint control. It has been stated in several different forms that Canada is willing to enter into a treaty if she secures those two objects in return for the suppression of her pelagic sealers. It is plainly intimated that Japan and Russia are willing to make treaties with us as soon as we have made one with Canada, but not before. The necessity for the suppression of our killing on land, for a period, surely is obvious.

Notes and Questions

1. William Hornaday was more than the representative of the Camp–Fire Club of America. He was also "the influential director of the New York Zoological Park. Hunter, taxidermist, polemicist, student of animal morals, advocate of the white man's burden, enthusiast for phrenology and temperance, Hornaday was a passionate man who got under a lot of people's skins.

Of all his enthusiasms, biodiversity was the foremost: he spent more than forty years railing against the loss of species." CHARLES C. MANN & MARK L. PLUMMER, NOAH'S CHOICE: THE FUTURE OF ENDANGERED SPECIES 119 (1994). Hornaday's 1913 book *Our Vanishing Wild Life* "was probably the first book devoted to the biodiversity crisis." *Id.* at 120. That book proclaimed that "Labor, Money, and Publicity" are the "three very slender threads" upon which "[t]he fate of wild life in North America hangs." WILLIAM T. HORNADAY, OUR VANISHING WILDLIFE: ITS EXTERMINATION AND PRESERVATION 393 (1913). Hornaday focused on biodiversity in the United States, but he also described the present and future of game in Asia and Africa, the game preserves and game laws of Canada, and British game preserves in Africa and Australia.

Hornaday recounted the battle to save the fur seals in a later book describing efforts to protect biodiversity. *See* WILLIAM T. HORNADAY, THIRTY YEARS WAR FOR WILDLIFE: GAINS AND LOSSES IN THE THANKLESS TASKS (1931). (As an aside, the book is probably the only one ever "dedicated to the Congress of the United States as a small token of appreciation of its generous services to wild life during the decade from 1920 to 1930, in new legislation to provide game sanctuaries, and to reduce excessive killing privileges." *Id.* at vii). Three years after the United States purchased Alaska from Russia in 1867, the United States entered into a 20 year lease that allowed the Alaskan Commercial Company to kill 100,000 fur seals–also known as Alaskan sea lions–on the Pribolof Islands. In 1880, pelagic sealing began to replace the killing of the seals on land. The resulting slaughter of the seals involved British, Canadian, Russian, and Japanese hunters. After earlier efforts to protect the seals failed, Hornaday testified before the Senate Committee on Natural Resources on February 26, 1910. As Hornaday described it, it was "a very odd and spectacular Hearing" in which eleven Senators listened to just one witness–Hornaday. "The Committee did not seem to think it worth while to notify or invite any other persons, possibly because no one outside of Congress, save [Henry Wood Elliott, whom Hornaday credited as doing 'by far the most to save the Fur Seal industry'] had manifested the slightest interest in the fate of the unhappy fur seals." *Id.* at 175, 179. The result of Hornaday's testimony was an unanimous resolution opposing the renewal of the lease permitting the killing of fur seals and advocating a treaty to prevent pelagic sealing. "And that action," reported Hornaday, "instantly killed the leasing system of seal killing, forever and a day." *Id.* at 176.

Next, "[i]n due course of time and diplomatic procedure, the Department of State completed an excellent treaty with England and Canada, Japan and Russia, which was fully ratified, and which at once put a stop to the wicked and wasteful killing of seals at sea." *Id.* at 181. The 1911 Treaty for the Preservation and Protection of Fur Seals banned pelagic sealing, empowered each country to seize vessels violating that ban, prohibited the importation of unapproved seal skins, and directed each country to supervise seal harvests on the islands' rookeries within their jurisdiction. The treaty also authorized the continuing harvest of the seals (albeit not by killing at sea) and mandated the payment of various sums for existing skins and in lieu of future hunting. Japan withdrew from the treaty on the eve of

World War II: in fact, it "may have telegraphed its intention to go to war with the United States when on October 23, 1941, it broadcast a message to the Pribolofs in English rejecting the treaty and announcing plans to 'wage war' on American fur seals wherever they were found." George Reiger, *Song of the Seal*, Audubon, Sept. 1975, at 19. Informal restrictions continued until the original parties entered into the Interim Fur Seal Convention of 1957, which essentially retained the same provisions while adding an explicit policy of "achieving the maximum sustainable productivity of the fur seal resources . . . so that the fur seal population can be brought to and maintained at the levels which will provide the greatest harvest year after year." That interim agreement was continually renewed until 1985. At that time, the dramatic increase in the Pribolof Islands fur seal population from 30,000 in 1910 to as many as 2,500,000 in the 1950's began to be reversed. The 1985 population of about 800,000, combined with listing of the fur seal as threatened under the Endangered Species Act, ended the commercial harvest of the fur seals. Since then, the fur seal population has grown slightly to about 1,000,000. *See generally* MICHAEL J. BEAN & MELANIE J. ROWLAND, THE EVOLUTION OF NATIONAL WILDLIFE LAW 471–78 (3d ed. 1997); SIMON LYSTER, INTERNATIONAL WILDLIFE LAW: AN ANALYSIS OF INTERNATIONAL TREATIES CONCERNED WITH THE CONSERVATION OF WILDLIFE 39–48 (1985).

2. Could the United States have saved the seals without a treaty? Was Hornaday right to insist that the fur seal could only be saved through the efforts of the "small group of men" that he named? What would have happened if the nations involved in pelagic sealing had been unable to reach an agreement? Consider the example of the fisheries of the high seas beyond the jurisdiction of any individual country, which were long treated as "common resources subject to the law of capture. Resources belonged to whoever took the property into possession first." HUNTER, SALZMAN & ZAELKE, *supra*, at 680. The many efforts to establish treaties to govern the use of such fisheries are described *id.* at 703–731.

3. The fur seal treaty was one of the first international agreements to protect biodiversity. The few agreements before then were even more limited in their focus, such as a 1781 convention that protected forests and game birds along the border between France and Basel. *See* P. VAN HEIJNSBERGEN, INTERNATIONAL LEGAL PROTECTION OF WILD FAUNA AND FLORA 9 (1997). The fur seal treaty was soon followed by the Migratory Bird Treaty of 1916, which established a number of protections for birds that migrated between the United States and Canada. The ensuing federal statute to enforce the treaty was sustained against a federalism challenge in Missouri v. Holland, 252 U.S. 416 (1920), in which Justice Holmes observed that "[b]ut for the treaty and the statute there soon might be no birds for any powers to deal with." *Id.* at 434–35.

Today there are hundreds of bilateral treaties, regional treaties, and other treaties involving a small number of countries that address certain species or ecosystems. They attempt to protect such biodiversity as:

● *Whales*–The hunting of whales has captured the public's attention at least since the publication of *Moby Dick*. International efforts to regulate the killing of whales began in the 1920's, culminating in the International

Convention for the Regulation of Whaling in 1946. The preamble to the convention explained that the parties sought "to provide for the proper conservation of whale stocks and thus make possible the orderly development of the whaling industry." That convention created the International Whaling Commission (IWC), which can designate protected species of whales and govern the hunting of whales. But "[w]ith no clear mandate for whale conservation, a directive to consider the needs of the whaling industry, and a regulatory structure that enables individual nations to frustrate conservation actions by registering objections to them, the IWC has for most of its life served as overseer of successive depletions of individual whale stocks." BEAN & ROWLAND, *supra*, at 480. Then, in 1982, the IWC imposed a moratorium on commercial whaling. Japan, Norway, and the Soviet Union filed objections to the moratorium and continued whaling until the threat of American trade sanctions convinced each nation to accept the moratorium. Those nations, and others, still kill some whales pursuant to a controversial "research" provision contained in the convention, and Alaskan Eskimos and other aboriginal groups engage in subsistence whaling. Moreover, Iceland withdrew from the convention in 1992, and Norway resumed its commercial whaling in 1993 based on the objection it had lodged against the moratorium. The competing perspectives on the proper role of international law in governing whaling are presented in the web sites of the IWC, <http://ourworld.compuserve.com/homepages/iwcoffice/>; the Japan Whaling Association, <http://wwww.jp-whaling-assn.com/english/index.htm>; and Greenpeace, <www.greenpeace.org.chio.html>.

• *Polar bears*–The decline in the number of polar bears living in the Arctic ecosystem prompted the Soviet Union to ban polar bear hunting in 1955 and the state of Alaska to prohibit sport hunting in 1972. An international agreement followed in 1973, when Canada, Denmark (acting on behalf of Greenland), Norway, the Soviet Union, and the United States entered into the Agreement on the Conservation of Polar Bears. The agreement is designed to limit the killing and capture of polar bears, to protect their habitat, and to facilitate research efforts. The agreement prohibits the "taking" of polar bears–defined as "hunting, killing and capturing"–but it contains a series of exceptions for scientific purposes, conservation purposes, "local people using traditional methods in the exercise of their traditional rights," and takings "wherever polar bears have or might have been subject to taking by traditional means by its nationals." The agreement also requires each party to "take appropriate action to protect the ecosystems of which polar bears are a part, with special attention to habitat components such as denning and feeding sites and migration patterns." That ecosystem provision poses a particular challenge for the United States because "[t]he most important polar bear on-land denning area in Alaska is in the coastal plain of the Arctic National Wildlife Refuge," which has been eyed for increased oil production. BEAN & ROWLAND, *supra*, at 492.

• *Wetlands*–The Convention on Wetlands of International Importance Especially as Waterfowl Habitat–better known as the Ramsar Convention after the Iranian city in which it was signed in 1971–seeks to better protect the most important of the wetlands that constitute four percent of the world's surface. It was "the first international treaty focused on conservation of a single type of ecosystem." HUNTER, SALZMAN & ZAELKE, *supra*, at 1029. Each of the 130 countries that are parties to the convention are required to designate at least one wetland for the List of Wetlands of International Importance. The list contains 1,111 wetlands, such as Karaginsky Island in the Bering Sea along a major bird migration route, a wetland formed during mining construction in the Transvaal region of South Africa in 1930, Kakadu National Park in northern Australia, and the Everglades National Park. (For the entire list, visit the convention's web site at <http://www.ramsar.org/profile_index.htm>). The convention's general guidance regarding the ecological, botanical, hydrological, and other international importance of a wetland has been supplemented by more specific quantitative criteria to determine whether a wetland enjoys the requisite importance. Listed sites that are threatened are included in the Montreux Record, which identifies "wetlands requiring urgent national and international conservation attention." HUNTER, SALZMAN & ZAELKE, *supra*, at 1031. Once listed, a party must encourage the "wise use" of a wetland. Parties are also required to establish nature reserves containing wetlands. Even though "[t]he inclusion of a wetland in the List does not prejudice the exclusive sovereign rights of the Contracting Party in whose territory the wetland is situated," Ramsar Convention, Art. 2, para. 3, the effect of listing has often been to block harmful development from occurring on or near a listed wetland. *See* HUNTER, SALZMAN & ZAELKE, *supra*, at 1035–36 (citing examples from Canada, Chile, Great Britain, Norway, Pakistan, and South Africa).

• *Antarctica*–The fragile and unique ecosystem of Antarctica lies outside the jurisdiction of any individual nation. Instead, 26 countries now play a role in governing the region pursuant to a treaty established in 1959. That treaty paid scant attention to biodiversity, though, simply authorizing recommendations for the "preservation and conservation of living resources." Antarctica's biodiversity gained international protection in 1980 when many of the same interested nations agreed to the Convention on the Conservation of Antarctic Marine Living Resources (CCAMLR). The convention's preemptive effort to protect the abundant, shrimp-like krill makes it "one of the few international treaties concerned with wildlife conservation to be concluded prior to heavy commercial protection of the species it was designed to protect." SIMON LYSTER, INTERNATIONAL WILDLIFE LAW: AN ANALYSIS OF INTERNATIONAL TREATIES CONCERNED WITH THE CONSERVATION OF WILDLIFE 157 (1985). The convention applies to all "marine living resources" in Antarctica, not just krill, by focusing upon the size of harvested populations, maintaining ecological relationships, and preventing irreversible changes in the marine ecosystem. The Commission for the Conservation of Antarctic Marine Living Resources is charged with establishing a marine reserves and reaching a consensus on any necessary regulations. The only such regulations issued by the commission to date permitted a krill harvest nearly three times greater than the largest actual harvest of just over a half million metric tons in 1982. Most recently,the countries responsible for Antarctica agreed to a Protocol on Environmental Protection to the Antarctic Treaty, better known as the Madrid Protocol, in

1992. Two features of the Madrid Protocol are of particular note for Antarctica's biodiversity. It states that "protection of the Antarctic environment and dependent and associated ecosystems, including its wilderness and aesthetic values ... shall be fundamental considerations in the planning and conduct of all activities" there. The protocol also prohibits the introduction of any exotic animal or plant species to the Antarctic without a permit.

• *Food crops*–In 2000, 56 countries entered into the first international agreement specifically designed to protect genetic diversity. The International Undertaking on Plant Genetic Resources emerged from a Rome meeting of the members of the United Nations Food and Agriculture Organization (FAO). That agreement seeks "to ensure the conservation and sustainable use of plant genetic resources used for food and agriculture and the fair and equitable sharing of the benefits arising from their use." Food and Agriculture Organization of the United Nations, *Agreement Reached on Protecting Plant Genetic Resources*, <http://www.fao.org/news/2001/010703–e.htm>. Two complementary provisions are designed to achieve these goals: a system that facilitates access to gene banks, actual farms, and other sources of agricultural information and technology; and a cost-sharing plan that requires mandatory payments when commercial benefits are gained from the use of the plant genetic resources covered by the system, with most of the collected funds invested in developing countries. The agreement also directs countries to "take measures to protect and promote Farmers' Rights" by protecting traditional knowledge about plant genetic resources, encouraging public participation by farmers in decisions about genetic resources, and promoting the ability of farmers to share equitably in the benefits produced by such resources. In November 2001, the FAO submitted the agreement for adoption by its member states, and by the end of 2001 the agreement had gained the approval of eight of the forty nations needed for it to become effective.

Are such agreements more or less desirable than multinational agreements that seek to protect all kinds of biodiversity? Can you identify the factors that should determine whether an issue is best addressed by a focused treaty with few parties, a broad treaty with many parties, or some other kind of agreement?

B. The Convention on International Trade in Endangered Species

Barnabas Dickson, CITES in Harare: A Review of the Tenth Conference of the Parties

Colorado Journal of International Environmental Law.
1997 Year Book 55.

The Convention on International Trade in Endangered Species of Wild Fauna and Flora (CITES) held its tenth Conference of the Parties in

Harare, Zimbabwe, from June 9 to June 20, 1997. There was a record attendance with ninety-six percent of the Parties registered by the start of the conference. Not for the first time at a CITES meeting, discussions were dominated by the African elephant, but this time a proposal was passed to loosen the restrictions on the international trade in ivory. There were other signs at the meeting that the Parties are moving toward a greater acceptance of the philosophy of sustainable use. The pro-trade advocates did not have everything their way, however, since Japan's proposals to weaken the moratorium on whaling were quashed. An inquiry into the effectiveness of the treaty was regarded by many as unsatisfactory and is unlikely to lead to significant reforms, due in part to the absence of consensus among the Parties about the relationship between trade and conservation.

A.   Background

CITES was originally signed in Washington, D.C., in March 1973 and came into force two years later. The guiding assumptions of the treaty are that international wildlife trade is a major threat to the continued existence of some wild species and that, where this is so, the appropriate response is to restrict or halt that trade. The Parties can limit trade by listing a species on either Appendix I or II of the treaty. Appendix II is for the less seriously threatened species, and it imposes some conditions on international trade in the species. Appendix I is for the more seriously threatened species, and the relevant clause states that "Appendix I shall include all species threatened with extinction which are or may be affected by trade." In contrast to Appendix II, an Appendix I listing involves an almost complete ban on international trade of the species in question. Decisions on the listing of species are made at the roughly biennial Conferences of the Parties, and they require a two-thirds majority of those voting to pass. Since the inception of the treaty, the Parties have endeavored to develop clearer and more detailed criteria for listing decisions than are contained in the original articles. Two resolutions passed at the 1976 first Conference of the Parties (CoP) in Berne, Switzerland, represent the first such attempt, and a resolution passed at the 1994 ninth CoP held in Fort Lauderdale, Florida, the latest. Alongside these developments, a growing number of Parties have also questioned the underlying premise of CITES, arguing, under the banner of "sustainable use," that trade can actually be a positive tool for conservation. One outcome of this has been an increasing tendency to qualify listing decisions in various ways, involving such things as split-listings for different populations and quotas for trade. This allows for a wider range of options than the limited choice between Appendix I (no trade) and Appendix II (partial regulation of trade) that is offered by the original treaty. The decision that was eventually reached in Harare regarding elephants illustrates this more flexible approach.

B.   Elephants

The fate of the African elephant (*Loxodonta africana*) has always been a major concern of CITES. The elephant was placed on Appendix II at the first CoP in 1976. With elephant populations continuing to decline precipitously in some countries, the fifth CoP, held in Buenos Aires in 1985, introduced the Management Quota System. When this quota system was perceived to have failed, elephants were placed on Appendix I at the seventh CoP, held in Lausanne, Switzerland, in 1989. Botswana, South Africa, and Zimbabwe were prominent among the opponents, but many critics of the decision now concede that, at least in the short term, the Appendix I listing had beneficial effects on some populations of elephants. With all legal trade banned, the market for illegal ivory was dramatically reduced and, with it, the amount of poaching. Elephant populations in countries such as Kenya began to recover. Nevertheless, the southern African countries continued to argue that their elephant populations had never been seriously threatened, that these populations were well managed, and that they were being punished for the failures of others.

At the eighth CoP, held in Kyoto, Japan, in 1992, some southern African countries tried to have their elephant populations transferred to Appendix II. They withdrew their proposals, however, after encountering a great deal of opposition. At the ninth CoP in 1994, a South African proposal to allow trade in elephant hide and meat (but not ivory) was also withdrawn before a vote was taken. By the time of the 1997 CoP in Harare, however, the mood had changed.

Botswana, Namibia, and Zimbabwe submitted proposals to transfer their populations of elephants to Appendix II. These proposals were all annotated to allow, in 1998 and 1999, direct export of specified amounts of their registered stocks of ivory to one customer, Japan. Critics of the proposals argued that even a limited reopening of the trade would lead to an escalation of poaching. Indeed, some contended that the submission of the proposals had itself caused an increase in poaching. These critics also pointed to the report of the CITES Panel of Experts. This panel, created by a resolution passed at the Lausanne CoP, was to review the proposals from the southern African countries prior to the conference. The report judged that the population status and the management of elephant populations in all three countries was adequate, but it identified weaknesses in the system of regulation in Japan. It stated that "the control of retail trade is not adequate to differentiate the products of legally acquired ivory from those of illegal sources." The report was even more critical of the situation in Zimbabwe. It stated that in that country, "law enforcement with respect to the ivory trade has been grossly inadequate. . . . Officials from the Customs Department declared that they had no interest in controlling ivory exports." "Between September 1995 and October 1996 [the Department of National Parks and Wildlife Management's] control over the carving industry appeared to have broken down. . . ."

However, these considerations were not decisive with the majority of Parties. When the debate opened, South Africa submitted an amendment that incorporated the three proposals for Appendix II listing into a single proposal and added several further restrictions, including the postponement of any resumption of trade until eighteen months after the transfer to Appendix II occurred. More than fifty Parties contributed to the debate.

Among the factors cited in the discussion were the limited nature of the proposed trade, the healthy elephant numbers in the countries concerned, the right of countries to use their natural resources as they wish, and the benefits that trade can bring to poor rural communities. These last two considerations do not formally come under the purview of CITES, but, nevertheless, they appear to have carried some weight in the voting. There was a clear majority for the proposal, with seventy-five in favor, forty-one against, and seven abstentions. This was a substantial turn around from the Lausanne vote of eight years before, but still three votes short of the two-thirds majority needed.

At this point, with no vote having been taken on the original proposals introduced by Botswana, Namibia, and Zimbabwe, the chairman of Committee I proposed that a drafting group be established to work on those proposals. The group was chaired by Norway and included representatives from North America, the European Union, and Africa. The group returned with three documents. The chairman ruled that there would be no further discussion of the elephant issue, and the three documents were then put to a vote. The first of these, Com. 10.34, concerned the conditions under which trade in ivory could resume. It stipulated that trade should not resume until the deficiencies identified by the Panel of Experts were remedied. It also included a clause empowering the CITES Standing Committee to halt trade and retransfer the elephant to Appendix I if there was an escalation of illegal trade or hunting due to the resumption of trade. Moreover, trade could not resume until the Standing Committee had agreed that all of the conditions had been met. This document was passed by a two-thirds majority, 76 to 21, with twenty abstentions. Compared to the vote on the South African amendment, the most notable features were the decrease in the number of votes against and the corresponding rise in the number of abstentions. This change was largely explained by a shift in the European Union's position.

The second document from the drafting group, Com. 10.33, consisted of a new set of annotations to the original proposals. These annotations included the requirement (as had the failed South African amendment) that there would be no resumption of trade until eighteen months after the transfer to Appendix II came into effect. This meant that there would be no trade until March 1999, and then only if the conditions set out in Com 10.34 had been met. Thus annotated, each of the original proposals was voted on and passed with the necessary two-thirds majority. In the case of Botswana, the voting was 74 to 21, with 24 abstentions. The proposals from Namibia and Zimbabwe received similar majorities.

The third and final document from the drafting group, Com. 10.35, concerned the disposal of ivory stocks from African elephant range states. This allowed range states that registered their ivory stockpiles with the CITES Secretariat to sell those stockpiles within a ninety-day period if they were purchased for noncommercial purposes. The revenue generated from these sales would have to be managed by boards of trustees and directed towards conservation, monitoring, capacity-building, and local community-based programs. The vote on this document was 90 to 18 in favor.

These decisions prompted Zimbabwean supporters to break into a rendition of the African anthem *Ishe Komborera Africa* ("God Bless Africa"). The United States and Australia remained the most prominent opponents of the transfer, and Don Barry, head of the US delegation, pledged continued support for elephant conservation, saying, "We all now need to work together for the benefit of the elephants." Some nongovernmental organizations (NGOs) were more outspoken, with Wayne Pacelle of the Humane Society of the United States saying, "It is premature and reckless to renew this trade once again."

C. Other Species

1. Southern White Rhinoceros

At the previous CoP, the South African population of the southern white rhinoceros (*Ceratotherium simum simum*) had been placed on Appendix II with an annotation to restrict trade to live animals and hunting trophies. This time South Africa submitted a proposal to amend the annotation to allow the possibility of establishing a legal trade in horn and other products. This trade would be subject to a quota, and the South Africans proposed that initially this quota should be set at zero. In effect, this would have maintained the ban for the present, while establishing the acceptability of trade in principle. The amendment to the annotation also included permission to investigate the establishment of bilateral trade in rhino horn. The vote was 60 to 32 in favor, one short of a two-thirds majority. Peter Mokaba, South Africa's deputy minister of environmental affairs, said that CITES had failed to recognize South Africa as a responsible sovereign state that had the capacity to manage its natural resources. He commented, "Once you take away these rights over natural resources you are certainly closing many developing options."

2. Whales

At the second CoP the Parties passed a resolution recommending that CITES not allow trade for any species that was protected by the International Whaling Commission (IWC). This time Japan submitted a proposal to repeal that resolution and make CITES no longer automatically linked to the IWC. This was interpreted by many as the first step in an attempt to undermine the IWC's moratorium on commercial whaling, and the proposal was easily defeated (27–51). Subsequent proposals from Japan to transfer one population of gray whale (*Eschrichtius robustus*) and two populations of minke whale (*Balaenoptera acutorostrata*) from Appendix I to Appendix II also failed to achieve even simple majorities. Japan withdrew a similar proposal concerning a population of Bryde's whale (*Balaenoptera edeni*). Norway's proposal to transfer another population of minke whale (in the Northeast Atlantic and North–Central Atlantic) received a majority of 57 to 51, but this was well short of two-thirds support. Nevertheless, this vote seems to have reflected the greater sympathy some Parties had for Norway's proposal.

3. Sturgeon

Since the breakdown of the Soviet Union, the sturgeon population in the Caspian region has declined rapidly. These fish are harvested for their eggs, better known as caviar, and the trade has become uncontrolled. A proposal was also submitted to list on Appendix II the twenty-three species of sturgeon (*Acipenseriformes spp.*) not already listed. This proposal was adopted by consensus.

4. Hawksbill Sea Turtle

Cuba submitted a proposal to transfer the Cuban population of the hawksbill sea turtle (*Eretmochelys imbricata*) from Appendix I to Appendix II, with an annotation to allow trade with one trading partner. The annotation allowed the export of a single shipment of Cuba's registered stock of shells, together with annual shipments of up to 500 shells taken from the traditional fishery. There was considerable support for Cuba on this issue but also some concern that the "Cuban" population of turtles actually included animals from other parts of the Caribbean that pass through Cuban waters. The proposal received a majority of 53 to 39 but was thus defeated, as it did not achieve a two-thirds majority.

5. Bigleaf Mahogany

Efforts at the two previous CoPs to list the bigleaf mahogany (*Swietenia macrophylla*) had failed. This time, the United States and Bolivia submitted a proposal to have it listed on Appendix II. They contended that it would enhance exporting nations' capacity to control trade without harming the interests of logging companies. Their opponents, including Cameroon, argued that the International Tropical Timber Association already ensured that the trade was sustainable. There may also have been fears among other timber exporting nations that a decision to list this species would be followed by attempts to list other commercially traded trees. The vote was 67 to 45 in favor, but the proposal failed, by eight votes, to get the necessary two-thirds majority. The United States subsequently successfully proposed that a working group be established to look at the advantages of a listing on Appendix II. This proposal was supported by the range states and the group is to report before the next CoP.

D. *Secret Voting*

The previous CoP had made it easier for Parties to call for a secret vote by stipulating that such a motion simply needs to be seconded by ten Parties. Nevertheless, some environmental NGOs were critical of secret voting. Gordon Shepherd of the World Wide Fund for Nature expressed concern that it would no longer be possible to know how individual Parties had voted. When Japan successfully called for a secret vote on its proposal to de-link CITES from the IWC, he described this as "a disastrous precedent" and "a blow for transparency." What heightened feelings on the issue were allegations that Japan had engaged in a vote-buying exercise, providing financial support to small Caribbean states in return for their votes at international conferences. Secret ballots were also called for in the debates on elephants, the hawksbill turtle, and the bigleaf mahogany. A Zimbabwean delegate defended the practice, saying, "A secret ballot is needed to allow poor African countries to vote freely without fear of pressure from the wealthy Western donor countries...." The precedent of secret votes now appears to be securely established and it has proved easy to implement....

G. Next Conference of the Parties

The conference decided that the next CoP will take place in Indonesia. The only other candidate was the United Kingdom, and although the results were not announced, the vote was rumored to have been close. There was some speculation that many developing countries preferred Indonesia because there they would be less likely to face environmental NGOs critical of their policies.

H. Conclusions

Insofar as the decisions on the African elephant are a barometer of opinion within CITES, the Harare CoP signaled that those who favor the sustainable use of wild species as a means of conservation are winning more support. The closeness of the vote on the South African white rhino, the sympathy with which the Cuban proposal on the hawksbill turtle was received, and the unwillingness of the Parties to put more emphasis on enforcement measures add weight to this interpretation. Nevertheless, the potential benefits to poor rural communities of a reopened ivory trade, and the fact that the elephant populations of Botswana, Namibia, and Zimbabwe are not under obvious threat, were also relevant factors in the elephant decision. The decisions on Japan's whale proposals, however, show that where these factors are not present, a majority for more strict protectionist measures remains. It would therefore be a mistake to assume that CITES has swung irrevocably in favor of promoting sustainable use, for the Parties remain divided on the fundamental issues of trade and conservation.

On the one hand, there are those who cleave closely to the guiding assumptions of the treaty. The Convention is premised on the assumption that international wildlife trade can be a serious threat to wild species and that the appropriate way to respond to that threat is by restricting or halting that trade. According to this perspective, decisions about listing particular species should hinge solely on the degree to which the species is threatened. This motivates the attempt to devise increasingly detailed criteria so that listing decisions become "technical" decisions, determined by whether the given criteria are satisfied in particular cases. Other considerations, such as whether trade may benefit either people or the species itself, and the rights of nations to manage their own resources, are held to have no legitimate place in CITES' deliberations.

On the other hand, there are those Parties who favor sustainable use and argue that, in the right circumstances, trade can be a useful conservation tool. What has allowed CITES to move in their direction is the way in which listing decisions are made. These are based on voting by the Parties at the CoPs, and there is no guarantee (and no means of guaranteeing) that the Parties will make their decision solely on the basis of the technical criteria developed by CITES. Indeed, as the elephant debate in Harare illustrates, many Parties are quite happy to claim publicly that their vote is motivated by considerations such as the benefits to poor communities and the role of trade in promoting sustainable conservation policies. It is this disjuncture between the original rationale of CITES and its own decision-making procedures that has made it possible for CITES to evolve toward the philosophy of sustainable use, as was seen in Harare.

This tension may also be responsible for some of the dissatisfaction with the way CITES is currently working. The fact that the review of the treaty's effectiveness led nowhere suggests that there is no consensus among the Parties either for a radical overhaul or for a return to a more strict adherence to the original rationale. In the meantime, the short-term evolution of CITES is likely to be shaped most by what happens with the limited reopening of the ivory trade. Therefore, the African elephant provides a test case for the proponents of sustainable use.

Notes and Questions

1. Dickson (joined by Jon Hutton) has written that "CITES has proved to be the most controversial of the international environmental conventions." Jon Hutton & Barnabas Dickson, *Introduction*, in ENDANGERED SPECIES, THREATENED CONVENTION xv (Jon Hutton & Barnabas Dickson eds. 2000). Another treatise describes CITES as "perhaps the most successful of all international treaties concerned with the conservation of wildlife." SIMON LYSTER, INTERNATIONAL WILDLIFE LAW: AN ANALYSIS OF INTERNATIONAL TREATIES CONCERNED WITH THE CONSERVATION OF WILDLIFE 240 (1985). Who is right? Or could CITES be both successful and controversial? What would you need to know to judge both claims? Additional, thorough overviews of CITES include WILLEM WIJNSTEKERS, THE EVOLUTION OF CITES (6th ed. 2001); DAVID S. FAVRE, INTERNATIONAL TRADE IN ENDANGERED SPECIES: A GUIDE TO CITES (1989); and the CITES web site at <http://www.cites.org>.

2. The first two CITES appendices contain the lists of species that are "threatened with extinction" and that "may become" threatened with extinction, respectively. Species listed on the third appendix do not require a vote of the parties because they are simply placed there by their host nation. As of 2001, about 5,000 species of animals and 25,000 species of plants were listed on the three appendices. The contested votes concerning the listing of elephants, whales, sturgeon, the Hawksbill sea turtle, and bigleaf mahogany reveal the difficulty in deciding the appropriate treat-ment of a species. CITES itself does not define the operative terms "threatened with extinction" and "may become" threatened with extinc-tion. The Berne Criteria, adopted at the first CITES conference in 1976, sought to provide guidance regarding the biological and trade information needed to judge the status of a species. Widespread objections to those guidelines resulted in the adoption of the Fort Lauderdale Criteria at the ninth convention of the parties in 1994. The new criteria provide that a species should be included in Appendix I if the species "is or may be affected by trade" and if (1) its wild population is small, (2) its population is limited to a small area, (3) the population is dropping, or (4) either of the first three biological factors will be satisfied within five years. The criteria further specify the quantitative measurements that judge whether a population is small, restricted, or declining. The criteria thus represent an effort to employ objective scientific standards in listing decisions, rather than relying upon ad hoc political reasons. One recent study described the Fort Lauderdale Criteria as "a small but useful attempt to increase the influence of science in the decisionmaking process," but it contended that the criteria "will improve the deliberations of the parties to CITES not so much by increasing the influence of scientific data as by cooling expressive disputes among the parties, encouraging compromise, and promoting further scientific research." Note, *The CITES Fort Lauderdale Criteria: The Uses and Limits of Science in International Conservation Decisionmaking*, 114 Harv. L. Rev. 1769, 1792 (2001).

3. The substantive provisions of CITES regulate trade in a species once it is listed. Species listed on Appendix I may be traded only if the exporting and importing countries permit the trade after they determine that the individual animal or bird was obtained properly, the individual will not be used "primarily for commercial purposes," and the survival of the species would not be threatened. Appendix II and III species require an export permit to be traded, but not an import permit. CITES also contains a number of exceptions for animals that were bred in captivity for commercial purposes and plants that were artificially propagated for commercial purposes, specimens that are personal or household effects, specimens collected by scientists and scientific institutions, and specimens that were acquired before the species was listed.

Enforcement of these regulations depends upon the efforts of each country that is a party to the treaty. CITES Article VIII provides that parties "shall take appropriate measures to enforce the provisions of the present Convention and to prohibit trade in specimens in violation thereof." Each party must designate a Management Authority that is responsible for issuing permits and a Scientific Authority that monitors those permits. Note, too, that CITES Article XIV ensures that countries that are party to the treaty are allowed to adopt stricter domestic regulations regarding the trade of species protected by CITES.

The success of CITES depends upon domestic statutes prohibiting trade in the species protected by CITES. The United States Congress, for example, enacted the Endangered Species Act in 1973 in part to enforce the obligations imposed by CITES. How much CITES adds to the efforts of the United States to protect endangered species is questionable. One observer suggests that CITES "is best treated as an extra facility under which any Party can invoke the assistance of law enforcement agencies of other Parties in improving the implementation of its own policies." R.B. Martin, *When CITES Works and When It Does Not*, in Endangered Species, Threat-

ENED CONVENTION, *supra*, at 31. By contrast, "CITES is unlikely to work . . . where control of wildlife is not centralized or popularly accepted or where the state bureaucracy is weak and inefficient." *Id.* at 32.

The ability of CITES to control the international trade in endangered species also depends upon the treatment of those countries that are not parties to the treaty. CITES Article X states, "Where export or re-export is to, or import is from, a State not a Party to the Convention, comparable documentation issued by the component authorities in that State which substantially conforms with the requirements of the present Convention for permits and certificates may be accepted in lieu thereof by any Party." This provision anticipates a problem created by the very popularity of CITES, which now has 154 parties: "There are so few countries that are not members of CITES that considerable money can be made from the channeling of illegal trade through these few countries. There is now an economic incentive to not be a member, if you are willing to deal in illegal goods." FAVRE, *supra*, at 256. Article X responds to that concern by requiring all parties to CITES to trade with non-parties only when there has been substantial compliance with the CITES provisions. What else can be done about countries that are not parties to CITES that countenance trade in endangered species?

4. Contrary to Dickson's report that the next conference would take place in Indonesia, the eleventh conference of the CITES parties occurred in Nairobi in 2000. Many of the issues debated in the tenth conference resurfaced in Nairobi. Voting with secret ballots, proposals to downlist gray and minke whales and the hawksbill sea turtle gained the support of most of the parties, but they failed to reach the necessary two-thirds majority to pass. A compromise agreement maintained the ban on trading elephant ivory for three years while allowing a limited trade in live elephants, elephant hides, and leather products. A new proposal to waive the permitting scheme for scientists engaged in research on endangered species failed because of concerns about its effects on domestic controls and the ability to regulate bioprospecting. *See* Wendy Williams, *CITES Puts Off Plan to Hasten Shipments; Convention on International Trade in Endangered Species*, SCIENCE, Apr. 28, 2000, at 592.

The parties participating in the eleventh conference also approved a strategic plan that seeks to guide the implementation of CITES through 2005. That plan contains seven goals: (1) enhance the ability of each party to implement the treaty; (2) strengthen the scientific basis of the decision-making processes; (3) contribute to the reduction and ultimate elimination of illegal trade in wild fauna and flora; (4) promote greater understanding of the treaty; (5) increase cooperation and conclude strategic alliances with international stakeholders; (6) progress toward global membership; and (7) provide the treaty with an improved and secure financial and administrative basis. The plan lists numerous specific objectives to achieve these goals, and it is complemented by an action plan that details which entities are responsible for taking the steps necessary to reach those goals. *See* Convention on International Trade in Endangered Species of Wild Fauna and

Flora, Strategic Vision Through 2005, *available at* <http://www.cites.org/eng/decis/11/annex1.shtml>. Should Dickson, who worried that there was little agreement about the long-term direction of CITES, be satisfied with that plan? Are the goals so generally stated that they are unobjectionable but meaningless?

5. How important is CITES given that habitat destruction is a greater threat to biodiversity than trade? David Favre defends CITES as follows:

> Controlling land use decisions within a particular country would be very difficult to do in an international convention, but controlling the movement of animals is possible because the flow of international commerce already goes through identifiable locations which can be monitored. Controlling trade may not provide all of the protection necessary for a species' survival, but it would be impossible to protect most species without controlling trade. Also, the process of listing an animal or plant under this international document will hopefully bring public attention and government action in protecting the habitat of the species listed.

Favre, *supra*, at 30. Conversely, can the international wildlife trade be "a positive tool for conservation," as the sustainable use proponents described by Dickson contend? Is the contrary premise of CITES outdated? Or is it more necessary than ever?

C.   The Convention On Biological Diversity

United Nations Conference on Environment and Development: Convention on Biological Diversity

31 International Legal Materials 818 (1992).

Preamble

The Contracting Parties,

Conscious of the intrinsic value of biological diversity and of the ecological, genetic, social, economic, scientific, educational, cultural, recreational and aesthetic values of biological diversity and its components,

Conscious also of the importance of biological diversity for evolution and for maintaining life sustaining systems of the biosphere,

Affirming that the conservation of biological diversity is a common concern of humankind,

Reaffirming that States have sovereign rights over their own biological resources,

Reaffirming also that States are responsible for conserving their biological diversity and for using their biological resources in a sustainable manner,

Concerned that biological diversity is being significantly reduced by certain human activities,

Aware of the general lack of information and knowledge regarding biological diversity and of the urgent need to develop scientific, technical and institutional capacities to provide the basic understanding upon which to plan and implement appropriate measures.

Noting that it is vital to anticipate, prevent and attack the causes of significant reduction or loss of biological diversity at source.

Noting also that where there is a threat of significant reduction or loss of biological diversity, lack of full scientific certainty should not be used as a reason for postponing measures to avoid or minimize such a threat,

Noting further that the fundamental requirement for the conservation of biological diversity is the in-situ conservation of ecosystems and natural habitats and the maintenance and recovery of viable populations of species in their natural surroundings,

Noting further that ex-situ measures, preferably in the country of origin, also have an important role to play,

Recognizing the close and traditional dependence of many indigenous and local communities embodying traditional lifestyles on biological resources, and the desirability of sharing equitably benefits arising from the use of traditional knowledge, innovations and practices relevant to the conservation of biological diversity and the sustainable use of its components,

Recognizing also the vital role that women play in the conservation and sustainable use of biological diversity and affirming the need for the full participation of women at all levels of policy-making and implementation for biological diversity conservation,

Stressing the importance of, and the need to promote, international, regional and global cooperation among States and intergovernmental organizations and the non-governmental sector for the conservation of biological diversity and the sustainable use of its components,

Acknowledging that the provision of new and additional financial resources and appropriate access to relevant technologies can be expected to make a substantial difference in the world's ability to address the loss of biological diversity,

Acknowledging further that special provision is required to meet the needs of developing countries, including the provision of new and additional financial resources and appropriate access to relevant technologies,

Noting in this regard the special conditions of the least developed countries and small island States,

Acknowledging that substantial investments are required to conserve biological diversity and that there is the expectation of a broad range of environmental, economic and social benefits from those investments,

Recognizing that economic and social development and poverty eradication are the first and overriding priorities of developing countries,

Aware that conservation and sustainable use of biological diversity is of critical importance for meeting the food, health and other needs of the growing world population, for which purpose access to and sharing of both genetic resources and technologies are essential,

Noting that, ultimately, the conservation and sustainable use of biological diversity will strengthen friendly relations among States and contribute to peace for humankind,

Desiring to enhance and complement existing international arrangements for the conservation of biological diversity and sustainable use of its components, and

Determined to conserve and sustainably use biological diversity for the benefit of present and future generations,

Have agreed as follows:

Article 1. Objectives

The objectives of this Convention, to be pursued in accordance with its relevant provisions, are the conservation of biological diversity, the sustainable use of its components and the air and equitable sharing of the benefits arising out of the utilization of genetic resources, including by appropriate access to genetic resources and by appropriate transfer of relevant technologies, taking into account all rights over those resources and to technologies, and by appropriate funding.

Article 2. Use of Terms

For the purposes of this Convention:

"Biological diversity" means the variability among living organisms from all sources including, inter alia, terrestrial, marine and other aquatic ecosystems and the ecological complexes of which they are part; this includes diversity within species, between species and of ecosystems.

"Biological resources" includes genetic resources, organisms or parts thereof, populations, or any other biotic component of ecosystems with actual or potential use or value for humanity.

"Biotechnology" means any technological application that uses biological systems, living organisms, or derivatives thereof, to make or modify products or processes for specific use.

"Country of origin of genetic resources" means the country which possesses those genetic resources in in-situ conditions.

"Country providing genetic resources" means the country supplying genetic resources collected from in-situ sources, including populations of both wild and domesticated species, or taken from ex-situ sources, which may or may not have originated in that country.

"Domesticated or cultivated species" means species in which the evolutionary process has been influenced by humans to meet their needs.

"Ecosystem" means a dynamic complex of plant, animal and micro-organism communities and their non-living environment interacting as a functional unit.

"Ex-situ conservation" means the conservation of components of biological diversity outside their natural habitats.

"Genetic material" means any material of plant, animal, microbial or other origin containing functional units of heredity.

"Genetic resources" means genetic material of actual or potential value.

"Habitat" means the place or type of site where an organism or population naturally occurs.

"In-situ conditions" means conditions where genetic resources exist within ecosystems and natural habitats, and, in the case of domesticated or cultivated species, in the surroundings where they have developed their distinctive properties.

"In-situ conservation" means the conservation of ecosystems and natural habitats and the maintenance and recovery of viable populations of species in their natural surroundings and, in the case of domesticated or cultivated species, in the surroundings where they have developed their distinctive properties.

"Protected area" means a geographically defined area which is designated or regulated and managed to achieve specific conservation objectives.

"Regional economic integration organization" means an organization constituted by sovereign States of a given region, to which its member States have transferred competence in respect of matters governed by this Convention and which has been duly authorized, in accordance with its internal procedures, to sign, ratify, accept, approve or accede to it.

"Sustainable use" means the use of components of biological diversity in a way and at a rate that does not lead to the long-term decline of biological diversity, thereby maintaining its potential to meet the needs and aspirations of present and future generations.

"Technology" includes biotechnology.

Article 3. Principle

States have, in accordance with the Charter of the United Nations and the principles of international law, the sovereign right to exploit their own resources pursuant to their own environmental policies, and the responsibility to ensure that activities within their jurisdiction or control do not cause damage to the environment of other States or of areas beyond the limits of national jurisdiction.

Article 4. Jurisdictional Scope

Subject to the rights of other States, and except as otherwise expressly provided in this Convention, the provisions of this Convention apply, in relation to each Contracting Party:

(a) In the case of components of biological diversity, in areas within the limits of its national jurisdiction; and

(b) In the case of processes and activities, regardless of where their effects occur, carried out under its jurisdiction or control, within the area of its national jurisdiction or beyond the limits of national jurisdiction.

Article 5. Cooperation

Each Contracting Party shall, as far as possible and as appropriate, cooperate with other Contracting Parties, directly or, where appropriate, through competent international organizations, in respect of areas beyond national jurisdiction and on other matters of mutual interest, for the conservation and sustainable use of biological diversity.

Article 6. General Measures for Conservation and Sustainable Use

Each Contracting Party shall, in accordance with its particular conditions and capabilities:

(a) Develop national strategies, plans or programmes for the conservation and sustainable use of biological diversity or adapt for this purpose existing strategies, plans or programmes which shall reflect, inter alia, the measures set out in this Convention relevant to the Contracting Party concerned; and

(b) Integrate, as far as possible and as appropriate, the conservation and sustainable use of biological diversity into relevant sectoral or cross-sectoral plans, programmes and policies.

Article 7. Identification and Monitoring

Each Contracting Party shall, as far as possible and as appropriate, in particular for the purposes of Articles 8 to 10:

(a) Identify components of biological diversity important for its conservation and sustainable use having regard to the indicative list of categories set down in Annex I;

(b) Monitor, through sampling and other techniques, the components of biological diversity identified pursuant to subparagraph (a) above, paying particular attention to those requiring urgent conservation measures and those which offer the greatest potential for sustainable use;

(c) Identify processes and categories of activities which have or are likely to have significant adverse impacts on the conservation and sustainable use of biological diversity, and monitor their effects through sampling and other techniques; and

(d) Maintain and organize, by any mechanism data, derived from identification and monitoring activities pursuant to subparagraphs (a), (b) and (c) above.

Article 8. In-situ Conservation

Each Contracting Party shall, as far as possible and as appropriate:

(a) Establish a system of protected areas or areas where special measures need to be taken to conserve biological diversity;

(b) Develop, where necessary, guidelines for the selection, establishment and management of protected areas or areas where special measures need to be taken to conserve biological diversity;

(c) Regulate or manage biological resources important for the conservation of biological diversity whether within or outside protected areas, with a view to ensuring their conservation and sustainable use;

(d) Promote the protection of ecosystems, natural habitats and the maintenance of viable populations of species in natural surroundings;

(e) Promote environmentally sound and sustainable development in areas adjacent to protected areas with a view to furthering protection of these areas;

(f) Rehabilitate and restore degraded ecosystems and promote the recovery of threatened species, inter alia, through the development and implementation of plans or other management strategies;

(g) Establish or maintain means to regulate, manage or control the risks associated with the use and release of living modified organisms resulting from biotechnology which are likely to have adverse environmental impacts that could affect the conservation and sustainable use of biological diversity, taking also into account the risks to human health;

(h) Prevent the introduction of, control or eradicate those alien species which threaten ecosystems, habitats or species;

(i) Endeavour to provide the conditions needed for compatibility between present uses and the conservation of biological diversity and the sustainable use of its components;

(j) Subject to its national legislation, respect, preserve and maintain knowledge, innovations and practices of indigenous and local communities embodying traditional lifestyles relevant for the conservation and sustainable use of biological diversity and promote their wider application with the approval and involvement of the holders of such knowledge, innovations and practices and encourage the equitable sharing of the benefits arising from the utilization of such knowledge, innovations and practices;

(k) Develop or maintain necessary legislation and/or other regulatory provisions for the protection of threatened species and populations;

(*l*) Where a significant adverse effect on biological diversity has been determined pursuant to Article 7, regulate or manage the relevant processes and categories of activities; and

(m) Cooperate in providing financial and other support for in-situ conservation outlined in subparagraphs (a) to (*l*) above, particularly to developing countries.

Article 9. Ex-situ Conservation

Each Contracting Party shall, as far as possible and as appropriate, and predominantly for the purpose of complementing in-situ measures:

(a) Adopt measures for the ex-situ conservation of components of biological diversity, preferably in the country of origin of such components;

(b) Establish and maintain facilities for ex-situ conservation of and research on plants, animals and micro-organisms, preferably in the country of origin of genetic resources;

(c) Adopt measures for the recovery and rehabilitation of threatened species and for their reintroduction into their natural habitats under appropriate conditions;

(d) Regulate and manage collection of biological resources from natural habitats for ex-situ conservation purposes so as not to threaten ecosystems and in-situ populations of species, except where special temporary ex-situ measures are required under subparagraph (c) above; and

(e) Cooperate in providing financial and other support for ex-situ conservation outlined in subparagraphs (a) to (d) above and in the establishment and maintenance of ex-situ conservation facilities in developing countries.

Article 10. Sustainable Use of Components of Biological Diversity

Each Contracting Party shall, as far as possible and as appropriate:

(a) Integrate consideration of the conservation and sustainable use of biological resources into national decision-making;

(b) Adopt measures relating to the use of biological resources to avoid or minimize adverse impacts on biological diversity;

(c) Protect and encourage customary use of biological resources in accordance with traditional cultural practices that are compatible with conservation or sustainable use requirements;

(d) Support local populations to develop and implement remedial action in degraded areas where biological diversity has been reduced; and

(e) Encourage cooperation between its governmental authorities and its private sector in developing methods for sustainable use of biological resources.

Article 11. Incentive Measures

Each Contracting Party shall, as far as possible and as appropriate, adopt economically and socially sound measures that act as incentives for the conservation and sustainable use of components of biological diversity.

Article 12. Research and Training

The Contracting Parties, taking into account the special needs of developing countries, shall:

(a) Establish and maintain programmes for scientific and technical education and training in measures for the identification, conservation and sustainable use of biological diversity and its components and provide support for such education and training for the specific needs of developing countries;

(b) Promote and encourage research which contributes to the conservation and sustainable use of biological diversity, particularly in developing countries, inter alia, in accordance with decisions of the Conference of the

Parties taken in consequence of recommendations of the Subsidiary Body on Scientific, Technical and Technological Advice; and

(c) In keeping with the provisions of Articles 16, 18 and 20, promote and cooperate in the use of scientific advances in biological diversity research in developing methods for conservation and sustainable use of biological resources.

Article 13. Public Education and Awareness

The Contracting Parties shall:

(a) Promote and encourage understanding of the importance of, and the measures required for, the conservation of biological diversity, as well as its propagation through media, and the inclusion of these topics in educational programmes; and

(b) Cooperate, as appropriate, with other States and international organizations in developing educational and public awareness programmes, with respect to conservation and sustainable use of biological diversity.

Article 14. Impact Assessment and Minimizing Adverse Impacts

1.   Each Contracting Party, as far as possible and as appropriate, shall:

(a) Introduce appropriate procedures requiring environmental impact assessment of its proposed projects that are likely to have significant adverse effects on biological diversity with a view to avoiding or minimizing such effects and, where appropriate, allow for public participation in such procedures;

(b) Introduce appropriate arrangements to ensure that the environmental consequences of its programmes and policies that are likely to have significant adverse impacts on biological diversity are duly taken into account;

(c) Promote, on the basis of reciprocity, notification, exchange of information and consultation on activities under their jurisdiction or control which are likely to significantly affect adversely the biological diversity of other States or areas beyond the limits of national jurisdiction, by encouraging the conclusion of bilateral, regional or multilateral arrangements, as appropriate;

(d) In the case of imminent or grave danger or damage, originating under its jurisdiction or control, to biological diversity within the area under jurisdiction of other States or in areas beyond the limits of national jurisdiction, notify immediately the potentially affected States of such danger or damage, as well as initiate action to prevent or minimize such danger or damage; and

(e) Promote national arrangements for emergency responses to activities or events, whether caused naturally or otherwise, which present a grave and imminent danger to biological diversity and encourage international cooperation to supplement such national efforts and, where appropriate and agreed by the States or regional economic integration organizations concerned, to establish joint contingency plans.

2. The Conference of the Parties shall examine, on the basis of studies to be carried out, the issue of liability and redress, including restoration and compensation, for damage to biological diversity, except where such liability is a purely internal matter.

Article 15. Access to Genetic Resources

1. Recognizing the sovereign rights of States over their natural resources, the authority to determine access to genetic resources rests with the national governments and is subject to national legislation.

2. Each Contracting Party shall endeavour to create conditions to facilitate access to genetic resources for environmentally sound uses by other Contracting Parties and not to impose restrictions that run counter to the objectives of this Convention.

3. For the purpose of this Convention, the genetic resources being provided by a Contracting Party, as referred to in this Article and Articles 16 and 19, are only those that are provided by Contracting Parties that are countries of origin of such resources or by the Parties that have acquired the genetic resources in accordance with this Convention.

4. Access, where granted, shall be on mutually agreed terms and subject to the provisions of this Article.

5. Access to genetic resources shall be subject to prior informed consent of the Contracting Party providing such resources, unless otherwise determined by that Party.

6. Each Contracting Party shall endeavour to develop and carry out scientific research based on genetic resources provided by other Contracting Parties with the full participation of, and where possible in, such Contracting Parties.

7. Each Contracting Party shall take legislative, administrative or policy measures, as appropriate, and in accordance with Articles 16 and 19 and, where necessary, through the financial mechanism established by Articles 20 and 21 with the aim of sharing in a fair and equitable way the results of research and development and the benefits arising from the commercial and other utilization of genetic resources with the Contracting Party providing such resources. Such sharing shall be upon mutually agreed terms.

Article 16. Access to and Transfer of Technology

1. Each Contracting Party, recognizing that technology includes biotechnology, and that both access to and transfer of technology among Contracting Parties are essential elements for the attainment of the objectives of this Convention, undertakes subject to the provisions of this Article to provide and/or facilitate access for and transfer to other Contracting Parties of technologies that are relevant to the conservation and sustainable use of biological diversity or make use of genetic resources and do not cause significant damage to the environment.

2. Access to and transfer of technology referred to in paragraph 1 above to developing countries shall be provided and/or facilitated under fair and most favourable terms, including on concessional and preferential terms where mutually agreed, and, where necessary, in accordance with the financial mechanism established by Articles 20 and 21. In the case of technology subject to patents and other intellectual property rights, such access and transfer shall be provided on terms which recognize and are consistent with the adequate and effective protection of intellectual property rights. The application of this paragraph shall be consistent with paragraphs 3, 4 and 5 below.

3. Each Contracting Party shall take legislative, administrative or policy measures, as appropriate, with the aim that Contracting Parties, in particular those that are developing countries, which provide genetic resources are provided access to and transfer of technology which makes use of those resources, on mutually agreed terms, including technology protected by patents and other intellectual property rights, where necessary, through the provisions of Articles 20 and 21 and in accordance with international law and consistent with paragraphs 4 and 5 below.

4. Each Contracting Party shall take legislative, administrative or policy measures, as appropriate, with the aim that the private sector facilitates access to, joint development and transfer of technology referred to in paragraph 1 above for the benefit of both governmental institutions and the private sector of developing countries and in this regard shall abide by the obligations included in paragraphs 1, 2 and 3 above.

5. The Contracting Parties, recognizing that patents and other intellectual property rights may have an influence on the implementation of this Convention, shall cooperate in this regard subject to national legislation and international law in order to ensure that such rights are supportive of and do not run counter to its objectives.

Article 17. Exchange of Information

1. The Contracting Parties shall facilitate the exchange of information, from all publicly available sources, relevant to the conservation and sustainable use of biological diversity, taking into account the special needs of developing countries.

2. Such exchange of information shall include exchange of results of technical, scientific and socio-economic research, as well as information on training and surveying programmes, specialized knowledge, indigenous and traditional knowledge as such and in combination with the technologies referred to in Article 16, paragraph 1. It shall also, where feasible, include repatriation of information.

Article 18. Technical and Scientific Cooperation

1. The Contracting Parties shall promote international technical and scientific cooperation in the field of conservation and sustainable use of biological diversity, where necessary, through the appropriate international and national institutions.

2. Each Contracting Party shall promote technical and scientific cooperation with other Contracting Parties, in particular developing countries, in implementing this Convention, inter alia, through the development and implementation of national policies. In promoting such cooperation, special attention should be given to the development and strengthening of national capabilities, by means of human resources development and institution building.

3. The Conference of the Parties, at its first meeting, shall determine how to establish a clearing-house mechanism to promote and facilitate technical and scientific cooperation.

4. The Contracting Parties shall, in accordance with national legislation and policies, encourage and develop methods of cooperation for the development and use of technologies, including indigenous and traditional technologies, in pursuance of the objectives of this Convention. For this purpose, the Contracting Parties shall also promote cooperation in the training of personnel and exchange of experts.

5. The Contracting Parties shall, subject to mutual agreement, promote the establishment of joint research programmes and joint ventures for the development of technologies relevant to the objectives of this Convention.

Article 19. Handling of Biotechnology and Distribution of its Benefits

1. Each Contracting Party shall take legislative, administrative or policy measures, as appropriate, to provide for the effective participation in biotechnological research activities by those Contracting Parties, especially developing countries, which provide the genetic resources for such research, and where feasible in such Contracting Parties.

2. Each Contracting Party shall take all practicable measures to promote and advance priority access on a fair and equitable basis by Contracting Parties, especially developing countries, to the results and benefits arising from biotechnologies based upon genetic resources provided by those Contracting Parties. Such access shall be on mutually agreed terms.

3. The Parties shall consider the need for and modalities of a protocol setting out appropriate procedures, including, in particular, advance informed agreement, in the field of the safe transfer, handling and use of any living modified organism resulting from biotechnology that may have adverse effect on the conservation and sustainable use of biological diversity.

4. Each Contracting Party shall, directly or by requiring any natural or legal person under its jurisdiction providing the organisms referred to in paragraph 3 above, provide any available information about the use and safety regulations required by that Contracting Party in handling such organisms, as well as any available information on the potential adverse impact of the specific organisms concerned to the Contracting Party into which those organisms are to be introduced.

Article 20. Financial Resources

1.   Each Contracting Party undertakes to provide, in accordance with its capabilities, financial support and incentives in respect of those national activities which are intended to achieve the objectives of this Convention, in accordance with its national plans, priorities and programmes.

2.   The developed country Parties shall provide new and additional financial resources to enable developing country Parties to meet the agreed full incremental costs to them of implementing measures which fulfil the obligations of this Convention and to benefit from its provisions and which costs are agreed between a developing country Party and the institutional structure referred to in Article 21, in accordance with policy, strategy, programme priorities and eligibility criteria and an indicative list of incremental costs established by the Conference of the Parties. Other Parties, including countries undergoing the process of transition to a market economy, may voluntarily assume the obligations of the developed country Parties. For the purpose of this Article, the Conference of the Parties, shall at its first meeting establish a list of developed country Parties and other Parties which voluntarily assume the obligations of the developed country Parties. The Conference of the Parties shall periodically review and if necessary amend the list. Contributions from other countries and sources on a voluntary basis would also be encouraged. The implementation of these commitments shall take into account the need for adequacy, predictability and timely flow of funds and the importance of burden-sharing among the contributing Parties included in the list.

3.   The developed country Parties may also provide, and developing country Parties avail themselves of, financial resources related to the implementation of this Convention through bilateral, regional and other multilateral channels.

4.   The extent to which developing country Parties will effectively implement their commitments under this Convention will depend on the effective implementation by developed country Parties of their commitments under this Convention related to financial resources and transfer of technology and will take fully into account the fact that economic and social development and eradication of poverty are the first and overriding priorities of the developing country Parties.

5.   The Parties shall take full account of the specific needs and special situation of least developed countries in their actions with regard to funding and transfer of technology.

6.   The Contracting Parties shall also take into consideration the special conditions resulting from the dependence on, distribution and location of, biological diversity within developing country Parties, in particular small island States.

7.   Consideration shall also be given to the special situation of developing countries, including those that are most environmentally vulnerable, such as those with arid and semi-arid zones, coastal and mountainous areas.

Article 21. Financial Mechanism

1. There shall be a mechanism for the provision of financial resources to developing country Parties for purposes of this Convention on a grant or concessional basis the essential elements of which are described in this Article. The mechanism shall function under the authority and guidance of, and be accountable to, the Conference of the Parties for purposes of this Convention. The operations of the mechanism shall be carried out by such institutional structure as may be decided upon by the Conference of the Parties at its first meeting. For purposes of this Convention, the Conference of the Parties shall determine the policy, strategy, programme priorities and eligibility criteria relating to the access to and utilization of such resources. The contributions shall be such as to take into account the need for predictability, adequacy and timely flow of funds referred to in Article 20 in accordance with the amount of resources needed to be decided periodically by the Conference of the Parties and the importance of burden-sharing among the contributing Parties included in the list referred to in Article 20, paragraph 2. Voluntary contributions may also be made by the developed country Parties and by other countries and sources. The mechanism shall operate within a democratic and transparent system of governance.

2. Pursuant to the objectives of this Convention, the Conference of the Parties shall at its first meeting determine the policy, strategy and programme priorities, as well as detailed criteria and guidelines for eligibility for access to and utilization of the financial resources including monitoring and evaluation on a regular basis of such utilization. The Conference of the Parties shall decide on the arrangements to give effect to paragraph 1 above after consultation with the institutional structure entrusted with the operation of the financial mechanism.

3. The Conference of the Parties shall review the effectiveness of the mechanism established under this Article, including the criteria and guidelines referred to in paragraph 2 above, not less than two years after the entry into force of this Convention and thereafter on a regular basis. Based on such review, it shall take appropriate action to improve the effectiveness of the mechanism if necessary.

4. The Contracting Parties shall consider strengthening existing financial institutions to provide financial resources for the conservation and sustainable use of biological diversity.

Article 22. Relationship with Other International Conventions

1. The provisions of this Convention shall not affect the rights and obligations of any Contracting Party deriving from any existing international agreement, except where the exercise of those rights and obligations would cause a serious damage or threat to biological diversity.

2. Contracting Parties shall implement this Convention with respect to the marine environment consistently with the rights and obligations of States under the law of the sea. . . .

Notes and Questions

1. The Convention on Biological Diversity resulted from years of international negotiations. In 1972, the Stockholm Convention became the first effort to state binding principles of international environmental law. Discussions about biodiversity continued through the 1980's and culminated at the United Nations Conference on the Environment and Development held in Rio de Janeiro in June 1992. The Convention emerged from Rio with the signatures of nearly every country that attended the conference, and it became effective at the end of 1993. It has become the primary framework for the protection of global biodiversity. As of 2001, 182 countries have ratified the Convention.

But not the United States. The Bush Administration declined to send representatives to Rio in 1992 because of concerns about the nature of the evolving agreement. It emphasized both that it "strongly supports the conservation of biodiversity" and that "international cooperation in this area [is] extremely desirable," but it found "particularly unsatisfactory the text's treatment of intellectual property rights; finances, including, importantly, the role of the Global Environmental Facility (GEF); technology transfer and biotechnology." United States: Declaration Made at the United Nations Environment Programme Conference for the Adoption of the Agreed Text of the Convention on Biological Diversity, 31 I.L.M. 848 (1992). One year later, President Clinton signed the Convention. He then dispatched State Department official (and former Colorado Senator) Timothy Wirth to defend the Convention before a House of Representatives committee. Wirth testified:

> The Administration strongly supports the Convention on Biological Diversity as an important vehicle for the conservation and sustainable use of biodiversity worldwide. It sets in place a global commitment to promote biodiversity conservation and sustainable use for the benefit of this and future generations. U.S. adherence to the Convention is critical for global efforts to conserve and utilize biodiversity and for maintaining our position as the world leader in environmental protection.
>
> As noted above, no implementing legislation will be required. The existing assemblage of Federal, State and private sector biodiversity programs–comprising numerous State and Federal laws and programs, an extensive system of Federal and State wildlife management areas, marine sanctuaries, parks and forests, and research and education programs–is considered sufficient in meeting our responsibilities under the convention. The Administration does not intend to disrupt the existing balance of State and Federal authorities through this Convention and, indeed, is committed to expanding and strengthening these productive partnerships....
>
> As you know, Mr. Chairman, industry has expressed concerns about the way some of the language in the Convention is drafted, particularly on the issues of intellectual property rights protection, terms of trans-

fer of technology, and participation in U.S. research projects. We share those concerns, but feel we can best protect U.S. interests by:

1. Sending clear messages to the rest of the world as to how the U.S. expects these provisions to be implemented by all Parties;

2. Participating actively in the Convention to protect U.S. intellectual property rights interests; and

3. Depositing with our instrument of ratification statements of understanding on specific issues raised in articles of the convention.

Specifically, we propose to deposit with our instrument of ratification statements of U.S. understanding that make clear our positions on the issues of technology transfer and intellectual property rights, research, funding, sovereign immunity, and the Article 3 Principle. . . .

Finally, many have asked what mechanisms are available to the United States to ensure that any decisions of the Conference of the Parties (COP) accord with U.S. interests. First, we will ensure that the COP's rules of procedure, which will be adopted by consensus, are fully acceptable to the United States. Second, we have supported a proposal to require that all decisions concerning the financial mechanism are also made by consensus. Most importantly, however, the Conference of the Parties could not legally bind the United States to a legal interpretation of the Convention that the U.S. did not accept. The United States will have an opportunity to formally accept or reject any amendment or protocol to the Convention.

Timothy E. Wirth, *Ratification Sought for the Convention on Biological Diversity*, 5 Dept. of State Dispatch 213, 214–15 (Apr. 18, 1994). But Congress has not been persuaded. The Senate has not ratified the Convention, and there is little prospect that it will in the foreseeable future.

Meanwhile, some environmentalists object to the Convention as well. Professor Lakshman Guruswamy contends that the Convention "may halt the advance of international environmental law on three fronts. . . ." First, it rejects the concept of sustainable development by prioritizing economic growth over environmental protection, allowing international resources earmarked for the protection of biodiversity to be expended on economic growth that could destroy biodiversity. Second, it denies state responsibility for damage to the global commons. Finally, it repudiates the idea that the plant, animal, insect, and genetic resources of the world (our biodiversity) are the common heritage of humankind, and that it is the responsibility of the community of nations to protect this heritage. Lakshnan D. Guruswamy, *The Convention on Biological Diversity: A Polemic*, in Protection of Global Biodiversity: Converging Strategies 351 (Lakshman D. Guruswamy & Jeffrey A. McNeely eds. 1998).

Notwithstanding these objections, nearly every nation has ratified the Convention. The discussion of the key issues raised by the Convention that follows draws from numerous sources, though it is especially indebted to David Hunter, James Salzman & Durwood Zaelke, International Environmen-

TAL LAW AND POLICY 932–76 (2d ed. 2002); and to the Convention's web site at <http://www.biodiv.org>.

2. The Preamble to the Convention states that the conservation of biological diversity is a "common concern of humankind." Can that be disputed? The most frequent complaint with that assertion is that it does not go far enough. As Professor Guruswamy notes, the alternative formulation would have proclaimed biodiversity to be "the common heritage of humankind." How are the two descriptions different? Which is more accurate? The rejected "common heritage of humankind" description was opposed by developing countries who "did not want common heritage applied to all biodiversity, because the concept implies a requirement to allow all States free access to the resource." It was also opposed by industrial countries who worried that "the common heritage principle if applied too far might require benefit sharing or open transfers of technology derived from biodiversity." HUNTER, SALZMAN & ZAELKE, *supra*, at 934.

The Preamble also seeks to balance state sovereignty with state responsibility for the protection of biodiversity. Does it do so successfully? Opponents of the Convention answer no, though for much different reasons. Professor Guruswamy sees the Convention as tilted in favor of national economic development instead of global biodiversity. American complaints, by contrast, insist that the Convention will deprive the United States (and other concerned countries) of the freedom to balance biodiversity and economic development themselves, even if that balance is more favorable toward biodiversity. Still, "the primary rights and responsibilities regarding biodiversity conservation remain at the national level." *Id.* at 936. Is that an accurate assessment of the Convention? If so, why did the signatories adopt an approach that emphasizes domestic laws rather than international law? What does the Convention add that interested countries could not accomplish themselves?

Is there anything surprising in the steps that each country is asked to take to protect its biodiversity? Articles 6 through 11, 14, and 26 each set forth specific duties. Do you agree with Assistant Secretary of State Wirth that the United States already satisfies those requirements? Does the experience of the United States in protecting species and ecosystems suggest any appropriate duties besides those listed in the Convention? The strategies adopted by different countries can be discerned from plans, required by Article 6, that each signatory is required to prepare. *See* Convention on Biological Diversity, All National Reports, *available at* <http://www.biodiv.org/world/reports.asp?t=all>.

3. The Convention regulates the biotechnology trade. The nexus between that trade and the preservation of biodiversity is provided by the concern that genetic diversity could suffer if foreigners are allowed to exploit a nation's native plant and animals resources. This became a North–South conflict centered on the appropriate scope of intellectual property rights. The South "wanted to retain the right to control the access of northern industries to 'prospect' for biodiversity in their countries," and "they saw intellectual property rights (IPR) regimes, which protected the patents of biotechnology firms, as a major obstacle to benefit-sharing and ultimately to biodiversity conservation." The South also objected to "biopiracy": the exploitation of traditional knowledge about local biodiversity by multinational bioprospecting companies who seek to patent their "discoveries." For its part, "the North wanted to ensure free and open access to biodiversity so that pharmaceutical and agricultural research firms could expand their efforts to identify potentially valuable plants and animals," and to protect the intellectual property rights of those firms in order to reward and encourage their investments in research and product development. HUNTER, SALZMAN & ZAELKE, *supra*, at 943–44.

Is the Convention's treatment of these concerns–embodied in Articles 15 through 19–satisfactory? Some developing nations object that "the Convention does not explicitly address the global and national roots of biodiversity destruction, and pays mostly lip-service to the genuine needs of disprivileged people everywhere." Anshish Kothari, *Beyond the Biodiversity Convention: A View from India* 67–72, in BIODIPLOMACY: GENETIC RESOURCES AND INTERNATIONAL RELATIONS (V. Sanchez & C. Juma eds. 1994). The United States reads the Convention quite differently. Assistant Secretary Wirth insisted that "the Convention cannot serve as a basis for any Party unilaterally to change the terms of existing agreements involving public or private U.S. entities." He continued:

> On the critical issue of technology transfer, our understandings make clear that any access to and transfer of technology that occurs under the convention must recognize and be consistent with the adequate and effective protection of intellectual property rights (IPR). The U.S. understandings also make clear that the term "fair and most favorable terms" for technology transfer contained in Article 16(2) of the Convention means terms that are voluntarily agreed to by all Parties to the transaction. This lays down the clear marker to all other countries that the Convention cannot be used by any Party to unilaterally impose terms or obligations on any other Party regarding technology transfer.

Wirth, *supra*, at 215. Is Wirth's fear of unilateral actions legitimate? Should the Convention permit such actions?

4. Biodiversity can be harmful. The traditional illustrations of that maxim involve the smallpox virus, predators who eliminate their prey, and alien species that wreak havoc when found outside their native ecosystems. Now biotechnology presents far more ominous concerns about the potential effects of genetically modified organisms (GMOs) that are released into the environment. Developing countries have been particularly worried about GMOs because of their concerns that foreign corporations will be insufficiently sensitive to the risks of creating such organisms within their borders. Again, the United States sees the issue differently. Testifying in 1995, Assistant Secretary Wirth explained:

> Industry representatives have also inquired as to U.S. intentions regarding the negotiation of a biosafety protocol under the Convention. We stated at the time we signed the Convention that the need for a

protocol must be demonstrated before further steps are considered. At the first intergovernmental meeting relating to the Convention we restated this position, adding that based on our experience, we did not feel that a biosafety protocol to this Convention is warranted. However, should negotiations on a protocol eventually proceed, we will be in a better position to protect U.S. interests if we have a seat at the table. And the United States, in cooperation with U.S. industry and other interested groups, would work to ensure that any protocol is scientifically based and analytically sound.

Wirth, *supra*, at 215.

In 2000, the parties to the Convention adopted the Cartagena Protocol on Biosafety. That protocol allows countries to decide whether or not they are willing to accept agricultural imports that contain GMOs. Commodities that might contain GMOs must be clearly labeled when exported, with heightened notice requirements imposed upon seeds, live fish, and other GMOs that are intentionally released into the environment. The protocol thus embodies the precautionary principle, a common yet contested feature in environmental law that allows regulators to err on the side of caution when faced with uncertain scientific risks. *See* Michael Pollan, *The Year in Ideas: A to Z.; Precautionary Principle*, N.Y. TIMES, Dec. 9, 2001, § 6, at 92 (discussing the role of the precautionary principle in the Biosafety Protocol). The protocol also establishes a Biosafety Clearing House that collects and shares information about GMOs. The protocol will become effective once it has been ratified by 50 countries; only nine countries had approved the protocol by the end of 2001.

5. So who should pay for all of these efforts? The South agreed to emphasize biodiversity conservation as an international goal in exchange for the North's promise to help pay for that conservation. Was that trade-off necessary? Should developing countries treat biodiversity as important even if that means that they have pay for its protection themselves? Should developed countries pay for the protection of biodiversity even if the host country does not regard it as especially important? Consider, moreover, the financial obligations contained in Articles 20 and 21. What does Article 20 require of developed nations? Developing nations? What examples are there of "countries undergoing the process of transition to a market economy" who may voluntarily assume the financial responsibilities imposed upon developed countries?

Management of the requisite funds has generated questions of its own. Not surprisingly, the dispute has centered on control of the distribution of the money collected pursuant to the Convention. The American position, as explained by Assistant Secretary Wirth, notes that "U.S. funding for convention-related activities will be handled through periodic contributions to the Global Environment Facility (GEF). The United States has committed to provide $430 million over the next four fiscal years toward the replenishment of the recently restructured GEF; total pledges by all countries will come to slightly more than $2 billion over the replenishment period. In order to make clear how the United States interprets key funding provisions of the convention, the Administration has recommended several understandings to be deposited with our instrument of ratification. . . ." Wirth, *supra*, at 214–15. By contrast, many developing countries have objected to the role of the GEF because of its relationship to the World Bank, which those countries criticize as insufficiently democratic and responsive to the needs of developing nations. The Convention itself does not specify the entity that will oversee the funding efforts, and Article 21 mandates that the chosen entity possess a "democratic and transparent system of governance." The parties to the Convention selected GEF to perform the role described in Article 21, and as modified, GEF continues to be responsible for the funds promised by the Convention. *Cf.* HUNTER, SALZMAN & ZAELKE, *supra*, at 1501–05 (providing an overview of the GEF).

6. Numerous other issues are addressed in the provisions of the Convention not reprinted above. Articles 23 establishes a Conference of the Parties that meets periodically and oversees the implementation of the Convention, while Article 24 creates the Secretariat that performs the daily work of the Convention. Article 27 addresses the resolution of any disputes between parties to the Convention, as complemented by the arbitration procedures detailed in Annex II of the Convention. Article 37 states simply, "No reservations may be made to this Convention," thereby preventing a common strategy by which countries have signed treaties even as they refuse to be bound by key provisions of them. Article 38 permits a country to withdraw from the Convention after giving one year's notice.

D.   CUSTOMARY LAW

Beanal v. Freeport–McMoran, Inc.

United States Court of Appeals for the Fifth Circuit, 1999.
197 F.3d 161.

CARL E. STEWART, CIRCUIT JUDGE:

. . . This case involves alleged violations of international law committed by domestic corporations conducting mining activities abroad in the Pacific Rim. Freeport–McMoran, Inc., and Freeport–McMoran Copper & Gold, Inc., ("Freeport"), are Delaware corporations with headquarters in New Orleans, Louisiana. Freeport operates the "Grasberg Mine," an open pit copper, gold, and silver mine situated in the Jayawijaya Mountain in Irian Jaya, Indonesia. The mine encompasses approximately 26,400 square kilometers. Beanal is a resident of Tamika, Irian Jaya within the Republic of Indonesia (the "Republic"). He is also the leader of the Amungme Tribal Council of Lambaga Adat Suki Amungme (the "Amungme"). In August 1996, Beanal filed a complaint against Freeport in federal district court in the Eastern District of Louisiana for alleged violations of international law. Beanal invoked jurisdiction under (1) 28 U.S.C. § 1332, (2) the Alien Tort Statute, 28 U.S.C. § 1350, and (3) the Torture Victim Protection Act of 1991, sec. 1, et seq., 28 U.S.C. § 1350 note. In his First Amended Complaint, he alleged that Freeport engaged in environmental abuses, human rights violations, and cultural genocide. Specifically, he alleged that Freeport mining operations had caused harm and injury to the Amungme's environment and habitat. He further alleged that Freeport engaged in cultural genocide by destroying the Amungme's habitat and religious symbols, thus forcing the Amungme to relocate. Finally, he asserted that Freeport's private security force acted in concert with the Republic to violate international human rights. [The district court dismissed Beanal's claims for failure to state a claim. *See* Beanal v. Freeport–McMoRan, 969 F.Supp. 362 (E.D.La.1997)].

... We observed in 1985, "the question of defining 'the law of nations' is a confusing one which is hotly debated, chiefly among academics." Carmichael v. United Technologies Corp., 835 F.2d 109, 113 (5th Cir.1985). However, in Cohen v. Hartman, 634 F.2d 318, 319 (5th Cir.1981) (per curiam), we "held that the standards by which nations regulate their dealings with one another inter se constitutes the 'law of nations.'" These standards include the rules of conduct which govern the affairs of this nation, acting in its national capacity, in relationships with any other nation. The law of nations is defined by customary usage and clearly articulated principles of the international community. One of the means of ascertaining the law of nations is "by consulting the work of jurists writing professedly on public law or by the general usage and practice of nations; or by judicial decisions recognizing and enforcing that law." Courts "must interpret international law not as it was in 1789. but as it has evolved and exists among the nations of the world today." Although Beanal's claims raise complex issues of international law; nonetheless, the task before us does not require that we resolve them. We are only required to determine whether the pleadings on their face state a claim upon which relief can be granted. Although the day may come when we will have to join other jurisdictions who have tackled head-on complex issues involving international law, "this case, however, does not require that we stand up and be counted." ....

Environmental Torts and Abuses

Next, Beanal argues that Freeport through its mining activities engaged in environmental abuses which violated international law. In his Third Amended Complaint, Beanal alleges the following:

FREEPORT, in connection with its Grasberg operations, deposits approximately 100,000 tons of tailings per day in the Aghwagaon, Otomona and Akjwa Rivers. Said tailings have diverted the natural flow of the rivers and have rendered the natural waterways of the plaintiff unusable for traditional uses including bathing and drinking. Furthermore, upon information and belief, the heavy metal content of the tailings have and/or will affect the body tissue of the aquatic life in said rivers. Additionally, tailings have blocked the main flow of the Ajkwa River causing overflow of the tailings into lowland rain forest vegetation destroying the same.

FREEPORT in connection with its Grasberg operations has diverted the aforesaid rivers greatly increasing the likelihood of future flooding in Timika, the home of the plaintiff, TOM BEANAL.

FREEPORT, in connection with its Grasberg mining operations has caused or will cause through the course of its operations 3 billion tons of "overburden" to be dumped into the upper Wanagon and Carstensz creating the likely risk of massive landslides directly injurious to the plaintiff. Furthermore, said "overburden" creates acid rock damage which has created acid streams and rendering the Lake Wanagon an "acid lake" extremely high in copper concentrations, ...

However, Freeport argues that Beanal's allegations of environmental torts are not cognizable under the "law of nations" because Beanal fails to show that Freeport's mining activities violate any universally accepted environmental standards or norms. Furthermore, Freeport argues that it would be improper for a United States tribunal to evaluate another county's environmental practices and policies. The district court conducted a thorough survey of various international law principles, treaties, and declarations and concluded that Beanal failed to articulate environmental torts that were cognizable under international law.

Beanal and the amici refer the court to several sources of international environmental law to show that the alleged environmental abuses caused by Freeport's mining activities are cognizable under international law. Chiefly among these are the Principles of International Environmental Law I: Frameworks, Standards and Implementation 183–18 (Phillip Sands ed,. 1995) ("Sands"),[5] and the Rio Declaration on Environment and Development, June 13, 1992, U.N. Doc. A/CONF. 151/5 rev.1 (1992) (the "Rio Declaration").

Nevertheless, "it is only where the nations of the world have demonstrated that the wrong is of mutual and not merely several, concern, by means of express international accords, that a wrong generally recognized becomes an international law violation in the meaning of the [ATS]." Thus, the ATS "applies only to shockingly egregious violations of universally recognized principles of international law." Beanal fails to show that these treaties and agreements enjoy universal acceptance in the international community. The sources of international law cited by Beanal and the amici merely refer to a general sense of environmental responsibility and state abstract rights and liberties devoid of articulable or discernable standards and regulations to identify practices that constitute international environmental abuses or torts. Although the United States has articulable standards embodied in federal statutory law to address environmental violations domestically, *see* The National Environmental Policy Act (42 U.S.C. § 4321 et seq.) and The Endangered Species Act (16 U.S.C. § 1532), nonetheless, federal courts should exercise extreme caution when adjudicating environmental claims under international law to insure that environ-

---

5. Sands features three environmental law principles: (1) the Polluter Pays Principle; (2) the Precautionary Principle; and (3) the Proximity Principle.

mental policies of the United States do not displace environmental policies of other governments. Furthermore, the argument to abstain from interfering in a sovereign's environmental practices carries persuasive force especially when the alleged environmental torts and abuses occur within the sovereign's borders and do not affect neighboring countries.[6] Therefore, the district court did not err when it concluded that Beanal failed to show in his pleadings that Freeport's mining activities constitute environmental torts or abuses under international law.

Notes and Questions

1.   Is there a customary international law norm requiring the protection of biodiversity? The court explains that "the law of nations"–*i.e.*, customary international law–"is defined by customary usage and clearly articulated principles of the international community." More specifically, the statute governing the International Court of Justice identifies five sources of customary international law: (1) treaties, (2) international custom that provides "evidence of a general practice accepted as law," (3) "the general principles of law recognized by civilized nations," (4) certain judicial decisions and the writings of "the most highly qualified publicists of the various nations," and (5) equity. Statute of the International Court of Justice, June 26, 1945, 1945 U.S.T. Lexis 199, Art. 38. United Nations resolutions, the resolutions of other international organizations, intergovernmental communiqués, and international consensus, are among the other acts and documents that can become relevant to the determination of customary international law. *See generally* LAKSHAM D. GURUSWAMY ET AL., INTERNATIONAL ENVIRONMENTAL LAW AND WORLD ORDER: A PROBLEM–ORIENTED COURSEBOOK 67–170 (2d ed. 1999) (discussing the sources of international law).

Extending these principles to the protection of biodiversity has proved to be difficult. In 1990, Michael Glennon described the state of customary international law related to biodiversity as follows:

> It is now possible to conclude that customary international law requires states to take appropriate steps to protect endangered species. Customary norms are created by state practice "followed by them from a sense of legal obligation." Like highly codified humanitarian law norms that have come to bind even states that are not parties to the instruments promulgating them, wildlife protection norms also have become binding on nonparties as customary law. Closely related to this

---

6. Although Beanal cites the Rio Declaration to support his claims of environmental torts and abuses under international law, nonetheless, the express language of the declaration appears to cut against Beanal's claims. Principle 2 on the first page of the Rio Declaration asserts that states have the "sovereign right to exploit their own resources pursuant to their own environmental and developmental policies," but also have "the responsibility to ensure that activities within their jurisdiction or control do not cause damage to the environment or other States or areas beyond the limits of national jurisdiction." Beanal does not allege in his pleadings that Freeport's mining activities in Indonesia have affected environmental conditions in other countries.

process of norm creation by practice is that of norm creation by convention: customary norms are created by international agreements "when such agreements are intended for adherence by states generally and are in fact widely accepted." Several such agreements are directed at wildlife protection, and CITES is one of them. It is intended for adherence by states generally and is accepted by the 103 states that have become parties. In addition, some nonparties comply with certain CITES documentary requirements so as to trade with parties. CITES is not "rejected by a significant number of states"; only the United Arab Emirates has withdrawn from the agreement. In such circumstances, the International Court of Justice has observed, international agreements constitute state practice and represent law for nonparties.

Moreover, customary norms are created by "the general principles of law recognized by civilized nations." Because CITES requires domestic implementation by parties to it, and because the overall level of compliance seems quite high, the general principles embodied in states' domestic endangered species laws may be relied upon as another source of customary law. Even apart from the CITES requirements, states that lack laws protecting endangered species seem now to be the clear exception rather than the rule. That there exists opinio juris as to the binding character of this obligation is suggested by the firm support given endangered species protection by the UN General Assembly and various international conferences.

While the existence of a norm requiring the protection of endangered species thus seems likely, its scope remains uncertain. To the extent that the norm derives from CITES and laws implementing CITES, that scope would be fairly narrow, for the norm would cover only species in international trade, not those taken for domestic consumption or those endangered by threats to their habitat. Even if it could be shown that major legal systems generally comprise endangered species legislation, more work needs to be done to determine exactly what elements those laws have in common. What constitutes an "endangered species," for example, is debatable. Is it one that is endangered in every state, or only in the state making the assessment? And to what lengths must a state go in protecting a species it finds "endangered"? Must it do everything necessary to protect that species, notwithstanding the cost or the ecological significance of the species?
. . .

It thus appears doubtful that a customary norm concerning the elephant or any other endangered species can yet play any significant role in its protection. But the trend cannot be doubted; and once its contours are more clear, the customary norm requiring states to protect endangered species ought to take on the character of an obligation erga omnes. Ordinarily, claims for the violation of an international obligation may be made only by the state to which the obligation is owed. Obligations erga omnes, however, run to the international community as a whole; thus, their breach is actionable by any state since such matters are "[b]y their very nature ... the concern of all States...." [T]hey are obligations erga omnes.

Michael J. Glennon, *Has International Law Failed the Elephant?*, 84 AM. J. OF INT'L L. 1, 30–33 (1990). Do any events since 1990 provide sufficient evidence that customary international law now requires nations to prevent the destruction of biodiversity? Why did the Fifth Circuit conclude that the operation of the Grasberg mine does not violate customary international law?

Suppose that the court had determined that the mine did violate customary international law. What remedies would have been available to the plaintiffs? What obligations would the Fifth Circuit's decision have imposed on the government of Indonesia? What other forum has jurisdiction to render a decision that Indonesia has acted in violation of international law? If there is no such forum, or if Indonesia could ignore without sanction any holding that it had violated international law, what would it mean to say that international law protects biodiversity? *Cf.* GURUSWAMY, *supra*, at 49–66 (reviewing the debate concerning the character of customary international law as "law").

2.   Tom Beanal has many responsibilities besides lead plaintiff in this case. He is the head of the tribal council of the Amunge community which used to live on the land now occupied by the mine. He has served as deputy chairman of the Papuan People's Congress and is a leading supporter of West Papua's (formerly Irian Jaya) bid for independence from Indonesia. And during a speech in New Orleans–the corporate home of Freeport in 1996–Beanal described the environmental consequences of the Grasberg mine: "Our environment has been ruined and our forests and rivers polluted by waste. The sago forests which serve as our primary food source have become dry, making it hard for us to find food.... The animals we have hunted in the past have disappeared so we no longer know what to hunt.... Our water is contaminated by chemicals so we can no longer drink it." *Tom Beanal's Speech at Loyola University in New Orleans* (May 23, 1996), *available at* <http://www.corpwatch.org.issues/PID.jsp?/articleid=987>.

Other environmentalists and human rights groups have been even more scathing in their criticism of the mine. *See, e.g.,* PROJECT UNDER-GROUND, RISKY BUSINESS: THE GRASBERG GOLD MINE (May 1998), *available at* <http://www.moles.org/ProjectUnderground/downloads/risky_business.pdf>; Robert Bryce, *Spinning Gold; Freeport–McMoRan Copper & Gold's Mining Operations in Indonesia*, MOTHER JONES, Sept. 1, 1996, at 66. Yet one travel agency serving Timika–Beanal's hometown–boasts of *both* "the unparalleled bio-diversity" of the area and of the Grasberg mine. "A major attraction of Irian Jaya is the wide variety of flora and fauna with steamy mangrove swamps in the lowland coastal areas rising to tropical rainforests in the island's heart.... Wildlife diversity is incredible with 640 species of birds including the cassowary and 36 types of Birds of Paradise," along with 800 species of spiders and 2,700 species of orchids, pitcher plants, and giant anthouse plants. At the same time, the agency proclaims that the

Grasberg mine "has been called one of the world's great engineering marvels," adding that "[a] virtual tour of the mining operations is available on interactive touch-screen videos in the Irian Jaya Room of the Sheraton Timika." *Welcome to Timika–Irian Jaya–Indonesia*, <http://www.timika-chaters.com.au>. The mine contains the largest reserves of gold, and the third largest reserves of copper, in the world. Does that explain why, in the words of the Rio Declaration, Indonesia has exercised its "sovereign right to exploit [its] own resources pursuant to [its] own environmental and development policies?"

3. How *should* Indonesia balance its environmental and economic needs? Consider the September 2001 announcement by Freeport that it had agreed to pay $2.5 million and then $500,000 annually to a trust benefitting the Amungme and Kamoro tribal communities who were native to the land now containing the Grasberg mine. Tribal leaders plan to use some of the trust's money to purchase shares of Freeport stock, thus obtaining an ownership interest in the mine. Beanal applauded the agreement, saying, "We used to be on the outside, but now we stand together. We have a stake in this mining operation and we will work hard so that we can share in its success." *Freeport to Establish Trust for Indigenous Communities*, JAKARTA POST, Sept. 26, 2001.

4. Is the court correct that "the argument to abstain from interfering in a sovereign's environmental practices carries persuasive force especially when the alleged environmental torts and abuses occur within the sovereign's borders and do not affect neighboring countries?" One student commentator disagrees because (1) the ATCA's restriction to "shockingly egregious" violations of international law provides sufficient caution in all cases, (2) the ATCA does not require concern for the domestic policies for foreign nations, and (3) there is nothing unique about environmental claims that justifies greater judicial caution than any other violation of international law. *See* Russell Unger, Note, *Brandishing the Precautionary Principle Through the Alien Tort Claims Act*, 9 N.Y.U. ENVTL. L. REV. 638, 646–47 (2001). Should a foreign nation's unwillingness to regulate the destruction of biodiversity within its borders affect the success of an ATCA claim? Does it matter that the Indonesian government has been labeled one of the most corrupt in the world? Or could a foreign nation authorize a lawsuit against an activity that harms biodiversity within the United States?

5. A lawsuit brought by the Indonesian Forum for the Environment claiming violations of Indonesia's own law proved to be more successful than Beanal's customary international law litigation in the United States. In 2001, the South Jakarta District Court held that Freeport had spread misleading information about its environmental activities and violated an Indonesian law concerning environmental management. *Experts Allay Fears Over Freeport Ruling*, JAKARTA POST, Aug. 31, 2001. Does that suggest that domestic law will remain a more fruitful means of protecting biodiversity than customary international law? Would there be any legal cause of action against such habitat destruction if the mine were located in the United States? What is the ideal legal regime to prevent the loss of biodiversity attributed to an activity like the Grasberg mine?

*

INDEX

References are to pages

†